Dynamic Anatomy and Physiology

Dynamic

and

Anatomy

Physiology

Ben Pansky

Ph.D., M.D., Professor of Anatomy, Medical College of Ohio at Toledo.

Formerly, Director of Basic Sciences, School of Nursing, New York Medical College

Macmillan Publishing Co., Inc.
New York

Collier Macmillan Publishers
London

Copyright © 1975, Ben Pansky

Printed in the United States of America

All rights reserved. No part of this book may be reproduced or transmitted in any form or by any means, electronic or mechanical, including photocopying, recording, or any information storage and retrieval system, without permission in writing from the Publisher.

ACKNOWLEDGMENT: The electron micrographs appearing on the first page of each unit were supplied by Dr. G. Colin Budd, Medical College of Ohio at Toledo. The subjects are as follows:

Unit I (p. 1): Hepatocyte in the liver of a mouse.
Unit II (p. 123): Striated muscle of a mouse.
Unit III (p. 245): Hypothalamic neurons of a mouse.
Unit IV (p. 325): Small capillary close to two acinar cells in the pancreas of a mouse.
Unit V (p. 571): Ovarian follicle of a mouse.
Unit VI (p. 611): Aging pigment (lipofuscin granules) in hepatocytes from an elderly patient.

Macmillan Publishing Co., Inc.
866 Third Avenue, New York, New York 10022

Collier-Macmillan Canada, Ltd.

Library of Congress Cataloging in Publication Data

Pansky, Ben.
  Dynamic anatomy and physiology.

  Includes bibliographies.
  1. Human physiology. 2. Anatomy, Human.
I. Title. [DNLM: 1. Anatomy. 2. Physiology.
QS4 P196d 1974]
QP34.5.P36   612   73-13170
ISBN 0-02-390740-1

Printing: 2 3 4 5 6 7 8   Year: 6 7 8 9 0

This book is dedicated to
my wife, Julie, and son, Jon,
without whose constant
encouragement, confidence,
and support this book might
never have been.

Preface

This textbook has been written to meet the needs of undergraduate students of nursing and allied health professions. It provides both verbal and visual descriptions of structure and function and demonstrates the beauty, harmony, and complexity of the human organism. The facts are centered about the structure of cells, tissues, and organs, and their interactions with each other in the performance of daily functions. The hundreds of newly created illustrations—many of which are reproduced in color—depict major concepts and pertinent facts in a simple manner, and are closely interwoven with and inseparable from the accompanying text discussions.

The early chapters of *Dynamic Anatomy and Physiology* are concerned with cellular morphology and physiology, for here is where life begins and ends—with the cell. It is the cell that varies in its structure and function in each of the systems of the body. Indeed, each cell changes and operates to perform a specific chore within a particular system, whether the task be conducting fluids or impulses, receiving or emitting messages, or altering metabolism. In each system, however, the life of the cell is finite, not infinite. Not only do individual cells change, but whole systems are modified, both normally and pathologically, with time and age.

The reader follows the living organism throughout the life cycle—from the early chapters on cell morphology, function, and viability (Chapters 1–4), through each of the systems (Chapters 5–18), and finally to the aging process and ultimate death of the cell and the organism itself (Chapters 19–20). Along the way the student learns how both male and female pass on some cells to a new generation (Chapters 17–18). Thus, the perpetual cycle begins over again.

Developmental anatomy is included in areas where knowledge of change from the simple form enhances understanding of the more complicated adult organ. Disturbances of basic morphology and function are also discussed because they are a natural sequence of events. Such discussions may encourage further study into the effects of injury, disease, and aging. However, the book does not concentrate on pathology, since its major aim is to convey a broad understanding of normal anatomy and physiology.

Repetition is employed to emphasize and, at times, to view a concept from another perspective. In addition, detailed descriptions and discussions are occasionally necessary because of the intricate nature of the organs involved. Thus, the text is rich and complete and presents a full, readable, and comprehensive story rather than an uneven, abbreviated version. The anatomic terminology used throughout is an anglicized version of the standard nomenclature (*Nomina Anatomica*) approved at the Seventh and Eighth International Congresses of Anatomists, held respectively at New York in 1960 and at Wiesbaden in 1965.

Pertinent questions follow each chapter. These are intended to test knowledge and comprehension and to provoke discussion. An up-to-date bibliography also accompanies each chapter and offers a source of continuing education. The back matter of *Dynamic Anat-*

omy and Physiology includes a four-color atlas of regional anatomy as well as a listing of prefixes, suffixes, and combining forms.

The contributions of many friends, colleagues, teachers, and students are gratefully acknowledged. The author is particularly indebted to Dr. G. Colin Budd, for supplying many excellent electron micrographs and for reviewing the manuscript of Chapter 1; to Dr. Dale E. Bockman, for providing additional electron micrographs; to Dr. Pietro Motta, for making available scanning micrographs of the ovarian surface; to Drs. Bruce A. Gastineau, Leonard Nelson and Jyotsna Chakraborty, and David D. Cherney, for supplying photographs of the fetal skull, sperm, and ovum, respectively; and to the late Dr. A. Gorman Hills, for reading Chapter 14 and making helpful comments and suggestions. John Doman photographed the bones shown in Chapter 5 and the surface anatomy illustrated in Chapter 7. Kenneth G. Flora, of the Audio-Visual Service, Medical College of Ohio at Toledo, provided the photograph of the attractive young lady shown in Figure 9-1. The author wishes to give special thanks to Ronald M. Watterson, librarian of the Medical College of Ohio at Toledo, and Roberta H. Raeder, research technologist at the same institution, for assisting with the arduous task of proofreading.

B. P.

Contents

Dynamic Anatomy
and Physiology

The Body: Its Structure and Organization

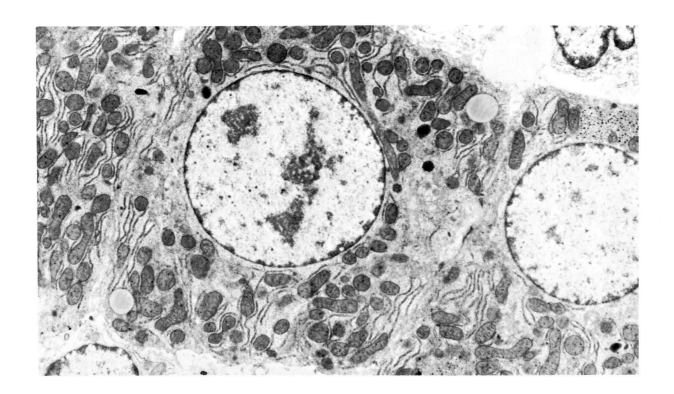

UNIT I

THE CELL

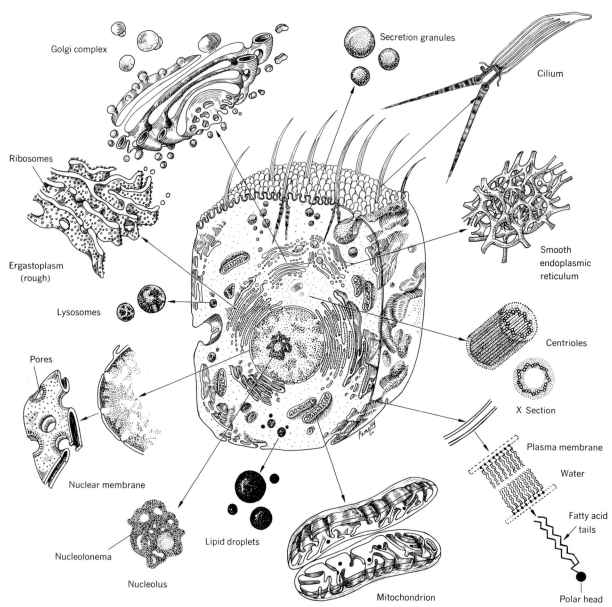

Golgi complex

Secretion granules

Cilium

Ribosomes

Smooth
endoplasmic
reticulum

Ergastoplasm
(rough)

Lysosomes

Centrioles

X Section

Pores

Plasma membrane

Water

Nuclear membrane

Fatty acid
tails

Nucleolonema

Lipid droplets

Nucleolus

Mitochondrion

Polar head

Figure 1-1. Diagram of a cell, illustrating the form of its organelles and inclusions as they appear by light and electron microscopy. Around the periphery are representations of the finer structure of these components.

Cell Structure

All life's activities, such as seeing, breathing, moving, and thinking, need energy, which is derived from chemical reactions taking place in each individual cell. The activities of cells, to be understood, require a knowledge of the basic structures that they comprise: in breathing, the lungs and the diaphragm; in movement, the bones and muscles; etc. All these structures are composed of cells, and it is the arrangement of the molecular components that determines their appearance, activity, and function (Fig. 1-1).

A cell, regardless of its nature, must possess (1) the necessary functioning parts to make its products or perform its work, (2) a source of energy to supply its capabilities, (3) an organization to enable the parts to function with efficiency, and (4) a central control to direct and time it. Thus, all cells, although variable in dimension and function, must share basic features.

Every living organism has individuality and uniqueness that can be explained in molecular terms, with the cell as the single structural and functional unit of all true living systems. The cell is far greater in size and complexity than an atom or molecule and has a definite boundary within which chemical activity takes place. It is usually an integral part of a more elaborate organ and organ system in a multicellular plant or animal. There is no life without the cell; and as life has its variations of form and function, so does the cell vary. Some cells live alone as freely independent moving entities; others belong to organized communities (that may also move about); and some remain relatively fixed and immobile in the tissue of a larger organism. In any form, whatever function, the cell is the fundamental unit of all living matter, containing all the essential ingredients for survival in an environment that is constantly undergoing change.

We recognize the complexity of the world around us by touch, taste, hearing, smell, or sight, and even these unaided senses can differentiate solid from liquid, living from nonliving. We can refine this to differentiate rough from smooth, intensity and color shades, and a variety of taste sensations (salt, sweet, or bitter). Yet even these senses are limited. We see only that portion of the light spectrum in the visible range and hear only a limited range of sound. Beyond these we need instruments to penetrate beyond the "natural" sphere and act as extended sensors-telescopes to bring the macrocosm to view, microscopes to probe the microcosm, and photosensitive surfaces to detect the long infrared rays and short ultraviolet. Thus, we strive continually to determine the basic units of structure and function.

The cellular divisions of organic matter were first identified and called cells in 1663 by the English scientist Robert Hooke. However, the modern cell theory, "that all animal and vegetable tissues are made up of cell collections," was not established until the early part of the nineteenth century. Matthias Schleiden, the German botanist, in 1838 stated that the cell was the basic structural unit of all vegetable matter. In 1839, Theodor Schwann, also a German, declared that "cells are organisms and that entire animals and plants [are] aggregates of these organisms arranged according to definite laws." Twenty years later, Rudolf Virchow further stated that "cells come only from pre-existing cells." Thus was born the foundation upon which all

3

subsequent investigations into the world of the cell were undertaken. It involves three basic concepts: (1) Life exists only in cells; organisms are made up of cells and the activity of an organism depends on cellular activity, both individually and collectively. (This breaks down in the case of viruses, which lack the usual cellular organization.) (2) The continuity of life has a cellular basis; genetic continuity in a very exact sense includes not only the cell as a whole but some of its smaller components (genes and chromosomes). (3) A relation exists between structure and function (principle that complementary biochemical activities of the cell occur within and are determined by structures organized in a specific way).

Figure 1-2. Three kinds of microscopes are compared in schematic diagrams that emphasize their similarities and show their differences. The transmission electron microscope resembles a conventional light microscope except for the use of magnetic lenses instead of glass lenses and an electron beam instead of incandescent light. In the scanning scope the source is imaged onto the sample by magnetic lenses. The small spot of electrons thus formed is then deflected by scanning coils (not shown) which moves the beam across the sample in a particular pattern to illuminate the sample, point by point at sequential times. Secondary electrons, generated where the primary beam strikes the sample, are drawn to a detector and produce an electrical signal that modulates the intensity of an electron beam inside a television tube, scanned in step with the spot. It is these secondary electrons, collected by the detector that are normally used to produce a scanning electron micrograph. Thus, the scanning scope is essentially a closed-circuit television system.

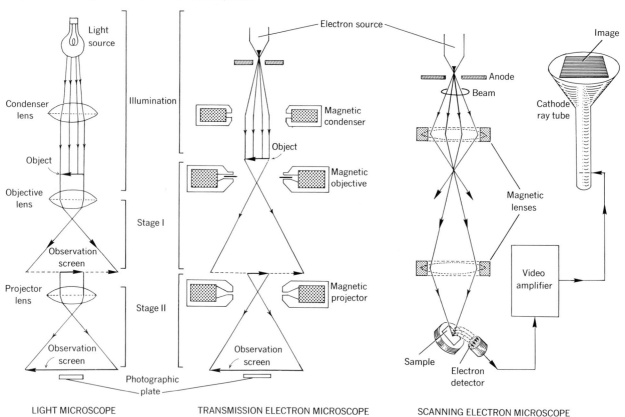

The Cell: History and Study

The study of biology tends to take place in an ever-narrowing field, having begun with that of the body as a whole in the days of Leonardo da Vinci (1452–1519) and Vesalius (1514–1564), who detailed bone and muscle anatomy with surprising accuracy. Harvey's (1578–1657) blood circulation theory, a century later, described how blood makes its way through the body—with no microscope available and guessing at the existence of small connecting vascular channels since he could only see the arteries and veins with the naked eye. It was the microscope that revealed the connecting capillaries and later was to reveal the cellular nature of all living matter.

Until only a few years ago, man's conception of the cell was limited by what he could see in a microscope under visible and ultraviolet light and with magnifications of 100 to 2000 times (Fig. 1-2). He saw the cell outline, cytoplasm, and nucleus. Smaller entities ap-

Figure 1-3. Cell fractionation. Cells are disrupted (homogenized) by a pestle. The space between pestle and tube wall can be varied to achieve cell breakage and minimal organelle damage. Inevitably the plasma membrane and endoplasmic reticulum fragment. Ultracentrifugation of 50,000 to 125,000 G forces per minute permits sedimentation of the smallest structures and many macromolecules.

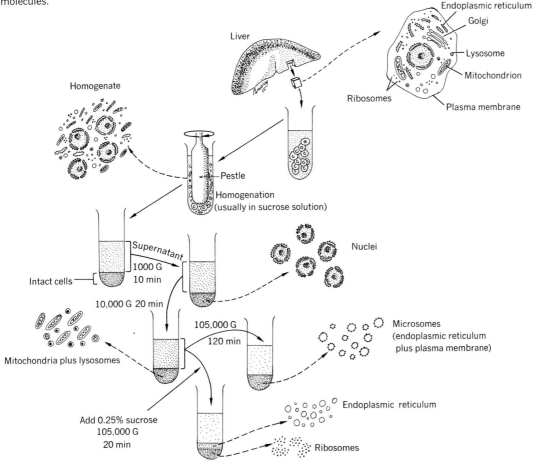

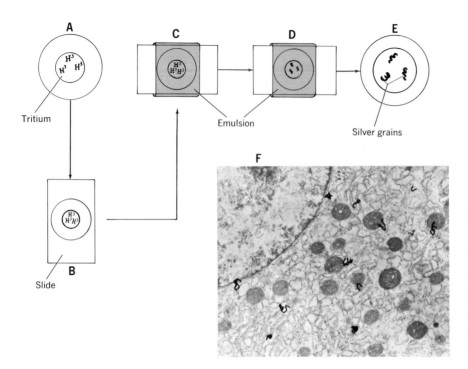

Figure 1-4. Autoradiography procedure. **A.** Cell is grown in radioactive precursor (thymidine) labeled with tritium (H^3), a radioactive isotope. **B.** Cell is fixed and placed on slide. **C.** Cell is covered with photographic emulsion in the dark and exposed (several days to several months) and developed as with photographs. **D** and **E.** Silver grains are located in the emulsion over the areas where radioactive atoms have emitted radiation (affects emulsion). The number of grains is proportional to the number of radioactive atoms. **F.** Actual electron micrograph of silver grains developed. ×6,514. (Electron micrograph courtesy of Dr. G. Colin Budd, Medical College of Ohio at Toledo.)

peared as strands or dots or were entirely invisible. Using phase-contrast optics, light microscopes made various parts of cells "stand out" in sharp contrast when light just out of phase passed through them. Progress, however, particularly in cytology, has depended on refined tools and techniques of analysis.

If one takes a group of cells and uses ultracentrifugation at carefully regulated high speeds, cell "parts" settle out in different layers, depending on their weight, and once separated each can be analyzed (Fig. 1-3). Living cells can also be cultured and then tested for behavior, probed with microneedles, dissected, and tagged with radioactive substances. If a cell is suspended in solution containing a substance labeled by a radioactive isotope, which the cell is capable of incorporating, it is often possible to determine where in the cell incorporation takes place and to what extent. It has been shown that ^3H thymidine is selectively incorporated into DNA, etc. Thus, autoradiography has been of value in localizing the sites of a variety of syntheses in the cell (Fig. 1-4). About 20 years ago,

with the application of the electron microscope the interior of the cell was further revealed.

The electron microscope unit of measurement is the Angstrom (Å) (named after a Swedish physicist), which is equal to about $\frac{1}{250}$ millionth of an inch or $\frac{1}{10,000}$ of a micrometer (μm). A micrometer is also referred to as a micron (μ). Investigators embed the cells in transparent plastics, cut them into very thin sections (a few millionths of an inch) by glass or diamond knives on a special microtome, and examine them at magnifications of up to 200,000 times or more.

The unaided eye has a resolving power (the property in an optical system of distinguishing between objects lying very close together) of about 0.1 mm (it can see two objects as separate entities when they are as near to each other as 0.1 mm). Objects having a diameter smaller than 0.1 mm will be invisible or blurred. The eye has no power of magnifying and thus we calculate sizes mentally. Microscopes, on the other hand, resolve and magnify but are limited by their light sources (illumination). Objects smaller than one half the wavelength of the illuminating light cannot be seen in a light microscope. Thus, since white light has a wavelength of about 5500 Å, the visible light microscope cannot resolve a diameter less than 2750 Å or 0.275 μm. Since many cell parts have smaller dimensions, they cannot be seen.

Having exhausted the optical (light) microscope's power to reveal the structure of life's smaller particles, scientists turned to the electron microscope for greater magnification of individual cells. The electron microscope makes use of a different source to increase its resolving power. It uses high-speed electrons (rather than light). As the electrons pass through the specimen, parts of the cell absorb and scatter them differentially. Electrons passing through the specimen form an image of the specimen on an electron-sensitive photographic plate or fluorescent screen (Fig. 1-2). The human eye is not stimulated by electrons, and thus we need the plate or screen to visualize the image. The optics are the same as for the light microscope except that focusing is by magnetic rather than glass lenses. When electrons pass through the scope, they have a wavelength of about 0.05 Å (with 50,000 volts) and thus theoretically one could resolve objects of $\frac{1}{2}$ of 0.05 Å, or 0.025 Å (less than the diameter of a hydrogen atom, which is only 1.06 Å). Unfortunately this resolution is not usually achieved owing to lens construction and other difficulties. About 5 to 10 Å is usual. Thus, the eye resolves to 100 μm, the light microscope to 0.2 μm, and the electron microscope to 0.001 μm (or if the human eye resolving power is 1, the resolution with a light microscope is 500 times greater and with an electron microscope is 10,000 times greater) (Fig. 1-5).

Although the electron microscope seems to have answered the problems of cellular biologists and revealed details of cellular structure never before realized, the images of various cell parts seen have been frequently criticized since one is looking at fixed and stained tissues. Stains are likely to alter structure, and different fixatives often give different images. Furthermore, in preparation and processing, there is also a removal of salts and water, which may lead to artifact formation. The question of whether or not we are looking at the "real thing" or artifacts is a real one.

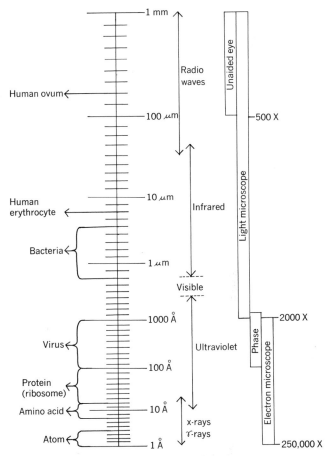

Figure 1-5. Schematic diagram of the approximate size relationships of certain specimens and useful magnification ranges of the light, phase, and electron microscope.

Recently it has been possible to study the cell organelles without the use of fixatives in a method called freeze etching. One places the object to be studied in 20 percent glycerol (antifreeze), and then it is frozen at −100° C, mounted on a chilled holder, and "splintered" with a knife along natural cleavage planes, usually along membrane surfaces. The structure may be cross-fractured, giving cross sections of organelles. Such replicas give the outline of various cell structures and verify the structures seen in thin, fixed, and stained sections. The views of structures are somewhat three-dimensional.

Another instrument used recently is the scanning electron microscope. The outstanding characteristic of the image seen with this instrument is its three-dimensional quality (Fig. 18-1, p. 588). The scanning scope is able to provide images of three-dimensional objects because in its normal mode of operation it records not the electrons passing through the specimens (as with conventional electron microscopy) but the secondary electrons that are released from the sample by the electron beam impinging on it. The sample then can be of any size and thickness that fits into the instrument's evacuated sample chamber. The secondary electrons need not be focused but simply collected, and since the beam used is like a sharp needle, there is great depth of focus not limited by the thinness of the specimen. Since thin slices are not needed, preparation of electrically conducting specimens is much simpler. In the scanning electron micrograph the specimen is scanned by a focused electron beam and the image that results is formed by a time-sequence technique similar to that used in commercial television. There is no lens aberration, no limit to practical resolution, no limitation due to specimen damage as a result of electron beam illumination, and no drift of electrons. The specimen image is viewed as a whole at any one time. The scanning scope is capable of a wide range of magnifications, from about 15 to 100,000 diameters with a resolution of 100 to 200 Å. This broad range of magnification (along with the ease of changing magnification) permits easy "zooming" from a gross image to an image showing great detail. As one increases the magnification, however, one does not decrease the depth of focus, and the resolution of detail is good.

Modern biologic study and research lie in the single cell, the unit of life. Successful living depends on the integrity and "normal" function of every cell. The body's structure and function have arisen as a result of the limitations, requirements, and infinite variety of cells. Magnified hundreds of thousands of diameters, the cell is revealed as containing an integrated molecular world, a world of perpetual chemical interchange and reaction known as metabolism.

The microscope has shown us a great variation in

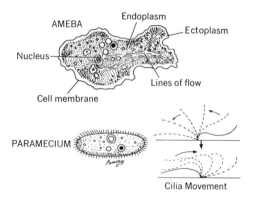

Figure 1-6. Ameba and paramecium. Shown, too, are cilia and ciliary movement.

both the function and structure of cells. They are shaped like spirals, rods, squares, spheres, stalks, and everything imaginable. Their shape is often dictated by their environment or related to their function (human red blood cells are saucer shaped and flat, allowing for easy transfer of the O_2 and CO_2 they carry [see Chap. 11]; nerve cells, on the other hand, have long, thin processes to carry messages [see Chap. 8]). Cells can be self-sufficient, independent, and free-living single-celled creatures like the ameba or paramecium (Fig. 1-6). Each has its own individuality, shape, size, and pattern of behavior, living in and on its environment with the ability not only to react to external stimulation but also to reproduce itself. These simple forms, under favorable conditions, reproduce by simple division of one cell into two, called mitosis (Fig. 2-1, p. 54). The cell enlarges somewhat; its nucleus divides into two (each part forming a complete new nucleus); the nuclei separate, taking a share of the cytoplasm and cell membrane with them; and now we have two cells, replicas of the original except in size, each complete, independent, and having the same traits, characteristics, and abilities of its parent before it. The ameba moves by bulging itself out in one direction and drawing itself in at the opposite end, a movement of ectoplasm and endoplasm. The shape of this organism therefore changes constantly and is produced by molecular movement in the cytoplasm. In contrast, the paramecium has a relatively constant shape, moving about with the help of hairlike protrusions from the cell membrane called cilia (Fig. 1-6).

Cells can be specialized to do particular tasks, as in higher organisms, and depend on the integration with other cells for survival. They can vary in size from $\frac{1}{250,000}$ of an inch, as in some microbes, to the size of a bird's egg. Yet, with all this variety and diversity, all adhere to a fundamental design, a basic pattern apparently essential for life.

The human body is a mass of billions of microscopic living cells. Each cell has the ability to reproduce and a limited span of life; yet none are independent. Each cell has predetermined functions, interrelated, interdependent, and integrated with each other. Animal cells of comparable function show little variation in size from one animal to another, and though, in the various animals, the component cells may vary greatly in number, the cells of each particular part are often similar in dimensions.

There are two major kinds of cells, whose differences are such as to suggest that they represent a major event in evolutionary time. The eucaryotic cell (eucell or true cell) has a complex internal organization and is found in all plants and animals as well as in fungi, protozoa, and most algae. The procaryotic cell or procell is found in bacteria and blue-green algae. Both have an external cellular membrane that is a complicated chemical mosaic of protein and lipids. Their intracellular organization does differ. The eucell is distinguished by the grouping of its principal hereditary material, DNA (a form of nucleic acid), into a membrane-bound nucleus. The DNA occurs in association with protein in structures called the chromosomes (whose number and form are characteristic for the species). The nuclei of the eucell also contain distinct bodies, the nucleoli, which are rich in RNA (another type of nucleic acid). In the procell, all the DNA exists as one large molecule and is not incorporated into any subcellular particle.

Eucells and procells also differ in other ways. The eucells contain several types of membrane-bound particles of specialized function. The largest is the mitochondrion (see p. 39), a spherical or cigar-shaped structure enclosed by a membrane. Mitochondria are the chief sites of oxygen-consuming respiratory metabolism and produce ATP, essential in the use and transfer of energy. There are also internal membranes that delimit the surface of cytoplasmic vacuoles and channels (cisternae) and constitute the endoplasmic reticulum (see p. 36). The membranes contain a complex enzyme system important in oxidative metabolism. Seen, too, are semispherical granules called ribosomes that are attached to the outer surface of the endoplasmic reticulum, or dispersed in the cytoplasmic matrix, which contains RNA and functions in protein synthesis; lysosomes, which contain high concentrations of enzymes responsible for degradation of biopolymers into smaller fragments; Golgi apparatus (see p. 38), a complicated network of cisternae and vesicles important in secretory activity; and others. In plants and higher algae one finds chloroplasts, which contain chlorophyll used in photosynthesis. The procells lack these structures but do have ribosomes, which, however, are not membrane associated.

A logical sequence of evolution can be postulated with the bacteria being the least complex, having a "nuclear" area (not membrane bound) of low electron density that contains filaments of DNA-protein (hereditary material) and a "cytoplasm" with membranous structures. The blue-green algae are more highly organized and also have no nuclear envelope, but there are numerous membranes found in its cytoplasm representing photosynthetic lamellae (not really chloroplasts). Last, we have the true cell or eucell of the advanced plants and animals.

The Cell: Shape and Size

All cells share certain structural features. All are enclosed within a membrane (thin film or wall) with selective permeability and all show a localization of specific functions in certain structural elements. Within the cell membrane is a semifluid jellylike material, the cytoplasm, in which the activities of the cell take place; within the cytoplasm various organelles regulate the synthesis of materials and mobilize energy that is essential for cell life. Near its center is found the nucleus (see p. 46) for central control necessary for cellular existence and containing the cell's reproductive and hereditary material enabling the cell to survive. A nuclear membrane separates the cytoplasm and the nucleus. Besides these cellular and nuclear membranes, there are other cytoplasmic membranes that provide surfaces on which cellular activity can take place.

The numerous organs and structures in the body (muscles, nerves, skin, etc.) apparently have little in common in appearance and function, and yet all are made up of cells and cell products. The varied character of these structures results from the varied character of the individual cells and cell products, which is acquired during the course of growth and development, since each has arisen from a single fertilized egg. Thus, the cell is remarkably plastic and is capable of modifying itself with the maturing organism.

Cell Shape

The general shape of a unicellular organism is mainly an inherent and inherited feature of the organism rather than a product of surface tension or other external forces. In multicellular organisms, the shape is more often but not entirely the result of the reaction of the cell to its environment, cell shape depending on contacts with other cells and on the external tensions exerted on the cell. When spherical cells are packed together, they tend to become faceted at points of contact. Cells, furthermore, are not always packed similarly. Some are layered in flat sheets (epidermis, blood vessel lining, etc.) and are wider than they are deep since the mechanics of stretching force them into such a shape. Muscles and bones, on the other hand, are elongated structures with their cells tending to run parallel to the long axis of each organ, oriented to some degree by the stresses placed on them.

Besides mechanical forces one must consider the cell's function. Red blood cells that function to transport O_2 and CO_2 are spherical from one view but appear flattened and concave from another. Their thin dimension permits rapid gas exchange, and their rounded contours allow them to slide easily through the vascular capillaries. Muscle cells contract and relax rhythmically or spasmodically when in action, and nerve cells can reach over 90 cm (3 ft) in length for relaying messages in the body. Thus cellular shape must be compatible with cellular function.

Cell Size

Cells vary widely in size. The smallest cells visible under the light microscope are bacteria (0.2 to 5.0 μm). The largest cell of all is the egg of the ostrich (about 15 cm [6 in.]). The smallest free-living organism, seen only in the electron microscope, is the causative agent of pleuropneumonia, Mycoplasma (about 0.1 μm). In size, the latter are like viruses and pass through fine filters but can, like bacteria, grow in a nonliving medium.

In the human, cells range from the small white blood cells (leukocytes) with a diameter of 3 to 4 μm to nerve cells whose body size is only 100 μm but have axons up to 90 cm (3 ft) long. The ova is about 0.1 mm in diameter. Nerve, muscle, and blood cells differ in size, related to the particular function each performs.

Table 1-1. **Comparative Cell Sizes**

Hen's egg	60 × 45 mm
Human egg	0.1 mm
Liver cell	20 μm
Red blood cell	7 μm
Typhoid bacillus	2.4 × 0.5 μm
Bacterial virus	80 mμ
Hemoglobin molecule	7 mμ

The size of cells of similar function is controlled by (1) the ratio of nucleus to cytoplasm; (2) the ratio of cell surface area to cell volume; and (3) the rate of cellular activity (or metabolism).

1. The nucleus regulates growth, development, and the existence of the cell. The nucleus cannot exercise control over an indefinitely large amount of cytoplasm, and thus most cells of a particular type maintain a relatively constant nucleocytoplasmic ratio.
2. Metabolism occurs continuously throughout the entire mass of a cell but substances needed for

metabolism can only pass in and out through the surface membrane (Fig. 1-7). If the cell volume is abnormally too large, normal cell function is impaired because the exchange across the membrane is inadequate. These limitations are somewhat varied by cells altering their shape: flattening, folding, elongating, etc. (the digestive tract is a long, coiled tube with an inner surface of convoluted surfaces that increase its absorptive surface). Thus, cells may increase their surface area without changing their volume (Fig. 1-8).

3. The rate of activity is related to cell size. The rapidly metabolizing cells of organisms such as bacteria, bees, etc., are smaller than in man, and the surface exchanges in the cells of smaller, more active animals are more rapid. The cells must be

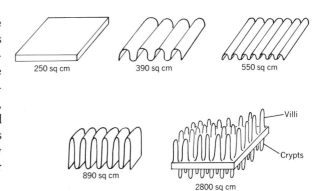

Figure 1-8. Diagram to illustrate the increase in surface area as a result of folding and inclusion of villi and crypts.

smaller to attain the greatest amount of surface area to relative volume. Thus, man's maximum size is limited by structural considerations and the need to maintain adequate flow of substances and energy in and out of the cell.

There is, however, a limit beyond which a cell cannot grow without endangering survival. This is reached when the surface area of the cell's membrane ceases to be big enough to meet the cell's needs for normal metabolism. Thus, when a cell reaches its appropriate size, it either remains that way or grows bigger and divides into two bringing its surface-to-volume ratio back into line with its needs.

The internal processes of each living cell in the body except for certain specializations are essentially the same as the smallest single-celled, free-living organism. Man's billions of cells, having no access to external environment, are not independent but are subsidiaries of a huge biologic entity. Size, therefore, is the result of the cell's primary need for an efficient surface-to-volume ratio.

The Cell: Chemistry

The human body is composed of atoms and molecules and its function depends on "millions" of interactions taking place between them. The great varia-

Figure 1-7. Diagram of a hepatocyte with its plasma membrane and structures associated with control of entry and exit of materials. Arrows indicate interchange between blood and cell. The junctional complex restricts intercellular movement.

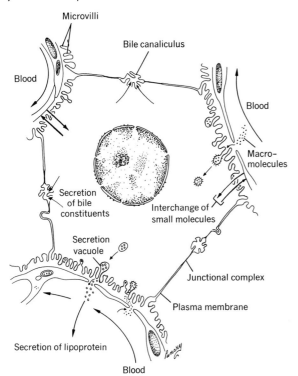

tion of chemical structure reflects the body's wide diversity of structure and function.

The development of biochemistry over the past decades and the use of radioactive agents to "tag" atoms and molecules forming the basic structure of the cell and follow them in their changing and adapting cellular world have led to many advances in cellular study and understanding.

The amount of a chemical compound present in a cell or in the total body is represented by: the number of grams, the number of molecules, or the number of moles. One mole of a compound is the weight of the compound in grams equal to its molecular weight. (The number of moles of a compound present in a system is equal to the weight of the compound in grams divided by its molecular weight.) Thus, 40 gm of a compound whose molecular weight is 40 would represent 1 mole of the compound.

Only a few chemical elements (of more than 103) play an important role in the living organism. In fact, 96 percent of the body weight and over 99 percent of the body atoms consist of four elements: hydrogen, carbon, oxygen, and nitrogen. Other elements (Table 1-2) also play an important role in living organisms, but account for only 1 percent of the total atoms and only 4 percent of the total body weight.

Molecules are formed when atoms join together by chemical bonds. It is during the process of forming and breaking these chemical bonds that the chemical potential energy of a molecule is absorbed or released. With the formation of the covalent chemical bond, both atoms share electrons and the "electric" interaction created holds the atoms together.

Since the atoms of oxygen and hydrogen are electrically neutral, the molecule of water formed from them is also neutral. On the other hand, some interactions between atoms result in the complete transfer of an electron from one atom to another producing atoms having a net charge and known as ions. One of the most important ions in the body is the hydrogen ion (H^+). The acidity of a solution is determined by the concentration of hydrogen ions present (see Chap. 13 for detailed discussion).

When a number of molecules come together, they form larger aggregates of matter. Matter exists as a gas, a liquid, or a solid. All three states of matter are found in the body, and in all three states the molecules are in constant motion. The air we breathe is a mixture of gases. Bones, teeth, and skin are solids and consist primarily of the liquid water. The major portion of the body is in a liquid or colloidal (particles dispersed in a liquid) state.

Water

Sixty percent of the body weight is water. Eighty percent of the weight of a living cell is water, and 82 percent of all the oxygen atoms and 67 percent of all the hydrogen atoms in the body combine to form water. Thus, 99 of every 100 molecules in the body are water molecules. A molecule that dissolves in a liquid is called a solute and the liquid in which it dissolves is the solvent. Solutes are usually divided into two classes. The first class consists of whole molecules such as the gas oxygen or the compound sugar, which dissolve (by merely "falling apart" into the smallest units or molecules). The second class consists of compounds such as acids, alkalis, or salts. In water, these break up and dissociate into two kinds of electrically charged particles called ions. Salt, NaCl, for example, consists of one sodium (Na) atom linked to one Cl (chloride) atom, but in solution splits up into a cation of sodium (Na^+), positively charged, and an anion of chloride (Cl^-), negatively charged. These ions move about as individual units, and either can travel across the cell membrane without the other. Although most molecules can dissolve in water, fats are not soluble, which is an important aspect of their functional role in living systems. Since water is such a good solvent, it is an ideal medium for chemical reactions.

The body must maintain a proper water balance to function normally. The circulation of blood and body fluids requires a liquid medium such as water, and even the regulation of body temperature depends partly on the evaporation of water from the skin. (See Chap. 13 for detailed discussion.)

Despite the importance of water, life is actually centered around the chemistry of the element *carbon*. Although hydrogen is the most numerous atom in the

Table 1-2. The Inorganic Molecules

P, S, Ca, Mg, K, Na, and Cl ions account for about 75 percent of the inorganic matter in the body and are thus considered most frequently in attempting to regulate mineral balance. Iron may be required as a supplement under abnormal conditions (hemorrhage or pregnancy).

Inorganic Molecules	Function
Sulfur (S)	Occurs mostly as a component of the amino acids cystine and methionine Component of certain vitamins (thiamine and biotin) Component of coenzyme A
Calcium (Ca)	Important in formation of bones and teeth Controls nerve and muscle action, blood clotting, and cell permeability Important in formation of acetylcholine (nerve impulse transmission)
Phosphorus (P)	Phosphate is essential in metabolic schemes Important in formation of bones and teeth
Magnesium (Mg)	75% found in teeth and bones Activator for a variety of enzymes
Manganese (Mn)	Needed for proper bone formation and development and function of the reproductive system Needed by several enzymes
Iron (Fe)	0.001% of animal body is composed of iron and $\frac{2}{3}$ of that occurs as a component of hemoglobin (blood pigment that functions in oxygen transport in bloodstream) Component of myoglobin, a protein found in muscle tissue
Sodium (Na)	Operates with potassium in regulation of nerve and muscle function 25% found in the skeleton in an insoluble, inert form Important in maintaining proper osmotic concentration of cells and tissues
Chloride (Cl)	Important in maintaining proper osmotic concentration of cells and tissues Indirectly involved in transport of CO_2 by blood A major electrolyte in controlling protein solubility (particularly globulins) Amounts to $\frac{2}{3}$ of anions of blood
Potassium (K)	Found in all cells, particularly in muscle and cartilage Important in physiologic fluids Enters into electrolyte and water balance with Na and Cl ions An activator for metabolic enzymes Participates in the actomyosin system
Cobalt (Co)	Supplied as vitamin B_{12} Effective enzyme activator
Copper (Cu)	Needed for the formation of hemoglobin Component of the prosthetic group of some respiratory pigments (hemocyanins), which serve as major oxygen-transporting proteins in the blood of lower forms
Zinc (Zn)	Constituent of a number of enzymes Insulin appears to be associated with it
Molybdenum (Mo)	Required component of certain enzymes
Iodine (I)	Important in the function of the thyroid gland and formation of thyroxine
Fluoride (Fl)	Status in doubt Found in teeth, bone, and other tissues
Others	
Selenium (Se)	May be essential in tissue respiration as a component of the electron transfer system in cells
Chromium (Cr)	?
Vanadium (V)	Mineralization of bones and teeth

body, on the basis of weight, carbon (not oxygen) is the second most numerous atom.

Molecules in Living Systems

Water and the mineral elements make up the inorganic molecules of the body (Table 1-2), and proteins, lipids, carbohydrates, intermediates, and nucleic acids make up the organic molecules. The minerals are inorganic substances that exist as ions such as Na^+, K^+, Cl^-, Ca^{++}, and Mg^{++}. The concentration of each of these substances is regulated within very fine limits.

The Organic Molecules. PROTEINS. Proteins (Greek *proteios*, of the first rank) make up the bulk of the cell (besides water). They account for 17 percent of the body weight and 50 percent of the organic body material. They take part in many body functions. They give the organism its specific properties and are the basic structural units of cellular architecture. They control the cell's activities and give shape and form to cells and their organelles. The proteins serve structural and transport functions and include enzymes, a class of biochemical catalysts that make the many chemical processes of synthesis and breakdown of organic molecules possible. The connective tissue of the body (in-cluding the skin, hair, ligaments, and tendons) is composed mostly of protein molecules. Proteins help regulate the body's chemistry and its ability to derive energy. The ability of muscle to contract depends on "contractile" proteins. Many of the chemical messengers in the body, the hormones (including insulin), are proteins. Antibodies are proteins. Hemoglobin, the red pigment in blood, is a protein and is involved in the carrying of oxygen in the blood. These are but a few functions of proteins.

Proteins consist of up to about 24 different subunits called amino acids (Table 1-3), which are strung together in chains with linking branches and complicated folds. All amino acids have one structural property in common: an amino group (NH_2) and a carboxyl or acid group (COOH) are attached to the terminal carbon atoms of the molecule (Fig. 1-9). The R, or remainder, of the molecule can have many different molecular forms and is called the amino acid "side chain." About 20 different amino acids (20 different R groups) have been found in proteins but not all need be found in any one protein.

Protein molecules are formed by linking the amino group of one amino acid to the carboxyl group of

Table 1-3. Essential Amino Acids

Basic Amino Acids	Hydroxylic Amino Acids
1. Lysine	14. Serine
2. Hydroxylysine	15. Threonine
3. Arginine	
	Sulfur Amino Acids
Aromatic (Agreeable, Pungent Odor) Amino Acids	
	16. Cysteine
4. Phenylalanine	17. Cystine
5. Tyrosine	18. Methionine
Heterocyclic (Closed-Chain) Amino Acids	*Acidic Amino Acids*
6. Histidine	19. Aspartic acid
7. Tryptophan	20. Asparagine (β amide)
8. Proline	21. Glutamic acid
	22. Glutamine (α amide)
Aliphatic (Fatty) Amino Acids	
9. Glycine	
10. Alanine	
11. Valine	
12. Leucine	
13. Isoleucine	

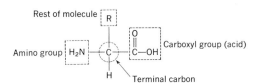

Figure 1-9. General structure of amino acids.

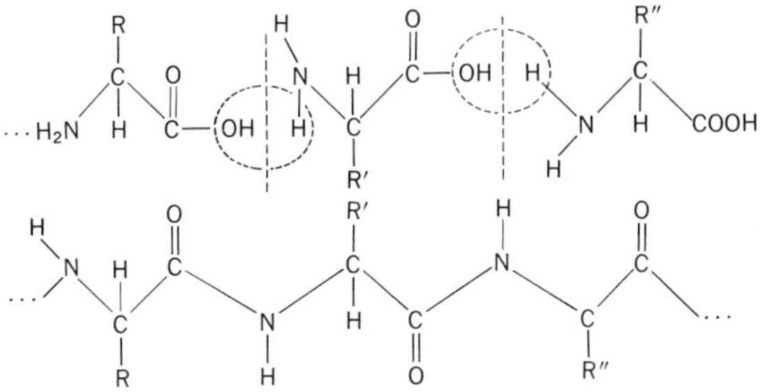

Figure 1-10. A peptide bond. Linking of the amino group of one amino acid to the carboxyl group of another amino acid.

another amino acid through a special bond, a peptide bond (Fig. 1-10). It is this bonding that forms the "backbone" of the protein. Thus a protein molecule is a chain of amino acids held together by peptide bonds. Starting with 20 amino acids an almost unlimited variety of protein molecules can be created by just rearranging the sequence and altering the total number of amino acids, which changes the physical and chemical properties of the molecule. Because the peptide bonds are regularly spaced along the protein "backbone," a regular repeating pattern of attraction occurs along the protein chain and produces a coiled configuration, the alpha helix (Fig. 1-11). Actually, however, we know relatively little about the specific structure of most protein molecules in the body and only a few have been analyzed in detail. Insulin is one of the few.

The assembly of proteins in orderly sequences is biologically regulated. This regulation of cell structure and reproduction is called a code, which passes unchanged into new cells at mitosis and is transmitted from generation to generation in an identical form. This code is packaged in chromosomes (threadlike bodies in the nucleus). Each of man's cells has 46 chromosomes, except for the sex cells, each of which has only 23 chromosomes necessitating their combination and interplay to determine the appearance and characteristics of the new generation.

Deoxyribonucleic Acid (DNA). The cell's ability to synthesize protein depends on the presence of DNA in the nucleus (Fig. 1-11). Practically all the cell's DNA is found in the nucleus. The DNA protein, deoxyribonucleic protein, has a great affinity for dyes and thus has been called a chromosome or colored body. The nucleus also contains about 10 percent of the total cellular ribonucleic acid (RNA), which is associated with the nucleoli. Apparently RNA is formed in the nucleus and passes into the cytoplasm. Thus, chromosomes are made of DNA in combination with RNA and protein.

It is the nucleic acids that serve as a cellular repository of genetic information to direct synthesis of proteins and govern the ultimate structure and function of the cell. DNA forms the pattern on which other molecules are molded since coded information built in the DNA molecules of the cell nucleus is used by the cell to synthesize proteins of all kinds.

Energy must be available and in a compact form for the above to take place. These energy-rich packages are made of a chemical substance called adenosine triphosphate (ATP). The latter easily changes to adenosine diphosphate (ADP) and, in so doing, emits energy. Most of the ATP is formed in the cytoplasmic particles called mitochondria (see p. 39). The major source of such energy is glucose, which is broken down bit by bit, and at certain specific stages, energy is created and transferred to ATP. The latter then leaves the mitochondria to go where needed.

The DNA molecule is the largest molecule in the cell and consists of a sequence of molecular subunits linked together into a long chain. The subunits are called nucleotides. Each nucleotide consists of three parts: phosphate, a sugar, and a ring-shaped organic molecule, the base (nucleoside), which contains nitrogen in addition to carbon, hydrogen, and oxygen (Fig. 1-11). The nucleotides in DNA contain deoxyribose as the sugar. Four different bases may be linked to the

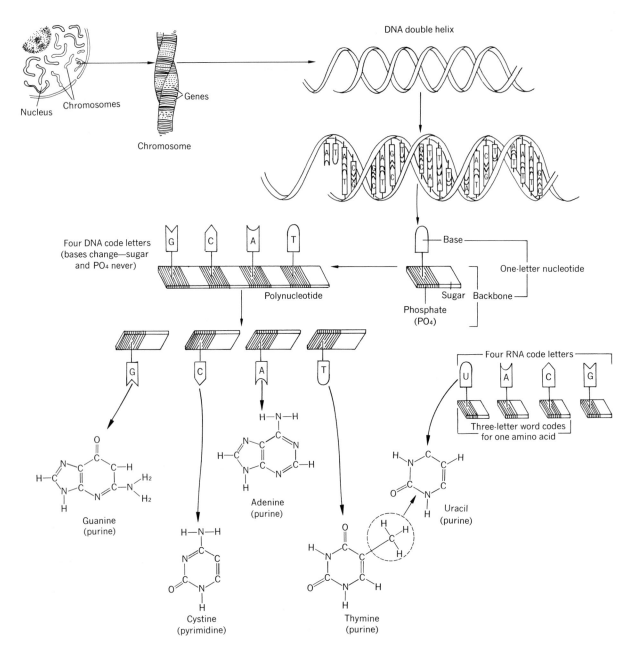

Figure 1-11. DNA passes hereditary information from generation to generation and directs the manufacture of proteins. Shaped like a spiral staircase, it spells out messages in a complex code. Each of the steps that join DNA's twin spiral consists of two complementary chemical bases, each base strand unit forming a nucleotide (single letter in the code). Three letters make up a word. When a protein is needed, this is communicated to the nuclei. Here an enzyme unwinds and separates the section, or gene, of the DNA molecule that contains the coded instructions for the protein. As the DNA strands unwind, their paired nucleotide links come apart. Other nucleotides floating freely in the cell fluid attach themselves to complementary nucleotides on one of the DNA strands to form a single-stranded molecule of messenger RNA. Once imprinted with the DNA's message, it detaches and leaves the nucleus, and the open end of the DNA molecule rewinds.

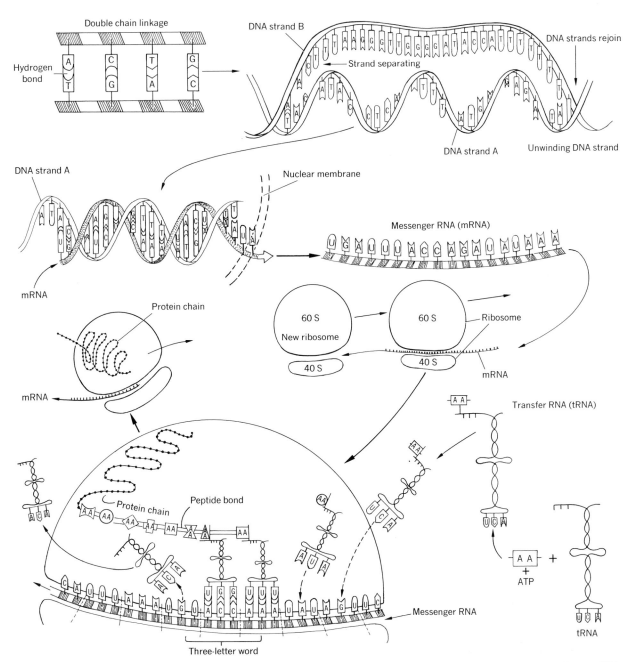

After leaving the nucleus, the mRNA is picked up by a ribosome, which runs the mRNA through itself and reads off the mRNA three-letter words, each with a specific amino acid necessary to create the protein molecule. As each is "read," another type of RNA, transfer RNA, picks up the appropriate amino acid from the cell cytoplasm and arrives with it, carrying the amino acid at one end and nucleotides at the other. The nucleotides are drawn to their complements on the mRNA, and thus the amino acids are brought to the ribosome in proper sequence to form a long protein chain. When finished, the chain is released to the cytoplasm, where it organizes into protein molecules. The free end of the mRNA is picked up by another ribosome for the formation of another protein molecule.

sugar to form four different nucleotides and these are found in two classes: purine bases containing the bases adenine or guanine (having two rings) and pyrimidine bases containing the bases cytosine or thymine (having but one ring) (Fig. 1-11). Linkages take place between a phosphate group of one nucleotide and the sugar of the next, producing a repeating pattern along the DNA molecule. Attached to each sugar is a base located to the side of the sugar-phosphate chain (Fig. 1-11).

The first three-dimensional structure of the DNA molecule was proposed by Watson and Crick in 1953 and won them together with Wilkins the 1962 Nobel Prize for Medicine. It made possible the correlation of genetic information with chemical structure in the cell, and the story of genetic coding as well as protein synthesis became known. They described a DNA molecule consisting of two chains of nucleotides, coiled about each other to form a double helix (Fig. 1-11). It is seen that a purine base is paired with a pyrimidine base (adenine is paired with thymine and guanine with cytosine). The purine-pyrimidine pairs form six ridges between the two sugar-phosphate chains, and the hydrogen bonds formed between the repeating nucleotide units produce the helical configuration of the molecule. The specific pairing of adenine with thymine and guanine with cytosine has implications for replication of the DNA molecule and allows for transfer of information from DNA to RNA during protein synthesis.

RNA resembles DNA. It too consists of nucleotides made up of phosphate, sugar, and base in repeated units, but its sugar is ribose instead of deoxyribose. The nucleotides in both DNA and RNA contain adenine, guanine, or cytosine. In RNA, however, thymine is replaced by the pyrimidine base uracil (Fig. 1-11). RNA unlike DNA has predominantly only a single chain of nucleotides rather than a double chain, and thus the purine and pyrimidine bases are not equal, as they are in DNA. There are several classes of RNA molecules with different functions.

Genetic Coding. The genetic code or language relies on the sequence of bases in DNA, for it is this order that determines the sequence of amino acids in proteins (Fig. 1-11). The code word or three-base sequence for a single amino acid is called a codon, and the codon sequence of a gene determines the sequence of amino acids in a protein. Since there are 20 different essential amino acids there are at least 20 different codons. The evidence now indicates that several different codons may specify the same amino acids. Thus, the codons CGA, CGG, CGT, and CGC all refer to the amino acid alanine. Codes in which several different code words describe the same thing are called degenerate codes. Most of the 64 codons appear to specify specific amino acids but a few perform the function of "punctuation marks" in the genetic message, merely denoting the beginning and ending of genes.

Protein Synthesis. Although the DNA molecules contain the information needed to synthesize specific proteins, they do not take part directly in assembling a protein molecule (Fig. 1-11). The molecular arrangement of the DNA molecule, however, does determine (is the template for) the arrangement of molecular subunits that link with it and relay the information to the cytoplasm since DNA molecules are found in the nucleus, yet most protein is formed in the cytoplasm.

A single DNA molecule carries information required for the synthesis of several thousand different protein molecules. This information is "spread out" into a number of separate units of the DNA molecule called genes. The gene is the functional unit of heredity and occupies a special place on a chromosome. Genes usually occur in pairs in all cells except the gametes since all chromosomes are paired except the sex chromosomes (X and Y) of the male. The gene is able to reproduce itself exactly at each cell division and can direct the formation of an enzyme or other protein. As a functional unit, it probably consists of a discrete segment of a giant DNA molecule containing the proper number of bases in the correct sequence to code the sequence of amino acids needed to form a specific peptide.

Coded information on the DNA, with the gene unit acting as a template, is transferred from DNA in the nucleus to protein-synthesizing sites (ribosomes) in the cytoplasm by molecules of RNA called messenger RNA (mRNA). The latter are synthesized on the surface of the DNA, where the sequence of DNA nucleotides determines the sequence of RNA nucleotides.

During the synthesis of mRNA, the bonds between the base-pairing DNA chains are broken and the two chains unwind and separate. Bonds are then formed between the free nucleotide triphosphates of RNA and the bases in one of the DNA chains. Thus, the base adenine of RNA pairs with the base thymine of DNA and uracil with adenine of DNA, etc. If the DNA sequence is C-G-T, the corresponding sequence of mRNA is G-C-A.

The formation of mRNA does not involve the entire DNA molecule, but only selected parts of the molecule are copied at any one time. The synthesis of mRNA on the DNA surface results in a molecule with a linear sequence of bases that is the mirror image of the base sequence in DNA. There is a one-to-one transfer of codons in the three-base sequence.

Once a molecule of mRNA is synthesized, it leaves the nucleus and enters the cytoplasm where it attaches to small particles called ribosomes and the actual process of protein formation occurs. The ribosomal particle moves along the mRNA strand, and some of the proteins in the ribosome act as enzymes to catalyze the linking procedure. The enzymes for fatty acid and steroid synthesis (as well as those for metabolizing foreign molecules such as drugs) are located in the membranes, the endoplasmic reticulum, but they do not synthesize proteins. It is the ribosomes that synthesize protein. Some ribosomes lie free in the cytoplasm whereas others are attached to membranes (see p. 36). Cells that synthesize much protein have many ribosomes. Cells that synthesize and secrete protein actively have ribosomes attached to membranes. Rapid-growing cells that make protein for themselves yet do not "export" it have many free ribosomes but few attached to membranes. The ribosomes produce proteins and release them into the lumens of the endoplasmic reticulum, from which they are eventually secreted.

Free amino acids do not have the ability to bind to particular bases on the mRNA by themselves. Orientation of the amino acids on the mRNA involves another class of RNA called transfer RNA (tRNA) made in the cytoplasm (Fig. 1-11). Thus, a free amino acid first undergoes a reaction with a molecule of tRNA, becoming bound to one end of it. There are a number of different tRNA molecules, each specific for a different amino acid. Each tRNA molecule contains, in its nucleotide sequence, a specific three-base sequence that is able to pair with a special codon in the mRNA. The binding of a specific tRNA and its specific amino acid to mRNA provides a means for arranging the amino acids in sequence. The three-base sequence in tRNA that pairs with a codon in mRNA is called an anticodon. There are 64 mRNA codons and also 64 different types of tRNA anticodons.

The amino acid combines with a molecule of tRNA in the cytoplasm helped by a specific enzyme, amino acyl synthetase. There are as many enzymes as there are amino acids. The energy used is obtained from splitting ATP.

The final interactions between mRNA, ribosomes, and tRNA occur on the surface of the ribosomal particles. Synthesis is begun by the attachment of one end of the mRNA to a ribosome particle. The various tRNAs, with their specific amino acids, then base-pair through their anticodons with the corresponding codons in mRNA. Enzymes in the ribosomes catalyze the formation of a peptide bond between two adjacent amino acids while they are attached to their tRNA molecules. With the formation of a peptide bond between the protein chain and the next amino acid, the initial tRNA is released from the mRNA and the peptide chain is transferred to the next tRNA. The ribosome now moves one codon space down the mRNA to make room for the binding of the next amino acid–tRNA molecule. This procedure is repeated over and over as each amino acid is added in succession to the growing peptide chain. When the ribosomes reach the end of the mRNA (end of a coded sequence for a single protein), the completed protein is released from the ribosome. Thus, a protein of about 150 amino acids can be synthesized in about 10 to 60 seconds.

The same strand of mRNA can be used many times to synthesize many molecules of the same protein since the mRNA is not destroyed during assembly. In fact, the same strand can be used simultaneously, for as a ribosome moves down the mRNA a new ribosome may become attached and begin synthesis following along behind the first. Therefore, one can find a number of ribosomes attached to the same strand of mRNA to form a polyribosome.

Protein Secretion. Gland cells synthesize as well as secrete protein molecules. Dense cytoplasmic granules called zymogen or secretory granules, surrounded by a membrane, are found in many types of gland cells. They contain high concentrations of protein. It is suggested that these granules are released during cellular secretion processes.

The Golgi apparatus (see p. 38) also appears to be directly involved in protein secretion from the cell. Its membranes, however, do not carry ribosomes, and it is suggested that it plays a role in the secretory process. A sequence of events suggested is that proteins to be secreted are synthesized on the surface of the ribosomes attached to the endoplasmic reticulum. They then pass into the lumen of the endoplasmic reticulum. As proteins gather here, portions of the reticulum "break away," forming small membrane-bound vesicles containing the protein. The vesicles accumulate near the Golgi apparatus, fuse with other vesicles, and thus form the "stacks" of Golgi cisternae (membranes). Once in the Golgi complex, the protein solution in the vesicles becomes concentrated and may even be modified by the addition of materials to it such as carbohydrate groups. Last, the protein leaves the Golgi in concentrated form as a secretory granule, the latter usually migrating to the cell membrane where its boundary membrane fuses with the plasma (cell) membrane and its contents are released from the cell (Fig. 1-12).

Enzymes. Among the most important of all proteins are substances called enzymes, which speed vital biochemical processes without themselves being destroyed (Table 1-4). The chemistry of our bodies is governed by the complicated work of the enzymes, of which about 10,000 are needed for the body's chemistry. Without them, we could not digest our food, breathe the air, move our muscles, or perform any of the many functions that maintain life.

Simply defined, enzymes are biologic or organic

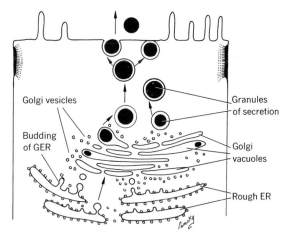

Figure 1-12. Pathway of protein secretion, with amino acids formed into proteins on ribosomes, carried to the Golgi apparatus for storage and alteration, and then secreted. Note lamellar pattern of Golgi membranes and merocrine type of secretion at cell surface.

"accelerators" whose role is to trigger vital chemical changes efficiently and quickly. They are placed in the family of chemical substances called catalysts, named such because no matter how often they take part in a chemical reaction, they themselves do not change in any way. Their very presence apparently is all that is needed to set things in motion although they make no direct contribution to the nourishment of the body.

Enzymes are large protein molecules that have unique characteristics. Each enzyme present in our body does one specific job and nothing else. They are sensitive to heat and work best at body temperature. They are readily destroyed by boiling and in both animals and plants can be found in very small amounts.

Table 1-4. Six Major Enzyme Categories

Class	Type of Reaction Catalyzed
1. Oxidoreductases	Oxidation-reduction
2. Transferases	Transfer of groups
3. Hydrolases	Hydrolysis
4. Lyases	Elimination and addition reactions
5. Isomerases	Isomerizations
6. Ligases (synthetases)	Energy-dependent condensation

Man unknowingly has had experience with enzymes since primitive times. He knew that meat allowed to age a few days was more flavorful than that eaten just after the kill. He learned also that he could make wine and cheese and beer but never suspected that catalysts found in or produced by living organisms were helping him do these things. And it remained for Louis Pasteur and others to finally explain the processes of fermentation and enzymatic action. Yeast was perhaps the first microorganism used to benefit man. The word *yeast* (Greek for zyme) is really responsible for the name now used for many of nature's biologic catalysts.

Enzymes are produced by all body cells by linking together amino acids that have penetrated the cell walls. Cells are able to take the amino acids, put them together in a correct order, and create a specific enzyme. The majority of enzymes remain in the cells in which they are formed and command the varied chemical reactions that take place in that particular cell. Each enzyme somehow appears at the exact place in the chain of events and at the precise time. Other enzymes leave the cells of their origin for specialized jobs elsewhere in the body (e.g., the digestive enzymes, which are secreted into the digestive tract, where they speed up the breakdown of complex proteins, carbohydrates, and fats into fragments and small molecules to be absorbed and used by cells of the body).

Only very small amounts of enzymes are needed to do their jobs. For example, one molecule of the enzyme catalase (found in the blood) is able to split five million molecules of hydrogen peroxide into molecules of water and oxygen in 60 seconds. Invertase, an intestinal enzyme, can break up a million times its own weight of sugar. And 30 gm (1 oz) of rennin will curdle almost three million liters of milk.

The mechanisms of enzyme action are very intricate. A theory, in simplified form, states that a specific enzyme fits into a specific molecule like a key (lock-and-key theory). Turn the key and the molecule it fits into is converted into something else. For example, the enzyme ptyalin in saliva is the key that fits the starch molecules. When we chew food that is composed of starch, the enzyme quickly converts the starch to sugar.

The enzymes do not always do their work alone. They are sometimes assisted by substances called co-enzymes, small, nonprotein molecules that give the enzymes a boost or enhance their activity. Just as important, there are inhibitors that put the brakes on the enzymes when necessary.

Over the years, scores of enzymes have been purified and their properties and activities determined, making the use of many of these enzymes possible in medicine and industry.

In medicine, their most familiar use is as "digestants," preparations taken mostly by elderly people to make up for shortages of their own peptic and pancreatic enzymes. Less well known are those used in diagnosis and treatment of disease. For example, after a suspected heart attack, tests can be made to determine whether certain enzymes known as transaminases have risen in the blood. If they are present in abnormal amounts, it is an indication that they have gotten there by seeping into the blood from injured heart cells and the physician uses this along with other information to test his diagnosis. Blood enzyme levels may also indicate the severity of the attack. Unfortunately, this test is not specific for heart attacks alone since many diseases cause the enzyme to rise in the blood. It is known that the prostate is rich in the enzyme, called acid phosphatase. If unusually large amounts of it appear in the blood, cancer of the prostate is often diagnosed. Furthermore, blood levels of this enzyme can be repeatedly taken during treatment, and if they decline this is a good indication that the cancer's growth has been retarded. Hopefully, specific enzyme tests will be perfected for many diseases including those that affect the unborn child. It is known that 30 or more defects can be detected between the twelfth and fourteenth weeks of pregnancy, by taking cells from the amniotic fluid (the liquid bathing the developing fetus) and studying them for specific enzyme deficiencies. Among the conditions that can be detected this way is cystic fibrosis (a lethal, inherited disease of white children).

Enzymes also play a role in the treatment of many diseases. They are used to clear up "black eyes," to speed healing of wounds, to reduce inflammation and fluids in local infections, to dissolve blood clots, to aid in cataract surgery, to make injections less painful and more effective, to enhance the action of drugs, and, in the case of penicillin, to fight serious reactions to the drug. The application of enzymes in wound healing apparently outranks them all and stems from the time when surgeons serving under Napoleon found that maggot-infested wounds healed more readily and with fewer complications than those that were maggot-free. It turned out that the maggot's digestive enzymes cleansed the wounds of festering material. Years later, scientists combined two highly selective enzymes, streptokinase and streptodornase, to help clear away waste material in wounds while not harming the healthy tissue. These two enzymes are derived from the very deadly streptococci bacteria. Many enzymes now being used in medicine and industry have strange origins. Some come from human urine, others from the organs of various animals, and some even from the pineapple.

The enzyme chymotrypsin, used in cataract surgery, when injected, selectively dissolves the small protein fibers (ligaments) that hold the lens of the eye in place. With the lens freed, the surgeon can remove it easily without the usual delicate cutting procedure previously used. Hyaluronidase has a liquefying effect on the "cement" or connective tissue between the cells and makes injections more effective and less painful. It can also be used with intravenous feedings to eliminate swelling or blocking near the point of needle entry. When used with certain anesthetics, hyaluronidase causes them to "spread out" rapidly, work sooner, and last longer.

Many gaps exist in our knowledge of how drugs act in our bodies. There is, however, no doubt that their function is greatly influenced by the enzymes of the body. Some drugs attain their effectiveness when changes are created in their molecular structure by body enzymes. Many and possibly most drugs would undoubtedly be too toxic for the body except for the enzymes produced by the liver. As detoxifiers, they break down compounds into inactive forms for excretion by the kidneys. An example of this is: when allergic reactions to penicillin occur, prompt use of the enzyme penicillinase will inactivate enough of the drug to stop or alleviate the reaction.

Researchers have long felt that enzyme disturbances must contribute in some way to the uncontrolled growth of cancer cells. Recently, the enzyme l-asparaginase has been used, with limited results, in the treatment of a type of leukemia. Since the growth of the leukemic cells requires the amino acid asparagine and since the enzyme destroys this acid in the body fluids, the growth of the leukemic cells is affected.

What happens if enzymes are defective or even missing, since they are essential to all life's processes? The real answer lies with the enzyme and its importance in body function. One that is missing may not be missed at all because others of close chemical identity may substitute for it. Generally, however, if something goes wrong with an enzyme a noticeable defect occurs. Examples of this are: (1) Albinism is characterized by pink eyes, white hair, acute sensitivity to light, and lack of ability to make dark pigment that gives color to the hair or skin. This is due to an inborn enzyme-gene defect and thus melanin fails to develop. (2) Two other enzyme-gene inherited defects have been examined, and science has made great strides against the inherited diseases caused by their lack, phenylketonuria (PKU) and galactosemia. PKU, the more prevalent of the two, occurs in babies born without an enzyme needed for the utilization of an amino acid found in most proteins. The acid thus accumulates in the body and damages the nervous system. If the disease is not detected early and promptly treated, severe mental retardation results. If the urine of babies is tested and the acid is found, a diet in which the amino acid is completely eliminated will enable the child to grow and develop normally in most cases. Galactosemia occurs when the baby lacks an enzyme needed to make use of milk sugar or galactose. The disease can cause serious brain damage as well as damage to the eyes and liver leading to early death or feeblemindedness and physical abnormalities (dwarfism). Tests are now available in the blood and urine to spot this amino acid and it can then be eliminated from the milk or milk products.

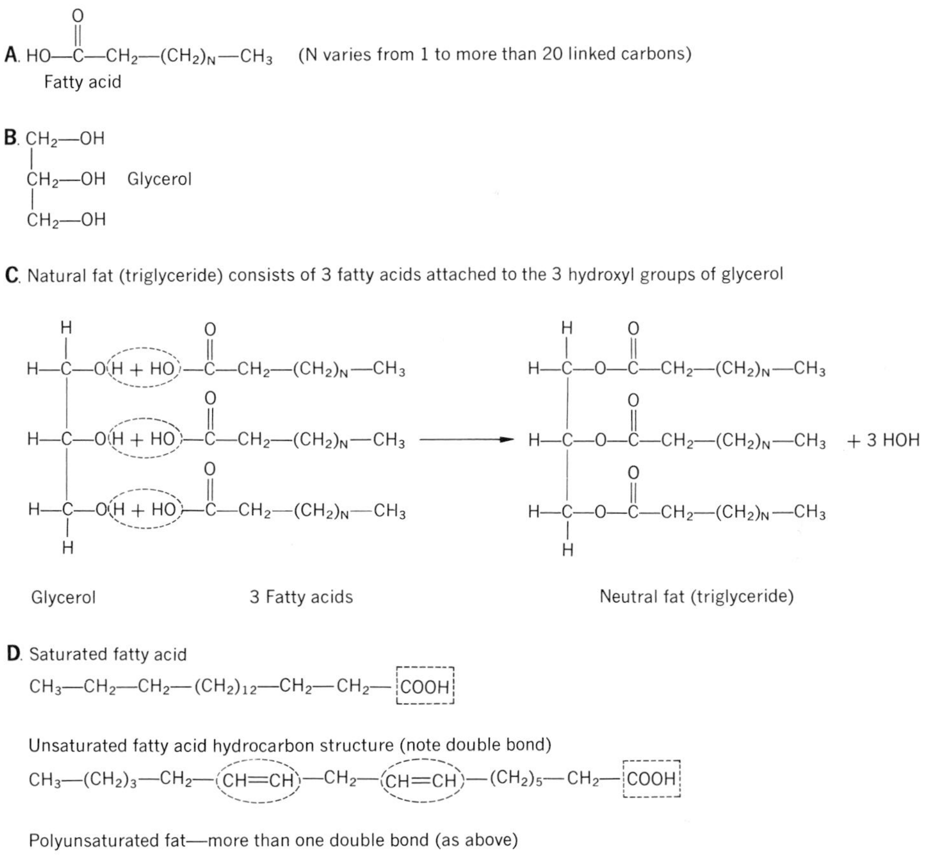

A.

$$\overset{\displaystyle O}{\underset{\displaystyle \|}{}}$$

HO—C—CH$_2$—(CH$_2$)$_N$—CH$_3$ (N varies from 1 to more than 20 linked carbons)
  Fatty acid

B. CH$_2$—OH
  |
  CH$_2$—OH   Glycerol
  |
  CH$_2$—OH

C. Natural fat (triglyceride) consists of 3 fatty acids attached to the 3 hydroxyl groups of glycerol

H—C—O(H + HO)—C—CH$_2$—(CH$_2$)$_N$—CH$_3$ H—C—O—C—CH$_2$—(CH$_2$)$_N$—CH$_3$

H—C—O(H + HO)—C—CH$_2$—(CH$_2$)$_N$—CH$_3$ ⟶   H—C—O—C—CH$_2$—(CH$_2$)$_N$—CH$_3$ + 3 HOH

H—C—O(H + HO)—C—CH$_2$—(CH$_2$)$_N$—CH$_3$ H—C—O—C—CH$_2$—(CH$_2$)$_N$—CH$_3$

  Glycerol   3 Fatty acids   Neutral fat (triglyceride)

D. Saturated fatty acid

CH$_3$—CH$_2$—CH$_2$—(CH$_2$)$_{12}$—CH$_2$—CH$_2$—[COOH]

Unsaturated fatty acid hydrocarbon structure (note double bond)

CH$_3$—(CH$_2$)$_3$—CH$_2$—(CH=CH)—CH$_2$—(CH=CH)—(CH$_2$)$_5$—CH$_2$—[COOH]

Polyunsaturated fat—more than one double bond (as above)

Figure 1-13. Neutral fats and fatty acids.

Much remains to be learned about the enzymes and their chemical function. Recently, for the first time, scientists have produced a man-made enzyme. Researchers at Rockefeller University and Merck Sharpe and Dohme Laboratories have synthesized ribonuclease, an enzyme that consists of 19 amino acids linked to form a 124-unit molecular chain and triggers many reactions essential to cell life.

Industry, too, benefits from enzymes even more extensively than medicine since these are used to produce better textiles, improve quality of leather, do away with sediments that cloud beer and wine, and give better texture to bread as well as keeping qualities, to name a few.

LIPIDS. Lipids (Greek *lipos*, fat) are molecules relatively insoluble in water and make up about 15 percent of the total body weight. The fats belong to this class. In fact, the majority of lipids in the body are fats. Fats, however, make up only one of several subclasses of lipid molecules. Lipids are composed largely of hydrogen and carbon. The simplest are hydrocarbons. Few simple hydrocarbons exist in living organisms. Oils and gasoline are examples of mixtures of hydrocarbons.

Lipids in the body exist in three subclasses: neutral fats, the phospholipids, and steroids. All are relatively insoluble in water.

Neutral Fats. These constitute the majority of lipids in the body. The base component of the neutral fat molecule consists of a straight-chain hydrocarbon with a carboxyl group at one end and is known as a fatty acid (Fig. 1-13). Neutral fats are synthesized in the body by linking two-carbon fragments, and thus, most fatty acids have an even number of carbon atoms. A variable is the presence or absence of double bonds. When a fatty acid contains a double bond, it is said to be unsaturated. If more than one double bond is present, it is said to be polyunsaturated (Fig. 1-13). Animal fat contains a high proportion of saturated fatty acids, and vegetable fat contains more polyunsaturated fatty acids. There is some evidence linking saturated fatty acids to fatty deposits in the blood vessels.

In addition to fatty acids, neutral fat molecules contain *glycerol*, a three-carbon molecule to which fatty acids are attached (Fig. 1-13). Glycerol itself is not a lipid. Thus, neutral fat (triglyceride) consists of three fatty acids attached to three hydroxyl groups of glycerol.

Neutral fat functions as a reservoir of potential energy. If one eats too much, the excess is stored as neutral fat. If needed, the reserve can be called on to release energy. The fat deposits of the body also provide thermal insulation to protect against cold.

Phospholipids. These are similar to neutral fat but

Figure 1-14. Phospholipid (lecithin).

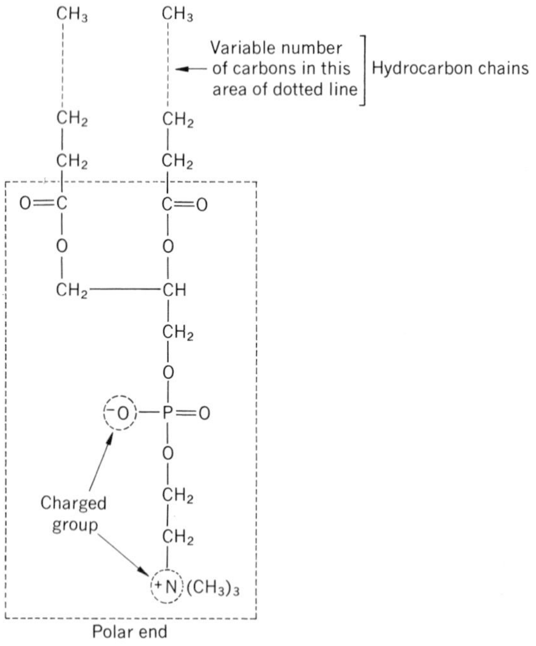

CH₃ ··· CH₃
Variable number of carbons in this area of dotted line ← Hydrocarbon chains
CH₂   CH₂
CH₂   CH₂
O=C   C=O
O   O
CH₂———CH
       CH₂
       O
(O)—P=O
       O
Charged group
       CH₂
       CH₂
(⁺N)(CH₃)₃
Polar end

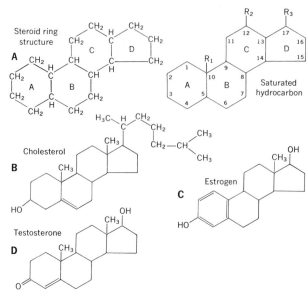

Figure 1-15. Carbon ring structure of various steroids. In female sex hormones the A ring is aromatic and R_1 is absent at C_{10}. Estrogens and androgens are devoid of carbon atoms at R_3. The R_3 substituent containing two carbons is characteristic of adrenocortical steroids and progesterone. Bile acids have a five-carbon unit in the R_3 position, whereas sterols have eight, nine, or ten carbon atoms here. Unsaturation usually occurs between C_4 and C_5.

contain only two fatty acids. The third hydroxyl group of the glycerol is linked to a phosphate group (Fig. 1-14). A small nitrogen-containing group is also usually attached to the latter. Phospholipids form one of the major components of cell membranes, the hydrocarbon part permitting the membrane to function as a barrier between two water compartments (see p. 30).

Steroids. Steroids (Greek word meaning solid) consist of four interconnecting rings of carbon atoms (Fig. 1-15). Few polar groups are attached to the "steroid ring" making it insoluble in water. The steroids include such molecules as cholesterol (a component in cell membranes), hormones (male and female sex hormones, testosterone and estrogen, respectively), and bile salts, which aid in digestion.

CARBOHYDRATES. Carbohydrates (Fig. 1-16) account for only 1 percent of the total body weight. Most of the chemical energy used by the cell is obtained by

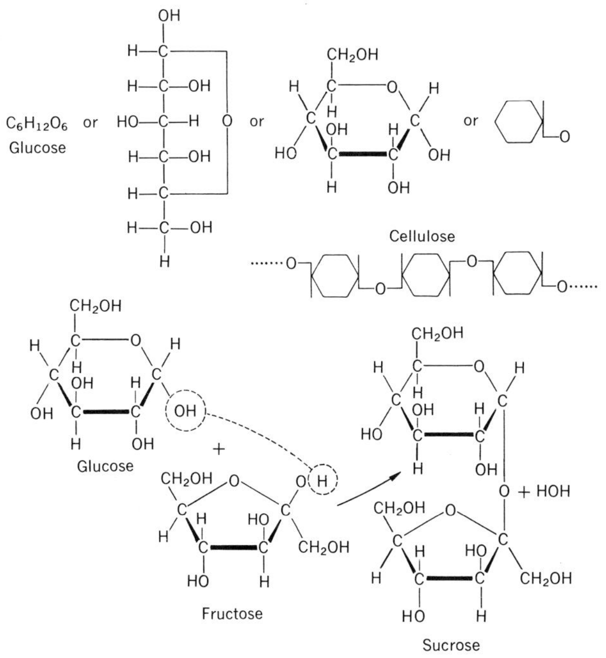

Figure 1-16. Carbohydrates.

the breakdown of the carbohydrate molecule into water and carbon dioxide. Carbohydrates are the most readily available source of chemical energy.

Carbohydrates are hydrated (contain water) carbon chains. The simplest carbohydrates are the sugars, e.g., glucose ($C_6H_{12}O_6$). Sugars are water-soluble. Larger carbohydrate molecules can be formed by linking sugars together. Sucrose (table sugar), a disaccharide, is a combination of the monosaccharides glucose and fructose. When more than two monosaccharides are linked, we call it a polysaccharide. The most important polysaccharides are starch, glycogen, and cellulose, which are composed of thousands of repeating units of the monosaccharide glucose. They differ in the way the glucoses are linked. Cellulose is found in plants. Man cannot use it as an energy source in food because his digestive tract cannot break it up. Yet starch, also of plant origin, is digestible. Glycogen can be stored in the body for future energy needs.

INTERMEDIATES. In a cell, molecules of carbohydrate, lipid, and protein are continually undergoing reactions that break them down into smaller units and new molecules are being created. These chemical reactions are referred to as metabolism (Greek word meaning change). The so-called intermediates represent the great variety of molecules formed during synthesis and breakdown of the many body molecules.

NUCLEIC ACIDS. Nucleic acids (Fig. 1-11) determine whether a cell is a muscle cell or liver cell or skin cell, etc. The nucleic acids, as has been shown, are the molecules that carry the genetic information and provide the "blueprints" for the organism. They are of two types: deoxyribonucleic acid (DNA) and ribonucleic acid (RNA). The former has the primary genetic information coded in its molecular structure; the latter serves in transcribing the information contained in the DNA molecule into a form that the cell can use to build structures to carry out its functions.

The Inorganic Molecules. The major inorganic molecules and their function are summarized in Table 1-2. Further discussion is also found in other chapters.

Cellular Energy

The functioning of a cell depends on its ability to extract and use the chemical potential energy bound up in the structure of its organic molecules. The cell obtains energy from reactions that degrade and release energy by coupling them to energy-carrier molecules. The major energy-carrier molecule in man is adenosine triphosphate (ATP) consisting of three parts: a chemical ring called adenine (composed of carbon, hydrogen, and nitrogen), a five-carbon sugar molecule called ribose, and three phosphate groups linked in series to the sugar (Fig. 1-17). When the bond between the terminal phosphorus atom and oxygen is broken in the presence of water, 7 kcal per mole is released, and the product is adenosine diphosphate (ADP) and inorganic phosphate. In order to synthesize ATP from ADP and inorganic phosphate, 7 kcal per mole must be added to the molecules. Thus, energy has been added to make ATP and is released from it in the process of transferring phosphate from one molecule to another.

The energy stored in ATP is used to perform work by the cell: mechanical (contraction of muscle), active

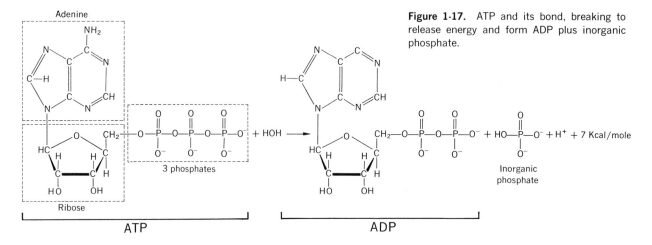

Figure 1-17. ATP and its bond, breaking to release energy and form ADP plus inorganic phosphate.

transport of molecules across cell membranes and assembly or chemical biosynthesis of nucleic acids, proteins, etc. (Fig. 1-18). Energy, too, is continually cycled through ATP molecules in the cell. The use of ATP provides the cell with flexibility. The energy released in the breakdown of organic molecules can be coupled to the synthesis of ATP, which is then used by the cell for energy. Furthermore, the cell depends on energy in the form of ATP, but the source of this supply may

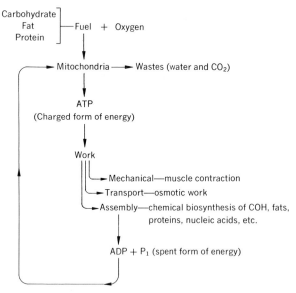

Figure 1-18. Cell energy cycle (use of ATP).

vary and the cell may utilize not only glucose but lipids or proteins as well.

Most of the ATP (95 percent) is synthesized in the cell's mitochondria from glucose breakdown and 100 percent of that from fatty acid breakdown. The total number of mitochondria in a cell varies with the size and energy expenditure of the cell. The ATP synthesis in the mitochondria occurs via a highly specialized coupling process called oxidative phosphorylation. The latter requires molecular oxygen and consists of the addition of free inorganic phosphate to ADP to form ATP. ATP usually cannot be synthesized by this method without oxygen,° and thus in the absence of oxygen, the cells die since they cannot synthesize ATP to meet energy needs.

In the breakdown of organic molecules, many reactions involve the removal of hydrogen atoms from intermediate steps of the degradation. These reactions need a coenzyme molecule such as nicotinamide adenine dinucleotide (NAD) to which the hydrogen atoms are transferred: $BH_2 + NAD \longrightarrow B + NADH_2$. Thus, the potential energy in the BH_2 molecule is transferred to the $NADH_2$ along with two atoms of hydrogen. As a result of similar shifts, much of the chemical energy in molecules of fat, protein, and carbohydrate is transferred to a coenzyme molecule like $NADH_2$. Oxidative phosphorylation uses this energy

° There is some anaerobic production of ATP.

to synthesize ATP. Thus, energy is obtained from many sources and transferred to the common energy carrier, ATP.

In order to transfer the energy in $NADH_2$ to ATP, a complex series of reactions takes place in the membranes of the mitochondria where the potential energy of $NADH_2$ is released in small amounts associated with the transfer of electrons from hydrogen to oxygen through a series of proteins called cytochromes. The latter are found in the inner mitochondrial membranes. When electrons are transferred from $NADH_2$ to a cytochrome, the total reaction releases a small amount of energy (1 to 12 kcal per mole):

$$NADH_2 + 2CyTFe^{3+}(\text{cytochrome}) \longrightarrow NAD + 2H^+ + 2CyTFe^{++} + \text{energy}$$

The energy gained by the cytochrome molecule from the transfer of electrons from the hydrogen carrier can now be given to a second cytochrome by passing on the electrons obtained from hydrogen and returning the original cytochrome to its original low-energy state (Fig. 1-19A). Thus, electrons pass through a series of cytochromes, which eventually transfer the electrons to an oxygen atom producing ionized oxygen, which combines with hydrogen ions to form water, a process of electron transport that, with each carrier cytochrome, releases small amounts of energy (Fig. 1-19B). If $NADH_2$ reacted directly with oxygen ($NADH_2 + \frac{1}{2}O_2 \longrightarrow NAD + HOH + 52$ kcal per mole), all calories would be released as heat and not be available for ATP synthesis. A cell cannot use heat to do work since it functions at a low and constant temperature.

Figure 1-19. **A.** Changes in chemical potential energy accompanying electron transfer from H_2 to cytochromes and cytochrome to cytochrome. These are the major steps of mitochondrial electron transfer and associated oxidative phosphorylation. Each component in the chain receives electrons from previous components and passes them on to the next. In the case of cytochromes, the receipt of electrons reduces iron ions in the molecules to Fe^{2+}, and the subsequent loss of energy (e^-) to the next component regenerates the oxidized Fe^{3+} form. **B.** Transfer of electrons from $NADH_2$ through the cytochromes with terminal transfer of electrons to an atom of oxygen (electron transport system). Results in NAD + HOH + 52 kcal per mole.

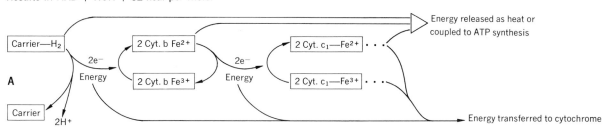

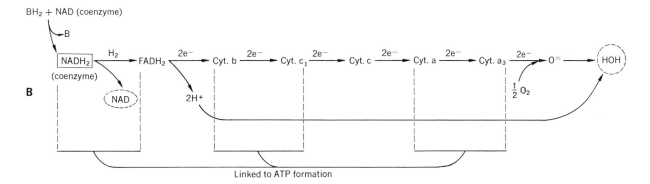

A cell's wastes are CO_2 and HOH, and the chemical energy of the fuels is repackaged as chemical energy and not dissipated as heat. Taken through the transport cytochrome system, the energy is released in small steps along with electron transfer and is coupled to the synthesis of ATP.

The transfer of energy from the cytochromes to ATP involves other reactions (see any standard textbook of biochemistry for details). The cytochromes and enzymes involved in oxidative phosphorylation are organized into structured units built into the membranes of the mitochondria. The proteins are bound to the membrane in a pattern that allows electrons to be transferred easily from molecule to molecule. A single structural unit with all the cytochromes and enzymes is called a respiratory assembly, and each mitochondria has about 15,000 such units in its cristae.

Figure 1-20. Emden-Meyerhoff glycolytic pathway. Reactions are numbered to correspond to individual steps. This summarizes the reactions that take place in the cells and is probably the major route for glucose metabolism by animal muscle tissue. Glucose, a six-carbon molecule, is broken down into two molecules of three-carbon lactate. In the process two molecules of ATP are used up but four are formed, and thus there is a net gain of two molecules of ATP.

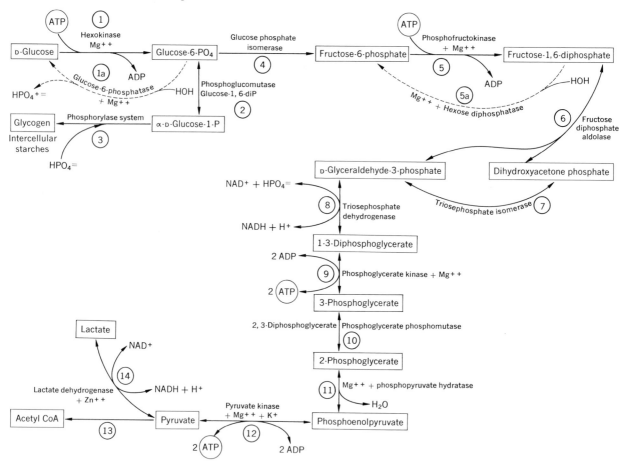

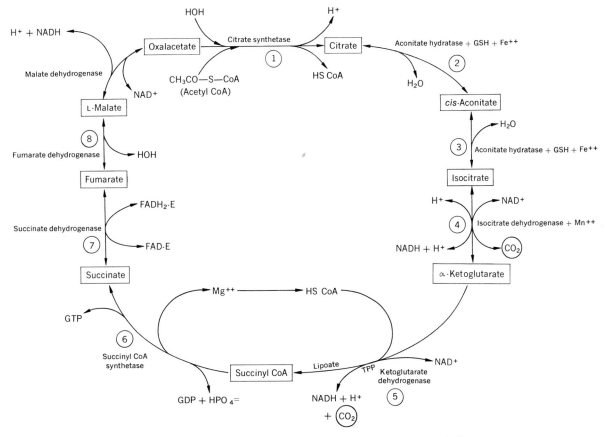

Figure 1-21. Citric acid (tricarboxylic acid or Krebs) cycle, a universal pathway for aerobic metabolism. It is a cycle because it generates CO_2, carries out oxidations and reductions, and continually restores the necessary intermediates—all at the expense of acetyl CoA. It is coupled with electron transport and oxidative phosphorylation systems, in most cells, and yields energy in the form of ATP. This cycle occurs in mitochondria, and certain of the enzymes are localized in the outer membrane, which controls mitochondrial permeability.

Pathways of Metabolism

The total metabolism of a cell can be divided into a number of metabolic pathways (sequence of enzyme-mediated reactions leading to the formation of a particular product) involving the major classes of molecules: carbohydrates, lipids, proteins, and nucleic acids. The metabolism of a cell is further divided according to location of enzymes in the cell, and many metabolic pathways and their enzymes are confined to specific organelles. As indicated, carbohydrates are most intimately related to the generation of ATP, but cells can couple the breakdown of lipids and proteins to synthesize ATP by converting them into intermediate compounds that then enter the carbohydrate sequence of reactions (Figs. 1-20, 1-21, 1-22). Cells also synthesize carbohydrates, lipids, and proteins in addition to using their chemical energy to synthesize ATP. (For a detailed description of metabolic pathways see Chap. 13. The reader is also referred to any standard textbook of biochemistry.)

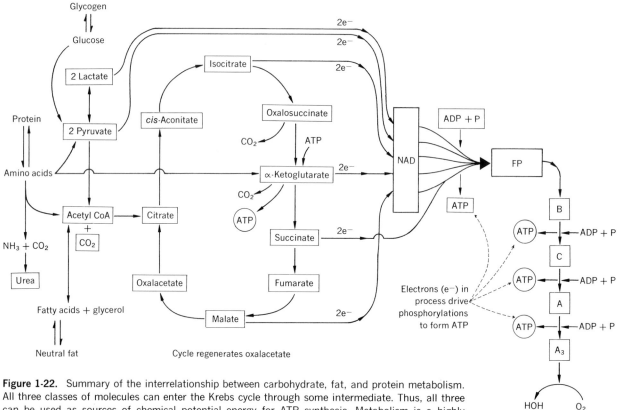

Figure 1-22. Summary of the interrelationship between carbohydrate, fat, and protein metabolism. All three classes of molecules can enter the Krebs cycle through some intermediate. Thus, all three can be used as sources of chemical potential energy for ATP synthesis. Metabolism is a highly integrated process in which all classes of molecules are used to provide energy for the cell through ATP synthesis and each class of molecules can provide the raw materials for synthesis. Thirty-six molecules of ATP are produced by two molecules of lactate formed from the original glucose molecule. The electrons removed (right side of figure) pass down a respiratory chain of electron carriers. NAD is nicotinamide adenine dinucleotide, and FP is flavoprotein enzyme. B, C, A, and A_3 are iron-containing enzymes that ultimately reduce O_2 to HOH.

A molecule of ATP in an active cell is recycled every 50 seconds. As a charged form of energy, ATP can yield its energy to other parts of the cell for work, such as mechanical, for muscle contraction; transportation, for movement; and assembly (biosynthesis and structure). Energy is used and ATP is degraded into ADP and inorganic phosphorus (P_1), which are spent forms of energy. To be recharged, they pass back into the mitochondria and emerge again as ATP to be recycled every 50 seconds.

Biochemistry has, thus, provided us with information about the chemical pathways in the cell but electron microscopy has identified the organizational cellular morphology.

Cellular Morphology

Cell Surface (Areas of Exchange)

Studies have confirmed the existence of a plasma membrane (outer cellular membrane) of about 60 to 100 Å (0.01 μm) thick (Figs. 1-1, 1-7, 1-23, 1-30). It not only acts as a barrier between the cell's contents and

Figure 1-23. Model of cell membrane with pores "lined" with unrolled protein. Attraction between polar areas of protein keeps pores expanded. Pores facilitate diffusion and active transport. The lipids serve to isolate the cell from its environment, with the liquid lipid bilayer acting as a two-dimensional solvent for macromolecules. The proteins serve to reinforce the lipid bilayer, insulating it from external physical and chemical stresses, promoting selective interchange with the environment, and providing polar pathways through the thickness of the membrane. (After studies by Danielli and Davson.)

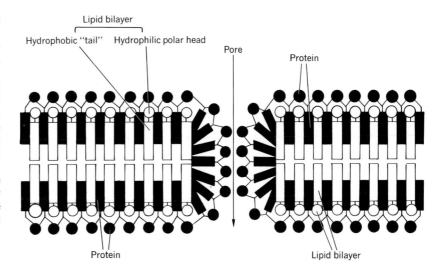

its surrounding environment but also serves as a "sieve" to allow select molecules to pass in and out of the cell. It permits intake of nourishment and oxygen from the cell's surroundings and allows removal of waste materials that have accumulated in the cell. It has been described as being made up of layers of complex molecules particularly a lipoprotein. In general, most membranes share the following common features. The protein portion gives the cell flexibility, wettability, elasticity, strength, and communication between cell and environment, whereas the phospholipid assists in the passage of both water- and fat-soluble substances. The phosphate end of the phospholipid molecule is electrically charged and admits water-soluble substances. The purely lipid or "fatty" end is uncharged and accepts fat-soluble compounds. Plasma membranes, which have a high degree of enzymatic activity, show approximately an equal amount of protein and lipid. Intracellular organelles, on the other hand, such as mitochondria or bacterial membranes have a protein content as high as 75 to 80 percent enabling them to conduct a considerable amount of biochemical activity on their surfaces.

It is thought by some that defective membranes, as seen in some muscle and nervous diseases, play an essential role in malignancy and offer a potential theory for the appearance of aberrant cells. Membrane function can be altered by radiation, chemical carcinogens, mutant or aberrant genes (that can code for inappropriate membrane constituents), and viruses that may transmit directly or indirectly through cellular genes. It is the membrane of cellular surface that makes the cell what it is and tells it where to go—like a computer directing cell travel and recognition. Altered membranes, due to surface changes, can interfere with normal cell function and disrupt the transport of nutrients across them altering cellular volume.

The cell must obtain substances from without and must discharge or eliminate either waste products or useful nutrients into an environment of tissue fluids that surround it. The environment contains the needs of the cell at a concentration adjusted before they reach the cell, and yet the concentration in the tissue fluids, at the cellular level, varies continuously. The cell membrane acts as a selective barrier as well as an agent of active export and import. The materials that arrive at the cell membrane, or leave it, do so in solution.

Since the materials that arrive at the cell membrane, or leave it, are in solution, they present a problem in their passage across the cell membrane. They accomplish their passage usually by diffusion—the movement of a substance from a region of high solute concentration to one of lower solute concentration until the

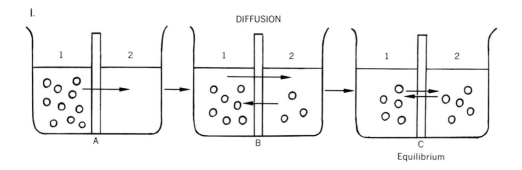

I.

DIFFUSION

1   2   1   2   1   2

A   B   C
Equilibrium

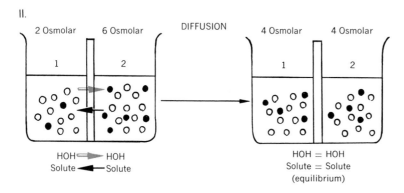

II.

2 Osmolar   6 Osmolar   DIFFUSION   4 Osmolar   4 Osmolar

1   2   1   2

HOH ⟹ HOH   HOH = HOH
Solute ⟸ Solute   Solute = Solute
               (equilibrium)

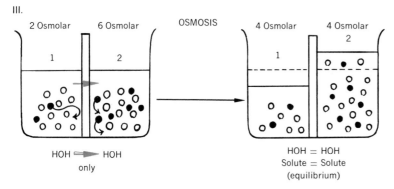

III.

2 Osmolar   6 Osmolar   OSMOSIS   4 Osmolar   4 Osmolar

1   2   1   2

HOH ⟹ HOH   HOH = HOH
only   Solute = Solute
               (equilibrium)

Figure 1-24. I. Diffusion of glucose between two compartments of equal volume. **A.** No glucose in one compartment. **B.** Some glucose moves across and some randomly back. **C.** Diffusion equilibrium with movement back and forth equally. **II.** Diffusion where separating membrane is permeable to water and solute. **III.** Only water can diffuse across this membrane separating two compartments of different solute concentration. Water moves from higher concentration of water to lower concentration of water; however, since solute does not move, the volume of compartment 2 increases. The loss of water raises the solute concentration in compartment 1 until equilibrium is reached.

final concentration is the same in both regions (Fig. 1-24).

In the body, gases and some solutes cannot enter cells independently of the water in which they are dissolved. The greater the difference between the two concentrations (a difference called the concentration gradient), the faster is the diffusion. Although this is a passive, automatic occurrence, it is very important and of practical value to the cell. Human cells live at a rapid pace and continuously pump ions (electrically charged particles) in and out of the cell against a concentration gradient. They transport them across the cell membrane usually with the help of a protein carrier and ATP. Furthermore, cells use up a solute quickly and must replace it before there is any cellular damage.

When a cell membrane permits things to pass through by diffusion, it is said to be permeable. Only a few substances enter or leave the cells by diffusion alone (oxygen and CO_2 pass through by diffusion and some waste materials leave the cells in that fashion). The cell often needs more of a substance whose concentration inside the cell is higher than outside, and the cell may need to remove a substance under the same circumstances: namely, against the concentration gradient. Clearly, this must involve an active process. The cell membrane must "work" to force materials into and out of the cell. This form of movement is called active transport or active absorption.

The tissue fluid that bathes all the different cells in the body contains approximately the same solutes wherever it may be in the body and yet different parts of the body need different substances in varying amounts. For example, each gland has special requirements. The need of muscle cells differs from that of nerve cells and the liver operates at a different tempo than does connective tissue and so on. Each cell membrane not only moves materials through against a gradient but must also be selective of the particular solute it needs at any one time.

Solutes are one problem, but the flow of water must also be kept in mind. This involves a process called osmosis—the diffusion of only the solvent through a semipermeable membrane (semipermeable because it lets through the water but not the large solute molecules). If, for example, there is more sugar inside the cell than in the tissue fluid and more water in the fluid than in the cell, then water cannot help but diffuse into the cell. The sugar, on the other hand, attempts to leave the cell but cannot because its molecules are too large to pass through the membrane. Osmosis is the diffusion of only solvent through a semipermeable membrane, and due to this process, most cells take up water passively.

With diffusion, the strength of the solution is important. The greater this is, the more water will tend to enter the cell. The force with which it does so is referred to as osmotic pressure. If too much enters, the cell bursts, and if too much leaves, the cell collapses (Fig. 1-25). Thus, the body spends much energy to prevent osmotic gradients from occurring between the cells and the tissue fluids.

Some cells have another way of taking in materials, acting like the ameba. The membrane engulfs fluids and good-sized particles. These substances are enclosed in an expandable portion of the cell membrane (engulfed by an involution of the membrane) and a vacuole is formed. Gradually the membrane closes over the pocket, and the vacuole separates and travels into the cell's interior. The nutrients then pass through the vacuole membrane or it dissolves. This process is called pinocytosis (Greek word meaning drink + cell) and occurs in part of the intestine, in the kidneys, in some white blood cells, and possibly elsewhere (Fig. 1-26). On the other hand, the cytoplasm may engulf "solid material" and take it inside into the cytoplasm where the membranes break down (in bacteria) and enzymes digest the particles. This is called phagocytosis (Greek *phagem*, to eat).

The membrane then can be permeable, semiperme-

Figure 1-25. Comparison of red blood cell in solutions of variable concentrations of NaCl. The volume of the red cell is surrounded by body fluids.

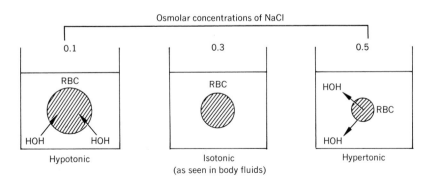

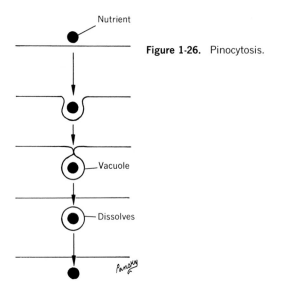

Nutrient

Figure 1-26. Pinocytosis.

Vacuole

Dissolves

both osmosis and diffusion and work selectively. How this is done has not been fully explained. We do know, however, that although the membrane is very thin, it is complex and a dynamic region of the cell and alters its properties from time to time and somehow manages to function and maintain the body's normal (and often abnormal) activities.

Since long and complex protein molecules can fold or unfold, the membrane can expand or contract and provide, by molecular spacing, a means of control over the molecules that pass in and out through this membrane—selective permeability. The membrane also permits growth and movement for the cell as a whole or selective parts of the cell. The lipid portion of the membrane is important since it allows fat solvents to readily penetrate the cell from its environment, but neither the structure nor the functional role of these lipids is completely understood. The surface molecular structure also provides a surface on which specific cellular reactions can occur.

able, and selective. Diffusion requires that the membrane let through small solutes, but osmosis prevents solutes from passing at all. Furthermore, as a result of both active and passive transport processes taking place at the same time, the membrane may "inhibit"

The membrane is not a simple wrapping surrounding the cell but its contour can be modified in specialized cells. In absorption in the digestive tract, the membrane is often greatly convoluted, and there

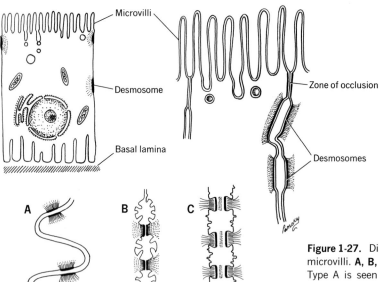

Microvilli

Desmosome

Basal lamina

Zone of occlusion

Desmosomes

A B C

Figure 1-27. Diagrammatic representations of desmosomes and microvilli. **A, B,** and **C** represent various forms of desmosomes. Type A is seen in ependymal cells; type B, in liver cells; type C, in flat epithelium. Larger drawing is typical of intestinal epithelium.

are regions of intimate connections with other cells called the desmosomes (Fig. 1-27). The upper surfaces can form a so-called brush border (Fig. 1-27) in which microvilli (projections) provide an absorptive area. Each cell may have up to 3000 individual microvilli and a square millimeter of the intestine could have about 200 million. The relationship between nerve cell and associated Schwann or satellite cell is another example where the plasma membrane is very elaborate (Fig. 1-28). Here the cytoplasm of the Schwann cell is squeezed to the outside leaving the axon surrounded by a multilayered membrane system to protect the nerve and assist in nerve impulse transmission as well as nutrition of the long axon.

Thus, the membrane of the cell is a major portion of the living cell and is intimately connected with its other internal systems. It has the ability of limited repair if torn and the capability of mobility and engulfing materials. It is elastic, changeable, pliable in certain cells and rigid and unyielding in others. It

Figure 1-28. A. Peripheral nonmyelinated nerve fiber. **B.** Peripheral myelinated nerve fiber. **C.** Formation of the myelin sheath. **D.** Relation of a Schwann cell to several unmyelinated fibers. **E.** Myelin sheath, neurolemmal sheath, and node of Ranvier (as seen in electron micrograph). **F.** Closeup of the myelin sheath.

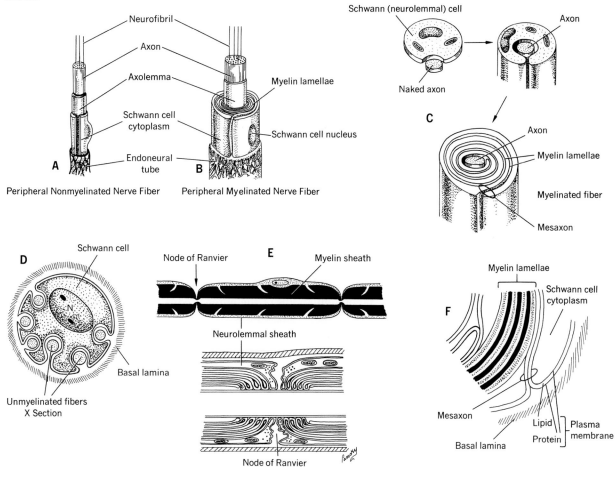

varies in thickness and its surface texture is variable (smooth in ameba, ciliated in paramecium). Its surface differs physiologically in different cells. Furthermore, cells from the same tissue have the ability to recognize one another when in a liquid suspension, and thus, the cell membranes possess discrete properties of movement and adhesiveness and yet their appearance is often deceptively similar morphologically under even the electron microscope.

Endoplasmic Reticulum

Submerged in the fluids of the cell and seen only as stringy structures under the light microscope are a maze of segmented microtunnels (hollow ducts) known as the endoplasmic reticulum (Fig. 1-1). In most cells it exists as an important part of the cell for the manufacturing of cellular products and is designed to carry materials from one part of the cell to the next. This is a membrane-limited saclike system enclosing interconnecting cavities or cisternae extending from the nuclear membrane (on the inside) to the plasma membrane (on the outside) of the cell. It is a part or extension of the outer membrane of the nucleus. The endoplasmic reticulum also provides the cell with an increased surface area, and if enzymes are part of the membrane systems, this allows for local patterns of synthesis. Although each cell has characteristic endoplasmic reticulum, it has a variable morphology and can alter its nature rapidly. It can be loosely arranged or tightly packed and can be either rough (granular) or smooth (agranular).

Rough or Granular Type. The rough or granular type of endoplasmic reticulum (Fig. 1-29) is found in great abundance in cells and is involved in protein synthesis. This capacity rests in the electron dense particles, the ribosomes that cover the endoplasmic reticulum and are rich in RNA (the site of cytoplasmic protein synthesis). The ribosomes are approximately 150 Å (less than a millionth of an inch) in diameter and are distinctive, small, dense granules. They are fundamental subcellular particles and are seen in every type of cell. They may exist in groups or singly in the cytoplasm, often arranged in clusters or rosettes and even spirals. The nucleolus contains an abundant amount of similar particles and they are also seen in the nucleus. Their name reflects the presence in the granules of a significant amount of RNA.

The granular endoplasmic reticulum is a dynamic system. Connections between different cisternae and between the endoplasmic reticulum and the perinuclear cisterna are thought to be continuously made and broken due to the activity of the cytoplasm. Areas of organized granular endoplasmic reticulum are sometimes called different things in different cells, for example, Nissl bodies in nerve cells. Biochemical studies of fractionated (broken-up) pancreatic cells show that protein synthesis is carried out by the so-called microsome fraction of the cell whose composition is mostly ribosomes and fragments of membranes of endoplasmic reticulum. However, since ribosomes may be free in the cytoplasm, they need not be associated with the endoplasmic reticulum (e.g., bacteria have little to no endoplasmic reticulum and yet are rich in ribosomes). The association of ribosomes with membranes does provide the cell with a means of compartmentalizing specific chemical reactions. Thus, the ribosomes, though looking alike, must differ in protein-forming capacity since each cell must be able to produce a variety of protein materials for structure and function.

Smooth or Agranular Type. The smooth endoplasmic reticulum (Fig. 1-29) is prevalent in cells that have membrane cisternae arranged in patterns, related to those of the granular reticulum, but have no associated ribosomes and appear to be engaged in synthesis of fatty substances (lipids, as in sebaceous glands, or steroid hormones in certain endocrine glands). The cells of the adrenal cortex and interstitial cells of the testis both have a well-developed smooth endoplasmic reticulum and both share the function of steroid hormone secretion. Under normal circumstances, liver cells have a variable amount of smooth endoplasmic reticulum (which can be increased by giving barbiturate drugs and many other substances that are detoxified by liver enzyme activity). The smooth endoplasmic reticulum may be associated with glycogen metabolism in liver cells, since there is an apparent close topographic relationship between smooth endo-

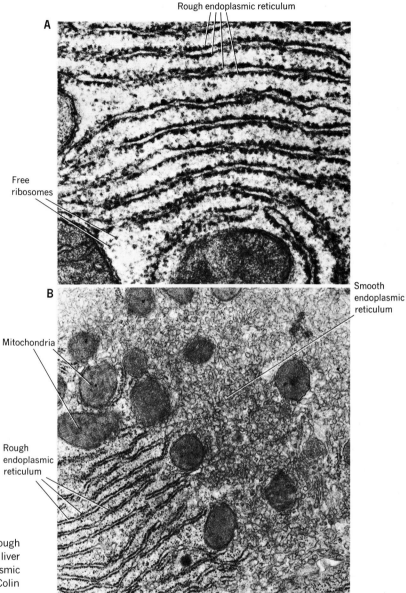

Rough endoplasmic reticulum

A

Free
ribosomes

B

Smooth
endoplasmic
reticulum

Mitochondria

Rough
endoplasmic
reticulum

Figure 1-29. **A.** Rat liver hepatocyte. Rough endoplasmic reticulum. ×55,500. **B.** Rat liver hepatocyte. Rough and smooth endoplasmic reticulum. ×20,000. (Courtesy of Dr. G. Colin Budd, Medical College of Ohio at Toledo.)

plasmic reticulum and glycogen granules. The amount of smooth endoplasmic reticulum in the intestinal absorptive cell may vary, but it is regular and suggests a possible metabolic function relating to digestion or absorption. An extreme specialization of smooth endo-

plasmic reticulum is seen in skeletal and cardiac muscle, where the smooth endoplasmic reticulum membranes are referred to as sarcoplasmic reticulum and may link membrane excitation of the muscle cell to the myofibril contraction. Although there is no sharp morphologic discontinuity between rough and smooth endoplasmic reticulum, it is more common for the two types to occupy different cell areas.

Thus, the endoplasmic reticulum makes surfaces available for chemical reactions, allows pathways for transport from one area of the cell to the other, and creates collection depots for synthesizing materials.

Golgi Apparatus

Another cellular unit seen in the cytoplasm is the Golgi apparatus or complex (Figs. 1-1, 1-30), named after Camillo Golgi, an Italian physicist who discovered it about 80 years ago.

Morphology. From electron microscopy, the Golgi apparatus appears to be a membrane system that forms a distinct structural component of the cytoplasm. It consists of four to five flat, hollow disks that can have

Figure 1-30. Mouse pituitary Golgi complex (arrow). ×22,500. (Courtesy of Dr. G. Colin Budd, Medical College of Ohio at Toledo.)

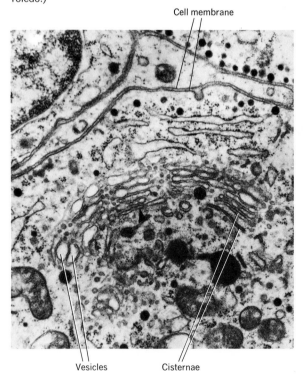

Cell membrane

Vesicles   Cisternae

swollen or bulbous edges and are piled like "saucers" one above the other. Its three-dimensional nature is not seen in electron microscopy since its interconnecting strands appear as isolated units. The Golgi apparatus generally lies at the apical pole of the nucleus occupying an oval or horseshoe-shaped area of cytoplasm. Its position and morphology appear to be fairly constant in a cell type. It is small in muscle cells and lymphocytes but well developed in goblet cells, secretory cells of the exocrine pancreas, and absorptive cells of the intestine.

Three major parts forming the Golgi are membrane-bound cisternae, dilated membrane-bound vacuoles, and small vesicles (Fig. 1-30).

The membranes often appear slightly thicker than the endoplasmic reticulum and the usual trilaminar structure is present. The cisternae, which can make up most of the complex, form closed sacs with opposing walls lying close together or widely separated.

The vacuoles often appear to consist of dilated cisternae and are thus continuous with the cisternae. It is usual for several typical Golgi cisternae to lie in a parallel arrangement. Thus, the lamellar structure of the cisternal component of the Golgi complex is built up.

Around these main components there lies a population of small membrane-bound vesicles, each a few hundred Angstroms in diameter. They may be very numerous and sometimes contain dense material.

The Golgi complex differs from the endoplasmic reticulum by its circumscribed appearance, position in the cell, thickness of component membranes, and distinctive architecture of its different elements.

Function. The Golgi complex has been linked with the process of secretion in certain cells. In some cells, granules of secretion form in the cytoplasm close to the Golgi complex, or even by accumulation of material in components of the Golgi complex itself. These granules subsequently become detached from the complex and pass toward the apex of the cell where maturation takes place and where the granules are stored prior to release. A similar relation between secretory granules and Golgi complex is seen in cells with an endocrine function—islet cells of the pancreas, chromaffin cells of the intestine, and cells of the anterior pituitary.

The Golgi thus participates in secretion by concentrating or packaging material made in the cell and transforming it into units that can be distributed outside the cell (Fig. 1-12). It may also function in the modification of lipid substances. In protein-secreting cells (e.g., cells of the exocrine pancreas), the major part of protein synthesis takes place in the endoplasmic reticulum, which then appears to transport this material in its cisternae to the region of the Golgi complex—there to be packaged into secretion granules. It is possible that vesicles filled with newly synthesized contents of the cisternae bud off from the endoplasmic reticular membranes adjacent to the Golgi. These vesicles could carry material to the Golgi cisternae releasing it into the cavities of the Golgi sacs. The secretory substance would then pass through this complex, undergoing concentration and even chemical change, and finally accumulate as a secretion granule at the far side of the complex. Since the final granule is surrounded by a membrane derived from the complex, it appears feasible that the enzymes attached to the membrane may continue to act on the granule even after its release from the complex.

In addition to segregation, concentration, and "packaging" of secretions elsewhere synthesized, the Golgi has independent synthetic functions of its own. It appears that the polysaccharide component of certain types of secretion may be manufactured here and become conjugated with a protein moiety synthesized in the endoplasmic reticulum. In the intestinal tract, the Golgi has been shown to be the site of synthesis of the carbohydrate component of the fuzzy coat of the microvilli. It also takes part in the process of absorption, and in certain protozoa, it may be concerned with the regulation of cellular fluid balance.

Mitochondria

Larger cellular structures seen in the cytoplasm and mainly concerned with energy problems of the cells are the mitochondria (the cell's powerhouse and energy source). All are made up of membranes arranged in a similar organized pattern (Figs. 1-1, 1-31, 1-32).

The membranes are thinner than the cell surface membranes but have a trilaminar pattern. The mitochondria are found in every type of cell (bacteria, blue-green algae, and the mammalian red blood cells are exceptions). In living cells, they are in constant motion and range in size from 0.2 to 0.5 μm and their shape varies from rodlike to spherical (shape, size, and internal structure may be characteristic of a given cell type or even organism). They appear to gather where cellular activity is highest (Fig. 1-33): (1) at the junction where nerve cells meet and transmit impulses across membranes; (2) near beating sperm tails; (3) just under the brush border of the intestinal epithelial cells where absorption is taking place; and (4) in areas of muscle cells that are in contraction.

Cells with a high rate of oxidative metabolism have large mitochondria (adrenal cortical cells) and some cells of the testes have cristae that are tubular rather than shelflike. The significance of all of this is unknown but it may reflect the spatial arrangements of the mitochondrial enzymes.

Morphology. Each mitochondrion is limited by an outer, double lipoprotein membrane, similar to the cell and nuclear membranes (Figs. 1-31, 1-32). This outer membrane is "elastic" and has the ability to swell or contract easily. Within this limiting membrane is the inner mitochondrial membrane separated from the outer by a constant narrow gap measuring about 80 Å in width. The inner membrane is thrown into folds or shelves called cristae that extend from the side walls of the mitochondria into the center, adding an increased interior surface area. Each cristae has two parallel layers of inner membrane separated by an extension of the outer space. The cristae are variable in length. The space into which the cristae project is completely enclosed by the inner membrane and is referred to as the inner mitochondrial space or matrix (this is the "liquid phase" of the mitochondria versus the "membrane phase"). The matrix is often finely granular and varies in density. In some cells, there are irregular dense intramitochondrial granules in the matrix about 500 Å in diameter, and one may even see dense droplets (apparently lipid) as in some cells of the adrenal cortex.

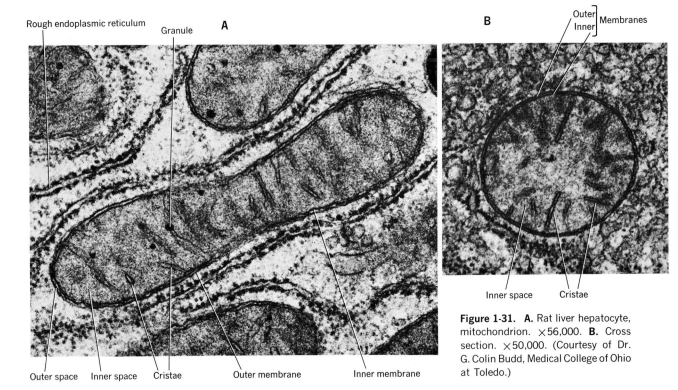

Rough endoplasmic reticulum
Granule
A
B
Outer
Inner
Membranes

Outer space   Inner space   Cristae   Outer membrane   Inner membrane

Inner space   Cristae

Figure 1-31. A. Rat liver hepatocyte, mitochondrion. ×56,000. **B.** Cross section. ×50,000. (Courtesy of Dr. G. Colin Budd, Medical College of Ohio at Toledo.)

Function. The mitochondria are the respiratory centers of the cell. They contain most of the enzymes of the citric acid (Krebs) cycle that are concerned with the breakdown of simple molecules from which the cell obtains energy (Fig. 1-21). This cycle is the final common path of catabolism for carbohydrates, fatty acids, and many amino acids. (As the molecule passes through the cycle, hydrogen atoms are removed by enzyme action and eventually combined with O_2 to form HOH. During these oxidative reactions, energy stored in chemical molecular structure is released for the cell's use.) The carbohydrates, proteins, and fats used as metabolic fuel are broken down by enzymes outside the mitochondria into smaller fragments (pyruvic, amino, and fatty acids). These then can pass through the membrane into the matrix where enzymatic reactions "chip off" carbon atoms one at a time from pyruvate and amino acids and two at a time from fatty acids. This is oxidation and the end products are CO_2 and HOH. The energy is not liberated as heat but is passed on in repackaged chemical form and ultimately into adenosine triphosphate (ATP). The mitochondria form ATP as part of an energy-trapping process, and these packets of energy can then be moved to the parts of the cell where needed. The breakdown of the metabolic fuels occurs in the liquid matrix where enzymes exist in soluble form. ATP production is the primary function of the mitochondrial membrane, and it has been shown that part of the inner membrane consists of assemblies of enzymes that transfer electrons from one site to another until incorporated into ATP. About 70 enzymes are known to exist. The oxidative ones occur in the matrix but the phosphorylating ones are hard to separate from the membranes, which are like a structural and functional mosaic. The enzymes may even be attached like ribosomes on the membranes.

The process of energy transfer is an essential factor in oxidative phosphorylation. By disrupting mitochondria, structural subunits have been isolated that contain

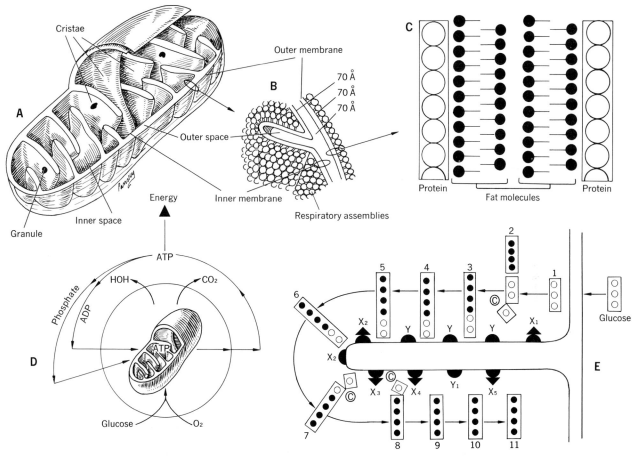

Figure 1-32. **A.** A diagram of mitochondrial structure. **B.** An enlarged view of the mitochondrial membranes. The globules are believed to be respiratory assemblies where ATP is produced. **C.** A still larger view of one of the mitochondrial walls, composed of two layers of fat molecules sandwiched between two layers of protein. **D.** The processes that occur in the mitochondrion. **E.** A detail of what is shown in **D.** Glucose breakdown to pyruvate takes place outside the mitochondrion. Note that the enzymes lie on the mitochondrial shelf (hydrogen transfer systems), whereas other enzymes may be suspended in the mitochondrial fluid. *1*, pyruvate; *2*, acetyl coenzyme A + oxaloacetate; *3*, citrate; *4*, cis-aconitate; *5*, isocitrate; *6*, oxalosuccinate; *7*, alpha-ketoglutarate; *8*, succinate; *9*, fumarate; *10*, malate; *11*, oxalo-acetate; *Y*, aconitase; X_1, pyruvic dehydrogenase; X_2, isocitric dehydrogenase; X_3, alpha-ketoglutarate dehydrogenase; X_4, succinic dehydrogenase; Y_1, fumarase; X_5, malic dehydrogenase; *C*, carbon dioxide.

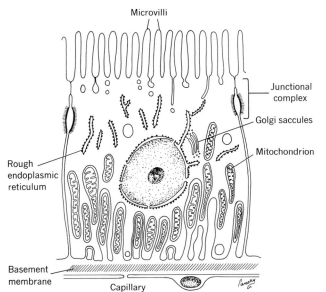

Figure 1-33. Schematic diagram of a kidney tubule cell, illustrating the many basal mitochondria.

the complete energy transport chain. These enzyme assemblies are called electron transport particles (ETPs), and from 10,000 to 50,000 may be present in a single mitochondrion. It has been suggested that the surfaces of the mitochondrial cristae are covered with closely packed subunits that represent the ETPs. Thus, theoretically, the more ETPs, the greater the capacity to produce energy from simple molecules. In cells that produce large amounts of energy by oxidative phosphorylation, the high metabolic rate of cells is reflected in the number, size, and complexity of the mitochondria. For example, prominent mitochondria are seen in the acid-secreting gastric parietal cells and in cardiac muscle cells.

Since gastric juice has hydrochloric acid (pH 1) and the tissue fluid surrounding the base of the gastric gland cell is at a pH greater than 7, the acid-secreting

Figure 1-34. Rat liver peribiliary dense bodies, lysosomes in hepatocyte. ×51,000. (Courtesy of Dr. G. Colin Budd, Medical College of Ohio at Toledo.)

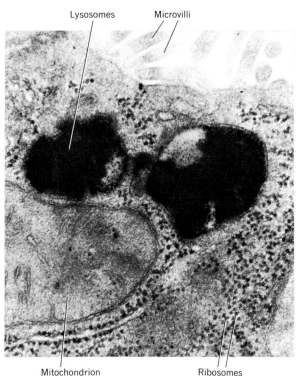

Lysosomes   Microvilli

Mitochondrion   Ribosomes

parietal cell can concentrate hydrogen ions by a factor of one million. The secretion of electrolyte against such a concentration gradient can be accomplished only with large amounts of energy produced by many large mitochondria. The heart "pumps" for many years, and the muscle cells are in active contraction for nearly one half of the animal's lifetime and can even be called upon to increase its output by a factor of 10 for prolonged periods of time. The energy for these constant efforts must come from the continued oxidation of simple molecules in the mitochondria.

The mitochondria, the source of the energy-rich ATP, often gather in a particular region of the cell in which energy is being used to produce metabolic work. In striated muscle, they lie in contact with the myofibrils, the contracting units of the cell. In the kidney tubule, they are sandwiched between complex basal infoldings of the cell membrane where large-scale water transfer and solute occur. The midpiece of the spermatozoon consists of a tightly wound spiral of mitochondria around a central core of functional units that have a probable contractile function and give the sperm motility.

There is no doubt that the mitochondria are the source of the oxidative energy in cells of all types. A puzzling fact is the finding of DNA in the mitochondria, thought previously to exist only in the cell nucleus. Its functional significance here is unknown.

Thus, within the membrane folds of the mitochondria, the conversion of food to energy takes place—the energy needed for growth, reproduction, and other functions. It "burns" the food absorbed by the cell and partly broken down in the cytoplasm to release its energy and then loads this energy onto a chemical molecule called ATP for the raw power to run the plant.

Lysosomes

Lysosomes (Figs. 1-1, 1-34) are also found in the cytoplasm. The term *lysosome* was used to describe a distinctive group of subcellular particles. They are of the same general size as mitochondria but can be separated from the large mitochondrial fraction of the cell by differential centrifugation. They contain hydrolytic enzymes. The name was really intended to convey the biochemical concept of a discrete cytoplasmic particle containing these enzymes.

The first enzyme described in the lysosome was acid phosphatase. Others to follow were beta galactosidase, beta glucuronidase, cathepsin, deoxyribonuclease, and ribonuclease. The number of enzymes varies in different tissues and there is a broad spectrum of enzyme activity.

Morphology. The lysosomes have a limiting mem-

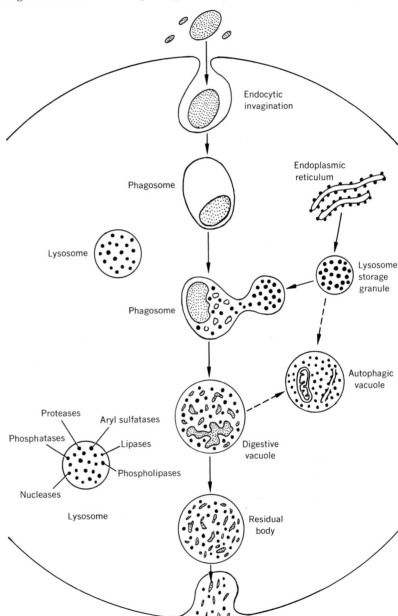

Figure 1-35. Lysosomes are involved in intracellular digestion. Four types are seen: storage granules, digestive vacuoles, residual bodies, and autophagic vacuoles. The first three are directly involved in the digestive process. The storage granule is the original form of the lysosome. The ribosomes are seen as black circles and produce enzymes. When the cell ingests a substance, a food vacuole (phagosome) is formed. A storage granule fuses with the phagosome to form a digestive vacuole, which continues digestion. Gradually a residual body results, which is then eliminated by fusion with the cell membrane.

brane but lack internal cristae and characteristic enzymes since the contents are variable. The protective membrane is essential to protect the cytoplasm from its contents, which if uninhibited would destroy the cell. They are pleomorphic dense bodies with membranes and contain patches of dense material, membrane lamellar structures, and pale amorphous areas. Acid phosphatase has become a major cytologic criterion for identifying them.

Function. The lysosomes are the intercellular digestive systems and function to destroy large molecules (Fig. 1-35). Most cells contain recognizable lysosomes; yet certain cells (the macrophage and polymorphonuclear cell) have become specialized to make and store them for use in the process of phagocytosis. In these and other cells, the probable site of lysosome synthesis is the Golgi apparatus. In the liver, the lysosomes are called peribiliary dense bodies since they lie close to the bile canaliculi. In damaged or injured cells, they may increase in number and complexity. They have been called cytolysosomes in unhealthy or degenerating cells.

Figure 1-36. Centrioles in mouse lymphoblast. ×10,912. (Courtesy of Dr. G. Colin Budd, Medical College of Ohio at Toledo.)

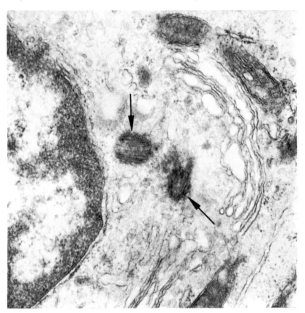

The lysosomes represent a safe method for storing a number of potent and potentially destructive enzymes in the cytoplasm. The enzymes may be used in the process of phagocytosis in specialized cells in the disposal of damaged cytoplasmic components.

The lysosome may take part in bringing about cellular death. Cell death is a part of growth and development since remodeling of tissues and organs can take place only by the planned death and removal of unwanted cells. The death of cells may result from the "purposeful" breakdown of the lysosomal membranes with the release of enzymes into the cytoplasm. The release of these enzymes from leukocyte granules when bacteria are ingested also leads to cellular death. The acrosome of the spermatozoa may be a type of large lysosome concerned with the penetration of the ovum. It is suggested that the lysosome rupture may be the stimulus to cell division and that abnormal lysosome behavior could be a factor in the production of certain forms of disease such as cancer.

Other Cell Components

Vacuoles. Vacuoles are fluid-filled, membrane-enclosed spaces in the cytoplasm. They probably arise by expansion of the open spaces or cisternae in the endoplasmic reticulum. They tend to force the cytoplasm to the outside of the cell where a change of gases can occur. Vacuoles also serve to maintain the proper internal pressure in the cells and may provide a means of getting rid of noxious materials no longer needed by the cell.

Centrioles and Centrosome. With the light microscope, an area is seen that appears more homogeneous than its surroundings, is free of mitochondria, and has two tiny spots at its center. The area is often associated with the Golgi and is known as the centrosome and the two central spots are the centrioles (Figs. 1-1, 1-36). On electron microscopy, each centriole is a cylindrical structure about 5000 Å long and 1500 Å in diameter. The two seen in the resting cell lie with their long axes at right angles to each other. In transverse section, the wall is composed of nine sets of microtubular structures that run the length of the centriole in a gentle spiral. Each set has three subunits termed subfibrils ab, b, and c arranged in a characteristic pattern. Dense satellites that may be attached to the outside of the centrioles could serve as points of insertion for microtubules related to the mitotic spindle.

They are best known in relation to cell division. Before cell division occurs, they reproduce themselves and take up positions at opposite ends of the nucleus where they form the two poles of the mitotic spindle. In ciliated cells, the centrioles divide repeatedly and form the basal bodies that give rise to the cilia, acting as their anchoring areas in the cell and possibly as the centers for their control. The mode of replication and significance of their fine structure are associated with the chromosomes at mitosis, and the relationship to the microtubules of the spindle is not fully understood.

The association of the centrioles in relation to cilia and flagella suggests that they are concerned with the organization of subcellular movement. The reason for the universal occurrence of nine subunits in the centrioles is unknown. The constancy of this pattern may reflect the preservation through evolution of a structural formula developed for an essential function at an early stage of development.

Cytoplasmic Fibrils and Microtubules. These are seen by electron microscopy in the cytoplasm of many cells (Fig. 1-37A). They consist of aggregates of elongated protein molecules believed to form a diffuse skeleton extending through the cytoplasm, giving resilience and even rigidity to the cell structure and perhaps forming a framework for attachment of enzyme groups and cytoplasmic structures. In other cases, the fibrils form an important part of the cell structure and are related to function (see the following).

In muscle, the contracting mechanism in the cytoplasm is made up of protein fibrils with special chemical nature and anatomic arrangement. In the skin, the fibrillar material that accumulates in the cell is called keratin and has a protective function. Prominent fibrils are seen in nerve cell processes and take on a definite longitudinal arrangement. However, the presence of microfibrils in a cell on electron microscopy may not be of specific functional significance since many differ-

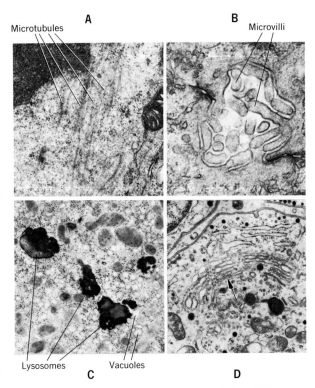

Figure 1-37. **A.** Microtubules forming part of the spindle apparatus of a mitotically dividing cell in a mouse lymph node. They are attached to a chromosome. ×23,215. **B.** Microvilli in a bile canaliculus between two liver cells. Mouse liver; ×25,348. **C.** Lysosomes (residual bodies) in cytoplasm of a hepatocyte. Human liver; ×5,510. **D.** Golgi complex and maturing secretion granules; rough endoplasmic reticulum. Mouse pituitary; ×11,305. (Courtesy of Dr. G. Colin Budd, Medical College of Ohio at Toledo.)

ent proteins with different biologic functions can aggregate to produce a fibrillar pattern.

With the use of special fixative and the electron microscope, fine tubules of about 230 Å in diameter and indefinite length have been identified in the cytoplasm. These microtubules have no true membranes and do not have the characteristic trilaminar structure. On cross section, they appear as a circular profile, and in longitudinal section are narrow enough for their entire thickness to be seen followed for distances through the cytoplasm, despite their irregular course.

They have been found to form the mitotic spindle

and radiate from the centrioles to the chromosomes during division. The tubules appear to run between the centrioles and the chromosomes suggesting a role in controlling movement of the chromosomes during metaphase and anaphase. Prominent tubules called neurotubules run parallel to the long axis of the axon and dendrite in the nerve cell. Tubules are also relatively plentiful in the irregular-shaped cells such as the podocyte of the renal glomerulus. It has been suggested that the tubules are concerned either with the maintenance of cell shape (a structural role) or with the establishment of diffusion channels through the cytoplasm. Since they appear in the central core of motile structures such as cilia and sperm tails, they may be simple contractile units. The true significance of the tubules is not clear.

The Nucleus

The control of the cell is located in the nucleus—the computer, the designer, and the "life center" (Figs. 1-1, 1-38). Almost everything that the cell does is here supervised. This is the most prominent feature of the cell. Its major function is seen in cell division. It is the controlling center for reproduction and heredity, for here are the genes and chromosomes that determine the cell's individuality. It is the necessary organelle providing information or parts to the cyto-plasm to keep it functioning properly for an indefinite period of time, and it also is the source of information that governs the cell's morphology.

Morphology. The nucleus is generally a rounded body (it may, however, be flattened or even lobed). It is bounded by a double-layered nuclear membrane, the outside layer being continuous with the membranes of the endoplasmic reticulum. The outer membrane is peculiar in that it has pores that vary with the cell type. It is not certain whether these pores permit the passage of large molecules in and out of the nucleus.

When stained with basic blue dyes (hematoxylin), the central portion shows a network of fine threads in which coarser lumps or patches of heavier material stand out—the chromatin (which is really the chromosomes in an extended or diffuse state permitting maximum surface contact with nuclear sap. This is necessary because the diffuse state in the nondividing cell is the time when the most active metabolic phenomena occur). The pattern of chromatin varies from cell to cell (it may be diffuse or lumped). The patches as seen by the light microscope have a granular dense appearance, whereas in the electron microscope they form discrete clumps in the nuclear ground substance. Their position varies in the nucleus. The relationship between the dense areas in the chromatin and the presence of DNA or nucleoprotein is not yet clear. Be-

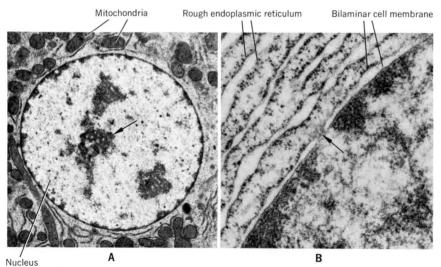

Mitochondria   Rough endoplasmic reticulum   Bilaminar cell membrane

Nucleus   **A**   **B**

Figure 1-38. A. Nucleus with nucleoli (arrow). Mouse hepatocyte; ×5195. **B.** Nuclear pore (arrow). Mouse pancreatic acinar cell; ×41,895. The double-layered nuclear membrane is also clearly visible. (Courtesy of Dr. G. Colin Budd, Medical College of Ohio at Toledo.)

tween the chromatin clumps there is the pale, finely granular ground substance, the euchromatin, with little obvious interrelationship between the dense particles seen. Scattered throughout the nucleus are larger discrete dense granules, 150 Å in diameter, with a slight angular appearance, similar in structure to the ribosomes of the cytoplasm and particles in the nucleolus. These may be newly formed ribosomes, synthesized in the nucleolus and not as yet having passed on into the cytoplasm. Correlation between areas of electron density and chemical constituents of the chromosomes is not completely understood as yet.

When an analysis is made of the nucleus, it is found to be made up of DNA, RNA, a protein of low molecular weight called histone, and a more complex protein. How they are grouped is unknown. It is known, however, that DNA is the molecule of heredity within whose structure is coded the information of inheritance to give the cell its unique qualities. The role of the others is uncertain.

Function. The nucleus is essential not only for cell division (a cell may live for years without dividing) but in daily control of the cell. When the nucleus of a protozoon dies, the protozoon loses the ability of motility and power to feed. The human red blood cell, once it loses its nucleus, can no longer divide or reproduce itself and its life-span is also limited. The nucleus directs the cell's activities according to the information (genetic) carried in the nuclear DNA.

Shortly before mitosis, the DNA of the nucleus doubles itself in a short synthetic phase. The subsequent cell division results in the distribution of a full complement of genetic information to each daughter cell (thus the code is carried intact to every body cell [Fig. 1-11]). The germ cells (sperm and ova) are the only exceptions since in meiosis (specialized type of cell division) there is a halving of the genetic information and each cell carries only one half of the full code and is thus haploid (see Chap. 2). Fertilization leads to the diploid state with full renewal of the complement. If at any time the sequence of bases is altered in the DNA, a mutation is said to take place. This can be ever so slight as to go unnoticed or so severe as to cause cell death or prevent any future successful

division. It is a permanent change since it involves altering the DNA molecule and thus is transmitted to all progeny of the mutant cell (assuming mitosis continues).

Day-by-day activity of the cell is maintained by the nucleus predetermined through the genetic code. Thus the nature of every protein molecule synthesized by the cell, and this includes enzymes in every step of metabolism, is nuclear controlled. It is now believed that the dense patches of chromatin seen in the interphase nucleus represent the inert portions of the chromosomes and are referred to as heterochromatin. The paler area of the nuclear ground substance is called the euchromatin and probably contains the extended metabolically active portions of the chromosomes engaged in the synthesis of specific messenger RNA.

The Nucleolus. Also seen in the nucleus are highly electron-dense, rounded bodies, the nucleoli (Figs. 1-1, 1-38), which are formed by particular chromosomes that have an active region called the nucleolar organizer, which in some way accumulates or manufactures nucleolar material and organizes it into a compact body. (It is not separated from the nuclear substance by a definite specialized structure such as a membrane.)

Chemically the nucleolus is rich in proteins and RNA, and it seems likely that this RNA passes into the cytoplasm to attach to the ribosomes to participate in protein synthesis. Two major components of the nucleoli have been described: a coiled or spongelike, coarse, dense material called the nucleolonema and a paler substance between the loops called the pars amorpha. Patches of chromatin are often associated with the rim of the nucleolus. Fibrils have also been described in the nucleolus in association with the granules seen there.

FUNCTION. The nucleolus appears to be an essential link in the chain of communication between the nucleus and the cytoplasm. The primary step in the biosynthesis of protein is the formation of messenger RNA. It is felt that the granules of the nucleolonema may be newly synthesized ribosome subunits that pass out of the nucleolus into the nuclear substance and finally into the cytoplasm, combining at some point with the messenger RNA already formed by the DNA of the nucleus. Thus, the nucleus can direct cell function through the messenger RNA using ribosomes, synthesized by the nucleolus, as a means of transporting the mRNA into the cytoplasm where protein synthesis actually takes place. This passage of material from the nucleus to the cytoplasm is of great importance in the communication of genetic information to the cell. Tumor cells, engaged in rapid growth and marked activity, have prominent nucleoli, perhaps another indication of the importance of the nucleolus in protein synthesis.

Extracellular or Intercellular Substances

The plasma membrane is generally believed to be the cell's outer living unit but not its outer boundary. Most extracellular substances are proteins or polysaccharides (macromolecules formed by linking together smaller repeating molecular units). They can combine with lipids and minerals to produce complex structures. The functions of these substances are varied: (1) HOH retention; (2) protection (tough chitin covering of the insects); (3) support (plant cellular walls as well as cartilage and bone); (4) rigidity and hardness (from mineralized bone to dentin and tooth

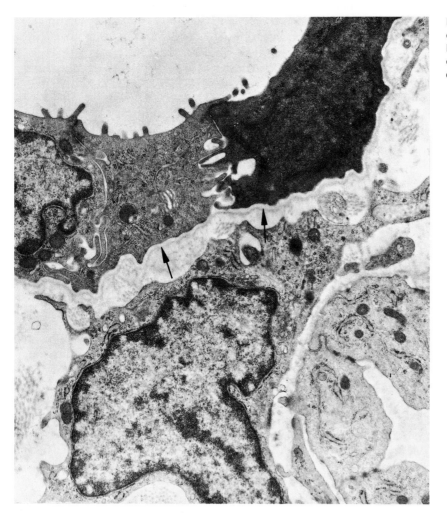

Figure 1-39. Rat liver, portion of bile ductule in portal tract to show basal lamina (arrow). ×16,800. (Courtesy of Dr. G. Colin Budd, Medical College of Ohio at Toledo.)

enamel to chitin); (5) elasticity (as seen in the elastic fibers of the skin or arterial walls); (6) adhesiveness (adhesiveness is critical because multicellularity depends on it and the outer bacterial cell walls often determine their virulence and immunology); and (7) permeability (also governed in part by extracellular substances).

Intercellular "Glue." The intercellular "glue" is of two kinds: hyaluronic acid and chondroitin sulfate, together known as ground substance. Combined with protein and lacking sulfur, hyaluronic acid is a jelly-like, amorphous, viscous polysaccharide of high molecular weight that retains water tenaciously. It functions as a "glue" by binding cells together yet permitting flexibility; in the fluid of joints it acts as a lubricant and shock absorber; and in the fluids of the eye it retains water and helps keep the shape of the eye uniform. Its viscosity is determined by the amount of calcium present. It can be dissolved by the enzyme hyaluronidase, which is present in sperm cells and formed readily by certain bacteria. In the sperm, it permits the sperm to penetrate the jellylike coat around the egg. Bacteria use it to effectively invade tissues and spread infection. Chondroitin sulfate, on the other hand, is a firmer "gel" but is also a polysaccharide combined with protein. One sees it in cartilage and where it is associated with fibrous elements such as collagen to provide a matrix in which the cartilage cells are nestled. It gives support and adhesiveness while preserving flexibility (found in the ear, nose, rib ends, new bone, joints, respiratory tract, etc.). In bone some is laid down in the original cartilage and is later calcified.

Basement Membrane. Basement membrane (Figs. 1-39, 1-40) is a definitely organized, special ground substance. It is seen in organs where it both binds cells together and helps shape the organ. It is prominent in the skin and lies between the epidermis and dermis as a condensation of intercellular substances. It may be highly laminated, which is not typical of the general ground substance.

Fibrous Elements of the Ground Substance: Elastin, Collagen, and Reticulin. These are all proteins of high molecular weight. About one third of the proteins of

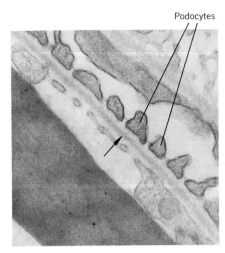

Podocytes

Figure 1-40. Basal lamina (arrow), mouse kidney. ×33,858. (Courtesy of Dr. G. Colin Budd, Medical College of Ohio at Toledo.)

an animal is collagen and is located where firmness or rigidity is needed (e.g., in muscles, bone, skin, and tendons).

Elastin and reticulin unlike collagen show no periodicity (periodic structure). Elastin is a stringy protein with the ability to stretch and "snap" back to original state and is seen in the skin and tissues around blood vessels. Reticulin consists of finer fibers than the other two but is related to collagen in everything but aggregation. It consists of finally branched fibers found in the ground substance and abundant in the basement membrane.

The Aging Process: Cell Death and the Intercellular Substances

Every individual has a life-span characteristic of its species—a few days for certain insects, months for plants, three score and ten for man, 250 to 300 years for an oak tree, and even several thousand years for some trees.

Cells, too, have a life-span to complete, and cell strains even in the same organism have variable life-spans. Some cells are apparently immortal, for when a unicellular organism divides, the original cell becomes part of two new cells, and as long as the species survives, the chain goes back to the original cell. Among sexually reproducing organisms, only the germ cells can lay claim to immortality, for they, like the unicellular organisms, span the generations. But among the cells of the body, death is a necessary process since to alter it would drastically alter the functioning of the organism. Cellular death is a result of (1) the wear and tear of existence, which is countered by equivalent cell replacement, and (2) the result of the normal process of development.

The human organism has a "new body" about every seven years, for it takes about that long for the old cells to be replaced by new. Some areas of the body require a constant replacement of cells and others are incapable of any replacement. At birth, all the nerve cells have been formed and function as long as the body lives. If a nerve cell is destroyed, it is *not* replaced by another for it cannot divide once fully differentiated, and no nerve cell replacement center exists. Muscle is capable of some very limited replacement. An organ remaining at a constant size is not an indication of constant replacement, for unless the cellular conditions are known, the size constancy only indicates no net gain or loss of cells. Death of cells equals the production of new cells.

The human body loses 1 to 2 percent of its cells through death per day. If the body weight remains constant these must be replaced, amounting to billions of cells per day. Since none are produced in muscle or nervous tissue, the centers of activity must be elsewhere. These exist in the protective layers (skin), blood, digestive, and reproductive systems. A liver cell, for example, has a life-span of 18 months.

The outer surface of the body is covered with a protective layer, mostly skin, which also lines all openings and even the cornea of eye. Other skin derivatives are nails and hair. The cells of these are constantly lost through death. Skin sloughs off and the growing nails and hair are composed of dead cells. The process of replacement is a rapid one with the underlying cells constantly dividing and pushing up toward the skin surface. The outer layer becomes hard and "kerati-

nized or cornified" as they die. It takes 12 to 14 days for skin of the forearm to move "up and out." Calluses, for example, are nothing more than thickened areas of dead cells and contain no pain fibers or blood supply. The eye cornea is a special skin modification in which the rate of cell death and replacement is exceptionally high.

The blood cells are the red blood cells from the bone marrow and the white blood cells from lymph nodes, spleen, and thymus. Cells and plasma form the blood with an average ratio of 1 white blood cell to every 400 to 500 red blood cells. The blood-forming areas manage to maintain the ratio and there is a relatively high cell loss to offset the new cell formation. Each type of cell "dies off" at a relatively constant rate (e.g., the red blood cell in 120 days). Since the normal wear and tear on the cells, as they pass through the vascular system, cannot be repaired, they grow fragile and finally break apart. Their life-span can be shortened by a variety of illnesses. In pernicious anemia, for example, the life-span is reduced to 85 days and in sickle cell anemia to only 42 days. Since the rate of replacement in disease often cannot keep up, the total blood count falls and the patient develops a so-called anemic state. The digestive tract is another organ system with a very high cell death rate.

A high rate of cell death is a pattern of existence as is the continuation of living cells. A plant preserves itself by the continual process of cell division and differentiation, paralleling these processes with an equally continuous succession of cell death. It discards these cells (leaves) or converts them into dead supporting tissue (wood). The "younger" the living cells, the less aged is the tree. A mammal on the other hand, achieves longevity by preserving the majority of cells in a functioning state. A cell that does not divide is destined to die, and thus the life-span of a mammal is short compared to that of a tree.

Cell death plays two major roles in development. The first is metamorphosis involving a change in shape and a change in organs when one way of life is exchanged for another (tadpole to frog; caterpillar to pupa to butterfly). The second is the role of cell death in the shaping of organs and body contour. Form can be achieved by relative rates of cell death as well as growth. As organs develop during metamorphosis, excess cells, which are transient and used by the embryo but not by the adult, are removed; e.g., when an organism forms secretory ducts, cells die instead of merely pulling apart to provide the lumen.

The ground substance and associated fibers serve several purposes: cells aggregate into organs of specific shape and size, and organ systems are held together to function. Adhesiveness, lubrication, rigidity, and elasticity are all necessary functions and are affected by the quality and quantity of the ground substance. If they change, the organism changes.

Aging, in part, must be related to changes in the intercellular substances—changes such as loss of skin elasticity, joint stiffness, arterial hardening, etc. There is some uncertainty about the actual meaning of changes seen in the ground substance, but one does note an increase in the amount and thickness of collagen fibers, elastic fibers become thicker and less springy with increasing binding of calcium ions, and reticular fibers become more brushlike and heavy.

If these fibers, necessary in the organization of living systems, during the course of development, continue to be formed after mature growth is reached, then aging is truly a process where development has gone beyond the needs of the system for proper functioning. Thus, multicellularity through adhesiveness and cell preservation has introduced into the life cycle an inevitable consequence of the aging process wherever there is a determinative or limited type of growth. Single-celled organisms are potentially immortal, and the long life of plants is due to cell preservation and the discarding of old cells and the continued production of new ones. (See Chap. 20 for detailed discussion of aging.)

Summary

The living cell is a very complex, restless, dynamic, changing center of life. It grows, reacts, moves, protects, and reproduces itself. It has a tightly organized system of many parts. It is a mass enclosed in a thin, flexible, highly active membrane. It is ever changing in shape.

Partly hidden at its center is the nucleus. The cellular interior, too, is not stagnant but is in continuous motion since the endoplasmic reticular network not only shifts but actually breaks up into segments and re-forms and breaks up again. The mitochondria, in the living state, move and twist and are also constantly splitting into equal active pieces and fuse end to end to form elongated moving systems. Moving along with them, in the cellular currents, are fatty globules and fragments of endoplasmic reticulum.

Somehow, all exist in a delicate homeostatic balance in the normal state, and though the total organism is made up of 60,000 billion or more ever-changing active cells, there is an order built in. This homeostasis is often disrupted, and the individual cells or groups of cells may go on to destroy and interfere with normal function, as in cancer. In the aging process, there are still many unanswered questions regarding alterations that vary the cellular processes. Yet, each and every second, millions of cells die, new cells are born, and the "biologic clock" is kept in balance, a balance that is somehow always active and never the same.

Review Questions

1. What were the contributions of the following men to the development of modern cellular thought?
   a. Robert Hooke (1663)
   b. Matthias Schleiden (1838)
   c. Theodor Schwann (1839)
   d. Rudolf Virchow (1859)

2. What methods have been used to study the cell and how do they compare?
3. What is the difference between a procell and a eucell?
4. About 96 percent of the body weight and over 99 percent of the body atoms consist of what four elements?
5. What is a protein and how does it function?
6. Relate the steps in protein synthesis, beginning with nuclear DNA.
7. What is an enzyme? List a few.
8. What are lipids and carbohydrates?
9. How does the cell acquire energy for its use? Relate the ATP story.
10. List the major elements of the cell and describe their functions.

References

Bernhard, S. A.: *The Structure and Function of Enzymes*, W. A. Benjamin, Inc., New York, 1968.

Bloom, W. B., and Fawcett, D. W.: *A Textbook of Histology*, 9th ed. W. B. Saunders Co., Philadelphia, 1968.

Clark, B. F. C., and Marcker, K. A.: How Proteins Start. *Sci. Am.*, **218**(1):36–42, 1968.

Crick, F. H. C.: The Genetic Code. *Sci. Am.*, **207**(4):66–74, 1962; **215**(4):55–62, 1966.

Danielli, J. F.: The Bilayer Hypothesis of Membrane Structure. *Hosp. Prac.*, **8**:63–71, 1973.

Dick, D. A. T.: *Cell Water*. Butterworths, Washington, 1966.

Everhart, T. E., and Hayes, T. L.: The Scanning Microscope. *Sci. Am.*, **226**(1):54–69, 1972.

Hayes, T. L., and Pease, R. F. W.: The Scanning E. M.: Principles and Applications in Biology and Medicine. *Adv. Biol. Med. Phys.*, **12**:85, 1968.

Kornberg, A.: The Synthesis of DNA. *Sci. Am.*, **219**(4):64–78, 1968.

Lenz, T. L.: *Cell Fine Structure*, W. B. Saunders Co., Philadelphia, 1971.

Lowey, A. G., and Sickevitz, P.: *Cell Structure and Function*, 2nd ed. Holt, Rinehart & Winston, New York, 1969.

Mirsky, A. E.: The Discovery of DNA. *Sci. Am.*, **218**(6):78–88, 1968.

Neutra, M., and Leblond, C. P.: The Golgi Apparatus. *Sci. Am.*, **216**(5):80–94, 1967.

Novikoff, A. B., and Holtzman, E.: *The Cells and Organelles*. Holt, Rinehart & Winston, Inc., New York, 1970.

Porter, K. R., and Bonneville, M. A.: *Fine Structure of Cells and Tissues*. Lea & Febiger, Philadelphia, 1968.

Singer, S. J.: Biological Membranes, *Hosp. Prac.*, **8**:81–90, 1973.

Sjostrand, F. S.: *Electron Microscopy of Cells and Tissues, Vol. 1*. Academic Press, Inc., New York, 1967.

Stein, W. D.: *The Movement of Molecules Across Cell Membranes*. Academic Press, Inc., New York, 1967.

Watson, J. C.: *Molecular Biology of the Gene.* W. A. Benjamin, Inc., New York, 1965.
———: *The Double Helix.* Atheneum Publishers, New York, 1968.
Yanofsky, C.: Gene Structure and Protein Structure. *Sci. Am.,* **216**(5):80–94, 1967.

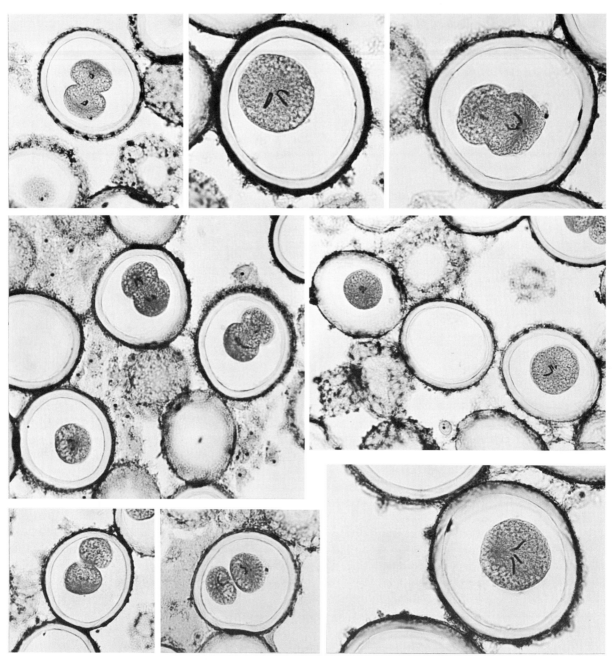

Figure 2-1. Oogenesis in the uterus of *Ascaris*. Note the various stages of cell division. Some chromosomes and centromeres are also clearly seen. The clear limited zone around the ovum is the zona pellucida surrounded by the darker granulosa layer. ×450–750. (Courtesy of Dr. Dale E. Bockman, Medical College of Ohio at Toledo.)

Cell Development

The development of the human fetus from its origin as a single fertilized egg to its completion as a baby with billions of cells and the phenomenon of cellular differentiation is complex and involves cell growth, cell replication, and cell differentiation.

Man's development passes through the following stages: fertilized egg $\longrightarrow$ prenatal life $\longrightarrow$ childhood $\longrightarrow$ adolescence $\longrightarrow$ sexual maturity $\longrightarrow$ physical maturity $\longrightarrow$ middle age $\longrightarrow$ senility (?) $\longrightarrow$ death!! (See Chaps. 18–20 for detailed discussion.) Though most obvious and prominent in early life, development also includes the continuous formation of blood cells, gametes (sperm and ova), and wound-healing tissue that daily occurs up until death, even into advanced age. The aging process is also associated with development of an excess formation of collagen in extracellular spaces and calcification of joints. All are "normal" developmental processes, with some continuing beyond a normal optimum. Yet this does not influence the mechanism of development or its phenomenon as a biologic process. For this we must look at development from a cellular level since the cell is the basic block or "bud" of life of all development. How do the basic potentials of the fertilized egg, located as coded information in the egg's DNA and the organization of its cytoplasm, actually become a total human being with all its symmetry and parts in their correct places and develop into perfect organs and organisms of size with all their proper cells to function where and how and when they should? This question is too immense to fully discuss here, but can be somewhat understood when one considers the major aspects of development—namely, growth, differentiation, and integration—and the role of the cell in the complex process.

Growth

Growth is an increase in mass and can result from enlargement of cells or more often by an increase in the number of cells through mitotic divisions. It is in most instances a process of replication, with the original cell using its environmental raw materials and converting them into more substance and cells like itself. If the rate of synthesis exceeds the rate of degradation, the cell grows. The growth of cells can continue only in association with repeated cell divisions during which time a parent cell splits into two daughter cells, each about half as large as the parent (mitosis—described later).

The egg weighs approximately one millionth of a gram and the sperm (at fertilization) adds only another 5 billionths more. During the nine-month gestation period there is an increase of about one billion times to a child of 3200 gm (about 7 lb). The final product (child) obviously is not just a mass of cells of comparable size and character as the original cell nor has its growth rate been uniform in its entire prenatal development. There has been a "relative growth rate" that determines the individual's form. Parts of the body grow at a faster or slower rate than others and some features appear earlier or later than others. Thus, in early development, the head and neck increase in size rapidly, the arms grow faster and at an earlier stage than the legs, but the trunk in contrast apparently grows at a steady rate until maturity. Growth, therefore, is not simply the generalized multiplication of cells but rather a pattern of multiplication with varying growth centers active at different times and at different rates during the course of development and these centers are coordinated to produce the correct form.

Differentiation

Where growth is an unending process of multiplication of similar units, differentiation makes the unit distinctive. In the process of differentiation, further multiplication of the cell is prevented. The more differentiated a cell becomes, the less it is likely to divide. Thus, when a cell nears its final form as a mature, differentiated structure, its potential for further change narrows with its specialization. Differentiation is the process that "stamps" each cell with its specific uniqueness of structure and takes place as the generalized cell is transformed into one that is specialized. Variation is introduced into the functioning organism, and differentiation provides variety.

Differentiation begins long before any visible change in the cells occurs and must be preceded by chemical alterations that are set into an unyielding pattern. There is no doubt, although little is known about all this, that the differentiation of cells can occur either "through a loss of old functions" or "through acquisi-

tion of new ones." For example, a nerve cell traces its origin back to the original fertilized egg, but in development irreversible changes occur and the early versatility of the egg cell is lost for the special property of conduction; yet if a nerve is cut, the axon regenerates and the power of repair is not lost, but the nerve cell cannot divide to form new nerve cells and if a nerve cell dies it cannot be replaced. Also, from studies it has been shown that differentiation of cells is essentially a change in cellular proteins.

The nucleus of a cell also undergoes a degree of differentiation. It has a set of genes with coded information in its DNA that exercises control over the cell. The nuclei from differentiated or "committed cells" are restricted in their developmental control and lack the versatility of nuclei of undifferentiated cells. Thus differentiation can be cytoplasmic and nuclear.

The cell's divisions are synchronous, the cells splitting in unison. With time, this synchrony gets out of phase and some cells divide more rapidly than others. Initially all the cells appear alike but soon differentiation occurs. As the generations of cells pile up, selective cells go their specific way to serve their specific function.

Integration

The fertilized egg, a single cell, initially gives rise to many cells by repeated divisions of quick succession

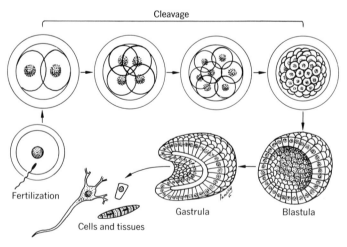

Figure 2-2. Diagram of early mammalian development, from fertilization to gastrula stage.

(Fig. 2-2; see Chap. 18). The fertilized egg is primarily an organism in its most simple state and differs from other cells in its potentiality for total development. Growth and differentiation are processes by which development is achieved. However, they are not enough to account for all development since there is a unity of structure and behavior, and cells develop as a whole or unit and not just a mass or collection of groups of cells; this unity is integration. It depends on chemical stimuli (hormones), cell movements, cell interactions, and the process of differentiation. Integration determines (1) why animals reach a certain size and stop growing; (2) the varying normal life-spans; (3) the relation of one body part to another; and (4) one phase of development leading to and affecting another. It involves a system of chemical checks and chemical balance.

Regulation of Development

Beginning with a single fertilized egg cell, the first division creates two cells, the next four cells, and so on. If the development proceeded this way, it would require only about 43 division cycles to produce the five trillion body cells from but one original. However, the process of differentiation marks certain cells to develop along specific lines, and all is regulated by the DNA molecules present in the single fertilized egg.

When the cell divides to form two independent cells, the genetic information in its DNA must duplicate itself and pass on to each daughter cell. DNA is the only molecule of the cell capable of duplicating itself without instructions from another source. Messenger RNA needs DNA to be formed. Proteins require messenger RNA for sequencing the amino acids (Fig. 1-11, p. 16).

DNA replicates as follows: The two strands of DNA separate and the exposed bases in each strand act as a template with which free nucleotides can base-pair to form a complementary strand. Once the free nucleotides have paired with the bases in each of the two strands of DNA, the enzyme DNA polymerase joins the nucleotides together. Thus, both strands of DNA act as templates for the synthesis of new DNA molecules. When two identical molecules of DNA have been formed, one copy is passed on to each of the two

Figure 2-3. Diagram of the karyotype of human chromosomes in metaphase, in accordance with size and centromeric constriction. It is a systematic way of detecting aberrations in the number and structure of chromosomes. Karyotype is the name given to a human complement of chromosomes arranged into pairs according to an international convention.

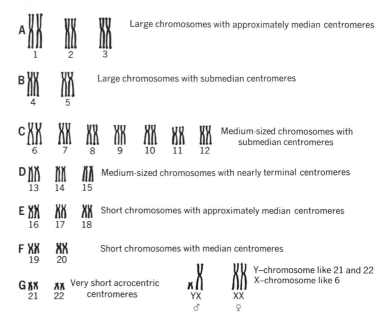

daughter cells during division so that each receives the original genetic code (Fig. 1-11, p. 16; Fig. 2-3).

Cell Division

The way new cells originate has been under investigation since the cellular nature of organisms was discovered and the cell theory recognized the cell as the basic unit of biologic function and organization. It was in 1858 that Rudolf Virchow stated that "cells originate through the division of pre-existing cells." The theory of biogenesis, credited to Pasteur (1822–1895) and the Italian Redi (1626–1698), noted that cells are the products of ages of evolution and that life spawned from preexisting life. "Each of us developed from a single cell—this cell came from our parents, each of whom was formed from a single cell and so back to the beginning of cellular life"—a slender series of perfectly formed individual cells that, though they vary in themselves, can live, reproduce, and give rise to new cells from the past and extending into the future.

The continuation of any species depends on an unending succession. Thus, its population must reproduce. In unicellular organisms (e.g., ameba), cell division leads to the continued formation of new individuals; since this is achieved by mitosis, a constant number of chromosomes is maintained in the processes. The ameba is asexual and has no eggs and sperm.

Man is made up of about 10^{14} cells that are formed and differentiated and integrated as the body matures and in many instances constantly replaced even through old age and up until death. In any healthy organism that grows and requires repair, the process of cell division produces new cells according to the need of the organism in its particular environment. Thus, all cells do not divide simultaneously or at the same rate and some cannot even be replaced. The process of cell division is uniquely similar in all organisms, and if once described in any particular cell, it can be said that it operates similarly in most living organisms. This process centers largely in the nucleus; yet one must not forget that the cytoplasm also undergoes changes.

Stages of Nuclear Division (Mitosis)

Interphase (Resting Phase). The more rapidly growing cells divide about once every 24 hours (the time between cell divisions varies in different cell types in the body). In a "24-hour" cell, visual changes in cell structure begin to appear 23 hours after the last division, and one hour later the cell has formed two cells. The 23-hour period between the end of one division and the beginning of the next is called interphase (Fig. 2-4). Thus, since division lasts about one hour, the cell spends most of its time in this period.

In interphase, the nucleus is large in comparison to a nondividing cell. During this period nucleic acids and proteins are synthesized as a preparation for division. There is little definable structure except for nucleoli and a chromatin fine network (nucleic acids of chromosome are either too diffused to absorb dye or are so hydrated that dye does not accumulate to stain). Therefore, chromosomes are not distinguishable here. The replication of DNA takes place in interphase and takes 10 to 12 hours. At the end of this period, each DNA thread has been duplicated but the duplicate threads are joined together in a region called the centromere. Each of the duplicate threads itself is a complete double-helix DNA molecule.

Prophase. Prophase (Fig. 2-4) begins when the chromosomes become visibly distinct as long thin threads. They are longitudinally divided into identical halves or chromatids. The increased visibility is caused by the loss of HOH; so stainable parts are more densely packed. The duplicated DNA threads become highly coiled and condense to form rod-shaped bodies called chromosomes. They shorten throughout prophase and the coils decrease in number as they increase in diameter. Isolated chromosomes consist of DNA, protein, and RNA. DNA from the nucleus is always associated with specific proteins. The combination of nucleic acid protein forms a nucleoprotein called chromatin. The protein may control gene activity as well as function in the coiling of chromatin threads to form chromosomes. The coiling and condensation of chromatin to form chromosomes, before cell division, are a means of transferring the chromatin threads to daughter cells.

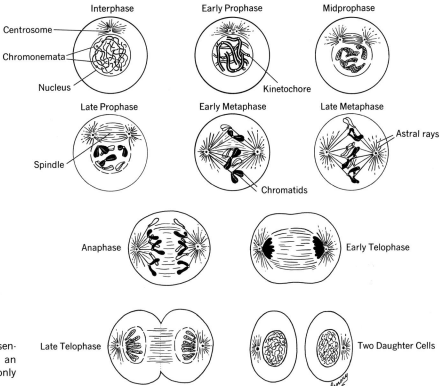

Interphase

Centrosome

Chromonemata

Nucleus

Early Prophase

Midprophase

Kinetochore

Late Prophase

Spindle

Early Metaphase

Late Metaphase

Astral rays

Chromatids

Anaphase

Early Telophase

Figure 2-4. Diagrammatic representation of the stages of mitosis in an animal cell. For the sake of clarity only six chromosomes are shown.

Late Telophase

Two Daughter Cells

Each chromosome consists of two identical chromatin threads attached at the centromere.

The nucleoli, which are formed by particular chromosomes, during prophase are initially prominent but decrease in size and vanish near the end of this phase and disappear. The nuclear membrane disintegrates in late prophase, and chromosome contraction ceases and the next period begins. In prophase, the cell also shows a radiating structure adjacent to the nuclear membrane. This is the centrosome with astral rays radiating from it and a central body or centriole contained within it. During prophase or even before, the centriole divides and each half migrates along the membrane to lie opposite the other. When the nuclear membrane breaks down, the cellular proteins are organized into the spindle so that the centrioles are at the poles with the spindle between.

Prometaphase. In prometaphase (Fig. 2-4) the spin-dle apparatus forms and the nuclear membrane begins to vanish.

Metaphase. In metaphase (Fig. 2-4) there is disappearance of the nuclear membrane coincident with the appearance of a new cytoplasmic structure, the spindle apparatus (chemically this is a long-chain protein molecule oriented longitudinally between the two poles). The fibers of the spindle are really fine tubules (not just protein threads).

The fibrous proteins of the spindle, in part, are synthesized during interphase. Chemical analysis of spindles shows a single type of protein combined with 5 percent RNA. With beginning prophase, spindle proteins are present but are unassembled. Most of the proteins are of cytoplasmic origin (not nuclear) although both can contribute. Just before metaphase, the spindle proteins are oriented in a linear fashion (longitudinally) between centrioles (or poles) with some organized into distinct filaments connecting centrioles and centromeres. Some of the fibers pass "unbroken" from one side of the cell to the other between two small cylindric bodies called centrioles. Some pass from the centrioles and are attached to the centromere of the chromosome. These project as astral rays. This organization of spindle proteins brings the centromeres onto the metaphase plate. With beginning anaphase, the fibers shorten separating chromatids and moving them to poles.

The centromere always contracts the spindle at the equator, and the chromosome arms are in random array. The centromere is the organ of movement (without it a chromosome cannot orient on the spindle and chromatids do not separate from each other).

Anaphase. In anaphase (Fig. 2-4) the centromeres divide so each chromatid has its own centromere and the sister chromatids then move apart from each other toward opposite poles toward the opposite centriole region. The spindle fibers act as if they are pulling the chromosome segments toward the poles, although the actual mechanism of movement is unknown. As the chromosome segments move toward opposite poles of the cell, the cell begins to constrict along a plane perpendicular to the spindle apparatus (constriction continues until the cell has been completely separated into two halves). Termination occurs when chromosomes form a densely packed group at two poles.

Telophase. Telophase (Fig. 2-4) is the reverse of stage 1. The nuclear membrane forms, the chromosomes uncoil and form threads again, and the nucleoli and chromocenters appear. The nuclei once again take on the interphase look. The division into two daughter cells is by furrowing, beginning at the outer edges of the cell near the equator and cleaving the cell. The spindle then disintegrates and division is complete.

The process of cell division just described is called mitosis (Greek *mitos*, thread). It is a mechanism for distributing equal amounts of DNA from cell to cell during most cell divisions in the body.

The cell cycle is really a delicate balance between nuclear and cytoplasmic events and their constituent parts and not simply a description of events from one phase to the next. Mistakes lead to abnormal cells. A cell really prepares to divide long before actual visible signs of division occur, but whether it does occur depends on later events; in some cells—e.g., nerve and muscle—once the division cycle is permanently interrupted they never divide again. Others, as in blood-forming tissues, divide throughout life. Replication of chromosomes and the genes they contain is related to the DNA molecule. Chromosomes also contain RNA, histones, and a (more) complex protein. The histones are bound to the DNA and are synthesized at the same time but little is really known of the chemical association of the four molecules.

Other Events in Cell Division

Shortening of chromosomes (by coiling) in prophase, uncoiling at telophase, disappearance and reappearance of nucleoli and nuclear membrane, formation of cell plate or furrow, and disappearance of the spindle are all chemical events, but little is known of the basic biochemistry except replication of DNA (see p. 18).

Duration of Cell Division

The rate of cell division is characteristic of the organism and governed by varying factors such as nutrition and temperature. For a human fibroblast (generalized type of connective tissue), the entire cycle takes 18 hours from interphase to interphase. Yet from beginning of prophase to end of telophase takes only 45 minutes. The cell spends approximately 17 hours preparing for division and then proceeds rapidly. Other cells take varying amounts of time, even much longer.

The effect of temperature is striking. A rise in temperature and the reaction is faster. A drop in temperature and the time is lengthened. Cell division in warm-blooded animals ceases at temperatures below 24° C, some even as high as 46° C.

Significance of Cell Division

1. Cell division is part of the process of growth. It involves the assimilation of outside materials and their transformation through breakdown and synthesis into new cellular parts and the utilization of

Table 2-1. Summary of Cell Division: Sequence of Events

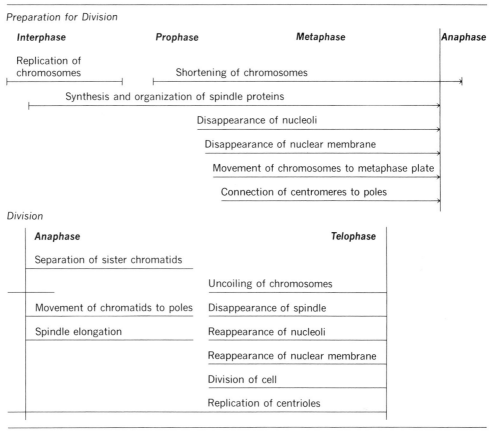

Preparation for Division

Interphase	Prophase	Metaphase	Anaphase
Replication of chromosomes			
	Shortening of chromosomes		→
Synthesis and organization of spindle proteins			→
	Disappearance of nucleoli		
	Disappearance of nuclear membrane		
		Movement of chromosomes to metaphase plate	
		Connection of centromeres to poles	

Division

Anaphase	Telophase
Separation of sister chromatids	
	Uncoiling of chromosomes
Movement of chromatids to poles	Disappearance of spindle
Spindle elongation	Reappearance of nucleoli
	Reappearance of nuclear membrane
	Division of cell
	Replication of centrioles

energy. Only the fertilized egg divides into half-size cells and then quarter-size cells. In most there are periods of growth in the process, and cell enlargement occurs.

2. Cell division ensures a continuous succession of similarly endowed cells necessary for species preservation. Accidents and variations do occur but are few in evolutionary history.

3. The chromosomes are regulators of cellular metabolism and structural characteristics of the cell.

4. Cell division is an attempt to keep a fairly constant ratio between the amount of the nucleus and cytoplasm. If the cytoplasm exceeds a certain amount, the power over it by the nucleus becomes less severe (the nucleus can control only a certain amount of cytoplasm). Both cytoplasm and nucleus must, however, be in a "state of maturity" or "prepared" for division.

5. Cell division is an act of survival since cells die eventually if they do not divide. Division accompanied by growth brings fresh substance into the cell, preventing aging and imparting a form of potential immortality. In a multicellular organism, division provides added cells to "divide the labor" and is really the first step to differentiation and specialization even though this is an antithesis of survival. Differentiation is a first step to eventual death as cells too specialized can no longer divide.

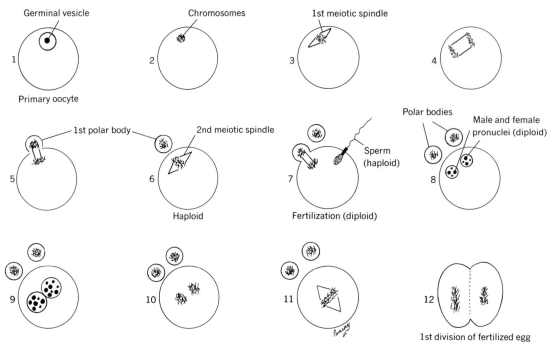

Figure 2-5. Primary oocyte changes prior to fertilization. Just before ovulation the "envelope" around the germinal vesicle disappears. The chromosomes arrange on a spindle and separate into two groups to complete the first meiotic division. The chromosomes remain in the oocyte and start the second meiotic division (completed at fertilization). After the second meiotic division both egg and sperm have only one half the number of chromosomes (haploid). They fuse to complete the number (diploid).

Fertilization and Chromosome Number

Most multicellular organisms produce by "sexual" means and during their life cycle produce gametes (any type of sexual cell—e.g., ovum and sperm) that unite in pairs to form a single new cell, the zygote. The union of gametes is called fertilization or syngamy (Fig. 2-5).

In fertilization, the major event is fusion of the gametic nuclei. In terms of chromosome number: the cells of all adult humans contain 46 chromosomes except the male and female reproductive cells. The female egg cell and male sperm cell contain only 23

chromosomes, which are brought together at fertilization. The zygote formed, if it were only a matter of mitosis and fusion, would have 92 chromosomes and so would the gametes of this individual. The next generation would have 184 and so on, with the tenth generation having about 23,332 chromosomes per person. This would prove absurd and unwieldy and obviously such an increase cannot go on indefinitely. Thus, during the life cycle there must be some compensatory mechanism to reduce this number since we know that all individuals possess a constancy of chromosome numbers with man having 46. There is a numerical constancy in all species from generation to generation. Thus, each of the gametes must have one half the number of chromosomes found in the zygote to keep the final number constant. The reduction in number of chromosomes is accomplished by a special cell division called meiosis.

The chromosomes in the gamete nuclei are called haploid, reduced, or N number, while in the zygote, and all cells derived from mitosis are called zygotic, diploid, 2N number, or unreduced. Prior to fertilization an egg has 23 chromosomes whereas a zygote has 46. There are, however, not 46 individually different chromosomes but 23 pairs with the pair members similar in shape, size, and genetic content—thus homologous to each other. In a zygote every pair of homologous chromosomes (homologs) consists of one member from the sperm and one from the egg (Fig. 2-5).

Meiosis

The exception to cell division by mitosis is concerned with sex determination. The reproductive cells that give rise to the egg cells and sperm cells undergo a slightly altered form of division called meiosis (Greek word meaning diminution) (Fig. 2-6). Meiosis is that process of cellular division in the gonads (testes or ovaries) resulting in the formation of germ cells or gametes (sperm or ova) and reduces the number of chromosomes distributed to the daughter cells from 46 to 23 (diploid to haploid). Meiosis consists of two cell divisions in succession. Before the first division there are 46 chromatin threads, each of which is duplicated,

Figure 2-6. Meiosis. Schematic representation showing the fate of two homologous chromosomes during the first and second meiotic divisions. The chromosomes are near each other **(2).** The chromosomes become double stranded **(3).** Recombination of homologous strands occurs **(4–5).** The chromosomes then move to opposite poles, and two cells are formed **(7–11).** The double-stranded chromosomes in each new cell divide at the centromere **(13).** Completion of the second division results in four cells **(14),** each with a different chromatid.

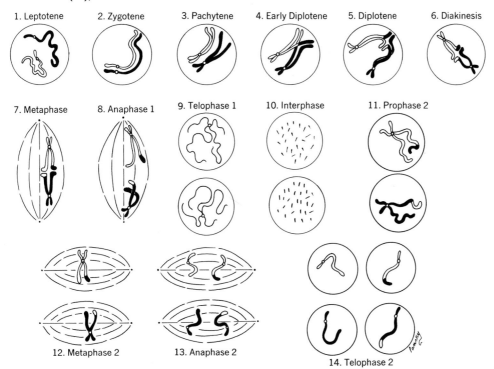

as in other cells. During the condensation of the chromatin threads into chromosomes, the duplicated maternal and paternal chromatin threads pair with each other and form a four-stranded chromosome. Thus there are 23 four-stranded chromosomes instead of 46 two-stranded chromosomes. When the cell divides, the two maternal strands of a chromosome pass into one of the daughter cells and the paternal strands into the other and the chromosome number in the daughter cells is 23. Since the four-stranded chromosomes attach to the spindle fibers at random, the two daughter cells receive a random distribution of the maternal and paternal chromosomes. Only by chance will all 23 maternal or paternal chromosomes end up in one cell. Since each chromosome in the daughter cells still has two strands, the next cell division proceeds without a DNA replication stage and resembles a normal mitotic division except that only 23 chromosomes are involved and one of the duplicate strands is passed to each daughter cell. The end result of the two meiotic divisions produces four cells each with half the number of chromosomes as the original parent cell. Each of the four contains a mixture of paternal and maternal chromosomes. Over eight million different combinations of parental chromosomes can take place during meiosis. Since each parent can produce any one of the eight million different sex cells, the random combination of a sperm and egg cell during fertilization can produce a fertilized egg having about 70 trillion possible combinations. With crossing-over during chromosomal pairing this number is further multiplied. Meiosis is complicated, but the nuclear events in all species are similar and except for the type of cell resulting, the process is the same in both sexes.

The X chromosome is involved in female determinant and the Y chromosome in male. Thus, the human female is XX (the paired chromosomes are similar and homologous), but the male is XY and the two chromosomes are nonhomologous.

Stages of Meiosis

Following chromosome duplication during interphase, there are several special prophase I stages.

Prophase is longer (than in mitosis) and modified in character. Five stages are recognizable in prophase meiosis.

Stage 1: Leptotene (Thin Thread Stage). This stage starts meiosis. There is the beginning condensation of chromosomes. These cells and nuclei are larger than the surrounding cells. Chromosomes (in diploid number) are thinner and longer than in mitosis and hard to tell apart. The chromosomes differ from mitotic prophase in that (1) they appear to be longitudinally single rather than double (although DNA synthesis timing suggests they are double), and (2) there is a more definite structure with dense chromomeres (granules) at irregular intervals along the chromosome length. The chromomeres of any organism are characteristic in number, size, and position. Chromosome movement starts stage 2.

Stage 2: Zygotene (Joined-Thread Stage). In this stage there is a movement of chromosomes as a result of attraction force between homologous chromosomes. The homologous pairing or synapsis begins at one or more points along the chromosome's length and proceeds to unite the pairs along their entire length (like a zipper). This is an exact not a random process.

Stage 3: Pachytene (Thick Thread Stage or Stable Stage). This stage involves complete pairing. The nucleus looks like a haploid number of chromosomes is present. However, each is a pair of homologous chromosomes referred to as bivalents. The paired chromosomes of each bivalent are easily seen. The chromosomes have shortened and thickened and are readily distinguished from each other. Even the chromomeres are visible with high magnification. This stage ends when the synaptic forces lapse and the homologous chromosomes separate.

Stage 4: Diplotene (Double-Thread Stage). In this stage the pairs begin to separate and each chromosome now consists of two chromatids. The bivalent, therefore, is composed of four chromatids (longitudinal division of each chromosome except in the region of the centromere took place prior to this stage but was not evident until now). Separation of homologs is not complete. Contact is retained by means of chiasmata. Each chiasma results from an exchange of chromatids between the two homologs. If only one chiasma is formed, the bivalent in this stage appears as a cross. If three or more are formed, they assume a looped appearance. In different cells, the number and position of the chiasmata vary (even for the same bivalent). As a rule, long chromosomes have more chiasmata (although even short ones have at least one). The chiasmata are evidence of exchange of homologous parts of maternal and paternal chromosomes (crossing-over).

Stage 5: Diakenesis: This stage is hard to distinguish from stage 4. Here the nucleolus becomes detached from its bivalent and disappears and the bivalent becomes more contracted. As contraction proceeds, the chiasmata tend to lose their original position and move toward the ends of the chromosome. This terminates the prophase of meiosis.

First Metaphase. The breakdown of the nuclear membrane and the appearance of the spindle end the prophase and start the first metaphase of meiosis. (The paired homologous chromosomes align at the equatorial plate in the form of paired bivalents or tetrads.)

The bivalents orient on the spindle, but instead of all centromeres being on the equatorial plate (in mitosis), each bivalent is located so that its centromeres lie on either side of and equidistant from the plate.

First Anaphase. Each tetrad chromosome separates into two dyad chromosomes, and migration of chromosomes to the poles begins. The two centromeres of each bivalent remain undivided, and their movement to opposite spindle poles causes the remaining chiasmata to slip off and free the homologs from each other. When movement ceases, a reduced or haploid number of chromosomes is seen at each pole. Unlike mitotic anaphase (here the chromosomes appear longitudinally single) each chromosome now has two distinctly separated chromatids united at their centromeres. The nucleus then forms, the chromosomes uncoil, and the meiotic cell is bisected by the membrane wall; this is the first telophase of meiosis.

First Telophase. This phase finds two daughter nuclei formed. At this point of the first division, homologous paternal and maternal chromosomes have separated at random into two daughter cells. This is now followed by an interphase period.

Interphase. This phase may be short or long, depending on the species, or may be absent altogether. The chromosomes are diffusely arranged, but no duplication takes place here. Then we begin the second meiotic division.

Prophase II. There is chromosome contraction resulting in wide chromatid separation.

Metaphase II. This phase finds the chromosomes arranged at the equatorial plate. There is no chromosomal reproduction and the centromere remains undivided.

Anaphase II. A spindle forms in each of the two cells, and at anaphase II the centromeres divide, the chromosomes split longitudinally, and the chromatids move to the opposite poles as in mitosis.

Telophase II. The nuclei are reorganized, the nuclear membrane is completed, and we have two daughter cells during telophase II. Haploid nuclei result that segregate into individual cells by cytoplasmic segregation.

Summary of Meiosis

The chromosomes remain unchanged in longitudinal structure from the diplotene stage to the end of the second meiotic division. The reproduction of each chromosome occurs during pachytene or earlier stages and this is followed by two divisions. In the first, the homologs separated from each other to reduce chromosome number (made possible because synapses joined them), and in the second, the two chromatids of each chromosome separated.

No explanation can be given why two divisions are needed in meiosis instead of one, but where only a single meiotic division takes place (as happens during sperm formation in the normal haploid male honey bee) and where a reduction in chromosome number is not necessary in the life cycle, it is essentially like the second rather than the first division. In all organisms with diploid chromosome numbers, the two divisions make more sense particularly when considered from the point of view of heredity.

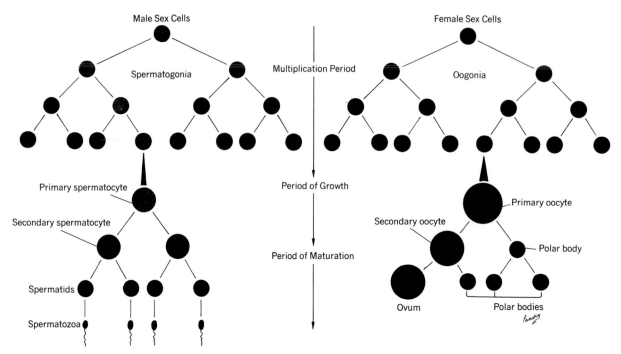

Figure 2-7. Diagram of the development (maturation) of the male (left) and female (right) sex cells.

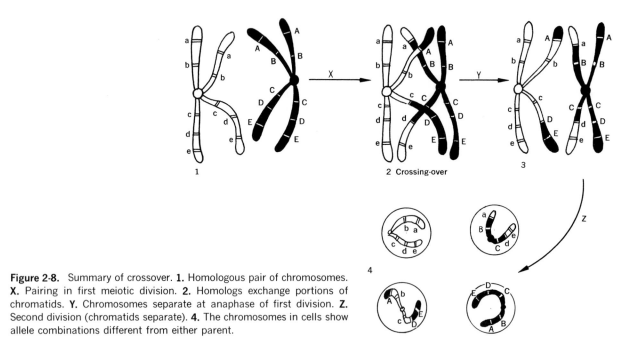

Figure 2-8. Summary of crossover. **1.** Homologous pair of chromosomes. **X.** Pairing in first meiotic division. **2.** Homologs exchange portions of chromatids. **Y.** Chromosomes separate at anaphase of first division. **Z.** Second division (chromatids separate). **4.** The chromosomes in cells show allele combinations different from either parent.

The process of meiosis is characterized by one DNA replication, the exchange of material between homologous chromosomes by crossing-over, and two-cell division. This results in the reduction of the chromosome number in the resulting gamete to one half the diploid number or a haploid cell. The combination of the two haploid germ cells at fertilization restores the original diploid number to form the zygote.

In man, one sees oocytes or immature cells in various stages of development—from their origin in the germinal epithelium, the outermost layer of cells of the ovary, to their eventual release prior to fertilization. The oocytes are formed early in life and do not increase in number thereafter. The female enters reproductive life with the ovaries provided with a finite number of oocytes. This number in humans is reached by the fifth or sixth month of fetal life, and by birth, the oocytes have entered meiosis and have reached the diplotene stage. Diplonema can, therefore, persist from about 12 to 50 years—the times when the first and last eggs are shed. During the remainder of the period the growth of the oocytes is essentially enlargement and nutrient storage. The time sequence varies according to length of the reproductive span. At puberty, one or more ova resume meiosis before ovulation into the oviduct at the middle of the menstrual cycle. Meiosis II is usually completed only after the ovum has been fertilized. Since the ovum needs many cytoplasmic nutrients, meiosis results in only one mature ovum from each oogonium while the other three daughter cells are lost as polar bodies (Fig. 2-5).

In the male, spermatogonia proceed to primary spermatocytes by a series of mitoses. Primary spermatocytes become secondary spermatocytes by meiosis I. In meiosis II each secondary forms two spermatids (four in all), which become mature spermatozoa without further dividing (Fig. 2-7).

Genetic Significance of Meiosis

Meiosis is a necessary part of the life cycle of sexually producing organisms. It is the opposite of fertilization in regard to the chromosomes.

When paternal and maternal gametes fuse, each bivalent at the first meiotic phase consists of two homologs, one from each parent. Orientation of all bivalents to a spindle is entirely random; so segregation at anaphase leads to random distribution of chromosomes. The haploid cells resulting would, therefore, have a mixture of maternal and paternal chromosomes. When four pairs of chromosomes are involved, 16 different combinations of four are passed, (calculated as $2n$, with n being the number of chromosomes). In man, with 23 pairs, the possible gametic chromosome combination $= 2^{23}$ or 8,388,608. The chance of any single sperm or egg containing only paternal or maternal chromosomes is negligible.

Chiasmata formation further complicates the distribution of paternal and maternal chromatin. Chiasmata result from an exchange between chromatids in the two homologs, one from maternal chromosome and the other from paternal. If a chromosome consists of genes strung along its length and the genes in one homolog are slightly different from those in others as a result of mutations, many changes can occur (Fig. 2-8).

Chiasmata formation breaks up linkage groups and alters the set of genes the chromosomes possessed before meiosis. Therefore, since the chromatids are distributed to the four haploid cells, each gamete is genetically different.

Mutation

To form the many cells of the body, a minimum of five trillion cell divisions take place and the DNA molecules in the fertilized egg cell are replicated at least five trillion times. During all this replication "errors" occur resulting in an altered sequence of bases and a change in the genetic message. Any alteration in the genetic message carried by DNA is called a mutation. It is during the replication of DNA, when free nucleotides are being incorporated into new strands of DNA, that incorrect base substitution most commonly takes place. Factors in the environment that increase the mutation rate are chemical agents and forms of ionizing radiation (x-rays, cosmic rays, or atomic radiation). They usually cause the breakage of chemical bonds, which results in alterations in the structure of DNA such that incorrect pairings between bases takes place or the wrong base is incorporated when the broken bonds are reassembled. There are many other possibilities for mutation, but the subject is far too extensive and complicated for detailed analysis here.

Mutations may have one of three effects on a cell: (1) they may cause no noticeable alteration in the function of the cell; (2) they may modify cell function but still be compatible with cell reproduction and growth; or (3) they may lead to the death of the cell. The consequence of the mutation depends on the type of cell in which it occurs. Thus, if a liver cell mutates causing it to function abnormally or die, the effect on the total organism is usually negligible since there are thousands of similar liver cells to perform the liver's function. However, if a mutation occurs during the early stages of development, it is passed on to most of the cells of the developing organism. If it is in an egg or sperm cell, all the cells descending from them inherit the alteration and affect the offspring.

Some types of abnormal functions in the body are the result of genetic mutations transmitted from generation to generation and are often due to a single gene that produces an abnormally active protein or fails to produce one. We refer to such inherited diseases as errors of inborn metabolism. The example usually cited is that of phenylketonuria, a disorder resulting in mental retardation in children due to a single abnormal enzyme. The individual is unable to convert the amino acid phenylalanine into the amino acid tyrosine at a normal rate. Thus, phenylalanine is diverted into other biochemical pathways resulting in abnormal activity of the nervous system. The by-products are secreted in the urine accounting for the name of the disease. One prevents the disease by restricting phenylalanine from the diet.

All mutations, however, are not harmful. This is the mechanism that permits organic evolution and may produce cells that function better or worse. This is the principle underlying natural selection. It is this process that has selected those mutations that have best enabled the cells to survive and propagate during man's evolution of life on earth.

Viruses

In a discussion of DNA, synthesis of protein, and cell growth one should mention viruses and their relation to cell function and disease.

Viruses are living organisms that approach molecular size but are not cells. Most elementary virus particles consist of nothing more than a molecule of nucleic acid (DNA or RNA) surrounded by several molecules of protein. The isolated virus particles are inert, but when entering a living cell, they can reproduce thousands of copies of themselves. They are the smallest known agent acting on cells and directly attack the genetic machinery essential for building cellular material. Over 300 different viruses are known with many being infective agents in diseases such as polio, smallpox, mumps, measles, rabies, etc.

The metabolism and structure of viruses differ from those of animal cells. They are not free-living organisms and can grow and multiply only in another cell they have parasitized. They change the metabolic machinery of the cell so that it now makes more virus instead of doing its usual job. Otherwise they resemble all characteristics of life in that they grow, multiply to produce exact replicas of themselves, and have a form of inheritance similar to our own. Viral nucleic acid carries the coded information for a limited number of proteins. Some genes code for structural proteins and others code for protein enzymes involved in nucleic acid synthesis. The nucleic acid in some viruses is RNA rather than DNA (here RNA carries genetic information). Thus, they have similar key molecules of protein and nucleic acid found in all living organisms, and though they are like cells with cytoplasm, they possess little more than the nuclear hereditary apparatus.

When the virus infects a cell, it is the nucleic acid (DNA or RNA core) part of the virus that usually enters the cell since it discards its protein shell as it enters. The viral nucleic acid then uses the biochemical machinery of the cell to synthesize virus-type proteins and to replicate itself using the nucleotides and energy sources of the cell. The virus wins the battle for the cell's hereditary material; 80 percent of the infected cell's DNA may be broken down and resynthesized into viral DNA. Thus, instead of producing its own hereditary material and proteins, the infected cell's nucleus produces viral substances. After 20 minutes or so, the cell can burst and release hundreds of new viruses.

Viral replication in cells may affect the cell function in several ways. (1) It may eventually kill the cell if it multiplies too rapidly. (2) The viral nucleic acid may become associated with the cell's own DNA and replicate along with the cell DNA and be passed on to daughter cells in division and may pass from generation to generation without any obvious deleterious effect on the cell. This dormant virus may later enter a stage of rapid replication and kill the cell.

The body responds to viruses by making antibodies, this time to the proteins of the virus shells.

Antibodies

Although the major objective in fighting viruses and germs is to create and enhance immunity, this may not always be desirable. Treatment may depend on inhibiting certain immune reactions since these natural systems may work against the organism. This is particularly true in transplanting tissues and organs. This is limited by the automatic efficiency of the body's antibody mechanism. Skin grafts are thus limited. If the skin is from the patient's own body, healing is prompt, and blood vessels grow in to replace the tissues and merge. The graft "takes" in weeks and lasts for life, but if it comes from another individual it usually, after an early good beginning, is rejected in 10 to 20 days. The body rejects this "foreigner" by producing antibodies that react specifically against the unique protein antigens of the graft, obeying its law of exclusion. This holds for tissues and organs. Only in identical twins is there an exception since they have similar genes and contain similar proteins in the graft and the body somehow recognizes "its own" and permits it to grow.

Fighting Antibodies

One technique involves intense irradiation and special drugs to kill the patient's antibody-producing cells. Unfortunately, this suppresses all the antibody processes and the patient is prone to infection of all kinds. This has been used in connection with organ transplants when they were lifesaving. Most attempted organ transplants are failures though a few successes of varying lengths of time have been reported.

It is known that the body usually distinguishes its own cells from foreign cells, but this often fails and the body "turns on itself" and destroys its own cells, which it mistakes for foreign cells. Certain blood, kidney, and thyroid diseases have been traced to this phenomenon. Experiments are under way to attempt to create immunologic paralysis by using killed donor cells (or purified antigens from donor cells) that are the specific proteins that in normal concentrations stimulate production of antibodies against grafts and in a massive concentration prevent such a reaction. Thus, heavy concentration of antigen molecules in dead cells could "paralyze" the antibody produced prior to operation and prepare the patient for his graft. (See Chap. 16 for detailed discussion of antibodies.)

Cancer

Abnormal types of cell growth are spoken of as cancer. A cell becomes cancerous when growth is no longer regulated by the processes that control normal cell growth. Any cell in the body has the potential of becoming a cancer cell. The underlying mechanism for this process is unknown. Ionizing radiation as well as various chemical agents can lead to the production of cancer cells. Viruses have been isolated that induce cancer in some animal species, but no virus has been isolated that has been specific for human cancer. Apparently the mechanisms that control normal cell growth cannot control cancer cell growth. The cancer cell grows and divides and forms a mass of cells called a malignant tumor, which tends to invade surrounding tissues and disrupt the normal structure and function of the body organs. It eventually leads to the death of the organism.

Many tumors, when detected early, can be surgically removed. However, cancer cells lack adhesiveness, and they break off from the original tumor and spread via the circulatory system throughout the body (metastases) where they grow and multiply to form new tumor sites. When cancer is spread, surgery is impossible.

Various drugs and ionizing radiation damage rapidly dividing cells and have been used in treating cancer. These agents, however, also damage normal cells, particularly the rapidly dividing cells of the blood-forming tissues and intestinal tract. The treatments prolong life but seldom cure. As we learn more of molecular and biochemical mechanisms of normal cell growth, perhaps we may find the answer to the cancer phenomenon.

Review Questions

1. Define growth, differentiation, and integration.
2. Correlate the happenings in the left column with the phases of mitosis in the right column.

   a. There are shortening and condensing to form distinct chromosomes.

   b. This is the period between the end of one division and the beginning of the next. The nucleus is large, but the chromosomes are not distinguishable.

   c. The nuclear membrane forms, the chromosomes uncoil, the nucleoli reappear, and the spindle vanishes.

   d. The nuclear membrane disappears, but the spindle apparatus becomes visible. Nucleoli disappear, and the centromeres connect to the poles.

   e. The sister chromatids separate and move to the poles, and the spindle elongates.

   Interphase
   Prophase
   Metaphase
   Anaphase
   Telophase

3. What is the actual significance of cell division?
4. What does fertilization contribute to the zygote?
5. Define meiosis.
6. Relate the stages of meiosis and their significance.

References

Baserga, R.: *The Biochemistry of Cell Division*. Charles C Thomas, Publisher, Springfield, Ill., 1969.

Franz, C. H.: *Normal and Abnormal Embryological Development*. National Research Council–National Academy of Sciences, Washington, D.C., 1967.

Haines, R. W., and Mohiuddin, A.: *Handbook of Human Embryology*. E. & S. Livingstone, Ltd., London, 1968.

Levitan, M., and Montagu, A.: *Textbook of Human Genetics*. Oxford University Press, New York, 1971.

Mazia, D.: The Cell Cycle. *Sci. Am.,* **230**(1): 54–64, 1974.

Thomas, J. B.: *Introduction to Human Embryology,* Lea & Febiger, Philadelphia, 1968.

Williams, P. L.; Wendell-Smith, C. P.; and Treadgold, S.: *Basic Human Embryology.* J. B. Lippincott Co., Philadelphia, 1966.

Wolfe, S. L.: *Biology of the Cell.* Wadsworth Publishing Co., Inc., Belmont, Calif., 1972.

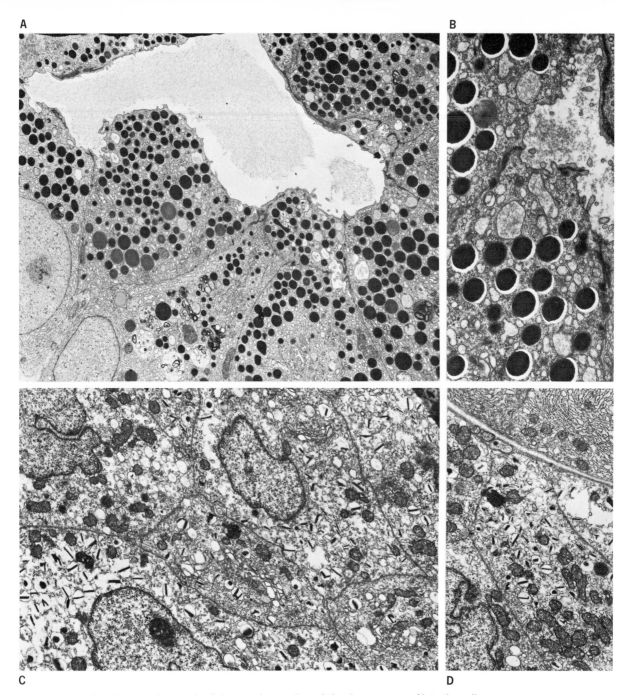

Figure 3-1. A. An electron micrograph of the exocrine portion of the dog pancreas. Note the cells and the exocrine duct. ×4900. **B.** An enlarged view of **A** showing the cells more clearly. ×12,474. **C** and **D.** Electron micrographs of the endocrine portion of the same dog pancreas. Beta cells with insulin granules are clearly seen. **C,** ×8,316; **D,** ×7,425. (Courtesy of Dr. Dale E. Bockman, Medical College of Ohio at Toledo.)

Cell Function

Patterns of cell structure form the machinery devoted to special jobs in special cells. There is a structural and functional relationship to be described. This will be done predominantly by example rather than in just a scattered survey. Greater detail follows in subsequent chapters of the book. In general, cells perform the following functions: (1) secretion, (2) absorption, (3) permeability, (4) phagocytosis, (5) storage and carriage, (6) protection, (7) support, (8) movement, (9) contraction, (10) communication, and (11) photoreception.

Secretion

Secretion is the elaboration by a cell, from precursor materials, of a new substance that is then released from the cell. Special groups of cells gathered together for such a purpose are known as glands. Although gland cells are particularly concerned with secretion, other types of cells also show this activity.

A secretory product may be stored in the cell (for some time) in the form of secretory granules, which often have characteristic morphology. Eventually these are discharged for use elsewhere. The fine structure of cells involved in secretion varies in characteristic ways according to the secretion.

Glands fall into two major classifications: endocrine and exocrine (Fig. 3-1; Fig. 4-1, p. 92). The exocrine gland passes its secretion through a duct onto a surface or into a hollow organ in the body (e.g., the salivary glands secrete into the mouth and the exocrine portion of the pancreas secretes digestive enzymes into the small intestine). Different exocrine glands form secretions of different chemical natures. Endocrine glands, on the other hand, secrete a chemical messenger or hormone that passes from the gland into the bloodstream and is carried to its destination (the pituitary, adrenal, thyroid, gonads, etc., are examples of endocrine glands). The hormones, too, have different chemical composition and produce specific biologic effects.

Protein Secretion

Protein molecules perform many metabolic and mechanical functions. All enzymes, including those of digestion released in the gut, are protein molecules. The secretions of the zymogenic cells of the gastric glands and the pancreas are protein secretions. Note that protein secretion is not confined to the exocrine glands. The endocrine thyroid gland manufactures a protein conjugated with its specific iodine-containing hormone. Cells that secrete proteins are not always found specifically in glands. The fibroblast makes the structural protein molecules of collagen, and the plasma cells produce the antibody protein as part of the defense mechanism of the body against foreign material. Protein secretion is widespread and a very important cell function. The secreting cell is specifically equipped for its job.

The essential feature of the protein-secreting cell is the presence of an elaborate system of granular endoplasmic reticulum, which may fill most of the cytoplasm. The cisternae are long, often extensively interconnected, and often so very closely packed that the

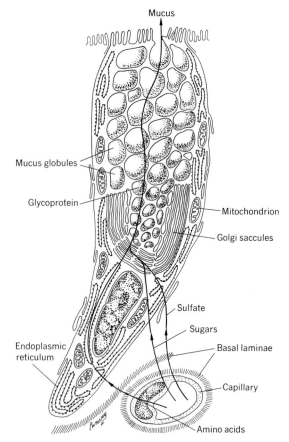

Mucus

Mucus globules

Glycoprotein

Mitochondrion

Golgi saccules

Sulfate

Sugars

Basal laminae

Capillary

Endoplasmic
reticulum

Amino acids

Figure 3-2. Goblet cell and mucus formation. Amino acids are synthesized on ribosomes into proteins, which then move through the endoplasmic reticulum to enter the Golgi saccules. Simple sugars are taken up by the saccules and combine with the protein to form glycoprotein. Sulfate is added to the latter. In the saccules the glycoprotein is converted into mucus globules, which migrate to the luminal side of the cell and are released to coat the intestinal surface. (After studies by LeBlond.)

cytoplasmic space containing ribosomes between adjacent cisternae is narrower than the width of the individual cisternae. Interconnections between cisternae may allow a widespread continuity throughout the system and allow for a dynamic unity of the components (Fig. 3-2).

The contents of the cisternae vary in appearance in different cells but are usually of low density as compared with cytoplasmic ground substance with its many ribosomes. The material in the cisternae may be flocculent and the cisternae may be dilated, suggesting the accumulation and storage of material in the cisternae. In certain cells (e.g., the pancreatic zymogenic cells) of a few species, intracisternal zymogenic granules are seen but the material is usually not organized and the secretion granules actually originate in the Golgi apparatus. In a cell with substantial endoplasmic reticulum the membranes of the cisternae divide the cytoplasm into two phases: the interconnecting cavities of the endoplasmic reticulum, related to the inner nuclear membrane through the perinuclear cisternae; and the ground substance, with ribosomes and other components related to the nucleus through the nuclear pores.

In the cells that secrete protein in granular form, discharge takes place at the cell surface and the Golgi is prominent. The lamellae of the Golgi are long and the vacuoles often prominent. Endoplasmic reticulum usually lies close to the outer aspect of the Golgi apparatus. Transfer of newly synthesized protein material from the cavities of the cisternae of the Golgi apparatus probably takes place by production of small blebs or buds, similar to micropinocytic vesicles, which form from the membrane of the cisterna and which fill its contents, finally pinching off to form a free vesicle. The vesicles then pass to the Golgi apparatus and fuse with its outer aspect, releasing their contents into the Golgi sacs. The components of the Golgi apparatus are functionally and structurally polarized since secretion granules tend to occur on the opposite side of the Golgi apparatus from the endoplasmic reticulum.

The final secretion granule seems to form by an accumulation of material within the Golgi sac. The formed granules that lie closest to the Golgi apparatus are often paler than those released from the Golgi apparatus and have passed to their storage area in the cell apex suggesting that the granules may continue to mature by withdrawal of fluid or by action of enzyme systems associated with Golgi membranes from which the membrane around the granule originates.

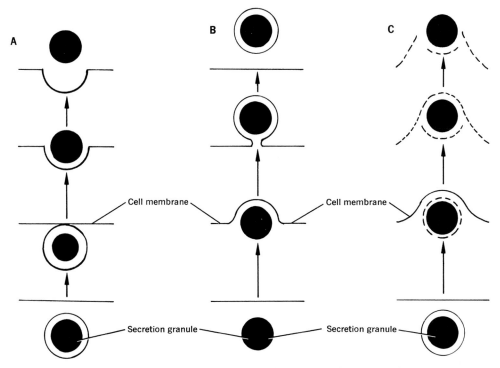

Figure 3-3. Schematic diagram showing forms of secretion. **A,** merocrine; **B,** apocrine; **C,** holocrine.

The mature granule is commonly released at the cell surface by fusion of the surrounding membrane with the cell surface permitting discharge without a breakdown of structural integration of the cell. This is called merocrine or eccrine secretion (Fig. 3-3). As the granulated dense material is exposed to the extracellular fluid, it changes to a pale, flocculent material probably owing to the uptake of H_2O. The apical surface of the protein-secreting gland usually has sparse microvilli, the significance of which is not clear since they do not seem to be associated with absorption.

Examples. The most extensively studied protein secretion cell is the zymogenic cell of the exocrine pancreas (Fig. 3-4), which is similar in fine structure to the chief cell of the stomach glands. The paneth cell of the intestinal crypt also has all the morphologic characteristics of a protein secretion cell, but its precise secretory function is uncertain. The cells of the

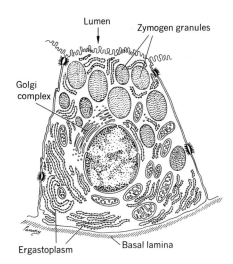

Figure 3-4. Diagram of a zymogen cell (as seen in electron microscope) of the exocrine pancreas.

seminal vesicle have similar ultrastructural as well as characteristic large secretion granules. The thyroid has a prominent and widely, often dilated, granulated endoplasmic reticulum and behaves like an exocrine cell, since initially it discharges its protein-conjugated hormone into storage follicles. The endoplasmic reticulum of the fibroblast varies morphologically, depending on how actively it is laying down collagen. The plasma cell (producer of antibody protein) clearly shows the peripheral distribution of endoplasmic reticulum, surrounding the centrally placed Golgi apparatus. In the plasma cell and fibroblast, the secretory product is probably released in molecular form at the cell surface and not as secretory granules.

Acid Secretion

This secretion is remarkable since the very physical nature of the secretion is far removed from normal body constituents. The parietal cells (Fig. 3-5) of the gastric glands of the stomach secrete HCl in sufficient concentration to produce gastric contents of pH 1.0, an acid level that could cause damage and death to most cells. The cell's function is the concentration of

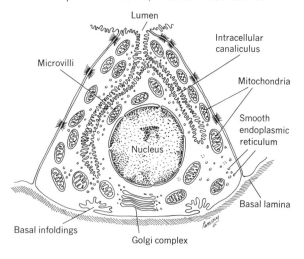

Figure 3-5. Gastric parietal cell (acid secreting). Note the intracellular canaliculus that almost surrounds the nucleus, the many large mitochondria (a sign of rapid oxidative metabolism), and smooth endoplasmic reticulum, involved in acid secretion.

hydrogen ions (H^+). Since the pH of blood and tissue fluids is maintained at slightly above pH 7.0, the parietal cell can concentrate H^+ by almost a million times. The gastric cells have several structural adaptations related to their function.

Surface specializations are important since the transport of ions seems to depend on the available area of surface membrane of the cell. The intracellular canaliculus, a tubular invagination of the apical surface, is its most characteristic feature. Branches of this canaliculus ramify in the cytoplasm around the nucleus and often even reach close to the cell base, forming an elaborate system that greatly increases the potential secretory cell surface and ensures that no part of the cytoplasm is far removed from the effective secretory surface. Since the lumen of this system is in continuity with the lumen of the gastric gland, acid secretion passed into the system at any part is discharged into the gland. The surface area of the canaliculus system itself is increased by the presence of numerous club-shaped microvilli that project into the canaliculus lumen and may even almost fill it. Through these specializations of the apical surface, an adequate area for secretory transfer is attained, despite the limited apical portion of the cell that is able to reach the restricted lumen of the gastric gland. It would otherwise be impossible to have the full acid-secreting potential within such a restricted space.

The base of the parietal cell has a further specialization to increase available membrane surface. Basal infolds (Fig. 3-5) of the membrane occur at several points (the basal lamina does not follow these folds). The extent and complexity of these folds vary with the functional state of the cell.

In the cytoplasm there are two major specializations involving the mitochondria and smooth endoplasmic reticulum. Mitochondria are large and numerous, occupying a large proportion of the cytoplasmic volume indicating the considerable demands that acid secretion places on the energy source of the cell. The rate of oxidative activity is well above average. The smooth endoplasmic reticulum is also extensively developed. The Golgi complex and the rough endoplasmic reticulum are poorly developed in the parietal cell and are not important in acid secretion.

Although in the mammalian stomach acid secretion and enzyme secretion take place in different cells (parietal and chief), certain lower species (e.g., bird) combine both in one type of gland cell. In the stomach of the hen, the gastric gland cell has a well-developed granular endoplasmic reticulum, prominent Golgi, and typical membrane-bound secretion granules. In addition, there is a well-developed smooth endoplasmic reticulum, and the mitochondria are characteristic of acid-secretion cells with the surface membrane specializations present. Deep clefts between adjacent cells replace the intracellular canalicular system that allows the sides as well as apex to be used for ion transfer. The junctional complex is displaced near the base of the cell. During active secretion, the cell surface and clefts become covered with the typical club-shaped microvilli, the basal infoldings become elaborate, and there is a decrease of smooth endoplasmic reticulum in the cytoplasm.

Endocrine Secretion

The typical endocrine gland has no lumen or duct and discharges its secretion into the capillaries at the base of the cells (Fig. 3-1). Thus, a rich blood supply is an essential part of an endocrine gland, and capillaries must come into close contact with these cells. Every cell in such a gland has at least one and often two surfaces in close reach of a capillary. The capillary endothelial wall is thin and fenestrated, and the vessels are separated by minimal amounts of connective tissue from the secreting cell surface. Several barriers do exist despite this close relationship. After discharge, the hormone must cross the epithelial basal lamina, the connective tissue space, the capillary basal lamina, and the capillary endothelium. Fenestrations commonly seen in the capillaries of endocrine glands may ease the passage. The hormone probably is dissolved in the tissue fluids after leaving the cell and then by diffusion, assisted by a concentration gradient, passes from the cell base to the capillary lumen.

The difference in cell structure of the various endo-

crine glands can be explained on the basis of the wide range of hormone secretions produced, and yet there is a common structural link between a number of glands including: the anterior pituitary, adrenal medulla, pancreatic islet, and the enterochromaffin cells of the intestinal epithelium. In all, the secretory product is formed into granules in the Golgi apparatus. The small dense granules surrounded by limiting membranes derived from the Golgi membranes accumulate in the cytoplasm. In the case of the enterochromaffin or argentaffin cells of the intestine, they are predominantly distributed toward the base of the cell and may even be discharged at the cell's base just as individual secretion granules leave the apex of the cell.

By careful study of the granular morphology, the various types of endocrine cells that contain small cytoplasmic granules can be separately identified. The different cells of the anterior pituitary each have distinctive patterns of granulation.

The beta cells of the pancreas islet have granules that may contain crystals with a species-specific morphology (Fig. 3-6). In the adrenal medulla, the granules contain epinephrine. In certain cases the relationship of the granules to the hormone of the gland can be confirmed when secretion is stimulated, and one observes degranulation of cells and a rise in level of hormone in the blood.

The secretion of steroid hormones by the adrenal cortex and by interstitial cells of the testis is associated with several fine structural features. The mitochondria of these cells often have tubular cristae instead of the typical shelf configuration, and a moderately well-developed smooth endoplasmic reticulum is seen. Accumulation of lipid material in the form of irregular cytoplasmic inclusions is often seen (Fig. 3-7).

The thyroid cell is unusual since it has features of an exocrine and an endocrine gland. The initial product of the cell is thyroglobulin, a protein conjugated with the thyroid hormone. This is not released into the circulation but is stored in a follicle (a blind vesicle lined by thyroid gland cells). The cytoplasm of these cells contains a prominent granular endoplasmic reticulum with dilated cisternae and a large Golgi appara-

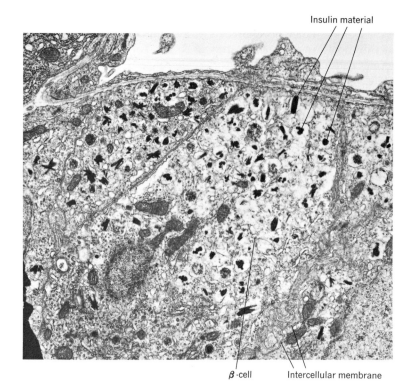

Insulin material

β-cell

Intercellular membrane

Figure 3-6. Several beta cells in an islet of Langerhans are seen. Note the black crystal-like "insulin" material in the beta cell. Dog pancreas; ×10,800. (Courtesy of Dr. Dale E. Bockman, Medical College of Ohio at Toledo.)

Figure 3-7. Histologic section of adrenal cortex indicating areas of lipid accumulation. **A,** ×75; **B,** ×200.

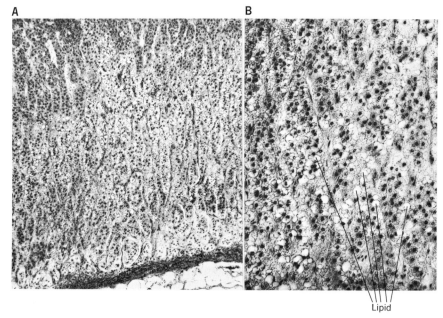

A

B

Lipid

tus. The thyroglobulin is released into the follicle by a process morphologically identical to exocrine secretion, but the follicles are not drained by ducts and act as storage spaces for reserve hormone. The cell also functions in the opposite direction as an endocrine gland. Thyroglobulin can be removed by pinocytosis from the follicle by the thyroid gland cells, which then detach the hormone component and discharge it from the base of the cell into adjacent capillaries. The final secretory activity of the gland is the result of these two opposing aspects of cell function.

Endocrine secretion is carried out by many different cells with widely different products, but common structural features are seen. Many aspects of endocrine function have not as yet been fully investigated owing to limitation of methods.

Absorption

When one thinks of absorption, one simultaneously thinks of the digestive tract, particularly the small intestine. Here, foodstuffs are broken down by the action of digestive enzymes secreted into the lumen of the intestine. The columnar cells of the intestinal villi were thought to absorb the simple molecules resulting from extracellular, intraluminal hydrolysis. The structural specializations of the intestinal cell were interpreted as being purely absorptive. However, recently it has been shown that many enzymes that were thought to conduct hydrolysis in the lumen and supposed to originate in the crypt cells of the intestine are not present in the lumen in concentration sufficient to account for the rapid disappearance of their substrates during normal digestion and absorption. It is now believed that these enzymes (such as the carbohydrases and peptidases), which are located in or on the cell surface and appear free in the lumen, are there as a result of the normal physiologic desquamation of cells from the villus. Thus, the structural specialization of the intestinal cell is both absorptive and digestive in nature with digestive hydrolysis occurring at the surface of the cell.

The distinctive feature of the intestinal epithelial cell is the presence of a striated border, a refractile zone about 1 to 2 μm thick at the surface of the cell. It forms the interface between the cell and contents of the lumen. As seen by electron microscopy, the surface of the intestinal cell is covered by highly organized, parallel fingerlike projections called microvilli that project into the lumen from the cell surface. Each consists of a cytoplasmic core with a surface formed by the surface membrane of the cell. As many as 1700 may be seen on the surface of a single cell, increasing the surface area more than 20-fold. The microvilli are longer and more numerous toward the tip of the intestinal villus where absorption is more rapid (Fig. 3-5).

When seen in longitudinal section, a central core of fibrils or narrow tubules from 30 to 60 Å in diameter is seen extending from the tip of the microvillus to the cytoplasm at its root. On cross section, the central fibrils are found randomly spaced in the core with about 40 in number (unlike the organized pattern seen in cilia, which are larger, more complex, and have a clearly defined constant internal pattern). The cores of the villi are linked together at their bases by a transverse network of "felted fibrils" extending across the cell apex between the junctional complexes, forming a region of cytoplasm—the terminal web. The fibrillar network of the terminal web gives rigidity to the apex of the cell and prevents deformation of the individual cell and of the epithelial surface as a whole in an otherwise very delicate mucosa.

The most significant part of the microvillus border is the apical surface membrane of the intestinal cell through which all absorbed material must pass. The membrane over the microvilli shows the usual trilaminar structure but is about 105 Å thick (thicker than most other biologic membranes). The membrane at the apex is thicker than the membrane that forms its lateral and basal surface. It is suggested that the membrane covering the villi has a mosaic of enzymes built into its structure. The increased surface at the apex can be considered as a specialization making available a greater surface area to accommodate larger numbers of digestive enzyme assemblies at the absorptive surface between cell and lumen.

The intestinal cell has a moderately well-developed smooth endoplasmic reticulum with strands of poorly organized granular reticulum and some free ribosomes. The Golgi apparatus forms a complex system in which dilated Golgi vacuoles are often prominent. The mitochondria have no special features. The nucleus is basally placed, moderate in size, and oval with diffuse chromatin. Like all epithelial cells, they are joined at their apices by junctional complexes and along their contact surfaces by intermittent adhesion plaques of desmosomes.

Variations occur, depending partly on functional state and on cell maturity. The cells of the villus are constantly replaced by division of precursor cells in the crypts and migration of these newly formed cells to the villus tips where, after a "working life" of only two to three days, they are extruded. Thus, the cells at the tip are more mature than those in the crypt and at the villus base. The crypt cell cytoplasm has numerous free ribosomes and little formed endoplasmic reticulum (characteristic of a "progenitor" cell). As the crypt cell passes on to the villus, it acquires the features of the mature cell while losing its potential for division; with this maturation there is a development of enzyme systems and an increase in absorptive capacity.

Fat Absorption

Early work suggested that fat droplets could be taken up directly from the lumen in micropinocytic vesicles formed at the roots of villi. It is now thought that this pathway is not significant in quantitative terms.° Fat absorption occurs to a greater extent in the form of lipid micelles less than 100 Å in diameter held in suspension by the bile salts and fatty acids in the intestinal lumen (Fig. 3-8). The lipid is probably broken down at the cell surface and resynthesized in the endoplasmic reticulum, which has the necessary enzymes. The lipid droplets then appear in the cisternae of the endoplasmic reticulum, through which they are transported in the cell and pass to the Golgi apparatus. Chylomicrons are made in the intestinal cell by the production of a lipoprotein envelope for the lipid droplets, which then pass out of the cell to the lacteal in the core of the villus.

Malabsorption

Abnormalities of the highly specialized intestinal epithelium may cause impairment of foodstuff absorption. If a single enzyme (e.g., lactase) is absent from the surface membrane of the microvillus (a normal abnormality), there is a failure to absorb and digest lactose from the diet, leading to malnutrition and diarrhea in infancy. The absence of a single enzyme, however, may cause little significant alteration in the ultrastructural change in the epithelium cells. On the other hand, certain people are sensitive to wheat protein, gluten, which damages the intestinal epithelium causing severe structural abnormalities such as disorganization and shortening of the microvilli. This is called idiopathic steatorrhea in adults and celiac disease in children and is called a gluten enteropathy. It is characterized by a malabsorption syndrome, in which all digestive and absorptive activities of the intestine are reduced.

Renal Absorption

In the renal tubule aspects of absorption and secretion are occurring simultaneously during urine formation. There are fine structural differences in the cells in the different segments of the nephron.

Urine formation begins in the glomeruli with filtration of plasma resulting in a dilute urine that collects in Bowman's space and first passes into the first part of the tubular nephron. In man 120 ml of filtrate are formed per minute, of which 99 percent is reabsorbed, along with many solutes, by the renal epithelium (mostly in the proximal convoluted tubule). The epi-

---

° Although pinocytosis is not now thought to be particularly important in fat absorption in adults, it does apparently form a pathway for uptake of antibodies by young mammals. The antibodies ingested in maternal milk are transferred unaltered to the infant circulation, which results in passive immunity to disease.

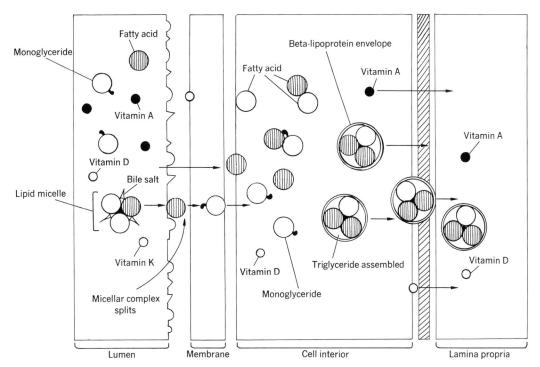

Figure 3-8. Normal fat transport. Monoglycerides, free fatty acids, phospholipids, and bile salts combine to form water-soluble micelles, essential for normal absorption into the cell. Enzymes in the cell reassemble fatty acids and glycerol and coat the molecule with a beta-lipoprotein envelope, permitting it to pass to the vessels in the lamina propria.

thelium of this tubule is columnar or cuboidal and has a striated border. As in intestinal epithelium the striated border consists of many closely packed microvilli. These are embedded in a dense surrounding layer. Tubular invaginations of the cell surface between the villi are common in this cell type and provide channels for the uptake of electron-dense materials from the nephron. The morphology of these apical channels is affected by osmotic variation.

The transport of HOH and ions across the proximal tubule is reflected in the basal membrane specializations. The cell base is infolded to a great degree, the invaginated paired membranes extending from the cell base deep into the cytoplasm. Marked ATPase activity has been shown at the cell base, suggesting pressure in the membrane on an energy-consuming active transport system. The large and numerous mitochondria of these cells are packed between the infolding membranes with orientation mostly parallel to the invaginations of the cell base. (See Chap. 13 for detailed discussion of renal absorption.)

Permeability

The rapid passage of nutrient or waste products of metabolism across cellular boundaries in and out of cells is of great importance. Governing this rate of diffusion across the cell barrier are the thickness and permeability of the barrier itself. In the capillaries of the lung and kidney, where materials cross cell barriers, the obstruction to diffusion is decreased to a minimum by thinning of the cells that form the barrier.

Capillaries

The blood capillaries (smallest vessels of circulation) form the major barrier throughout the body between circulating blood and cells of the various tissues. The oxygen and essential metabolites must pass across the capillary wall, and waste products of metabolism must move from the tissues into the capillary.

The capillary (Fig. 3-9) is a delicate tubule composed of flattened endothelial cells joined together by adhesion specializations along their contact or meeting edges. Each cell has a thickened area with a nucleus along with a few common cytoplasmic components (e.g., small packet of Golgi membranes, few vesicles, and occasional mitochondria). The endothelial cell is surrounded by a closely applied basal lamina around which are connective tissue elements with collagen fibers tending to fuse with the basal lamina. Thus, there is an extracellular connective tissue space around the capillaries that separates it from the cells of the organ through which it passes.

In the thinnest part of the capillary wall (only a few hundred Angstroms thick), two specializations occur to aid in passage of material across the endothelial barrier. (1) There are flask-shaped invaginations of the inner and outer surface of the endothelial cell called caveolae (probably micropinocytotic vesicles). These contain small amounts of extracellular or intravascular fluid that pinch off from the cell surface and form isolated vesicles in the endothelial cytoplasm. It is thought that they pass across the cell and fuse once again with the opposite surface membrane, discharging contents on the opposite side. Thus, the endothelial cells actively assist transport across the capillary wall. (2) Seen in some capillaries and not in others are capillary pores or fenestrations, which appear (in the endothelial lining) as discontinuities of several hundred Angstroms in diameter. These pores are bridged by a diffuse diaphragm that appears thinner than the sur-

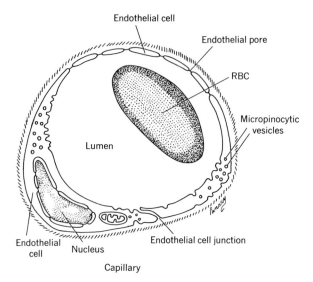

Figure 3-9. Schematic drawing of a capillary.

face membrane of the endothelial cell but is a significant structural barrier. The pores may appear at frequent intervals in the capillary wall.

Although the caveolae and pores play a role in vascular permeability, it is not certain to what extent they give either active or passive assistance to the passage of different substances. The normal vascular permeability may be due mainly to a physiologic leakage at the junction between the endothelial cells, since small molecules can pass between the cells, suggesting that their contact specializations are not designed to completely seal off the vessel lumen. During inflammation, when there is an increase in vascular permeability, this leakage between cells is apparently increased.

In different sites, the endothelium of the capillaries varies in structure. In the endocrine glands and intestine, where they are receiving hormones or absorbed food materials in addition to allowing normal metabolic interchange, fenestrations are commonly present. In the brain there are no fenestrations, but there are the perivascular foot processes of neuroglia cells closely applied to blood vessels, possibly providing a basis for the physiologic blood-brain barrier (Fig.

4-11, p. 115). Other differences may also exist with relation to specific endothelial function.

The capillary wall is not the only barrier to diffusion between tissue and blood. Between the circulation and the cytoplasm of an epithelial cell one finds the capillary endothelial cell membrane and cytoplasm, the endothelial basal lamina, and the membrane of the epithelial cell itself. Diffusion can be influenced at any of these points by selective processes. (See Chap. 10 for detailed discussion of the capillaries.)

The Lungs

The lung is designed to create a large interface between air in the alveoli and blood in the highly complex lung capillary bed. Across this interface pass oxygen, for metabolism of the body, and, in an opposite direction, carbon dioxide, the end product of metabolism. The interface must resist leakage of plasma from vessels that are poorly supported and must withstand mechanical stress during respiration and coughing and yet be thin enough to permit gas exchange at a sufficient rate.

The pulmonary capillaries have no fenestrations and show no other distinctive feature. At places, the wall shows flasklike caveolae, suggestive of micropinocytosis. The endothelial cells have the usual basal lamina investment with a surrounding but often very narrow connective tissue space where pulmonary macrophages are often seen. The capillaries are further reinforced by a delicate framework of fine collagen fibers and lie within the cavity of the common wall between adjoining alveoli. The epithelial lining is often so closely applied to either side of the capillaries that practically no loose connective tissue can intervene.

It is now known that there is a complete but tenuous epithelial cell layer continuous throughout the pulmonary air spaces. The nuclei of the main cell type, the flattened alveolar lining cells, tend to lie in the corners and angles of the alveoli leaving wide areas of wall covered by their flat and featureless cytoplasm. Since it is a continuous sheet there is an underlying continuous basal lamina closely applied to the cells. Thus, although the entire barrier between air and blood may

be less than 100 Å, it always consists of alveolar lining epithelium, its basal lamina, a connective tissue space, endothelial basal lamina, and endothelial cell.

The alveolar pattern of the lung creates certain problems. Most of the epithelial surfaces of the alveoli must be separated from each other to ensure aeration; yet the great surface tension forces acting at the alveolar level tend to produce adhesions between the alveolar walls and try to collapse them. The surface tension forces in the alveoli are apparently reduced by the secretion of a surface-acting substance (surfactant) by a cell in the alveolar epithelium, the great alveolar cell. These cells, which usually lie in the angles of the alveoli, are interspersed between the alveolar lining cells and are joined to them by junctional complexes. The great alveolar cells share the same basal lamina as the lining cell and are clearly epithelial in type. They have a flat, cuboidal shape with dense cytoplasm, lamellated secretion inclusions, and characteristic short, irregular surface microvilli. (See Chap. 11 for detailed discussion of the lungs.)

The Kidneys

Here a complex relation between endothelial and epithelial cells exists at a site where controlled permeability is essential. In contrast to the lung, where retention of fluid in capillaries is essential, filtration of the circulating plasma occurs. The fluid from the capillary passes through the filtration barrier into the urinary space, and the first part of the tubular system of the nephron (functional unit) has a dilute form of urine, which is then concentrated and altered as it proceeds along to secretion.

The renal glomerulus consists of a tuft of capillaries projecting into Bowman's capsule, the dilated blind end of the nephron (see Chap. 14). The flat cells of the parietal epithelium lining of the capsule are continuous with the visceral epithelium of the capsule at the point where the capillaries invaginate the capsule. The visceral epithelium of the capsule is a system of cells that invest the individual capillaries of the tuft. Between visceral and parietal epithelium cells of capsule lies a narrow cleft, the urinary space, which is continuous with the lumen of the nephron and drains ultimately through the renal pelvis and ureter into bladder and urethra.

The glomerular capillaries are extensively fenestrated with pores in the endothelial cells, bridged by a diaphragm. A prominent basal lamina surrounds the capillary. Thus, each capillary of the glomerulus is covered by an epithelial sheath consisting of the podocytes and the visceral epithelial cells of Bowman's capsule. The epithelium does not form a continuous layer but a complex interdigitating covering of closely packed podocyte foot processes or pedicels sent out in different directions by the podocytes, which lie between the capillary loops. A single podocyte may have foot processes applied to the surface of several capillaries while its nucleus and surrounding cytoplasm are in a cleft between the vessels. Adjacent foot processes are joined by a delicate connection, the filtration slit membrane, which is the major barrier to free passage between the processes. It is possible that the podocytes do more than perform a simple mechanical filtration but may influence the composition of the glomerular filtrate by selective action.

Since the podocytes are epithelial in nature, they lie in contact with the continuous basal lamina intervening between foot process and capillary endothelium. Between the epithelial and endothelial cells there is in fact only a single, thick, basal lamina called the basement membrane by histologists. Much of its substance may be produced by the podocyte. It forms a selective barrier in glomerular filtration.

Electron microscopy of the glomerular basal lamina and its relation to endothelial and epithelial cells has recently appeared to be significant in certain renal diseases. Early changes, not formerly visible with light microscopy, can be seen in glomerular disease. These changes influence the selective filtration function of the glomerulus. (See Chap. 14 for detailed discussion of the kidneys.)

Phagocytosis

Certain cells have the ability to pick up particulate matter, which then becomes segregated in their cyto-plasm in specialized structures designed for intracellular digestion. The process of uptake is called phagocytosis. Theoretically pinocytosis is the uptake of fluids while phagocytosis implies particulate uptake, but this is really artificial separation. Phagocytes ingest bacteria and debris from damaged cells as well as a great variety of foreign materials. The sequel to phagocytosis is enzymatic hydrolysis or digestion within the cell. In complex animals this process of phagocytosis is a function of special cells. The most important are the leukocyte (white blood cell) and the macrophage (see Chap. 15), which are widely encountered in different organs and form part of the so-called reticuloendothelial system of the body.

The most important cytoplasmic components of the macrophage are the lysosomes (Fig. 1-34, p. 42; Fig. 1-35, p. 43). These inclusions contain the powerful hydrolytic enzymes necessary for controlled digestion of particles in the phagocyte. The enzymes are an essential product of the cell, a form of secretion that is not discharged but performs in the cell itself. The primary lysosomes are newly formed, simple, membrane-bound structures with homogenous or finely granular contents and are probably produced in their final form by the Golgi apparatus. Aside from the lysosome, the macrophage has few distinctive cytoplasmic features. When the cell is stimulated by materials that induce phagocytic action, the cell surface becomes ruffled and irregular, throwing out surface projections that cause the cell's outline to fluctuate. Electron microscopy shows irregular indentations and projections of the cell surface membrane. The foreign material adheres to the cell surface, which then becomes invaginated to form a cytoplasmic vesicle. Thus, by phagocytosis, the cell ingests extracellular material.

Digestion follows phagocytosis. The material phagocytized is brought into contact with the enzymes of a lysosome. If it is in small particles, they can be incorporated directly within the lysosome. If larger (e.g., bacteria) it is engulfed, and the lysosomes are discharged into the cytoplasmic vacuole that enclosed the material (thus forming a sort of intracellular stomach full of active digestive enzymes) but is isolated from the cytoplasm by the persisting membrane of the vacuole. This activity usually results in the death of a leukocyte, but the macrophage remains unharmed. The products of hydrolysis are finally absorbed in the cell's cytoplasm, while any indigestible residue accumulates in the lysosomal structure.

The complex bodies formed by combination of ingested material with lysosomal enzymes during active phagocytosis are the secondary lysosomes, while those with undigested residues are the residual bodies.

The macrophages of the reticuloendothelial system form a network throughout the body that reacts to the presence of foreign or unwanted material, attempting to remove and destroy it. Phagocytes act as scavengers, removing debris when cells die or are injured by disease, and attempt to keep the circulation clear of foreign matter. They form a defense mechanism against disease and infection in the body.

Carriage and Storage

Certain cells take the form of specialization to function in storage and carriage of materials for widely different purposes. In such cells the stored material may have its own specific fine structural appearances, while the formation of the material and the metabolic activity of the cell are often reflected in the structure of its cytoplasmic organelles.

Red Blood Cell. The major function of blood is to carry oxygen to the tissues. The red blood cells have become specialized to assist this function by accumulating in their cytoplasm large amounts of hemoglobin, which has a strong affinity for oxygen. In mammals, this is carried to extreme by the red blood cell, which loses its nucleus during late developmental stages, and the circulating mature red corpuscle becomes, not a true cell, but little more than a membrane surrounding a concentrated solution of hemoglobin within an apparently structureless matrix.

The red corpuscle precursors carry the biochemical apparatus for hemoglobin synthesis in their cytoplasm. Since hemoglobin is not used outside the cell but is stored in the cytoplasm, the cytoplasmic organelles are not those typical of a protein-secreting cell such as the plasma cell. Yet despite absence of a complex granular endoplasmic reticulum, the active protein synthetic function is "betrayed" by the many free ribosomes that give the early red blood cell precursors their typical basophilic cytoplasm. The iron component of the forming hemoglobin is reflected in the presence of ferritin molecules in cytoplasm. As hemoglobin accumulates in the cell (maturing), it replaces the cytoplasmic organelles. The concentration of hemoglobin in the mature red blood cell reaches as high as 33 percent. Thus, the specialization of the developing red blood cell results in production of an efficient carrier for oxygen. However, in the process, with the loss of the nucleus, the cellular identity of this carrier has been sacrificed in the interests of function and efficiency and the red blood cell cannot reproduce itself and has a limited span of life! (See Chap. 11 for detailed discussion of the red blood cell.)

Adipose Tissue. Fat storage is important in the body economy, in view of its value not only as insulation but also as a fuel rich in energy. The oxidation of fat produces over two times as much energy as oxidation of a similar amount of carbohydrate or protein.

Fat is normally stored in connective tissue cells and is characterized by the presence of a surrounding membrane and homogeneous consistency. The typical fat cell contains a single, large fat droplet that compresses the other cytoplasmic components to the extent that they are difficult to recognize. When mobilization of fat stores is needed, the lipid material is withdrawn from the cells and transported to be metabolized.

Brown fat is a specialized form of fat storage tissue particularly prominent in very young mammals and hibernating animals. The brown fat cell (in contrast) contains a number of smaller droplets and is characterized by the presence of many large and well-organized mitochondria, indicating an unusual rate of oxidative metabolism. Brown fat cells are thought to oxidize stored fat within their own cytoplasm for the purpose of heat production to maintain body heat in very young, poorly insulated animals or to restore the normal body temperature of the hibernating mammal prior to reawakening. (See Chap. 4 for detailed discussion of adipose tissue.)

Protection

Skin. The protection provided by the intact skin surface is so essential that the loss of a significant proportion of skin can prove fatal (e.g., burns). The skin protects the body from mechanical injury and from the loss of tissue fluids and proteins from deeper tissues by virtue of its continuously renewed layer structure and accumulation of keratin in its cells.

The epidermis is a stratified squamous, epithelial layer supported on an elastic, resilient but strong underlying layer of connective tissue. The adhesion between cells of the epidermis is very strong and adds to the ability of the skin to resist injury.

The imperviousness and resistance to attrition of the skin are due to the accumulation of the structural protein keratin in the maturing epidermal cell. As the cell passes toward the surface from the basal layer, where the population is renewed by continuous division of progenitor cells, they begin to synthesize and accumulate keratin in their cytoplasm. This synthesis is accomplished mainly by free ribosomes in the cytoplasm rather than by an elaborate cytoplasmic membrane system. The accumulated keratin eventually replaces the other cell components, and when the cell dies it forms an integral part of the protective horny surface layer of skin that is essential for its mechanical function. Again the price paid for this protective structural specialization is death of the cell. (See Chap. 16 for detailed discussion of skin.)

Support

The connective tissue in the body (cells and their products) has a predominant mechanical function. In connective tissue, the cells are relatively sparse while the intercellular material, fibers, and matrix are abundant. The mechanical strength of connective tissue lies in the fibers produced by the connective tissue cells and in the matrix around the cells. This is in contrast to epithelial tissues, in which the cells, the essential functional units, lie close together with little extracellular material.

Connective Tissue Fibers. COLLAGEN. Collagen, the major fiber of connective tissue, is synthesized by the fibroblast. In the light microscope it appears as dense eosinophilic bundles of coarse fibrous material, while in the electron microscope it is seen as an aggregate of finer dense fibrils each with a characteristic repeating pattern of fine cross-bandings every 640 Å. It is believed that the collagen is secreted by the fibroblast in the form of a tropocollagen molecule, a fibrous macromolecule with a characteristically high proportion of the amino acids glycine, hydroxyproline, and proline. These molecules appear to line up in chains that fit together in some way, with a regular overlap. Collagen is very resistant to both physical and chemical damage and is seen in a similar form in different species. Its properties make it ideal for its role as the mechanical framework of connective tissue.

One differentiates dense and easily recognizable collagen, which is eosinophilic in the light scope, from the more delicate framework of the so-called reticular fibers, which form around tissue elements. On electron microscopy fibers of collagen and reticulin differ only in size of aggregates of fibers formed but are essentially alike. Elastic fibers, on the other hand, appear thick and homogeneous in electron microscopy. They have irregular margins and show no periodic structure.

MATRIX. In loose connective tissue the matrix has no organized fine structure, but in bone and cartilage the matrix has taken on a rigid, mechanical function and it becomes more prominent owing to increased density. In bone, the crystals of calcium salts that are laid down in the matrix have a dense needle-shaped appearance. Collagen fibers in bone form the framework on which the dense components of matrix are deposited, and the repeating pattern can still be seen in places. (See Chap. 4 for detailed discussion of connective tissue.)

Movement

Cells have the ability to move. The ameba moves by bulging itself out in the direction it cares to go and

Figure 3-10. **A.** Diagrammatic reconstruction of a cilium and basal body; a median longitudinal section in the plane of the central axial fibrils. **B.** Transverse section at the level indicated. Note that there are nine subunits with two tubular components in each (subfibers a and b). Subfiber a may have small arms and is often more dense. The arrow indicates the direction of ciliary beat.

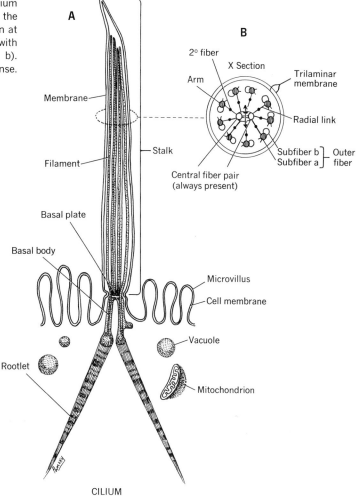

A

B

2° fiber

X Section

Arm

Trilaminar membrane

Membrane

Radial link

Stalk

Filament

Subfiber b — Outer
Subfiber a — fiber

Central fiber pair
(always present)

Basal plate

Basal body

Microvillus

Cell membrane

Vacuole

Rootlet

Mitochondrion

CILIUM

drawing itself in at the opposite end. The shape of this organism changes and is produced by molecular movement in the cytoplasm. In contrast, the paramecium (a protozoan) has a relatively constant shape but moves around with the help of hairlike protrusions from the cell membrane, called cilia. This movement, too, is a result of molecular movement. In higher forms of life, cilia are present in special sites where their action can be of value: the epithelial cells of the upper respiratory tract are ciliated, providing the surface of the trachea,

bronchi, and nasal passages with a sheet of cilia in constant motion. Mucus secreted by goblet cells forms a surface layer that traps inhaled particles (dust and bacteria), and the ciliary action sweeps the mucus back to the larynx where it can be removed by coughing.

Cilia. Each cilium is an elongated cytoplasmic extension covered by the apical surface membrane (Fig. 3-10) and measures 5 to 10 μm long and 0.25 μm in diameter and can barely be seen in the light scope.

The cilium has a central core of parallel subunits that extends from base to tip unbranched and is called the axial filament complex although in reality the subunits are tubular in structure. On cross section, one sees two tubules, each 240 Å in diameter, slightly apart, in the center of the complex; nine evenly spaced pairs of tubules with subfibrils; and two short arms extending from one doublet toward the adjacent pair. The components of the axial filament appear to arise from the basal body, a cylindric structure resembling a centriole that anchors the cilium in the base of the cell and may control and coordinate ciliary activity. The basal bodies arise during development from repeated division of the centrioles.

The cilia of a single cell or an entire region beat together with coordinated function. The direction of the beat is defined by the orientation of the central pair of the axial filament complex. The beat consists of a fast forward stroke and a slow recovery stroke. The mechanism is uncertain, but it appears that the nine peripheral parts bring about movement by contracting in a predetermined sequence with the two central tubules acting as a control or stabilizing force.

Flagella and Spermatozoa. The flagellum (a motile cell specialization) is closely related to cilia and is distinguished by its length (up to 150 μm) and its spiral action. The flagella provide spiral action for the sperm, which is carried over a considerable distance, to the ovum, with which it fuses to form the zygote and complete the genetic composition of the new individual produced.

The sperm (Fig. 17-11, p. 585) has two essential parts: the nucleus, which carries genetic information, and the propulsion unit, which carries it to its destination. The load that has to be carried has been streamlined to minimize it. Thus, during its development from precursor cells in the seminiferous tubules to its mature form, the spermatozoon sheds almost all its cytoplasm retaining only that part essential to motility. (The ovum, in contrast, has a considerable cytoplasmic accumulation of material available for cells of the early embryo.)

The head of the spermatozoon has a nucleus with highly condensed chromatin, existing predominantly in an inactive form. The restriction of nuclear size is clearly in the interest of efficiency and reduces drag. During development of sperm in the seminiferous tubules of the testis, a collection of material produced by the Golgi accumulates in the acrosome, a cap that becomes applied to the pole of the nucleus. It looks like a specialized lysosome that is concerned with penetrating the ovum at fertilization and starting division. All other components except those in the head are concerned with motility.

The sperm tail contains the axial filament complex (similar in cross section to that of cilia and flagella elsewhere). The axial filaments or tubules arise from a centriole at the posterior pole of the nucleus and run to the tip of the tail. In mammalian sperm, nine additional broad, dense components are arranged in a ring external to the nine tubular doublets of the axial filament complex. These are straight and unbranching and terminate at different levels along the sperm tail. They are asymmetric in arrangement and may be extracontractile elements reinforcing the axial complex.

Movement needs energy expenditure. This energy need must be met from mitochondria in the cell and by ATP produced by oxidative phosphorylation. The midpiece of the sperm consists of a tightly wound spiral of mitochondria that forms a coil through the core of which pass the components of the axial filament complex. The material for oxidation is gotten from the surrounding environment and broken down by mitochondrial enzymes to generate energy for conversion to mechanical work. Further along the tail, the mitochondrial sheath is replaced and the cell strengthened by a system of fibrous ribs linked longitudinally in the plane of the central filaments.

Muscle. Suffice it to say that complex animals composed of a myriad of cells cannot simply rely upon cilia or flagella. Specialized cells combine to form massive muscle tissue to enable the animal to move in a purposeful fashion. (See Chap. 7 for detailed discussion of muscle.)

Contraction

Undifferentiated protoplasm has primitive contractile powers and can contract down on itself slowly. In unicellular organisms, strands of specialized fibers have the ability to contract and do so quickly. Muscular tissue in the multicellular organism is the specialization that certain cells develop as their primary activity, the power of rapid contraction. Such cells contain numbers of intercellular contractile fibrils. The myofibrils and the cells tend to develop in an elongate manner to accommodate contractile fibrils and increase the extent of movement.

In the vertebrate, muscle is predominantly mesodermal in origin. In a few cases, such as some of the intrinsic eye muscles, it may be ectodermal. Associated with the epithelium of the glands of the skin and mouth, we find unusual contractile elements called the myoepithelial cells. These may represent an "old type" of contractile specialization as seen in the myoepithelial cells of the coelenterates.

Communication

Primitive cells respond to their environment. They possess irritability or the capacity to react to stimuli. Primitive cells can also conduct a "change of state" through their protoplasm. Nerve cells (see p. 114) represent, in man and higher forms, the elements in which a specialization has taken place for receptiveness and conduction of the changes initiated by the reception of a stimulus impulse.

Associated with the above is the enormous increase in total surface area of the cell membrane with a wealth of fine, branching, unsheathed dendritic processes for reception or input, the development in many nerve cells of long, often sheathed processes, the axons for output, the presence of special cytoplasmic Nissl granules, and the continual movement of molecules along the core of the axon fibers. (See Chap. 8 for detailed discussion of communication.)

Photoreception

Light is so familiar a part of the environment that its importance is often overlooked. Yet all of life depends (in different ways) upon the energy of light. In chloroplasts of green plants the reaction of photosynthesis occurs as the first essential link in the chain of life on which more complex animals depend. The animals, in turn, use light to provide information to assist their movement and have evolved photoreceptors, the eyes, which inform the nervous system about the environment. Although the chloroplast and eye use the energy of light for dissimilar purposes, the components concerned with trapping and transformation of light energy show ultrastructural specializations of a similar nature. (See Chap. 9 for detailed discussion of photoreception.)

Chloroplasts. The chloroplasts of plants are discrete, membrane-limited structures lying in the cytoplasm of plant cells and contain the green pigment chlorophyll. The internal structure of the chloroplast varies greatly in different species of green plant, but their essential common feature is the presence of parallel membrane lamellae associated structurally with chlorophyll molecules. In higher plants, the lamellae of the chloroplast form a number of discrete packages called grana in which the chlorophyll is located. Photosynthesis can take place only in the presence of chlorophyll associated with the membrane lamellae of the chloroplast, and thus the enzymes are probably spatially organized on the lamellar template. The reactions of photosynthesis use the radiant energy of light to promote the combination of the simple molecules of H_2O and CO_2, resulting in synthesis of CHO and release of O_2. Without this reaction, life, as we know it, would cease to exist.

The Eye. The photoreceptor cells of the eyes of animals show essential similarities in fine structural specialization. In all, the light-trapping cells contain numerous membrane lamellae or tubules with a high degree of spatial organization. The membranes are closely associated with molecules of photosensitive pigment. The similarity in principle to the chloroplasts is clear. In the vertebrate eye, the photoreceptor cells are the rods and cones each of which has an inner and outer segment joined by a narrow ridge. In the outer segment closely packed parallel membrane lamellae stacked like a pile of coins fill all the available space. Integrated with these lamellae are the molecules of the pigment retinene, closely related to vitamin A, in association with the protein component opsin. It is here that the radiant energy of light is trapped and transduced or converted into cellular activity leading to the initiation and propagation of a sensory nerve impulse. The inner segment of the receptor cell contains the other cytoplasmic components of the cell, including a significant amount of mitochondria. The ridge between the inner and outer segments is structurally similar to a cilium and is undoubtedly involved in the functional communication between the two parts of the cell. The efficiency of the light-trapping mechanism in the retina photoreceptor is so great that in some cases stimuli amounting to the simultaneous reception of only a few photons are sufficient to generate a visual sensation.

Review Questions

1. List ten major functions of a cell.
2. Differentiate between endocrine and exocrine secretion and give examples of each.
3. What is merocrine or eccrine secretion?
4. What cellular elements and modifications of cellular structure aid a cell in absorption?
5. If you were in the bloodstream and wanted to leave a capillary, what layers would you have to pass through?
6. What is phagocytosis?
7. What cells does the body use to carry and store materials?
8. Correlate the structure with the function:
   a. Protection   Connective tissue
   b. Support   Skin
   c. Movement   Cilia
   d. Contraction   Flagella
   e. Communication   Muscle
   f. Photoreception   Nerve cells
                           The eyes
9. Is each and every cell of the body endowed with only one function, or is it possible that some cells can have a duplicity of function? Discuss.

References

DeRobertis, E. D. P.; Nowinski, W. W.; and Saez, F. A.: *Cell Biology*, 5th ed. W. B. Saunders Co., Philadelphia, 1970.

DuPraw, E. J.: *Cell Physiology*, 3rd ed. W. B. Saunders Co., Philadelphia, 1968.

Frost, J. K.: *The Cell in Health and Disease*. Williams & Wilkins Co., Baltimore, 1969.

Giese, A. C.: *Cell Physiology*, 4th ed. W. B. Saunders Co., Philadelphia, 1973.

Goldsby, R. A.: *Cells and Energy*. Macmillan Publishing Co., Inc., New York, 1967.

Guyton, A. C.: *Textbook of Medical Physiology*, 4th ed. W. B. Saunders Co., Philadelphia, 1971.

Langley, L. L.: *Cell Function*, 2nd ed. Reinhold Publishing Corp., New York, 1968.

Levitan, M., and Montagu, A.: *Textbook of Human Genetics*. Oxford University Press, New York, 1971.

Winchester, A. M.: *Genetics*, 4th ed. Houghton Mifflin Co., Boston, 1972.

Wolfe, S. L.: *Biology of the Cell*. Wadsworth Publishing Co., Inc., Belmont Calif., 1972.

Woolridge, D. E.: *The Machinery of Life*. McGraw-Hill Book Co., New York, 1966.

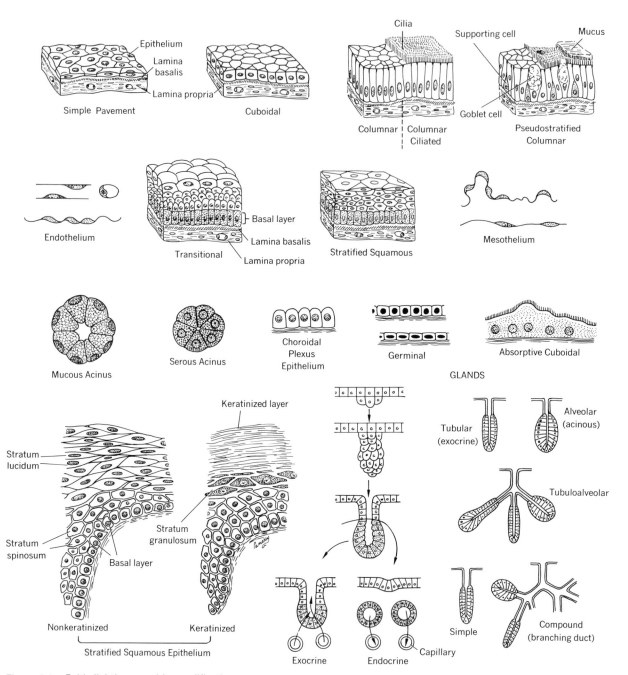

Figure 4-1. Epithelial tissue and its modifications.

Tissues of the Body

Identical cells multiply, diversify, and organize to form different tissues to make up the higher organism. The study of these changes and building processes is called embryology. Most cells in higher organisms belong to one of many specialist groups. They surround themselves by their own kind of specialist cells and limit themselves to a specific type of work in order to achieve the level of complexity found in multi-cellular organisms.

Much thought has been given to how cells become specialized and at what stage the irrevocable change occurs. It all begins with the original fertilized egg, which is no specialist as it produces many cell types in its development. In development the fertilized egg divides into two daughter cells, then into four cells, and so on. During all this early cleavage with rapid division the cells remain in a ball, which does not increase in size, and there are no signs of specialization (Fig. 2-2, p. 56).

In the human, the first step in specialization results in the production of three different types of cells. Ten to eleven days after fertilization, the embryo is a hollow ball, smaller than a pin's head, and contains these three cell types grouped into an inner, middle, and outer layer. These cells are now bound on a "no-turning-back course," and their descendants can only become more specialized with development. The outer layer forms the epidermis and the nervous system, the cells of the middle layer contribute to the formation of bone and muscle, cartilage, and blood vessels, and the inner layer to the digestive tract.

The chemical mechanism that directs the development of a cell into a specialized unit would be of little use if cells of similar character did not organize into tissue and the right cells did not manage to make their way to the right site at the right time. Movement is a cell's natural state and a fundamental fact of cell life. Yet when cells associate to form the tissues of an organism, they often give up their wandering and attach themselves exclusively to their own kind, ignoring dissimilar cells. They somehow "recognize their own." Although they are bound together, however, tissue cells (except for certain muscle) do not fuse or even come into actual contact with each other. The space between them is filled with a cementing substance or forms bridges to gap the intercellular space and bind the two cells together. This bonding is apparently ineffective in cancer cells that stick together less firmly than normal tissue and can migrate and spread. Furthermore, growth is apparently under strict control at all stages of development and cells of all types increase in number and size until structure is completed. Then, they suddenly stop as if a switch had been turned off.

Cell development does not end with the embryo or even with the adult organism that is taking shape. Cells change with age, and age change is part of the developmental process. As the cells age, man ages and death follows when a critical number of cells that are vital are no longer functioning. With age, the cell processes deteriorate due to imperfection or short supply, and though all cells carry the blueprints for replication and can divide and reproduce to replenish tissues for fresh, new parts at a rapid rate, this somehow fails and the cell does not live forever.

Varying cells have varying life-spans. An epithelial cell from the intestine lives about $1\frac{1}{2}$ days and is sloughed off; white blood cells live about 13 days; red blood cells die after 120 days. Liver cells rarely divide and nerve cells may live for 100 years.

The capacity of the cells to reproduce and repair decreases with age (thus an older person cannot repair a wound as readily). Sometimes the cell decreases in size, there is a slowdown in the amount of protein produced, and the Golgi complex fragments. The mitochondria break up or elongate. Even the genetic material and the enzymes can vary as a cell ages, to decrease the reproductive rate and threaten the very life of the cell.

Aging is a complex process involving the interaction of various cells and cell groups. There may even be a species, genetically determined limit to the life expectancy of both cells and the organism even with the modern medical means of lengthening the probable life-span to the seventh, eighth, or even ninth decade.

The Tissues

From a single cell, the egg (from the female parent), fertilization takes place by a sperm (from the male parent) to form the zygote and development begins. Cell division (mitosis) transforms the original cell into 2, 2 become 4, then 8, 16, 32, 64, 128, and onward until there are some 26 million cells in the newborn child, which may increase to about 100 million cells in the adult. With this rapidly accelerating division, a developing, growing cell mass called an embryo is formed. Within the embryo various cells take on different shapes, and these, in turn, produce cells similar to themselves. Thus differentiation takes place, which is an alteration in shape and structure of the cells, and groups of cells come to take special positions in the embryo. The cells not only increase but are responsible for the production of the extracellular elements that form most of the total body mass.

At birth most differentiation is complete, but mitosis continues until the child has grown to its full size. Mitosis never stops completely, for during childhood and maturity cells are continually wearing out and being replaced. Nerve cells, however, do not follow this pattern since we are born with a complete complement of these cells and they are not apparently replaced even should they die.

A large number of cells of similar character and function grouped together for the purpose of performing a single function is called a tissue. It is a convenient term, but few of the tissues, with the exception of the epithelia, are pure entities.

There are four fundamental tissues: (1) epithelia or lining tissues that cover surfaces inside and out; (2) the supporting or connective tissues; (3) the muscular tissues that form part of the structural framework as well as shorten and extend to do the body's work; and (4) the nervous tissue, the body's communicating system. As cells are grouped together to form tissues, different tissues group to form organs (e.g., liver, spleen, kidneys, etc.). The solid organs usually have a well-developed capsule or outer covering from which ingoing extensions form the basic framework. The hollow organs have their characteristic features on their inner or luminal surface, which is usually covered by a lining or mucous membrane.

In turn, the organs of the body are further organized into systems in which each individual organ is coordinated to assist in the functioning of the entire system (e.g., digestive system, urinary system, etc.). The various systems are finally coordinated to allow for interaction, balance, and integration.

The Epithelia

The epithelia (Fig. 4-1) form the lining membranes of the surfaces of the body, external and internal. Epithelial tissue may be derived from all three primitive layers of the embryo: ectoderm, endoderm, and mesoderm.

Epithelia are characteristic features of the special organs. Morphologically, they are composed of cells and a minimum amount of intercellular substances and thus are relatively pure tissues with few other elements in them. The cells are closely applied to one another with only a thin cementlike substance between them, which permits in many instances a gliding of cells over one another and offers little resistance to the passage of "wandering cells" of the connective tissue. In other instances the intercellular space is narrowed at the adjacent thickened cell membrane of the desmosomes (Fig. 1-27, p. 34). No blood vessels or lymph vessels penetrate the epithelia, and they maintain nutrition by absorption of materials through the basal or free surface (or both). Their regenerative powers are excellent.

The epithelia are predominantly protective. They may (1) act as cellular linings or cover surfaces; (2) protect against physical and chemical agents; (3) prevent excess fluid loss due to evaporation or loss of heat by radiation; (4) prevent the passage of molecules or permit such passage (called transudation and the fluid is a transudate) by osmosis if conditions are right; (5) aid filtration if there are actual differences in hydrostatic pressure; (6) help secretion if it is associated with cellular activity; and (7) function in the absorption of molecules that are selectively drawn into the cell.

Epithelial surface membranes, being exposed, are subject to normal wear and tear as well as injury but they are capable of a constant reepithelization. They are classified according to the arrangement of their cells in one or more layers and by the character of the individual cells. Note the presence of a thin layer of connective tissue just beneath the cell membrane at the base of the epithelial cells to form a basement membrane (see Chap. 1). It is a condensation of ground substance of the connective tissue with a supporting latticelike arrangement of fine reticular fibers.

Most epithelial cells produce some kind of secretion. The "tubes" carrying air from the nose to the lungs or those of the intestine (and many others) secrete mucus to keep the cell surface moist and protect the surface against mechanical injury. A modification of the "air passage" epithelium produces cilia (hairlike projections) on their surfaces that beat rhythmically in one direction in order to move fluid past the cell surface in a constant stream.

There are three major subtypes of epithelia based on the cellular arrangement as well as functional and morphologic characteristics of the cells.

Simple Epithelium. Simple epithelium (Fig. 4-1) consists of a single layer of cells that face the free or luminal surface on one side and the basal or subjacent tissue surfaces on the other. There are three types dependent on the individual cell shape: pavement, cuboidal, and columnar epithelium. These three forms of epithelia are further subdivided on a functional or positional basis (see below).

PAVEMENT EPITHELIUM. Pavement epithelium consists of thin, flattened cells that resemble diamond-shaped pieces of a puzzle; their luminal and basal surfaces are maximal but their intercellular contacts minimal (Fig. 4-1). These are lining structures, are not bound firmly, easily rub off, and have active secretory function.

Simple Squamous. This type of pavement epithelium is originally cuboidal but thins out as the area increases until it is like a "pavement." The cells are extremely thin, which allows substances to quickly pass through them. This epithelium is seen lining the anterior eye chamber.

Endothelium. Endothelium forms the essential lining surface of the blood- and lymph-filled cavities of the heart and vessels. The edges of cellular contact are sinuous and can withstand pressure without leaking. A white blood cell, however, can pass between two cells into the extravascular tissue and not rupture the wall. The luminal surface is normally nonwetting and thus blood platelets do not adhere. Mitochondria are few and small. Many clear vesicles and small openings into some of them are frequent suggesting that pinocytosis is common. The endothelial lining is a mobile, changing surface.

Mesothelium. Mesothelium is formed by mesoderm cells, which line the celomic cavity of the embryo and later the peritoneal, pleural, and pericardial cavities. The cells are broader than endothelial cells and the contacts less sinuate. They can swell up (abnormally), round off, leave their attachments, and wander as free, phagocytic cells. They have short microvilli and demonstrate both pinocytosis and phagocytosis.

CUBOIDAL EPITHELIUM. The cells are like cubes and

fit together like a mosaic (Fig. 4-1). They appear square in vertical section but may be pyramidal or even hexagonal when seen on their free surface.

Simple Cuboidal. This type of tissue is found lining the collecting tubules of the kidney and small bile ducts and is of mesodermal origin. The cells often bulge into the lumen and may have folding of the basal cell membrane possibly associated with water reabsorption and concentration of fluid in the lumen.

Germinal Cuboidal. These cells differentiate as "germ cells" (seen in the ovary).

Chorioidal Cuboidal. This tissue is engaged throughout life in producing cerebrospinal fluid. There are many microvilli on the free surface.

Secretory Cuboidal or Glandular Cuboidal. These cells elaborate special substances in their cytoplasm that may accumulate as secretion granules in the region near the nucleus and are later discharged in the secretion by various mechanisms (Fig. 3-3, p. 75; Fig. 4-1): eccrine (merocrine) secretion, where the colloidal secretory granules and other constituents pass through the luminal wall of the cell without its rupture; apocrine secretion, where the products and their surrounding cytoplasmic elements accumulate near the luminal part of the cell, which bulges, and the entire mass breaks off from the body of the cell and disintegrates to form the actual secretion (axillary sweat glands); and holocrine secretion, where the cytoplasm of the whole cell gradually fills with accumulated secretory products, the cell dies, and the whole cell disintegrates to form the secretion (sebaceous gland cell).

Another distinction is made on the nature of the secretion: serous or zymogen (enzyme-secreting) cells and mucous cells.

Absorptive Cuboidal or Brush Border Cuboidal. These cells show a highly specialized luminal "brush border." With this is associated the selective reabsorption of substances into the cell from the luminal fluid and later passed on to nearby blood vessels (renal convoluted tubules).

COLUMNAR EPITHELIUM. The cells resemble building bricks, but their long sides are placed vertically and are usually opposed; their cut surfaces appear hexagonal in shape (Fig. 4-1). They are of various heights and these cells show a wide variety of structure and function. In certain situations single cells in this epithelium may become "one-cell glands" or goblet cells.

Simple Lining Columnar. This tissue is the same as the simple lining cuboidal, but the cells are taller (excretory duct system of the kidney).

Mucus-Secreting Columnar. This tissue is composed of cells that have a few short microvilli and whose secretion granules form a clear homogeneous mass extending down from the luminal surface (free surface of the gastric mucosa and in the cervical and endometrial uterine glands).

Serous Columnar. These are cells whose secretion granules remain discrete. The basal part of the cells has many mitochondria (duct cells of the salivary glands).

Striated Border or Brush Border Columnar. These cells show a highly specialized border on the luminal surface with mitochondria just below the brush border. It is associated with increased absorptive ability. Very long microvilli are seen (epithelium of the villi of the small intestine).

Ciliated Columnar. The luminal surface shows numerous motile hairlike processes (cilia). Ciliated and nonciliated cells are seen side by side in the uterine tube epithelium.

Pseudostratified Epithelium. Pseudostratified epithelium (Fig. 4-1) is a modification of a simple epithelium in which additional shorter cells lie on the basal side but do not reach the luminal surface. The short cells are reserve cells at incompletely developed stages. Some develop fully and others undergo cell division to provide new basal cells.

PSEUDOSTRATIFIED COLUMNAR. This type of epithelium is seen in larger ducts of many glands.

PSEUDOSTRATIFIED COLUMNAR CILIATED. This tissue is found in the respiratory passage from the nose to the bronchioles. It often has goblet cells. One sees three cell types: small rounded basal cells with large nuclei; intermediate tall pyramidal cells; and fully developed columnar ciliated cells. The ciliary current moves the mucus toward the pharynx and impurities out of the respiratory system.

Stratified Epithelium. Stratified epithelium (Fig. 4-1) is many cells thick, covers surfaces most exposed, and is liable to be worn away. It carries a reserve layer of cells for continuity and rapid replacement.

TRANSITIONAL. This is the epithelium of the urinary passages. These types of cells are seen: basal, pear-shaped cells with their broad end toward the free surface and flattened or umbrellalike surface cells. This epithelium has three characteristics: (1) it can accommodate itself to cover large and small surface areas completely; (2) it has the ability to prevent leakage of the urinary constituents into the subjacent tissue; and (3) it lacks a clearly defined basement membrane on its basal surface. Since capillaries in the subjacent tissue lie very near the basal epithelial cells, injury can lead to hemorrhage.

STRATIFIED COLUMNAR. This is not a common form and is found in part of the male urethra and other small areas of mucous membranes. One sees several layers of small polyhedral basilar cells covered by a single layer of columnar cells.

STRATIFIED COLUMNAR CILIATED. This is a fetal type of epithelium found lining the fetal esophagus. The columnar ciliated cells lie superimposed on several layers of small cells.

STRATIFIED SQUAMOUS. Stratified squamous epithelium is the common form, the most resistant type of stratified epithelium, and is found where "wear and tear" occur (the surface of the skin, the lining of the mouth and anus, and the cornea of the eye where it is exposed and needs constant replacement) (see Chap. 16). It is thick and consists of many cell layers. The outer cells are continuously rubbed off and replaced by cells from below, by cell division. Two types are seen.

Nonkeratinized. The nonkeratinized type (Fig. 4-1) is composed of cells not easily divided into definitive layers. Nuclei are present throughout but vary in shape. The basal cells are columnar and are anchored by small processes to the subjacent connective tissue.

The cells in the superficial layer are flattened to thin scales and "flake" off the free surface. This epithelium is seen in the mouth, esophagus, and so forth. The cornea is a very special form of this epithelium.

Keratinized. In the keratinized type (Fig. 4-1) a more definitive layering exists. This epithelial type covers the exposed skin surfaces of the body. It varies in thickness and number of layers. Fully developed, its outer layer is a thick, compact layer of dead cells. It is thickest on the sole of the foot and palm of the hand. Its layers are named as follows: (1) stratum basale (basal cell layer); (2) stratum spinosum—broad, spindle-shaped cells with tonofibrils between cells (the latter are the so-called "prickle cells"; layers 1 and 2 are often referred to together as the stratum malpighii); (3) stratum granulosum; (4) stratum lucidum; and (5) stratum corneum (cornified or keratinized layer), the outer thick layer of dead cells. Cell division usually occurs in layers 1 or 2. The cell is gradually pushed farther up and accumulates granules in its cytoplasm. The granules later dissolve to form eleidin; the internal cell structure including the nucleus disappears to form a clear homogeneous mass and the cell dies. The eleidin solidifies to form keratin, and the cells remain as dehydrated scales that flake off the surface.

The Supporting or Connective Tissues

The supporting tissues (Fig. 4-2) are derived exclusively from mesenchyme. They are characterized by three major constituents: cells, homogeneous ground substance (matrix), and formed elements or fibers that lie in the matrix between the cells. There are many different types but they remain labile and can change from one to another easily.

The cells are the primary constituent and are widely separated from each other by matrix, which is referred to as intercellular (or extracellular). The matrix contains protein, collagen, elastin, and variable amounts of several mucopolysaccharides such as hyaluronic acid, chondroitin, chondroitin sulfate, and some glycoprotein. The cells are responsible for the production of the matrix and the modifications that occur in it. They are also essential for fiber production. The matrix

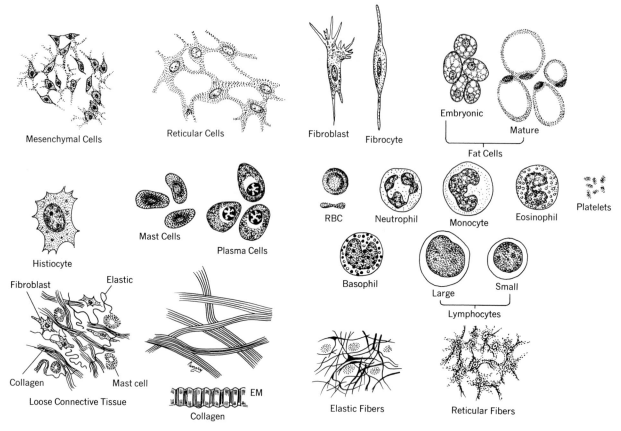

Figure 4-2. The supporting or connective tissues and their cells.

varies greatly in consistency and contains a variety of fibers. Blood and lymph vessels as well as nerve elements are seen in all subtypes except in cartilage. The degree of specialization is low, and thus nutritional needs are low and the matrix can withstand oxygen lack. For this reason, too, it has a high degree of regenerative ability.

The supporting tissues form the framework of the body giving mechanical strength. They build the skeleton and also serve as the binding, mortar, or packing material between parts. However, they are not just passive elements. They form coverings or capsules for protection, help support the soft tissues, and help maintain shape and size of organs. They serve as pathways for vessels and nerves. They even serve as storage depots and act as reservoirs of unspecialized, undifferentiated cells capable of many developmental lines.

Their classification is based on the modification of the matrix and the fibers in it: fluid (in blood), highly viscous or semisolid (fibrous or connective tissues), solid and elastic (cartilage), and solid and rigid (bone and dentin). Some tissues are compound tissues in which a subtype of connective tissue creates a framework and the intercellular spaces are large and filled with free cells. In the blood-forming or hemopoietic tissue the free cells are blood-cell-forming; in lymphatic tissue, the free cells are lymphocytes.

Undifferentiated Supporting Tissues. Mesenchyme. This is the earliest connective tissue to be formed. It is derived from mesoderm (in the embryo) and has an abundance of intercellular matrix with the cells widely separated. The cell cytoplasm is drawn out in fine processes. No fibrils are seen in the matrix or cells. This "primitive tissue" fills all the spaces between the ectoderm, mesoderm, and endoderm and their development. All supporting tissues develop from mesenchyme: blood cells, endothelium, mesothelium, reticular cells, fibroblasts, mast cells, plasma cells, reticuloendothelial cells, bone cells, cartilage cells, smooth muscle, and others.

Mucous Connective Tissue (Wharton's Jelly of the Umbilical Cord). This is one of the first developments from mesenchyme and resembles the primitive condition. The cells are "fibroblastic," and the intercellular matrix is a clear, homogeneous jelly with mucopolysaccharide in it. Some fibers are also seen.

Adult Supporting Tissues. There are four major types based on the matrix structure: (1) Intercellular matrix is a colloidal "sol" as seen in plasma (of lymph and blood) and contains proteins such as albumin, globulins, and fibrinogen. (2) The intercellular substance is more viscous, with the addition of mucopolysaccharides and mucoproteins to form a jellylike matrix in which lie formed elements (fibers); this is the group of connective fibrous tissues. (3) The intercellular substance is a solid, elastic matrix containing fibers (often hard to see); these are the cartilages. (4) The intercellular matrix is a solid, rigid material due to an increase of calcium and magnesium salts; these are the bones and dentin.

Fibrous Connective Tissues Proper. *Cellular Components.* These are derived from mesenchymal cells. The cells found here are: (1) undifferentiated cells; (2) reticular cells, which are primitive with long processes that form a three-dimensional network that contains fine reticular fibers (cells are potential phagocytes—i.e., ingest particles); (3) fibroblasts, associated with fiber formation (cells are flattened, irregular, with long cytoplasmic processes); (4) fibrocytes, inactive fibroblasts found surrounded by thick collagen fibers; (5) corneal cells, of the cornea lying flattened between layers of fibers; (6) tendon cells, modified to lie in rows between the thick parallel tendon fibers; (7) fat cells, with reduced cytoplasm and filled with coalesced fat globules of neutral fat and a flattened nucleus to one side; (8) pigment cells, seen in the connective tissue of the iris of the eye and containing yellow-brown pigment granules; (9) fixed macrophages (histiocytes or adventitial cells), which belong to the reticuloendothelial system and are the scavengers of the tissue spaces; (10) mast cells, filled with large granules that stain deeply with basic dyes and resemble blood basophilic granular cells (they contain histamine and may also liberate heparin); (11) plasma cells, seen in pathologic conditions but normally few in number (they are egg shaped with eccentric spherical nuclei and contain chromatin arranged in a sort of cartwheel); and (12) cells of circulating blood, consisting of all forms of white blood cells, most frequently monocytes and lymphocytes.

The Matrix Components. This is a colloid complex composed of tissue fluid plus substances such as albumin, mucopolysaccharides (hyaluronic acid), glycoproteins (mucopolysaccharides in protein complexes), and scleroproteins. The matrix has a smaller protein content than blood plasma. It is capable of storing and releasing substances. The degree of polymerization of substances creates a matrix of low or high viscosity. The basement membrane is a more highly polymerized layer of the ground substance supported by reticular fibers. Constant interchange takes place in the matrix.

Fibrous Components. These consist of three types of fibers: reticular, collagen, and elastic.

RETICULAR (ARGYROPHIL). This type of fiber is formed from protein reticulin and is fine, highly refractile, branching, and embedded in a matrix rich in carbohydrate.

COLLAGEN. Collagen appears as thin, wavy, unbranched fibers or in large bundles that may run long distances and have great tensile strength. Their constituent molecules are oriented in the fibril to create birefringent properties oriented longitudinally. As noted by electron microscopy there is a periodicity of 700 Å that creates a banding. Collagen fibers rapidly contract on heating to a critical temperature, which decreases their tensile strength and rubberlike elasticity. In dilute acetic acid they dissolve into aggregates of variable-sized individual collagen molecular units, procollagen or tropocollagen, which is a thin, rigid rod about 3000 Å long and 14 Å in diameter with molecular weight of 350,000 consisting of three intertwined polypeptide chains twisted into a helical formation. All known collagens are characterized by very high amounts of glycine, proline, and hydroxyproline. Collagen fibrils may be regenerated by neutralization of the acid solution of tropocollagen, and fibrils thus re-formed resemble natural collagen. Collagen makes up 95.99 percent of the organic matrix in ground substance and serves to bind the tissues.

ELASTIC. This type of fiber is composed of elastin. The fibers appear as fine, highly refractile branching structures that may become thick or even fuse to form elastic laminae. Elastin is a polypeptide containing glycine, alanine, and valine. Elastic fibers are formed in arteries, walls of the stomach, living tissues, and the bladder, as well as being the major component of the ligaments that bind bones together at the joints. They allow for stretch and rebound.

CONNECTIVE TISSUE CLASSIFICATION. *Reticular Tissue.* This is a three-dimensional net of reticular cells and fibers in which the interstices of the network are filled with blood-cell-forming cells or lymphocyte-forming cells.

Loose Fibrous (Areolar) Tissue. This is formed directly from mesenchyme, fills the spaces between developing tissues and organs, and is found under most skin. Most of its cells are fibroblasts. Reticular, collagen (singly or in bundles), and elastic fibers are present. Lymph capillaries, nerve endings, and fibers as well as a fine plexus of autonomic nerve fibers are found here. The elastic fibers in the tissue stretch to a limit determined by the inelastic collagen fibers. The areolar tissue prevents organs from "slipping about" when the body moves, and, in addition, the weight of man's erect body rests on the areolar tissue in the soles of the feet. One owes his youthful complexion to the consistency and elasticity of the areolar tissue beneath the skin.

Dense Fibrous Tissue. The amount of matrix is reduced, cells are fewer, capillaries are fewer, and there is a predominance of thick, coarse collagen bundles without any specific alignment. A few elastic fibers are seen.

Adipose or Fatty Tissue. There is an open framework of dense fibrous tissue with masses of fat cells. The matrix is minimal and the fat cells closely packed. In any fat depot, the condition is not static as new fat is continually being laid down and the old utilized. Most of the skin over the body has an underlying layer of adipose tissue that we refer to as subcutaneous tissue or fat. This fat not only is a long-term food store but also is a poor heat conductor and helps one maintain the body's temperature within a very narrow homeostatic range. Extra fat may help "fat" people keep warmer than thin ones in very cold climates.

Tendon, Ligament, or Fascial Membrane. This consists of a parallel alignment of coarse collagen bundles with a reduced matrix. Cells conform to the fiber alignment to form tendon cells. This makes them difficult to break, and they can withstand the loads needed to support the body as well as the stresses imposed by various activities. The tissue is relatively avascular. Few elastic fibers are present, and the cells are few in number compared to the amount of collagen they produce.

Elastic Tendons. This consists of thick elastic fibers bound together by collagen fibers. The ligamentum nuchae is the best example (found attached to the cervical spines and very prominent in horses and oxen).

Cartilage (Gristle). In this tissue, the matrix is a solid with marked elastic properties. Cartilage (Fig. 4-3) may persist throughout life or be a transitional stage to bone formation (in long bones). It serves as the major material of the endoskeleton before bone appears. Its elasticity is important and may be lost with age. Cartilage helps provide cushioning and flexibility. It is strong as well as flexible but may be cut or damaged by pressure. It is seen at the ends of bones where they form joints. Generally, cartilage has no blood or lymph vessels but is nourished by tissue fluids. Cartilage consists largely of collagenous material with living cells. The cells are surrounded by an amorphous mucopolysaccharide ground substance, giving it firmness and translucence. Calcium may be present but is usually associated here with cellular death, removal of the calcified tissue, and replacement by bone.

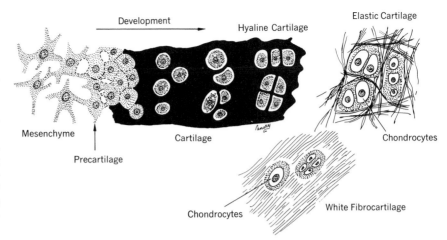

Figure 4-3. The origin of cartilage matrix. Simultaneous appositional and interstitial enlargement occurs. Mesenchymal cells form precartilage cells. The cartilage or chondrogenic cell is enveloped by new matrix and becomes a young chondrocyte. With growth the cell becomes successively more deeply enclosed as new cartilage is formed. The cell itself enlarges, thus contributing to interstitial growth, and it subsequently undergoes cell division to form groups of daughter chondrocytes. The cells become more separated as matrix is deposited.

The cells can divide and grow in cartilage, but this is limited by nutritional diffusion capabilities. Surface growth is attributed to the presence of chondrogen cells (chondrocytes) under a fibrous layer of perichondrium. The articular cartilage of synovial joints is different in that it has no perichondrium on its articular surface.

Classification of cartilage depends on the character of the matrix and the type of fiber present.

HYALINE CARTILAGES. These appear bluish-white, homogeneous, and translucent owing to the type of intercellular material. Such cartilages are easily cut and damaged by pressure. One sees them in articular cartilage of synovial joints in growth and in plates of cartilage between separately ossifying parts of a bone (costal cartilages). Hyaline cartilages are also found where one sees cartilage in the respiratory system: nose, trachea, bronchi, and larynx.

FIBROCARTILAGES. Fibers are seen in the matrix. Two forms are described: (1) white fibrocartilage and (2) yellow fibrocartilage.

White fibrocartilage has strong bundles of white fibrous tissue and is brighter, more flexible, less homogeneous, and less cellular than hyaline cartilage. It can be found as the cartilage in the knee joint; as the intervertebral disks between vertebrae; as sesamoid cartilage in some tendons; as the articular disk in the wrist joint as well as at the end of the clavicle; as the rim that deepens the sockets of the shoulder and hip joints; and as the disk that unites the two hip bones at the pubic symphysis.

Yellow fibrocartilage has bundles of yellow elastic fibers but little or no white fibrous tissue. One finds this type in the external ear, in some cartilages of the larynx, and in the auditory tube of the middle ear. It is very flexible, can bend, but springs back to its original form.

Synovial Membrane. Synovial membrane is a special type of connective tissue that covers the sides of any bony joint. It is attached to each articulating cartilage covering the ends of the bones and secretes synovial fluid into the joint cavity. It prevents the ends of the bones from touching each other and provides a perfect lubricant for these moving parts.

Bone. Fibers, matrix, and cells are seen in bone tissue (Fig. 4-4), but the matrix is solid and rigid. Magnesium and calcium salts impregnate the matrix between the fibers, and bone is laid down with the help of the cells (osteocytes). The adult skeleton is formed of bone. Bone gives rigidity to soft tissues, provides levers for movements while the joints supply the fulcrum, and also protects by covering certain areas. Bone is about one third organic and two thirds inorganic by weight.

Grossly bone is covered by a connective tissue membrane, the periosteum, which contains both elastic and collagen fibers. Some of the latter penetrate the bone. On its inner surface, lining the bone marrow cavity

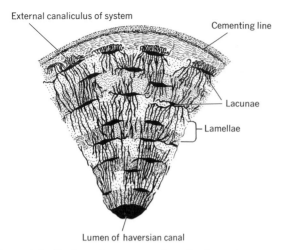

External canaliculus of system

Cementing line

Lacunae

Lamellae

Lumen of haversian canal

Figure 4-4. Portion of a cross section of a haversian system of a bone (macerated), showing lacunae, canaliculi, and lamellae.

Figure 4-5. Bone, showing haversian canals (systems) and Volkmann's canals. The difference between bone in the shaft and the head is noted.

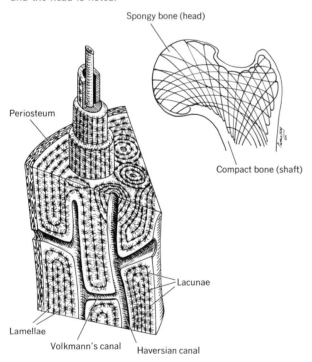

Spongy bone (head)

Periosteum

Compact bone (shaft)

Lacunae

Lamellae

Volkmann's canal

Haversian canal

is a finer connective tissue membrane, the endosteum. One sees a very dense hard form or compact bone and another form called spongy or cancellous bone. The latter has an open texture with bone spicules separated by wide channels of soft tissue, and usually oriented along specific lines and planes to follow the lines of stress. Compact bone is also penetrated by channels but they are narrower.

Histologically bone is composed of lamellae, and at points between them lie small lacunae (lakes), which are flattened ovoid spaces occupied by bone cells (osteocytes). The osteocytes are a modification of the bone-forming cells, the osteoblasts that have been trapped in the products of their own making. Their fine processes extend out through the matrix, between the fibers, in small channels, the canaliculi. The latter pass from one lacuna to another linking the cellular processes of the different cells. The lamellae lie in three series: (1) parallel with the outer or inner bone surfaces (the circumferential lamellae); (2) arranged in concentric patterns around a small central canal (the haversian lamellae); and (3) in the intervals between the haversian lamellae in an irregular interrupted series (the interstitial lamellae).

The central canal of the haversian lamellae and its contents (a small arteriole, venule, lymphatic vessel, and a few unmyelinated nerve fibers, all wrapped in connective tissue), the lamellae themselves (composed of an organic matrix of crystals of calcium carbonate, phosphate, fluoride, citrates, magnesium chloride, and regularly aligned fine collagen fibers [ossein] that add strength to the bone, but little weight), and an outer, denser limiting cementlike layer are spoken of as the haversian system (Fig. 4-5). These systems tend to run parallel to the bone's long axis but may branch off so that the systems actually communicate. There are other canals called Volkmann canals, which pass in from the outside of the bone and carry vessels that anastomose with those of the haversian canals. The Volkmann canals have no lamellae but merely pass through any of the above-named lamellae to reach their destination.

Bundles of collagen fibers penetrate the outer circumferential lamellae and are called Sharpey's fibers.

They originate from the periosteum and are embedded in the bone during development, thus binding the periosteum to the bone.

The lamellar arrangement as well as the haversian systems undergo constant remodeling during growth and even to a lesser degree in adult life.

Bone Formation (Osteogenesis). Bone never forms as a primary tissue but is a result of "conversion" of a preexisting tissue. Bone develops in mesoderm as a result of the deposition of mineral salts in the connective tissue or in a previously formed cartilage model. If the initial tissue is fibrous tissue we speak of intramembranous bone formation. This is seen in the flat bones of the skull (frontal, parietal, etc.) and in part of the clavicle (collarbone). If bone is laid down in cartilage, as is the case for the limb long bones, ribs, etc., we speak of endochondral or intracartilaginous bone formation. Note, however, that in the latter, the cartilage is actually removed and bone replaces it in the remaining space. The cartilage itself does not convert directly to bone. There is a common misconception that the terms *endochondral bone formation* and *intramembranous bone formation* refer to two types of osteogenesis. Actually the terms only refer to the formation of bones as organs. Essentially the formation of the bone tissue of the flat bones and long bones and the factors involved (formation, growth, and resorption) are similar since even in endochondral bone formation, part of the process takes place in the spaces once occupied by cartilage and bone formation occurs in the connective tissue that invades the spaces from the outside. The development and activity of the osteoblast are the most essential factors in bone formation. Bone may also form in abnormal sites in pathologic processes, in a wound, in tumors, in muscles, etc.

INTRAMEMBRANOUS FORMATION. One begins with an embryonic or simple connective tissue of fine collagen fibers and early fibroblasts (Fig. 4-6). Initially one sees a center or area where bone formation is to begin and from which it will extend outward. In this center there is a marked increase in vascularity with an outward extension (from the center) of wide, thin-walled blood vessels. Also noted here are new cells or osteoblasts that have formed from the cells of the embryonic connective tissue. A clear, jellylike matrix, the osteoid matrix, then forms around the osteoblasts and within it are seen irregularly arranged thin collagen fibers (ossein or osteogenic). Calcium salts are deposited in

Figure 4-6. Intramembranous bone formation. In the center of ossification the cells and matrix of undifferentiated connective tissue undergo changes to produce small bone spicules. Some cells **(a)** remain undifferentiated; others **(b)** develop into osteoblasts that lay down the first osteoid, which subsequently becomes mineralized. Original blood vessels are retained in close proximity to the forming bony trabeculae **(c).** As bone deposition continues by the osteoblasts some of these cells are enclosed in their own deposits (osteocytes). Some undifferentiated cells develop into new osteoblasts **(f),** and other osteoblasts undergo cell division to accommodate trabecular enlargement. The outline of the early spicule **(e)** is seen. Blood vessels are by now enclosed in the cancellous spaces, which also contain a scattering of fibers, undifferentiated cells of connective tissue, and osteoblasts.

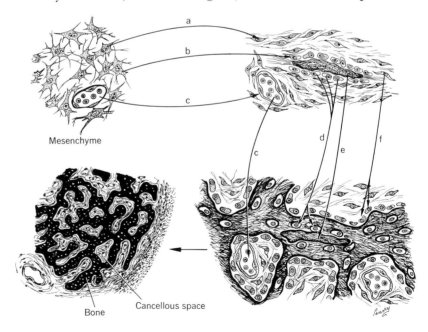

Mesenchyme

Cancellous space

Bone

the matrix about the osteoblasts and between the fibers to create a solid, rigid matrix that tends to trap the osteoblasts. They change name and are now called osteocytes. This results in the formation of spicules or long processes of true bone (primary bone). Primary bone contains a higher percentage of calcium salts than does bone formed later. Its fibers are not aligned. It has no lamellae or regular cellular arrangement nor any haversian systems. The lamellae appear only after birth with remodeling. The spicules thicken, the connective tissue spaces decrease in size, vascular connective tissue is incorporated in the spaces between the trabeculae, and as more bone is laid down, the spaces are reduced to capillary size and things get more compact.

After intramembranous development, the formation of new bone continues. This primary bone and its irregular disposition is later removed by resorption and replaced by a uniform lamellated bone with typical haversian systems. The surrounding connective tissue condenses to form a limiting layer of periosteum with osteogenic cells in its deeper part and blood vessels that allow the formation of haversian systems and continued reconstruction or growth as needed, during the period of skeletal development as well as in adult life.

ENDOCHONDRAL (CARTILAGE) BONE FORMATION. This is the commonest form of bone formation (Fig. 4-6). This process necessitates cartilage removal since it involves the presence of a model of the bone in hyaline cartilage surrounded by the perichondrium (a connective tissue membrane). There is really no direct conversion of cartilage to bone since the cartilage is almost completely removed, and in the space that remains, bone is formed.

One first sees the aggregation or condensation of the mesodermal cells where the bone is to develop. This aggregation is the anlage or rudiment of the bone, and in it, the cells become visibly modified by changing their shape to become rounded. An amorphous intercellular material containing collagen is laid down between the cells. Each cell enlarges and the surrounding material forms a "capsule" around it. The cells divide and a capsule is laid down around each daughter cell,

two new capsules being formed inside each older one, as the cells multiply and more intercellular material separates them. Thus, the cartilage enlarges by interstitial growth (growth of cells and intercellular substance). The surrounding mesenchyme forms a sheath of perichondrium with a superficial, more fibrous layer and a deeper cellular layer containing chondrogenic cells that provide for growth, on the surface of the cartilage, by apposition. Thus, there is the presence of a model of the bone in hyaline cartilage, surrounded by a connective tissue membrane, the perichondrium. The cartilage cells proliferate by mitosis and arrange themselves in rows parallel to the long axis of the bone. The cells then progressively enlarge with the formation of a large nucleus, vacuolization of the cytoplasm, and an increase in alkaline phosphatase in the cell and deposition of calcium in the matrix. The matrix thus becomes calcified and the cells degenerate, having been cut off from their nutrient source. The cartilage model of a long bone grows predominantly at its ends. The oldest part is near the middle and here, too, are the oldest cartilage cells.

The perichondrium also changes simultaneously. Cells on its inner surface differentiate to become osteoblasts and lay down long narrow spicules of bone that join together to form a cylindric "sleeve" around the weakened cartilage of the center of the bone shaft. This "sleeve" lies outside the cartilage and is laid down as in membranous bones by the perichondrium, which we now call the periosteum. Thus, the shaft of the model is encased in bone derived from perichondrium.

The ingrowth of connective tissue and vascularization of the weakened cartilage take place through an opening in the "sleeve" of bone carrying connective tissue, osteoclasts (that destroy bone), and vessels into the bone. Bone-forming osteoblasts, blood-forming cells, and wide, thin-walled blood vessels then follow into the spaces and channels thus formed; the calcified cartilage is excavated and largely removed; the widening spaces are filled with very vascular connective tissue; and the spaces are lined by osteoblasts. The osteoblasts begin to form bone on the surface of the remaining spicules of calcified cartilage to form trabeculae. In this process, some osteoblasts become

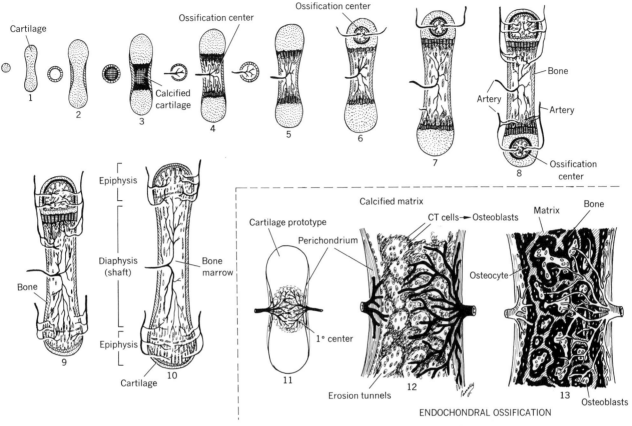

Figure 4-7. Diagram of development of a long bone (longitudinal sections). **1.** Cartilage model. **2.** Periosteal bone collar, seen before calcification of cartilage. **3.** Cartilage calcifying. **4** and **5.** Vascular mesenchyme enters matrix and divides area into two ossification zones. **6** and **7.** Blood vessels and mesenchyme enter upper epiphysial cartilage, and epiphysial ossification center develops. **8.** Ossification center is seen in lower epiphysial cartilage. **9.** As bone stops to grow in length, the lower epiphysial plate disappears. **10.** Upper epiphysial plate disappears. Bone marrow cavity is now continuous throughout bone, and all blood vessels communicate. **11–13.** Enlarged version of endochondral ossification.

trapped in the bone substance and become osteocytes lying in the small spaces or lacunae from which fine processes of the osteocytes extend into the bone to form canaliculi. Other cells multiply and fill the spaces between bone trabeculae with marrow. During growth, constant remodeling of the primary arrangement takes place involving bone resorption and formation.

Not only does bone formation occur in the center or shaft of the long bone but also separate centers of ossification appear in the cartilage at the ends of the bones. Thus, between the shaft (diaphysis) and the ends (epiphysis) (Fig. 4-7) is a line of cartilage, the epiphysial cartilage or plate, with bone formation taking place on either side of it, lengthening the bone. By age 18, this plate has usually been replaced by bone and no further growth is possible. Some bones do continue to grow until one is in one's twenties. Meanwhile, the middle of the shaft and extremities have hollowed out to accommodate the bone marrow.

Where resorption of bone is occurring, osteoclasts are often seen. They are large, multinucleated cells probably formed by cell fusion rather than division but their nature is uncertain. They may be converted osteoblasts, undifferentiated cells of connective tissue, macrophages, or reticular cells. Note that bone resorption can occur without them.

The cells on the outside of the shaft remain on "duty" throughout life in case the bone is injured. When this occurs, a substance called callus is laid down across the break to support the two halves while they grow together. The "join" is smoothed off by other cells that tend to consume any surplus bone. The same cells that "eat away" the bone (osteoclasts) after injury are also working while the bones are developing and growing. They help to sculpture the bones into adult shape; from the time we can walk until puberty bones are continually being sculptured into their correct shape as well as the arrangement of the trabeculae. In the shaft, very hard compact bone (Fig. 4-5) exists only on the outside for here most of the stress occurs.

The above process of sculpturing is not sufficient for bone's proper development. Of equal importance is the muscle pull on the bones, and that is why exercise is essential during the developmental periods. One must also be aware that the skeleton is not only of structural importance but is alive with nerves, veins, and arteries that permeate it, making it possible for the bone salts to be "called upon" at any time when the body content needs reinforcement. Furthermore, the bone marrow is the body's vital and essential source of red blood corpuscles as well as of some of its defensive white cells.

CENTERS OF OSSIFICATION. These are places where bone begins to be laid down (Fig. 4-7). The process spreads from such centers. The earliest (usually principal) center in the body (or shaft) of a bone is called the primary center.

Primary centers of ossification appear in different bones at different dates in an orderly sequence. Most of them occur before the end of the fourth intrauterine month (135 mm) about the seventh to twelfth week. All are present before birth.

Secondary centers occur at a much later time than primary and are formed in smaller areas of the bone that have remained cartilaginous. All long bones (and many others) have secondary centers in outlying parts. Nearly all secondary centers appear after birth. Ossification from these are the same as with primary centers except that it takes place in cartilage, which continues to enlarge by cellular multiplication until growth stops.

The part of the bone ossified by a 1° center is called the diaphysis and from a 2° center is called the epiphysis. A thin plate persists for a time between the diaphysis and epiphysis and is called the epiphysial or growth cartilage, or epiphysial plate, and its edge, at the surface of the bone, is called the epiphysial line. As long as this plate persists, the diaphysis can grow in length.

Extension of diaphysial ossification into the epiphysial plate finally obliterates it and fusion takes place preventing further growth in length. The part of the diaphysis where ossification is spreading is called the metaphysis. Extension of epiphysial ossification to the perichondrium converts it to periosteum and prevents further endochondral growth in this direction. With cessation of growth, no further cartilage cells proliferate toward the epiphysis from the articular surface and it ceases to grow toward the surface. Cartilage remaining is concerned with production of new material for the articulating surface, and if nonarticular, it is all replaced with bone.

OSSIFICATION AND EPIPHYSIAL FUSION. The dates at which epiphysial centers appear and fuse with each other or with the diaphysis may be of clinical importance, but one should not memorize such lists. The opposed epiphysial and diaphysial surfaces have depressions and elevations that fit together to help them withstand twists and knocks during growth. An epiphysis can be damaged or broken such that bone stops growing prematurely or unevenly and thus an ordinary fracture or an epiphysial separation must be recognized.

Postnatal changes in epiphyses of long bones occur mainly as follows:

1. Secondary centers of ossification appear from birth to age five.
2. Ossification spreads from these centers until the age of 12 (in girls) or 14 (in boys).
3. From 12 or 14 to 25 years, the epiphyses fuse with the diaphyses and growth ceases with fusion.

The process is speeded up in girls at five years and at ten years and thus at these ages they are in advance of the male.

As always, exceptions do occur: a center in the distal end of the femur occurs in the last month of fetal life as does one in the proximal end of the tibia and upper humerus (a child is considered "full term" with the appearance of one of these centers). The clavicle has an epiphysis at its medial end whose center appears late (18 to 20 years) and fusion occurs from 20 to 30 years. Generally, an epiphysis that appears early fuses late and vice versa, and if fusion is late, the bone grows longer and more rapidly. Most centers in the ribs and vertebrae, shoulder, and pelvic girdles do not appear until puberty and fuse about age 25. The ringlike epiphyses on the upper and lower surfaces of the vertebral bodies fuse at full adult age (25 or later), and, therefore, the back continues to grow after the limb bones have ceased (fuse before 21 years).

The date of appearance of any one epiphysis center in different children varies by weeks, months, or even one to two years. The date of union is even more variable. Thus some children reach their full height in their teens and others not until age 25. The sequence of dates of union, however, is remarkably constant and the intervals between them are proportionately the same in different people. Thus, if first union occurs early, growth is completed early.

Dentin. Dentin is a hard, yellowish tissue that forms a great portion of the hard substance of the teeth. It has a matrix and formed elements (fibers) but has no cell bodies or nuclei. The bodies of the cells (odontoblasts) lie in the tooth pulp, and their processes extend into the dentin. It has a higher concentration of mineral than bone.

Enamel. Enamel is really not a tissue since it has no cells or processes. It is really a material that is secreted on the ameloblast cell surface and the cells then disappear. It is the hardest of all body substances and consists of up to 97 percent mineral.

The Muscular Tissues

Muscle tissue (Fig. 4-8) is a specialization whereby certain cells develop the power of rapid contraction as a primary activity. It is derived from mesoderm, in part from the mesodermal somites (muscle masses) and partly from the mesenchyme (rarely from ectoderm—e.g., in a muscle of the eye). The fibers are

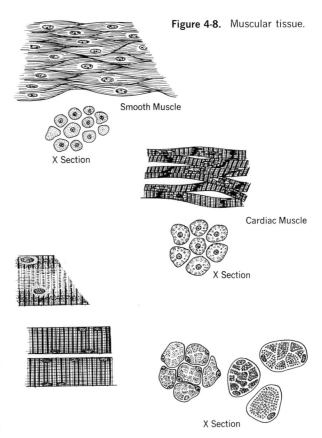

Figure 4-8. Muscular tissue.

Smooth Muscle

X Section

Cardiac Muscle

X Section

X Section

composed of elongated cells in which are found intracellular contractile fibrils called myofibrils or sarcostyles. Every fiber thus consists of many fibrils (many times thinner than red blood cells), each as long as the fiber itself yet no more than 1 to 2 μm in diameter. The cellular development tends for elongation to accommodate longer contractile fibrils and thus increase the extent of movement. The movement can be either rapid or of a sustained tension. Muscle tissue is penetrated by connective tissue elements, blood vessels, nerve fibers (sensory and motor), and lymph vessels (in the connective tissue). It has a high rate of metabolism, aided by its many blood vessels, but being so highly specialized, muscle tissue does not regenerate well. Owing to its high metabolic rate, muscle is a very important heat producer in the body.

Muscle tissue is classified on the basis of the type of myofibril present. Two forms are seen: (1) a less specialized, nonstriated, plain or smooth (visceral) muscle and (2) a highly specialized, striated or striped muscle. The latter is further divided, on the basis of its unit character (size, shape, and arrangement of cells) and the site in the body where it is found, into heart or cardiac muscle and skeletal muscle.

Types of Muscle. PLAIN OR SMOOTH MUSCLE. Skeletal muscle is designed for rapid, powerful but short-lived mechanical action and has a clearly defined rest period and a very narrow range of movement. Visceral muscle, on the other hand, is slow to respond, contracts with less force, has no clearly defined rest period, and may take up different ranges of movement depending on physiologic needs. Further, although skeletal muscle depends on nerve supply for normal activity and atrophies if this is interrupted, visceral muscle still retains reasonable function without a nerve supply. It is innervated by the autonomic nervous system and is not under voluntary control. There is an inherent cellular rhythmic activity not dependent on external innervation, but there is a response to certain hormone stimulation. These muscles are found in the walls of arteries and veins, uterus, and digestive tract, to name just a few. Their activity is a relatively slow and rhythmic relaxation and contraction.

The type in which the intracellular fibrils show no cross-striations (no alternating dark and light transverse bands) is also called visceral muscle (as to location) or involuntary muscle (as to control).

The cell is a long, spindle-shaped cell with its single nucleus lying near the middle of its length and almost centrally placed. Its chromatin is in fine granules, and two nucleoli are usually seen. The individual cells are usually shorter than striated muscle and are about 100 μm long (but may be larger as in the uterus) and about 4 to 7 μm wide. In the bundles, they are interdigitated with each other and may be linked by fine intercellular bridges. Blood and lymph vessels lie in the connective tissue between bundles. The myofibrils extend throughout the cell length, and the cell membrane has no distinctive thickened sarcolemma.

The cell membrane is covered on its outer surface by a delicate cell coat, composed partly of basal lamina material and partly of related collagen fibrils. This barrier exists between each cell and its neighbor except at certain limited areas where membranes come in close contact. The number of close contacts between smooth muscle cells varies in different sites. They may be areas of low electrical resistance allowing passage of excitation from cell to cell and accounting for characteristic spread of activity in sheets of visceral muscle. The term *nexus* has sometimes been used for this type of close junction between smooth muscle cells.

At other points on the surface of smooth muscle cells, numerous invaginations or caveolae form rows of flask-shaped structures whose significance is unknown but may represent micropinocytic activity.

There is no sarcomere pattern. The cytoplasm of the cell is filled with densely packed myofilaments generally arranged parallel to the long axis of the cell but that sometimes show whorls and spirals or variations in direction. The filaments are very thin in diameter but long and run the full length of the cell.

No meaningful molecular pattern has been attributed to filament arrangement of smooth muscle. At the poles of the nucleus are cones of cytoplasm free from filaments that contain such things as small Golgi apparatus, ribosomes, and occasional lysosomelike structures. Mitochondria are scattered at random and do not show features of high metabolic potential and occupy small proportions of the cell volume. There is no well-organized endoplasmic reticulum. (Although there may well be a common metabolic link between striated and smooth muscle at biochemical function, the two types have different forms of structure for the purpose of contraction.) Contraction presumably occurs as in striated muscle but is slow and gentle and less energy is needed, and thus the mitochondria are less conspicuous in size and number. In general, it can be said that mitochondria are more numerous in striated than smooth cells but in each type the number and size are related to the cell's demands for energy.

Striated or Striped Skeletal Muscle. Striated muscle contracts faster than either smooth or cardiac muscle but is incapable of the continuous contractions seen in either of the latter two. Thus, it works best intermittently. Skeletal muscle fibers are of two types: red and white. In mammals, the diaphragm is mostly of red fiber composition whereas the limb muscles contain a greater proportion of white fibers. Most skeletal muscles contain varying amounts of both. The differences in the two types are as follow: white muscles respond with a propagated action potential to a single nerve stimulus, red muscle to only repetitive stimuli; sustained contraction of the white requires a greater frequency of stimulation; white fibers develop more tension in a single twitch; white fibers are "fast" fibers while red are "slow" and tetanic contractions are more prolonged in the red; red fibers are smaller in diameter, mitochondria are more numerous and the sarcoplasmic reticulum less extensive; red fibers have no M bands; red fibers have more myoglobin. Heart fibers resemble red fibers. Thus, red fibers seem to be responsible for prolonged, sustained contraction (to maintain muscle tone) whereas the white are adapted for rapid, purposeful movement.

The cellular unit in skeletal muscle is the single muscle cell referred to as the myofibril owing to its elongated shape. Striated muscle cells can produce sudden contraction, and like the muscle, the cells are tapered and elongate and multinucleated (hundreds of nuclei may be present in one fiber). The cells vary from 1 to 40 mm long and 10 to 40 μm wide. Their nuclei lie at the outer edge (near the cell surface) of the cell and are needed to control the large cytoplasmic muscle. The nuclei are lined by double membranes with a few "pores" and often have nucleoli. A regular pattern of cross striations with a spacing of 2 to 3 μm is the most prominent feature of the cell and gives it its name (striated). These are due to alternate dark (birefringent) and light (monorefringent) bands that are opposite each other. Each fibril is individually cross striated. This muscle is under control of the spinal or cranial nerves and therefore is dependent on the integrity of its innervation for continued, voluntary function.

A connective tissue layer, the epimysium, encloses the entire muscle. In turn, each of the muscle bundles has a similar, but thinner covering, the perimysium. Reticular and fine collagen fibers, the endomysium, surround the individual muscle fibers (Fig. 4-9).

In the light microscope one notes that each cell contains a large number of intracellular longitudinally arranged, fine, threadlike myofibrils about 1 μm in diameter, packed together so that they are superimposed on each other in a histologic section and appear to have faint striations. A small amount of cytoplasmic material, the sarcoplasm, surrounds the myofibrils and separates them. The sarcoplasm, which is the substance filling the interorganelle space, is a continuous solution of protein, carbohydrate, and electrolytes containing lipid droplets, glycogen particles, lipofuscin granules, and ribosomes. The lipid droplets may serve in the oxidation of fatty acid as a source of energy to nearby mitochondria; the glycogen particles play a role in the energy economy of the cell as a storage form of glucose; the lipofuscin granules (seen in increased numbers in aging muscle) are frequent in the myocardium and uncommon in skeletal muscle. They are "wear-and-tear" pigments and probably are remnants of degenerated organelles. The ribosomes, which in normal adults are uncommon, are frequent in developing and regenerating muscle. Their major role in protein synthesis is well known. The smooth endoplasmic reticulum is a prominent component of the sarcoplasm. It has a complex regular segmental arrangement corresponding to the repeating sarcomere pattern of the fibril. It is called sarcoplasmic

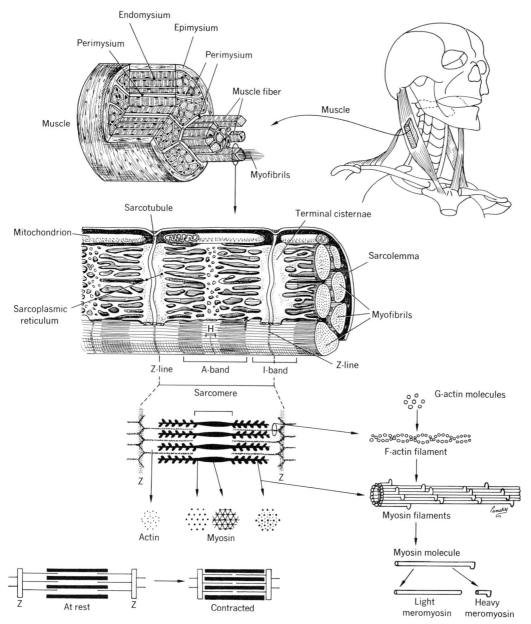

Figure 4-9. The histology and electron microscopy of skeletal muscle structure.

reticulum (Fig. 4-9) and has two structural elements: one transverse in relation to the muscle cell and the other longitudinal. The T tubules or transverse tubules (Fig. 4-9) cross the cell parallel to the cross striations and lie at a constant position in each sarcomere either at the Z line or at A-I junctions. They are closely related to the surface membrane of the cell, becoming continuous with it in some species so that the narrow lumen by communicating directly with extracellular space may contain an extracellular concentration of ions. The T tubule relates the whole thickness of the muscle cell to the surface membrane, ensuring that each sarcomere is kept in touch with the functional state of the cell surface.

The longitudinal component of the sarcoplasmic reticulum, the true smooth endoplasmic reticulum of the muscle cell, forms a network that winds between the myofibrils and invests each with a lacelike sleeve of intercommunicating cisternae (Fig. 4-9). The longitudinal tubules are also arranged in a segmental pattern, forming expanding foot processes that flank each side of the T tubules. The repeating system consisting of T tubule with two closely applied foot processes lying at constant position in each sarcomere is called the triad of the striated muscle. There is no obvious communication at the triad of the T system and longitudinal elements of the sarcoplasmic reticulum. Thus, the segmented sarcomere pattern produced by overlapping thin and thick filaments is accomplished by partial segmentation of mitochondria and a segmental arrangement of sarcoplasmic reticular components. Mitochondria can be detected in the sarcoplasm with special staining in the light scope. The nuclei lie in the sarcoplasm in close contact with outer myofibrils. The Golgi apparatus appears as a cluster or stack of parallel flat saccules near the nuclear pole. It is less frequent than in secretory cells. It is probably involved in internal cellular economy; it may play a role in protein synthesis and transport in muscle (in cells of secretory function it appears as a site of accumulation, segregation, and storage). A thin, tough membrane investment around the cell is called the sarcolemma, which consists of a dense inner plasma membrane, a narrow electron translucent intermediate zone, and a darker, irregular outer basement membrane. Across the dense, compact plasma membrane a bioelectric potential is maintained, and ionic changes take place during the propagation of the excitatory impulse (action potential that starts the muscle contraction). The basement membrane is essentially structureless (300 to 500 Å thick) and is a part of the cell membrane impermeable to Na$^+$. It serves a supportive role in maintaining cell shape, is resistant to trauma and persists after muscle injury, and may play a role as a matrix for regenerating myofibrils. Little is known of the translucent zone. The individual myofibril is composed of an organized longitudinal system of hundreds of regularly arranged parallel filaments of two types: thick or coarse (myosin) and thin or fine (actin) (Fig. 4-9).

The filaments are composed of bundles of long molecules. The alignment of the filaments, forming regular regions of overlap, is responsible for the distinctive repeating pattern of cross striations and forms the molecular basis for muscle contraction. The repeating unit of cross striation is called the sarcomere(s), and a complex nomenclature identifies its components.

The coarse filaments are 1.5 μm long and about 100 Å wide. Cut transversely, they appear arranged in a widely spaced hexagonal pattern, allowing the fine filaments to be placed between them. They are composed of myosin. The thin filaments are about 1.0 μm long and 50 Å wide and are composed of actin. There are twice as many fine actin filaments as coarse myosin filaments.

The region of the fibrils in which only thin filaments are found is the light I band or disk (isotropic) and consists mostly of actin, and that in which thick filaments are seen is the dark A band (consisting mostly of myosin) (Fig. 4-9). The fibril is made up of alternating I and A bands. The I band is bisected by the dense Z line or disk (contains both actin and myosin) that links together the midpoints of the light filaments. This Z line (disk) appears to be the point at which the parallel sarcostyles are bound to each other and may be the point of greatest excitability. The sarcomere extends from one Z line to the next and consists of two half I bands with a complete A band between. At rest, thick and thin filaments lie parallel to each other and interdigitate (Fig. 4-9). In the center of the A band is a thin, light H band (Hensen's line) within which lies an M line (Fig. 4-9).

At each end of the A band there is an overlap between the interdigitating thick and thin filaments. At intervals there appear to be minute cross linkages that join overlapping thin and thick filaments together.

The coarse filaments lie totally in the A band as they stretch from one edge of the A disk to the other, whereas the thin filaments extend from the Z line, pass across the I band, and pass a short distance between the ends of the coarse filaments when the muscle is relaxed. The H band is due to a lack of actin filaments in the middle of the A disk at the time of relaxation.

Function. Myosin is an asymmetric protein of molecular weight 500,000 and about 1500 Å long. It is thin, rigid, rod shaped with a 60 percent alpha-helical content. It is actually an enzyme that catalyzes the hydrolysis of ATP and ADP. It is stimulated by calcium and inhibited by magnesium. It can split into two fragments by hydrolysis with trypsin: a larger fragment H meromyosin and a smaller L meromyosin, with the former containing the ATPase activity of the original.

Actin is a globular protein of molecule weight 56,000. It has the ability to polymerize. At neutral pH, monomeric actin (G actin) is stable in the absence of added electrolyte. With salts polymerizing occurs to form extended filaments of F actin. Each G actin contains a molecule of bound ATP, and during polymerizing this is hydrolyzed to ADP. When myosin and F actin are mixed, they form actomyosin. With ATP added to actomyosin one gets dissociation to actin and myosin. Furthermore, with the addition of ATP (instant energy), the strands of actomyosin contract. Myosin can split ATP, releasing the energy stored in its high-energy phosphate linkage. This action is partly converted to mechanical work by the overlapping filament mechanism of the fibril. Mechanical force may be exerted in some way at the cross linkages between thick and thin filaments (actin), when energized by ATP, to slide with relation to the thick filaments, which themselves remain stationary, thus shortening the sarcomere. During contraction, the thick and thin filaments remain at constant length, but the sliding of the thin filament between the thick leads to an increase in their area of overlap and a corresponding decrease in width of the I band. The A bands remain of constant width. Since the thin filaments slide closer to the M line, the H zone narrows and may even disappear with contraction. Thus, during contraction the sarcomere becomes shorter and broader, shortening the cell. The mechanism of sliding filaments, repeated hundreds of times along the length of a single fibril, results in a shortening of the fibril's original length, and with every moment we make thousands of parallel fibrils carry on this action. (See Chap. 7 for detailed discussion of striated muscle.)

CARDIAC OR HEART MUSCLE. Cardiac muscle (Fig. 4-10) is found only in the heart and where the great veins enter the heart. It is required to provide "motive" force for the circulation by repeated contraction. It may rest for only a fraction of a second at a time and must be able to give rapid, powerful, sustained contraction. It needs wide resources of power to meet sudden increased demands, and the action of all parts must be coordinated so that all effort is directed to move blood and is not wasted by inefficiency.

Cardiac muscle is distinct from both striated and smooth muscle but has features of both. It is striated with some of the sarcomere pattern of skeletal muscle. It is a branching structure (Fig. 4-10), but the nuclei are centrally located as in smooth muscle. The cardiac cells are shorter than the long, cylindric skeletal fibers. Longitudinally the fibers contain irregular rodlike fibrils, which run the length of the fiber (cell). The fibrils are the regular, repeating sarcomere, which is the final unit of contraction in the heart, as in skeletal muscle (see earlier discussion). The sarcomere (contractile element) constitutes about 50 percent of the myocardial fiber, a much lower percentage than in skeletal muscle, where it is 80 to 90 percent of the total mass. This difference is due to the greater number of mitochondria in the heart muscle.

In contrast to the multinucleate appearance of skeletal muscle, the nuclei are separated by transverse partitions, the intercalated disks, now known to mark individual cell boundaries. This muscle closely resembles skeletal muscle in many respects. There are cross striations due to the presence of a repeating pattern of thick and thin filaments arranged in register and overlapping to form bands: the A band, where thick myosin filaments are found; the I band, formed by a portion of thin actin filaments not overlapped by myosin; the Z line, which bisects the I band; and the M line, which bisects the A band. The action of cardiac

Figure 4-10. The histology and electron microscopy of cardiac muscle structure. (After studies by Sonnenblick and Chidsey.)

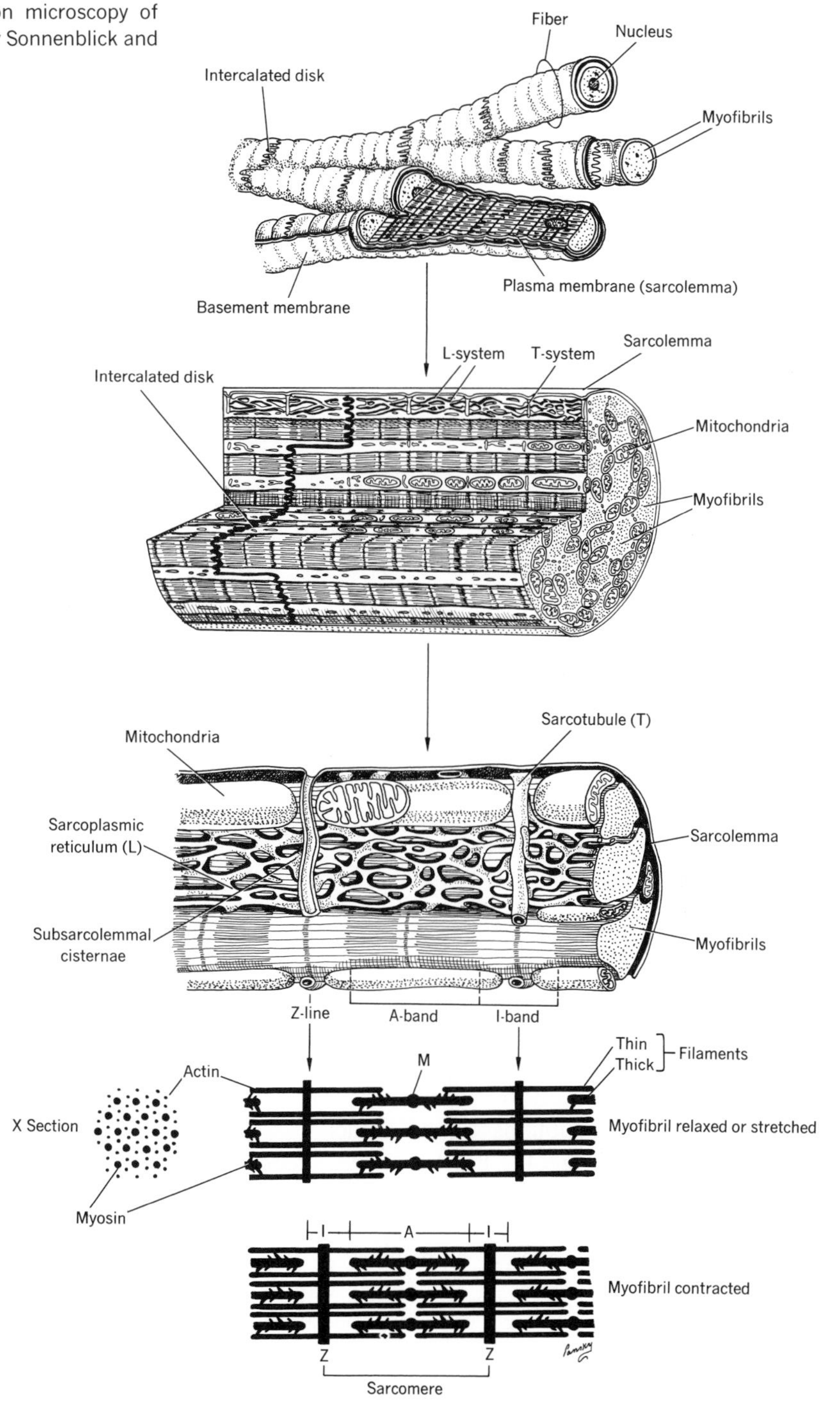

muscle is thought to depend on the same sliding mechanism, with energy released by ATP hydrolysis causing actin filaments to be drawn between myosin filaments with a resultant shortening of the sarcomere. We thus get powerful, rapid contraction with the consistency of rest period (provided by a striated pattern).

The organization of sarcoplasm is not significantly different from that of skeletal muscle. The T system and longitudinal components of the sarcoplasm reticulum have a similar arrangement, although they show less regularity in cardiac muscle. Small yellow-brown lipochrome pigment granules that increase with age are often seen in the sarcoplasm. Reticular fibers lie around and between the cells, and cell bundles are bound by collagen fibers that arise from the connective tissue skeleton of the heart.

The heart is in constant activity, and a high rate of energy production may be needed for prolonged exercise. The mitochondria (sources of energy for contraction) are correspondingly large and numerous (more per unit volume than any other tissue or organ) with cristae closely packed (importance of oxidation phosphorylation in cell economy). Glycogen and lipid droplets (potential fuel for oxidative processes) are found in the sarcoplasm near the mitochondria.

The intercalated disks (Fig. 4-10) are the most specific feature of the cell. Electron microscopy shows that each marks the points of contact between two separate cardiac muscle cells, each with its own nucleus. Thus cardiac muscle is composed of numerous separate cellular units joined closely at the disks. The disks serve two functions: (1) to provide a mechanical link between cells, preventing separation during contraction; and (2) to act as anchorage points for contracting fibrils of each cell. In addition, they allow a communication for possible passage of surface excitation from cell to cell, and the heart activities coordinate with mechanical efficiency. The zigzag form of the disk is characteristic.

The adjacent closely apposed cell membranes at the disks form specialized areas of contact at different points. There are extensive zones where membranes of adjacent cells appear to fuse. These areas are thought to allow permeability to ions between cells, permitting spread of surface excitation through the muscle mass.

Heart muscle cells function faster than smooth muscle and the time intervals are shorter, but there is a long refractory period. Rhythmic contraction is a definite feature. Physiologically, the entire heart responds in an "all-or-none" contraction.

The Nervous Tissue

Nerve cells are elements in which specialization has taken place for (1) receptiveness, (2) conduction of change initiated by the reception of a stimulus or impulse, and (3) manufacture of very highly potent chemical substances or neurohumoral agents that allow transmission across the spaces of no structural nervous continuity. The latter are called synapses. Associated with these functions are the increase in total surface area of the cell membrane to form fine, branching unsheathed processes, dendrites, for input or receiving; development of many nerve cells with long sheathed processes, the axon, for transmitting; the presence of special granules in the cytoplasm, Nissl granules, and the continual movement of molecules along the axon of the long nerve fiber.

Nervous tissue (Fig. 1-28, p. 35; Fig. 4-11) is derived from the cells of the neural plate, which also forms the neural groove and canal and primitive brain stem. Thus, it is ectodermal in origin. Morphologically the cells of the neural plate form the nerve cell proper or neurons and accessory cells. Nervous tissue contains connective tissue elements that penetrate through it. In the peripheral nervous system, found outside the bony case that holds the brain and spinal cord, one sees blood and lymph vessels. In the central nervous system, that part inside the bony case, blood vessels are many. There are no lymph vessels but rather a special fluid, called cerebrospinal fluid.

The neuron is a highly specialized cell and thus its regenerative powers have been lost. Mitosis (cell division) is never seen in fully developed nerve cells. The accessory cells, however, are less highly developed and have great reproductive ability. The metabolic rate is high in neurons, and they cannot stand an oxygen deficit for long. Oxygen lack can be fatal.

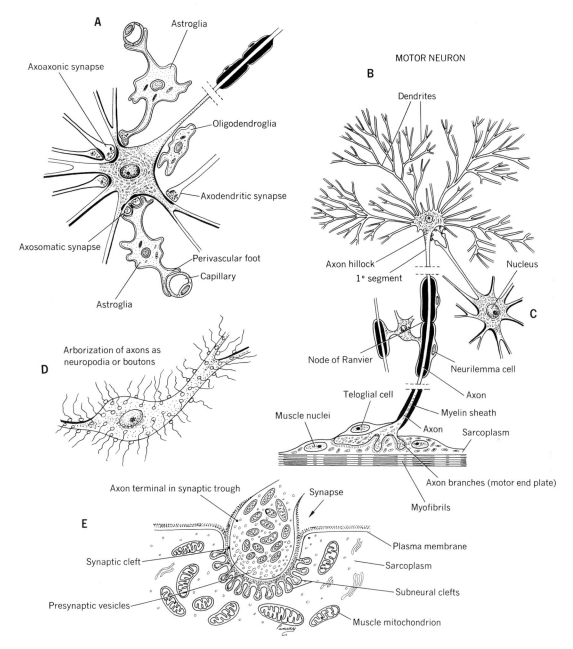

Figure 4-11. A. Relation of a neuron, astroglia, oligodendroglia, and nerve terminals. Astroglia are shown related to capillaries and neurons. A variety of synapses are also seen. **B.** Diagram of a motor neuron and its synapse with a voluntary muscle to form a motor end-plate. **C.** A neuron in the central nervous system. **D.** Arborization of axons as boutons on the surface of a nerve cell body. **E.** Relationship of the motor end-plate and the muscle fiber.

Neurons are highly receptive and have the ability of irritability and conductivity. They (1) receive stimuli in their sensory nerve receptors; (2) rapidly conduct along thin ingoing (afferent) nerve fibers; (3) maintain the excitable state in their cell bodies; (4) integrate and correlate changes in the excitable state in the arrangement of the central neurons in definite patterns; (5) conduct outward via efferent fibers to the effectors (glands, muscles, etc.); and (6) transmit across from the final nerve structure to the operative tissue or effector (across the synapse). The accessory cells are important in both the normal and abnormal state but little is known of their function.

Within the brain stem in the central nervous system, nervous tissue is divided into gray matter, which con-tains nerve cell bodies and nerve fibers, and white matter, which contains only nerve fibers. Outside the central nervous system, collections of nerve cells and fibers are called ganglia whereas bundles of fibers are referred to as nerves.

Cells Derived from the Primitive Neural Groove.
NERVE CELLS (NEURONS). Nerve cells consist of a cell body (perikaryon) composed of cytoplasm containing Nissl granules, neurofibrils, a large vesicular nucleus with little chromatin and a single conspicuous nucleo-lus; and two types of processes extending from the cell body. One type of process is the dendrites, which are numerous, short, and branch repeatedly like a tree. They contain neurofibrils. They are the receiving afferent or input fibers carrying impulses to the cell

Figure 4-12. Schematic drawings of motor neurons, anterior horn cells, and neuroglia.

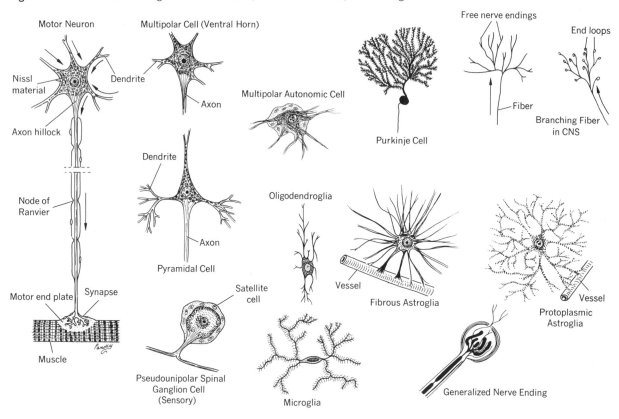

body. The other is a single process called the axon (Fig. 1-28, p. 35; Figs. 4-11, 4-12), which arises from the cell body in the form of a granule-free expansion, the axon hillock. Many neurofibrils enter the axon, which is usually very long, and, although a few branches may arise from it near its origin, it tends to pass for distances unbranched with little tapering and finally divides dichotomously near the structure it supplies. At its end, it may have a specialized end formation or have free tapering end-branches with end-loops or knobs. The axon carries impulses away from the cell body (efferent) to the structures to be innervated.

CHROMAFFIN CELLS. In early development these cells resemble neurons but they tend to wander into the peripheral tissues and organs (like the adrenal medulla). They can also be seen among the ganglion cells of the sympathetic ganglia. They retain a simple epithelial-like appearance and do not develop as do neurons. They are secretory and go on to produce and secrete epinephrine and norepinephrine (catecholamines). The enterochromaffin cells of the digestive tract also exhibit secretory activities and have wandered from the primitive neural canal.

MELANOBLASTS. Melanoblasts are cells that produce the black melanin pigment and have wandered from the neural crest. In higher vertebrates they pass to the skin only and become the melanoblasts or dendritic cells of the skin.

CHIEF CELLS OF THE PINEAL GLAND. Chief cells lie in the pineal gland, located near the posterior roof of the third ventricle of the brain. They have long processes with clublike ends and their function is as yet uncertain.

CELLS LINING THE CHOROID PLEXUS. These cells appear to be epithelial and are responsible for a great share of the formation of cerebrospinal fluid.

ACCESSORY CELLS OF THE NERVOUS SYSTEM. These are closely associated with both nerve cell bodies and axons but contain no Nissl granules or neurofibrils. Two forms are seen in the central nervous system and two in the peripheral nervous system.

Ependymal Cells. Ependymal cells line the central canal of the spinal cord and the ventricles of the brain. They are tall, ciliated cells and the base of each cell has a long glial-like fiber that extends to the surface of the brain stem. Its nucleus has large chromatin granules.

Neuroglial Cells. Neuroglial cells (Fig. 4-12) are scattered throughout the gray and white matter. They are very small and the nucleus takes up nearly the entire cell body. The cytoplasm is "drawn out" into many processes. One notes two types: astrocytes and oligodendroglia.

The neuroglial cells may control passage of materials between the circulation and nervous tissue (Fig. 4-12). The cerebral capillaries are surrounded by a cuff of neuroglia, mainly astrocyte foot processes that prevent the nerve cell from coming into extensive contact with blood vessels. Other processes of the neuroglial cells ramify through adjacent tissue contacting and often surrounding nerve cell processes.

The neuroglial cells may take the place of a connective tissue "space," providing a diffusion channel for metabolites under direct control by the cell. Through metabolic selectivity, the neuroglial cells may regulate the nutrition of the nervous system and constitute part of the physiologic blood-brain barrier. Certain neuroglial cells may participate in electrical activity of brain. The oligodendroglia participate in myelination in the central nervous system and are analogous to the Schwann cell of peripheral nerves. The microglial cells appear to act as phagocytes after injury.

Thus the neuroglial cells undoubtedly play more than a pure mechanical role in the brain (in view of large numbers and intimate relations to neurons).

Amphicytes or Satellite Cells. These are small cells of the peripheral nervous system that form capsules around ganglion cells in the autonomic and dorsal root ganglia (Fig. 4-12). They are closely related to the nerve cell bodies.

Schwann Cells (Sheath Cells). Schwann cells (Fig. 1-28, p. 35; Fig. 4-11) are similar to the satellite cells but are intimately associated with the axons of the peripheral nervous system and are a part of the peripheral nerve fiber (whether myelinated or not).

Autonomic Interstitial Cells. These cells are intermediary between postganglionic fibers and the effector units.

Neurons. LOCATION. Neurons can be found in several locations. These include:

1. Neurons that are entirely in the central nervous system. These are the intercalated neurons and neurons that give origin to the association, commissural, and projection fibers.
2. Neurons with cell bodies in the central nervous system but whose axons pass to peripheral tissues: lower motor arising in the spinal cord and going to the tissue of supply and preganglionic fibers of the autonomic system.
3. Neurons whose cell bodies wander outside the central nervous system but stay nearby, like the dorsal root ganglia.
4. Neurons that wander far from the central nervous system and lie in the autonomic ganglia.
5. Neurons in special sense organs or elsewhere.

STRUCTURE. The neuron is often a long cell with its axon up to 90 cm long. It is usually considered in three parts: the cell body (perikaryon) and its dendrites; the axon and its sheaths; and the nerve endings (telodendrites).

Cell Body and Dendrites. The cell body (Figs. 4-11, 4-12) varies in size (from 5 to 120 μm in diameter) and shape (spherical, ovoid, etc.). In the central nervous system multipolar forms are most common whereas they tend to be round or ovoid in the autonomic ganglia. The nerve cells (in the ganglia) are commonly encapsulated but rarely in the central nervous system. The dendrites of the cell body vary in number and in extent and type of branching.

The cytoplasm of the cell body consists of a cell membrane, cytoplasm (hyaloplasm), organelles, and inclusions. The cytoplasm of the cells in the central nervous system has a low viscosity, whereas that in the cells of the dorsal root ganglion is very high (almost gellike). The normal, fixed cytoplasm always appears to have Nissl granules (tigroid bodies, chromaphil), and these are variably disposed in angular masses, in fine granules, or are hardly visible. The amount and distribution of granules vary with the physiologic state. Exhausted cells show granule disappearance. This is also true if the axon of a motor cell is cut. Under the electron microscope the Nissl granules appear as areas of closely arranged flattened vesicles of the endoplasmic reticulum with large numbers of ribosomes on their cytoplasmic surfaces.

In fixed cells, very fine neurofibrils cross the cell body at all angles. They show no branching, and on injury to the cell or axon they break up and vanish. They are not seen in early neuroblasts but appear with development. The neurofibrils are not the actual basis of conduction.

Mitochondria are seen scattered throughout the nerve cell. Golgi apparatus is also seen. Its function may be associated with neurohumoral production in nerve cells. Other inclusions often seen are melanin pigment granules; lipochrome, which appears to increase with age; and protein masses, which are seen in some cells of the hypothalamus.

Nerve Fibers. Nerve fibers are generally long structures (Fig. 1-28, p. 35; Figs. 4-11, 4-12). The sheath elements that cover the axon are secondary structures but contribute to the fiber's normal stability. The diameter of the fiber is important in relation to its physiologic time constants. Thus, large-diameter fibers conduct rapidly and there is a slower rate in small-diameter fibers. In the central nervous system nerve fibers make up most of the white matter, and in the peripheral nervous system they form the essential part of the nerves of the body.

All nerve fibers have a myelin layer, which consists of closely packed regular lipoprotein lamellae, but the so-called nonmyelinated fibers have such a thin layer that it is not clearly seen with the usual histologic methods of preparation. The myelin is present in lamellae (see Fig. 4-14 for its formation). The number of myelin lamellae in the visibly myelinated fibers varies in different fibers. Thus, one still hears the terms *myelinated* or *medullated* and *nonmyelinated* or *unmyelinated*, but one should remember the above. Myelination is an active cellular process involving a dynamic relationship between axons and sheath cells.

The fully formed myelin sheath is still part of the living satellite cells and structurally distinct from the axon.

PERIPHERAL MYELINATED FIBERS. These fibers consist of the axon, a long cylinder that appears to have no visible internal structure but contains neurofibrils (Fig. 1-28, p. 35; Fig. 4-11). This is surrounded by a thin membrane, the axolemma. Outside of the latter is a series of myelin lamellae separated from each other by protein layers. Surrounding the lamellae is the Schwann or neurolemmal cell layer. The outer cell membrane of the Schwann cell forms a thin, homogeneous membrane (this used to be called the neurolemma but today we call the entire Schwann cell sheath the neurolemma). The Schwann cells are said to secrete the myelin sheaths. Overlying the Schwann cell complex are connective tissue elements (fibroblastlike cells, reticular fibers, and fine collagen fibers) called the endoneural tube.

Thousands of nerve fibers are held together in fibrous wrappings called the endoneurium. Many fibers are bundled together by a fibrous perineurium, and in turn these bundles are further wrapped in a common fibrous sheath, the epineurium. The bundles build up to eventually be seen by the naked eye as a "nerve." It should be noted that some nerves leading to and from the brain (cranial nerves) contain only sensory or motor fibers. Other cranial and *all* spinal nerves contain both sensory and motor fibers.

Circular constrictions interrupt the layering at intervals, the nodes of Ranvier (Fig. 1-28, p. 35; Fig. 4-11). At these constrictions, both the cytoplasm of the Schwann cell and the myelin lamellae are interrupted. Here the axolemma is unshielded and comes into contact with the connective tissue elements (tissue fluids) of the endoneural tube. A number of funnel-shaped narrow partitions are also seen in the internodal segments and are called the funnels of Schmidt-Lantermann. The axon itself, based on recent work, appears to be constricted at the nodes.

MYELINATED FIBERS OF THE CENTRAL NERVOUS SYSTEM. Here the axon with its axolemma is identical to that of the peripheral nerve. The myelin lamellae also form a myelin sheath that is covered by a thin homogeneous membrane, but no Schwann cells are seen.

Nodes of Ranvier are also not described or obvious. Surrounding the homogeneous membrane is a network of neuroglial fibers with neuroglial cells.

PERIPHERAL NONMYELINATED FIBERS. These are like the myelinated fibers but usually smaller (Fig. 1-28, p. 35). One or two lamellae of myelin are present but are not seen in a light microscope. The cytoplasm of the Schwann cell encloses the axon and membrane in a thin coat and the nuclei bulge giving the fiber a "knotted" appearance. A thin membrane covers the Schwann cell cytoplasm, which in turn is covered by an endoneural sheath.

NONMYELINATED FIBERS IN THE CENTRAL NERVOUS SYSTEM. Here the axon and its axolemmal covering are covered by an open network of neuroglial fibers and cells.

Nerve Terminals (Telodendrites). Nerve terminals (Fig. 4-11) are formed by the final axon branchings. These are points at which the neurohumoral agents (e.g., acetylcholine) are liberated and activate other neuronal units or effectors (muscle, glands, etc.).

IN THE CENTRAL NERVOUS SYSTEM. As the axon nears its termination, it divides several times until its branches are very fine, and on these fine fibrils, a small end-loop is found, in which the neurofibrils are seen to divide. These end-loops (end-feet or boutons terminaux) are numerous and lie close by or touch the dendrites or cell bodies of other neurons. Thus, hundreds of these may be associated with the surface of but one nerve cell body and its dendrites. The end-loop and the interval between it (about 200 Å) and the cytoplasm of the second neuron are called a synapse. Under the electron microscope one sees that the final axon branchings lose their sheaths and swell out to form a bulbous ending that lies isolated in a nest of neuroglial fibers. The bulb is flattened and its own membrane (presynaptic) is thickened as it faces the surface of the receiving cell membrane (postsynaptic). Numerous synaptic vesicles lie close to this flat surface and are said to contain a high concentration of neurohormone.

OUTSIDE THE CENTRAL NERVOUS SYSTEM. In synapses between preganglionic fibers and cells of the autonomic system that give rise to postganglionic fibers,

Table 4-1. Summary of Body Tissues

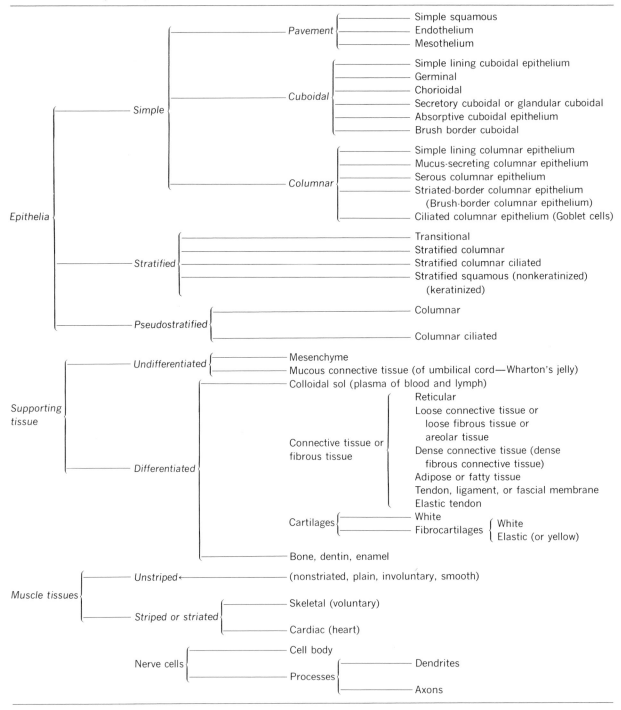

Epithelia
- Simple
  - Pavement
    - Simple squamous
    - Endothelium
    - Mesothelium
  - Cuboidal
    - Simple lining cuboidal epithelium
    - Germinal
    - Chorioidal
    - Secretory cuboidal or glandular cuboidal
    - Absorptive cuboidal epithelium
    - Brush border cuboidal
  - Columnar
    - Simple lining columnar epithelium
    - Mucus-secreting columnar epithelium
    - Serous columnar epithelium
    - Striated-border columnar epithelium (Brush-border columnar epithelium)
    - Ciliated columnar epithelium (Goblet cells)
- Stratified
  - Transitional
  - Stratified columnar
  - Stratified columnar ciliated
  - Stratified squamous (nonkeratinized) (keratinized)
- Pseudostratified
  - Columnar
  - Columnar ciliated

Supporting tissue
- Undifferentiated
  - Mesenchyme
  - Mucous connective tissue (of umbilical cord—Wharton's jelly)
- Differentiated
  - Colloidal sol (plasma of blood and lymph)
  - Connective tissue or fibrous tissue
    - Reticular
    - Loose connective tissue or loose fibrous tissue or areolar tissue
    - Dense connective tissue (dense fibrous connective tissue)
    - Adipose or fatty tissue
    - Tendon, ligament, or fascial membrane
    - Elastic tendon
  - Cartilages
    - White
    - Fibrocartilages
      - White
      - Elastic (or yellow)
  - Bone, dentin, enamel

Muscle tissues
- Unstriped ← (nonstriated, plain, involuntary, smooth)
- Striped or striated
  - Skeletal (voluntary)
  - Cardiac (heart)

Nerve cells
- Cell body
- Processes
  - Dendrites
  - Axons

Table 4-1 (Continued)

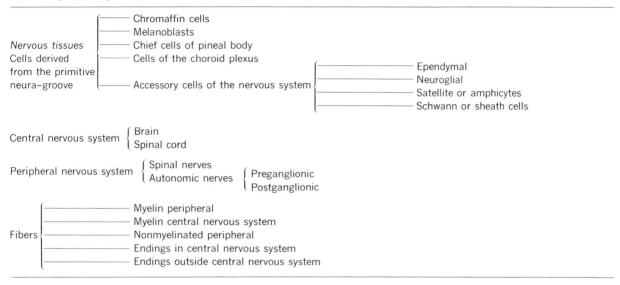

Nervous tissues
Cells derived
from the primitive
neura–groove
- Chromaffin cells
- Melanoblasts
- Chief cells of pineal body
- Cells of the choroid plexus
- Accessory cells of the nervous system
  - Ependymal
  - Neuroglial
  - Satellite or amphicytes
  - Schwann or sheath cells

Central nervous system { Brain / Spinal cord

Peripheral nervous system { Spinal nerves / Autonomic nerves { Preganglionic / Postganglionic

Fibers
- Myelin peripheral
- Myelin central nervous system
- Nonmyelinated peripheral
- Endings in central nervous system
- Endings outside central nervous system

the end-loop formation is most common. Where the connection is between a fiber and receptor structure there is a sensory nerve ending and between a fiber and an effector there is a motor nerve ending. One speaks of free nerve endings with simple, small, widely spread end points. Organized nerve endings are often complex, closely coiled, or discretely branched structures. The various sensory end-organs or receptors in areas of the skin are of this latter type.

In mammalian skeletal muscle, each branch of the motor axon terminates in a discrete, localized, compact structure, the motor end-plate (Fig. 4-11).

Central Nervous System. Brain tissue is complex. Supporting cells or neuroglia are clearly seen here as well as multiple synapses. In the brain there is no true connective tissue supporting the cells and no apparent significant intercellular space. The neuroglial cells take the place of connective tissue and surround the other elements of the nervous tissue. Nerve and neuroglial cells are so tightly packed in normal circumstances that intercellular spaces greater than 200 Å are rarely seen. The packing in the central nervous system is as close as compact epithelium.

Review Questions

1. What is a tissue? Give several examples.
2. What are some of the functions of epithelial tissue?
3. Endothelium, simple squamous epithelium, and mesothelium serve what purpose and what do they have in common?
4. Differentiate between cuboidal, columnar, and stratified epithelium. Can they be subdivided and, if so, into what general classifications?
5. Define stratified epithelium. What are its major functions?
6. What is a supporting or connective tissue? How is it constructed? How does its construction help in its function? List the kinds of connective tissue.

7. Where does one find cartilage? What function does it serve?

8. Compare intramembranous and endochondral bone formation.

9. Describe any typical long bone of the body and indicate how it manages to grow in width and length.

10. What is meant by centers of ossification? Primary and secondary?

11. What are the three major types of muscle tissue and where are they found?

12. The histology (and electron microscopy) of skeletal muscle is unique. Define the following terms:
    a. Myofibril
    b. Sarcoplasm
    c. T tubules
    d. Sarcolemma
    e. Myosin
    f. Actin
    g. Sarcomere
    h. Various bands and lines

13. How does skeletal muscle contract?

14. How does cardiac muscle differ from skeletal muscle?

15. What is the function of the nerve cell?

16. What makes up a neuron? Describe the function of each of its parts.

17. What is meant by a myelinated or nonmyelinated cell?

18. In summary, what are the major tissues of the body? Give an example of each and how it functions.

References

Bloom, W. B., and Fawcett, D. W.: *A Textbook of Histology,* 9th ed. W. B. Saunders Co., Philadelphia, 1968.

Copenhaver, W. M.; Bunge, R. P.; and Bunge, M. B.: *Bailey's Textbook of Histology,* 16th ed. Williams & Wilkins Co., Baltimore, 1971.

DeRobertis, E. D. P.; Nowinski, W. W.; and Saez, F. A.: *Cell Biology,* 5th ed. W. B. Saunders Co., Philadelphia, 1970.

Fallis, B. D., and Ashworth, R. D.: *Textbook of Human Histology.* Little, Brown & Co., Boston, 1970.

Leeson, T. S., and Leeson, C. R.: *Histology,* 2nd ed. W. B. Saunders Co., Philadelphia, 1970.

Levitan, M., and Montagu, A.: *Textbook of Human Genetics.* Oxford University Press, New York, 1971.

Porter, K. R., and Bonneville, M. A.: *Fine Structure of Cells and Tissues.* Lea & Febiger, Philadelphia, 1968.

Body Framework and Movement

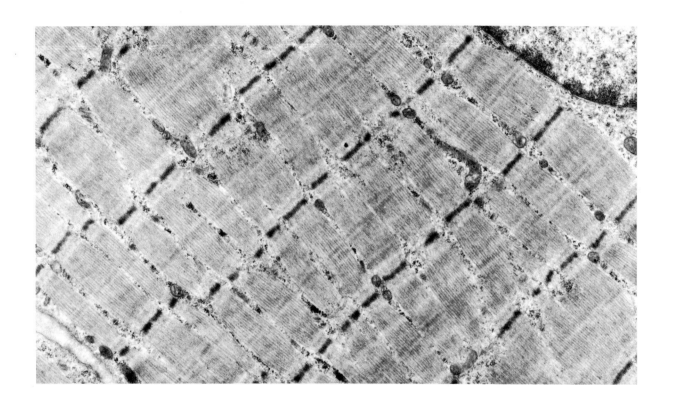

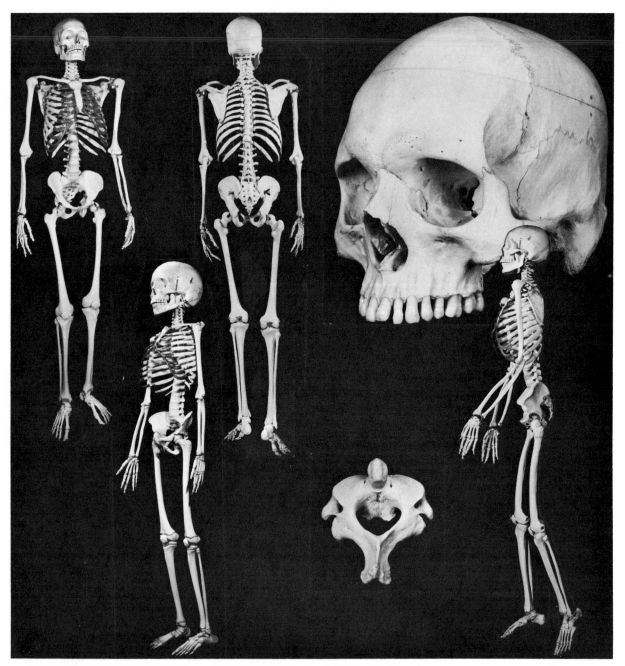

Figure 5-1. The human skeleton as seen from various views. The second from the left is a young skeleton; note the obvious epiphyses and lack of complete bone fusion.

Bony Framework: The Skeletal System

CHAPTER 5

Body Organization and Direction

A system of anatomic references has been used in this and subsequent chapters to facilitate description and discussion of the body as a unit. They are presented for organizational purposes and understanding and consist of direction, body planes, cavities of the body, and structural units.

Directions

The body is usually described as being in the anatomic position, which is erect, facing forward with arms at the sides and palms out with heels together. From this basic position, all directions are given (Figs. 5-2, 5-3).

Superior. *Superior* is uppermost or above; e.g., the chest lies superior to the abdomen.

Inferior. *Inferior* is lowermost or below; e.g., the abdomen is inferior to the chest.

Anterior. *Anterior* refers to the front; the umbilicus lies anterior to the rest of the body. The term *ventral* is also used for the same meaning; however, it probably makes more sense if one thinks of a four-footed animal, since his abdomen is truly ventral to his back.

Posterior. *Posterior* refers to the back; the vertebral column lies posterior to the rest of the body. As with ventral, the term here also used is *dorsal*.

Cephalad. *Cephalad* refers to the head; e.g., the shoulder is cephalad to the elbow and the elbow cephalad to the wrist.

Caudad. *Caudad* refers to the tail. The term usually refers to animals, but in man one can speak of the feet as being caudad to the hips, etc.

Medial. *Medial* is nearest the midline of the body; e.g., the umbilicus is medial to the side of the body or the ulna is medial to the radius in the forearm.

Lateral. *Lateral* is toward the side, e.g., away from the medial. The side is lateral to the umbilicus, the radius lateral to the ulna.

Proximal. *Proximal* is nearest the point of attachment; e.g., the elbow is proximal to the wrist and the wrist proximal to the fingers.

Distal. *Distal* is away from the point of attachment; e.g., the fingers are distal to the wrist, etc.

Visceral; Parietal. The term *visceral* refers to the covering of organs; one speaks of visceral pleura (covering lungs), visceral peritoneum (covering abdominal organs). The term *parietal* relates to the covering of cavity walls; examples are parietal pleura over the inside chest wall and parietal peritoneum over the inside abdominal wall.

Body Planes

The planes of the body (Fig. 5-3) are as follows:

Midsagittal. This is a plane vertically dividing the body into equal right and left sides, cutting through its midline.

Sagittal. This is any vertical plane dividing the body into right and left parts parallel to the midsagittal.

Horizontal or Transverse. This is any plane dividing the body into superior and inferior portions.

Frontal or Coronal. This is any plane dividing the

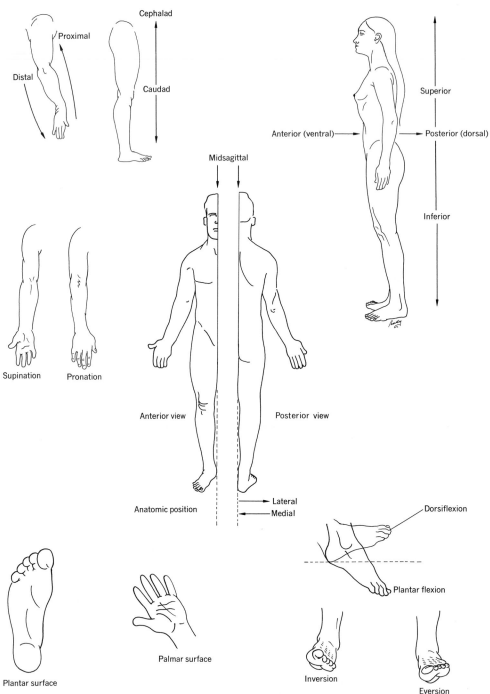

Figure 5-2. Body directions and surface views.

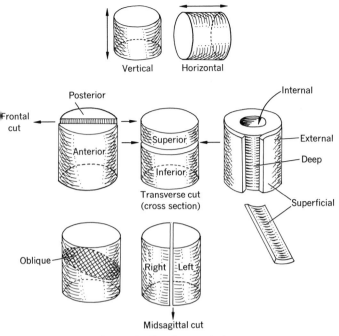

Vertical   Horizontal

Posterior

Frontal cut

Internal

Anterior

Superior

External

Deep

Inferior

Transverse cut
(cross section)

Superficial

Oblique

Right   Left

Midsagittal cut

Figure 5-3. Anatomic directions and planes.

body into anterior (or ventral) and posterior (or dorsal) parts. These are at right angles to the sagittal plane.

Major Body Cavities

The organs of the body are collectively called viscera. The major body cavities are as follows:

Dorsal Cavity. The dorsal cavity contains the structures of the nervous system. It is divided into a cranial part, containing the brain, and a spinal portion, which holds the spinal cord.

Ventral Cavity. This consists of several subdivisions.

THORACIC CAVITY. The thoracic cavity houses the lungs, pericardium, heart, and great vessels. The mediastinum is the area between the pleural cavities and contains the thymus, lymph vessels, esophagus, trachea, nerves, and heart.

ABDOMINOPELVIC CAVITY. The abdominopelvic cavity contains those organs between the respiratory diaphragm above and the pelvic diaphragm below. These include organs of the digestive system and the urogenital system.

Structural Units

The structural units of the body are as follows:

1. The cell: the basic unit of the body.
2. The tissues: composed of cells and intercellular substances or matrix.
3. The organs: comprised of cells grouped into tissues serving a common function.
4. The systems: cells grouped to form tissues, which combine to form organs, which, in turn, form a group of organs.
   a. Skeletal system.
   b. Nervous system.
   c. Circulatory system.
   d. Respiratory system.
   e. Digestive system.
   f. Urinary system.
   g. Endocrine system.
   h. Reproductive system.

Introduction to Osteology

The Skeleton

The body and its appendages (Fig. 5-1) are supported by a framework or skeleton of bones, cartilage, and fibrous tissue. The skeleton accounts for about 15 to 20 percent of the wet weight of the total body. The exoskeleton, prominent in many animals, is reduced and represented in man by the hair, nails, and teeth.

Bones and cartilage give support to the softer tissues and act as levers to which muscles and other tissues attach. The primary function of bone is mechanical: it affords stability and mobility as well as maximum strength with minimum weight and bulk. Where they surround structures like the central nervous system, heart, lungs, etc., bones provide protection. The interior of bones lodge and protect the bone marrow where blood cells develop. The bone substance also functions as an ionic reservoir and provides a storehouse for calcium and phosphates that can be drawn on by the body. It also has a facility for rapid ion exchange, important from a physiologic viewpoint.

Bone is a living tissue (bones are organs) made up

of protein components impregnated with mineral substances making it resistant to both tension and compression and still allowing for considerable elasticity. If bone is decalcified (in dilute acid), it retains its complete shape but becomes highly flexible since its minerals have been removed from the fibrous framework. If burned, on the other hand, to destroy the fibrous tissue, it also retains its shape but now is brittle, crumbles, and is inelastic. A bone is a cylinder of compact cortex covered by osteogenic periosteum and bound externally to muscle by fascia and ligaments that aid in support as well as providing stresses essential to remodeling, maturing, and maintaining structure (see Chap. 4). The mineral element that gives bone its hardness is calcium (65 to 70 percent). For a description of the architecture and structure of bone, see Chapter 4.

The final form and structure of any bone are well adapted to its function of resisting mechanical stresses. It is continually being modified to maintain this function as the stresses it is subjected to alter during life. Two factors create the strength and economy of material: (1) the intimate combination of fibrous tissue and mineral salts (already mentioned); and (2) the use of concentric microscopic tubular lamellae as basic structural units (all mature bone is laminated), which resist bending equally and, being "hollow," economize on material when it is not needed. The cylindric structure of the shaft of a long bone provides strength. When the shaft is curved, there is compensatory thickening of the compact bone in the concavity.

Adjacent bones and soft tissues, particularly muscles and ligaments, may cause or transmit compression or tension forces, and conditions vary when joints are moved to various positions and effectively arranged trabeculae are present to resist the normal forces.

Bone Classification (According to Shape)

Long Bones

Long bones are found in the limbs. When they are small (fingers and toes), they are called miniature long bones. Each has a shaft, two ends of compact bone, and a central, continuous medullary cavity.

The remaining bones consist of cancellous bone in a shell of compact bone with no shaft or cavity. The interstices of cancellous bone are filled with red marrow (e.g., kneecap).

Short Bones

The short bones include carpal, tarsal, and sesamoid bones. The latter are found in the substance of tendons and are so named because they resemble "sesame seeds."

Flat Bones

Flat bones are thin, rather than flat (all are usually curved somewhat) and include the bones of the skull vault, shoulder blade, and ribs. In the skull the spongy substance is sandwiched between plates of compact bone, has many venous channels, and is referred to as a diploe with veins called diploic veins.

Irregular Bones

Irregular bones include many skull bones and vertebrae.

Pneumatic Bones

Pneumatic bones are bones invaded by air-containing sacs. The spongy bone is absorbed and replaced by air-filled cavities with walls of compact bones lined by mucous membrane, which is continuous through small openings with the lining of the nose. The cavities are referred to as paranasal sinuses. Cavities are also seen in the mastoid bone (mastoid air cells). The sinuses lighten the bone, contribute to voice resonance, and often share in infections of the nasal mucous lining.

Fresh Bone

The articular parts of most bones are covered by articular hyaline cartilage, which is smooth and slippery (see Chap. 4). Many bones have no cartilage, as in the skull, where the articular margins are united by fibrous tissue.

All bones, except for the articular portion, are covered by a membranous periosteum, which consists of a fibrous layer containing connective tissue cells, with a second, more cellular layer on its deep surface directly on the bone. The deeper cellular or osteogenic layer gives rise to bone-forming cells, the osteoblasts, which are numerous only when bone is formed. They become incorporated in the developing bone substance as osteocytes (see Chap. 4). The periosteum is adherent to the bone because bundles of fibrous tissue penetrate the bone and are incorporated in it, plus the fact that blood vessels and nerves pass from the periosteum into the bone to supply both bone and marrow. If the periosteum is stripped off, by disease or accident, blood supply loss may cause death of underlying bone.

Muscles, tendons, ligaments, and intermuscular septa that are attached to the bone are really in continuity with it since the fibers blend with the periosteum and even join its constituent collagen fiber bundles.

The medullary cavity is filled with yellow marrow and the spaces in the cancellous bone with red marrow (see below).

The cavities in both the medullary cavity and the cancellous bone are lined with a layer of osteogenic cells that are continuous with the lining of the haversian (see Chap. 4) and nutrient canals and thus indirectly with periosteum. This lining is also called endosteum.

Bone Marrow

Red and white blood cells are formed in the bone marrow (Fig. 5-4), and after birth it is the only normal source of the red cells. In adults, the bone marrow is the major source of the granular leukocytes (white blood cells). In infants, red marrow is found in all bones, including the medullary cavities and haversian canals. It is replaced by yellow marrow, in time, and at puberty red marrow is seen only in cancellous bone. Red marrow persists throughout life in the sternum. With age, even this is replaced by yellow marrow.

Yellow marrow is composed mostly of fatty tissue. Some blood-forming cells remain and may even in-

Figure 5-4. Section of human bone marrow. ×200. (Courtesy of Dr. Dale E. Bockman, Medical College of Ohio at Toledo.)

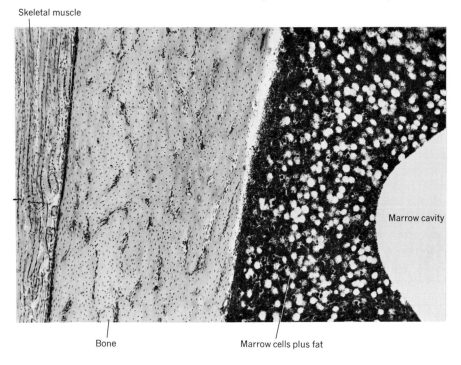

Skeletal muscle

Marrow cavity

Bone

Marrow cells plus fat

crease under abnormal conditions. Red marrow is an active blood-forming organ and consists of many cells in a mesh of a spongework of reticular tissue in which numerous blood vessels anastomose and ramify. The color is due to the red blood cells. Besides blood cells and reticular cells, the major group of cells in bone marrow are myelocytes, which give rise to white and red blood cells by division; nucleated red blood cells at various stages of development (not normally seen in the bloodstream); and megakaryocytes, which produce blood platelets. Large, multinucleated osteoclasts are seen on or near the bone surfaces and function in bone absorption. Gelatinous marrow is the degenerated type seen in skull bones of old people.

Nutrient vessels supply the marrow through canals in the compact bone of the shaft. They are few and small for yellow marrow; however, for red marrow, which is a more active marrow, they are larger and more numerous. The final vascular network consists of wide, thin-walled channels called sinusoids, which slow up flow, and the emergent veins are also correspondingly large. In the marrow cavity, vessels from the shaft run to the ends of long bones, except during growth when few, if any, pass through the epiphysial plate of cartilage.

Aspects of Bone Metabolism

The stability of bone is the result of regulatory mechanisms, especially by the kidneys. These mechanisms respond to changes in intake, and therefore, changes in blood composition affect bone. Bone is dependent on (1) mechanical factors such as stress, which causes growth disposition of the trabeculae and blood flow; (2) nutritional factors involving protein (starvation, kidney disease, and pregnancy can cause a decreased matrix and thus osteoporosis) and vitamins A, D, and C; (3) inorganic minerals such as calcium and phosphorus; (4) local enzymatic or metabolic controls (alkaline phosphatase, glycogen, etc.); and (5) endocrine control. At least six hormones affect bone: pituitary growth hormone stimulates bone growth, causing both chondrogenesis and osteogenesis; thyroid hormone increases both osteogenesis and osteolysis; an excess of thyrocalcitonin, and adrenal cortisol, leads to osteoporosis; androgens and estrogens of the gonads stimulate bone growth and closure of epiphyses; and parathormone (along with thyrocalcitonin and vitamin D) helps.

Bone Pathology

Bone is a complex and labile tissue that is subject to trauma, inflammation, and toxic and vascular phenomena, like all other somatic tissue. In addition, because of its close relationship to calcium and phosphorus metabolism, a unique group of diseases reflect alterations of this relationship. It is impossible to discuss all bone disorders, but a few prototypes will be mentioned.

Dysplasia
Dysplasia means "ill formed" or "disturbed form." In bone dysplasia the error is in form or modeling. It may be aplasia (lack of), hypoplasia (deficiency of), or hyperplasia (increased). Dysplasia may be generalized (affecting all growing bones), may be genetically determined, or may occur as an isolated happening. This is not dystrophy, which means "ill nourished" and involves a metabolic or nutritional involvement.

In this category we find achondroplasia, in which there is failure of cartilage growth and maturation and all growth centers are affected except for articular cartilage, resulting in a marked decrease in longitudinal bone growth. The individual is a dwarf who is squat, agile, and muscular with short, bowed extremities, deformed spine, and a large head. This is a mendelian-dominant transmitted disease.

Osteomalacia
Osteomalacia is literally "a softening of bone" as a result of calcium or vitamin D deficiency, to name but two major causes. It is a mineral deficiency. Rickets is a typical example. Osteomalacia is adult rickets.

Osteoporosis

Osteoporosis is literally "porosity of bone." There is bone atrophy, and the trabeculae become fewer and thinner than normal. Osteoporosis can lead to skeletal deformities. The skull is never affected, and the long bones much less than the spine. The cause of osteoporosis is unknown, but the disorder can, in some instances, be a result of medication.

Osteopetrosis

Osteopetrosis (Albers-Schönberg disease) is characterized by a failure of resorption. There is little or no marrow cavity in the diaphysis but, rather, a dense atypical bone with abnormal pattern. There is a history of frequent fractures in childhood since the bones are very fragile.

Osteogenesis Imperfecta

Osteogenesis imperfecta (brittle bone disease) is failure to form bone matrix. Multiple fractures also occur here. Of interest is the peculiar bluish scleral discoloration of the eye.

Bones of the Body

Vertebral Column

The spinal column (Fig. 5-5) serves many functions. It supports the weight of the head, acts as the central

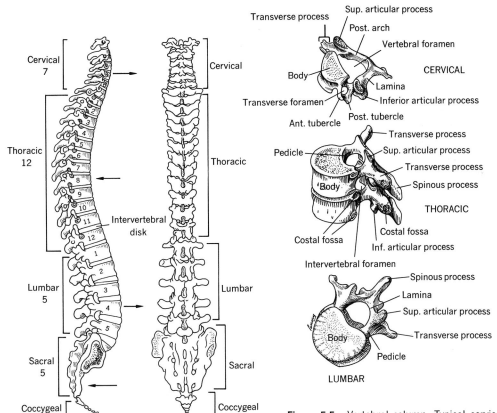

Figure 5-5. Vertebral column. Typical cervical, thoracic, and lumbar vertebrae.

pillar of the body, connects upper and lower segments of the trunk, gives attachment to the ribs, helps reduce "shock" transmitted from parts of the body, forms a tube for the spinal cord, and allows for a wide range of balances and movements.

The column consists of 33 vertebrae grouped according to region. The true or movable vertebrae are the seven cervical, 12 thoracic, and five lumbar. The false or fixed vertebrae are the five sacral (fused to form the sacrum) and the four coccygeal (fused to form the coccyx). Anteriorly the column forms a straight line, but when it is viewed from the side, several curvatures are seen.

Four curvatures are seen: two convex forward (cervical and lumbar) and two concave forward (the thoracic and sacrococcygeal). At birth only the thoracic and sacral curves are seen and are called the primary curves. Compensatory or secondary curves develop after birth (cervical and lumbar) as a result of elevation and extension of the head in infancy and assumption of the erect position in walking, respectively.

General Points of Interest. The structure of the vertebrae determines the spinal column mechanics. Each vertebra consists essentially of a body anteriorly and a vertebral arch posteriorly. The body and adjacent disks are weight-bearing. The size of each vertebra varies with the weight it supports and thus increases from the second cervical vertebra to the first part of the sacrum and then diminishes to the tip of the coccyx as the body weight is transmitted from the lower part of the column to the bony pelvis and lower extremities.

The arches enclose the spinal cord. Each arch is formed by a pedicle and a lamina, which extends from the pedicle. The latter fuse posteriorly in the midline to complete the arch. A bony spinous process and lateral transverse processes project from the laminae and are sites of muscle attachments. The spinous process is felt subcutaneously dorsally in the midline and is easily palpable except in the cervical (neck) region. In the thoracic region the tip of the process is at the level of the body of the adjacent lower vertebra, but in the lumbar area it is at a level with its own body.

The transverse processes (lateral) are not palpable. There are four articular processes per vertebra.

The intervertebral foramina increase in size down to L5 and diminish from above downward in the sacrum. In the midthoracic region, the tips of the spines are below the level of the bodies of corresponding vertebrae. The transverse processes in the cervical area are in front of the articular processes and in line with the intervertebral foramina; in the thoracic region they are behind both; and in the lumbar region they are behind the foramina but in front of the articular processes. As seen in front, the bodies of the vertebrae increase in breadth from C2 to T1, but diminish from T1 to T4 and increase from T4 to L1, below which they rapidly reduce in size. The transverse processes of C1 (atlas) are wide, but those of the next vertebrae are short and nearly equal in length. Those of C7 are nearly as long as T1. They diminish gradually to T12 where they are represented only by tubercles. They stand out in the lumbar region being longest at L3.

The vertebral column is about 70 cm (28 in.) long in men and 60 cm (24 in.) in women (but subject to great variation). The bodies of the vertebrae form a column for trunk and head support, the vertebral foramina provide a canal for the spinal cord and its coverings, and the intervertebral foramina transmit the spinal nerves. The spines and transverse processes provide attachments for muscles and form a longitudinal groove between them for muscles that move the column.

Certain surface landmarks of the column are commonly used. The spinous processes of C7 and T1 are prominent in the midline at the base of the neck. The inferior scapular angle normally lies at the level of and lateral to the interspace between T7 and T8. The iliac crests are palpable from anterior to posterior iliac spines. A line joining the top of the crests crosses the body of L4. The spinous processes are palpable from the external occipital protuberance to midsacrum. A line connecting the posterior superior iliac spines crosses the body of S2, and the sacroiliac joint lies medial and adjacent to the posterior-superior spine. A line drawn horizontally at the level of the ischial tu-

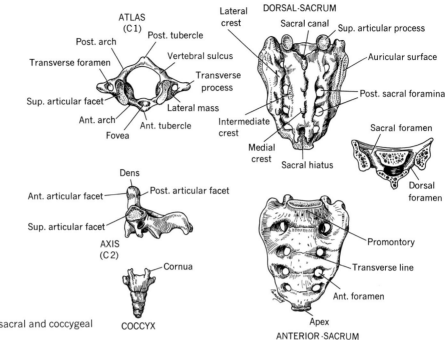

Figure 5-6. Atlas **(C1)**, axis **(C2)**, sacral and coccygeal vertebrae.

berosities crosses the femurs at the level of the lesser trochanters, and the tip of the coccyx lies above the level of the ischial tuberosities.

Cervical Vertebrae. The cervical vertebrae (Figs. 5-5, 5-6) are distinguished by their transverse processes, which are usually wide and contain a canal for the transmission of the vertebral vessels. The seventh cervical vertebra has a prominent spinous process, the vertebra prominens. The first two cervical vertebrae are different from the others. The first or atlas has no body and no spine, but consists only of a pair of lateral masses united by anterior and posterior arches resembling a large ring. Its cavity is divided by a transverse ligament into two, a larger posterior one for the spinal cord and a smaller anterior one for the odontoid process of C2. The upper surface of the vertebra has two smooth, oval areas for articulation with the under-surface of the skull. The second cervical vertebra or axis differs from the others in that its body is reduced in size and projecting upward from it is a prominent

elevation, the odontoid process or dens, which forms the axis of rotation for the skull and C1.

Thoracic Vertebrae. The thoracic vertebrae (Fig. 5-5) consist of a short cylindric, anterior portion, the body, the posterior aspect of which is attached to a bony arch enclosing the vertebral foramina within which is the spinal cord. On its external surface the arch has three processes: one pointing backward, the spinous process, and the other two extending laterally, the transverse processes. The spinous processes mark the midline of the back (tips can be felt beneath the skin) and the transverse processes give partial attachment to the ribs. Four small articular processes that interlock with similar elevations on the vertebrae above and below are also seen. Facets for the heads of the ribs are seen on the lateral parts of the bodies.

Lumbar Vertebrae. The lumbar vertebrae (Figs. 5-5, 5-6) are larger than the thoracic, and their transverse processes do not give attachment to ribs. They have no foramina in their transverse processes and have no costal facets. Their superior and inferior articular facets face laterally and medially, respectively, in contrast to those of the thoracic region, which face anteroposterior, and of the cervical area, which face superior and inferior. Thus, a vertebral dislocation without fracture can take place in the neck, whereas elsewhere in the column pure dislocation does not usually occur since the articular process must break off before the body can be moved.

Sacral Vertebrae. The sacral vertebrae (Figs. 5-5, 5-6) fuse (five vertebrae) to form the sacrum, which diminishes in size from above downward. It is triangular in shape and has a base (upper surface), an apex (lower end), and dorsal, anterior (pelvic), and two lateral surfaces. The sacrum is divided by paired rows of foramina on the dorsal and anterior surfaces into median and lateral parts. The median part is composed of portions of the five fused vertebrae. The lateral part, the lateral masses or alae, represents fused costal and transverse processes; these masses have articular surfaces for articulation with the ilium. The laminae of S4 and S5 do not usually meet in the midline and form the openings to the sacral canal, the sacral hiatus. The sacral canal contains the cauda equina (nerves), the filum terminale (of the spinal cord), and the meninges down to the middle of the third sacral vertebra (see Chap. 8). At that level, the meninges end and the lower portion of the canal has only the nerve roots of the lower sacral and coccygeal nerves with the coverings from the meninges. The base of the sacrum is formed by the upper surface of S1, which on its anterior border forms the bulging landmark called the sacral promontory. Four ridges cross the anterior surface of the sacrum to mark the site of fusion of the bodies before the twenty-first year of life. Lateral to these ridges (bilateral), the four anterior sacral foramina open and transmit the anterior rami of the upper four sacral nerves.

Coccyx. The coccyx (Figs. 5-5, 5-6) consists of three, four, or five rudimentary coccygeal vertebrae attached to the apex of the sacrum and represents the bony remnants of the tail of lower animals. It is the terminal segment of the spine.

Intervertebral Disks. The intervertebral disks (Fig. 5-5; Fig. 6-6, p. 173) are discussed in Chapter 6.

Skull

Chondrification begins in the base, but ossification begins in the calvarium (supraorbital part) before the former has gone too far. Membrane bones are the frontal, parietal, squamous temporals, greater wings of the sphenoid (except roots), and the occipital above the nuchal lines. Centers appear here in the seventh week of life. Generally the "older" basal bone is preformed in cartilage, but the facial and roofing bones are formed intramembranously. At birth there is considerable movement or "molding" between bony sutures to allow the head to pass through the mother's pelvis. The brain, within the skull, is protected by a series of membranes that are continuous with those of the spinal cord.

The Fontanelles. The fontanelles (Figs. 5-7, 5-8) are unossified spaces at the angles of the parietal bones. There are two midline fontanelles that help to determine position during birth since one can feel and see intracranial pulsations here. They are the anterior, which closes before two years of age, and the posterior, which closes in the first year of life. Fontanelles are

Figure 5-7. A lateral view of the skull at birth; the fontanelles.

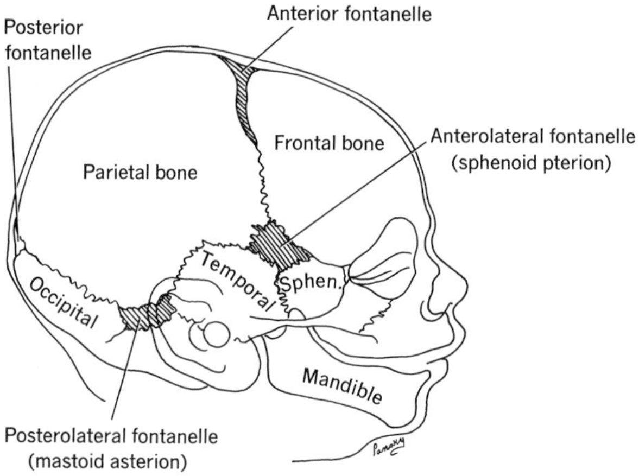

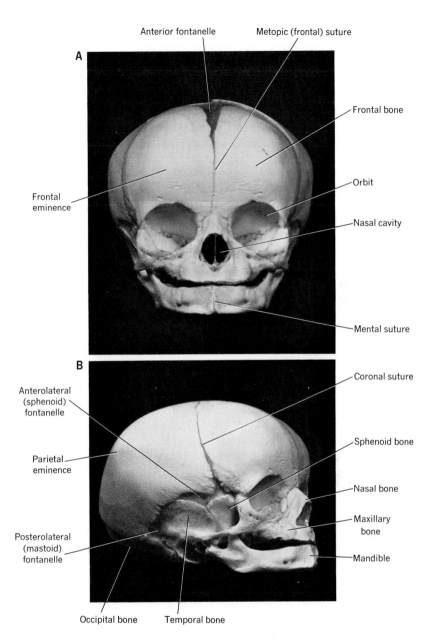

Anterior fontanelle   Metopic (frontal) suture

A

Frontal bone

Orbit

Nasal cavity

Frontal eminence

Mental suture

B

Coronal suture

Anterolateral (sphenoid) fontanelle

Sphenoid bone

Parietal eminence

Nasal bone

Maxillary bone

Posterolateral (mastoid) fontanelle

Mandible

Occipital bone   Temporal bone

Figure 5-8. Skull at birth, showing fontanelles. **A.** Anterior view. **B.** Semiside view.

also seen at the pterion (anterolateral) and asterion (posterolateral). The suture between the frontal bones disappears in the third year of life but is sometimes visible in the adult and is called the metopic suture.

The Skull Proper. The word *skull* refers to the entire head and face skeleton including the mandible. Cranium is the skull minus the mandible, and calvarium is the skull above the supraorbital ridges.

The skull is slightly flattened from side to side and appears smooth from above but uneven from below. It is made up of irregular bones that are immovably joined together (except for the mandible). It consists of 24 bones (including mandible). The bones consist of two compact plates that enclose a layer of spongy bone between them, the diploe. In some bones the diploe is absorbed, leaving cavities, referred to as air sinuses, which are found between the bony plates. The sinuses communicate with the nasal cavities and are lined and continuous with the same mucous membrane that lines the nasal cavities (see Chap. 10).

One views the exterior of the skull from five vantage points: from above (norma verticalis), from below (norma basalis), from in front (norma frontalis), from behind (norma occipitalis), and from the side (norma lateralis).

FROM ABOVE. The top of the skull (Fig. 5-9) shows four bones: the frontal, the occipital, and the right and left parietals, united by sutures with interlocking saw-like edges. The suture uniting the frontal to the parietal bones is called the coronal suture; that between the occipital and parietals is the lambdoid suture (inverted v and shaped like the Greek letter lambda). The lambdoid suture is connected to the coronal suture by the sagittal suture (sagitta meaning arrow), and they meet at the anterior fontanelle. The point of bony fusion at this fontanelle becomes the bregma. The point where the lambdoid and sagittal sutures meet is called the lambda. The vertex, the highest point of the skull, is near the middle of the sagittal suture. The parietal foramen is a small opening on either side of the sagittal suture and admits a small artery and vein. The latter connects the vein of the scalp with the

Figure 5-9. Skull as seen from above, in front, and from behind.

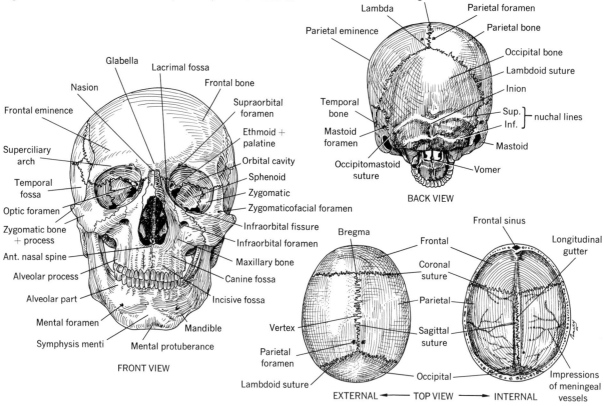

FRONT VIEW

BACK VIEW

EXTERNAL ◄── TOP VIEW ──► INTERNAL

Figure 5-10. External view of the skull as seen from below (its base).

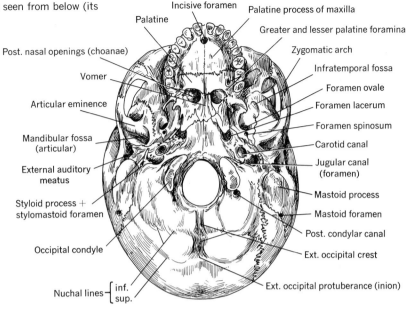

Incisive foramen
Palatine process of maxilla
Palatine
Greater and lesser palatine foramina
Post. nasal openings (choanae)
Zygomatic arch
Vomer
Infratemporal fossa
Articular eminence
Foramen ovale
Foramen lacerum
Mandibular fossa (articular)
Foramen spinosum
Carotid canal
External auditory meatus
Jugular canal (foramen)
Styloid process + stylomastoid foramen
Mastoid process
Mastoid foramen
Occipital condyle
Post. condylar canal
Ext. occipital crest
Nuchal lines { inf. sup.
Ext. occipital protuberance (inion)

BASAL VIEW–SKULL

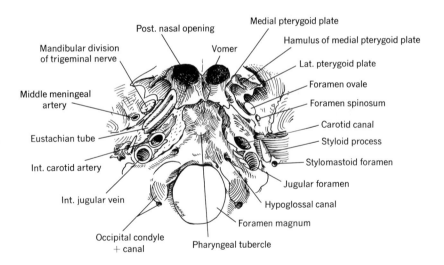

Post. nasal opening
Medial pterygoid plate
Vomer
Hamulus of medial pterygoid plate
Mandibular division of trigeminal nerve
Lat. pterygoid plate
Foramen ovale
Middle meningeal artery
Foramen spinosum
Carotid canal
Eustachian tube
Styloid process
Int. carotid artery
Stylomastoid foramen
Jugular foramen
Int. jugular vein
Hypoglossal canal
Foramen magnum
Occipital condyle + canal
Pharyngeal tubercle

superior sagittal dural sinus. Thus, an infection from the scalp can follow this very pathway.

FROM BELOW. The anterior aspect (Fig. 5-10) is occupied by the bony plate, which is formed by the palatine process of the maxillae and the horizontal plate of the palatine bones. Anteriorly, in the median plane, the incisive foramen receives the openings of the lateral incisive canals that transmit the terminal branches of the greater palatine vessels to the nose as well as the descending terminal branches of the long sphenopalatine nerves. The greater palatine foramina, which transmit the greater palatine vessels and nerves, are in the posterolateral corner near the last molars. The lesser palatine foramina lie just behind the greater. Behind and above the hard palate are the choanae (posterior openings of the nose), which are separated by the vomer and bounded laterally by the medial pterygoid plates.

The pterygoid plates (medial and lateral) are processes that project down from the roots of the greater wings of the sphenoid bones (Fig. 5-10). Between them is the pterygoid fossa. The free border of the medial plate ends below in a hook, the hamulus, which gives attachment at its tip to the pterygomandibular ligament (Fig. 7-9, p. 202) and by its posterior border to the fibers of the superior pharyngeal constrictor muscle. The tendon of the tensor palati muscle twists around the hamulus (Fig. 7-9, p. 202). The lateral pterygoid plate gives origin to the lateral pterygoid muscle on its lateral side and the medial pterygoid muscle on its medial side. The roof of the infratemporal fossa lies just lateral to both plates. The foramen ovale is posterior to the plates and transmits the mandibular nerve (V), accessory meningeal artery, lymph vessels, and small veins that connect the dural cavernous sinus (venous) with a venous plexus outside the skull. Posterolateral to the foramen ovale is the foramen spinosum for the middle meningeal artery. The foramen lacerum is a large opening at the base of the medial pterygoid plate and is closed by cartilage *in vivo*. The carotid canal, posterolateral to the foramen lacerum, is a tunnel in the petrous temporal bone through which the internal carotid artery passes into the cranial cavity. The canal is in relation to the middle and inner ear, and during periods of excitement or exertion one often hears a thumping sound in the head due to this artery beating against the bone near the inner ear.

The jugular foramen is a large opening directly behind the carotid canal. It conducts the internal jugular vein and cranial nerves (IX, X, and XI). The foramen is opposite the external auditory meatus. Part of the bone that forms the foramen also forms the floor of the middle ear. Thus, in diseases of the middle ear, infections may pass through the bone and involve the internal jugular vein.

Lateral to the jugular foramen is the styloid process, which gives attachment to three muscles (styloglossus, stylohyoid, and stylopharyngeus) and two ligaments (stylohyoid and stylomandibular). The stylomastoid foramen is located at the base of the process and transmits the facial (VII) nerve from the brain to the exterior. Stylomastoid branches of the posterior auricular vessels also pass through the foramen. The zygomatic arch is also a prominent feature of this aspect, and one sees the articular fossa for the mandible at its caudal end.

The mastoid process is felt behind the lobule of the ear and is not really recognized as a bony mass until the age of two. The mastoid foramen is posterior to the process and transmits a vein to the transverse dural venous sinus and a small branch of the occipital artery to the dura mater.

The foramen magnum is the largest bony foramen in the skull. Through it the medulla oblongata (lowest brain division), vertebral artery, and accessory nerve (XI) pass.

The occipital condyles are the large, smooth, and oblong protuberances lying at the margins of the foramen magnum. They articulate with the atlas (C1). Nodding movements of the head take place at this joint. The anterior condylar canals are above the lateral margins of the anterior part of the condyles (usually hidden by them). They are smaller than the jugular foramina and transmit the twelfth cranial nerve (hypoglossal). The posterior condylar canals (when present) transmit emissary veins connecting the suboccipital venous plexus and the dural sigmoid venous sinus.

Behind the foramen magnum is a crest of bone, the external occipital crest, ending in the elevated inion or external occipital protuberance. From here the inferior nuchal lines curve laterally but are poorly defined. The superior nuchal lines also curve laterally from the inion and separate the neck (nuchal) area from the scalp area above.

FROM IN FRONT. From in front (Fig. 5-9) one sees six areas: frontal (forehead), orbital, nasal, zygomatic, maxillary (upper jaw), and mandibular (lower jaw).

The Frontal Area. The frontal area is formed by the

frontal bone. The nasal root depression is the nasion and is opposite the anterior extremity of the brain.

Above the orbital margins are the superciliary arches, which give prominence to the eyebrows. The elevation between the arches (or eyebrows) is called the glabella. Behind the superciliary arch and in the frontal bone is a large air space called the frontal sinus. The frontal eminences are the most convex part of each frontal bone.

The supraorbital foramen or notch is immediately above the border of the orbital opening and transmits the supraorbital nerve and vessels. The supraorbital margin ends laterally in the zygomatic process of the frontal bone and is felt at the lateral end of the eyebrow. The anterior part of the temporal line curves backward from here.

The Orbital Area. Each orbit is a cavity resembling a pyramid with four walls, an apex, and a base (Fig. 5-9). The bones of this pyramid are the maxillary, zygomatic, sphenoid, frontal, palatine, ethmoid, and lacrimal. At the apex is the optic foramen. The roof is formed by the orbital plate of the frontal bone and the lesser wing of the sphenoid. It separates the orbit from the frontal sinuses and cranial cavity. The lacrimal gland sits in a fossa in the anterolateral part of the roof. At the medial angle the trochlea is found (a small fibrocartilage ring for the tendon of the superior oblique ocular muscle).

The floor is formed by the orbital plate of the maxilla. It separates the orbit from the maxillary sinus. Posteriorly one sees the inferior orbital fissure, which transmits the infraorbital branch of the maxillary nerve, the infraorbital vessels, the zygomatic nerve, and twigs from the sphenopalatine ganglion to the lacrimal gland and orbital periosteum. A vein is also seen, which connects the ophthalmic veins with the pterygoid plexus of veins in the infratemporal fossa via the infraorbital fissure.

The lateral walls (see Chap. 9) are at right angles to each other and are formed by the great wing of the sphenoid and zygomatic bone. The superior orbital fissure lies between the roof and lateral wall near the orbital apex. It transmits the third, the fourth, the ophthalmic division of the fifth, and the sixth cranial nerves accompanied by the orbital branches of the middle meningeal artery. The ophthalmic veins and recurrent branch of the lacrimal artery also pass back through this fissure. Only the lateral wall is not related to the paranasal sinuses.

The medial walls (see Chap. 9) are parallel, very delicate, and formed from anterior to posterior by the frontal process of the maxilla, the lacrimal bone, the ethmoid, and a small part of the body of the sphenoid. The wall contains the lacrimal fossa (for the lacrimal sac), the anterior ethmoid foramen (for the nasociliary nerve and anterior ethmoidal vessels), and the posterior ethmoidal foramen (for the posterior ethmoidal vessels and nerves) (see Chaps. 9 and 11).

The apex is at the back, corresponds to the optic foramen, and transmits the optic nerve and ophthalmic artery (not the ophthalmic vein).

Orbital infection may pass superiorly, to the frontal sinus or anterior cranial fossa (frontal lobe of the brain); inferiorly, to the maxillary sinus; medially, to the sphenoidal and ethmoidal sinuses; and laterally, to the middle cranial fossa (temporal lobe of the brain).

The Nasal Area. The nasal area (see Chap. 11) is formed by the two nasal bones in the bridge of the nose and on each side by the frontal process of the maxilla. The cavity is divided by a median partition, the septum, into right and left halves. The septum is formed by the perpendicular plate of the ethmoid, the nasal cartilage, and the vomer. Three curled, bony plates, the conchae, project downward from the side walls of the nasal cavity. There are superior and middle conchae from the ethmoid bone and an independent concha, the inferior one. The space below each concha is called a meatus. Thus, we have three: a superior, a middle, and an inferior.

The Zygomatic and Maxillary Areas. These are from the cheek bones and upper jaw, respectively. The latter is between the orbits and upper teeth. The anterior nasal spine projects forward from the two maxillae at the lower margin of the nose. The maxilla ends inferiorly as the alveolar border with ridges marking the roots of the teeth.

Near the lower margin of the orbit and just lateral to the side of the nose is the infraorbital foramen for

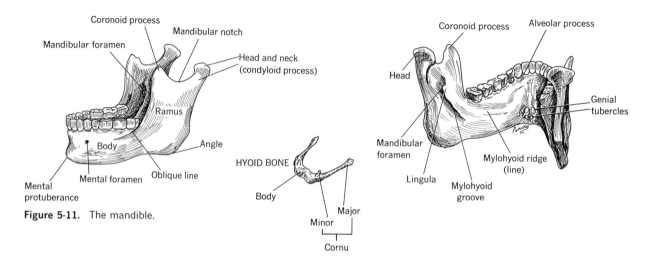

Coronoid process
Mandibular notch
Mandibular foramen
Head and neck (condyloid process)
Ramus
Body
Angle
Mental foramen
Oblique line
Mental protuberance

HYOID BONE
Body
Major
Minor
Cornu

Coronoid process
Alveolar process
Head
Genial tubercles
Mandibular foramen
Lingula
Mylohyoid groove
Mylohyoid ridge (line)

Figure 5-11. The mandible.

the infraorbital vessels and nerve. The zygomatico-facial foramen is an opening on the zygomatic bone and transmits a small branch of the lacrimal artery and the zygomaticofacial branch of the zygomatic nerve. The large air space in the maxilla is the maxillary sinus (see Chap. 11), which communicates with the middle meatus of the nasal cavity.

The Mandibular Area. This is mainly the mandible (Fig. 5-11), the largest and strongest of the facial bones, and contains the lower teeth. It consists of a body and a pair of flat, broad rami that stand up from the back of the body. Two processes project from each ramus: (1) an anterior, coronoid process for attachment of the temporalis muscle; and (2) a posterior, condyloid process, which articulates with the articular fossa. Articular cartilage covers its superior and anterior aspect but not the posterior. The parotid gland covers the condyloid process laterally. The process can be felt with a finger just in front of the ear when opening and closing the mouth. The notch between the coronoid and condyloid processes is the mandibular notch, which transmits the vessels and nerves to the masseter muscle. The outer surface of the body sometimes shows a faint median ridge, the symphysis menti, which divides to enclose a triangular eminence, the mental protuberance. The bone is bent forward here to form the chin. The upper border is the alveolar one and has a row of pits or sockets for 16 teeth. The mental

foramina, about 2.5 cm (1 in.) from the symphysis, transmit the mental vessels and nerves.

The internal surface of the body has the mylohyoid ridge or line for origin of the mylohyoid muscle. The angle of the mandible is where the posterior border of the ramus joins the lower border of the body and can be felt subcutaneously. The mandibular foramen leads into a canal in the bone and transmits the inferior alveolar vessels and nerves to the teeth. The lingula is a bony spur that overlaps the foramen.

LATERAL VIEW. On the lateral surface of the skull (Fig. 5-12) one sees the superior and inferior temporal lines, which curve backward and posteriorly into the supramastoid crest. They give attachment to the epicranial aponeurosis, the temporal fascia, and the temporalis muscle fibers.

The infratemporal crest on the great wing of the sphenoid is a ridge that separates the temporal fossa and the infratemporal fossa below. The former is outlined by the temporal lines and the zygomatic arch and contains the temporalis muscle, vessels, and nerves and the zygomaticotemporal nerve. This is the thinnest and weakest part of the skull and, since the middle meningeal artery passes through here, fractures injure this vessel. The infratemporal fossa is behind the maxilla, below the crest, and lateral to the pterygoid plates. It contains the pterygoid muscles, internal maxillary artery and its middle meningeal branch, the

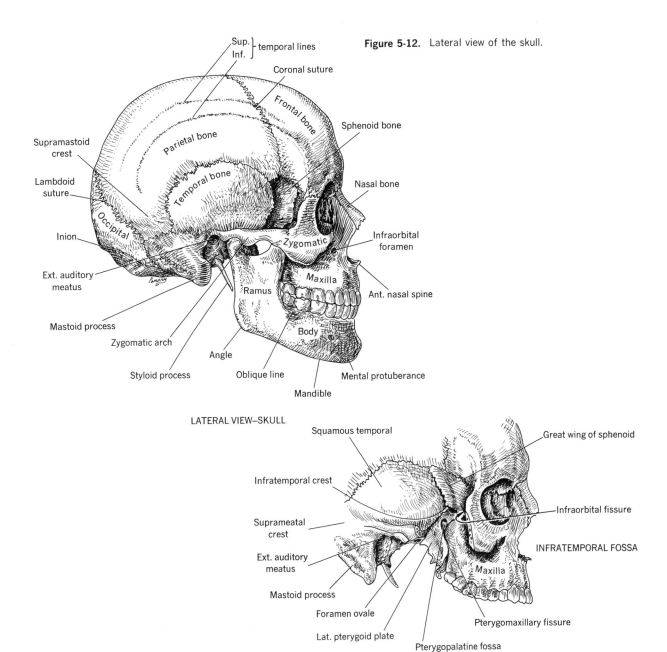

Figure 5-12. Lateral view of the skull.

Sup. ⎱ temporal lines
Inf. ⎰

Coronal suture

Frontal bone

Sphenoid bone

Parietal bone

Nasal bone

Supramastoid crest

Temporal bone

Lambdoid suture

Zygomatic

Infraorbital foramen

Occipital

Inion

Maxilla

Ext. auditory meatus

Ant. nasal spine

Ramus

Mastoid process

Body

Zygomatic arch

Angle

Styloid process

Oblique line

Mental protuberance

Mandible

LATERAL VIEW—SKULL

Squamous temporal

Great wing of sphenoid

Infratemporal crest

Infraorbital fissure

Suprameatal crest

INFRATEMPORAL FOSSA

Ext. auditory meatus

Maxilla

Mastoid process

Foramen ovale

Pterygomaxillary fissure

Lat. pterygoid plate

Pterygopalatine fossa

mandibular nerve and its branches, the chorda tympani nerve, the pterygoid plexus of veins, and the temporalis muscle. Two fissures can be seen in its depths: the infraorbital, connecting it to the orbit; and the pterygomaxillary, which transmits the terminal part of the maxillary artery. The latter fissure leads to the pterygopalatine fossa.

The zygomatic arch can be felt from the cheek

prominence to just in front of the ear. It is formed by the zygomatic process of the temporal and the temporal process of the zygomatic bone. Its upper border gives attachment to the temporal fascia and the lower border and medial surface to the masseter muscle. Its posterior root continues back above the external auditory meatus as the suprameatal crest. Below the posterior root of the arch is the external auditory meatus, and lateral to this is the cartilage of the external ear. The bony meatus is hardly wide enough to admit a pencil. It passes medially and forward and opens obliquely into the middle ear. The mastoid process is absent at birth and appears at age two and can be felt behind the ear lobule.

POSTERIOR VIEW. The posterior view (Fig. 5-9) is the back of the skull and consists of the two parietal bones, the occipital and the mastoid portions of the temporal bones (with the mastoid process). One sees the parietal eminences, the posterior part of the sagittal suture, the parietal foramina, the lambda and the lambdoidal suture, the occipitomastoid suture, and the mastoid foramina (which transmit an emissary vein). The inion is well marked and can be felt above the nape of the neck. The superior nuchal lines arch laterally.

The Skull Interior. THE SKULL CAP (CALVARIUM). The skull cap (Fig. 5-9) has depressions for cerebral convolutions and furrows for branches of the meningeal vessels. Along the midline is a longitudinal groove that lodges the superior sagittal sinus and at its margins gives attachment to the falx cerebri. Granular pits (which increase with age) are seen along the sides of the groove due to erosion by the arachnoid granulations of the arachnoid mater. Nutrient foramina are also seen.

CRANIAL FOSSAE AND BASE OF THE SKULL. The base of the skull (Fig. 5-13) is divided into three cranial fossae—anterior, middle, and posterior—with the first being on a higher plane than the second, and the second higher than the third (three terraces).

Anterior Cranial Fossa. The anterior cranial fossa is limited posteriorly by the lesser wings of the sphenoid bone and the edge of the optic groove. The roof of the frontal sinus and orbit are located in this area.

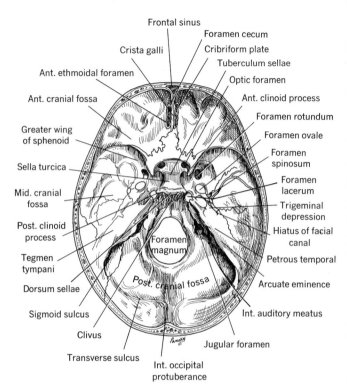

Figure 5-13. Interior base of the skull.

It also contains the frontal lobes of the brain and olfactory bulbs and tracts. Its midline is formed by the cribriform plate of the ethmoid bone, which roofs the nasal cavity and above which the crista galli (cock's comb) rises. The latter is an extension upward of the nasal septum and gives attachment to the falx cerebri. The foramen cecum is a small opening in front of the crista, which in early life transmits some veins but is closed in the adult. The cribriform plate is perforated by the olfactory nerves, which arise from the olfactory cells in the nasal mucosa. The posterior ethmoidal foramina are located at the side of the cribriform plate and lead into the nasal cavity, transmitting the anterior and posterior ethmoidal arteries and anterior ethmoidal nerve.

Fractures in this fossa may involve the cribriform plate and cause lacerations of the meninges and mucous membrane of the root of the nose, resulting in bleeding and/or discharge of cerebrospinal fluid. Loss of smell may result following laceration of the olfactory nerves; if the dura is injured, infection can lead to meningitis or abscess formation of the brain. A fracture of the orbital plate can produce bleeding that seeps into the orbit and produces bulging of the eyes. The frontal sinuses can also be involved.

Middle Cranial Fossa. The middle cranial fossa resembles a "butterfly" with a small median and two lateral expanded, concave parts. The median part is the upper surface of the body of the sphenoid bone. The sella turcica is the "saddle-shaped" area for the pituitary gland. Anteriorly there is a ridge, the tuberculum sellae, and on either side, the anterior clinoid processes. Anterior to the processes are the optic foramina at the end of the optic grooves. The posterior part of the sella is formed by a crest, the dorsum sellae, which ends laterally in the posterior clinoid processes. The carotid groove for the internal carotid artery is found on either side of the sella turcica. Three foramina run approximately parallel with this groove: the foramen rotundum, for the maxillary nerve (V); the foramen ovale, for the mandibular nerve (V), accessory meningeal artery, and lesser petrosal nerve; and the foramen spinosum, for the middle meningeal artery and a recurrent branch from the mandibular nerve. Medial to the foramen ovale is the foramen lacerum, in reality a canal whose lower part is filled with fibrocartilage and whose upper part transmits the internal carotid artery and a plexus of sympathetic nerves. The greater wing of the sphenoid forms part of the floor of the middle fossa, along with the petrous and squamous temporal. These lateral areas lodge the temporal lobes of the cerebrum. The superior orbital fissure enters this middle fossa but is not seen clearly from a direct top view.

The petrous temporal bone forms a large part of the middle fossa. The elevated part of this bone is the arcuate eminence, which overlies the superior semicircular canal of the inner ear. Lateral to the eminence and adjoining the squamous temporal is the tegmen tympani, which is a thin plate of bone that roofs middle ear and auditory tube. This bone is the only barrier between the middle ear and the temporal lobe of the brain and the meninges of this area.

The hiatus of the facial canal (for a branch of the seventh nerve) is really a small slit on the anterior surface of the middle fossa. The nerve of the hiatus originates from the facial nerve (VII) in the middle ear. The trigeminal impression, a hollowed-out area, is found at the apex of the petrous bone and lodges the trigeminal ganglion (cranial V).

The middle fossa is the most common site of skull fracture because of bone weakness as a result of all the canals and foramina. The tegmen tympani is frequently fractured and the tympanic membrane (eardrum) torn, so that blood and cerebrospinal fluid are discharged from the outer ear. The auditory (VIII) and facial (VII) nerves may be involved. The cavernous sinus walls may be lacerated and cranial nerves III, IV, and VI involved. Bleeding into the mouth may occur as a result of fractures of the middle fossa by passing through the sphenoid bone or base of the occipital.

Posterior Cranial Fossa. The posterior cranial fossa is the deepest and largest of the fossae and lodges the cerebellum, pons, and medulla oblongata. Its floor is formed by the occipital bone, its lateral wall by the petrous temporal bone, and its medial surface by the mastoid temporal.

The foramen magnum is at the lowest part of the fossa and is the most prominent feature of this fossa. At the anterolateral edge of the foramen are the anterior condylar canals for the hypoglossal (XII) nerves (not clearly seen). The foramen magnum transmits the medulla oblongata, the meninges, the vertebral arteries, and the ascending part of the accessory nerve (XI). The clivus is the sloping surface between the anterior margin of the foramen magnum and the root of the dorsum sellae and is related to the pons and medulla. The internal auditory meatus is located at the posterior aspect of the petrous temporal. It transmits the facial nerve (VII), the auditory nerve (VIII), the internal auditory branch of the basilar artery, and the auditory vein.

The jugular foramen is seen between the lateral part of the occipital and the petrous temporal bone. It transmits the inferior petrosal sinus (which becomes the internal jugular vein); a meningeal branch of the ascending pharyngeal artery; the ninth, tenth, and eleventh cranial nerves; the sigmoid sinus as it becomes the internal jugular vein; and a branch of the occipital artery. The transverse groove begins near the internal occipital protuberance and sweeps around laterally, where it joins the sigmoid groove, which in time curves downward in a medial direction to end at the jugular foramen. The right transverse groove usually receives the sagittal sinus and thus is larger than the left.

Fractures of the posterior fossa are important because small fractures can be fatal. The bone is thin in places and, since there are no outlets for either blood or cerebrospinal fluid, these fractures can be overlooked and it may be several days before one notices blood over the mastoid process. Fractures over the skull base involving the hypoglossal canal can lead to paralysis of one side of the tongue.

Upper Extremity

The Scapula (Shoulder Blade). The scapula (Fig. 5-14) is a flat, triangular bone on the posterolateral aspect of the thorax opposite ribs 2 to 7. It has three borders, three angles, two surfaces, and two processes.

BORDERS. The three borders are the superior, medial, and lateral. The superior is the shortest and in-

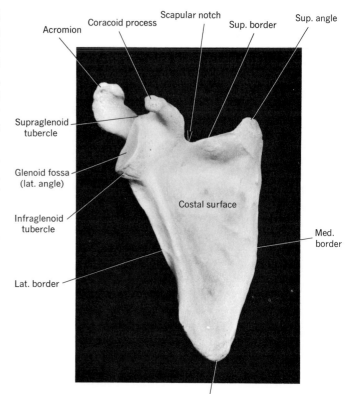

Figure 5-14. Above: Anterior aspect of the right scapula. **At left:** Scapula and clavicle.

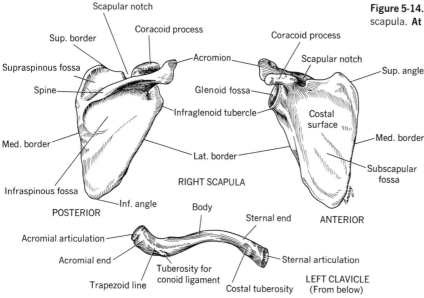

clines laterally and downward from the superior angle. The scapular notch is located on its lateral part and is converted into a foramen by the suprascapular ligament (transmits the suprascapular nerve). The border is thin and sharp and gives attachment to only the small omohyoid muscle. The medial (vertebral) border is the longest and is thick and gives insertion to the rhomboid major and minor as well as the levator scapulae muscles. The lateral (axillary) border is the thickest of the three. At its lower end is found a triangular impression, the *infraglenoid tubercle*, for attachment of the long head of the triceps muscle.

ANGLES. The three angles are the superior, inferior, and lateral. The superior is covered by the trapezius muscles and is difficult to palpate. The acute inferior angle is felt easily at about the seventh intercostal space, with the extremity hanging at the side, and is an important anatomic (and surgical) landmark. The lateral angle forms the glenoid cavity with a roughened area above it, the supraglenoid tubercle, for origin of the long head of the biceps brachii muscle.

SURFACES. The two surfaces are the dorsal and the ventral (costal). The former is subdivided by the scapular spine into a smaller supraspinous and larger infraspinous fossa. Grooves on its surface for the circumflex scapular artery can be seen. The latter forms the floor of the subscapular fossa and is covered by the serratus anterior and subscapularis muscles.

PROCESSES. The two processes are the acromion and the coracoid. The entire upper surface of the acromion is palpable subcutaneously, and just in front of the shoulder joint it gives attachment to the lateral end of the coracoacromial ligament. The coracoid process projects the lateral superior border of the scapula. Its lateral border gives attachment to the coracoacromial ligament, helping the acromion form an arch over the head of the humerus. Its tip gives origin to the coracobrachialis and short head of the biceps muscles and can be felt on deep pressure through the lateral boundary of the infraclavicular fossa about 2.5 cm (1 in.) below the clavicle.

The Clavicle (Collarbone). The clavicle (Fig. 5-14) joins the sternum to the scapula and thus connects the skeleton to the upper extremity. It is a bone that is commonly fractured. It consists of a sternal end, which articulates with the sternum; a body; and an acromial end, which articulates with the acromion of the scapula.

The Humerus (Arm, Brachial Region). The humerus (Fig. 5-15) is a long, cylindric bone that articulates with the scapula above and the radius and ulna below. The head is covered with cartilage and is directed upward, medially, and slightly backward. The anatomic neck is a constriction around the articular cartilage and gives attachment to the capsular ligament. Projecting lateralward and forward from the head is a bony mass divided into two unequal parts, the greater and lesser tubercles. The bicipital or intertubercular groove between them holds the long biceps tendon. The greater tubercle has been called the "point of the shoulder" and, although covered by the deltoid muscle, can be felt on deep palpation. The tubercle gradually becomes continuous with the shaft. The lesser tubercle is below the anatomic neck and on the front of the shaft can be felt on deep pressure during bone rotation. It provides attachment for the subscapularis muscle. The surgical neck is a fingerbreadth below the tubercles. Slightly above the level of the surgical neck is the epiphysial line of the bone.

The deltoid tuberosity is a roughened, inverted "delta" with its apex downward and is found halfway down the shaft of the bone on its lateral aspect. The spiral or radial groove lies behind this tuberosity and contains the radial nerve and profunda artery.

The shaft or body of the humerus is almost cylindric in its upper half and flattened or prismatic in its lower half. It has three borders (anterior, lateral, and medial) and three surfaces (anterolateral, anteromedial, and posterior). A nutrient foramen is found near the middle of the medial border and another frequently in the spiral groove.

The lower end of the humerus is divided into the capitulum (sphere shaped), which receives the head of the radius, and the trochlea (spool shaped) for the trochlear (semilunar) notch of the ulna. The radial fossa is located just above the capitulum and receives the radius in flexion. The coronoid fossa is just above the trochlea and receives the coronoid process of the ulna

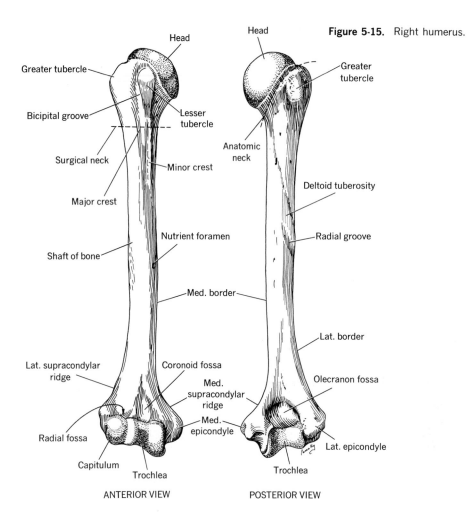

Head

Greater tubercle

Bicipital groove

Surgical neck

Major crest

Shaft of bone

Lat. supracondylar ridge

Radial fossa

Capitulum

Trochlea

Head

Lesser tubercle

Minor crest

Nutrient foramen

Coronoid fossa

Med. supracondylar ridge

Med. epicondyle

ANTERIOR VIEW

Head

Anatomic neck

Greater tubercle

Deltoid tuberosity

Radial groove

Med. border

Lat. border

Olecranon fossa

Lat. epicondyle

Trochlea

POSTERIOR VIEW

Figure 5-15. Right humerus.

during flexion. The olecranon fossa, the largest of the three, is also above the trochlea, posteriorly, and receives the olecranon of the ulna during extension at the elbow. The lateral epicondyle is just above and lateral to the capitulum, and its posterior aspect can be felt subcutaneously. The medial epicondyle is large and felt easily medial to and above the trochlea and has a groove posteriorly in which the ulnar nerve can be felt and rolled around.

Radius and Ulna (Forearm). These bones (Fig. 5-16) form the framework of the forearm. The radius lies lateral and the ulna medial in relation to the upper extremity. They both articulate above with the hu-merus and below with the carpal bones of the wrist. The radius and ulna also articulate with each other at their upper and lower ends and are united by the ligaments of the radioulnar joints and the interosseous membrane. The radius supports the entire hand and revolves on a longitudinal axis around the ulna at the radioulnar joints, providing for pronation and supination. The ulna does not enter into the wrist joint but is the more important bone at the elbow joint. The radius has a secondary function at the elbow.

RADIUS. Proximally there is a head, neck, and radial tuberosity. The head is a cup-shaped disk and articulates with the capitulum of the humerus. It is covered

Figure 5-16. Right radius and ulna.

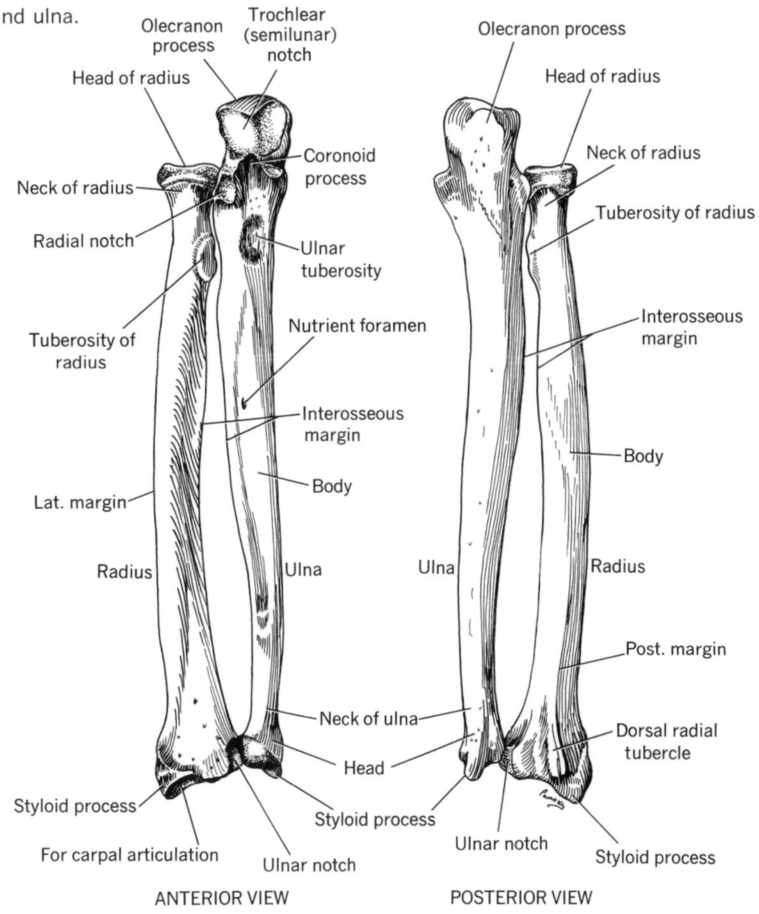

ANTERIOR VIEW   POSTERIOR VIEW

by hyaline cartilage, and its circumference articulates with the radial notch of the ulna. It is embraced by the annular ligament, but the ligament is not attached to it. The head can be felt from the back, where the lateral epicondyle of the humerus is also felt, as is the joint between them. The neck is the constricted part that supports the head and marks the point where the brachial artery divides into its ulnar and radial branches. The tuberosity, for the insertion of the biceps tendon, is on the medial side, below the neck. The shaft has a roughened area on its lateral aspect for insertion of the pronator teres muscle. The interosseous membrane (Fig. 6-4, p. 170) is attached to the interosseous border but does not extend up to the radial tuberosity.

The distal end of the radius is the widest and more bulky part of the bone. The *styloid process* is a downward prolongation on the lateral aspect of the bone and normally projects about 0.6 cm ($\frac{1}{4}$ in.) distal to the styloid process of the ulna. The dorsal radial tubercle is a very prominent ridge over this area and can be felt. It is grooved by the tendon of the extensor pollicis longus muscle. The medial surface of the bone is smooth and articulates with the ulna. The lateral surface extends to the styloid process. A series of grooves is seen on the distal end of the bone; the grooves lodge the tendons that cross the bone on their way to the hand. The carpal surface (distal) of the radius articulates with the scaphoid and lunate bones of the wrist.

A common fracture of the lower end of the radius with displacement of the hand backward and outward is called Colles' fracture.

Ulna. The ulna is medial and longer than the radius. Proximally one sees the olecranon, a coronoid process, and radial and trochlear notches. The olecranon is palpable posteriorly, and in extension its upper edge is on a level with the epicondyle of the humerus. It gives insertion to the triceps tendon. The coronoid process projects forward and gives insertion to the brachialis muscle. The radial notch is on the lateral side for articulation with the radial head. Inferiorly and posteriorly it gives attachments to the annular ligament. The trochlear notch is a wide concavity that articulates with the trochlea. The shaft of the ulna tapers distally and its posterior border can be felt subcutaneously from olecranon to styloid process.

The distal end has the head and styloid process. The former is small and round and articulates with the triangular articular disk (Fig. 6-5, p. 171). It can be felt subcutaneously when tendons are relaxed. The styloid process projects down from the posteromedial aspect of the head.

Carpal Bones (Part of Wrist). The wrist links the hand to the forearm and includes the carpus, distal parts of the radius and ulna, and the bases of the metacarpals (Fig. 5-17). One finds here the radiocarpal, midcarpal, and carpometacarpal joints. The tendons of the forearm cross here to insert below, being held by deep fascial thickenings.

The carpus is cartilaginous at birth with complete ossification between the twentieth and twenty-fifth years. The capitate begins to ossify during the first year and, except for the pisiform, which ossifies during the twelfth year, all the carpal bones are in the process of ossification by the eighth year. The carpal bones, although fitted closely together, do permit a certain amount of movement and afford some flexibility at the wrist. The carpus is generally concave from side to side on its palmar side, but convex on its dorsum. The extremities of this concavity give attachment to the flexor retinaculum (transverse carpal ligament) (Fig. 7-19, p. 218). The latter, with its bony attachments, creates an osteofascial tunnel, the carpal tunnel, for the flexor tendons of the fingers and the median nerve.

The carpal bones are eight in number and are arranged in a proximal and distal row of four bones each. Proximally, from lateral to medial, are the scaphoid (navicular), lunate (semilunar), triquetrum (cuneiform), and pisiform. The distal row, also from lateral to

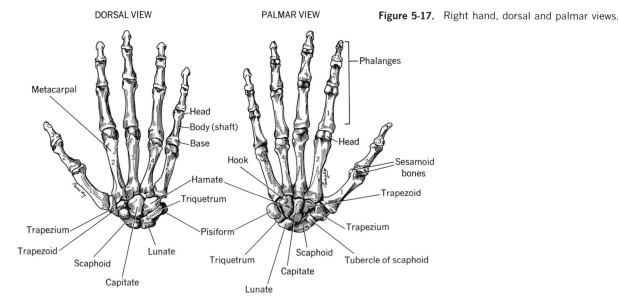

Figure 5-17. Right hand, dorsal and palmar views.

medial, has the trapezium (greater multangular), trapezoid (lesser multangular), capitate (os magnum), and hamate (unciform). The eight bones are roughly cubical in shape and thus have six surfaces: proximal, distal, lateral, medial, anterior, and posterior.

THE SCAPHOID (BOAT SHAPED). The scaphoid cannot tolerate an indirect blow well, such as might occur when falling on the palm of the hand in radial deviation. Thus, it is the most commonly fractured of all the carpal bones. The tubercle of this bone can be felt through the skin at the base of the thenar eminence in line with the radial side of the middle finger. The distal transverse skin crease at the front of the wrist crosses this tubercle as well as the pisiform bone.

THE LUNATE (MOON SHAPED). The lunate is the middle bone of the proximal group. It is the carpal bone that is most frequently dislocated.

THE TRIQUETRUM (TRIANGULAR SHAPED). The triquetrum articulates with the articular disk and with the ulnar collateral ligament, permitting it (with the pisiform) to glide toward the ulna in ulnar flexion. The rest of the proximal surface of this bone is nonarticular.

THE PISIFORM (PEA SHAPED). The pisiform is a small bone in the tendon of a muscle at the base of the hypothenar eminence on the medial side of the front of the wrist. It does not participate in the radiocarpal joint.

THE TRAPEZIUM (GREATER MULTANGULAR). The trapezium is the most radially placed bone of the distal carpal row. It can be felt along with the navicular in the "anatomic snuff box," which is found at the lateral side of the wrist between the styloid process of the radius and the base of the first metacarpal. Here the radial artery crosses these bones and pulsations can be felt easily.

THE TRAPEZOID (LESSER MULTANGULAR). The trapezoid is the smallest of the carpal bones, except for the pisiform.

THE CAPITATE (HEADLIKE). The capitate is the largest carpal bone and is centrally placed. Its head transmits the force of a fall on the hand to the radius through the navicular and lunate bones.

THE HAMATE (HOOKLIKE). The hamate has a projecting process, the hook of the hamate, which can be felt through the skin in the ball of the little finger about 2.5 cm (1 in.) below and lateral to the pisiform bone. The ulnar nerve and artery pass between the hook and the pisiform.

Metacarpal and Phalangeal Bones. The five metacarpal bones (Fig. 5-17) form the skeleton of the palm and articulate with the distal row of carpal bones as well as the phalanges. Each has a base (proximal), shaft (middle), and head (distal). The proximal ends articulate with the carpal bones as well as with each other. The heads of the medial four metacarpal bones articulate with the phalanges. At the metacarpophalangeal joints the prominences of the knuckles are formed by the distal aspects of the heads of the metacarpal bones.

The metacarpal bone of the thumb articulates with the trapezium by a joint entirely separated from the carpometacarpal joints. This bone is the shortest and most movable of all, and its base is "saddle shaped."

The construction of all five fingers is essentially the same except that the thumb has but two phalanges and the other fingers have three. There are 14 phalanges in all, three (proximal, middle, and distal) for each finger and two (proximal and distal) for the thumb. The distal ends are neither weight bearing nor force transmitting.

Bony Thorax

The thorax (Fig. 5-18) is oval shaped and flattened from front to back and contains the organs of respiration, circulation, and digestion. Its anterior wall is the shortest and is formed by the sternum and the anterior portions of the first ten pairs of ribs and their costal cartilages. The lateral walls are formed by the ribs, which slope downward and forward. The posterior wall consists of the 12 thoracic vertebrae and the ribs to their angles. At birth (and two years after) the thorax is circular rather than oval shaped. Thus, the adult thorax can be increased in chest breathing, but this is not possible in the child since the circumference remains constant. For this reason, breathing in the early years is almost entirely diaphragmatic (abdominal), and as one grows older it becomes more thoracic (intercostal). This makes the child more susceptible to postoperative pneumonia because when he restricts his

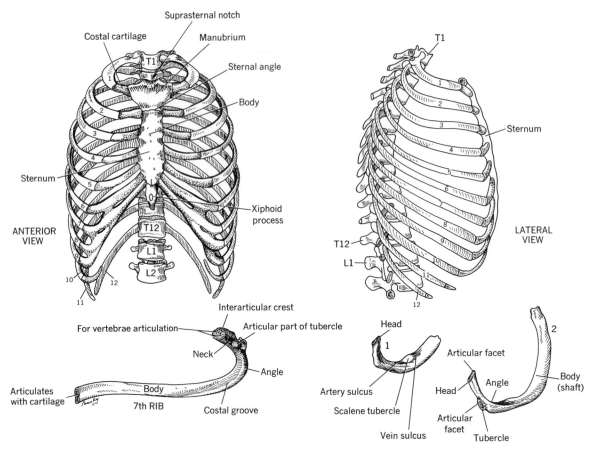

Figure 5-18. Bony thorax, ribs and sternum. Rib 7 is a typical rib and is seen from its inside. Ribs 1 and 2 are seen from above.

abdominal movements due to pain this interferes with diaphragmatic movements, resulting in poorly aerated lungs that fill with accumulated secretions.

The thorax has a small, "kidney-shaped" superior opening, about 5 to 10 cm (2 to 4 in.), and a large, irregular inferior one.

The Superior Aperture (Inlet). The superior aperture is bounded by the body of the first thoracic vertebra, the first ribs, the first cartilages, and the upper border of the manubrium of the sternum. Because of its downward slope, the lung apices rise above the inlet into the root of the neck.

The Inferior Aperture (the Outlet). The inferior

aperture is large and bounded by the twelfth thoracic vertebra, lowest ribs, seventh to twelfth costal cartilages, and the xiphisternal joint. It is occupied by the diaphragm.

Conventional Longitudinal Lines. The conventional longitudinal lines (Fig. 5-19) are used on the chest wall for description and orientation, and they run parallel to the axis of the body. They are: (1) midsternal, which bisects the sternum; (2) mammary line, which runs from the clavicle, passing through the nipple; (3) parasternal line, which is midway between (1) and (2); (4) anterior, middle, and posterior axillary lines, which go through the anterior, middle, and posterior axillary folds, respectively; (5) scapular line, which passes through the inferior angle of the scapula; and (6) paravertebral line, which runs opposite the tips of the transverse processes of the vertebrae.

The Ribs (Costae). There are 12 pairs of ribs (Fig. 5-18), and they form a series of oblique bony arches that form the greater part of the thoracic cage wall. They overhang the upper abdomen, articulate with the thoracic vertebrae posteriorly, and end in costal cartilages anteriorly. They increase in length from the first to the seventh and then shorten progressively. The first seven are called the true ribs because their cartilages articulate with the sternum. The cartilages of the last five pairs are known as false ribs. The eighth, ninth, and tenth ribs end by joining costal cartilages. The eleventh and twelfth ribs are free at their extremities and are spoken of as floating ribs.

The ribs, vertebrae, sternum, and skull diploe are all filled with red blood–forming marrow, and here (not in the limbs) is where those blood elements are formed after puberty. The first seven costal cartilages

are continuous, extending forward to join the sternum. The eighth through the tenth end anteriorly by joining with cartilages above. These form a cartilaginous ridge, the subcostal margin, that forms a part of the inferior thoracic aperture.

One can feel all the ribs through the soft tissues of the chest wall except for the first rib, which is hidden by the clavicle. The twelfth rib may be absent or too short to palpate. The second rib can be identified where its cartilage reaches the sternum at the sternal angle and is used as a guide to succeeding ribs by counting downward from above. The ribs vary in length, the shortest being either the first or twelfth and the longest usually the seventh or eighth. The typical ribs are the third to the ninth inclusive, whereas the first, second, tenth, eleventh, and twelfth have special characteristics.

TYPICAL RIB. A typical rib articulates with two vertebrae: the one to which it numerically corresponds and the vertebra above this. It has a head, neck, tubercle, and shaft. The shaft has an angle posteriorly and

Figure 5-19. **A.** Surface lines of the front of thorax and abdomen. **B.** Surface lines as related to underlying viscera—colon, liver, and stomach.

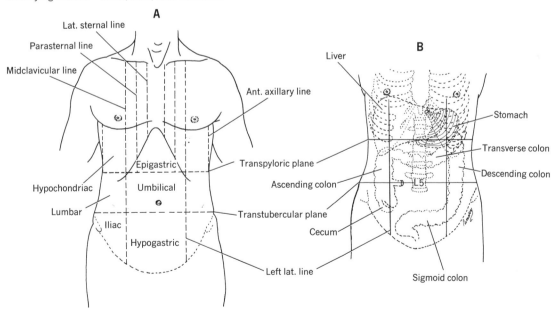

a costal groove inferiorly. The head has an articular surface divided by a crest into two facets, which articulate with contiguous facets on the lateral aspect of the bodies of the two adjacent vertebrae. From the intervening crest an interarticular (radiate) ligament passes to the intervertebral disk (Fig. 6-6, p. 173). The neck is constricted, about 2.5 cm (1 in.) long, and at its upper border is raised into a crest. The tubercle, found on the back of the junction of the neck and shaft, articulates with the corresponding transverse process. The body (shaft) is compressed laterally, giving it internal and external surfaces separated by superior and inferior borders. The rib is not only curved but looks twisted. The maximum area of curvature is spoken of as the angle and is marked by a ridge on its external surface. The internal surface is smooth and covered with pleura. The inferior surface has a groove, the costal groove, which contains the intercostal vessels and nerves. The sternal or anterior end of the rib shaft receives the costal cartilage.

THE SPECIAL RIBS. *First Rib.* The first rib is the highest, strongest, flattest, broadest, the most curved, and the shortest of the ribs. It lies almost horizontally and articulates with but one vertebra, the first thoracic. Its upper surface has a roughened area for insertion of the scalenus medius muscle and origin of the first digitation of the serratus anterior muscle. A wide shallow groove crosses the front of the rib and houses the subclavian artery. The groove is limited by the scalene tubercle where the scalenus anterior muscle inserts. Anterior to this tubercle another shallow groove lodges the subclavian vein. This rib has no costal groove.

Second Rib. The second rib is intermediate in shape and size and, although curved, is not twisted. The rib has a rough elevation near the middle of its outer surface where the lower part of the first digitation and the entire second digitation of the serratus anterior attach. The posterior scalene muscle attaches behind this point.

Tenth Rib. The tenth rib is like a typical rib, but shorter. Its head usually has but one facet for the body of the tenth vertebra. This rib is variable.

Eleventh Rib. The eleventh rib is short, has a single articular facet for vertebra 11, and has no neck or tubercle.

Twelfth Rib. The twelfth is very short, has but one facet on its head, and has no tubercle, angle, or costal groove.

BLOOD SUPPLY. The blood supply reaches each rib by way of its own nutrient vessel, which enters just beyond the tubercle and runs forward as far as the inner extremity of the rib. Periosteal vessels supply additional blood. It should be noted that there is no anastomosis between vessels of the diaphysis and epiphysis of the ribs and, therefore, the vessels are "end arteries."

OSSIFICATION. Ossification occurs as in the long bones of the extremities and takes place in the second fetal month beginning near the angle and proceeding in both directions, yet never reaching the sternal end resulting in the costal cartilages.

The Sternum (Breastbone). The sternum (Fig. 5-18) is an elongated flat bone in the midventral part of the thorax. It is shaped like an "old Roman sword" with a short handle, the manubrium; a longer blade, the body; and a small cartilaginous piece at its lower end, the xiphoid process. The bone ossifies late in fetal life; yet the constituent parts of the body do not fuse until puberty. Initially there are six segments, then in the adult the four middle segments are fused, resulting in three mature parts.

THE MANUBRIUM. This is the broadest and most massive part. There are three notches on its upper border, two called clavicular notches for the clavicle articulations and a centrally placed suprasternal or jugular notch that can be felt through the skin and lies on a horizontal plane with the disk between the second and third thoracic vertebrae.

THE BODY. The body is about 10 cm (4 in.) long. The sternal angle (angle of Louis) is the angle formed where the body and the manubrium meet and is easily felt beneath the skin (an important landmark since it overlies important thoracic structures and its surface also marks the attachment of the second costal cartilages). It is on a horizontal plane with the disk between the fourth and fifth thoracic vertebrae. The sternomanubrial joint is important in respiration as it allows

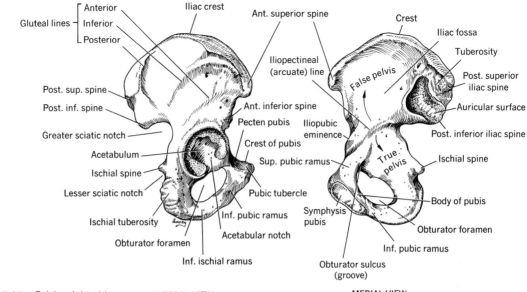

Gluteal lines { Anterior, Inferior, Posterior

Iliac crest

Ant. superior spine

Post. sup. spine

Post. inf. spine

Greater sciatic notch

Acetabulum

Ischial spine

Lesser sciatic notch

Ischial tuberosity

Obturator foramen

Inf. ischial ramus

Iliopectineal (arcuate) line

Ant. inferior spine

Pecten pubis

Crest of pubis

Sup. pubic ramus

Pubic tubercle

Inf. pubic ramus

Acetabular notch

Crest

Iliac fossa

Tuberosity

Post. superior iliac spine

Auricular surface

Post. inferior iliac spine

Ischial spine

Body of pubis

Obturator foramen

Inf. pubic ramus

Obturator sulcus (groove)

Symphysis pubis

Iliopubic eminence

False pelvis

True pelvis

Figure 5-20. Pelvis, right side.   LATERAL VIEW   MEDIAL VIEW

the body of the sternum to move forward and back-ward while the manubrium remains stationary. The pectoralis major muscles attach anteriorly, while the posterior surface is related to the pleurae and peri-cardium.

THE XIPHOID (ENSIFORM) PROCESS. The xiphoid process is the smallest of the three sternal parts and has a variable appearance; it may be bent or depressed, bifid, or perforated. It usually is incompletely ossified in the adult and gives attachment posteriorly to fibers of the diaphragm.

Pelvis

The pelvis (Fig. 5-20) is so named because it resem-bles a basin. It is composed of two hipbones (anteriorly and laterally) and the sacrum and coccyx (posteriorly). The cavity created is divided into a smaller inferior part called the true pelvis, formed by the pubis and ischium, and a larger superior one, the false pelvis (greater), formed by the ilia. The separation line is the iliopectineal (arcuate) line.

The male pelvis and the female pelvis show distinct differences. The female pelvis, in contrast to that of

the male, is marked by the fact that its muscular im-pressions are less marked, its bones are more delicate, its depth is less, the pelvis is less massive, the anterior iliac spines are more widely separated, the superior aperture of the lesser pelvis is larger and more circular, the obturator foramina are triangular, the coccyx is more movable, and the pubic arch is wider, more rounded, and forms an angle rather than an arch.

There are a variety of pelvic types based on meas-urements of various parts of the pelvis. These measure-ments are particularly essential to the female since the safe passage of the baby requires adequate pelvic room. Figure 5-21 gives one an indication of some of the variables. The gynecoid type is the most common female type, whereas the android is the most common male-type pelvis. Figure 5-22 illustrates where and how some of the pelvic measurements are taken.

The Hipbone (Innominate or Unnamed). The hip-bone consist of three parts: the ilium, the ischium, and the pubis. They all meet at the acetabulum. Ossifica-tion begins at 12 years of age and is usually complete by the age of 16 years.

THE ILIUM (FLANK). The ilium is the largest of the

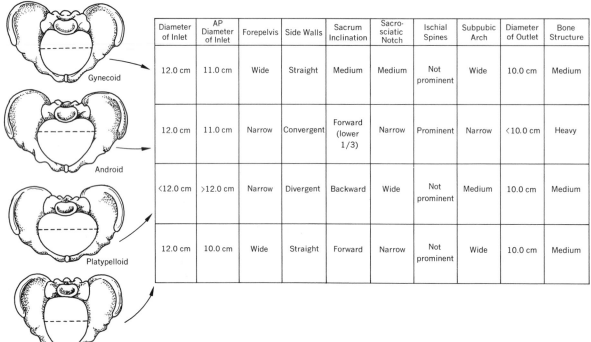

	Diameter of Inlet	AP Diameter of Inlet	Forepelvis	Side Walls	Sacrum Inclination	Sacro-sciatic Notch	Ischial Spines	Subpubic Arch	Diameter of Outlet	Bone Structure
Gynecoid	12.0 cm	11.0 cm	Wide	Straight	Medium	Medium	Not prominent	Wide	10.0 cm	Medium
Android	12.0 cm	11.0 cm	Narrow	Convergent	Forward (lower 1/3)	Narrow	Prominent	Narrow	<10.0 cm	Heavy
Platypelloid	<12.0 cm	>12.0 cm	Narrow	Divergent	Backward	Wide	Not prominent	Medium	10.0 cm	Medium
Anthropoid	12.0 cm	10.0 cm	Wide	Straight	Forward	Narrow	Not prominent	Wide	10.0 cm	Medium

Figure 5-21. The major pelvic types and how one arrives at their classification.

three parts and forms the upper expanded portion of the hipbone, being fan shaped with the handle of the fan placed downward where the ilium joins the pubis and ischium. Its arched upper border, the iliac crest, can be felt at the lower limit of the waist. The highest point of the crest is at its middle and toward the back, marking a level at the fourth lumbar vertebrae. The crest is crossed by cutaneous nerves, but no muscles cross it and fascia is attached to it. The crest ends anteriorly in a tubercle called the anterior superior iliac spine, which can easily be felt at the upper end of the groin fold. The crest terminates posteriorly in a sharp, posterior superior iliac spine, which is felt on a level with the second sacral spine in the upper and medial part of the buttock. The tubercle of the crest is a projection on its outer lip, palpable about 5 cm (2 in.) above and behind the anterior superior spine.

The anterior border of the ilium is concave and extends from the anterior superior spine to the iliopubic eminence (junction between pubis and ilium).

The iliopsoas muscle leaves the abdomen through the notch between the anterior inferior iliac spine and this eminence. The posterior border extends from the posterior superior spine to a point on the margin of the greater sciatic notch. It is subdivided into two concavities by the posterior inferior iliac spine. The lower concavity forms the upper part of the greater sciatic notch.

The ilium has an inner surface that is subdivided into the iliac fossa, tuberosity, auricular surface, and area in the true pelvis. The fossa is large and smooth and limited inferiorly by a ridge, the iliopectineal line. The tuberosity is the rough part of the inner surface, and the auricular surface articulates with the sacrum. The true pelvis is the lowest, smooth part. The outer (gluteal) surface shows three rough lines that converge on the greater sciatic notch, dividing the surface into three areas for gluteal muscles: the posterior gluteal line, the middle (anterior) gluteal line, and the inferior gluteal line (which is usually indistinct).

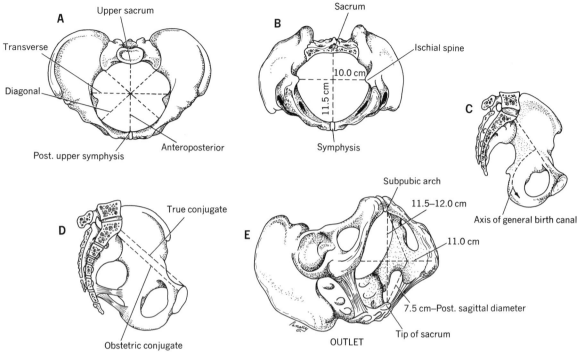

Figure 5-22. Pelvis. **A.** Boundaries of the pelvic inlet (greatest diameters). **B.** Boundaries of the midpelvis. **C.** The pelvic axis is the line that passes through the center of the planes of the inlet, midpelvis, and outlet. **D.** The obstetric conjugate, not the true conjugate, is the critical anteroposterior diameter through which the fetal head passes at the inlet. **E.** The outlet consists of two triangular planes with a common base, the intertuberous line (transverse diameter) measuring about 11.0 cm.

THE ISCHIUM (BUTTOCK). The ischium consists of three parts: a body that adjoins the ilium, a tuberosity that projects down from the body, and a ramus (inferior) that passes upward from the tuberosity, below the obturator foramen, to join the inferior ramus of the pubis. The ischial tuberosity is a mass of bone capping the posterior aspect of the lower body. It helps form part of the lesser sciatic notch and gives attachment to the sacrotuberous ligament and several muscles. The tuberosity's chief function is to support body weight when sitting. The ramus of the ischium projects forward and medially from the tuberosity and joins the inferior ramus of the pubis to help form the pubic arch. The greater sciatic notch is a "gap" between the posterior inferior iliac spine and the ischial

spine. The lesser sciatic notch is a smaller "gap" between the ischial spine and tuberosity. The notches are converted into foramina (greater and lesser) by the sacrospinous and sacrotuberous ligaments.

THE PUBIS (ADULT). The pubis is named from the region where hair develops in adulthood. It consists of a body and a superior and inferior ramus. The body is the wide, flattened medial part of the bone. The symphysial surface is joined to the opposite pubic bone by a fibrocartilage and ligaments. The lateral border helps form a boundary of the obturator foramen. The pubic crest forms the upper border of the body of the bone and terminates laterally in the pubic tubercle, which gives attachment to the inguinal ligament. In the erect position the tubercle and anterior superior iliac spines are in the same vertical plane. The medial end of the superior ramus is expanded and becomes the body, whereas the lateral end expands and fuses with the ilium and ischium to form part of the acetabulum. The inferior ramus passes down and lateral from the body to meet the ramus of the ischium.

THE ACETABULUM (LITTLE CUP FOR VINEGAR). The acetabulum forms the socket for the head of the femur. The ischium, ilium, and pubis all meet in the center of the cavity. The fusion lines can be seen in a child, but they obliterate with age. The acetabulum is directed downward and laterally, is horseshoe shaped, and is covered with cartilage, leaving a rough fossa in its floor that adjoins the acetabular notch in its lower part. The transverse ligament converts the notch into a foramen for the articular branches of the obturator and medial circumflex arteries. The obturator foramen is below the acetabulum and is closed by a membrane except above at the obturator groove through which the obturator vessels and nerves leave the pelvis to enter the thigh.

Lower Extremity

The Femur (Thighbone). The femur (Figs. 5-23, 5-24) is the longest, largest, and heaviest bone in the body and is about one fourth of the entire length of the individual. It consists of proximal and distal ends and a shaft.

PROXIMAL END. This consists of a head, neck, greater and lesser trochanters, trochanteric fossa and line, trochanteric crest, and quadrate tubercle.

Head. The head forms two thirds of a sphere and is directly medial, upward, and forward and is "gripped" firmly by the acetabular labrum. It is more secure in its socket than the humerus. The head is covered with hyaline cartilage and has a depression (pit), the fovea capitis, a little below and behind its center point that gives attachment to the ligament of the head (ligamentum teres).

Neck. The neck is about 1.25 cm ($\frac{1}{2}$ in.) long, is continuous with the shaft, and supports the head. It joins the shaft at about 125 degrees in the adult. This angle is more obtuse in the child. We speak of a "vertical neck-shaft angle." The neck is separated from the shaft

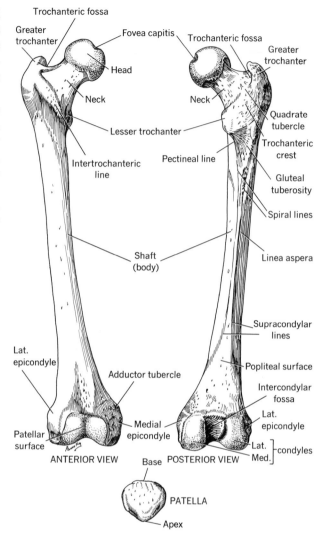

Figure 5-23. Right femur and patella.

by the intertrochanteric line in front and the trochanteric crest behind. The former is a roughened edge produced by the attachment of the iliofemoral ligament. The upper end of the line is at the front of the greater trochanter, while its lower end is continuous with a faint ridge, the spiral line, that winds around the lesser trochanter to the back of the shaft. The trochanteric crest crosses the posterior aspect of the

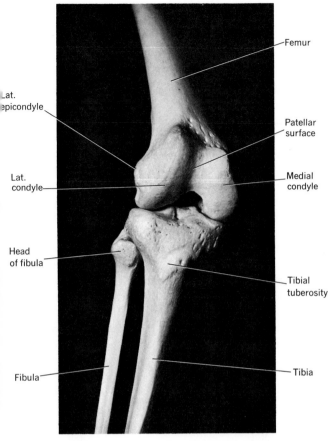

Lat. epicondyle

Lat. condyle

Head of fibula

Fibula

Femur

Patellar surface

Medial condyle

Tibial tuberosity

Tibia

Figure 5-24. The femur and its relation to the tibia and fibula in slight flexion (the patella is not seen).

bone and is continuous with the lesser trochanter below. The quadrate tubercle, midway along the crest, gives insertion to the quadratus femoris muscle. The trochanteric fossa is small and found at the junction of the posterior part of the neck and the medial side of the greater trochanter and serves for insertion of the obturator internus muscle.

Greater Trochanter. The greater trochanter is a fixed process that lies in a line with the lateral aspect of the shaft and can be felt beneath the skin about one handbreadth below the iliac crest. The muscles attached to it create the rotary movements of the thigh.

Lesser Trochanter. The lesser trochanter is a blunt-shaped process that is directed backward and medially from the lower posterior part of the neck of the femur. It is not palpable. It gives attachment to the tendon of the iliopsoas muscle and has been thought of as the traction epiphysis of that structure.

Quadrate Tubercle. The quadrate tubercle is an ill-defined protrusion found about the center of the trochanteric crest. It gives insertion to the quadratus femoris muscle.

DISTAL END. The distal end is larger than the proximal end. Found here are the lateral and medial condyles, the epicondyles, a patellar surface, the intercondylar fossa, the adductor tubercle, and the popliteal surface.

Condyles. The condyles make up almost all of the distal end. They are separated behind by the intercondylar notch. The top of the tibia and cartilages of the knee joint articulate with the posterior surfaces of the condyles when the knee bends and with their inferior surfaces when the knee is straightened.

The lateral condyle is broader than the medial one, and its epicondyle is situated on the lateral surface of the condyle.

The medial condyle is farther from the side of the shaft than the lateral. Immediately above and behind its center is the medial epicondyle, to which the medial ligament of the knee is attached. The adductor tubercle, for insertion of the tendon of the adductor magnus muscle, is just above this epicondyle.

The intercondylar fossa separates the condyles below and behind and is occupied by the cruciate ligament.

The patellar (trochlear) surface is situated on the anterior aspect of the femur, extending farther upward on the lateral condyle than on the medial condyle.

SHAFT (BODY). The shaft is thinnest in its middle and is bowed slightly forward. In the erect position the shaft is oblique, such that the distal ends of the femora touch each other while the proximal ends are separated by the pelvis. Each shaft has anterior, medial, and lateral surfaces with ill-defined lateral and medial borders and a well-marked posterior border, the linea aspera. The latter is a broad, rough line running from the middle of the shaft to bifurcate above and below into diverging lines. The upper lines are the spiral lines, which become continuous with the intertrochanteric line and greater tuberosity. The lower lines are the medial and lateral supracondylar lines, which descend to the epicondyles. The flat, triangular area between them is the popliteal surface. The medial aspect of the femoral shaft is devoid of any muscular attachment.

The Patella. The patella (Fig. 5-23) is a sesamoid bone in the tendon of the quadriceps femoris muscle. It is triangular in shape, with its inferior angle being the apex and the upper border the base. One can feel its anterior surface through the skin, from which it is separated by a subcutaneous prepatellar bursa. The bone begins to ossify between the third and fifth years, with ossification being complete at puberty. The rectus femoris and vastus intermedius muscles are attached to its base and on contraction pull obliquely upward. The vastus medialis is attached to the medial border of the patella and the vastus lateralis to its lateral border (see Chap. 7). If the quadriceps is relaxed (e.g., by placing the heel on a chair), the patella can be moved medially and laterally. As one proceeds from extension to flexion of the knee joint, the patella glides laterally onto the under aspect of the lateral femoral condyle.

The Tibia and Fibula. TIBIA (SHINBONE). The tibia (Figs. 5-24, 5-25) is the medial and larger of the two leg bones and has a proximal end, distal end, and a shaft. The proximal end is the larger of the two. It consists of a tuberosity, lateral and medial condyles, an intercondylar area and eminence, and fibular facets. The proximal end is massive and broad and provides a good weight-bearing surface for the femur. The tuberosity is below and in front of the condyles, has a rough surface, can be felt subcutaneously, and gives attachment to the ligamentum patellae. When one kneels, the body rests on the lower part of the tuberosity, the front of the condyles, the patellar ligament, and the lower patella.

The two condyles can be felt to the side of the bone and make up most of the proximal end of the tibia. Superiorly the condyles are covered with cartilage and articulate with the femoral condyles and semilunar cartilages. Between them is the intercondylar area, which is marked by an intercondylar eminence. The medial condyle is larger than the lateral. The latter has a facet for the fibular head.

The shaft is thick and strong. The anterior border can be felt as it extends from the tuberosity to the front of the medial malleolus below. The interosseus or lateral border gives attachment to the interosseous membrane. The medial border is traced from the medial condyle to the back of the medial malleolus. The shaft also has a lateral, medial, and posterior surface. The medial surface is subcutaneous and can easily be palpated. The soleal or popliteal line appears as a rough ridge that crosses the posterior border obliquely.

The distal end of the tibia has five surfaces, a medial malleolus, and a fibular notch. The medial surface is subcutaneous and continues as the medial malleolus; the anterior is covered by extensor tendons; the posterior is grooved for the tibialis posterior muscle; the inferior (tarsal) articulates with the talus; and the lateral has a depression, the fibular notch. The medial malleolus can easily be palpated; it articulates with the talus.

FIBULA. The fibula (Figs. 5-24, 5-25) is the lateral of two leg bones. It is slender and is attached above and below to the tibia. It has a proximal and distal end and a shaft.

The proximal end has a head, neck, and styloid process. The head articulates with the lateral condyle of the tibia. Projecting upward from the head is the apex or styloid process, to which the lateral knee ligament is attached. The head articulates with the lateral condyle of the tibia. The apex can be felt through the skin below the knee joint. The neck is a constricted portion below the head. The shaft has three borders: an interosseous border, which is ill defined and on the medial side; the anterior border, which is sharp and distinct; and the posterior border, which is blunt yet well defined.

The distal end of the bone is the lateral malleolus with a facet for articulation with the talus. The malleolus can be felt subcutaneously and forms the lateral prominence of the ankle. It lies at a lower level than

Figure 5-25. Right tibia and fibula.

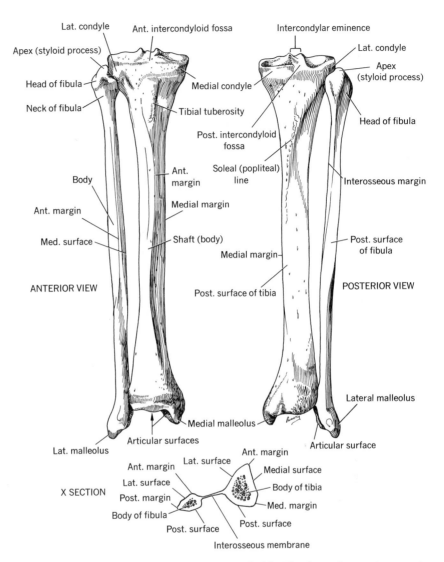

Lat. condyle
Ant. intercondyloid fossa
Intercondylar eminence
Apex (styloid process)
Lat. condyle
Head of fibula
Medial condyle
Apex (styloid process)
Neck of fibula
Tibial tuberosity
Head of fibula
Post. intercondyloid fossa
Body
Soleal (popliteal) line
Ant. margin
Interosseous margin
Ant. margin
Medial margin
Med. surface
Shaft (body)
Post. surface of fibula
ANTERIOR VIEW
Medial margin
POSTERIOR VIEW
Post. surface of tibia
Lateral malleolus
Medial malleolus
Articular surfaces
Articular surface
Lat. malleolus
Ant. margin
Ant. margin
Lat. surface
Medial surface
Lat. surface
Body of tibia
X SECTION
Post. margin
Med. margin
Body of fibula
Post. surface
Post. surface
Interosseous membrane

the medial malleolus. There are grooves on the posterior surface for the tendons of the peroneus longus and brevis muscles.

The Ankle. The bones that enter into formation of the ankle joint are the talus and the distal ends of the tibia and fibula. As the two malleoli project downward, they grasp the talus firmly at each side.

Foot. The foot (Fig. 5-26) is triangular in outline and extends from the heel to the root of the toes. It is divided into the tarsus (posterior half) and the metatarsus (anterior half). The lateral margin rests in contact with the ground over its entire extent. Near the middle of this lateral border, the tuberosity of the base of the fifth metatarsal bone presents a landmark for the tarsometatarsal joint. The medial aspect is arched and rests on the ground at the heel and on the ball of the great toe. The sustentaculum tali is found about 2.5 cm (1 in.) below the medial malleolus, and a tuberosity of the navicular bone is felt about 2.5 cm (1 in.) in front and slightly below the medial malleolus.

BONES. The tarsus consists of seven tarsal bones formed by the calcaneus and talus (posterior), the navicular (middle), and the three cuneiform bones and cuboid (anterior).

Talus (Astralgus). The talus rests on the anterior two thirds of the calcaneus and has a body, neck, and head. It lies below the tibia, sits on the calcaneus, and is grasped by the malleoli. The head is anterior and articulates with the navicular, while it rests below on the plantar calcaneonavicular (spring) ligament and calcaneus (Fig. 6-10, p. 178). The neck is the constricted part of the bone. The body is hidden by the tibia and is grasped by the malleoli.

Calcaneus. This is the heel bone. Its anterior two thirds supports the talus, while its posterior one third forms the heel prominence and rests on the ground. The anterior surface articulates with the cuboid bone. The lateral surface is almost all subcutaneous and can be felt below the lateral malleolus. On the medial side the sustentaculum tali (a horizontal shelf of bone) can be felt below the medial malleolus.

Navicular (Scaphoid). The navicular is found on the medial side of the foot and articulates with the head of the talus posteriorly and the three cuneiforms anteriorly. It, too, has a tuberosity (on the medial side) that can be felt about 2.5 cm (1 in.) below and in front of the medial malleolus midway between the root of the big toe and the heel.

Three Cuneiforms. These are named the first (medial), second (intermediate), and third (lateral) cuneiforms. They articulate with the navicular posteriorly and the first three metatarsals anteriorly. They are wedge shaped, lie side by side, and articulate with each other. The lateral one articulates with the cuboid and the fourth metatarsal, and the other two grip the base of the second metatarsal between them.

Cuboid. The cuboid lies on the lateral side of the foot, articulating with the fourth and fifth metatarsals anteriorly and the calcaneus posteriorly. Medially it articulates with the three cuneiforms and the navicular, and on its plantar surface it has an oblique groove for the tendon of the peroneus longus muscle.

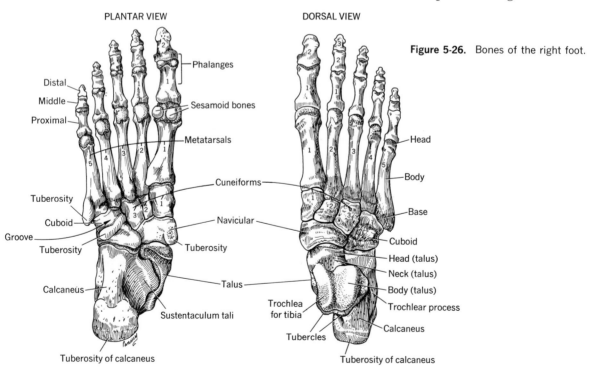

PLANTAR VIEW

DORSAL VIEW

Phalanges

Distal

Middle

Proximal

Sesamoid bones

Metatarsals

Cuneiforms

Tuberosity

Cuboid

Groove

Tuberosity

Navicular

Tuberosity

Calcaneus

Talus

Sustentaculum tali

Tuberosity of calcaneus

Head

Body

Base

Cuboid

Head (talus)

Neck (talus)

Body (talus)

Trochlear process

Calcaneus

Trochlea for tibia

Tubercles

Tuberosity of calcaneus

Figure 5-26. Bones of the right foot.

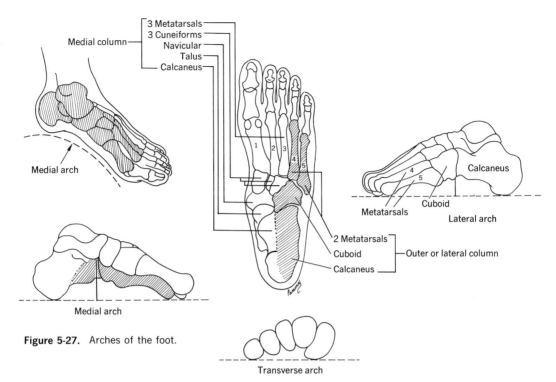

Medial column
- 3 Metatarsals
- 3 Cuneiforms
- Navicular
- Talus
- Calcaneus

Medial arch

Medial arch

Calcaneus

Metatarsals

Cuboid

Lateral arch

2 Metatarsals
Cuboid
Calcaneus
— Outer or lateral column

Figure 5-27. Arches of the foot.

Transverse arch

Metatarsals. These consist of five bones numbered 1 to 5 from medial to lateral. Each has a head (distal end), a body (midpart), and a base (proximal end). The bases of the first, second, and third articulate with the three cuneiforms and the bases of the fourth and fifth with the cuboid. The bases also articulate with each other, and the heads articulate with the proximal phalanges. The first is the shortest and stoutest. The base of the fifth metatarsal forms a prominent landmark on the outer foot margin.

Phalanges. These are the bones of the toes and are numbered from medial to lateral, with the first toe called the hallux and the little toe the small digit. The big toe has two phalanges; all the others have three. There is a proximal, middle, and distal phalanx, and each proximal one articulates with the head of a metatarsal bone to form the metatarsophalangeal joint. The middle phalanx articulates with the other two to form interphalangeal joints. Each phalanx has a base (proximal end) and a head (distal end).

Accessory Bones. There are two types: sesamoids, which are regular skeletal constituents; and true accessory bones, which are occasional bones seen in definite sites. These bones are usually bilateral and occur in 25 percent of feet.

ARCHES. The arches (Fig. 5-27) are important to (1) give elasticity and spring to the step; (2) distribute body weight properly; (3) break the shock from walking, running, and jumping; and (4) provide space for the soft tissues that lie in the arches and thus prevent excessive pressure on these structures. We speak of transverse and longitudinal arches.

Transverse Arches. These are a series of arches extending from that which is formed by the heads of the metatarsals backward to that formed by the navicular and cuboid bones. If these arches flatten, the digital vessels and nerves are pressed upon and pain results. The arches are supported by muscles, tendons, ligaments, fasciae, and bones.

Longitudinal Arches. Two columns exist, one medial and the other lateral, and both rest on the tuberosity of the calcaneus posteriorly. The medial or inner column consists of the calcaneus, talus, navicular, three cuneiforms, and the medial three metatarsals. It is a high arch, touching the ground behind at the calcaneus and in front at the head of the first metatarsal bone. It is absent in flat feet but increased in pes cavus. It has much elasticity and is the arch of movement. The lateral or outer arch is formed by the calcaneus, cuboid, and the two outer metatarsals. This arch is so low that the foot border touches the ground along its entire length. Thus, the body weight is borne by a tripod arrangement with the weight transmitted to the tuberosity of the calcaneus and head of the first and fifth metatarsals. In a child the plantar fat pad masks the arches of the foot.

The tendons that support the arches (Fig. 6-10, p. 178) are the peroneus longus and the tibialis posterior. The former passes down the lateral side of the foot, crosses the lateral side, turns at right angles on itself, and crosses the sole of the foot from lateral to medial to insert into the outer side of the first cuneiform bone and base of the first metatarsal. It acts as a sling for the longitudinal arch and bolsters the transverse arch. The tibialis posterior tendon inserts about the middle of the longitudinal arch to the undersurface of the navicular and sends a slip into every bone of the tarsus except the talus as well as the bases of metatarsals 2, 3, and 4. It balances the inner side of the longitudinal arch and supports the spring ligament.

The muscles support the arches by pulling the two pillars close together or upward. The muscles that adduct and invert the foot increase the longitudinal arch, and those that abduct and evert flatten it. The long toe flexors and short foot muscles pull the pillars together and increase the arch. The transverse arch is maintained mainly by the transverse head of the adductor hallucis muscle (and, to a lesser degree, by its oblique head).

The ligaments associated with arch support (Fig. 6-10, p. 178) are weak on the dorsum of the foot but powerful on the sole. All are important, but of special concern is the calcaneonavicular (spring) ligament, which is under the weakest part of the longitudinal arch (head of the talus) and prevents the sinking of the talus. The long plantar ligament runs from the calcaneus to the cuboid and to the bases of the second, third, and fourth metatarsals. The short plantar (calcaneocuboid) passes obliquely from the calcaneus to the cuboid and is concealed partly by the long plantar. The plantar intertarsal and tarsometatarsal ligament support the transverse arch.

The fascia of the plantar aponeurosis holds the extremities of the arches together.

The form and shape of the bones support the arches by being broader on the dorsum, thereby making less support necessary.

Review Questions

1. Select the correct answers for the statements on direction from the list below:

   Superior   Posterior   Medial   Proximal   Cephalad
   Inferior   Anterior   Lateral   Distal   Caudad

   a. Toward the head
   b. Nearest the midline of the body
   c. Uppermost or above
   d. Lowermost or below
   e. Refers to the back
   f. Refers to the front
   g. Toward the tail
   h. Toward the side
   i. Away from the point of attachment
   j. Nearest the point of attachment

2. Define the following body planes:
   a. Sagittal   **b.** Horizontal   **c.** Frontal

3. What are the major body cavities?

4. What are the major classifications of bone? Give an example of each.

5. Name the true or movable vertebrae and indicate how you would differentiate between each group.
6. Why are the cervical vertebrae unique? Describe the first and second.
7. What are fontanelles? Where are they? Why are they important?
8. Relate the following skull structures (or areas) to either the front, back, top, side, or bottom of the skull:

   a. Orbit
   b. Nasal cavity
   c. Sagittal suture
   d. Nasion
   e. Foramen ovale

   f. Hard palate
   g. Pterygoid plates
   h. Jugular foramen
   i. Styloid process
   j. Mastoid process

   k. Foramen magnum
   l. Zygomatic arch
   m. Infratemporal fossa
   n. External auditory meatus
   o. Occipital bone

9. Describe the mandible.
10. What are the cranial fossae?
11. Describe the location and essential features of the scapula and the clavicle.
12. What bones make up the upper extremity? What are the landmarks of each?
13. Name the wrist (carpal) bones. With what do they articulate both proximally and distally?
14. How many metacarpal and phalangeal bones are there?
15. How many ribs do you have? Are all of your ribs alike? If not, how do they differ?
16. What is meant by the sternal angle? Is it important?
17. What constitutes a pelvic bone? Name its parts and essential features. Do male and female pelvic bones differ? How?
18. List the bones of the lower extremity and give several features of each.
19. What makes up the ankles?
20. What bones of the foot form the medial and lateral plantar arches?

References

Fourman, P., and Royer, P.: *Calcium Metabolism and Bone.* F. A. Davis Co., Philadelphia, 1968.

Goss, C. M. (ed.): *Gray's Anatomy of the Human Body*, 29th ed. Lea & Febiger, Philadelphia, 1973.

Grant, J. C. B.: *Grant's Atlas of Anatomy*, 6th ed. Williams & Wilkins Co., Baltimore, 1972.

Johnson, L. C.: Kinetics of Skeletal Remodeling. In Bergsma, D., and Milch, R. A. (eds.): *Structural Organization of the Skeletal System.* Birth Defects Original Article Series, Vol. II, No. 1. National Foundation–March of Dimes, New York, 1966, pp. 66–142.

Pansky, B., and House, E. L.: *Review of Gross Anatomy*, 2nd ed. Macmillan Publishing Co., Inc., New York, 1969.

Woodburne, R. T.: *Essentials of Human Anatomy*, 4th ed. Oxford University Press, New York, 1969.

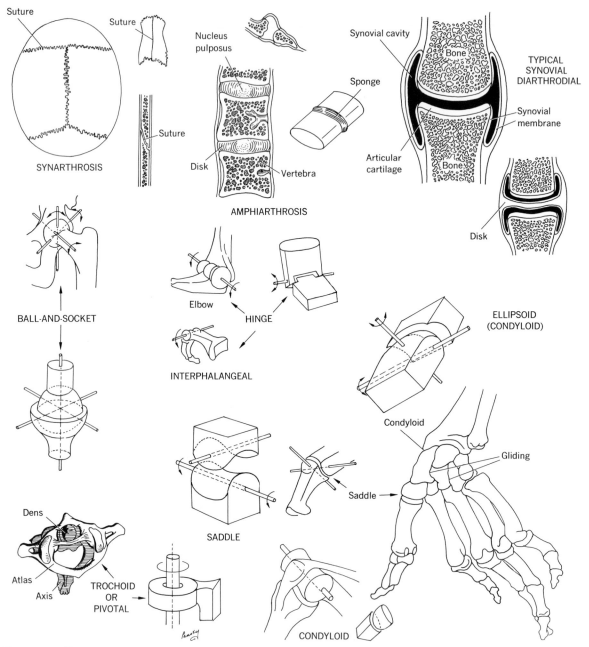

Figure 6-1. The various types of joints one sees in the body and how their shapes govern their classification.

The Body in Motion: Joints

A joint or articulation is the union between two or more adjacent bones. The surface of a bone that meets another at a joint is its articular surface and, when it is freely movable, a space is present between them. The articular surfaces are covered by an adherent, smooth, well-lubricated articular cartilage. Where less movement is needed, the joint is made by some form of connective tissue, such as fibrous tissue or fibrocartilage, and the union may even be completed by fusion of the bones so that no movement takes place. Joint movements are controlled and produced mainly by muscles attached to the bones; yet the bones are held together by all the tissues around them. Usually, fibrous tissue runs from bone to bone as thickened bands or ligaments. These bands and other soft tissues transmit tensile stresses and maintain the continuity of the skeleton. Compression stresses require a solid structure, and thus the ends of bones are enlarged and modified in shape to provide bearing surfaces and distribute the pressure.

Classification of Joints

Joints (Fig. 6-1) are of three major types: (1) synarthroses (immovable), (2) amphiarthroses (slightly movable), and (3) diarthroses (synovial, freely movable).

Synarthroses

Synarthrodial joints are joined and held together by cartilage or fibrous tissue, and the surfaces of the bones are in almost direct contact with each other, allowing for no appreciable motion. The joints between bones of the skull are examples of synarthroses.

Amphiarthroses

Amphiarthrodial joints are united by fibrocartilaginous disks (seen in the articulations between vertebral bodies) or are joined by a fibrous interosseous ligament (seen in the inferior tibiofibular or radioulnar articulations).

Diarthroses

Diarthroses are freely movable, synovial-lined joints in which a fluid-containing cavity separates the two opposing bones (seen in the knee, elbow, and shoulder joints). They are the most common body joints and are the joints most affected by rheumatism, arthritis, and other pathology. The articulating ends of the bones forming a diarthrodial joint are covered by hyaline cartilage and are enclosed in a capsule of dense fibrous tissue. The joint capsule is strengthened by strong ligaments extending between the bones of the joint and the muscles crossing it. The articular cartilage is normally resilient and acts as a cushion between the bones, the smooth surface allowing for ease of movement. A synovial membrane lines the inner surface of the fibrous capsule, forms the inner lining of the joint, and is attached at the margins of the articular cartilage. Under normal conditions only a small amount of synovial fluid is found in the confines of the synovial membrane (synovial cavity) to provide lubrication of the inner joint surfaces. The amount of fluid varies from joint to joint, from a thin viscous layer of fluid to about 3.5 ml of free fluid.

Diarthrodial joints are further classified according to the shapes of the articulating surfaces, which determine the type and extent of motion in the joint. Seven types are described: plane, spheroidal, ellipsoid (cotylic), hinge, condylar, trochoid or pivot, and sellar.

Plane Joints. Plane joints allow only gliding movements. This is the simplest type of joint movement and consists of one surface moving over another without any angular or rotary movement. The articular surfaces are usually flat or plane. The movement is limited by bony processes and ligaments around the articulation. Examples are the carpal joints (except those of the capitate with the navicular and lunate) and the tarsal joints (except between talus and navicular).

Spheroidal Joints. Spheroidal joints are ball-and-socket joints formed by the articulation of a rounded convex surface with a cuplike cavity. There is a wide range of movements possible, including flexion, extension, abduction, adduction, rotation, and circumduction. Examples are the shoulder and hip joints.

Ellipsoid Joints. Ellipsoid (cotylic) joints are similar to spheroidal except that the articular surfaces are elliptical rather than circular. They are more restricted in motion than spheroidal and do not permit axial rotation (but permit other movements). These joints are subdivided into (1) simple ellipsoid, where the articular capsule encloses only one pair of articulating surfaces, and (2) compound ellipsoid, where more than one pair of surfaces are enclosed. Examples are the metacarpophalangeal joints of the hand, which are of the simple type, and the radiocarpal (wrist) joint, which is of the compound type.

Hinge Joints. Hinge joints (ginglimus) permit motion in only one plane (flexion and extension). Strong collateral ligaments restrict lateral movement; yet the extent of flexion and extension can be considerable. The range of flexion usually exceeds that of extension. Examples are the interphalangeal joints of the hand and foot and the humeroulnar (elbow) joint.

Condylar Joints. One bone articulates with the other by way of two distinct articular surfaces, and each articular surface is referred to as a condyle. There is a condyloidlike ball and socket with no rotation. These joints resemble hinge joints but allow several kinds of movement. The condyles may be close together and enclosed in the same articular capsule (as in the knee joint) or widely separated and enclosed in separate articular capsules (as in the temporomandibular joints). Examples are the knee joint and the temporomandibular joints.

Trochoid or Pivotal Joints. The movement here is limited to rotation. The joint is formed by a bony pivotlike process turning in a ring or vice versa. Examples are the proximal radioulnar joint (the radial head rotates in a ring formed by the radial notch of the ulna and the annular ligament) and the atlantoaxial joint.

Sellar Joints. Sellar joints have saddle-shaped articular processes in which a convex surface articulates with a concave surface to allow flexion, extension, abduction, adduction, and circumduction but no axial rotation. An example is the carpometacarpal joint of the thumb.

Range of Motion

The range as well as the type of motion in a normal joint depends on the shape of the articular surfaces, the restraining effects of the supporting ligaments, and the control exerted by muscles acting on the joint. Limitation of motion may be a manifestation of articular disease. One should be aware of the normal type and range of motion in order to detect limitations due to abnormalities of the joint or adjacent structures.

Limitation may occur on either active or passive motion. In the former, motion is restricted when voluntary movement is attempted, and in the latter, it is limited when the examiner attempts to move the part with the muscles relaxed. When the two are not equal, the passive range is usually greater and a more reliable indication of actual range of movement, although one must observe the active range initially. A patient may restrict range of passive motion out of fear of being hurt or may not be able to move a joint fully due to pain from one tissue or another (e.g., swollen bursae, nodular tendon sheaths, torn tendons, etc.). Normally both active and passive range of motion should be approximately equal.

Limitation of motion may be transient (reversible) or permanent. The former may be due to muscle spasm from fear of pain, periarticular fibrous "gelling" that improves with movement, intra-articular fluid and inflammation "locking" secondary to loose bodies in the joint due to disorders of a cartilage or malposition of tendons, or fibrous proliferation that produces adhesions, inflammations, or contractures. Permanent limitation may be due to both intra- and extra-articular causes: ankylosis (stiffening of a joint), destruction of articular surfaces, bony spurs, or contractures.

Pathology (Brief Summary)

Almost all diseases affecting the articulations usually involve the synovial or diarthrodial joints. Few diseases affect the symphyses or fibrocartilaginous joints.

Arthritis

Arthritis is a general term used to describe disease "within" the joint itself, affecting the subchondral bone, articular cartilage, or synovial membrane. Inflammation of the membrane is usually synonymous with arthritis even though other parts may be involved. The most common signs are swelling, tenderness, and limitation of movement. The two most common types are rheumatoid and osteoarthritis.

Crepitation

Crepitation is palpable or audible grating or crunching sensation produced by motion (may not be accompanied by discomfort). It can be due to roughened articular or extra-articular surfaces and rubbing or slipping of ligaments and tendons over bony surfaces in motion.

Deformity

Deformity occurs as a bony enlargement, articular displacement, contracture, and ankylosis (stiffening).

Specific Joints of Body

Temporomandibular Joint (Condylar)

Anatomy. The temporomandibular joint (Fig. 6-2) is formed by the condyle of the mandible and the articular tubercle, mandibular fossa, and postglenoid tubercle of the temporal bone. An articular disk of fibrocartilage divides the joint into two cavities, each lined with synovial membrane. The synovial membrane is covered by a loose fibrous capsule strengthened laterally by the temporomandibular liga-

Figure 6-2. Temporomandibular joint.

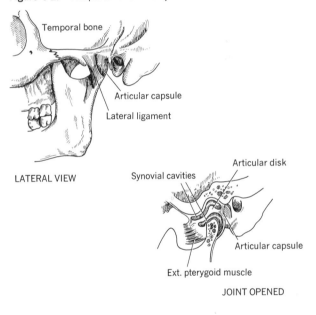

LATERAL VIEW

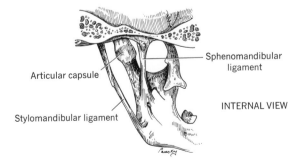

ment. It is located by placing the tip of the forefinger just in front of the external auditory meatus and opening the mouth.

Movement and Range of Motion. Three types of motion occur: opening and closing of jaws, protrusion and retrusion of mandible (anterior and posterior motion), and lateral (side-to-side) motion.

Movement at each joint has two components: (1) an inferior part, between the mandibular condyle and the articular disk that acts as a hinge joint; and (2) a superior part, between the temporal bone and the disk that acts as a sliding joint, allowing both the disk and mandible to glide forward, backward, and from side to side. Anteroposterior movements are mainly by a

gliding action of the superior compartments and lateral displacement of the jaw, causing one disk to glide forward while the other remains in position. The grinding or chewing is produced by alternate movements in both compartments.

Sternoclavicular Joint (Simple Spheroidal)

Anatomy. At the sternoclavicular joint (Fig. 6-3) the medial end of the clavicle articulates with the upper end of the manubrium of the sternum and with the first costal cartilage on each side. A fibrocartilaginous disk separates each joint into two separate cavities, both lined with synovial membranes. A fibrous tissue capsule surrounds each entire joint and is strengthened

Figure 6-3. Shoulder joint and sternoclavicular joint.

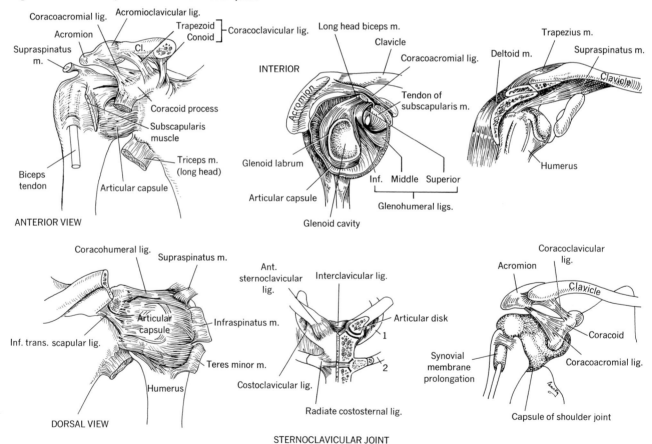

by anterior, posterior, and interclavicular ligaments. The joint lies just beneath the skin.

Movement and Range of Motion. The joints are the only points of articulation of the shoulder girdle with the trunk. Motion of the shoulder girdle results in a motion of this joint.

Acromioclavicular Joint (Simple Spheroidal)

Anatomy. The acromioclavicular joint (Fig. 6-3) is formed by the lateral end of the clavicle and the medial margin of the acromion of the scapula. The joint is enveloped by a fibrous capsule strengthened by superior and inferior acromioclavicular ligaments. A subcutaneous bursa is often located superficially over the joint but rarely connects with the cavity. Although this joint lies near the surface, it is difficult to visualize owing to its close proximity to the prominence of the shoulder.

Movement and Range of Motion. This joint permits the scapula to move vertically when the shoulder girdle is raised (shrugging shoulders) and falls. It enables the scapula to rotate backward and forward on the clavicle. When the arms are raised over the head, it participates in the scapular movement and elevation of the shoulder. Scapular movement accounts for most of the vertical movement in raising the arm over the head.

Shoulder Joint (Glenohumeral, Spheroidal)

Anatomy. The shoulder joint (Fig. 6-3) is formed by the articulation of the head of the humerus with the shallow glenoid fossa of the scapula. The joint permits considerable movement of the arm and is enclosed by powerful muscles and tendons to strengthen it. It is protected above by an arch formed by the coracoid process, the acromion, the coracoacromial ligament, and the clavicle. The shoulder girdle is formed by the scapula and clavicle, which articulate at the acromio-clavicular joint.

The fibrous capsule surrounding the glenohumeral joint is attached to the circumference of the glenoid fossa beyond the fibrocartilaginous rim of the glenoid labrum and to the anatomic neck of the humerus. The capsule is loose and allows the bones of the joint to separate from each other, permitting freedom of motion. There are two openings in the capsule: one allows the long tendon of the biceps muscle to enter the bicipital groove of the humerus; the other, under the subscapular tendon, permits an extracapsular pouching of the synovial membrane to act as a "subscapular bursa."

The synovial membrane is located beneath the fibrous capsule, tendons, muscles, and bursae and lines the inner capsular surface. It has two outpouchings: (1) the subscapular portion, which is extracapsular and acts as a bursa for the subscapularis muscle; and (2) a bicipital part, which extends along the bicipital groove to function as a sheath for the tendon of the long head of the biceps.

The rotator cuff of the shoulder consists of four muscle tendons: the supraspinatus, infraspinatus, and teres minor muscles, all inserting into the greater tuberosity of the humerus; and the tendon of the subscapularis muscle, inserting into the lesser tuberosity (Fig. 6-3). Before inserting, each of the tendons is incorporated in a fibrous capsule that encloses the joint and reinforces the articular capsule.

Overlying the tendinous and capsular cuff is the subacromial bursa. A lateral extension of it is called the "subdeltoid" bursa under the deltoid muscle. This bursa facilitates movement of the greater tuberosity under the acromion in abduction of the arm. The subcoracoid bursa lies between the shoulder capsule and the coracoid process and may communicate with the subacromial bursa. There may be several others about the shoulder. Inflammation or irritation of the bursae (usually subacromial) results in palpable swelling, tenderness, and warmth of the upper part of the arm and is referred to as bursitis.

Movement and Range of Motion. Motion of the upper extremity is a combination of shoulder-girdle and shoulder-joint movement. The shoulder joint permits wide movements: flexion (forward arm movement), extension (backward arm movement), abduction (elevation of arm from side), adduction (lowering arm to side), rotation, and circumduction. The shoulder girdle (as a unit) movements include elevation (shrugging or raising), depression (downward movement), protrusion (moving girdle anterior), retraction (moving girdle posterior), and circumduction.

Arm elevation from the side over the head takes place by the combined movement of the shoulder joint and rotation of the scapula on the chest wall associated with motion of the sternoclavicular and acromio-clavicular joints.

Relatively little external rotation of the humerus is needed when the arm is elevated over the head by forward flexion, yet external rotation of about 180 degrees occurs when elevation is performed by abduction. Thus, lesions of the rotating musculature or cuff cause pain, spasm, or limiting motion on abduction while range of flexion is relatively normal. Motions most valuable to evaluate the function of the gleno-humeral joint are internal and external rotation and abduction. The normal range is about 90 degrees with abduction and stabilization of the scapula.

Elbow Joint (Hinge Joint)

Anatomy. The elbow joint (Fig. 6-4) is formed by the humeroulnar, radiohumeral, and proximal radio-ulnar articulations. The major one is the humeroulnar. All bony junctions are enclosed in an articular capsule in a common synovial articular cavity, making the synovial membrane extensive. Three fat masses lie between the synovial membrane and the overlying capsule: over the olecranon fossa, coronoid fossa, and radial fossa. The articular capsule is thickened laterally by the radial collateral and medially by the ulnar collateral ligaments.

One large bursa, the olecranon, and several small ones are found about the joint. None normally communicates with the joint cavity. Swelling and redness of the olecranon bursa are easily seen because of the closeness of the bursa to the skin. The elbow is a common site of synovitis, which is associated with limitation of extension of the joint. The medial and lateral epicondyles of the humerus and the head of the radius and the tendinous attachment of muscles to these areas are sites of inflammation, pain, or localized tenderness ("tennis elbow").

Movement and Range of Motion. From full extension at 0 degrees, flexion of the joint (angles between

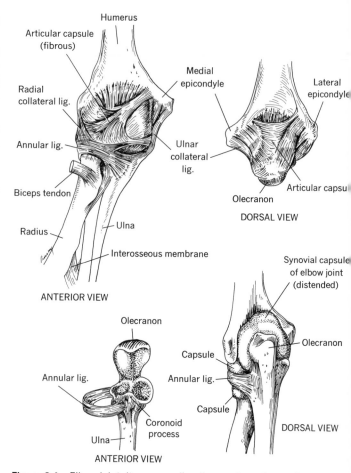

Figure 6-4. Elbow joint, its surrounding ligaments and capsule.

arm and forearm narrowed) is 150 to 160 degrees. The movement of flexion and extension occurs in the humeroulnar and radiohumeral joints. Pronation (palm down) and supination (palm up) of the forearm and hand require motion of both radioulnar joints as well as the radiohumeral joint, and they allow about 180 degrees of movement (90 degrees of either pronation or supination from a midway position between the two).

Wrist and Carpal Joints

Anatomy. WRIST (RADIOCARPAL) JOINT. The wrist joint (Fig. 6-5) is formed by the distal end of the radius

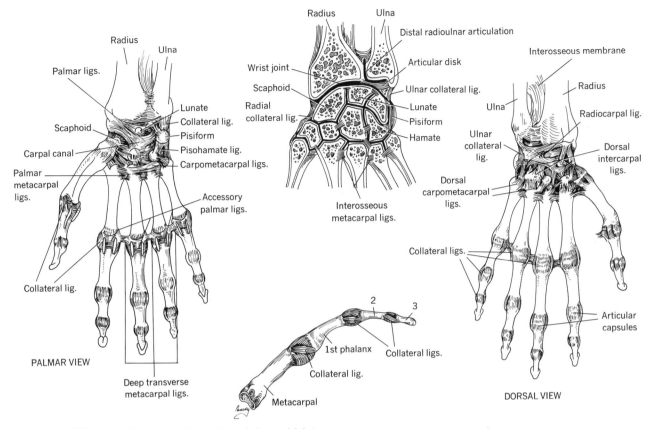

Figure 6-5. Wrist, carpal, metacarpal, and interphalangeal joints.

(and the articular disk) and by a row of carpal bones, namely, the scaphoid, lunate, and triquetrum. The disk joins the radius to the ulna and separates the ulna from the wrist joint proper. The joint is surrounded by a capsule (lined by a synovial membrane) and is supported by ligaments.

DISTAL RADIOULNAR JOINT. The distal radioulnar joint (Fig. 6-5) is near the wrist joint but not a part of it. The synovial membrane is loose and lines the deep surface of the capsule and internal ligaments.

MIDCARPAL JOINT. The midcarpal joint (Fig. 6-5) is formed by junction of the proximal and distal rows of carpal bones. It permits some flexion, extension, and very slight rotation. The midcarpal and carpometacarpal articular cavities often communicate.

Movement and Range of Motion. Wrist movement includes palmar flexion (flexion), dorsiflexion (extension), radial and ulnar deviation (abduction and adduction), and circumduction, which is a combination of the four movements. Pronation and supination of the hand and forearm occur at the proximal and distal radioulnar joints. The carpometacarpal joints move very little, except at the thumb, which permits flexion, extension, adduction, abduction, and medial and lateral rotation. In general, the range of wrist motion varies among individuals. The wrist usually can be dorsiflexed 70 degrees and palmar-flexed 80 to 90 degrees from a straight position. Ulnar deviation averages 50 to 60 degrees, while radial deviation allows only about 20 degrees. The most common loss or limitation (for many reasons) is dorsiflexion.

Metacarpophalangeal and Interphalangeal Joints (Hinge)

Anatomy. METACARPOPHALANGEAL JOINTS. The metacarpophalangeal joints (Fig. 6-5) are joints of flexion and extension with fibrous or fibrocartilaginous ligaments over the palmar surface (volar plate) that are reinforced by collateral ligaments on each side that become tight in flexion and loose in extension and prevent lateral motion. An extensor tendon crosses the dorsum of each joint to strengthen the thin articular capsule here.

INTERPHALANGEAL JOINTS. At the interphalangeal joints (Fig. 6-5) movement is restricted to flexion and extension. Each joint has a thin dorsal capsular ligament strengthened by the extensor hood and a dense palmar ligament (volar plate) as well as collateral ligaments on each side. The palmar and collateral ligaments prevent hypertension of the proximal and distal interphalangeal joints.

Movement and Range of Motion. A simple measure of overall function of the fingers is the ability to make a complete fist and extend the fingers fully. A normal complete fist produced by flexion of all fingers is a 100 percent fist, and a flat hand with no ability to flex fingers is a 0 percent fist. The lack of full extension of the fingers is described in degrees of full extension.

In range-of-motion observations, each digital joint should normally contribute to the total range seen. When range of motion is limited and voluntary or active motion equals passive motion, the limitation is attributed to involvement of the joint, to tightening of the articular capsule and periarticular tissues by distention or fibrosis, or to fixed muscle contractures. The metacarpophalangeal joints of the fingers flex about 90 to 100 degrees, while the thumb metacarpophalangeal joint flexes only about 50 degrees. The proximal interphalangeal joints flex 100 to 120 degrees, whereas the distal ones flex only 45 to 90 degrees. Each metacarpophalangeal joint may hyperextend as much as 30 degrees; the proximal interphalangeal joint rarely hyperextends more than 10 degrees, whereas the distal one hyperextends as much as 30 degrees. The interphalangeal joint of the thumb hyperextends 20 to 35 degrees and flexes to 80 to 90 degrees. Each finger can abduct (spread) and adduct (move toward third or middle finger) when the metacarpophalangeal joint is extended. The range of abduction-adduction at the metacarpophalangeal joint is about 30 to 40 degrees but may vary. Abduction at the thumb carpometacarpal joint measured at right angles to the plane of the palm is about 70 degrees.

Spinal Column

Anatomy. The structure of the spinal column (Fig. 6-6) allows flexibility of the trunk and helps retain the upright posture by means of coordinated action of bones, muscles, and ligaments. The shape of the column and resilient structure of the intervertebral disks help absorb much shock. A normal balanced spine can be maintained with less muscular effort than one that is unbalanced.

The vertebral arch supports articular processes originating from the junction of laminae and pedicles. There are four per vertebra, one from each side projecting upward and downward to form diarthrodial joints between vertebrae (apophysial joints or articular facets) and having articular cartilages, thin capsules, and synovial membranes. The angle of the articular surfaces in relation to the horizontal plane varies at different levels and determines the type and extent of the movement in that part of the column.

The bodies articulate by means of a fibrocartilaginous intervertebral disk and thin cartilaginous plates that cover the surfaces of the vertebral bodies with the plates between body and disk. The superficial layers of the disk are of tough fibers concentrically arranged, the annulus fibrosis, while the center or nucleus pulposus (relic of the notochord) is a soft, elastic semifluid tissue mass lying more posterior than central. The nucleus allows the vertebrae to rock about while the annulus prevents body displacement. The elasticity of the disk allows compression and compensatory expansion of various areas of the disk as well as upward, downward, and rotary motion between bodies. Movement is greatest in the cervical and lumbar areas (disk thickest here) and least in the thoracic region (disk thinner). The disks act as "shock absorbers," distribute body weight, and prevent weight concentration on any

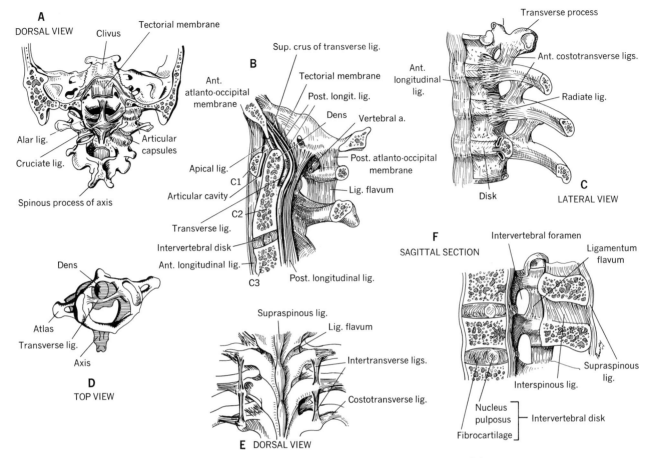

Figure 6-6. Ligaments of the vertebral column. **A, B,** and **D,** at the atlanto-occipital and atlantoaxial joints. **C** and **E,** in the thoracic region. **F,** in the lumbar region.

edge. The disks are thicker ventrally than dorsally in the cervical and lumbar area contributing to the curvatures. The disks make up about one fourth of the total length of the column above the sacrum. When normal stress and strain is placed on the column, the disks bulge in all directions but return to where they were. They are held in place predominantly by the anterior and posterior longitudinal ligaments. If the stress is too great, the disk proper or its nucleus may be extruded beyond normal limits and fail to return to its usual position. Posterior bulging of the disk may cause nerve root pain due to pressure on the spinal nerve. A common area affected is the lower lumbar region, and the symptoms are associated with sciatic nerve pain. After the age of 60, the disks begin to atrophy, giving rise to the "lower back" of old age.

The atlanto-occipital joints permit flexion and extension and slight lateral bending but little rotation.

The axis (C2) articulates with the atlas through paired lateral joints and a joint formed between the dens and the ring of the anterior arch and transverse ligament of the atlas (medial atlantoaxial joint). The latter is a pivot or trochoid joint with two synovial cavities between the dens and arch, and the dens and

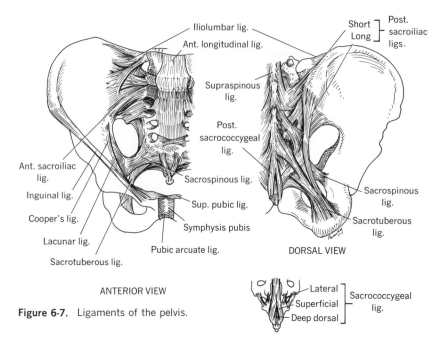

Figure 6-7. Ligaments of the pelvis.

Labels in figure:
Iliolumbar lig.
Ant. longitudinal lig.
Short / Long — Post. sacroiliac ligs.
Supraspinous lig.
Post. sacrococcygeal lig.
Ant. sacroiliac lig.
Inguinal lig.
Cooper's lig.
Lacunar lig.
Sacrotuberous lig.
Sacrospinous lig.
Sup. pubic lig.
Symphysis pubis
Pubic arcuate lig.
Sacrospinous lig.
Sacrotuberous lig.
ANTERIOR VIEW
DORSAL VIEW
Lateral / Superficial / Deep dorsal — Sacrococcygeal lig.

ligament. Each lateral joint has an articular capsule and a synovial membrane. Rotation takes place between the skull and the atlas on the dens.

The lumbosacral region (Fig. 6-7) is a point of junction between movable and nonmovable spinal column. There is a sharp anterior-posterior angulation at the lumbosacral junction, resulting in much leverage exerted in this area by the entire column above the sacrum. A tendency for the fifth lumbar vertebra to slip on the first sacral is countered by the posterior apposition of articular processes between L5 and S1. The disk here is thicker (especially anterior) and allows for extra compression and greater motion. This area is also a common site of congenital anomalies and disk abnormality.

The sacroiliac joints (Fig. 6-7) transmit the body weight to the bony pelvis and lower extremities. The upper sacrum is wider than the lower and extends farther forward as well. The joints have a capsule and a synovial membrane. A series of strong, short intra-articular fibers (interosseous sacroiliac ligament) connect the sacral tuberosities and ilium, filling the narrow space between the bones posteriorly. Strong extra-

capsular ligaments stabilize the joints and resist the tendency of the upper sacrum to rotate forward and the lower end backward. Thus, very little motion occurs here.

The vertebrae are bound by anterior and posterior longitudinal ligaments from the sacrum to the base of the skull. This increases the stability of the column. The anterior is stronger and limits extension of the column. The space between laminae of two adjacent vertebrae is filled by the ligamentum flavum (elastic tissue) that helps to restore the column to its erect position after bending. Supraspinous and interspinous ligaments between spinous processes limit flexion and lateral bending of the column.

Curvature of the spine to one side or the other is called scoliosis. A "hump" or gibbus (prominence) in the thoracic region is due to scoliosis and vertebral rotation on the side of the convexity. Posterior curvature is called kyphosis, and anterior curvature is known as lordosis.

Movement and Range of Motion. The column permits extension (back bending), flexion (forward bending), abduction (lateral bending), and rotation. Each

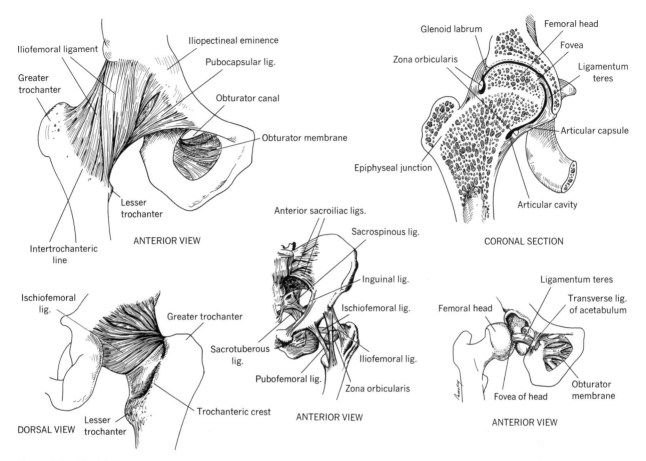

Figure 6-8. Hip joint.

varies in different regions. Movement is greatest in the cervical region, restricted in the thoracic and lumbar, and not even present in the sacral. Mobility depends on the elasticity and thickness of the intervertebral disks, interarticular facets, and limitations created by the ligaments. Forward bending usually includes flexion of the hips and column. There is a tendency for spinal flexibility to decrease with age.

The column as a whole flexes to about 90 degrees and involves the cervical, thoracic, and lumbar regions. Extension involves mainly the cervical and lumbar regions.

To maintain balance, the pelvis shifts in the opposite direction as the trunk moves forward or backward.

The column bends about 60 degrees to either side. Rotation is marked in the cervical area and restricted in others. If no neck motion occurs and the pelvis is stabilized, the column rotates about 30 degrees to each side.

Hip Joint (Enarthrodial, Spheroidal) (Ball-and-Socket)

Anatomy. The hip joint (Fig. 6-8) is formed by the articulation of the head of the femur with the acetabulum of the hip bone. It bears weight and has a wide range of motion. It has greater strength and stability, yet less mobility, than the shoulder joint due to the strong fibrous capsule and powerful muscles.

The acetabular cavity is strongest and deepest superiorly and posteriorly. A fat mass lies in the bottom of the acetabular fossa. The cavity is deepened by a circular fibrocartilaginous rim, the glenoid labrum, which forms a tight collar around the head of the femur and stabilizes the head. Inferiorly, the labrum is incomplete, forming a notch that is bridged by a transverse ligament, which converts the notch into a foramen to allow blood vessels to enter the joint.

The articular capsule is a dense, strong capsule attached to the edge of the acetabulum, glenoid labrum, and transverse ligament proximally and to the intertrochanteric line and neck of the femur distally. Thus, the anterior and medial half of the posterior surface of the femoral neck are intracapsular. The capsule is strong and thick over the upper and anterior part of the joint but is thinner and weak over the lower and posterior portions. The ligaments of the capsule include the iliofemoral, which is the strongest, crosses the joint anteriorly, and extends from the ilium to the neck and intertrochanteric line of the femur. The iliofemoral ligament is Y shaped. It relaxes in flexion of the thigh, tightens in extension, and prevents excessive hyperextension of the hip. In the upright position, it keeps the pelvis from rolling back on the femoral head and stabilizes the hip. The pubofemoral and ischiofemoral ligaments are weaker but reinforce the capsule anteriorly and posteriorly, passing from the pubic and ischial parts of the acetabular rim to the capsule attachments. The ligamentum teres is intracapsular and loosely attaches the femoral head to the lower part of the acetabulum and adjacent ligaments. It has no effect on normal motion or stability but is a channel for blood vessels to reach the femoral head. The iliotibial tract is a part of the fascia of the thigh that covers the joint laterally.

Synovial membrane lines the deep surface of the capsule. It covers the glenoid labrum (the mass of fat in the acetabular fossa), encloses the ligamentum teres in a sheath of synovial tissue, and is attached distally to the femur.

Several bursae are found here. The iliopectineal bursa lies between the iliopsoas muscle and the anterior surface of the joint. It communicates with the cavity in about 15 percent of cases. The trochanteric bursa is found between the gluteus maximus muscle, separating it from the ischial tuberosity.

Movement and Range of Motion. In normal gait the abductor muscles of the weight-bearing extremity contract and hold either both sides of the pelvis level or the side not bearing the weight slightly raised. A limp or gait abnormality may be associated with hip disease. The hip has a wide range of motion, allowing flexion, extension, abduction, adduction, circumduction, and rotation. The angulation of the neck and shaft of the femur converts the angular movements of flexion, extension, abduction, and adduction into rotary movements of the head in the acetabulum. With hip flexion and abduction, loss of stability occurs in the extended position since then only a part of the head is covered by the acetabulum and the rest only by the capsule.

Hyperextension of the hip is about 150 degrees when leg and thigh are straight and pelvis and spine are immobilized. With movement of the latter structures, extension is increased to 40 degrees. The greatest degree of flexion is possible when the knee is flexed and then the thigh can be flexed to 120 degrees from a neutral (extended) position (if the knee has been flexed to 90 degrees first). If the knee remains extended, hamstring muscle tension limits flexion of the hip to about 90 degrees. The amount of abduction increases with flexion and decreases with extension of the hip. Normally, with the leg and thigh extended, the hip abducts about 40 to 45 degrees from neutral. Adduction with the leg straight is limited by the legs and thighs coming in contact with each other; however, with flexion of the hip to permit leg crossing, one can adduct about 20 to 30 degrees. The hip normally rotates inwardly about 40 degrees and outwardly about 45 degrees, but this can vary. Rotation increases with flexion and decreases with extension. Limitation of internal rotation of the hip is an early sign of hip disease.

Knee Joint (Compound Condylar)
Anatomy. The knee joint (Fig. 6-9) is the largest joint in the body, formed by three articulations that

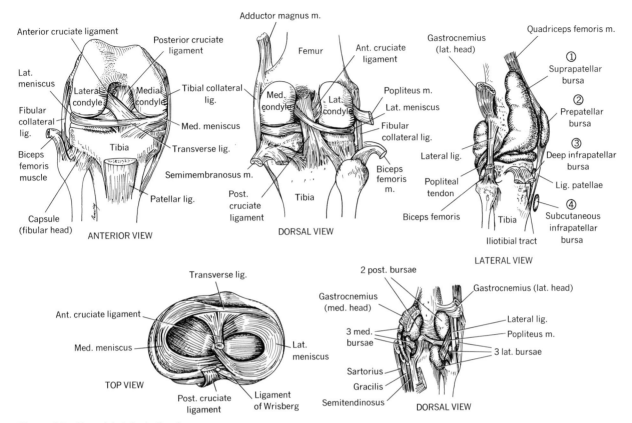

Figure 6-9. Knee joint, including bursae.

have only one articular cavity: between the lateral and medial condyles of the femur and tibia with their corresponding menisci and between the patella and femur. The fibula is not a part of this joint.

The bones are stabilized by the articular capsule, ligamentum patellae, medial (tibial) and lateral (fibular) collateral ligaments, and anterior and posterior cruciate ligaments. The medial and lateral menisci are fibrocartilaginous disks found in the knee joint interposed between the femoral and tibial condyles.

ARTICULAR CAPSULE. The articular capsule is a thick fibrous membrane strengthened by the fascia lata, tendons, and ligaments around the joint. Anteriorly and superiorly beneath the quadriceps tendon, it does not cover the synovial membrane. Posteriorly, fibers of the capsule enclose the membrane of the suprapatellar

pouch and the capsule here consists of fibers from the condyles and sides of the intercondylar fossa of the femur. Thus, the capsule lies on the side of and anterior to the cruciate ligaments, which in turn are outside of the joint cavity. Distally, the capsule attaches to the borders of the menisci and continues to the tibial condyles.

LIGAMENTS. The ligamentum patellae is an extension of the common tendon of the quadriceps muscle. A triangular fat pad, the infrapatellar pad, lies below the patella and between the ligament and the synovial membrane. The collateral ligaments lend medial and lateral support. The lateral is strong, round, and fibrous and is attached to the lateral femoral condyle and lateral side of the fibular head. The medial is broad, flat, and membranous and is attached to the medial

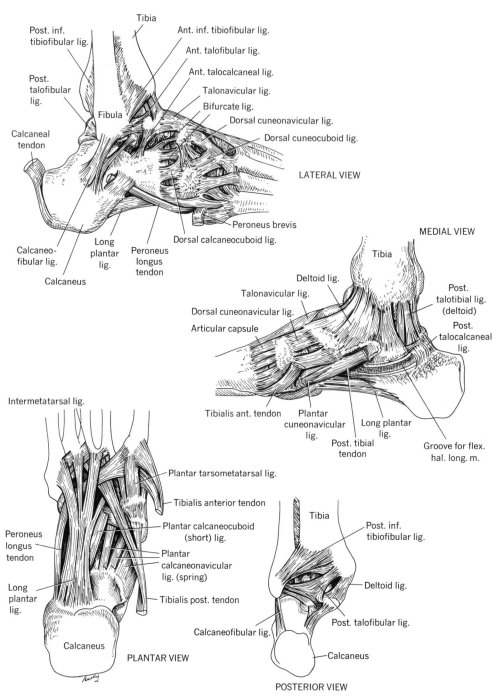

Figure 6-10. Ligaments of the ankle and foot.

The labels in the figure:

Tibia
Post. inf. tibiofibular lig.
Ant. inf. tibiofibular lig.
Ant. talofibular lig.
Ant. talocalcaneal lig.
Post. talofibular lig.
Talonavicular lig.
Bifurcate lig.
Fibula
Dorsal cuneonavicular lig.
Dorsal cuneocuboid lig.
Calcaneal tendon
LATERAL VIEW
Peroneus brevis
Calcaneo-fibular lig.
Long plantar lig.
Peroneus longus tendon
Dorsal calcaneocuboid lig.
Calcaneus

MEDIAL VIEW
Tibia
Deltoid lig.
Talonavicular lig.
Post. talotibial lig. (deltoid)
Dorsal cuneonavicular lig.
Post. talocalcaneal lig.
Articular capsule
Tibialis ant. tendon
Plantar cuneonavicular lig.
Long plantar lig.
Post. tibial tendon
Groove for flex. hal. long. m.

Intermetatarsal lig.
Plantar tarsometatarsal lig.
Tibialis anterior tendon
Peroneus longus tendon
Plantar calcaneocuboid (short) lig.
Plantar calcaneonavicular lig. (spring)
Long plantar lig.
Tibialis post. tendon
Calcaneus
PLANTAR VIEW

Tibia
Post. inf. tibiofibular lig.
Deltoid lig.
Post. talofibular lig.
Calcaneofibular lig.
Calcaneus
POSTERIOR VIEW

femoral condyle and to the medial surface of the tibia and tibial condyle. The two cruciate ligaments give anteroposterior support. The anterior (medial) is attached anteriorly to the intercondyloid tibial eminence and posteriorly and superiorly to the lateral femoral condyle. The posterior (lateral) ligament is attached to the posterior intercondyloid tibial fossa and to the posterior part of the lateral meniscus and attaches to the medial femoral condyle. They are both outside the synovial cavity. The medial and lateral menisci (disks) are wedge-shaped, crescentic, fibrocartilaginous disks. The outside edge of each is thick and attached to the capsule, whereas the inside edge is thin and unattached. They rest on the head of the tibia and articulate with the surface of the femoral condyles.

SYNOVIAL MEMBRANE AND BURSAE. The synovial membrane is the largest in the body. At the superior patellar border it forms a sac or pouch beneath the quadriceps femoris muscle. On each side, the membrane extends under the aponeuroses of the two vastus muscles.

The knee has many bursae: (1) a large prepatellar bursa, separating skin from patella; (2) a small superficial infrapatellar bursa, between skin and ligamentum patellae; (3) a deep infrapatellar bursa, beneath the ligamentum patellae; (4) a subpopliteal bursa, between the popliteal muscle and the lateral condyle of femur; (5) two gastrocnemius bursae, one between the lateral head of the muscle and articular capsule and one between the medial head of the muscle and capsule; (6) a large semimembranosus bursa, between that muscle and the medial head of the gastrocnemius; and (7) a bursa between the medial collateral ligament and tendons of the sartorius, gracilis, and semimembranosus muscles. Synovitis is seen as distention and swelling of the suprapatellar pouch. "Housemaid's knee" is a swelling of the prepatellar bursa (prepatellar bursitis).

Movement and Range of Motion. The knee normally extends to a straight line (0 degrees or 180 degrees) and may be hyperextended up to 15 degrees. The angle of flexion ranges from 135 to 150 degrees. The knee may "lock" or become fixed in partial extension while flexion may be unrestricted. Instability involves the collateral ligaments more than cruciates.

Normally there is practically no adduction or abduction of the leg on the femur.

Bowlegs or lateral angulation is called genu varum, whereas knock-knees or medial angulation is called genu valgum.

Ankle and Foot

Anatomy. ANKLE JOINT. The ankle joint (Fig. 6-10) is a true hinge joint with movements of plantar flexion and dorsiflexion and is formed by the distal ends of the tibia and fibula and proximal aspect of the body of the talus. The tibia is the weight-bearing part of the joint, and the fibula gives lateral stability. The fibula does not bear weight. The tibial (medial) and fibular (lateral) malleoli extend down beyond the roof or tibial portion of the joint and envelop the talus in a mortiselike fashion.

The capsule of the ankle joint is weak and lax anterior and posterior but is bound down by ligaments on both sides. It extends from the tibia to the neck of the talus. It is lined by synovial membrane. The cavity does not articulate with any other joints, bursae, or sheaths.

The medial or deltoid ligament is the only ligament on the medial side of the ankle and is a strong, triangular-shaped fibrous band that resists eversion of the foot and may be torn in eversion sprains. The lateral ligaments are three distinct bands: the posterior and anterior talofibular and the calcaneofibular. These ligaments are torn in inversion sprains.

All tendons crossing the ankle joint lie superficial to the articular capsule and are enclosed in synovial sheaths about 8 cm long.

INTERTARSAL AND SUBTARSAL JOINTS OF THE FOOT. *Intertarsal Joints.* The intertarsal joints (Fig. 6-10) provide increased mobility to the foot since ankle motion is limited to flexion and extension. Here the foot can be inverted and adducted (supinated) or everted and abducted (pronated). Since the foot is arched, body weight is transmitted to the head of the metatarsals and posteriorly to the calcaneus (both in contact with the ground). If the arch is to be preserved during weight bearing, the intertarsal joints must be braced and thus unusually strong intertarsal ligaments bind the tarsal bones and prevent collapse of the arch. The plantar aponeurosis (extending from the calcaneus and forming slips for each toe), the short foot muscles, and the long tendons crossing the ankle on their way into the plantar area of the foot all help to support.

Subtarsal Joint. The subtarsal joint (Fig. 6-10) is an important intertarsal joint since it permits and is responsible for most of the inversion and eversion of the foot. It is a functional unit, including not only the posterior talocalcaneal joint but also the talocalcaneal part of the talocalcaneonavicular joint and the talocalcaneal interosseous ligament that lies between these joints. The articular capsule and synovial membrane are tightly bound to the bones and allow little distention of the articular cavity.

METATARSOPHALANGEAL AND INTERPHALANGEAL JOINTS. The anatomy of these joints (Fig. 6-10) resembles that of the corresponding joints of the hand. Each has an articular capsule lined with a synovial membrane. The extensor tendon completes the capsule dorsally, the collateral ligaments strengthen the capsule on its sides, and the plantar ligaments support the plantar surface of the capsule. The metatarsophalangeal joints undergo little flexion.

Bursae are usually found over the first and fifth metatarsophalangeal joints and around the heel. Bursae (subcutaneous) usually develop in areas where most abnormal weight bearing or friction occurs.

Movement and Range of Motion. At the ankle joint, movement is limited to plantar flexion and dorsiflexion (about 20 degrees dorsally and about 45 degrees in the plantar direction).

Inversion and eversion of the foot take place mainly at the subtalar articulation. Inversion or supination exists when the sole is turned inward and eversion or pronation when the sole is turned out. The subtalar joint permits about 20 degrees of eversion and 30 degrees of inversion.

The metatarsophalangeal joint of the great toe flexes about 80 degrees and extends about 35 degrees, whereas the metatarsophalangeal joints of toes 2 to 5 move only 40 degrees in either direction. The proximal interphalangeal joints normally do not extend but do flex about 50 degrees. The distal interphalangeal joints may extend to 30 degrees and flex 40 to 50 degrees.

Muscle weakness and involved ligaments (stretched or inflamed) in the midfoot and intertarsal joints are associated with deformities and motion limitation.

ARCHES. See Chapter 5.

ABNORMAL FOOT POSITIONS. A lowering of the long arch is called pes valgo planus or flatfeet, while abnormal elevation of the same arch is called pes cavus. Talipes equinus is a foot in plantar flexion as a result usually of Achilles tendon contracture. The position of adduction and inversion is called varus while the position of abduction and eversion is referred to as valgus; these are often associated with knee abnormalities such as genu varum (bowlegs) or genu valgum (knock-knees). The most common deformity of the great toe is hallux valgus, a lateral or outward deviation resulting in an abnormal angulation and rotation at the first metacarpophalangeal joint. The typical hammertoe deformity consists of hypertension at the metacarpophalangeal joint, flexion at the proximal interphalangeal joint, and extension of the distal interphalangeal joint.

Review Questions

1. What are the three major types of joints? Define them and give examples of each.
2. What is meant by movement and range of motion?
3. What may limit range of motion?
4. For each of the following joints, name the articulating parts, indicate the type of joint, and discuss the movement and range of motion of each:
   a. Temporomandibular joint   **c.** Shoulder joint
   b. Sternoclavicular joint   **d.** Elbow joint

e. Wrist joint

f. Joints of the hand

g. Spinal column (what is a disk?)

h. Atlanto-occipital joint

i. Lumbosacral joints

j. Hip joint

k. Knee joint

l. Ankle joint

References

Basmajian, J. V.: *Grant's Method of Anatomy,* 8th ed. Williams & Wilkins Co., Baltimore, 1971.

Beetham, W. P., Jr.; Polley, J. F.; Slocumb, C. H.; and Weaver W. F.: *Physical Examination of the Joints.* W. B. Saunders Co., Philadelphia, 1966.

Goss, C. M. (ed.): *Gray's Anatomy of the Human Body,* 29th ed. Lea & Febiger, Philadelphia, 1973.

Grant, J. C. B.: *Grant's Atlas of Anatomy,* 6th ed. Williams & Wilkins Co., Baltimore, 1972.

Pansky, B., and House, E. L.: *Review of Gross Anatomy,* 2nd ed. Macmillan Publishing Co., Inc., New York, 1969.

Woodburne, R. T.: *Essentials of Human Anatomy,* 4th ed. Oxford University Press, New York, 1969.

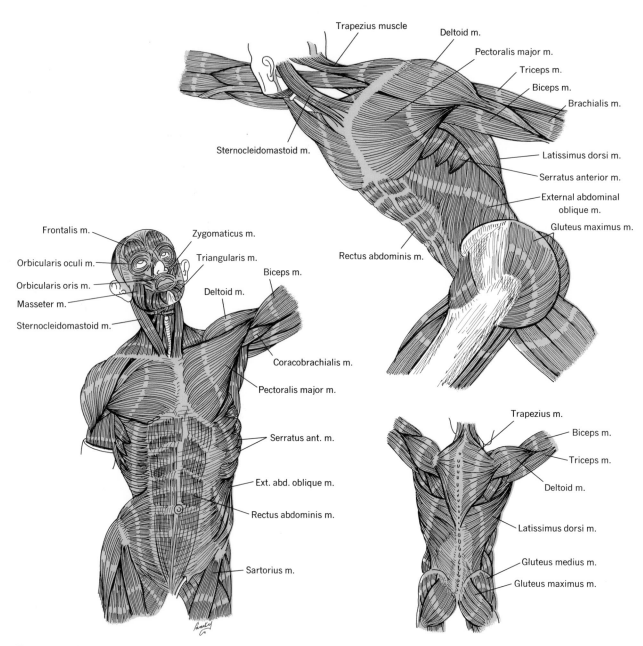

Figure 7-1. Muscles of the body.

The Body in Motion: Muscles

Life's processes consist of an external and internal environment of motion. The ability to sustain life is related to the ability to perform purposeful movement. Cells can move in a variety of ways. The property of contractility is widespread in biologic systems. Single cells, like the ameba, crawl along by extending their cell wall and internal cytoplasm. The paramecium moves by a hairlike extension called a cilia. Sperm cells use a whiplike flagellum. All this motion occurs at the molecular level. In man, muscle is one of the most remarkable of the specialized tissues. The chest muscles move air in and out of the lungs, the intestines move food from one end to the other, and the heart moves blood throughout the body—internal movements, done without changing the relative position of the body, in contrast to moving about from place to place.

Muscle tissue is very adaptable and has a reserve power built in. To pick up a pin a fraction of an ounce of exertion is needed; yet this group and similar groupings of muscle can exert a force of 7 lb in holding a heavy book, of 20 lb in lifting a suitcase, and even over 150 lb in a hard-forced squeeze.

Muscle structure is altered constantly during routine daily activity. In relaxation the muscle feels soft and pliable. However, in a fraction of a second it undergoes drastic and rapid change into a hard, tough, dynamic elastic material capable of contracting and exerting great force.

In terms of basic structure, muscle tissue is not unlike other body tissues. It contains about 78 percent H_2O, 20 percent protein, and 1 percent carbohydrate plus some fat and salts. Thus, it is the organization of its parts and not the actual basic ingredients that make it unique.

Muscle, by definition, is a tissue of cells (Fig. 4-8, p. 107) that enables the organism and its parts to move (Fig. 7-1). Contraction is one of the basic functions of protoplasm, and muscle is specialized for this function. Muscle is a source of movement or internal motility by its action on the bones of the skeleton or control of the blood vessel diameter and hollow organs. Contraction (mechanical work) can be sudden and forceful or gentle and continuous (walking versus rhythmic breathing). Muscle cells possess a form of contractile apparatus capable of converting chemical energy into mechanical work. The mechanical process is the contraction of many closely bound muscle cells or fibers. A muscle cell contracts when it receives a stimulus from an appropriate motor nerve. The degree of stimulus must be above a minimum level or threshold for the muscle fiber to react, but when it does contract it does so maximally—an "all-or-none" type of response. To avoid having the entire muscle respond, all the cells are not usually activated and just sufficient impulses are sent to achieve the force needed. The response of a muscle is matched to the load, and a feedback system operates to prevent expending useless energy.

Unlike many other cell types, muscle cells cannot divide and do not reproduce themselves in the usual manner. In man, however, new muscle cells can be formed to replace damaged cells. Replacement maintains a level of muscle ability but cannot offset the attrition that is the result of dying cells. The weight of some muscles between ages 30 and 75 can decrease by as much as 30 percent. The increase in size seen in "growth" in the above ages is due to the cell or fiber growing larger rather than an actual increase in cell number.

As to muscle growth and exercise, it is thought that growth is correlated to the level of activity of individual muscle fibers—a direct product of muscle use.

Man is an erect animal that operates on an elaborate system of passive structural tissues called bones, which are held together by ligaments and moved actively by *muscles*.

Every bone is a lever. A small movement of a muscle results in a large movement of the bone. The movements do not develop large forces because speed rather than strength is usually more essential to survival. Thus, every movement we make is the consequence of a muscle contraction, a muscle "pull." To push open a door, the triceps muscle in the upper arm pulls the forearm from a bent to a straight position. Thus, a muscle can contract and regain its original length only because of opposing muscles that work the lever in the opposite direction, forcing the first muscle to stretch again.

Myoneural Junction

Nerve-axon terminals fit into "gutters," the synaptic clefts, indenting the muscle-cell surface. Infoldings of the sarcolemma, the junctional folds, extend into regions beneath the synaptic gutter to form the postsynaptic membrane. The junctional folds and the space between the nerve endings and the muscle cell are filled with an amorphous substance. Nerve endings in the synaptic cleft are naked, and the Schwann cells are spread out above the endings and cover the synaptic region. Mitochondria are numerous in the nerve endings and in the subsynaptic area. Small, numerous membrane-bound vesicles, synaptic vesicles, are found in the nerve endings and in the muscle cell below the

postsynaptic membrane. The junction is specialized to achieve the transmission of neural impulses to the muscle-cell membrane (Fig. 4-11, p. 115).

Muscle fibers, like nerve fibers, show an "all-or-none" response to a single impulse and contract with all their force or not at all. However, a single muscle with thousands of fibers does vary its force of contraction. When a motor nerve meets a muscle, each of its axons branches over more than one muscle fiber. Each ends on a single muscle fiber to form a synapse called a motor end-plate. In some muscles a single axon may have branches over hundreds of muscle fibers. The combination of one nerve axon and its attached muscle fibers constitutes a motor unit. By varying the number of motor units in action, the muscle varies its force of contraction. Thus, the muscle can contract with variable force, depending on how many units are employed at one time. In the eye or fingertip, a muscle contains motor units in which one axon passes to ten fibers, whereas in a mass of leg muscle, a single axon usually feeds hundreds of fibers. This results in the leg muscle not being able to control the force of its movement as precisely as a muscle of the eye or finger. In either case, motor units work in relays or shifts to give a steady contraction without fatigue; some rest while others are active. Motor areas of the central nervous system control this activity and are arranged so that some of the fibers in the motor nerve are active at any given time; yet in emergencies impulses can be sent along all fibers, allowing for maximum muscle reaction for a short time.

A single impulse along a motor nerve fiber may cause the muscle to contract for a brief moment, a twitch. This does not normally occur in the body since a controlled muscle contraction is a response to a series of impulses. A series of such twitches produces a sustained contraction called tetanus. Thus, the force of contraction in muscle is "coded" by the frequency of nerve (motor) impulses.

The tension generated by a muscle can be varied by changing the frequency at which the individual motor units discharge or by altering the total number of motor units activated. If one increases the frequency of action potentials, one can summate the muscle cell contraction and develop an increased tension. Thus, many motor units must be activated or recruited if the combined tension is to create an effective movement or joint stabilization.

Muscle Action

Muscle is no exception to the rule that work needs energy, and here both electrical and chemical forces prevail. Contraction of a striated muscle cell is triggered by a wave of membrane depolarization and repolarization initiated or fired at the motor end-plate (at a point where the associated nerve is joined to the outer membrane of the fiber) (Fig. 4-11, p. 115), and spreads along the cell in milliseconds (excitation of the cell). The resting muscle cell, charged with energy, goes into action when the nerve delivers a signal "telling it" to contract. This involves two sequences of electrical activity with an intervening chemical event.

Electrical impulses are flashed along the various nerve fibers from the brain to the myoneural (nerve-muscle) junction on the muscle-fiber membrane. At this junction a chemical reaction releases a neurohumor that changes the property of the muscle cell's outer membrane so that the cell releases its "pent-up" electrical charge. This electrical discharge spreads over the cell surface, and with this discharge the muscle fiber is almost ready to contract when the second sequence of electrical action occurs. An action message is sent from the surface of the muscle fiber to the fibrils (the contracting elements) that takes about several thousandths of a second. The fibrils then contract. The total procedure from a single muscle cell twitch—stimulation to contraction to relaxation—takes about 0.1 second. Muscle contraction, the mechanical response to this surface excitation, is a slower process than the neurologic one, taking up to 100 milliseconds to be completed. Contraction involves the chemical and mechanical events that take place in the fibril and sarcoplasm.

Muscle Cell Function

The muscle cell functions in movement in the performance of work. In voluntary muscle, control of contraction is via the neural impulse and the cellular process of excitation as well as a mechanism to restore the contractile elements to their relaxation (resting) state. Energy is needed and is provided in the form of chemical-bond energy supplied by metabolism. The cell maintains and elaborates contraction, excitation, excitation-contraction coupling, relaxation, and energy metabolism by means of the processes of protein and lipid synthesis. The cell also has the ability to degrade and discharge damaged elements that cannot be restored and has the ability to transport elements into the cell and wastes out.

Muscles work in two ways. One way is isometric (Greek *isos*, equal; *metron*, measure), when the opposing forces are matched, their body parts barely move, and the muscles contract without shortening. The other is isotonic (Greek *tonos*, stretch), when opposing forces are unequal and contraction produces movement by shortening.

Contraction (Chemical Energy Converted to Mechanical Work)

Basic Mechanisms. Adenosine triphosphate (ATP) supplies the chemical energy. The contractile proteins are actin and myosin. The exact "key link" between the two is not clearly known. The properties of contraction are apparently the same in different types of muscle (and even animal species), and, therefore, the basic mechanisms are undoubtedly the same in all muscle cells.

Molecular Basis of Contraction. The contractile unit is the sarcomere, consisting of interdigitating thick myosin (A) and thin actin (I) filaments (Fig. 4-9, p. 110). It is felt that the two sets of filaments in the region of overlap interact by bridges that extend from the A filaments at regular intervals. The interaction of the filaments results in a "sliding" of the I filaments toward the center of the sarcomere, which thereby undergoes contraction. The energy for this is provided by the simultaneous splitting of ATP at the bridges (the sliding hypothesis of Huxley). As contraction continues, the H zone disappears and the I filaments extending from the two ends of the sarcomere begin to overlap. The Z line thus thickens, forming a "contraction band." On relaxation, the sarcomere returns to resting length and the filaments resume their rest positions. Contraction also results in some type of bridge movement, which is not clearly understood.

Excitation at the Synapse. Acetylcholine (ACh) is the mediator of the neural impulse across the myoneural junction (Fig. 4-11, p. 115). Its synthesis and storage take place in the terminal part of the motor neuron in the specialized "synaptic vesicles." Its synthesis is effected by the enzyme choline acetylase, which is also present in the vesicles. With the onset of the neural impulse, ACh is released into the synaptic space, where it diffuses and combines with receptors on the postsynaptic membrane. Inactivation of ACh is brought about by the enzyme acetylcholinesterase, present in the synaptic space.

In a resting nerve-muscle, ACh is released in "spontaneously" small packets or "quanta." The impulse increases the number of quanta released per unit of time. The frequency appears to be directly related to the extracellular concentration of Ca^{++} and reciprocally to that of Mg^{++}. The postsynaptic receptor is thought to be a triphosphate capable of binding ACh.

The muscle membrane is a complex of carbohydrate, fat, and protein separating intracellular and extracellular material. The osmotic pressure intracellularly and extracellularly is the same, but the ionic distribution varies. Na^+ and Cl^- concentrations are high in extracellular fluid and K^+ is low (reversed in intracellular fluid). (The unequal distribution of Na^+ and K^+ is said to be the basis for the difference in electrical potential across the membrane.) The resting potential is negative from outside to inside—the membrane is thus said to be polarized. The membrane serves as a barrier to ionic movement; yet Na^+ and K^+ are actively transported against this gradient with energy supplied by splitting ATP. The binding of ACh to receptors on the postsynaptic membrane results in

a movement of Na^+ and K^+ in the membrane, transiently depolarizing it. The passage of the impulse along the axon creates a wave of depolarization with movements of Na^+ and K^+ along their electrochemical gradients. The difference in potential between the peak of this wave and the resting potential is called the action potential. The rise of the action potential is due to the "inflow" of Na^+ and repolarization to a subsequent outflow of K^+. The return to a resting state occurs during the recovery period by the return of Na^+ to the outside and K^+ to the inside of the muscle. It has been suggested that when a nerve impulse reaches the myoneural junction, Ca^{++} enters the nerve terminal and may be responsible for disruption of the synaptic vesicle and the release of ACh. Furthermore, in the contraction of the fibril, calcium plays an important role (not fully understood). It is felt that the sarcoplasmic reticulum may bind calcium in its cisternae, releasing it on activation of the T system. The presence of Ca^{++} allows splitting of ATP and triggering of the sliding mechanism. The smooth endoplasmic reticulum then once again takes up the Ca^{++}, allowing a reestablished resting stage.

If interference occurs at this junction (e.g., toxin of botulism, poison strychnine, and cocaine or curare all work to block this junction), you get many effects. Myasthenia gravis, a disease of abnormal muscle weakness and fatigue, is a result of some yet-unknown defect in the operation at this myoneural junction.

Once a muscle fiber is stimulated it needs energy to work, and ATP supplies this energy. Immediate energy for muscle contraction comes from ATP breakdown. One fourth of the ATP energy is converted into mechanical energy, and the rest is lost as heat. Thus, our muscles are 25 percent efficient as energy converters. Although this seems like much waste, some of the waste heat from muscle activity is used to maintain the body's temperature. Only with violent exertion is an excess of heat produced and really wasted and dispelled to the outer air. If the exertion is beyond normal muscle activity, the ATP is used up more rapidly than it can be replaced and the muscle experiences oxygen debt. Since ATP synthesis depends on reactions between glucose and oxygen, the oxygen supply may lag behind cell need in strenuous exercise despite increased blood flow and respiratory rate.

When oxygen deficiency occurs, glucose metabolism no longer proceeds normally and lactic acid is produced in the muscle, resulting in fatigue and an aching sensation in the muscle. The lactic acid passes from the muscle into the bloodstream, increasing the acidity of the blood and helping to stimulate the respiratory center. The lactic acid also passes through the liver, where the energy in one fifth of the lactic acid is used to make ATP, which in turn provides the energy to rebuild the remaining four fifths of the lactic acid back into glucose that can be used again.

Summary of Muscle Activity

Contraction

1. As a result of synaptic events on the cell body and dendrites of the motor neuron in the central nervous system, an action potential is begun in a motor axon.
2. The action potential in the axon causes the release of acetylcholine from the axon terminals at the myoneural junction.
3. Acetylcholine is bound to the reactive sites on the motor end-plate membrane.
4. Acetylcholine causes a change in permeability in the motor end-plate, producing a motor end-plate potential.
5. The motor end-plate potential depolarizes the muscle membrane to its threshold potential, generating a muscle action potential that is propagated over the muscle membrane surface.
6. Acetylcholine is quickly destroyed by acetylcholinesterase on the end-plate membrane.
7. The muscle action potential depolarizes transverse tubules at the A-I junction of sarcomeres.
8. Depolarization of transverse tubules leads to the release of calcium ions from the lateral sacs of the sarcoplasmic reticulum surrounding the myofibrils.
9. Calcium ions bind to troponin-tropomyosin on the thin actin myofibrils, releasing the inhibition that prevented actin from activating the myosin ATPase.
10. Actin combines with myosin ATP in the presence of magnesium ions:
$$A + M + ATP \longrightarrow Mg^{++} + A + M + ATP.$$
11. Actin activates the myosin ATPase, which splits ATP, releasing energy needed to produce a movement of the myosin cross bridge:
$$A + M + ATP \longrightarrow A + M + ADP + P.$$
12. ATP exchanges with ADP on the myosin bridge, breaking the actin-myosin bond and allowing the cross bridge to relax:
$$A + M + ADP + ATP \longrightarrow$$
$$A + M + ATP + ADP.$$
13. Thus, cycles of cross-bridge contraction and relaxation continue as long as the calcium concentration is high enough to inhibit the action of the troponin-tropomyosin system.
14. Movements of the cross-bridges lead to relative movement of the thick and thin myofilaments past each other.
15. Concentration of calcium ions falls as they are moved into the lateral sacs of the sarcoplasmic reticulum by an energy-requiring process that splits ATP.
16. Removal of calcium ions restores the inhibitory action of troponin-tropomyosin; in the presence of ATP and magnesium, actin and myosin remain dissociated and in a relaxed state.

For neuromuscular transmission to take place, the following must occur:

1. There must be synthesis and storage of ACh.
2. Sufficient ACh must be released after nerve stimulation (Ca^{++} facilitates, Mg^{++} inhibits).
3. The transmitter substance must diffuse across the synaptic space.
4. The postsynaptic membrane must be structurally and functionally intact.
5. The muscle-cell membrane must also be structurally and functionally intact. (There must be an adequate depolarized membrane, which depends on the in-

tegrity of membrane Na$^+$- and K$^+$-activated ATPase.)

Relaxation

Relaxation of the contracted myofibrils is brought about by withdrawal of Ca^{++} from its binding sites in the overlapping myofilaments, which is accomplished by the "Ca^{++} pump" of the sarcoplasmic reticulum. (Ca^{++} taken from the myofibrillar spaces returns to the lateral cisternae of the sarcoplasmic reticulum, restoring the system to a preexcitation state.) As Ca^{++} is removed from the filaments, its concentration in the intramyofibrillar space drops to a subthreshold level, ATPase activity in the myofibrils decreases, tension falls, and relaxation follows.

The brain and other parts of the nervous system are always adjusting and responding to sliding filaments in millions of muscle fibrils. Every movement involves a continual balance of forces. Certain muscles, if not controlled, would flex the arms and legs in arcs and others would extend them stiffly. Thus, there are many pairs of opposing muscles in equilibrium and all are monitored automatically by the cerebellum. Other centers control sequences of movements in walking, tying shoes, and undertaking other tasks of unconscious thought.

Muscular dystrophy, which affects mostly children, is a hereditary disease characterized by the wasting away of muscle tissues. Recent studies show that one important factor may be changes in membranes of the muscle cells. These diseased membranes allow protein loss from cell. There is also an increased activity in the muscles of enzymes that break down proteins. Much research is being done in this area.

Movement and Posture

Any effective movement involves the motor units. As one picks up an object, the fingers are at first straightened (extended) and then bent (flexed). The coordination resulting in the degree and force of flexion and extension depends on the size of the object as well as its weight and consistency. At the same time, activity is occurring in the wrist, elbow, and shoulder (which are extending) and the trunk (which is leaning forward), related to the direction and distance where the object lies. Thus, it can be seen that most movements require the coordinated action of many motor units and many muscle groups.

The actions we "think about" are called voluntary and utilize voluntary or striated muscles. Voluntary movement is a movement of conscious awareness, our attention is directed to the movement, and the actions are usually the result of learning. One must, however, keep in mind that most muscle activity involves both conscious and unconscious components. If you are threading a needle, you might be consciously concentrating on the needle opening and the held thread but there is also the unconscious use of the arm and hand and inhibition of antagonistic muscles, let alone body and head movements and position. Thus, through learning, a complicated pattern of muscle movement is shifted from a conscious state to an involuntary reflex.

Movement and its control can take place only by controlling the individual motor neuron, which receives thousands of synaptic endings converging on it from many sources. The output from this neuron depends on both an excitatory and inhibitory synaptic balance. If either predominates, the response can be a rapid firing of action potentials, a mild excitation and facilitation, no firing at all, or a state of rest. Thus, the precision of coordinated movement depends on a balance of impinging influences. If one system of incoming stimuli becomes hyperactive, the balance can be destroyed, and one speaks of spasticity, rigidity (stiff and awkward), or flaccidity (weak, soft), depending on the abnormality developed.

Higher brain centers and various subcenters are programmed to transmit information to motor neurons and on to muscle. During any movement, receptors in the muscles, tendons, and joints send information back to the central nervous system, providing minute information about the movement in progress and allowing for adjustments to be made in that movement. Thus, we have a cerebral cortex with neurons that makes decisions as well as cortex and subcortical areas that program the input. The subcortical areas include the basal ganglia, the brain stem nuclei, and the brain stem reticular formation. The peripheral receptors include the muscle stretch receptors, the Golgi tendon organs, the joint proprioceptors (position receptors), skin receptors, vestibular receptors of the inner ear that detect the position of the head in space, and the rods and cones of the eyes. Errors between reception and programming are compensated for by the cerebellum.

Stretch Receptors

The stretch receptors of skeletal muscle are made up of afferent nerve endings wound around modified muscle cells (spindle fibers), both of which are encased in a fibrous capsule (Fig. 7-2). Together they make up the muscle spindle. These spindles occur in muscles of the fingers and eyes that perform fine movements as well as in larger muscles that maintain body posture. The receptors constantly work, particularly when we sit erect or stand (standing makes us top-heavy). Anterior and posterior trunk muscles working in opposition continuously contract and relax, preventing our swaying from a vertical position. The spindles play their part here, since a slight pull to one side tilts the body but simultaneously muscles on the other side are stretched a little, which stretches the spindle muscle fibers and thereby leads to the discharge of sensory

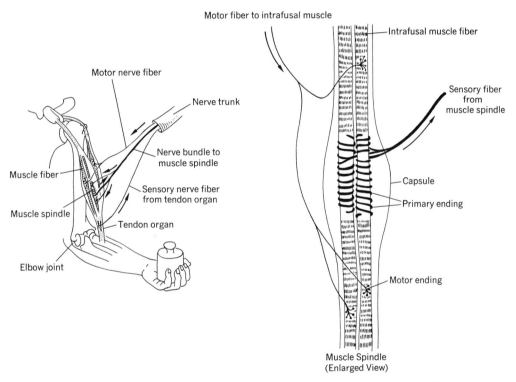

Figure 7-2. Sense organ arrangement in a typical muscle (diagrammatic). The muscle spindle is thin and consists of specialized structures called intrafusal muscle fibers. The terminations of the sensory nerve fiber are wrapped around the intrafusal muscle fibers and respond to mechanical deformation by causing nerve impulses to be sent up the sensory nerve. The cross striations are absent in the equatorial region. Thus, when the intrafusal fibers contract, this region is extended and excites the sensory endings just as if the region had been extended by stretching the entire muscle and the spindle in it.

impulses of the spindle. The nerve fiber travels to the spinal cord and synapses with a motor nerve of the same muscle, and the muscle contracts to maintain the erect posture.

Both fine and posture movements work on a "feed-back" mechanism. Every movement causes a change of state that is reported to the nervous system, and adjustments follow that in turn generate new information.

The neurons to the spindle are not the same as to the muscle fibers. The motor neurons to the muscle fiber are large and called alpha motor neurons. Those to the spindle fibers are smaller and called gamma motor neurons (Fig. 7-3). The latter cause the spindle fibers to shorten. The spindle muscle fibers are found at the ends of the spindle, and their shortening stretches and distorts the stretch receptors. Thus, the joint itself is not moved; only the receptor is involved. The afferent fibers from the stretch receptors, however, synapse directly on the alpha motor neurons of the same muscle and are excitatory. They activate the

stretch reflex via the alpha neuron to the skeletomotor muscle fibers. This is the gamma loop (Fig. 7-3).

The gamma motor neuron receives most of its synaptic input from pathways originating in the brain. The alpha fibers are stimulated by reflex pathways or descending neural pathways or via the gamma loop.

The gamma motor neuron activation maintains continual tension on the spindle stretch receptors by contraction of the spindle muscle fibers. This, in turn, coordinates alpha motor activity, indicating how much effort is needed for a job or how much a muscle needs to be shortened. The number of muscle spindles in a muscle is related to the precision by which it is controlled. It is difficult to specifically distinguish between the action of the individual alpha and gamma components of movement. Both are needed for coordination and effectiveness.

The stretch reflex is one in which information about the change in length of a muscle is fed back to the motor neurons controlling the same muscle in both its voluntary and involuntary movement. Thus, if a mus-

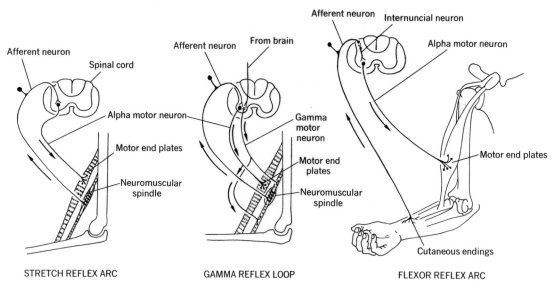

STRETCH REFLEX ARC   GAMMA REFLEX LOOP   FLEXOR REFLEX ARC

Figure 7-3. Examples of spinal reflexes. The stretch reflex is based on a two-neuron or monosynaptic reflex arc; the gamma reflex loop brings muscle tension under control of descending motor fibers from higher centers; and the flexor reflex is protective, as in withdrawal of hand in response to a painful stimulus (polysynaptic).

cle is stretched, it reflexly contracts. For example, the knee jerk is solicited by tapping on the patellar tendon, which stretches over the knee and connects a thigh muscle (quadriceps femoris) to the leg bone, the tibia. As the tendon is depressed, the muscle is stretched and the receptors in the muscle spindles are activated. The information regarding rate and magnitude of stretch is relayed to the central nervous system via afferent nerve fibers. The afferent nerves enter the central nervous system where branches from each of the fibers take a number of different pathways. One group synapses directly with motor neurons going back to the stretched muscle, are excitatory, and shorten the muscle, thereby raising the patient's foreleg in extension (the knee jerk). This stretch reflex activity is a monosynaptic reflex. All other reflexes (besides the stretch reflex) tend to be polysynaptic and involve at least one interneuron between the afferent and efferent pathways. To complete the above picture one must realize that a second group of afferents terminate on interneurons that inhibit the motor neurons controlling antagonistic muscles, which could interfere with the stretch reflex response. The excitation of one muscle and inhibition of its antagonist is called reciprocal innervation. Furthermore, a third group of afferents end on interneurons that activate synergistic muscles assisting the reflex motion (in this case involve other muscles that help extend the leg). Lastly, a group of afferents synapse with interneurons that go to the brain to convey information regarding the muscle length, and here coordination of muscle movement takes place. The latter is subconscious, coming from the muscles. The conscious awareness of the position of a limb or joint is relayed from the joint and ligament by afferent proprioceptor fibers. The latter relay information regarding the state of the muscles, their position, degree of tension, position of the limbs in relation to each other, and position of the body in relation to its surroundings. They are located in the ligaments, tendons, joints, and muscles. Reflex action tells us about the afferent limb of the reflex, the balance of synaptic input to the motor neuron, the motor neuron's function, the myoneural junction, and, finally, the muscle action itself. The stretch reflex permits us to maintain a muscle at a particular length and thus an upright posture.

Tension Monitors

Positioning is fundamentally a spinal cord phenomenon in which opposing pairs of muscles are coordinated into a working unit by reciprocal innervation. It involves the afferent and efferent neurons, which are discussed above. Positioning is also modified by the Golgi tendon organs, which respond to forces or tension created by the contracting muscle. The activity of these tendon organs results in both inhibition and excitation of postsynaptic potentials in the neurons of the contracting and antagonistic muscles. We find the organs responding to both high and low thresholds of tension, the latter supplying the motor nerves with continuous information about tensions developed. Thus, when an extensor muscle contracts, the flexor is passively stretched and both Golgi tendon organs of the extensor and muscle spindles of the flexor muscle are excited. This leads to inhibition of the extensor motor neurons and facilitation of the flexor motor neurons. These reflex inputs then oppose the original extension of the joint, and their combined activity reflexly decreases extension force (the converse is also true). Thus, there is a balance and stabilization of the joint, and a balance between flexion and extension is maintained.

Higher Control

Muscle activity is coordinated by the cerebral cortex, subcortical areas, the cerebellum, and afferent or sensory information relayed to these centers (see Chap. 8). Descending motor pathways in the spinal cord contain motor neurons and can affect the stretch and other reflexes.

Many areas of the cortex influence skeletal muscle activity, and a large number of motor fibers arise from the motor areas of the frontal cortex (particularly its posterior part). Adjacent neurons in this area are also functionally organized in groups. Neuron function in the motor cortex varies with position in the cortex. The relative size of the cortical area related to each part is proportional to its control and degree of refined movement of the part in question. Thus, the areas of the hand and face are the largest, accounting for the fine degree of motor control of these areas.

Many of the cortical neurons found in the major motor cortex area project to the adjacent frontal cortex and thus have some direct control over them. Furthermore, the axons of the motor cortical neurons may also pass directly, without synapse, and end near the motor neurons of the cord. The fibers innervating areas of the eye, face, tongue, and neck relay to the brain stem and there contact motor neurons of the cranial nerves. Others descend to terminate in the spinal cord, to innervate motor neurons controlling skeletal muscle associated with fine movements. Many of the latter cross from one side to the other in the medulla oblongata, and thus skeletal muscles on the left side of the body are controlled by neurons from the right half of the brain. These fibers make up the direct corticospinal pathway or pyramidal system (tract), which is so named because of their origin from large pyramidal neurons of the cortex. Generally speaking (not completely true since cutting this tract does not totally obliterate its function), the pyramidal system activity results in execution of delicate, skilled, voluntary muscle movements. The pyramidal system also influences ascending (afferent) systems, and, therefore, the two are functionally tied.

In the anterior part of the motor cortex are neurons largely responsible for the multineuronal pathways as well as for exerting some control over the motor area itself. The subcortical centers are the basal ganglia, brain stem nuclei, and reticular formation. The multineuronal motor paths descending from the cortex, as stated, synapse in these centers. Here muscle movement is controlled and patterns of neuron interaction into motions are relayed to lower centers such as the basal ganglia, brain stem, and cord. Thus, the walking mechanism lies at subcortical levels but the cortex sets the motion off. Neurons of the anterior motor cortex and subcortical centers are grouped together under the multineuronal or extrapyramidal system (see Chap. 8). They particularly influence interneurons in the cord and affect motor nerves on both the same and opposite sides of the body, thereby enabling reciprocal activity in movement. Thus, normal movements, especially the fine voluntary ones, require coordination between both pyramidal and extrapyramidal systems.

Posture

Postural mechanisms are needed to "hold" the mass of bones and spine and muscle erect against the forces of gravity. Disorders of posture may affect the head, the trunk, the limbs, or the entire organization. One holds the head erect without effort or thought or fatigue, and man supports his weight and general stability, keeping himself vertical, by reflex postural mechanisms. The basal ganglia, brain stem nuclei, reticular formation, thalamic nuclei, and cerebellar areas of the brain are involved in attention, wakefulness, sleep, and posture.

Posture is maintained by the continual balance of afferent and cerebellar input to the motor pathways. Input from general body skin receptors leads to facilitation of flexor motor neurons, and input from vestibular receptors in the ear causes reaction to head position and motion and leads to facilitation of extensor motor neurons. These inputs and outputs are continually active and shifting, at least while we are awake and part of us is in movement.

Locomotion

We must maintain equilibrium while in motion. Effective "stepping" or the flexion of one foot with the simultaneous extension of the other carries the body along as well as maintains an antigravity position. This requires the functioning of the upper brain stem. The center of gravity is relayed by afferent impulses from muscles, joints, skin, ears, and eyes. In walking, the weight of the body is supported on each leg alternately, and the body moves from side to side so that the center of gravity is shifted from right to left leg, etc.; only when the center of gravity shifts can the leg be raised from the ground and advanced. The body simultaneously sways to the firm side to counterbalance the weight of the raised leg. Thus, a fine control of the center of gravity is essential. If the center shifts, balance is lost.

As we walk, the body leans forward slightly and the weight of the body is shifted to one foot and the other is lifted from the ground. During this process of the body falling forward and losing its equilibrium (until it is caught on the leg that swings forward), the center of gravity has moved forward and sideways over one leg to a similar position over the other and this rhythmic movement depends on the forward shift of the center of gravity. The components involved are antigravity support of the body, stepping, control of the gravity center for equilibrium, and a means of acquiring forward motion (muscle movement)—all continuous and simultaneous. Disease of the basal ganglia leads to abnormalities in gait, posture, and locomotion. Parkinson's disease is a typical example associated with involuntary and continuous tremor at rest.

Cerebellar Control

The cerebellum is involved in muscle control of both posture and coordination (see Chap. 8). It receives input from the spinal cord, cerebral cortex, vestibular (ear) system, and reticular formation. It acts by influencing other areas of the brain and subcortical areas responsible for motor activity and does not initiate movement. Damage to the cerebellum is associated with a general inadequacy of movement and not the loss of any specific movement. With damage, movements lack smoothness and are jerky. Tremors appear when reaching for an object (intention tremor) rather than when one is resting. Walking is awkward and feet are held apart, and the gait is reeling and like a "drunk." It is difficult to either start or stop movements, and combined movement of several joints is difficult. Speech, too, is affected since this also requires an intricate coordination of many muscles. There is, however, no sensory or intellectual deficiency with cerebellar injury.

Vestibular Control

For details of vestibular control, see Chapter 8. Suffice it to say that this system relays information about head movements, to a degree controls eye muscles, and reflexly helps to maintain an upright posture where vision cannot be used.

Proprioceptors and Skin Receptors

Unconscious control of posture and movement, as well as awareness of position and joint movement, is associated with receptors in the joints, ligaments, and tendons. Input from these receptors and the skin receptors is integrated with vestibular and visual input to provide information about the body in space.

Muscles of the Body

Muscles of Facial Expression

The skin of the face is very vascular, thin, highly mobile, and movable and contains numerous sweat and sebaceous glands. Anteriorly there is no deep fascia so that muscles arising from the bones insert into the skin itself. Since the glands of the skin lie in the loose areolar connective tissue unsupported by deep fascia, edema spreads rapidly. Over the lower nose the skin is firmly tied to the cartilage, making any inflammation quite painful. The skin over the chin is like that over the scalp, being dense and adherent to the underlying structures.

The skin of the face is adaptable to healing and plastic surgery owing to its extreme vascularity and mobility. The area of the face around the root of the nose is important because its venous drainage enters the angular vein by the side of the nose and runs into the cavernous sinus of the skull via the ophthalmic veins. This area is often referred to as a "danger area" since carbuncles, boils, and other infections may create a cavernous sinus thrombosis.

Muscles. The muscles of facial expression (Fig. 7-4, Table 7-1) are placed around the eyes, ears, nose, and mouth and act as sphincters or dilators. They are all innervated by the facial (VII) nerve. One can consider all the muscles as being arranged around the two orbicularis muscles, the oculi and oris. In addition, there are two muscles associated with the nose, two are associated with the zygoma, two are levators of the lips, two are at the angle of the mouth, two are at the lower lip, and two are related to the chin and cheek. The buccinator is deeply set and forms the fleshy part of the cheek, running horizontally to the

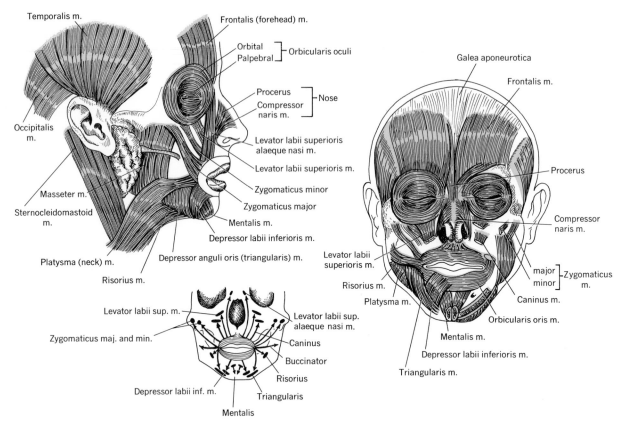

Figure 7-4. Muscles of facial expression.

angle of the mouth. Its inner surface is lined by the mucous membrane of the cheek and lips. The upper and lower fibers pass to the upper and lower lips, respectively, while the middle fibers decussate. The buccinator, which blends with the orbicularis oris at the mouth angles, acts to retract the mouth angle, and thus is antagonistic to the orbicularis. It is a muscle of facial expression but is used during mastication to press the cheek against the teeth and prevents food from escaping into the vestibule of the mouth (between the cheeks and teeth); it also helps in sucking and blowing. The buccal fat pad or suctorial pad is located on the muscle, tucked in between it and the masseter muscle (see below). It thickens the cheek and aids in reducing atmospheric pressure in sucking, being large

in infants and accounting for the full and round cheeks of the baby.

Muscles of Mastication (Temporal and Infratemporal Region)

The muscles of mastication (Fig. 7-5, Table 7-2) include the temporal, masseter, external and internal pterygoids, mylohyoid, geniohyoid, and anterior belly of the digastric.

The temporal, masseter, and internal pterygoid produce great grinding and biting force, while the muscles that depress the jaw (external pterygoids, digastric, mylohyoid, and geniohyoid) are very weak. Thus, when spasm occurs, the strong muscles prevail. If spasm is clonic, the teeth chatter; however, if it is

Table 7-1. Muscles of the Scalp and Face (Muscles of Expression): All Innervated by the Facial (VII) Nerve

Muscle	Origin	Insertion	Action
1. Orbicularis oculi (eye sphincter)			Generally closes eyes strongly, wrinkles forehead vertically, draws lid medially since it has no lateral attachment
a. Orbital part	Medial palpebral ligament and medial orbital margin	Same as origin	Gives "crow's feet" laterally, with age
b. Palpebral part (in eyelids)	Medial palpebral ligament	Lateral palpebral raphe	Gently closes lids in sleeping and blinking
c. Lacrimal part (behind lacrimal sac)	Posterior lacrimal ridge of lacrimal bone	Joins palpebral part	Keeps lids loosely applied to eyeball and wrinkles skin around eye to protect against light or wind
2. Orbicularis oris (mouth sphincter) (unpaired)	From neighboring muscles, chiefly the buccinator	Skin around the lip rim	Closes, opens, purses, inverts, points, and twists mouth
3. Procerus (unpaired)	Fascia over lower nose	Frontalis muscle and skin above and between eyebrows	Draws skin at root of nose down—transverse wrinkling
4. Nasalis	Maxilla, below wing of nose	Wing of nose; common tendon on crest of nose	Lowers and compresses wing of nose
5. Depressor septi	From orbicularis oris	Septal cartilage of nose	Depresses septum
6. Zygomaticus major	Zygomatic bone	Muscles at angle of mouth	Draws upper lip upward—smiling muscle
7. Zygomaticus minor	Zygomatic bone	Orbicularis oris of upper lip	Draws upper lip upward—and outward
8. Levator labii superioris	Maxilla, below infraorbital foramen	Orbicularis oris of upper lip	Elevates upper lip
		Ala of nose and orbicularis oris of upper lip	Elevates upper lip and wing of nose
9. Levator labii superioris alaque nasi	Root of nasal process of maxilla	Orbicularis oris and skin at mouth angle	Raises angle of mouth
10. Levator anguli oris (caninus)	Canine fossa of maxilla	Blends with muscles in lower lip near angle of mouth	Pulls down corners of mouth
11. Depressor anguli oris (triangularis)	Lower border of mandible		
12. Risorius	Platysma, skin, and fascia of masseter	Orbicularis oris and skin at corner of mouth	Draws corner of mouth outward; causes "dimple"; the "grinning" muscle
13. Depressor labii inferioris	Anterior part of lower border of mandible	Orbicularis oris and skin of lower lip	Depresses lower lip
14. Mentalis	Incisor fossa of mandible	Skin of chin	Raises and wrinkles skin of chin and pushes up lower lip
15. Buccinator	Buccinator ridge of mandible, posterior part of alveolar process of maxilla, and pterygomandibular ligament	Orbicularis oris at angle of mouth	Flattens cheek, retracts angle of mouth
16. Corrugator	Nasal part of frontal bone	Skin of eyebrow	Draws skin of forehead medially, wrinkles it vertically
17. Epicranius (scalp muscles)			
a. Frontalis	Galea aponeurotica	Into muscles and skin of eyebrows and nose	Draws scalp backward, eyebrows upward, wrinkles forehead
b. Occipitalis	Occipital bone	Galea aponeurotica	Draws scalp backward
18. Platysma	Fascia and skin of breast and shoulder regions	Fascia of face, overlying jaw; corners of mouth	Wrinkles skin of neck and upper part of chest
19. Auricula (anterior, superior, posterior)	Fascia in temporal region	Into auricle (ear)	Moves ear anterior, superior, and posterior

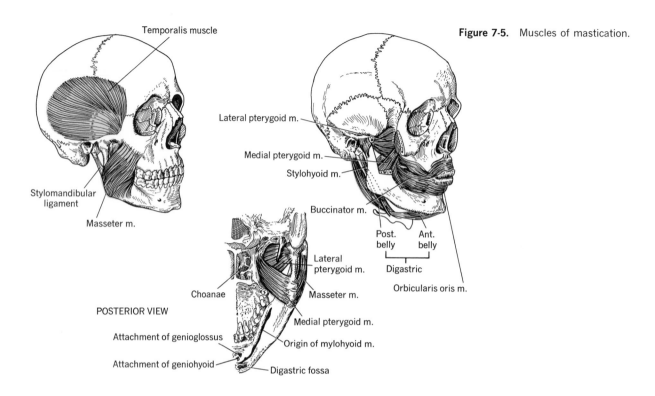

Temporalis muscle

Figure 7-5. Muscles of mastication.

Stylomandibular ligament

Masseter m.

Lateral pterygoid m.

Medial pterygoid m.

Stylohyoid m.

Buccinator m.

Post. belly   Ant. belly

Digastric

Orbicularis oris m.

Lateral pterygoid m.

Masseter m.

Medial pterygoid m.

Choanae

POSTERIOR VIEW

Attachment of genioglossus

Attachment of geniohyoid

Origin of mylohyoid m.

Digastric fossa

Table 7-2. Muscles of Mastication: All Innervated by the Mandibular Division of the Trigeminal (V) Nerve

Muscle	Origin	Insertion	Action
1. Masseter	Deep surface and anterior two thirds of zygomatic arch	Outer surface of ramus and coronoid process of mandible	Closes jaw; protrudes jaw (felt when jaws are clenched)
2. Temporalis	Temporal line and fossa of cranium	Anterior border of ramus and apex of coronoid process of mandible	Powerful closer of jaw; retracts jaw
3. Internal (medial) pterygoid	Medial surface of the lateral pterygoid plate and tuberosity of maxilla	Inner surface of lower jaw between angle and mylohyoid groove	Raises mandible, closing jaw
4. External (lateral) pterygoid	Lateral surface of the lateral pterygoid plate and greater wing of the sphenoid	Pterygoid fossa of head of mandible and articular disk	Brings jaw forward (protrudes) and opens jaw; one muscle pulls jaw to opposite side
5. Mylohyoid (deep to anterior belly of digastric)	Mylohyoid line (inner surface) of mandible	Upper border of hyoid bone and meets its partner in a midline raphe	Elevates floor of mouth and tongue; depresses jaw when hyoid bone is fixed
6. Digastric (anterior belly)	Intermediate tendon (fastened by loop of fascia to hyoid bone)	Lower border of mandible near symphysis	Depresses jaw if hyoid bone is fixed; raises hyoid if jaw is fixed
7. Geniohyoid (innervated by C1 and C2)	Mental spine of mandible	Body of hyoid bone	Draws hyoid forward or depresses mandible when hyoid is fixed

Table 7-3. Muscles of the Soft Palate: All Innervated Through the Pharyngeal Plexus Except for the Tensor Veli Palatini (Trigeminal V via Otic Ganglion)

Muscle	Origin	Insertion	Action
1. Levator veli palatini	Apex of petrous portion of temporal bone and cartilage of eustachian tube	Aponeuroses of soft palate	Raises soft palate
2. Tensor veli palatini	Scaphoid fossa of sphenoid bone and cartilage of eustachian tube	Posterior border of hard palate and aponeuroses of soft palate	Stretches (tenses) the soft palate
3. Palatoglossus	Undersurface of soft palate	Side of tongue	Raises back of tongue and narrows fauces during stage 1 of swallowing (forms anterior pillar of fauces in front of faucial tonsil)
4. Palatopharyngeus	Soft palate	Posterior border of thyroid cartilage and aponeurosis of pharynx	Narrows fauces and shuts off nasopharynx (forms the posterior pillar of fauces behind the faucial tonsil), aids in elevation of pharynx in stage 2 of swallowing
5. Musculus uvulae	Posterior nasal spine and palatine aponeurosis	Into uvula to form its chief bulk	Shortens and raises uvula

tonic, the mouth is rigid and closed and in a state of trismus or lockjaw.

Muscles of the Soft Palate

The muscles of the soft palate (Table 7-3; see Chap. 12) are five in number. Two descend from the palate to the tongue and pharynx (glossopalatine and glossopharyngeus); two descend to the palate from the skull (levator and tensor veli palatini); and one lies in the substance of the palate itself (musculus uvulae).

Several nerves supply the soft palate. The levator palatini, palatopharyngeus, and musculus uvulae are supplied (along with the pharyngeal muscles) through the pharyngeal plexus, by fibers derived from the internal branch of the spinal accessory (XI) nerve. The palatoglossus is supplied by the pharyngeal plexus and the tensor palatini by the third division of the trigeminal (V) nerve. The mucous membrane of the soft palate is supplied by the maxillary division of the trigeminal (V) nerve.

Muscles of the Tongue

The muscle substance of the tongue (Fig. 7-6, Table 7-4) is divided into a right and left half separated by a median septum, each half made up of extrinsic and intrinsic muscles. The extrinsic muscles alter its shape and move the tongue, whereas the intrinsic muscles only change its shape.

The motor nerve of the tongue is the hypoglossal (XII), which supplies all the intrinsic and extrinsic muscles of the tongue except for the glossopalatine (pharyngeal plexus). If the twelfth nerve is injured on one side, the protruded tongue deviates toward the paralyzed side (indicating the side of the lesion). If both nerves are cut, the muscles are paralyzed and the tongue may fall back and cause suffocation.

The glossopharyngeal (IX) nerve supplies taste and sensory fibers to the posterior one third of the tongue. The lingual (V) nerve supplies sensory fibers to the anterior two thirds of the tongue, but it is the chorda tympani (VII) nerve that "rides" along with it, on its way to the tongue, that receives the taste impulses from the anterior two thirds of the tongue.

Muscles of the Neck

The upper limits of the neck (Figs. 7-7, 7-8; Table 7-5) are the lower border of the jaw, a line from the angle of the jaw to the mastoid process, and the supe-

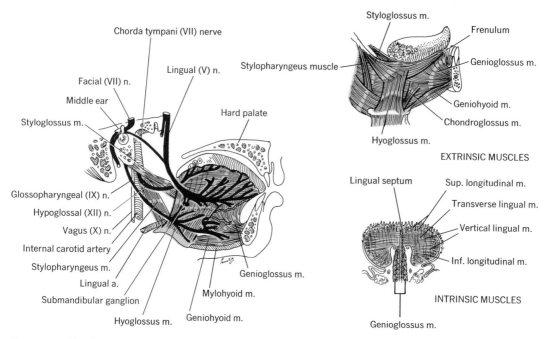

Figure 7-6. Muscles (intrinsic and extrinsic) of the tongue and their nerve supply.

Figure 7-7. Muscles of the neck.

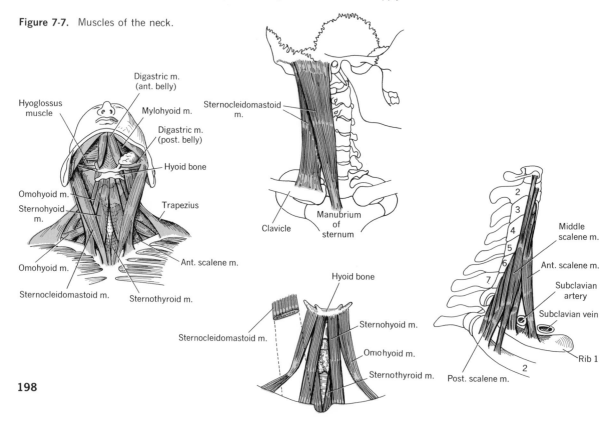

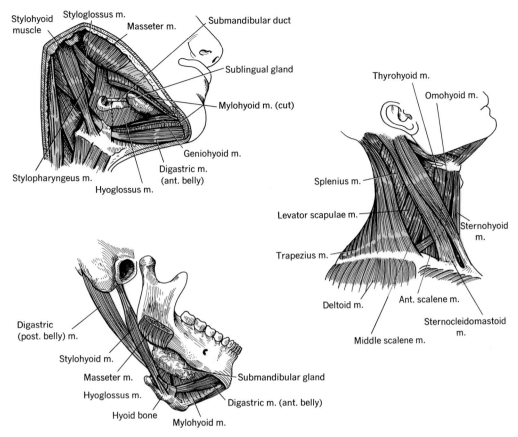

Figure 7-8. Muscles of neck and submandibular region.

Table 7-4. Muscles of the Tongue: All Innervated by the Hypoglossal (XII) Nerve Except Palatoglossus

Muscle	Origin	Insertion	Action
1. *Extrinsic Muscles* (4)			
a. Genioglossus (paired)	Mental spine of mandible	Undersurface of tongue and epiglottis	Depresses and protrudes tongue
b. Styloglossus (paired)	Styloid process	Side and undersurface of tongue	Retracts tongue
c. Hyoglossus (paired)	Body and great cornu of hyoid bone	Side of tongue	Retracts and pulls down side of tongue
d. Palatoglossus (paired)	Undersurface of soft palate	Side of tongue	Raises back of tongue and narrows fauces
2. *Intrinsic Muscles* (4) Superior and inferior longitudinal, transverse, and verticalis			

Table 7-5. Muscles of the Neck*

Muscle	Origin	Insertion	Action	Nerve
1. Sternocleidomastoid	By two heads, from the anterior surface of the manubrium sterni and sternal end of the clavicle	Mastoid process and outer half of superior nuchal line of occipital bone	Each turns head obliquely to opposite side; when acting together, they pull head downward and forward	Spinal accessory (XI)
2. Infrahyoid muscles or strap muscles (4)				
a. Sternohyoid	Posterior surface of manubrium sterni and first costal cartilage	Body of hyoid bone	Depresses hyoid bone	Upper cervical via ansa cervicalis
b. Omohyoid	By inferior belly from upper border of scapula between notch and inner angle	By superior belly into hyoid bone	Depresses hyoid bone	Upper cervical via ansa cervicalis
c. Sternothyroid	Posterior surface of manubrium sterni and first or second costal cartilage	Oblique line of the thyroid cartilage	Depresses larynx	Upper cervical via ansa cervicalis
d. Thyrohyoid	Oblique line of thyroid cartilage	Body of hyoid bone	Approximates hyoid bone to larynx	Upper cervical via ansa cervicalis
3. Digastric				
a. Posterior belly	Digastric groove near mastoid process	Central tendon on hyoid bone	Fixes hyoid bone	Facial (VII) nerve
b. Anterior belly	Central tendon on hyoid bone	Lower border of mandible near symphysis	Depresses jaw	Mylohyoid nerve from third division of trigeminal (V) nerve
4. Mylohyoid	Mylohyoid line of mandible	Upper border of hyoid bone and raphe separating muscle from its fellow	Elevates floor of mouth and tongue, depresses jaw, when hyoid is fixed	Mylohyoid from third division of trigeminal (V) nerve
5. Hyoglossus	Body and great cornu of hyoid	Side of tongue	Retracts and pulls down side of tongue	Hypoglossal (XII) nerve
6. Pharyngeal constrictors				
a. Superior	Medial pterygoid plate, pterygomandibular raphe, mylohyoid line of mandible, and mucous membrane of floor of mouth and side of tongue	Posterior wall of pharynx (raphe)	Narrows pharynx	Pharyngeal plexus
b. Medial	Stylohyoid ligament, lesser and greater cornua of hyoid bone	Middle of posterior wall of pharynx (raphe)	Narrows pharynx in act of swallowing	Pharyngeal plexus
c. Inferior	Outer surfaces of thyroid and cricoid cartilages	Posterior portion of wall of pharynx (raphe)	Narrows lower pharynx in swallowing	Pharyngeal plexus
7. Laryngeal muscles	(see Chap. 11)			

Table 7-5 (Continued)

Muscle	Origin	Insertion	Action	Nerve
8. Levator scapulae	Posterior tubercles of transverse processes of four upper cervical vertebrae	Superior angle of scapula	Raises scapula	Dorsal nerve of scapula from brachial plexus
9. Splenius capitis	Spine of the last four cervical and first three thoracic vertebrae	Outer half of superior nuchal line of occipital bone and mastoid	Rotates head; together, they draw head backward in extension	Cervical nerves II-VIII
10. Scalenus muscles (anterior, middle, and posterior)	Tubercles of the transverse processes of second to sixth cervical vertebrae	Upper two ribs	Elevate first and second ribs; bend cervical spine lateralward and together bend it forward	Cervical plexus

* The other deeper muscles of the neck will be considered with the muscles of the vertebral column.

rior curved line of the occipital bone. The lower limits are the sternal notch, clavicles, and a line from the acromioclavicular joint to the spinous process of the seventh cervical vertebra. The contour of the neck varies with age and sex. Anteriorly it contains the respiratory tube (larynx and trachea) and alimentary tube (pharynx and esophagus) (Fig. 7-9); on the side are the great vessels and nerves; and posteriorly is the cervical segment of the spine and its musculature.

The skin of the neck is loosely attached, especially anteriorly. It is very vascular and favors plastic surgery. The posterior area is thick and adherent and contains many sebaceous glands, making it prone to carbuncles and faruncles. The superficial fascia has fat and connective tissue and is not well defined. Lying in this fascia is the platysma muscle (see Muscles of Facial Expression).

Deep Cervical Fascia of the Neck. The deep cervical fascia consists of three layers: a superficial or investing layer, a middle or pretracheal layer, and a deep or prevertebral layer (Fig. 7-10). All layers protect structures in the neck.

Muscles of the Torso (Thorax, Abdomen)

The Thorax. The thorax (Figs. 7-11, 7-12, 7-13; Table 7-6) is flattened from front to back and contains the organs of respiration, circulation, and digestion. Its anterior wall is short and is formed by the sternum,

the first ten ribs, and cartilages. Its lateral walls are formed by ribs, and its posterior wall consists of the 12 thoracic vertebrae and ribs as far as the angles.

The diaphragm (Greek *dia*, in between; *phragma*, fence) (Fig. 7-14) is a dome-shaped, musculoaponeurotic partition found between the thorax and abdomen. When relaxed and seen from below (abdominal cavity), it forms a dome-shaped roof for the abdomen. Its circumferential portion is muscular, and these fibers curve upward and inward from every side to join the edges of an aponeurotic area called the central tendon, which acts as a site of insertion for the structure.

The diaphragm is "broken" by three large openings and several smaller ones. The large ones are for the (1) aorta, thoracic duct, and azygos vein at the level of T12; (2) esophagus, right and left vagus, and esophageal branches of the left gastric artery and veins at the level of T10; and (3) inferior vena cava, right phrenic nerve, and lymph vessels at the level of T8. The smaller openings transmit superior epigastric vessels, musculophrenic vessels, lower five intercostal nerves, last thoracic (subcostal) nerve and vessel, sympathetic trunk (behind the medial arcuate ligament), lesser and greater splanchnic nerves (pierce the crura), and the inferior hemiazygos vein (pierces the right crus).

Abnormal openings through the diaphragm that permit herniation of abdominal viscera into the tho-

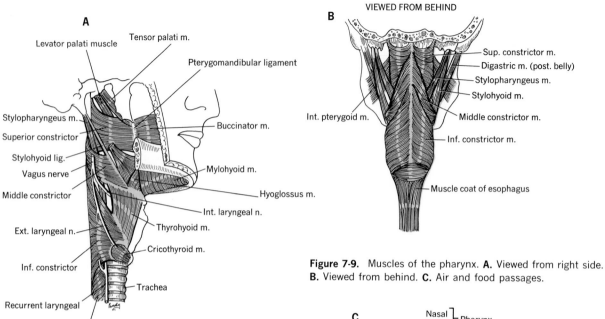

A

Levator palati muscle
Tensor palati m.
Pterygomandibular ligament
Stylopharyngeus m.
Buccinator m.
Superior constrictor
Stylohyoid lig.
Vagus nerve
Mylohyoid m.
Middle constrictor
Hyoglossus m.
Int. laryngeal n.
Ext. laryngeal n.
Thyrohyoid m.
Cricothyroid m.
Inf. constrictor
Trachea
Recurrent laryngeal
Esophagus
VIEWED FROM RIGHT SIDE

B VIEWED FROM BEHIND

Sup. constrictor m.
Digastric m. (post. belly)
Stylopharyngeus m.
Stylohyoid m.
Int. pterygoid m.
Middle constrictor m.
Inf. constrictor m.
Muscle coat of esophagus

Figure 7-9. Muscles of the pharynx. **A.** Viewed from right side. **B.** Viewed from behind. **C.** Air and food passages.

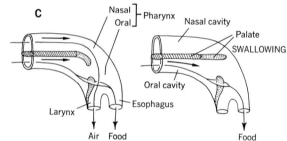

C

Nasal
Oral } Pharynx
Nasal cavity
Palate
SWALLOWING
Oral cavity
Larynx
Esophagus
Air Food
Food

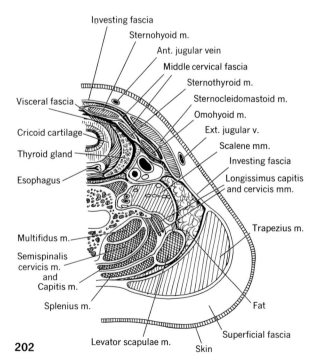

Investing fascia
Sternohyoid m.
Ant. jugular vein
Middle cervical fascia
Sternothyroid m.
Sternocleidomastoid m.
Visceral fascia
Omohyoid m.
Cricoid cartilage
Ext. jugular v.
Thyroid gland
Scalene mm.
Investing fascia
Esophagus
Longissimus capitis and cervicis mm.
Trapezius m.
Multifidus m.
Semispinalis cervicis m. and Capitis m.
Splenius m.
Fat
Levator scapulae m.
Skin
Superficial fascia

Figure 7-10. Fascia of the neck.

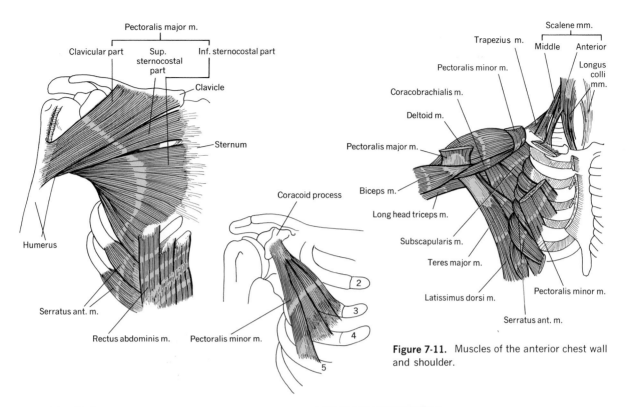

Pectoralis major m.

Clavicular part   Sup. sternocostal part   Inf. sternocostal part

Clavicle

Sternum

Humerus

Serratus ant. m.

Rectus abdominis m.   Pectoralis minor m.

Coracoid process

2

3

4

5

Scalene mm.

Trapezius m.   Middle   Anterior

Pectoralis minor m.   Longus colli mm.

Coracobrachialis m.

Deltoid m.

Pectoralis major m.

Biceps m.

Long head triceps m.

Subscapularis m.

Teres major m.

Latissimus dorsi m.   Pectoralis minor m.

Serratus ant. m.

Figure 7-11. Muscles of the anterior chest wall and shoulder.

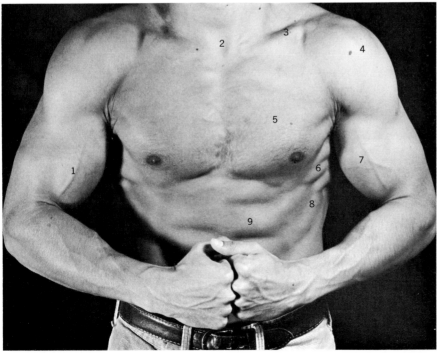

Figure 7-12. Muscles of the thorax (anterior view). Included, too, is the upper extremity. **1.** Cephalic vein. **2.** Suprasternal notch. **3.** Clavicle. **4.** Deltoid muscle. **5.** Pectoralis major muscle. **6.** Serratus anterior muscle. **7.** Biceps muscle. **8.** External oblique muscle. **9.** Rectus abdominis muscle.

203

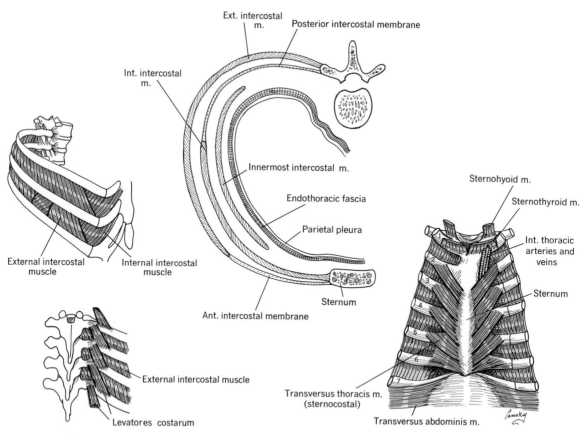

Figure 7-13. Deep muscles of the chest wall.

Table 7-6. Muscles of the Torso (Thorax)

Muscle	Origin	Insertion	Action	Nerve
1. Pectoralis major	Medial half of clavicle; anterior surface of manubrium and body of sternum and cartilages of first to sixth ribs	Crest of greater tubercle of humerus	Adducts and rotates arm; lowers arm when raised vertically	Anterior thoracic
2. Pectoralis minor	Third to fifth rib at the costochondral junctions	Top of coracoid process of scapula	Draws scapula down or raises ribs	Anterior thoracic
3. Transversus thoracis (sternocostalis)	Back of xiphoid cartilage	Costal cartilages of second to sixth ribs	Narrows chest	Intercostal

Table 7-6 (Continued)

Muscle	Origin	Insertion	Action	Nerve
4. Subcostal	Lower margins of the ribs near their angles	Upper margins of the second or third rib	Elevates ribs	Intercostal or thoraco-abdominal nerves
5. External intercostal*	Lower border of one rib from tubercles to rib cartilage	Upper border of rib below	Contracts during inspiration to elevate ribs and maintain tension in intercostal space	Intercostal
6. Internal intercostal*	Lower border of one rib from sternum to rib angle	Upper border of rib below	Contracts during expiration but also keeps tension in intercostal space	Intercostal
7. Innermost intercostal (part of 6)	Costal groove above	Upper margin of rib below	Uncertain	Intercostal
8. Levatores costarum	Transverse processes of last cervical and upper 11 thoracic vertebrae	Ribs next and second between angle and tubercle	Raises ribs	Intercostal
9. Serratus anterior	Center of external aspect of first eight or nine ribs	Superior and inferior angles and intervening vertebral border of scapula	Rotates scapula and pulls it forward; elevates ribs	Long thoracic from brachial plexus
10. Diaphragm			Muscle of respiration: descends when it contracts to increase thoracic pressure, increase abdominal pressure, and reduce abdominal volume	Phrenic nerve
a. Sternal part	Back of xiphoid process	Central tendon		
b. Costal part (right and left domes)	Inner surface of the lower six costal cartilages and lower four ribs	Central tendon		
c. Lumbar (vertebral) part	Medial and lateral arcuate ligaments; right crus: upper three to four lumbar vertebrae; left crus: upper two to three lumbar vertebrae	Central tendon		

*External intercostal: becomes membranous anteriorly between rib cartilages—the anterior intercostal membrane.
Internal intercostal: becomes membranous posteriorly beyond rib angles—the posterior intercostal membrane.

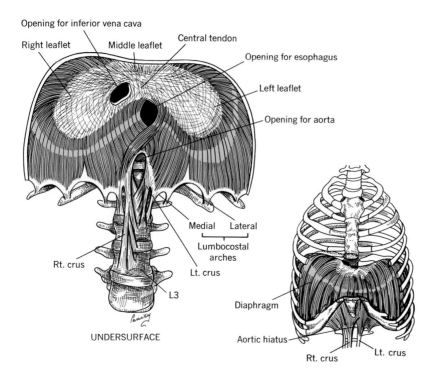

Opening for inferior vena cava
Right leaflet
Middle leaflet
Central tendon
Opening for esophagus
Left leaflet
Opening for aorta
Medial   Lateral
Lumbocostal arches
Rt. crus
Lt. crus
L3
UNDERSURFACE

Diaphragm
Aortic hiatus
Rt. crus   Lt. crus

Figure 7-14. Diaphragm as seen from below and anterior (slightly above).

racic cavity are the most common lesions at this site and often require surgery. Diaphragmatic hernias can be congenital, acquired, or traumatic.

The Abdominal Wall. The abdominal wall (Figs. 7-12, 7-15, 7-16; Table 7-7) is bounded above by the costal margins and xiphoid process of the sternum, and below by the iliac crest, inguinal ligaments, pubis, and pubic symphysis.

The lines of "tension" of the abdominal skin are almost transverse and, therefore, vertical scars tend to stretch, but transverse incisions are less conspicuous with time. The abdominal skin is loosely attached to the underlying structures except at the umbilicus. The linea alba is a fibrous raphe in the midline from the xiphoid to the symphysis pubis and is formed by the decussation of the three lateral abdominal muscles. Few to no blood vessels are found here. The superficial fascia appears as a single layer in the upper abdomen but consists of two layers between the umbilicus and pubis: (1) a superficial, fatty layer called Camper's fascia; and (2) a deeper, membranous layer called Scarpa's fascia, which is in contact with the deep fascia. The superficial layer contains small vessels and nerves and gives roundness to the body. The deep layer is devoid of fat and blood vessels and blends with the deep fascia. In the area of the pubic bone the deep fascia passes over the spermatic cords, penis, and scrotum and into the perineum, where it is called Colles' fascia. The skin is supplied by the lower six thoracic and first lumbar nerves.

The lower abdomen consists of nine layers. From superficial to deep, they are skin, superficial fascia of Camper, superficial fascia of Scarpa, external oblique muscle, internal oblique muscle, transversus abdominis muscle, transversalis fascia, extraperitoneal fat, and peritoneum.

The rectus abdominis muscle is a broad, long muscular band running from pubis to thorax on each side of the linea alba. The anterior part of the muscle is crossed by three tendinous intersections: one at the costal margin, one at the umbilicus, and one between the other two. Thus, a long muscle is divided into a

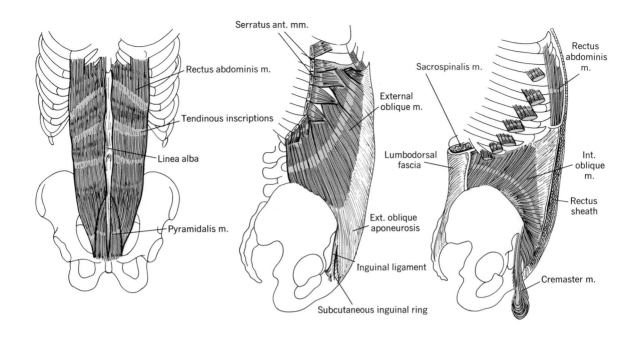

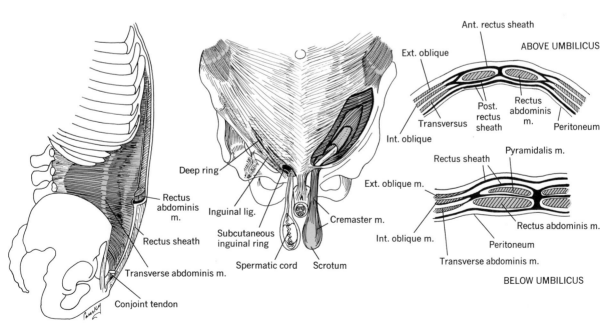

Figure 7-15. Muscles of the abdominal wall.

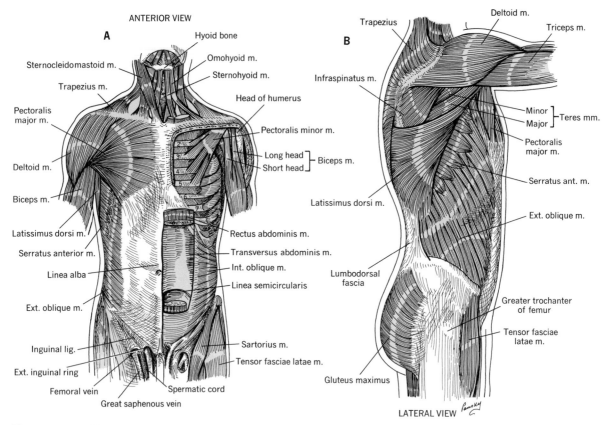

ANTERIOR VIEW

A

Hyoid bone

Sternocleidomastoid m.

Omohyoid m.

Sternohyoid m.

Trapezius m.

Head of humerus

Pectoralis
major m.

Pectoralis minor m.

Deltoid m.

Long head ⎤
Short head ⎦ Biceps m.

Biceps m.

Latissimus dorsi m.

Rectus abdominis m.

Serratus anterior m.

Transversus abdominis m.

Int. oblique m.

Linea alba

Linea semicircularis

Ext. oblique m.

Inguinal lig.

Sartorius m.

Ext. inguinal ring

Tensor fasciae latae m.

Femoral vein

Spermatic cord

Great saphenous vein

B

Deltoid m.

Trapezius

Triceps m.

Infraspinatus m.

Minor ⎤
Major ⎦ Teres mm.

Pectoralis
major m.

Serratus ant. m.

Latissimus dorsi m.

Ext. oblique m.

Lumbodorsal
fascia

Greater trochanter
of femur

Tensor fasciae
latae m.

Gluteus maximus

LATERAL VIEW

Figure 7-16. **A.** Muscles of the trunk, anterior view. **B.** Superficial muscles of the trunk as seen from the right side.

number of shorter ones and its strength and efficiency are increased. Besides its action on the pelvis and spine, it shares functions with the other abdominal muscles: they efficiently protect the abdominal viscera; by their tonicity they maintain intra-abdominal pressure and help keep the viscera in place; they function as respiratory muscles, since by contraction they press on the viscera, forcing them up and elevating the diaphragm; and they function in defecation, since contraction increases intra-abdominal pressure, which helps the rectum evacuate.

The aponeuroses of the abdominal muscles that approach the rectus from the sides form a sheath around the rectus, called the rectus sheath. It is usually divided into three parts:

1. Above the costal margin, where it is incomplete since its anterior wall is made up of external oblique aponeurosis while its posterior wall is absent and the rectus muscle lies directly on the cartilages.

2. From the rib margin to midway between the umbilicus and pubis. Here it is complete. The internal oblique aponeurosis forms an anterior and posterior layer of the sheath, having split around the rectus. Thus, anteriorly we have the aponeurosis of the external and internal oblique and posteriorly we have the internal oblique, transversus abdominis and transversalis fascia.

3. Midway between the umbilicus and pubis. Again, the internal oblique does not split but lies completely anterior to the rectus. The anterior wall consists of

the aponeuroses of external, internal, and transversus muscles, and posteriorly only the transversalis fascia is found. The inferior epigastric artery enters the sheath here, posteriorly; and where the internal layer is missing posteriorly, one sees an arched border called the linea semicircularis.

The rectus sheath contains rectus and pyramidal muscles, the superior and inferior epigastric vessels, and the terminal parts of the lower five intercostal thoracic nerves and their vessels.

The muscle fibers of the external oblique becomes tendinous below the line, joining the anterior superior iliac spine and the pubic tubercle. Here the aponeurosis forms a free border called the inguinal ligament (Poupart's ligament). The muscles, vessels, and nerves from the abdomen to the thigh pass under it.

The superficial or subcutaneous inguinal ring is a triangular, thinned-out portion of the external oblique aponeurosis, through which the spermatic cord (in the male) and the round ligament (in the female) pass. The "ring" is not an open defect but is covered by fibers. The descending testes push the fibers of the external oblique aponeurosis ahead of them to form a covering for the cord called the external spermatic fascia.

The lower fibers of the internal oblique aponeurosis become tendinous and arch over the spermatic cord to insert conjointly with the transversus abdominis into the pubic crest as the conjoined tendon (inguinal aponeurotic falx). The testicle in its embryonic descent "drags" some of the muscle loops of the internal oblique with it, as the cremasteric muscle (middle spermatic fascia). This muscle draws the testicle upward.

The transversus abdominis is the deepest of the abdominal muscles. The nerves in the region are found in the interval between the internal oblique and transversus and then enter the rectus sheath. The abdominal muscles are supplied by the seventh to twelfth thoracoabdominal nerves.

The transversalis or endoabdominal fascia is a connective tissue layer covering all of the internal surface of the abdomen, lying between the extraperitoneal layer and the transversus abdominis. Its thickness varies and when torn or weakened, particularly in the lower medial abdomen, predisposes to the development of a direct inguinal hernia. This layer is prolonged into the scrotum with the descent of the testes and becomes the internal spermatic fascia. The place

Table 7-7. Muscles of the Torso (Abdominal Walls): All Innervated by Branches of the Lower Thoracic Intercostal Nerves

Muscle	Origin	Insertion	Action
1. Rectus abdominis	Crest of symphysis of pubis	Xiphoid process and fifth to seventh costal cartilages	With pelvis fixed: draws thorax downward and bends spine forward. With thorax fixed: elevates pelvis
2. Pyramidalis	Crest of pubis	Lower portion of linea alba	Tenses linea alba
3. External oblique	Fifth to twelfth ribs	Anterior half of outer lip of iliac crest, inguinal ligament, and anterior layer of rectus sheath	Diminishes capacity of abdomen; draws thorax downward
4. Internal oblique	Iliac fascia deep to lateral part of inguinal ligament; anterior half of iliac crest; lumbar fascia	Tenth to twelfth ribs and sheath of rectus; some terminate in falx inguinalis	Diminishes capacity of abdomen, bends thorax forward
5. Transversus abdominis	Seventh to twelfth costal cartilages; lumbar fascia; iliac crest; inguinal ligament	Xiphoid cartilage, linea alba and through the falx inguinalis to spine and crest of pubis and iliopectineal line	Compresses abdominal contents

where the testicle meets the fascia at the abdominal wall and pushes it forward is called the internal or deep abdominal inguinal ring (really a thinned-out area of the fascia).

The inguinal canal is really a cleft through the inguinoabdominal region that measures about 4 to 5 cm. It extends from the deep inguinal ring, located a little above the center of the inguinal ligament, to the subcutaneous or superficial inguinal ring. Its boundaries are (1) anterior wall—aponeurosis of the external oblique plus fibers of the internal oblique laterally; (2) posterior wall—transversalis fascia in the entire length and the conjoined tendon medially; (3) roof—arched lower border of the internal oblique and to a lesser extent the transversus abdominis; and (4) floor—groove formed by a fusion of the inguinal ligament and transversalis fascia. Its contents include the spermatic cord (ductus deferens or round ligament), deferential vessels, testicular artery, pampiniform plexus of veins, lymphatics and autonomic nerves, ilioinguinal nerve, genital branch of the genitofemoral nerve, cremasteric muscle and artery, and internal spermatic fascia.

Hernias are prevalent in this area. A direct hernia passes into the scrotum (or labia) medial to the inferior epigastric artery, whereas an indirect (most common) passes lateral to this vessel. Umbilical hernias, too, are common in the abdominal region.

Muscles of the Upper Extremity

The Shoulder. This area (Fig. 7-17, Table 7-8) is divided into the axillary, pectoral, deltoid, and scapular regions.

THE AXILLARY AND PECTORAL REGIONS. The axilla is found between the medial side of the upper arm and the upper lateral chest wall. It has four walls, an apex, and a base, being pyramidal in shape. The anterior or pectoral wall is made up of the pectoralis major muscle and its fascia (superficially), the pectoralis minor and subclavius muscles, and the enveloping clavipectoral fascia (inner or deeper layer). The posterior or scapular wall is formed by the scapula, and the subscapularis, latissimus dorsi, and teres major muscles. The medial or costal wall consists of ribs 2 to 6 and the serratus anterior muscle. The lateral or humeral

wall consists of the humerus. The apex is bounded by three bones: the clavicle, anteriorly; the upper border of the scapula, posteriorly; and the first rib, medially. The base is made up of the skin, subcutaneous tissue, and axillary fascia.

The key to the axilla and pectoral region is the pectoralis minor muscle because of its relationship to the axillary vessels and brachial plexus. The vessels and nerves beneath the minor are covered by a connective tissue sheath known as the axillary sheath, which is a prolongation of the prevertebral fascia from the neck into the arm and extends almost to the elbow.

THE DELTOID AND SCAPULAR REGIONS. The scapula is the center of the area, giving muscle attachments and presenting many surgical and anatomic landmarks. The skin over the deltoid region is thick, and fibrous septa pass into the fibrous investment of the deltoid muscle. The deltopectoral groove, in which the cephalic vein runs, lies between the deltoid and pectoralis major muscles.

Three muscles in the scapular region are attached to the greater tuberosity of the humerus. The three "SIT" muscles are the supraspinatus, infraspinatus, and teres minor. The shoulder joint itself is bounded by the supraspinatus muscle above, the long head of the triceps below, the infraspinatus and teres minor behind, and the subscapularis in front. The tendons of the supraspinatus, infraspinatus, teres minor, and subscapularis converge and fuse with the capsule of the shoulder joint to form a common tendon, the cuff of the capsule. The subdeltoid bursa lies on the cuff and greater tuberosity. The subscapularis muscle is also separated from the shoulder joint by a "subscapular" bursa, which facilitates its movement on the front of the head and neck of the humerus.

The Arm or Brachium. The arm (Figs. 7-17, 7-18; Table 7-8) appears flattened from side to side because of the anterior and posterior muscles: the biceps brachii (anterior) and the triceps brachii (posterior). The medial bicipital groove (sulcus) separates the biceps and coracobrachialis muscles in front from the triceps behind and indicates the course of the brachial vessels, median nerve, and basilic vein. The lateral bicipital groove (sulcus) is not as clearly demarcated and in its

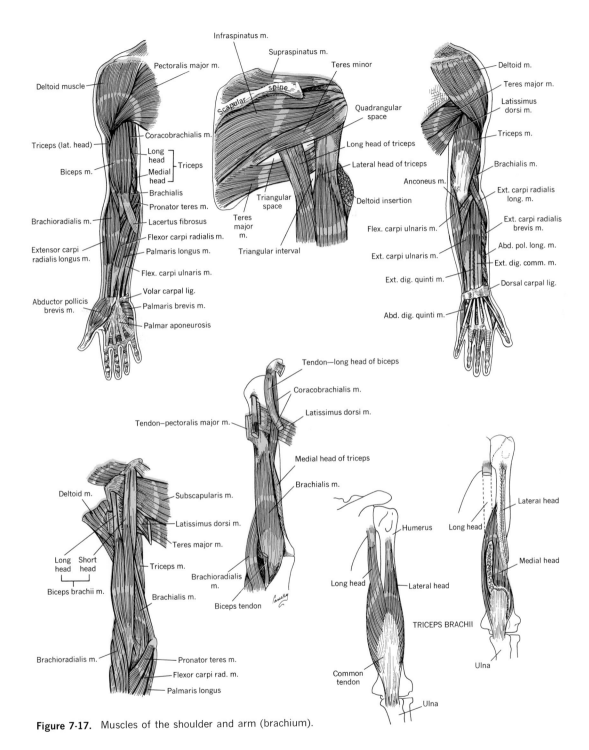

Figure 7-17. Muscles of the shoulder and arm (brachium).

Table 7-8. Muscles of the Upper Extremity

Muscle	Origin	Insertion	Action	Nerve
A. Shoulder and Arm Muscles				
1. Pectoralis major (see Table 7-6)				
2. Pectoralis minor (see Table 7-6)				
3. Deltoid*	Lateral third of clavicle; lateral border of acromion process, lower border of scapular spine	Lateral side of shaft of humerus just above its middle	Abduction, flexion, and rotation of arm	Axillary from fifth and sixth cervical via the brachial plexus
4. Biceps brachii	Long head: supraglenoid tuberosity of scapula Short head: coracoid process of scapula	Bicipital tubercle of radius (crosses both radius, shoulder and elbow joint)	Flexes and supinates forearm (puts a "corkscrew in and pulls a cork out")	Musculocutaneous
5. Coracobrachialis	Coracoid process of scapula	Middle of medial border of humerus	Raises arm	Musculocutaneous
6. Supraspinatus	Supraspinous fossa of scapula	Greater tuberosity of humerus	Abducts arm	Suprascapular from fifth and sixth cervical
7. Infraspinatus	Infraspinous fossa of scapula	Middle facet of great tuberosity of humerus	Extends arm and rotates it laterally	Suprascapular from fifth and sixth cervical
8. Teres minor	Posterior aspect of axillary border of scapula	Lower facet of great tuberosity of humerus	Adducts arm and rotates it laterally	Axillary from fifth and sixth cervical
9. Teres major	Lower third of posterior aspect of axillary border of scapula	Medial border of bicipital groove of humerus	Adducts and extends arm and rotates it medially	Lower subscapular from fifth and sixth cervical
10. Subscapularis	Subscapular fossa	Lesser tuberosity of humerus	Rotates arm medially	Subscapular from fifth and sixth cervical
11. Brachialis	Lower two thirds of anterior surface of humerus	Coronoid process of ulna	Most powerful flexor of forearm	Musculocutaneous and (usually) radial
12. Triceps brachii	Long head: axillary border of scapula below glenoid fossa Lateral head: lateral and posterior surface of humerus below greater tubercle Medial head: posterior surface of humerus below radial groove	Olecranon process of ulna	Extends forearm	Radial

B. *Forearm Muscles*
13. Anterior (volar) muscles (4)

	Origin	Insertion	Action	Nerve
a. Superficial group				
(1) Pronator teres	Medial condyle of humerus (superficial head), medial side of coronoid process of ulna (deep head)	Middle of outer surface of radius	Pronates forearm, flexes forearm	Median
(2) Flexor carpi radialis	Medial condyle of humerus	Anterior surface of base of second and third metacarpal bones	Flexes and abducts wrist	Median
(3) Palmaris longus	Medial condyle of humerus	Flexor retinaculum of wrist and palmar fascia (superficial to transverse carpal ligament)	Tenses palmar fascia and flexes forearm and hand	Median
(4) Flexor carpi ulnaris	Medial epicondyle, medial olecranon, posterior border of ulna	Pisiform, hamate, and fifth metacarpal	Flexes and adducts hand, flexes forearm	Ulnar
b. Middle group (1)				
(1) Flexor digitorum superficialis (sublimis)	Humeral head from medial condyle of humerus; Ulnar head from medial border of coronoid process of ulna; Radial head from oblique line and middle third of the lateral border of radius	Four split tendons, passing to either side of the profundus tendons into the sides of the second phalanx of each finger	Flexes middle phalanges of the fingers, flexes hand and forearm	Median
c. Deep plane (3)				
(1) Flexor pollicis longus	Anterior surface of the middle third of radius	Terminal phalanx of thumb	Flexes terminal phalanx of thumb	Median (anterior interosseous)
(2) Flexor digitorum profundus	Anterior surface of the upper third of ulna	By four tendons, piercing those of the superficialis, into the base of the terminal phalanx of each finger	Flexes terminal phalanges of fingers; Flexes hand	Ulnar and median

Table 7-8 (Continued)

Muscle	Origin	Insertion	Action	Nerve
(3) Pronator quadratus	Lower fourth of anterior surface of ulna	Lower fourth of anterior surface of radius	Pronates forearm	Anterior interosseous of median
14. Posterior (dorsal) muscles a. Superficial group (6)				
(1) Brachioradialis	Lateral supracondylar ridge of humerus	Base of styloid process of radius	Flexes forearm and assists slightly in supination	Radial
(2) Extensor carpi radialis longus	Lateral supracondylar ridge of humerus	Back of base of second metacarpal bone	Extends and abducts wrist, slight forearm flexor	Radial
(3) Extensor carpi radialis brevis	Lateral epicondyle of humerus	Back of base of third metacarpal bone	Extends and abducts wrist	Radial
(4) Extensor digitorum communis†	Lateral epicondyle of humerus	By four tendons into the backs of the first and second and base of terminal phalanges	Extends fingers	Radial (posterior interosseous)
(5) Extensor carpi ulnaris	Lateral epicondyle of humerus, oblique line and posterior ulna	Base of fifth metacarpal	Extends and adducts wrist	Radial (posterior interosseous)
(6) Anconeus‡	Back of lateral condyle of humerus	Olecranon process and posterior surface of ulna	Extends forearm and abducts ulna in pronation of wrist	Radial
b. Deep group (5) (1) Supinator	Lateral epicondyle of humerus and supinator ridge of ulna	Anterior and lateral surface of radius	Supinates forearm (with radial pronation, it tightens and then acts to "untwist" itself by supination)	Radial (posterior interosseous)
(2) Abductor pollicis longus	Interosseous membrane and posterior surfaces of radius and ulna	Lateral side of base of first metacarpal bone	Abducts and assists in extending thumb	Radial
(3) Extensor pollicis brevis	Dorsal surface of radius	Base of first phalanx of thumb	Extends and abducts first phalanx of thumb	Radial
(4) Extensor pollicis longus	Dorsal surface of ulna	Base of second phalanx of thumb	Extends all joints of the thumb and helps initiate forearm supination	Radial

Muscle	Origin	Insertion	Action	Nerve
(5) Extensor indicis (*proprius*)	Dorsal surface of ulna	Dorsum of first phalanx of index finger	Extends all joints of the index finger	Radial
C. Hand Muscles 15. Lumbricales (4)	The two lateral, from the radial side of the flexor tendons of the digitorum profundus (to fingers one and two); the two medial, from the adjacent sides of the second and third and third and fourth tendons (to fingers three and four)	Radial side of the first phalanx and extensor tendon on the dorsum of each of the four fingers	Flexes the first and extends the second and third phalanges	The two radial by the median, the two ulnar by the ulnar
16. Interossei a. Palmar (3)	First from ulnar side of second metacarpal; second and third from radial sides of fourth and fifth metacarpals	First into ulnar side of index, second and third into radial side of ring and little finger	Adducts fingers toward axis of middle finger and flexes the first phalanx	Ulnar
b. Dorsal (4)	By two heads, each from the shafts of the adjacent metacarpal bones	First phalanges and extensor expansion, first and second on radial side of index and middle finger, third and fourth on ulnar side of same fingers	Abducts all four fingers in relation to axis of middle finger and flexes first phalanx	Ulnar
17. Adductor pollicis a. Transverse head	Shaft of the third metacarpal	Medial side of base of first phalanx of thumb	Adducts thumb	Ulnar
b. Oblique head	Front of base of the second metacarpal, the trapezoid, and capitate bones			
18. Thenar eminence muscles a. Abductor pollicis brevis	Ridge of trapezium and flexor retinaculum	Lateral side of first phalanx of thumb	Abducts thumb	Median
b. Flexor pollicis brevis	From flexor retinaculum and ulnar side of first metacarpal	Base of first phalanx of thumb	Flexes first phalanx of thumb	Median and ulnar

Table 7-8 (Continued)

Muscle	Origin	Insertion	Action	Nerve
c. Opponens pollicis	Ridge of trapezium and flexor retinaculum	Anterior surface of first metacarpal bone	Opposes thumb to other fingers	Median
19. Hypothenar eminence muscles				
a. Abductor digiti V	Pisiform bone	Medial side of base of first phalanx of little finger	Abducts little finger	Ulnar
b. Flexor digiti V	Hamulus of hamate bone	Medial side of first phalanx of little finger	Flexes first phalanx of little finger	Ulnar
c. Opponens digiti V	Hamulus of hamate bone	Shaft of fifth metacarpal	Draws ulnar side of hand toward center of palm	Ulnar

* The deltoid is the great abductor of the arm although the first 15° is done by the supraspinatus.
† The tendons of this muscle are connected to each other by oblique bands so that extension of one finger at the metacarpal phalangeal joint is impossible if the other fingers are kept fixed.
‡ Placed here although not really part of the superficial extensor group.

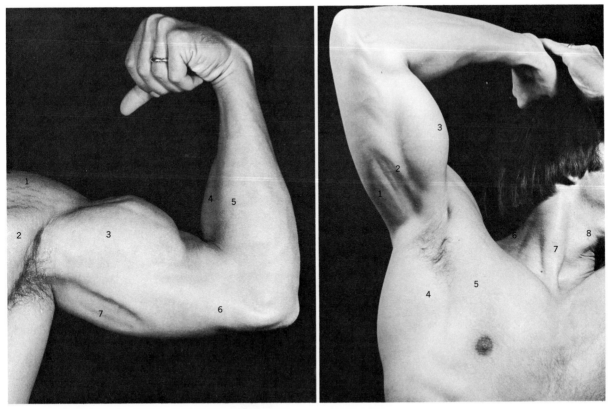

Figure 7-18. Muscles of the brachium (arm). **Left:** *1.* Deltoid m. *2.* Pectoralis major m. *3.* Biceps m. *4.* Flexor muscle group. *5.* Brachioradialis m. *6.* Triceps (medial head) m. *7.* Triceps (lateral head) m. **Right:** *1.* Triceps m. *2.* Coracobrachialis m. *3.* Biceps m. *4.* Subscapularis m. *5.* Pectoralis major m. *6.* Trapezius m. *7.* Sternocleidomastoid m. *8.* Thyroid cartilage.

lower portion separates the biceps from the brachioradialis and radial extensors. The cephalic vein is found here.

The deep fascia of the arm is continuous with that over the forearm. Intermuscular septa, on each side, are attached to the inner and outer margins of the humerus. Thus, the arm is divided into anterior and posterior compartments that limit inflammations, exudates, hemorrhage, and effusion. Fluids, however, can pass into the compartments by following the vessels and nerves that pierce the septa. The anterior or flexor compartment consists of three muscles: the biceps brachii, coracobrachialis, and brachialis. The posterior or extensor compartment is filled by the triceps brachii muscle.

The Elbow Region. Anteriorly the brachialis muscle covers the entire anterior aspect of the fibrous capsule of the elbow joint and separates the joint from the brachial artery and accompanying veins, the biceps tendon, and the median nerve. Laterally the common tendon of the extensor muscles overlies the lateral ligament of the elbow, and the extensor carpi radialis brevis and supinator muscles arise from it. Medially the common tendons of the flexor muscles lie over the anterior band of the medial ligament and the flexor superficialis muscle arises from it. The flexor carpi

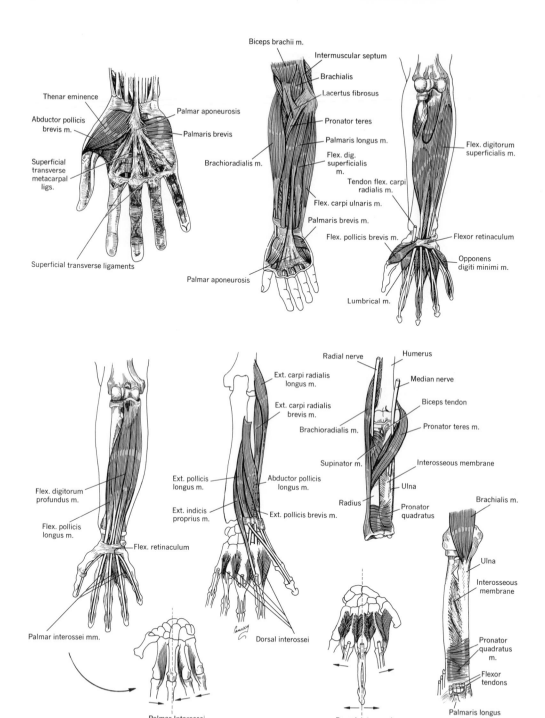

Figure 7-19. Muscles of the hand and forearm.

218 *Body Framework and Movement*

ulnaris muscle overlies the posterior band of the medial ligament of the joint. Posteriorly the triceps muscle and olecranon bursa lie over the olecranon of the ulna, with the anconeus muscle just to the lateral side of the olecranon. (See Table 7-8.)

The Forearm (Between Elbow and Wrist). The anterior (volar) area lies anterior to the radius and ulna and includes the internal and external muscle groups arising from the medial and lateral epicondyles of the humerus, respectively; the posterior (dorsal) region contains the extensor muscle group (Figs. 7-19, 7-20;

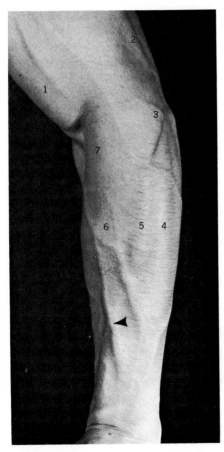

Figure 7-20. Superficial veins of the forearm and other landmarks. **1.** Biceps m. **2.** Triceps m. **3.** Lateral epicondyle. **4.** Extensor carpi ulnaris m. **5.** Extensor digitorum m. **6.** Extensor carpi radialis m. **7.** Brachioradialis m. The arrow points out a major tributary of the cephalic vein.

Table 7-8). Muscle masses that arise from the epicondyles increase the transverse diameter near the elbow, while distally bulk is lost because the muscles become tendons. The shafts of both radius and ulna can be felt superficially in the distal forearm.

The deep fascia is continuous with that of the arm. It is strengthened about the olecranon by expansions of the triceps muscle and reinforced anteriorly by the lacertus fibrosis. At the wrist it is continuous with the transverse and dorsal carpal ligaments (extensor retinaculum). From its deep side there are intermuscular septa to the radius and ulna forming muscle compartments.

Note that all of the muscles of the anterior forearm are supplied by the median nerve except the flexor carpi ulnaris, and the ulnar (medial) one half of the flexor digitorum profundus (to fingers 4 and 5). The latter two muscles are supplied by the ulnar nerve. On the other hand, all muscles of the dorsal region are innervated by the radial nerve. Furthermore, no motor nerves are found on the dorsum of the hand, no muscles originate from the dorsum of the hand, and no tendons insert into the dorsum of the carpal bones.

The long flexor tendons of the forearm muscles are enclosed in a common flexor tendon sheath or ulnar bursa, beginning at the wrist and extending to mid-palm. The sheath of the flexor pollicis longus usually separates and is called the radial bursa (Fig. 7-21). Part of the sheath is in a fibro-osseous canal (the carpal tunnel) bounded anteriorly by the transverse carpal ligament and posteriorly by the carpal bones and ligaments on the canal floor. The ligament is crossed anteriorly by the ulnar nerve, artery, and vein and tendon of the palmaris longus. The median nerve and flexor tendons run under the ligament and through the tunnel. Beyond the carpal ligament, the flexor tendons continue without a synovial sheath until they near the metacarpal heads, where they are individually covered by digital synovial sheaths. It is in this interval where no sheath is present that the lumbrical muscles arise from the deep flexor tendons.

The extensor retinaculum (dorsal carpal ligament, posterior annular ligament) (Figs. 7-22, 7-23) is a specialized part of the deep fascia of the forearm. It is

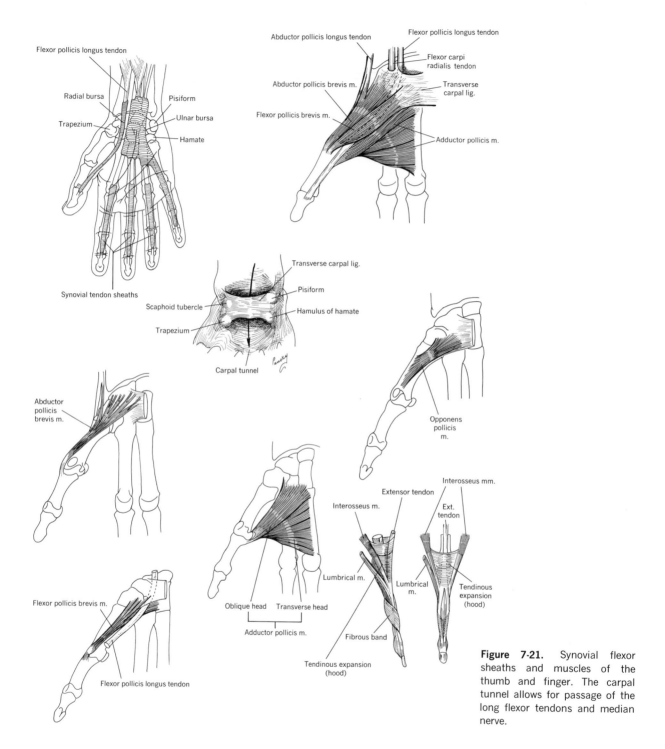

Flexor pollicis longus tendon

Radial bursa

Pisiform

Trapezium

Ulnar bursa

Hamate

Synovial tendon sheaths

Abductor pollicis longus tendon

Flexor pollicis longus tendon

Flexor carpi radialis tendon

Abductor pollicis brevis m.

Transverse carpal lig.

Flexor pollicis brevis m.

Adductor pollicis m.

Transverse carpal lig.

Pisiform

Scaphoid tubercle

Hamulus of hamate

Trapezium

Carpal tunnel

Abductor pollicis brevis m.

Opponens pollicis m.

Interosseus mm.

Extensor tendon

Interosseus m.

Ext. tendon

Lumbrical m.

Lumbrical m.

Fibrous band

Tendinous expansion (hood)

Flexor pollicis brevis m.

Oblique head   Transverse head

Adductor pollicis m.

Flexor pollicis longus tendon

Tendinous expansion (hood)

Figure 7-21. Synovial flexor sheaths and muscles of the thumb and finger. The carpal tunnel allows for passage of the long flexor tendons and median nerve.

220 *Body Framework and Movement*

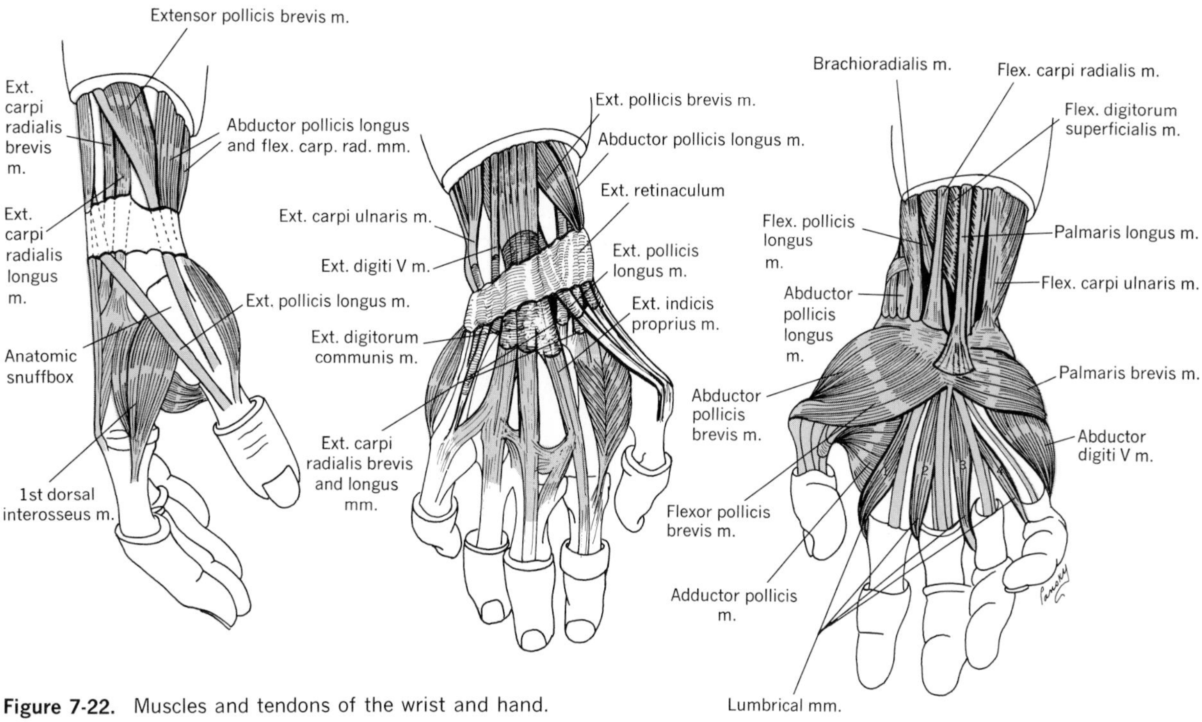

Extensor pollicis brevis m.
Ext. carpi radialis brevis m.
Abductor pollicis longus and flex. carp. rad. mm.
Ext. carpi radialis longus m.
Ext. pollicis longus m.
Anatomic snuffbox
1st dorsal interosseus m.
Ext. pollicis brevis m.
Abductor pollicis longus m.
Ext. retinaculum
Ext. carpi ulnaris m.
Ext. digiti V m.
Ext. pollicis longus m.
Ext. indicis proprius m.
Ext. digitorum communis m.
Ext. carpi radialis brevis and longus mm.
Brachioradialis m.
Flex. carpi radialis m.
Flex. digitorum superficialis m.
Flex. pollicis longus m.
Palmaris longus m.
Abductor pollicis longus m.
Flex. carpi ulnaris m.
Palmaris brevis m.
Abductor pollicis brevis m.
Abductor digiti V m.
Flexor pollicis brevis m.
Adductor pollicis m.
Lumbrical mm.

Figure 7-22. Muscles and tendons of the wrist and hand.

about 2.5 cm (1 in.) wide and acts as a strap that binds the extensor tendons down. Being obliquely placed, it does not interfere with pronation or supination at the wrist. Its inferior border is continuous with the deep fascia of the hand. It is attached medially to the triquetrum and pisiform bones and laterally to the lower 2.5 cm of the anterior border of the radius, forming an annular ligament for the head of the ulna. It is not attached to the lower end of the ulna and thus does not interfere with radial movements around the ulna. Five septa arise from its deep surface and attach to the head of the ulna and ridges on the back of the lower radius, forming six compartments. (The flexor retinaculum, in contrast, has only one compartment for the flexor tendons.) Each compartment transmits one or two tendons and is lined by a synovial sheath about the tendons (Fig. 7-22). Each tunnel is lined with a synovial sheath that extends about 2.5 cm (1 in.) proximally and distally from the retinaculum. The long abductor and short extensor tendons of the thumb pass through the most radial of the six tunnels and go beyond the prominence of the radial styloid process, and are subject to trauma and friction. A triangular depression over the dorsolateral aspect of the wrist, visible when the thumb is extended and abducted, is called the "anatomic snuffbox," whose radial border is formed by the long abductor and short extensor tendon of the thumb, medial border by the long extensor tendon of the thumb, and radial styloid process in the floor. The radial artery crosses the wrist in the "box" and the cephalic vein of Treves lies superficial to it. The long extensor is particularly prone to wearing, fraying, and rupture since it functions over bony prominences.

The Wrist. The distal skin crease is a good landmark, crossing the tip of the styloid process of the radius and lower part of the lunate bone. It marks the proximal border of the transverse carpal ligament and proximal row of carpal bones and is bisected by the palmaris longus tendon (clench the fist and see it). The median nerve lies beneath the tendon of the palmaris longus.

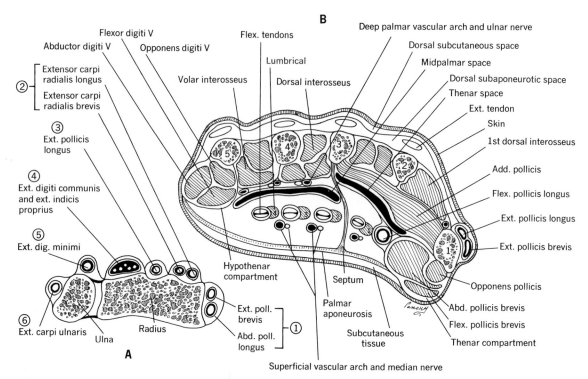

Figure 7-23. **A.** Dorsal wrist, tendon compartments. **B.** Cross section of hand through metacarpals.

The flexor carpi radialis tendon crosses its lateral third, and the flexor carpi ulnaris is at its extreme medial end (Figs. 7-19, 7-21, 7-22, 7-23; Table 7-8).

Lying proximal to the distal skin crease are layer 1 (lateral to medial), radial artery, flexor carpi radialis tendon, palmaris longus tendon, median nerve, flexor carpi ulnaris, and ulnar artery and nerve. The radial artery lies on the skeleton plane (lower end of the radius), covered here only by skin and superficial and deep fascia; therefore, it can be felt in taking the pulse. The radial artery is not related to the radial nerve at this site. The flexor carpi radialis tendon lies between the radial artery and median nerve and makes a private tunnel for itself as it pierces the flexor retinaculum. It travels over the scaphoid bone and can be used as a guide to it. The palmaris longus tendon (when present) bisects the crease and crosses anterior to the flexor retinaculum, continuing into the palm as the palmar

aponeurosis. It protects and overlies the median nerve. The flexor carpi ulnaris, at the medial extreme of the crease, roofs and protects the ulnar artery and nerve and acts as a guide to the pisiform bone. (Layer 2 consists of the flexor digitorum sublimis, four tendons lying between the median and ulnar nerves; layer 3 includes the flexor digitorum profundus and flexor pollicis longus tendons.)

Swelling of the wrist is diffuse or localized (and on the back of the hand may involve the synovium). The term *ganglion* is commonly used to describe a cystic enlargement, usually occurring on the back of the hand and related to a synovial lining inflammation or injury.

The Hand. The hand (Figs. 7-19, 7-21, 7-22, 7-23, 7-24; Table 7-8) is capable of a great variety of movements due to the coordinated actions of its extrinsic and intrinsic muscles and many joints. The thumb, with its ability of opposition, is all-important.

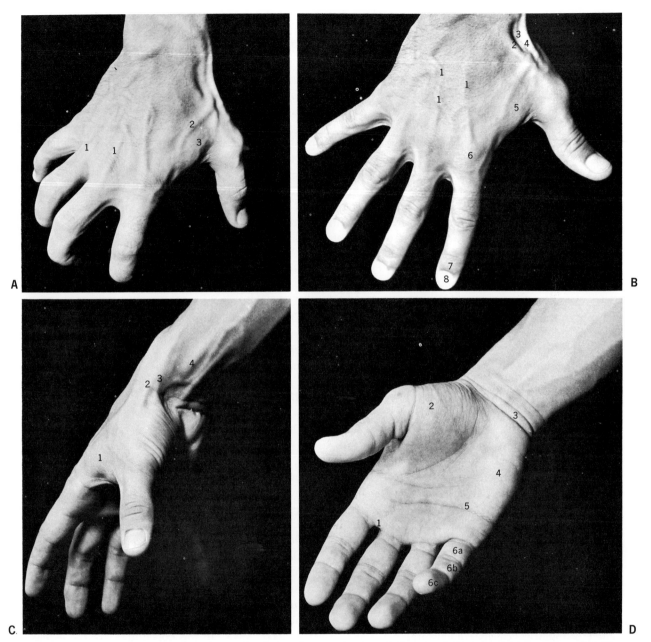

Figure 7-24. Surface anatomy of the wrist and hand. **A.** *1.* Extensor tendons. *2.* Dorsal v. *3.* First dorsal interosseus m. **B.** *1.* Dorsal venous plexus. *2.* Extensor pollicis longus. *3.* Anatomic snuffbox. *4.* Extensor pollicis brevis. *5.* First dorsal interosseus. *6.* Metacarpal head. *7.* Eponychium. *8.* Nail body. **C.** *1.* First dorsal interosseus. *2.* Extensor pollicis longus m. *3.* Extensor pollicis brevis m. *4.* Flexor digitorum m. **D.** *1.* Web. *2.* Thenar eminence. *3.* Distal wrist crease. *4.* Hypothenar eminence. *5.* Distal skin crease of palm. *6a.* Proximal phalanx. *6b.* Middle phalanx. *6c.* Distal phalanx.

PALMAR AREA. The "hollow" of the hand is bounded on the radial side by the thenar eminence and on the ulnar side by the hypothenar eminence. The center consists of nine layers (from superficial to deep): skin, subcutaneous tissue, palmar aponeurosis, superficial volar arterial arch and median nerve, flexor tendons (their sheaths and lumbrical muscles), deep palmar spaces, deep volar arterial arch and ulnar nerve, adductor pollicis, and interosseous muscles with the metacarpal bones.

Skin. The skin in the palm is thicker, coarser, and more vascular than that of the dorsum of the hand. There are no sebaceous glands or hairs but many sweat glands. One sees a proximal and distal transverse skin crease. A vertical crease ("lifeline") is also apparent.

Subcutaneous Tissue. The skin is bound down by fibrous septa with granular fat between. Over the thenar and hypothenar eminences the fat is sparse. Thus, where the skin creases are seen, little fat is found and penetrating wounds can be quite serious. The palmaris brevis muscle is a superficial muscle supplied by the ulnar nerve and lies over the hypothenar area. It raises the skin and fascia (wrinkles) in that area and enables the hand to grasp more firmly. The webs of the fingers are found on the palm side but not on the back of the hand. They are created by the superficial transverse ligaments.

Palmar Aponeurosis. The palmar aponeurosis is the deep palmar fascia that, opposite the distal skin crease, divides into four slips, one to each finger. A spontaneous contracture of the palmar fascia results in flexion of the fingers and is known as Dupuytren's contracture. The superficial transverse palmar ligament is in the webs of the palmar side of the hand and is often incomplete; the superficial transverse metacarpal ligament holds the metacarpal heads together anteriorly, while the dorsal transverse metacarpal ligament holds them together dorsally.

The intervals between the slips of palmar fascia are called the lumbrical canals (commissural spaces or web spaces) and are surgically important because infections from the subcutaneous tissue of the finger to the palm must pass this way. Their contents are fatty tissue, lumbrical muscles, and digital vessels and nerves.

Superficial Volar Arterial Arch and Median Nerve. See Figure 7-23 and Chapters 9 and 10.

Flexor Tendons, Sheaths, and Lumbrical Muscles. As the flexor tendons (sublimis and profundus) pass into the palm, they receive two sheaths, a fibrous and a synovial (Figs. 7-19, 7-21, 7-22, 7-23). The fibrous sheaths of the thumb and little finger are similar to those of the other three fingers and form, with the phalanges, a strong osteofibrous (fascial) tunnel to keep the tendons in place. Both superficial and deep tendons enter each tunnel, which is anchored to the lateral and medial borders of the phalanges. The tendons are thin and pliable over the joints but strong over the middle of the bones. Thus, the flexor retinaculum, palmar aponeurosis, and fibrous flexor sheaths form a continuous "fibrous plate" to hold the tendons in position and increase their efficiency. The palmar aponeurosis gives no fibrous slip to the thumb (in contrast to the large toe), thus allowing a wide range of thumb motion.

The synovial sheath, in contrast to the above, is a lubricating structure in a tubular form that surrounds a tendon and acts as a "bursa" to protect the tendon from the bones in hand motions. The synovial sheaths are sacs, closed at both ends and made up of endothelial-lined membranes.

See Table 7-8 for a description of lumbrical muscles.

Deep Palmar Spaces. A fibrous septum exists from the palmar aponeurosis to the third metacarpal, dividing the palm into a medial midpalmar space and a lateral thenar space. These spaces lie deep to the tendons of the hand and are limited by surrounding structures (Fig. 7-23). Infections are limited by these spaces, but they can also follow the structures within them such as the lumbricals to the dorsum of the hand.

Deep Volar Arterial Arch and Ulnar Nerve. See Figure 7-23 and Chapters 9 and 10.

Adductor Pollicis. See Table 7-8.

Interosseous Muscles and Metacarpal Bones. See Table 7-8 for muscles.

DORSAL AREA. In the dorsal area are the extensor tendons, palpable and visible as are the metacarpal bones. The skin is fine with many short hairs and sebaceous glands. Directly under the skin and above the tendons is a dorsal subcutaneous space filled with areolar connective tissue; if the space is infected, pus spreads easily and readily over the entire dorsum of the hand. Beneath the tendons is the dorsal subaponeurotic space, which is also filled with loose connective tissue; pus here is limited distally by the tendon attachments and proximally at the bases of the metacarpal bones and to either side by the tendon attachments.

THE FINGERS (PHALANGES). The hand has a thumb and four fingers, or five digits beginning with the thumb as number 1 (Figs. 7-19, 7-21, 7-22, 7-24). They are all the same except that the thumb has but two phalanges (not three). The skin of the flexor surface is thick, has some subcutaneous fat, is slightly mobile, and has no hair follicles. Over the dorsum it is thinner, is mobile, and has hair with very little subcutaneous fat. The creases on the palmar side are deceiving with regard to the joints since the proximal one is distal to the metacarpophalangeal joint, the middle one is a guide to the joint, but the distal one is somewhat proximal to the interphalangeal joint. The amount of fat at the creases is minimal, and, therefore, a wound at the crease may penetrate the deeper synovial membrane. The subcutaneous tissue over the flexor surface consists of fibrous tissue enclosing small amounts of fat; it connects the skin to the fibrous sheaths and to the terminal phalanx. The digital vessels run in the subcutaneous tissue.

The distal closed space of the finger is important since trauma here is common. The subcutaneous tissue consists of strong fibrous septa radiating to the periosteum with fatty tissue between. The distal four fifths of the phalanx becomes a closed space. Thus, edema or exudates in the space are limited, tension rises, the blood supply shuts off, and necrosis of the bone can occur. The arrangement is also important for understanding a felon (infection of this closed space) and a paronychia (an acute infection of the subepithelial tissue at the side of the nail).

Muscles of the Posterior and Posterolateral Body Wall (Including Posterior Neck)

This region is a quadrilateral area located between the lowest rib, iliac crest, vertebral column, skull, and

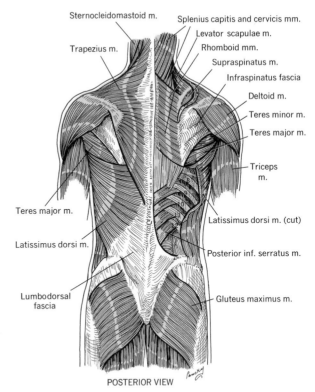

POSTERIOR VIEW

Figure 7-25. Superficial muscles of the trunk.

anterior superior iliac spine (Figs. 7-25, 7-26, 7-27, 7-28, 7-29; Table 7-9). The superficial fascia has two layers with much fat between. The area is related to the lumbodorsal fascia. The latter is made up of three layers that fuse laterally to give origin to the transversus abdominis muscle. Between the posterior and middle layers is the sacrospinalis muscle group (erector spinae), and between the middle and anterior is the quadratus lumborum muscle.

The muscles consist of several groups. In the neck, from superficial to deep, we have the trapezius, splenius, semispinalis, and suboccipital muscles. The deep anterior muscles of the neck are the longus capitis and cervicis. In the thoracic and lumbar region the superficial group includes the latissimus dorsi and external abdominal oblique; the middle group consists of the serratus posterior superior and inferior, the sacrospinalis, and the internal abdominal oblique; and the

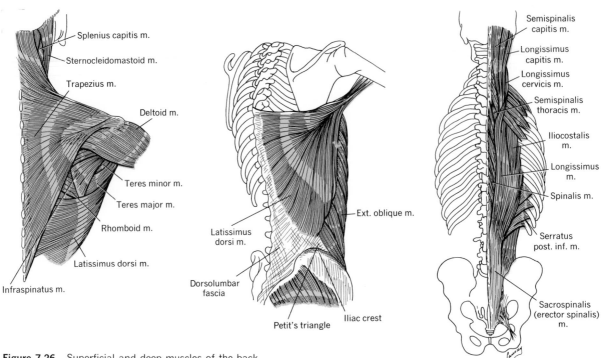

Figure 7-26. Superficial and deep muscles of the back.

Table 7-9. Posterior and Posterolateral Muscles of Neck and Body Wall

Muscle	Origin	Insertion	Action	Nerve
1. Latissimus dorsi	Spinous processes of lower five or six thoracic and the lumbar vertebrae; median ridge of sacrum and outer tip of iliac crest	Into posterior lip of bicipital groove of humerus	Adducts arm, rotates arm medially and extends it	Thoracodorsal from brachial plexus
2. External oblique (see Table 7-7)				
3. Serratus posterior superior	Spines of two lower cervical and two upper thoracic vertebrae	Outer side of angles of second to fifth ribs	Depresses ribs to which it inserts	First to fourth intercostal
4. Serratus posterior inferior	Spines of lower two thoracic and two upper lumbar vertebrae	Lower borders of last four ribs	Depresses ribs to which it inserts	Ninth to twelfth intercostal

Table 7-9 (Continued)

Muscle	Origin	Insertion	Action	Nerve

5. Sacrospinalis
  a. Iliocostalis
   cervicis
	Angles of middle and upper ribs	Transverse processes of middle cervical vertebrae	Extends thoracic spine	Branches of thoracic spinal nerves

Iliocostalis thoracis	Medial side of angles of seventh to twelfth ribs	Angles of first to sixth ribs and transverse processes of seventh cervical vertebra	Extends thoracic spine	Branches of thoracic spinal nerves
Iliocostalis lumborum	With erector spinal (sacrum, ilium, and lumbar vertebrae)	Angles of fifth to twelfth ribs	Extends lumbar spine	Branches of thoracic and lumbar spinal nerves
b. Longissimus capitis	Transverse processes of upper thoracic and transverse processes and articular processes of lower and middle cervical vertebrae	Mastoid process	Keeps head erect; bows head back or to one side	Posterior branches of cervical spinal nerves
Longissimus cervicis	Transverse processes of upper thoracic vertebrae	Transverse processes of middle and upper cervical vertebrae	Extends cervical spine	Posterior branches of lower cervical and upper thoracic spinal nerves
Longissimus thoracis	With iliocostalis and from transverse processes of lower thoracic vertebrae	Into most or all of ribs between angles and tubercles; tips of transverse process of upper lumbar and thoracic vertebrae	Extends spinal column	Thoracic and lumbar spinal nerves
c. Spinalis capitis	Spines of upper thoracic and lower cervical vertebrae	Blends with semispinalis capitis (into occipital bone)	Draws head back	Posterior branches of cervical spinal nerves
Spinalis cervicis	Spines of sixth and seventh cervical vertebrae	Spines of axis and third cervical vertebra	Extends cervical spine	Posterior branches of cervical spinal nerves
Spinalis thoracis	Spines of upper lumbar and two lower thoracic vertebrae	Spines of middle and upper thoracic vertebrae	Supports and extends vertebral column	Posterior branches of thoracic spinal nerves
6. Splenius capitis	Spines of last four cervical and first three thoracic vertebrae	Outer half of superior nuchal line of occipital mastoid process	Rotates head; together draw head back	Second to eighth cervical nerves
Splenius cervicis	Spines of third to fifth cervical vertebrae	Posterior tubercles of transverse processes of first and second cervical vertebrae	Rotates head; together draw head back	Second to eighth cervical nerves
7. Suboccipital muscles a. Rectus capitis posterior major	Spine of axis	Middle of inferior nuchal line of occipital bone	Rotates and draws head backward	Posterior branch of first cervical (suboccipital)

Table 7-9 (Continued)

Muscle	Origin	Insertion	Action	Nerve
b. Rectus capitis posterior minor	Posterior tubercle of atlas	Medial third of inferior nuchal line of occipital bone	Rotates and draws head backward	Suboccipital
c. Obliquus capitis superior	Transverse process of atlas	Outer third of inferior curved line of occipital bone	Rotates head	Suboccipital
d. Obliquus capitis inferior	Spine of axis	Transverse process of atlas	Rotates head	Suboccipital
8. Deep anterior neck				
a. Longus capitis	Anterior tubercles of transverse processes of third to sixth cervical vertebrae	Basilar process of occipital bone	Twists and bends neck forward	Cervical plexus
b. Longus colli (cervicis)	(1) From bodies of third thoracic to fifth cervical vertebrae	(1) Bodies of second to fourth cervical vertebrae	Twists and bends neck forward	Anterior branches of cervical spinal nerves
	(2) Anterior tubercles of transverse processes of third to fifth cervical vertebrae	(2) Anterior tubercle of atlas		
	(3) Bodies of first to third thoracic vertebrae	(3) Anterior tubercles of the transverse processes of the fifth and sixth cervical vertebrae		
9. Transversus spinalis				
a. Semispinalis capitis	Transverse processes of five or six upper thoracic and four lower cervical vertebrae	Occipital bone between superior and inferior nuchal lines	Rotates head and draws it backward	Suboccipital, greater occipital, and branches of cervical
b. Semispinalis cervicis	Transverse processes of second to fifth thoracic vertebrae	Spines of axis and third to fifth cervical vertebrae	Extends cervical spine	Branches of cervical spinal nerves
c. Semispinalis thoracis	Transverse processes of fifth to eleventh thoracic vertebrae	Spines of first four thoracic and fifth to seventh cervical vertebrae	Extends vertebral column	Branches of cervical spinal nerves
d. Multifidus	Sacrum, sacroiliac ligament, mammillary processes of lumbar vertebrae, transverse process of thoracic vertebrae, and articular processes of last four cervical vertebrae	Spines of all vertebrae up to and including axis	Rotates vertebral column	Posterior divisions of spinal nerves

Table 7-9 (Continued)

Muscle	Origin	Insertion	Action	Nerve
10. Quadratus lumborum	Iliac crest, iliolumbar ligament, and transverse processes of lower lumbar vertebrae	Twelfth rib and transverse processes of upper lumbar vertebrae	Flexes trunk laterally, fixes the last rib, and helps in inspiration	Upper lumbar spinal nerves
11. Psoas major	Bodies of vertebrae and intervertebral disks from the twelfth thoracic to fifth lumbar; transverse processes of the lumbar vertebrae	Lesser trochanter of femur (with iliacus, as the iliopsoas)	Flexes thigh and rotates it slightly medially	Lumbar plexus

Figure 7-27. Deep muscles of the neck, anterior and posterior. **A, B,** and **C** indicate depth, from superficial to deep.

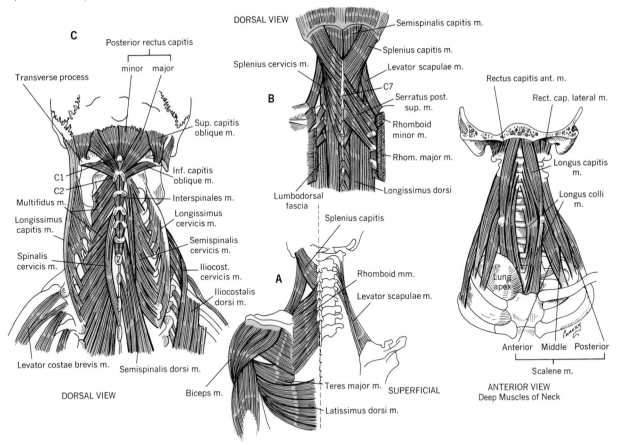

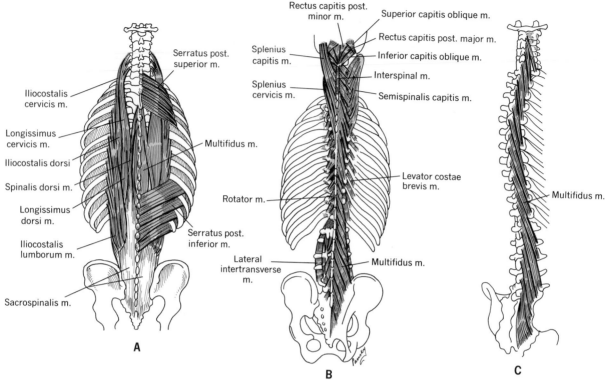

Figure 7-28. Deep muscles of the back. **A, B,** and **C** indicate approach, from superficial to deep.

deep set are the quadratus lumborum, psoas major (Fig. 7-30), transversus abdominis, and multifidus.

Pelvic Muscles

The pelvic muscles are discussed with the male and female reproductive systems (see Chaps. 17 and 18).

Muscles of the Lower Extremity

The Hip. DEEP FASCIA. The deep fascia of the gluteal region is strongly attached to the iliac crest (Fig. 7-31). Where it overlies the gluteus medius muscle (Table 7-10), it appears as a dense sheet. When it reaches the upper border of the gluteus maximus muscle, it splits into two layers to enclose that muscle. It appears thin and transparent over the maximus and sends septa into that muscle.

MUSCLES. The gluteus maximus muscle is the most massive muscle in the body as well as the strongest, heaviest, and coarsest. It is especially well developed in man because of his upright position. The lower border of the muscle is not parallel with the gluteal fold but crosses it obliquely, extending from the coccyx and across the ischial tuberosity to the shaft of the femur. The tuberosity is covered when one stands erect but is uncovered when sitting.

This muscle is the great extensor of the thigh, bringing the bent thigh in line with the body. It is important in walking, going up an incline, or rising from the sitting position by pulling the pelvis backward.

Three subgluteal synovial bursae are found below the muscle: (1) a small one between the muscle and ischial tuberosity, which may enlarge in those who have sedentary occupations (weavers' bottom); (2) an-

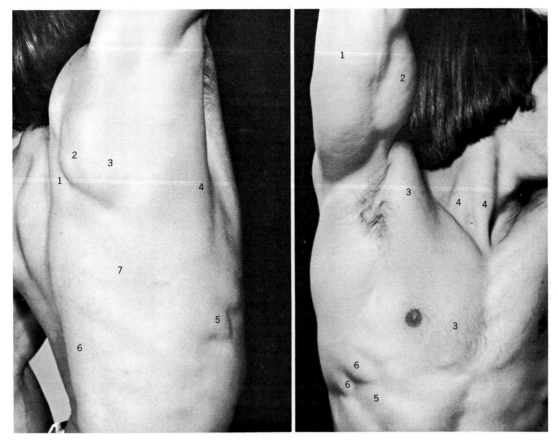

Figure 7-29. Surface anatomy of some chest and back muscles. **Left:** *1.* Rhomboid m. *2.* Scapula. *3.* Infraspinatus m. *4.* Latissimus dorsi m. *5.* Serratus ant. m. *6.* Erector spinae m. *7.* Latissimus dorsi. **Right:** *1.* Triceps m. *2.* Biceps m. *3.* Pectoralis major m. *4.* Sternocleidomastoid m. *5.* Ext. oblique m. *6.* Serratus ant. m.

other between the muscle and greater trochanter (usually large); and (3) still another between the muscle and the upper part of the vastus lateralis muscle.

The gluteus medius and minimus have powerful actions on the pelvis when the thigh is fixed and are especially important in walking. When walking, the muscles of the limb on the ground abduct the pelvis (hold or tilt it, so that the side of the swinging [free] limb is prevented from sagging and thus can clear the ground). Paralysis of the medius leads to a lurching or waddling gait.

The subcutaneous tissue of the buttock is usually thick and fatty and encloses the gluteus maximus, continues forward as a strong aponeurotic sheet over the gluteus medius, and splits around the tensor fasciae latae. Its deep layer fuses with the capsule of the hip joint. Below, it continues distally as the iliotibial tract of the fascia lata. It is bound to the skin along the gluteal fold, below the lower gluteus maximus border.

The arrangement of nerves and vessels is such that the upper and lateral quadrant of the buttock and anterior part of the gluteal region (part with the tensor fasciae latae) are relatively avascular and free of major nerves and thus are regions commonly used for intramuscular injections.

The Thigh. Deep Fascia. The deep fascia (fascia lata) attaches above and below to all bony and ligamentous structures available around the limb (Figs. 7-32, 7-33). It fuses with the iliotibial tract laterally and is very strong there. The fascia lata provides for septa that separate the various muscle groups of the thigh (Table 7-10), each inserting at the linea aspera and each enclosed in a separate fascial compartment. Three muscle groups need three septa: the lateral intermuscular septum separates the extensors from the flexors; the medial intermuscular septum separates the extensors from the adductors; and the posterior intermuscular septum separates the adductors from the flexors.

The fascia lata has several small openings for the passage of vessels and nerves and one large opening, the fossa ovalis, for the greater saphenous vein. The opening is about 2.5 cm (1 in.) wide and is found 2.5 cm below the medial end of the inguinal ligament. The fossa is covered by loose areolar and fatty tissue that fills the space (cribriform fascia).

Muscles. *The Adductor or Medial Group (Obturator Nerve).* These are situated on the medial aspect of the thigh and consist of six muscles: the adductor magnus, longus, and brevis; the pectineus; the gracilis; and the obturator externus. These muscles are interposed between the extensor group in front and the flexor group behind. They arise from the bones around the obturator foramen and membrane and insert from the trochanteric fossa of the femur above to the medial surface of the tibia below.

The Extensor or Anterior Group (Femoral Nerve). These are referred to as either the flexors of the hip or extensors of the knee and include the sartorius, the quadriceps femoris, the iliopsoas, and the pectineus muscles.

The adductor or subsartorial canal of Hunter (named after John Hunter, who here ligated the femoral artery) is an intermuscular canal situated on the medial aspect of the middle third of the thigh. It extends from the apex of the femoral triangle (of Scarpa) to the opening in the adductor magnus muscle (the adductor hiatus).

The canal is bound laterally by the vastus medialis and posteriorly (its floor) by the adductor longus and magnus along with the medial intermuscular septum; it is roofed by the sartorius muscle and its fascia.

The femoral triangle (of Scarpa) lies directly below the inguinal ligament, which forms its base. It is bounded laterally by the medial border of the sartorius and medially by the medial border of the adductor longus. The floor (lateral to medial) is formed by the iliacus, pectineus, and adductor longus muscles. It is roofed by fascia lata, and its contents include the femoral vessels and nerve.

The Flexor or Posterior Group (Sciatic Nerve). Here we find the three hamstring muscles: the semimembranosus, the semitendinosus, and the biceps femoris. A hamstring muscle is one that arises from the ischial tuberosity, inserts into one of the two bones of the leg, and is supplied by the tibial division of the sciatic nerve. It covers the femur but does not attach to it. Only the short head of the biceps does not follow the "rules" and is an exception. It really belongs to the muscle plane of the gluteus maximus and is supplied by the peroneal division of the sciatic nerve. Note, too, that the adductor magnus fulfills the criteria, but it has been considered with the adductor group.

The Knee. The popliteal region is the posterior aspect of the knee (Fig. 7-32). The popliteal fossa or space is "lozenge shaped." Its roof is formed by deep fascia that acts to restrain the hamstrings. The roof is pierced, near its center, by the small saphenous vein that passes between the two heads of the gastrocnemius muscle (Table 7-10). The floor is formed by the lower end of the femur, the posterior part of the knee joint capsule, and the popliteus muscle with its fascia. The upper borders of the space are bounded medially by the semimembranosus overlaid by the tendinosus and laterally by the biceps femoris muscle. The lower borders of the space are bounded by the two heads of the gastrocnemius muscle along with a "variable" plantaris muscle laterally. The contents of the space are the lateral and medial popliteal nerves and branches, the popliteal vein and its tributaries, the popliteal artery and its branches, the posterior cutaneous nerve, fat, and lymph glands.

[*Text continues on p. 237.*]

Table 7-10. Muscles of the Lower Extremity

Muscle	Origin	Insertion	Action	Nerve
A. *The Hip*				
1. Gluteus maximus (most massive body muscle)	Ilium behind posterior gluteal line; posterior surface of sacrum and coccyx and sacrotuberous ligament	Iliotibial band of fascia lata and gluteal ridge of femur	Extends thigh	Inferior gluteal (L5, S1, S2)
2. Piriformis (see Table 17-1)				
3. Obturator internus (see Table 17-1)				
4. Gluteus medius	Ilium between anterior and posterior gluteal lines	Outer surface of greater trochanter of femur	Abducts and rotates thigh medially	Superior gluteal
5. Gluteus minimus	Ilium between anterior and inferior gluteal lines	Greater trochanter of femur	Abducts thigh and helps medial rotation	Superior gluteal
6. Tensor fasciae latae	Anterior superior spine and adjacent surface of the ilium	Iliotibial band of fascia lata	Tenses fascia lata, extends knee via the fascia lata, and flexes thigh with iliopsoas	Superior gluteal
7. Gemellus superior and inferior	Ischial spine and margin of lesser sciatic notch	Tendon of obturator internus muscle	Rotates thigh laterally	Sacral plexus (L5, S1, S2)
8. Quadratus femoris	Lateral border of tuberosity of ischium	Intertrochanteric ridge	Rotates thigh laterally (may adduct)	Sciatic
9. Obturator externus	Lower half of margin of obturator foramen and adjacent part of external surface of obturator membrane	Digital fossa of greater trochanter	Rotates thigh laterally (may adduct)	Obturator
B. *The Thigh*				
1. Adductor or medial group				
a. Pectineus	Crest of pubis	Pectineal line of femur	Adducts thigh and assists in flexion	Obturator and femoral
b. Adductor longus	Symphysis and crest of pubis	Middle third of medial lip of linea aspera	Adducts thigh	Obturator
c. Adductor brevis	Superior ramus of pubis	Upper third of medial lip of linea aspera	Adducts thigh	Obturator
d. Adductor magnus*	Ischial tuberosity and ischiopubic ramus	Linea aspera and adductor tubercle of femur	Adducts and extends thigh	Obturator
e. Gracilis	Ramus of pubis near symphysis	Shaft of tibia below medial tuberosity	Adducts thigh; flexes knee; rotates leg medially	Obturator

Table 7-10 (Continued)

Muscle	Origin	Insertion	Action	Nerve
2. Extensor or anterior group (2)				
a. Sartorius	Anterior superior spine of ilium	Medial border of tibial tuberosity	Flexes thigh and leg; rotates leg medially; rotates thigh laterally	Femoral
b. Quadriceps femoris				
(1) Rectus femoris	Anterior inferior spine of ilium and upper margin of acetabulum	Inserts into patella and then by the ligamentum patellae into the tibial tuberosity	Extends leg; flexes thigh by action of rectus femoris	Femoral
(2) Vastus medialis	Medial line of linea aspera			
(3) Vastus intermedius	Upper three fourths of anterior surface of shaft of femur			
(4) Vastus lateralis	Lateral lip of linea aspera			
3. Flexor or posterior group (3)				
a. Biceps femoris				
(1) Long head	Tuberosity of ischium	Head of fibula	Flexes knee and rotates it laterally; long head also extends hip	Via tibial
(2) Short head	Lower half of lateral lip of linea aspera	Head of fibula	Flexes knee and rotates it laterally; long head also extends hip	Via peroneal
b. Semitendinosus	Ischial tuberosity	Side of shaft of tibia below medial condyle	Flexes leg and rotates it medially	Tibial
c. Semimembranosus	Ischial tuberosity	Medial condyle of tibia and to tibial collateral ligament of knee joint	Flexes leg and rotates it medially; tenses capsular ligament of knee joint	Tibial
C. *The Leg*				
1. Anterior or extensor group (4)				
a. Tibialis anterior	Upper two thirds of lateral surface of tibia; interosseous membrane and intermuscular septum	Medial cuneiform and base of first metatarsal bone	Dorsal flexion and supination (invertor) of foot	Anterior tibial
b. Extensor hallucis longus	Front of tibia and interosseous membrane	Base of terminal phalanx of big toe	Extends big toe and aids in dorsiflexion of foot	Anterior tibial

Table 7-10 (Continued)

Muscle	Origin	Insertion	Action	Nerve
c. Extensor digitorum longus	Lateral tuberosity of tibia; upper two thirds of anterior surface of fibula	By four tendons into the dorsal surface of second to fifth toes	Extends four lateral toes; dorsiflexes foot	Anterior tibial
d. Extensor tertius (part of c) (may be absent)		Along dorsum of fifth metatarsal	Extends little toe	Anterior tibial
2. Lateral or peroneal group (2)				
a. Peroneus longus	Upper two thirds of outer surface of fibula and lateral tibial condyle	Tendon passes behind lateral malleolus and crosses sole of foot to medial cuneiform and base of first metatarsal	Plantar flexes and everts foot	Peroneal (superficial)
b. Peroneus brevis	Lower two thirds of outer fibular surface	Base of fifth meta-tarsal	Everts foot	Peroneal (superficial)
3. Posterior or flexor group				
a. Superficial muscles (3)				
(1) Gastroc-nemius	Lateral head and medial heads from lateral and medial condyles of the femur	With soleus into tendo calcaneus into lower posterior surface of calcaneus	Plantar flexion of foot; flexor of knee	Posterior tibial
(2) Soleus	Posterior surface of head and upper third of fibula shaft; oblique line; middle third of medial tibia and tendinous arch between tibia and fibula	With (1) into tendo calcaneus	Plantar flexion of foot	Posterior tibial
(3) Plantaris	Lateral supracondylar ridge of femur	Medial margin of tendo calcaneus and deep fascia of foot	Plantar flexion of foot	Posterior tibial
b. Deep muscles (4)				
(1) Popliteus	Lateral condyle of femur	Posterior surface of tibia above oblique line	Flexes leg and rotates it medially	Posterior tibial
(2) Flexor hallucis longus	Lower two thirds of posterior surface of fibula	Base of terminal phalanx of big toe	Flexes big toe; inverts ankle	Posterior tibial

Table 7-10 (Continued)

Muscle	Origin	Insertion	Action	Nerve
(3) Flexor digitorum longus	Middle third of posterior surface of tibia	By four tendons into the bases of terminal phalanges of four lateral toes	Flexes second to fifth toes	Posterior tibial
(4) Tibialis posterior	Shaft of fibula and tibia	Navicular, three cuneiforms, cuboid, sustentaculum tali and second, third, and fourth metatarsal bones	Plantar flexion and supination of foot (evertor)	Posterior tibial

D. *The Foot*

1. Layer I (3)

Muscle	Origin	Insertion	Action	Nerve
a. Abductor hallucis	Medial process of calcaneal tuberosity, flexor retinaculum, and plantar aponeurosis	Medial side of first phalanx of big toe	Abduction of big toe	Medial plantar
b. Flexor digitorum brevis	Medial tubercle of calcaneus and plantar fascia	Second phalanx of four outer toes	Flexes toes	Medial plantar
c. Abductor digitorum V	Lateral process of calcaneus	Lateral side of proximal phalanx of fifth toe	Abducts and flexes little toe	Lateral plantar

2. Layer II (4)

Muscle	Origin	Insertion	Action	Nerve
a. Flexor digitorum longus (see above)				
b. Quadratus plantae	Lateral and medial borders of inferior surface of calcaneus	Tendons of flexor digitorum longus	Assists long flexor	Lateral plantar
c. Lumbricales (4)	First from tibial side of tendon of second toe of flexor digitorum longus; second, third, and fourth from adjacent sides of all tendons of the flexor digitorum longus	Bases of first phalanx of the outer toes (like hand)	Flexes the first and extends the second and third phalanges	First by medial plantar; second, third, and fourth by lateral plantar
d. Flexor hallucis longus	(See above)			

3. Layer III (3)

Muscle	Origin	Insertion	Action	Nerve
a. Flexor hallucis brevis	Medial surface of cuboid and middle and lateral cuneiforms	Sides of the base of first phalanx of big toe	Flexes big toe	Medial plantar

Table 7-10 (Continued)

Muscle	Origin	Insertion	Action	Nerve
b. Adductor hallucis				
(1) Transverse head	Capsules of the lateral four metatarso-phalangeal joints	Lateral side of base of first phalanx of big toe	Adducts big toe	Lateral plantar
(2) Oblique head	Lateral cuneiform and bases of third and fourth metatarsals	Lateral side of base of first phalanx of big toe		
c. Flexor digitorum V brevis	Base of metatarsal of little toe	Lateral surface of base of first phalanx of little toe	Flexes little toe	Lateral plantar
4. Layer IV (4)				
a. Plantar interossei (3)	Medial side of the third to fifth metatarsal bones	Corresponding side of first phalanx of same toe	Adducts three outer toes	Lateral plantar
b. Dorsal interossei (4)	By two heads each from the shafts of adjacent metatarsal bones	First into medial, second to fourth into lateral sides of phalanges	Abducts second to fifth toes	Lateral plantar
c. Peroneus longus tendon (see above)				
d. Tibialis posterior (see above)				

* At its insertion, a series of *osseoaponeurotic openings* are formed by tendinous arches that attach to the bone. The upper openings, usually four in number, are small and give passage to the perforating branches of the deep (profunda) femoral artery. The lowest, the *adductor hiatus*, is the largest and transmits the femoral vessels from the anterior thigh to the popliteal fossa (behind the knee joint).

The Leg. DEEP FASCIA. The deep fascia does not completely invest the leg, for it is absent over the anterior (subcutaneous) part of the tibia. It sends two partitions inward that attach to the fibula, and divides the leg into three compartments: lateral, anterior, and posterior. The anterior compartment contains the extensor muscles and the anterior tibial artery and nerve; the lateral compartment, the peroneal muscles and the superficial peroneal nerve; and the posterior compartment, the flexors of the ankle and the posterior tibial vessels and nerve. Still another layer subdivides the posterior compartment into a superficial and deep one. The fascia thins out distally but thickens in the area

of the ankle to form fascial bands called retinacula, which retain the tendons in position when the muscles that move the joint are active. The superior extensor retinaculum (transverse ligament) is such a band about 3.75 cm ($1\frac{1}{2}$ in.) wide that crosses the leg from tibia to fibula and splits medially to enclose the tibialis anterior muscle. The tendons of the extensor hallucis longus, extensor digitorum longus, and peroneus tertius pass behind this retinaculum in a common compartment. There is no synovial sheath here. The inferior extensor retinaculum (cruciate ligament) lies distal to the ankle joint. It is Y shaped, extending from the lateral part of the calcaneus, and splits into an upper limb that

[*Text continues on p. 241.*]

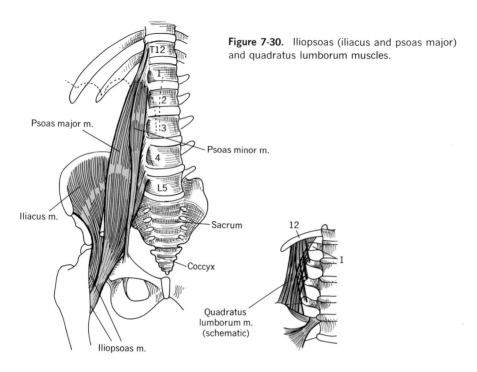

Figure 7-30. Iliopsoas (iliacus and psoas major) and quadratus lumborum muscles.

T12
1
2
Psoas major m.
3
Psoas minor m.
4
L5
Iliacus m.
Sacrum
Coccyx
Iliopsoas m.

12
1
Quadratus
lumborum m.
(schematic)

Figure 7-31. Muscles of the gluteal and posterior femoral regions.

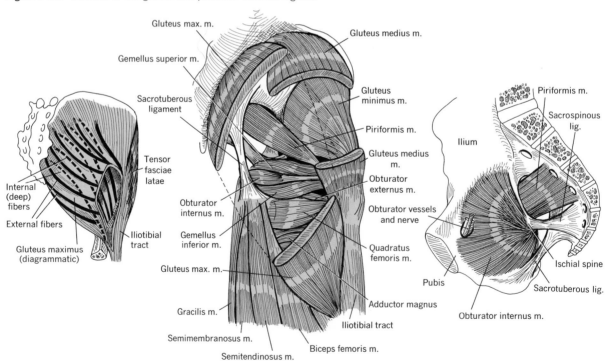

Gluteus max. m.
Gluteus medius m.
Gemellus superior m.
Gluteus
minimus m.
Sacrotuberous
ligament
Piriformis m.
Piriformis m.
Sacrospinous
lig.
Tensor
fasciae
latae
Gluteus medius
m.
Ilium
Internal
(deep)
fibers
Obturator
externus m.
External fibers
Obturator
internus m.
Obturator vessels
and nerve
Gluteus maximus
(diagrammatic)
Iliotibial
tract
Gemellus
inferior m.
Quadratus
femoris m.
Ischial spine
Gluteus max. m.
Adductor magnus
Sacrotuberous lig.
Gracilis m.
Iliotibial tract
Pubis
Obturator internus m.
Semimembranosus m.
Biceps femoris m.
Semitendinosus m.

238

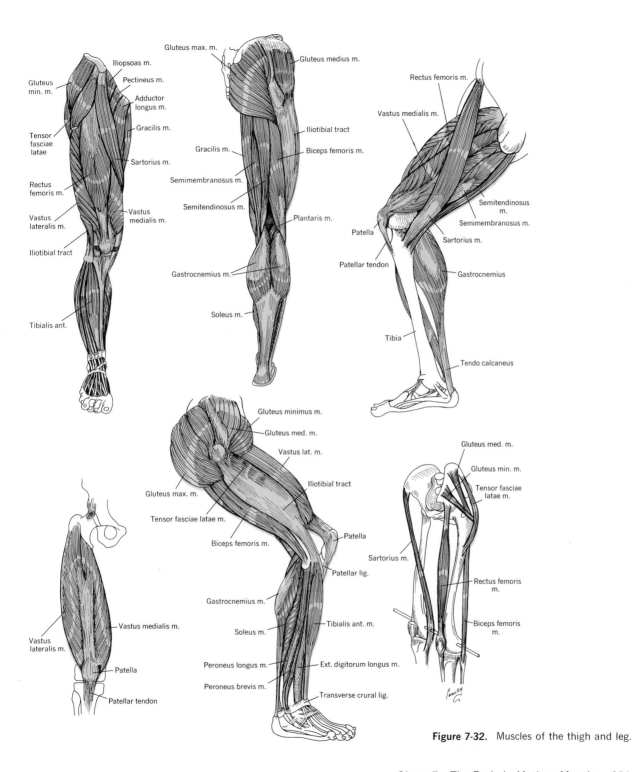

Figure 7-32. Muscles of the thigh and leg.

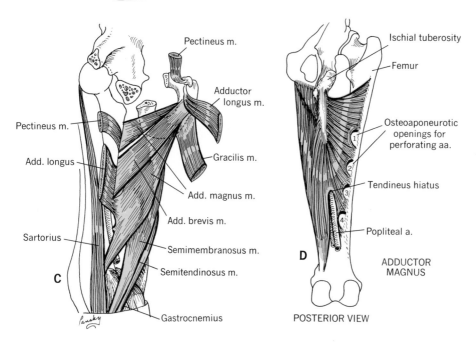

Iliacus m.

Inguinal lig.

Femoral nerve

Iliopsoas m.

Sartorius m.

Rectus femoris m.

Saphenous nerve

Opening of adductor canal

Vastus medialis m.

Patella

Psoas major m.

Femoral artery and vein

Pectineus m.

For 1st perforating artery

Adductor magnus m.

Add. longus m.

Gracilis m.

Femoral a.

Vastoadductor fascia

Popliteal a.

A

Pectineus m.

Add. longus m.

Gracilis m.

Add. magnus m.

Semimembranosus m.

Semitendinosus m.

Gastrocnemius m.

Sartorius m.

B

Pectineus m.

Adductor longus m.

Pectineus m.

Add. longus

Gracilis m.

Add. magnus m.

Add. brevis m.

Sartorius

Semimembranosus m.

Semitendinosus m.

Gastrocnemius

C

Ischial tuberosity

Femur

Osteoaponeurotic openings for perforating aa.

Tendineus hiatus

Popliteal a.

D

ADDUCTOR MAGNUS

POSTERIOR VIEW

Figure 7-33. Muscles of the medial side of the thigh. From **A** to **D** the muscles are seen at deeper levels of dissection.

240

passes to the medial malleolus of the tibia and a lower limb that passes to the deep fascia on the medial side of the foot.

MUSCLES. Three groups of muscles (Figs. 7-32, 7-34, 7-35; Table 7-10) are spoken of: the anterior or extensor group, supplied by the anterior tibial (deep peroneal) nerve; the posterior or flexor group, supplied by the posterior tibial nerve; and the lateral peroneal or evertor group, supplied by the superficial peroneal nerve.

With the foot off the ground, the peroneal muscles produce eversion. The peroneus longus also maintains the transverse arch of the foot. Since the muscles draw the foot to the lateral side, they balance the medial pull of the tibialis posterior and long flexors of the toe.

The Ankle. The tendo calcaneus stands out prominently at the back of the ankle (Figs. 7-34, 7-35, 7-36). Between it and the malleoli are two hollow grooves.

Over the front the extensor muscle tendons stand out (particularly when the joint is flexed). Above and behind the medial malleolus are the tendons of the posterior tibial and flexor digitorum muscles, and behind the lateral malleolus are the short and long peroneal tendons. The interval between the medial malleolus and calcaneus is crossed by the laciniate ligament, which forms a canal for the tendons of the deep posterior muscles. The tendons lie in close relationship to the joint, but the calcaneal tendon is separated from the joint by an interval filled with fatty areolar tissue. The deep fascia of the ankle is strong and continuous with that over the leg and foot. It forms five bands, in front and at each side of the ankle, that keep the tendons in contact with the bones and form osteoaponeurotic tunnels for the synovial sheaths and tendons. Anteriorly there is an upper transverse crural ligament between the fibula and tibia just above the ankle joint.

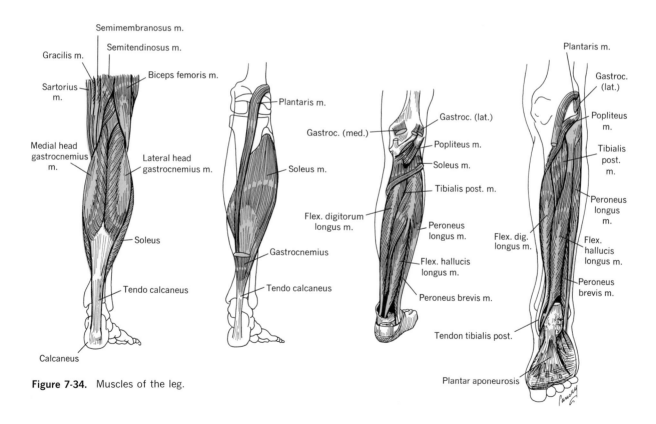

Figure 7-34. Muscles of the leg.

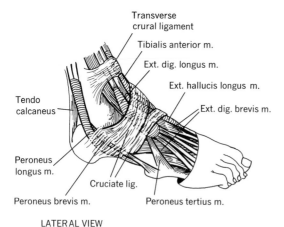

Transverse
crural ligament

Tibialis anterior m.

Ext. dig. longus m.

Ext. hallucis longus m.

Ext. dig. brevis m.

Tendo
calcaneus

Peroneus
longus m.

Cruciate lig.

Peroneus brevis m.

Peroneus tertius m.

LATERAL VIEW

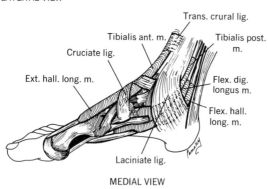

Trans. crural lig.

Tibialis ant. m.

Tibialis post.
m.

Cruciate lig.

Ext. hall. long. m.

Flex. dig.
longus m.

Flex. hall.
long. m.

Laciniate lig.

MEDIAL VIEW

Figure 7-35. Synovial sheaths of the tendons around the ankle.

Anterior-inferiorly is the cruciate ligament or inferior extensor retinaculum, which is the more important of the two. It is shaped like the letter Y, with the stem being lateral and the Y running across the top of the foot.

The peroneal retinacula are two lateral thickenings that bridge the area between the calcaneus and lateral malleolus. They hold the tendons of the peroneal muscles close to the outside of the ankle.

The tendon sheaths around the ankle are synovial sheaths and are associated with the tendons as they cross the joint. They extend about 5 cm (2 in.) above the malleoli and about 5 cm below. As in the hand, new sheaths cover the tendons, distally, beginning near the metatarsal bones.

The Foot. The skin is thicker on the sole than over its dorsum and particularly at the weight-bearing points such as the heel, ball of the big toe, and lateral sole margins. It is sensitive with many sweat glands. The superficial fascia is thick and tough laterally on the heel and ball of the foot, and it is transversed by small, tough, fibrous bands that divide the fatty tissue into lobules and connect the skin to the deep fascia.

DEEP FASCIA. The deep plantar fascia (aponeurosis) lies superficial to the nerves, vessels, muscles, and tendons and consists of a thin medial and lateral portion and a very strong, dense middle area spoken of as the plantar aponeurosis (Fig. 7-37). It runs from the calcaneus and divides anteriorly into five slips, one for each toe. The big toe, in contrast to the hand, is thus tied to the others, and its mobility is diminished when compared with the thumb. The five slips are connected to the fibrous flexor sheaths and to the sides of the metatarsophalangeal joints. The sheaths are strong (and dense) opposite the phalanges but thin and weak at the joints so as not to interfere with motion. They are attached to the margins of the phalanges and to the plantar ligaments and form tunnels for the tendons. Each tunnel is lined with a synovial sheath that encircles the tendons.

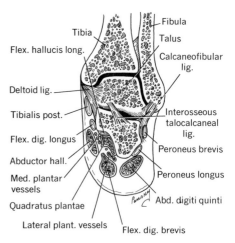

Tibia

Fibula

Talus

Flex. hallucis long.

Calcaneofibular
lig.

Deltoid lig.

Tibialis post.

Interosseous
talocalcaneal
lig.

Flex. dig. longus

Peroneus brevis

Abductor hall.

Peroneus longus

Med. plantar
vessels

Abd. digiti quinti

Quadratus plantae

Lateral plant. vessels

Flex. dig. brevis

Figure 7-36. Muscles related to the ankle joint. Bone and ligament relations are also seen.

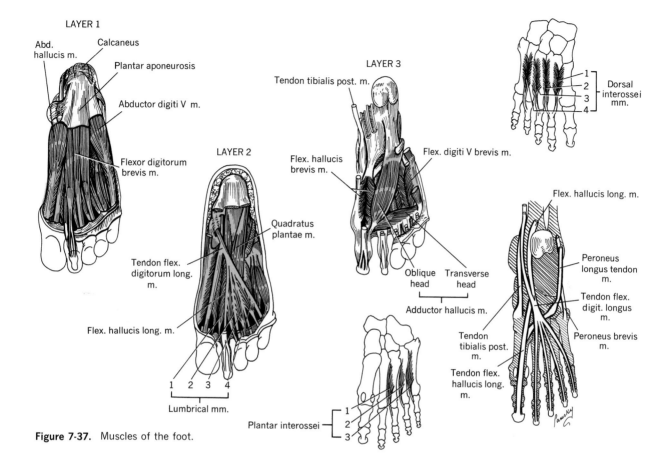

Figure 7-37. Muscles of the foot.

MUSCLES AND TENDONS. Muscles and tendons of the sole are in four layers (Table 7-10), separated by fascial partitions in which the plantar vessels and nerves are found. Only the first layer of muscles covers the entire sole. All muscles of the second layer attach to the flexor digitorum longus, forming an X-shaped figure that allows the first and third layers to come in contact with each other on the foot borders.

Review Questions

1. What is the difference between isometric and isotonic contraction?
2. What causes a muscle to contract?
3. What is necessary for neuromuscular transmission?
4. Define a stretch receptor, a spindle fiber, and a muscle spindle.
5. How do alpha and gamma motor neurons function?
6. What is a stretch reflex?

7. How do we control muscle activity?
8. For each of the following, list the muscles and indicate their action (and nerve supply):
   a. Muscles of facial expression
   b. Muscles of mastication
   c. Muscles of the palate
   d. Muscles of the tongue
   e. The neck muscles
   f. Muscles of the thorax and abdominal walls
   g. Muscles of the upper and lower extremities
   h. Muscles of the back
9. Inadvertently you stick a needle into the palm of your hand. Name the respective layers you pass through if the needle goes from palm to back.
10. What is the significance of knowing about the "spaces" in the hand?
11. What is the significance of the synovial tendon sheaths in the hand and foot?

References

Basmajian, J. V.: *Grant's Method of Anatomy*, 8th ed. Williams & Wilkins Co., Baltimore, 1971.

Goss, C. M. (ed.): *Gray's Anatomy of the Human Body*, 29th ed. Lea & Febiger, Philadelphia, 1966.

Grant, J. C. B.: *Grant's Atlas of Anatomy*, 6th ed. Williams & Wilkins Co., Baltimore, 1972.

Huxley, H. E.: Recent X-ray Diffraction and E.M. Studies of Striated Muscle in the Contractile Process. In *Proceedings of a Symposium of the New York Heart Association.* Little, Brown & Co., Boston, 1967, pp. 71–81.

——: The Mechanism of Muscular Contraction. *Sci. Am.,* **213**(6):18–27, 1965; *Science,* **164**:1356, 1969.

Pansky, B., and House, E. L.: *Review of Gross Anatomy,* 2nd ed. Macmillan Publishing Co., Inc., New York, 1969.

Porter, K. R.: The Sarcoplasmic Reticulum: Its Recent History and Present Status. *J. Biophys. Biochem. Cytol.,* **10**(Suppl):211–26, 1969.

External Integration, Correlation, and Coordination

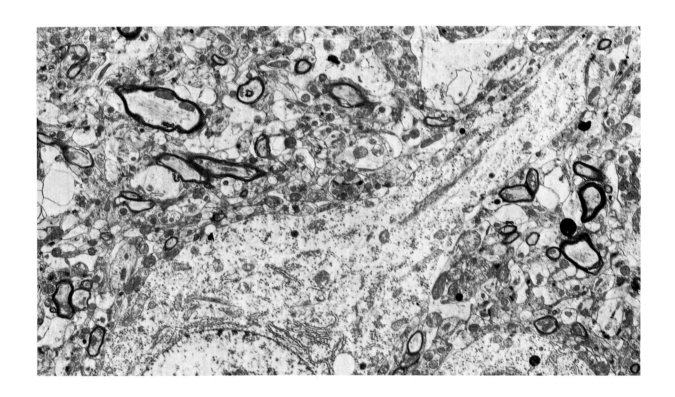

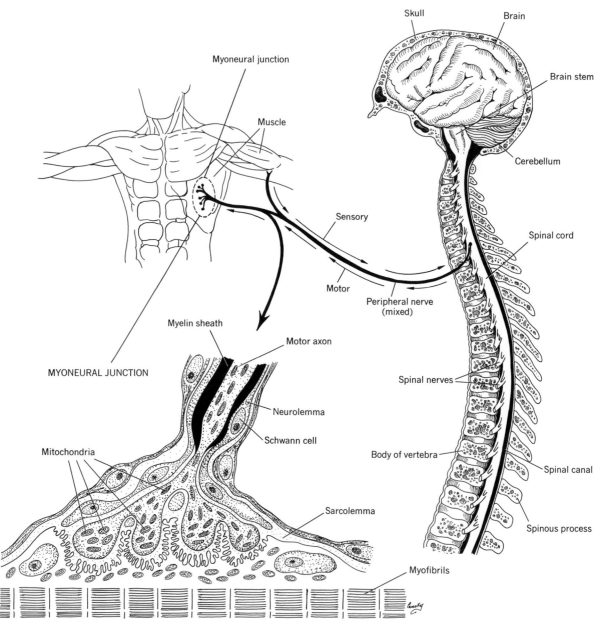

Figure 8-1. Diagram of some major components of the central and peripheral nervous systems. At the lower left is a schematic drawing of a myoneural junction.

Neural Control Mechanisms

Every movement is a response to a stimulus, whether it be conscious or subconscious. Living matter reacts to changes in both internal and external environment—i.e., to stimuli (agents or conditions that act or impinge on it). The billions of cells in our bodies continuously face the problem of seeking food and avoiding danger. Each cell, however, is involved in a communal existence with all the others. Many of the functions of the single-celled animal have, in man, been taken over by specialized organs. One of the most elaborate systems that enables the body's billions of cells to function as a single organism and coordinate all other systems is called the nervous system.

There is an intimate relation between nerve and muscle, brain and brawn. As an animal increases in size, its nervous system must do likewise. The nerve cell is thus the critical factor in the constant development of more complex life organisms.

In the multicelled organism, the nervous system is the primary network of communication. The brain is the site of the mental processes concerned with learning, reasoning, dreaming, imagination, and the planning capabilities associated with man. Man's capacity for thought depends on the workings of many millions of nerve cells.

Nerve cells utilize the same mechanisms as other cells for the synthesis of proteins and energy-rich ATP, and they have similar cellular structures. Nerve cells are like muscle cells in being electrical units that discharge in response to appropriate stimuli. Like all cells, nerve cells are centers of electrical activity and what really distinguishes these cells is that much of their electrical activity is in the form of pulsed signals that flow through the body.

When one reacts to a stimulus in the form of light, sound, smell, temperature, or touch (to name just a few), you check this information with earlier learned or experienced events by memory and then make a response. All elements in the complexity of the nervous system respond.

The human nervous system is very complex, and we are far from understanding everything regarding the way it works and can here discuss only its major features. Although facts are discussed in some detail, it is hoped that the reader will come away with a general concept of its morphology and function and appreciate its great complexity.

The nervous system is divided into two parts: (1) the central nervous system (CNS) and (2) the peripheral nervous system (PNS). The CNS consists of the spinal cord, which runs up inside the vertebral column and expands into the brain inside the skull. It receives information (afferent impulses), makes decisions (variety of interconnections and relays), and transmits its information (efferent impulses). The PNS is made up of a network of nerves and sense organs that gathers information from all over the rest of the body and feeds it into the CNS. It also transmits instructions away from the CNS so the body may respond and react to the sensations it receives. Thus, the peripheral nervous system carries messages but makes no decisions (Fig. 8-1).

Anatomy of the Nervous System

The Brain and Its Coverings

General Considerations. A primitive neural groove along the dorsal surface of the embryo is converted into a tube by the elevation and eventual fusion of the neural folds. The cranial or head end of the tube forms the brain. It expands and constricts to form a series of three primary brain vesicles (communicating sacs): the forebrain (prosencephalon), the midbrain (mesencephalon), and the hindbrain (rhombencephalon). The tube's cavity becomes the ventricular system (lateral and third ventricles, aqueduct of Sylvius, and fourth ventricle) and the central canal of the spinal cord (Fig. 8-2). The ventricular system "bathes" the inner CNS and carries cerebrospinal fluid.

The forebrain (prosencephalon) consists of the telencephalon, which becomes the cerebral hemispheres; the optic vesicles, which help form the eye; and the diencephalon. The latter includes: the thalami (way stations, relaying stimuli from the spinal cord to the brain), the medial and lateral geniculate bodies (associated with vision), the pineal gland, and the hypothalamus. The midbrain (mesencephalon) forms the brain stem and consists of ventral cerebral peduncles (fiber pathways connecting higher and lower centers) and a dorsal tectum with its corpora quadrigemina (related to visual and auditory pathways). The hindbrain (rhombencephalon) consists of the medulla oblongata, the pons, and the cerebellum (related to balance). (See later discussion of these brain parts.)

The brain weighs about 1400 gm in the adult male

Figure 8-2. Development of the brain and its major subdivisions.

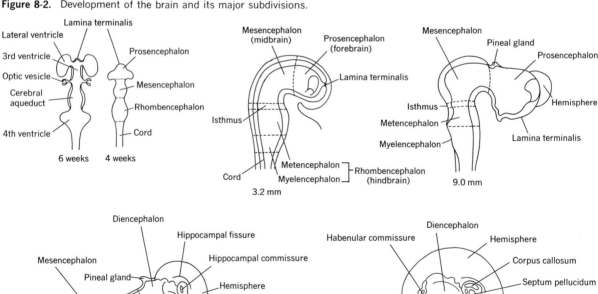

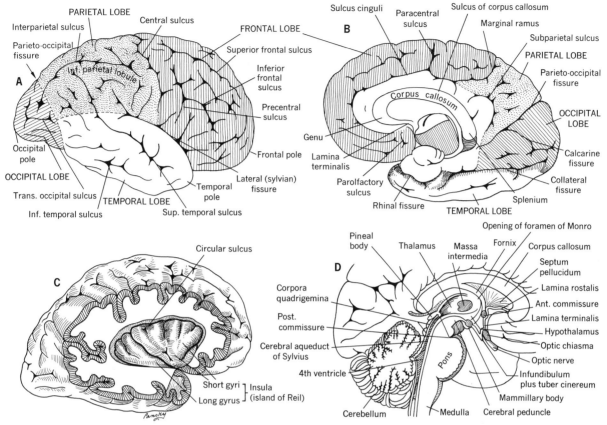

Figure 8-3. The brain. **A.** Lateral view showing lobes and sulci. **B.** Medial view showing lobes and sulci. **C.** View showing insula. **D.** Medial sagittal section.

and 1200 gm in the adult female. It reaches its maximum weight at the age of 20 and then actually decreases slowly with age.

The brain contains about nine tenths of the body's ten billion nerve cells. Within its convolutions is a tangle of dendrites and axons in a mass of integrated, complicated, interconnecting circuits; a single neuron may communicate with up to 270,000 other neurons. Thus, an unlimited number of variations and combinations are possible. Brain and computer both use electricity. A computer operates on 70,000 watts; the brain makes its own electricity and needs only the power of a 10-watt bulb. Once a computer has been programmed it can only perform as built; the brain is constantly reprogramming.

Lateral Surface of Cerebral Hemispheres. From the side, only the cerebral hemispheres can be seen (Figs. 8-3, 8-4). The longitudinal fissure separates the right and left hemispheres, except for the area deep in the central portion, where the corpus callosum forms the floor of the fissure as it runs between the hemispheres. The fissure contains a fold of dura (one of the brain coverings), which projects into it. The surfaces of the hemispheres are irregular due to the many sulci or fissures that separate the many convolutions or gyri from one another. The sulci are created by infoldings of the cerebral cortex, and thus the amount of cortical substance is increased without increasing the actual surface of the hemispheres. The convolutions have been associated with a variety of functions, and the

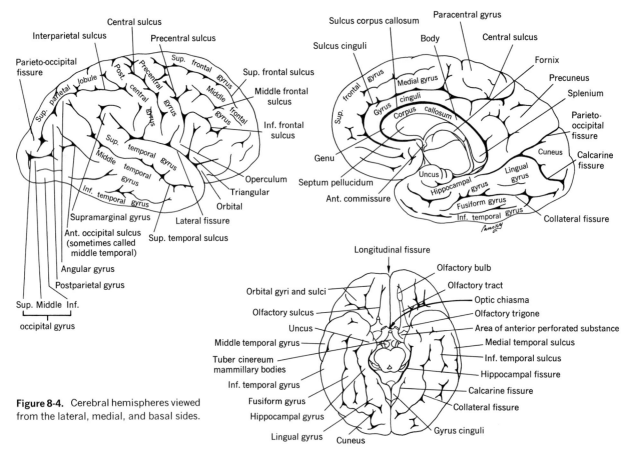

Figure 8-4. Cerebral hemispheres viewed from the lateral, medial, and basal sides.

various areas have shown structural differences (Figs. 8-3, 8-4, 8-5).

HEMISPHERES AND LOBES. The cerebral hemispheres lie in the anterior and middle cranial fossae, and each consists of five lobes (Fig. 8-3): frontal, parietal, temporal, occipital, and central (insula or the island of Reil).

Frontal Lobe. As seen on the lateral surface the frontal lobe is bounded posteriorly by the central sulcus (the fissure of Rolando). In front of this sulcus is the precentral gyrus, which contains the higher centers for motor control of movements of the opposite side of the body. Inferiorly, the lateral fissure of Sylvius separates the frontal lobe from the temporal lobe.

In the cortex, the body is represented in an inverted position, with the centers for the lower limbs occupying the upper portion of the gyrus. Note the arrangement of body parts to position on the gyrus (Fig. 8-6). It is essential to realize that corresponding muscles of the opposite sides of the body are connected with the cortex of both hemispheres and, therefore, movements (e.g., those of the eye) are not totally affected by unilateral lesions or brain damage. Note, too, that movements are represented in the cortex and not individually named muscles.

The superior and inferior frontal sulci run from the precentral sulcus and help delineate the superior, middle, and inferior frontal gyri. Broca's area, which is related to the function of speech, consists of an orbital portion, a triangular part, and an opercular part in the

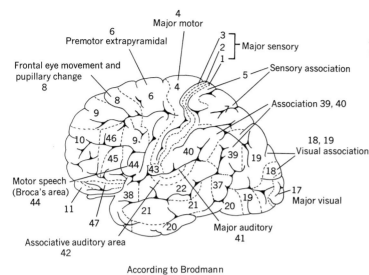

4
Major motor
6
Premotor extrapyramidal
3
2 } Major sensory
1
Frontal eye movement and
pupillary change
8
5
Sensory association
4
8
6
7
Association 39, 40
9
10
46
9
40
39
19
18, 19
Visual association
45
44
43
18
Motor speech
(Broca's area)
44
38
22
37
17
Major visual
11
21
21
20
19
47
20
Major auditory
41
Associative auditory area
42

According to Brodmann

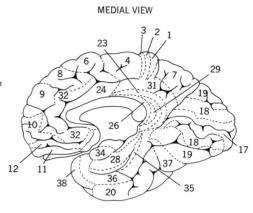

Figure 8-5. Cytoarchitectural maps of motor areas and primary and secondary sensory areas of the cerebral cortex according to Brodmann and von Economo. Functional localizations are also shown.

MEDIAL VIEW

3 2 1
23
6
4
8
7
29
9
32
24
31
19
10
18
32
26
34
18
12 11
28
19
17
37
38
36
35
20

According to Brodmann

LATERAL VIEW

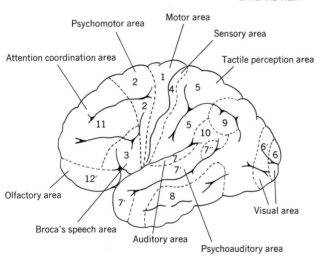

Psychomotor area
Motor area
Sensory area
Attention coordination area
Tactile perception area
2
1
4
5
2
5
9
11
10
3
7
6 6
12'
7'
Olfactory area
7'
8
Visual area
Broca's speech area
Auditory area
Psychoauditory area

According to von Economo

MEDIAL VIEW

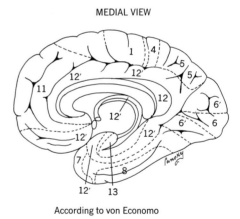

1
4
12'
5
12'
11
5
12
6'
12'
6'
6
7'
12'
8
12'
13

According to von Economo

lower portion of the frontal lobe. In right-handed people, the left side of this region is better developed.

Parietal Lobe. The parietal lobe is bounded anteriorly by the central sulcus, posteriorly by the boundary of the occipital lobe and the lateral parieto-occipital sulcus, above by the cerebral border, and below by the sylvian fissure.

Behind the central sulcus is the postcentral gyrus, in which the sensory projection area is located (Figs. 8-4, 8-5). If this area is involved by a tumor, there

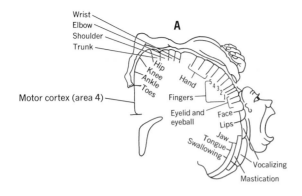

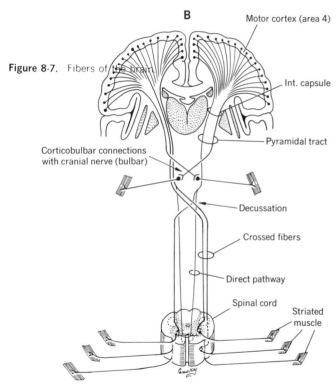

Figure 8-7. Fibers of the brain

Figure 8-6. **A.** Homunculus of motor representation in area 4. (After studies by Penfield and Rasmussen.) **B.** Corticospinal (pyramidal) tract.

results a diminution in various sensations and the individual loses the ability to localize a painful stimulus or even to determine its intensity. There may also be difficulty in determining size and weight as well as the

texture of things. From the area of the postcentral gyrus, the interparietal sulcus passes backward and demarcates a superior parietal lobule from an inferior (some texts refer to these as gyri rather than as lobules). Lesions of the inferior parietal gyrus produce astereognosis (the inability to correlate and interpret a variety of sensory impressions). This lobule is divided from below by three sulci: the posterior ramus of the lateral fissure, the superior temporal, and the anterior occipital. As a result, each sulcus is capped by an arched gyrus: the supramarginal, angular, and postparietal gyri. Aphasia or the inability to understand written words (word blindness) is characteristically associated with a lesion of the angular gyrus of the left side in a right-handed person. Lesions in the upper parietal lobe, on the other hand, are associated with inability to recognize the form or nature of objects. With lesions of the postcentral gyrus (the sensory brain area) the patient loses his ability to localize a painful stimulus or even measure its intensity. If, on the other hand, this area is irritated, a numbness or needle-pin sensation occurs in his extremities. Such sensory attacks may actually initiate symptoms of a type of motor convulsions. Since fibers associated with vision pass through a part of the parietal lobe on their way to the visual center, a deep tumor can produce visual field defects.

Temporal Lobe. The temporal lobe is bounded above by the lateral fissure and an arbitrary line drawn posterior from the fissure to the front of the occipital lobe (Fig. 8-3). The temporal lobe has three gyri, a superior, middle, and inferior, which are separated by two sulci, a superior and inferior. The superior sulcus runs parallel to the lateral fissure and is "capped" by the angular gyrus. The superior temporal gyrus contains the higher auditory centers (Fig. 8-5). The temporal lobe occupies part of the middle cranial fossa and is susceptible to injury with basal skull fractures. Lesions of the superior temporal gyrus have been associated with word deafness (inability to understand the spoken word). The middle temporal gyrus lies between superior and inferior temporal sulci, and the inferior gyrus is below the inferior sulcus.

Occipital Lobe. The occipital lobe is marked off by a line drawn downward from the parieto-occipital

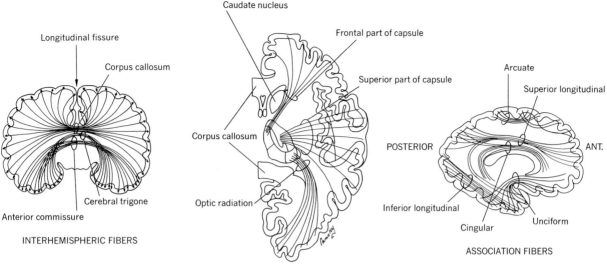

Longitudinal fissure

Corpus callosum

Caudate nucleus

Frontal part of capsule

Superior part of capsule

Arcuate

Superior longitudinal

Corpus callosum

POSTERIOR

ANT.

Optic radiation

Cerebral trigone

Anterior commissure

INTERHEMISPHERIC FIBERS

Inferior longitudinal

Cingular

Unciform

ASSOCIATION FIBERS

INTERNAL CAPSULE

Figure 8-7. Fibers of the brain.

fissure to the inferolateral border of the cerebral hemisphere. Its surface is very variable. The transverse and lateral occipital sulci divide the lobe into superior, middle, and inferior gyri. It is believed that the chief function of the occipital lobe is concerned with vision. Visual hallucinations may result due to irritation of the lobe, and the patient claims to see rainbow colors, bright and flashing light, or even brilliant lightning patterns.

Central Lobe (Insula). The central lobe (Fig. 8-3) is seen only when the edges of the lateral sulcus are pulled apart, and even then part of it remains buried. It is surrounded by a circular sulcus and overlapped by parts of the cortex called opercula (Fig. 8-4).

Medial Surface of Cerebral Hemisphere. If one divides the two cerebral hemispheres by a vertical incision between them, the white commissure (fibers connecting the two parts) known as the corpus callosum becomes very evident since it lies at the bottom of the longitudinal fissure in the midline (Figs. 8-3, 8-4). It is about 10 cm (4 in.) long and extends to about 3.75 cm ($1\frac{1}{2}$ in.) of the anterior and about 6.25 cm ($2\frac{1}{2}$ in.) of the posterior extreme ends of the hemispheres. The corpus callosum is subdivided into an anterior part, the genu, which ends below in a point

called the rostrum. The latter is connected to the upper part of the optic chiasma by a sheet of gray matter, the lamina terminalis. The posterior part of the corpus callosum is called the splenium, which covers the dorsal part of the midbrain. The major part between the ends is called the body of the corpus. Fibers in the corpus callosum travel predominantly transversely and are mainly association fiber tracts that connect the two halves of the hemisphere (Fig. 8-7).

Below the middle third of the corpus is found the body of the fornix. As the latter passes forward, it arches downward, terminating in the mammillary body of its own side (Fig. 8-3). The fornix forms the efferent tract from the hippocampus of one hemisphere to that of the other and to the brain (Fig. 8-4) and is involved in olfaction (smell). A thin two-layered membrane, the septum pellucidum, stretches from the anterior portion of the fornix to the underside of the corpus callosum. The septum forms a partition between the anterior horns of the lateral ventricles.

Behind the fornix are the interventricular foramina of Monro, which appear as openings and provide a communication between the lateral and third ventricles. The two lateral ventricles thus open into the third ventricle (Fig. 8-8).

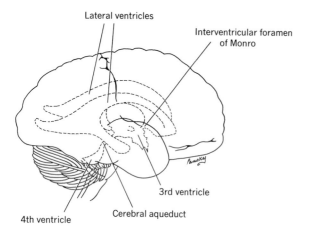

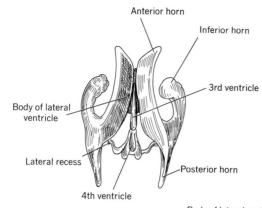

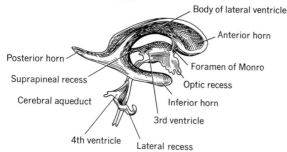

Figure 8-8. Ventricular cavities of the brain.

The anterior commissure found below the foramen and in front of the fornix (Fig. 8-3) is a small bundle of white fibers seen in the anterior wall of the third ventricle; it relays olfactory fibers. Also seen here is the thalamus (an association center for impulses to and from the cerebrum). The thalamus is joined to its counterpart by the so-called massa intermedia, which is actually gray matter and not really interconnecting fibers. The hypothalamus is the area below the thalamus and is an autonomic center controlling temperature, breathing rates, and secretion of some hormones. It includes the mammillary bodies, tuber cinereum, infundibulum, optic chiasma, and many nuclei (see Chap. 15).

The posterior commissure (Fig. 8-3), which is a layer of white fibers connecting the two thalami posteriorly, forms the posterior boundary of the third ventricle. It can be found just above the openings of the aqueduct of Sylvius. The pineal body (gland) is directly above and posterior to the posterior commissure.

The sulcus cinguli (Fig. 8-4) follows the curve of the corpus callosum but is separated from it by the cingular gyrus. The sulcus turns up just behind the upper end of the central sulcus. The medial gyrus lies in front of the upper end of the central sulcus and forms the medial surface of the frontal lobe and contains the higher motor centers that control lower limb movements of the opposite side of the body. The latter part of the gyrus is called the paracentral gyrus, behind which is the precuneus. The precuneus forms the medial side of the parietal lobe and is separated from the occipital lobe by parieto-occipital sulcus.

Also seen on the medial brain surface is the calcarine fissure, which runs forward from the occipital lobe to enclose an area of cortex that is called the cuneus, which forms the medial part of the occipital lobe. The lingual gyrus lies below the calcarine sulcus.

Inferior Brain Surface, Cerebellum, and Medulla Oblongata. Inferior Brain Surface. The anterior portion of the frontal lobe rests on the floor of the anterior cranial fossa and is thus related to the orbit, nasal cavity, and frontal sinus (Figs. 8-3, 8-4). The orbital gyri and sulci are found here. Very little is known of the function of this part of the brain.

The posterior surface lies on the floor of the middle cranial fossa. This relates to the middle ear. Medially, one notes the hippocampal gyrus, uncus (which receives incoming olfactory fibers and is associated with olfactory impressions), rhinal fissure, collateral fissure, lingual gyrus, fusiform gyrus, and inferior temporal gyrus (all related to olfaction).

Anteriorly, one notes the olfactory bulb, which continues backward as the olfactory tract. The olfactory bulb rests on the cribriform plate of the ethmoid bone, through which the olfactory nerves from the nasal mucous membrane pass into the olfactory bulbs on their way to the brain. The olfactory tract divides into medial and lateral olfactory roots at the trigone. Directly behind these roots one sees the anterior perforated substance, which transmits numerous arteries and veins to and from the inside of the brain.

The optic chiasma results from the crossing of fibers from the optic nerves (which come from the eye). The optic tracts continue backward from the chiasma and are lost from view under cover of the uncus. Immediately behind the optic chiasma, the tuber cinereum is located (it is poorly developed in man but well developed in fish, whose sense of taste dominates the sense of smell). The stalk of the pituitary or infundibulum is at the summit of the tuber cinereum. Behind the tuber, two small white mammillary bodies are seen (their nuclei form centers of smell). These structures have previously been mentioned as belonging to the hypothalamus.

If the optic tract is followed backward around the lateral side of the cerebral peduncle (seen best if the temporal lobe is retracted), it divides into medial and lateral roots. The lateral root ends in the lateral geniculate body (Fig. 8-9) and the superior colliculus (Fig.

Figure 8-9. Lateral view of the brain stem, cranial nerves, and central nuclei.

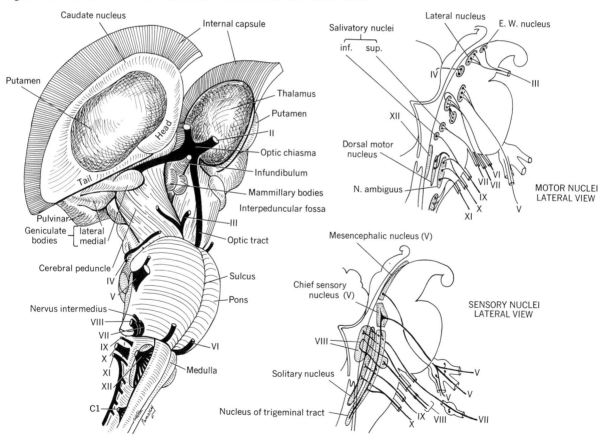

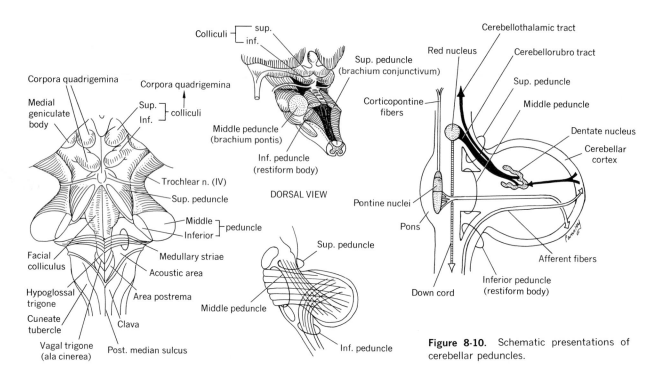

Colliculi — sup.
— inf.

Corpora quadrigemina

Corpora quadrigemina

Medial
geniculate
body

Sup. ┐
Inf. ┘ colliculi

Sup. peduncle
(brachium conjunctivum)

Red nucleus

Cerebellothalamic tract

Cerebellorubro tract

Sup. peduncle

Middle peduncle

Corticopontine
fibers

Dentate nucleus

Middle peduncle
(brachium pontis)

Cerebellar
cortex

Inf. peduncle
(restiform body)

DORSAL VIEW

Trochlear n. (IV)

Sup. peduncle

Pontine nuclei

Middle ┐
Inferior ┘ peduncle

Pons

Facial
colliculus

Medullary striae

Acoustic area

Sup. peduncle

Afferent fibers

Hypoglossal
trigone

Area postrema

Inferior peduncle
(restiform body)

Down cord

Cuneate
tubercle

Clava

Middle peduncle

Vagal trigone
(ala cinerea)

Post. median sulcus

Inf. peduncle

Figure 8-10. Schematic presentations of
cerebellar peduncles.

8-10). The brain nuclei here form the lower visual centers. The medial root ends in the medial geniculate body (Fig. 8-9) and is not associated with vision. Behind and above the geniculate bodies is the posterior end of the optic thalamus, the so-called pulvinar (Fig. 8-9), which is also associated with the lower visual centers.

The corpora quadrigemina consist of four "rounded" bodies found on the dorsal aspect of the midbrain just below and medial to the pulvinar. The superior corpora (superior colliculus) are larger than the inferior and have the pineal gland located just above them (Fig. 8-10). Each has a brachium running laterally from it. The lateral geniculate body, the superior colliculus, and the pulvinar of the thalamus have been referred to (collectively) as the "lower visual centers," with the lateral geniculate body being the most important since

it is a relay station on the visual pathway from the retina of the eye to the visual cortex. The inferior colliculi are associated with auditory pathways.

The dorsal part of each peduncle is known as the tegmentum and contains afferent pathways and the red nucleus. The tegmentum and the basis pedunculi (which is the ventral part of the pedunculi) are separated on each side by a band of pigmented gray matter known as the substantia nigra. The aqueduct of Sylvius appears in the midline near the dorsal side (Fig. 8-11).

The cerebral peduncles (Figs. 8-9, 8-10) are two large bundles of white substance lying together at the upper margin of the pons. They function to connect the cerebral hemispheres with structures below. The cerebral peduncles connect the pons to the cerebrum and by this means relay messages to the spinal cord.

Figure 8-11. [*Opposite*] Sections of the brain stem illustrating general appearance at various levels. A few structures have been labeled for identifying level. Most other structures and details can be found in any standard neuroanatomy text.

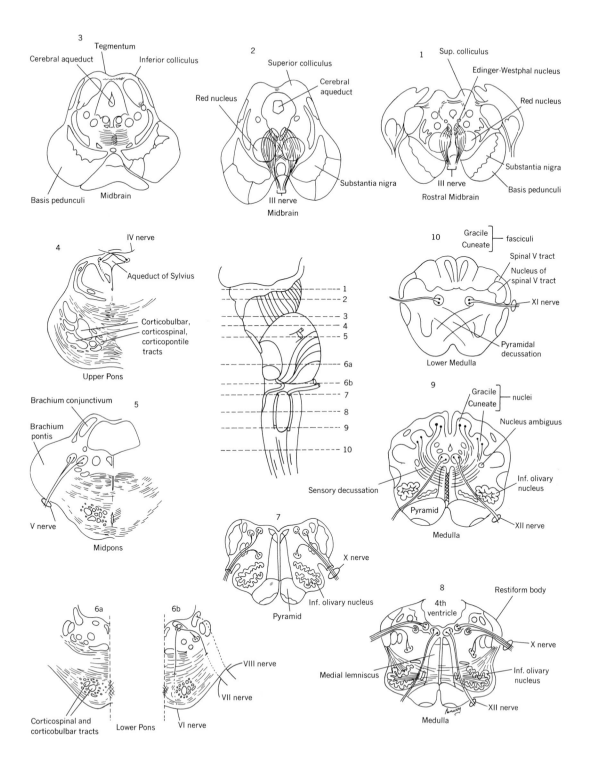

3

Cerebral aqueduct
Tegmentum
Inferior colliculus

Basis pedunculi
Midbrain

2

Superior colliculus
Cerebral aqueduct
Red nucleus
Substantia nigra
III nerve
Midbrain

1

Sup. colliculus
Edinger-Westphal nucleus
Red nucleus
Substantia nigra
III nerve
Basis pedunculi
Rostral Midbrain

4

IV nerve
Aqueduct of Sylvius
Corticobulbar,
corticospinal,
corticopontile
tracts
Upper Pons

5

Brachium conjunctivum
Brachium pontis
V nerve
Midpons

6a

Corticospinal and
corticobulbar tracts
Lower Pons

6b

VIII nerve
VII nerve
VI nerve

7

X nerve
Inf. olivary nucleus
Pyramid

8

4th ventricle
Restiform body
X nerve
Medial lemniscus
Inf. olivary nucleus
XII nerve
Medulla

9

Gracile
Cuneate } nuclei
Nucleus ambiguus
Inf. olivary nucleus
XII nerve
Sensory decussation
Pyramid
Medulla

10

Gracile
Cuneate } fasciculi
Spinal V tract
Nucleus of spinal V tract
XI nerve
Pyramidal decussation
Lower Medulla

1
2
3
4
5
6a
6b
7
8
9
10

The cerebral peduncles gradually diverge to form the interpeduncular fossa and seem to disappear under the optic tracts (Fig. 8-9). The fossa is bounded in front by the optic chiasma, anterolaterally by the optic tracts, and posterolaterally by the cerebral peduncles. The contents of the fossa form the floor of the third ventricle.

The pons (Fig. 8-9) is found above the medulla, below the cerebral peduncles, and between the halves of the cerebellum. It is about 5 cm (2 in.) wide and 3.75 cm ($1\frac{1}{2}$ in.) long and has a basilar sulcus (ventrally) in which the basilar artery rests. It forms part of the floor of the fourth ventricle and is continuous posteriorly with the medulla.

CEREBELLUM. The cerebellum (Fig. 8-3) occupies the posterior cranial fossa and is related to the pons and medulla in front, from which it is separated by the fourth ventricle. It consists of a middle strip called the vermis and two hemispheres. Gray matter forms a layer of cortex on its surface (not centrally as in the cord or medulla). The cerebellum maintains balance and also coordinates skillful hand movements.

MEDULLA OBLONGATA. The medulla oblongata (Figs. 8-3, 8-9) extends from the foramen magnum of the skull to the lower border of the pons, connecting the cord below to the pons above. The posterior part of the medulla is divided into an upper part, forming the floor of the fourth ventricle, and a lower part, which is directly continuous with the spinal cord. It is pyramidal in shape and about 3.125 cm ($1\frac{1}{4}$ in.) long.

The medulla is an important sensory and motor area as well as an autonomic center that, in conjunction with the hypothalamus, controls the heart and breathing rates as well as digestion.

Vascular Supply of the Brain

Arterial Supply. Four vessels supply the brain: two vertebrals and two internal carotids (Fig. 10-19, p. 356). The vertebral artery is a branch of the subclavian, ascends in the transverse foramina of the cervical vertebrae, and enters the skull via the foramen magnum. After ascending and perforating the dura, it unites with the same vessel of the opposite side to form the basilar artery. The branches of the basilar artery are distributed to the pons, cerebellum, internal ear, and cerebral hemispheres. They include the pontine, paired anterior inferior cerebellar, paired superior cerebellar, and terminal posterior cerebral arteries.

The internal carotid ascends in the neck and enters the skull via the carotid foramen. After penetrating the dura, it reaches the base of the brain at the angle between the optic nerve and optic tract, gives off an anterior choroidal artery to the choroid plexus of the lateral ventricle, and divides into two terminal branches, the anterior and middle cerebral arteries. These are interconnected by the anterior and posterior communicating arteries, respectively, to form the arterial circle of Willis (Fig. 10-21, p. 358). Two separate sets of branches originate from the cerebral arteries: (1) central branches, which pierce the surface of the brain to supply the internal parts of the cerebrum, including the basal nuclei, and do not anastomose with each other; and (2) cortical branches, which ramify over the cerebral surface and supply the cortex and anastomose in the pia mater.

The anterior cerebral arteries supply the superior and middle frontal convolutions and the entire medial surface of the hemisphere as far back as the parieto-occipital fissure. They also supply a part of the para-central lobule (leg center) and the highest point of the precentral gyrus.

The middle cerebral arteries are a direct continuation of the internal carotid artery. They run into the sylvian fissure and supply most of the exposed surface of the hemisphere, insula, and internal capsule. They also supply the major portion of the motor area of the brain, part of the visual center, and the motor speech area of the left hemisphere. They give off the lenticulo-striate vessels (most frequently involved in cerebrovascular accidents).

The posterior cerebral arteries supply much of the inferior and medial surfaces of temporal and most of the occipital lobes. Among their branches are the posterior choroid branches, which supply the choroid plexuses of the third and lateral ventricles.

Veins (of Brain and Head). The veins of the head and brain consist of emissary veins, which connect the veins of the inside of the skull with those of the neck, face, and head; diploic veins, which form venous plexuses found between the inner and outer tables of the skull; cerebral and cerebellar veins, which drain venous blood from the cerebrum and cerebellum; and the venous dural sinuses, which are located between the layers of the dura mater (Fig. 10-28, p. 367).

EMISSARY VEINS. Emissary veins connect intracranial and extracranial veins. Blood can travel in either direction. So can infection. This double direction of blood flow also equalizes the venous pressure in the superficial veins and sinuses. Some of the emissary veins include the following: the parietal vein goes from the occipital vein to the superior sagittal sinus; emissary veins of the foramen cecum connect the superior sagittal sinus with veins of the frontal sinus and root of the nose; the mastoid emissary veins connect the occipital vein with the transverse sinus; and the ophthalmic veins, which can also be considered emissary veins, drain into the cavernous sinus. In relation to these veins, it should be noted that the anterior facial vein communicates with the cavernous sinus by way of the ophthalmic veins. The part of the anterior facial vein that passes along the side or angle of the nose is called the angular vein. Boils and carbuncles in this area if opened may carry infection to the cavernous sinus, resulting in death. Since the anterior and posterior facial veins join in the neck to form the common facial vein that ends in the internal jugular, the latter may also become infected. Since these veins have no valves, infected thrombi may cause spread of the infection.

DIPLOIC VEINS. The diploic veins form venous plexi between the inner and outer skull tables. The veins form a plexus that is drained by four diploic venous trunks on each side: frontal diploics drain into the supraorbital vein; anterior parietal or temporals drain into the sphenoparietal sinus; posterior parietal or temporals drain into the lateral sinus; and occipital diploics drain into the lateral sinus. There are no diploic arteries since arterial blood supply comes via meningeal and pericranial arteries.

CEREBRAL AND CEREBELLAR VEINS. The cerebral and cerebellar veins are veins of the brain proper. They do not follow the arteries. They have no valves

and no muscle around them and are very thin walled. They lie on the brain surface and are covered by arachnoid. The superior veins run upward to end in the superior sagittal sinus. There are both external and internal cerebral veins, depending on whether they drain the surface or inner part of the hemispheres. The external veins are the middle, superior, and inferior cerebral veins. The internal veins that drain the deep areas of the brain are the terminal veins and the great cerebral vein of Galen.

VENOUS DURAL SINUSES. The venous sinuses of the dura (Fig. 8-12) are spaces between dural layers that collect blood and return it to the internal jugular veins. Spinal fluid also drains here from the subarachnoid space via the arachnoid villi and granulations. Their walls are a single layer of endothelium and do not easily collapse. There are seven paired sinuses and five unpaired. The paired sinuses are the sphenoparietal, cavernous, superior and inferior petrosal, occipital, transverse, and sigmoid. The unpaired sinuses are the superior and inferior sagittal, the straight, the intercavernous (circular), and the basilar.

Superior Sagittal Sinus. The superior sagittal sinus begins in front of the crista galli at the foramen cecum, and passes upward and backward in the falx cerebri as far back as the internal occipital protuberance. Here it forms a dilatation, the confluence of sinuses, at which point the superior sagittal, transverse, occipital, and straight sinuses all meet. Here, too, it becomes continuous with the transverse sinuses. The superior sagittal sinus receives emissary veins, diploic veins, and veins draining the cerebral hemispheres.

Inferior Sagittal Sinus. The inferior sagittal sinus passes back in the lower border of the falx cerebri. It unites with the great cerebral vein of Galen at the free edge of the straight sinus.

Straight Sinus. The straight sinus passes back along the attachment of the tentorium cerebelli to the falx cerebri. It terminates usually in the left transverse sinus and receives tributaries from the posterior cerebrum, cerebellum, and falx cerebri.

Transverse or Lateral Sinus. The transverse sinus is paired and begins at the internal occipital protuberance. The right usually is a continuation of the superior

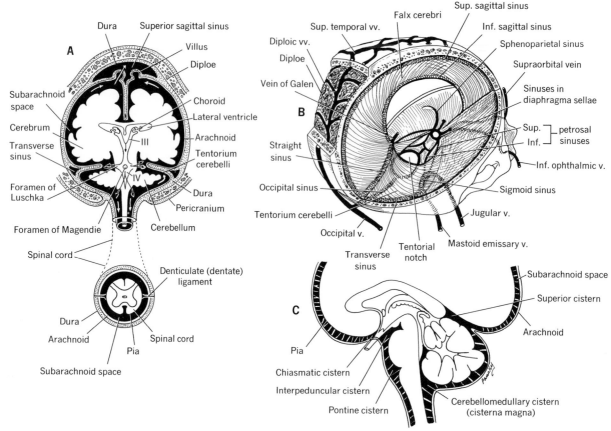

Figure 8-12. **A.** Diagram of essential points related to production, course, and absorption of cerebrospinal fluid; meningeal layers are also seen. **B.** Cranial dural sinuses; emissary, diploic, and superficial cranial veins. **C.** Diagram of major subarachnoid cisterns, lateral view.

sagittal, and the left, of the straight sinus. It receives the superior petrosal sinus and a few inferior cerebral and cerebellar veins. It is bounded by tentorium and outer layer of dura and runs lateral and then forward.

Sigmoid Sinus. The sigmoid sinus is a continuation of the transverse and is S shaped. It continues from where the transverse ends to the jugular foramen, where it becomes the internal jugular vein. The superior petrosal sinus joins it at its first bend and the inferior petrosal at its termination. It is related to the tympanic antrum of the ear, and, therefore, diseases of the middle ear can form a suppurative process and involve the sinus (sinus thrombosis).

Cavernous Sinus. The cavernous sinus is found on either side of the body of the sphenoid bone and is continuous with the ophthalmic veins in front. Posteriorly it leads into the superior and inferior petrosal sinuses. It surrounds the internal carotid artery, and an injury here can be very dangerous. Cavernous sinus thrombosis may be a result of inflammatory lesions of the upper lip or face that have extended through the facial, nasal, and ophthalmic veins. Infections here often result in meningitis.

Other Sinuses. The circular sinus connects the cavernous sinuses. The sphenoparietal sinus runs along the lesser wing of the sphenoid to the superior sagittal.

The occipital sinus lies in the border of the falx cerebelli. The basilar sinus lies behind the dorsum sellae and unites the cavernous and inferior petrosal sinuses of opposite sides and communicates with spinal veins.

The Cranial Meninges

The brain and spinal cord are surrounded by three envelopes of membranes, between which flows the cerebrospinal fluid (Figs. 8-12, 8-13). From inside out, they are the pia mater, arachnoid mater, and dura mater. The dura is firm and tough, the arachnoid is like a spider's web, and the pia is a thin, clinging, skinlike layer that "hugs" the brain surface and intimately follows all its irregularities. Unlike the pia, the dura and arachnoid do not dip into the brain's sulci and fissures but cover it more like a "mitten."

Pia Mater. The pia mater is the inner layer and is really the membrane of nutrition. It is closely adherent to the brain surface and dips into the depths of all sulci, carrying branches of the cerebral arteries with it. The larger blood vessels of the brain lie in the subarachnoid space, but the smaller ones ramify in the pia and proceed into the substance of the brain proper. At certain places, the pia sends strong vascular duplications into the brain, which spread over the cavities of the third and fourth ventricles and extend into the lateral ventricles as the tela choroidea. The vessels accompanying the tela form a plexus and project into the ventricles as the choroid plexus. From the choroid plexus, most of the cerebrospinal fluid is formed.

Arachnoid Mater. The arachnoid mater is a very delicate membrane that lies between the pia internally and the dura externally. It does not dip with the pia but follows the dura. Over the brain's convolutions, the arachnoid and pia are in close contact but they are separated at the sulci by the subarachnoid space, which contains cerebrospinal fluid and circulates around the brain and cord. Thus, the CNS is immersed in fluid. The subarachnoid space is crossed by a webby, filmy retinaculum of "fibers" that connects the pia and arachnoid. At the base of the brain, this cobweblike network is reduced and the pia and arachnoid are widely separated to form the subarachnoid cisternae.

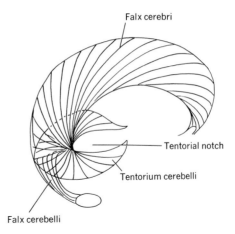

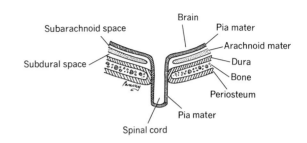

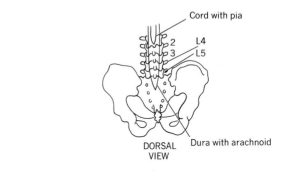

Figure 8-13. Cranial and spinal meninges.

SUBARACHNOID CISTERNAE. There are three major cisterns (Fig. 8-12).

Cisterna Magna (Cerebellomedullary Cistern). This is a cavity resulting from the arachnoid bridging the inferior surface of the cerebellum and the dorsal surface of the medulla oblongata. It continues below with the spinal subarachnoid space. Cerebrospinal fluid enters this cistern from the fourth ventricle by way of the foramen Magendie.

Pontine Cistern. This is found in front of the pons and medulla oblongata and is continuous with the subarachnoid space about the medulla. It forms a "water cushion" for the brain.

Interpeduncular Cistern. This is formed by the arachnoid that extends across and between the two temporal lobes, encloses the cerebral peduncles, and contains the circle of Willis.

ARACHNOID GRANULATIONS. The arachnoid granulations are best seen in old age, when they "pit" the parietal bone. When hypertrophied they are called the pacchionian bodies. They are villous processes of the arachnoid that push the dura ahead of them and serve as channels for the passage of cerebrospinal fluid from the subarachnoid space into the venous sinus system.

Dura Mater. The dura mater (Fig. 8-12) is the external membrane and consists of two layers firmly blended with each other except in certain areas. The more superficial is the endocranium (the endoperiosteum). Through the openings in the skull this layer is continuous with the pericranium (external periosteum). The endoperiosteum is intimately related to the bones of the skull and does not take part in the formation of the falx cerebri or tentorium cerebelli. The middle meningeal vessels ascend in the dura and produce grooves in the parietal bone. The inner layer of dura is smooth and lined by endothelial cells, resembling a serous membrane, and is separated from the outer layers by a small amount of fibrous tissue. The venous sinuses and meningeal vessels separate the two dural layers. By infolding and reduplicating itself, the inner dural layer forms four major membranes that subdivide the cranial cavities. The membranes are the falx cerebri, tentorium cerebelli, falx cerebelli, and diaphragma sellae.

The falx cerebri (Fig. 8-13) is sickle shaped and placed vertically between the two hemispheres as a reduplication of inner serous dura. It is attached to the crista galli in front and extends back to the internal occipital protuberance. Its lower border is attached to the tentorium cerebelli behind but remains free anteriorly and projects between the cerebral hemispheres.

The superior sagittal sinus appears in its upper border, and the inferior sagittal sinus is found in its lower border and continues back into the straight sinus. The falx cerebelli runs vertically from the tentorium to the foramen magnum and separates the two cerebellar hemispheres. It attaches to the internal occipital crest and encloses the occipital sinus. The tentorium cerebelli is tentlike and forms a partition between the cerebellum and cerebral hemispheres. It forms a roof for the cerebellum and floor for the occipital lobe and posterior temporal lobe. A wide space, anteriorly, between the two tentoria is known as the tentorial notch and allows passage for the midbrain. Thus, the tentorium has a free inner edge and is attached laterally to the margins of the transverse sinus, to the margins of the superior petrosal sinus, and to the posterior clinoid process of the sphenoid bone and runs forward to the anterior clinoid process. The upper layer is continuous with the falx cerebri in the midline. The diaphragma sellae is a fold with a foramen in its center. It forms the boundary of the foramen around the infundibulum and thus is related to the pituitary gland. Laterally, the dura extends from the diaphragma sellae to form the cavernous sinus.

Intracranial Spaces

There are several intracranial spaces (Fig. 8-13). One refers to an extradural space, which is only a potential one since the outer dura is closely adherent to the skull. The meningeal vessels are found here and, if they are injured, arterial bleeding occurs between dura and skull, which may peel the dura away from the skull. Such bleeding is rapid and can be fatal. The subdural space is between dura and arachnoid. Hemorrhage here may result from injury to large arteries, but this is rare. Subdural bleeding is usually venous bleeding since the large sinuses may be torn. The subarachnoid space is between arachnoid and pia, and the cerebrospinal fluid is found here. The fluid is extensive at the base of the skull, where the cisterns are found and where all but the anterior part of the brain "floats." The so-called intracerebral space (subpial space) is a potential one between brain and pia. One cannot strip pia off the brain without tearing brain tissue. Bleeding here usually is traumatic or due to the spontaneous rupture of an artery in its interior. Blood, in this case, would appear in the cerebrospinal fluid.

Cerebrospinal Fluid and the Ventricular System

Cerebrospinal fluid involves the ventricular system and the subarachnoid space (Figs. 8-8, 8-12). It is formed in the ventricular system by the choroid plexi, absorbed in the subarachnoid space, and released into the venous dural sinuses by the arachnoid villi, thus maintaining a pressure level. It takes the place of lymph and nourishes the neurons.

The ventricular system (Fig. 8-8) consists of four ventricles: two lateral, the third, and the fourth ventricles. These spaces normally communicate with each other through well-defined openings. Each lateral is located in a cerebral hemisphere and is divided into an anterior horn (frontal lobe), a body (parietal lobe), a posterior horn (in the occipital lobe), and an inferior horn (in the temporal lobe). Each communicates with the third ventricle via the foramina of Monro. The third ventricle empties into the fourth by way of the foramen of Sylvius, which passes through the midbrain and is the only source of exit for the lateral and third ventricles. It is narrow and small and is the weakest point in the system. The floor of the fourth ventricle is formed by the pons and medulla, and it is roofed by the cerebellum. The fourth ventricle connects with the subarachnoid space by three openings: two lateral foramina of Luschka and a median foramen of Magendie. The two lateral open into the lateral cistern and the median one into the cisterna magna, thus connecting the ventricular system with the subarachnoid space. The fluid from the latter circulates freely around the cerebrum and cerebellum, finally passing down the spinal subarachnoid space.

Cerebrospinal fluid is formed by the choroid plexus (mainly in the lateral ventricle but in all). It passes to the subarachnoid space, where it contacts the arachnoid villi, which absorb and return it to the venous system in the dural sinuses. The total amount of fluid is between 90 and 150 ml in adults. If the ventricular system is blocked, a condition called hydrocephalus results with distention of the ventricular system proximal to the block. One can determine the position of the block by the use of ventriculography, in which some cerebrospinal fluid from the sinuses is replaced by air and the head is x-rayed. Encephalography consists of withdrawing cerebrospinal fluid by means of a lumbar puncture needle and introducing air, which ascends and produces an outline of the ventricular system seen in x-ray. By this procedure one can also measure cerebrospinal pressure as well as sample the fluid for bleeding and other changes.

The Cranial Nerves

There are 12 pairs of nerve trunks, arranged symmetrically and attached to the base of the brain (brain stem). They leave the skull by way of foramina in the skull and are predominantly distributed to the head. They contain both sensory (afferent) and motor (efferent) fibers and constitute the cranial portion of the parasympathetic system. On leaving the brain, each nerve is ensheathed by a layer of pia mater, pierces the arachnoid, and acquires a sheath from that layer as well. It then passes through dura on its way to the skull foramina and is covered with dura as well, which is continuous with the epineurium of the nerve trunk proper. The cranial nerves are designated numerically from before backward but also have specific names based on either distribution or function (Figs. 8-9, 8-14; Table 8-1).

Cranial I. The olfactory nerves (Figs. 8-14, 8-15) are 15 to 20 in number. They arise in the upper third of the nasal mucous membrane and pass up through the cribriform plate of the ethmoid bone into the skull, where they enter the olfactory bulb. The olfactory tract runs back from the bulb to end in the brain, in areas associated with olfaction (the uncus of the temporal lobe being one). Lesions in the latter area may initiate with unpleasant olfactory prodromes followed by a state of dreaminess, the syndrome spoken of as uncinate gyrus fits.

Cranial II. The optic nerve (Figs. 8-9, 8-14, 8-15; see Chap. 9) runs from the posterior aspect of the eyeball and passes through the optic foramen (accompanied by the ophthalmic artery). In the skull, it converges toward the nerve of the opposite side, which

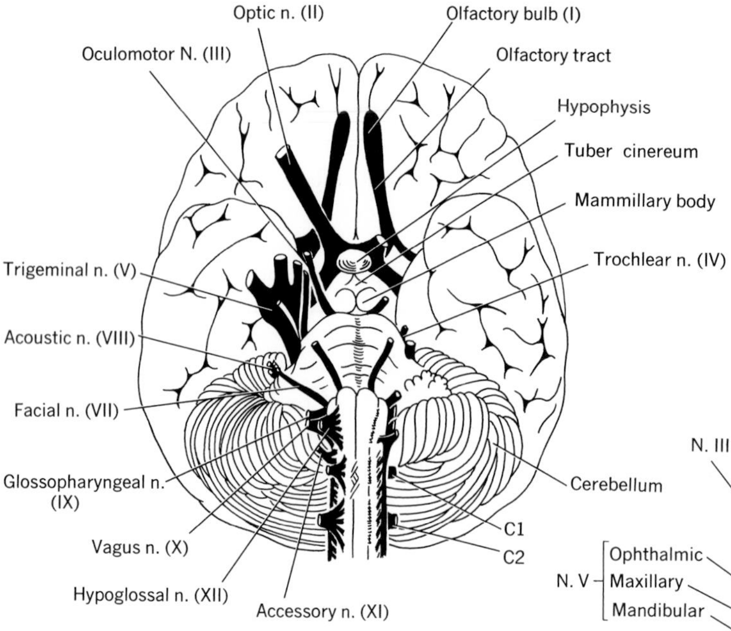

Optic n. (II)
Olfactory bulb (I)
Oculomotor N. (III)
Olfactory tract
Hypophysis
Tuber cinereum
Mammillary body
Trochlear n. (IV)
Trigeminal n. (V)
Acoustic n. (VIII)
Facial n. (VII)
Glossopharyngeal n. (IX)
Vagus n. (X)
Hypoglossal n. (XII)
Accessory n. (XI)
Cerebellum
C1
C2

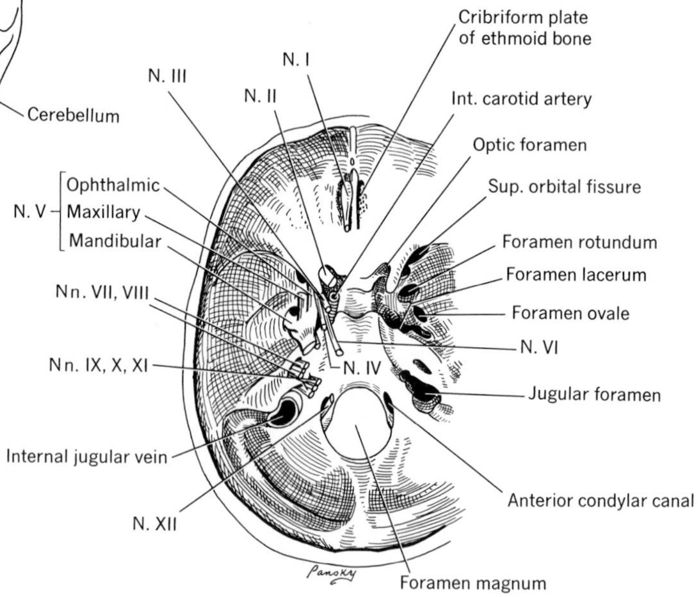

Cribriform plate of ethmoid bone
N. I
N. III
N. II
Int. carotid artery
Optic foramen
Sup. orbital fissure
N. V
Ophthalmic
Maxillary
Mandibular
Foramen rotundum
Foramen lacerum
Foramen ovale
Nn. VII, VIII
N. VI
Nn. IX, X, XI
N. IV
Jugular foramen
Internal jugular vein
Anterior condylar canal
N. XII
Pansky
Foramen magnum

Figure 8-14. Cranial nerves and their cranial foramina of exit (and entrance).

Table 8-1. Summary of Cranial Nerves

Nerve	***Component****	***Major Cell Body***	***Course***	***Termination***
Olfactory (I)	VS	Olfactory epithelium	Through roof of nasal cavities	Olfactory epithelium
Optic (II)	SSS	Retinal ganglionic layer	Orbit to optic chiasma to optic tracts	Bipolar cells of retina—rods and cones
Oculomotor (III)	SM	Oculomotor nucleus	Orbit	Rectus superior, inferior, medial; inferior oblique; levator palpebrae muscles
	VM	Edinger-Westphal nucleus	Ciliary ganglion to ciliary nerves	Constrictor pupillae and ciliary muscles of eyeball
Trochlear (IV)	SM	Trochlear nucleus	Orbit	Superior oblique muscle of eyeball

Table 8-1 (Continued)

Nerve	Component*	Major Cell Body	Course	Termination
Trigeminal (V)	BM	Masticator nucleus	With mandibular	Muscles of mastication
	GSS	Semilunar ganglion	Ophthalmic, maxillary, and mandibular branches	Face, nose, mouth
	GSS	Mesencephalic nucleus	With mandibular and maxillary branches	Proprioceptive to jaw muscles and teeth
Abducens (VI)	SM	Abducens nucleus	Orbit	Lateral rectus of eyeball
Facial (VII)	BM	Facial nucleus	Temporal bone, side of face	Muscles of facial expression and elevators of hyoid bone
	VM	Superior salivatory nucleus	Greater superficial petrosal nerve to sphenopalatine ganglion	Glands of nose, palate, and lacrimal gland
			Chorda tympani to the submandibular ganglion	Submandibular and sublingual glands
	VS	Geniculate ganglion	Chorda tympani	Anterior taste buds of tongue
Acoustic (VIII) 1. Vestibular	SSS	Vestibular ganglion	Internal acoustic meatus	Cristae of the semicircular canals; maculae of the utricle and saccule
2. Cochlear	SSS	Spiral ganglion	Internal acoustic meatus	Organ of Corti
Glossopharyngeal (IX)	BM	Nucleus ambiguus	Jugular foramen to sides of larynx	Superior constrictor of pharynx and stylopharyngeus muscle
	VM	Inferior salivatory nucleus	Lesser superficial petrosal to otic ganglion to auriculotemporal nerve	Parotid gland
	VS	Petrosal ganglion	Side of pharynx	Taste buds of vallate papillae of tongue
	GSS	Superior ganglion	Side of pharynx	Auditory tube
Vagus (X)	BM	Nucleus ambiguus	Recurrent and external branches of superior laryngeal nerve	Pharyngeal and laryngeal muscles
	VM	Dorsal motor nucleus	Along carotid artery, esophagus, and stomach	Viscera of thorax and abdomen
	VS	Nodose ganglion	With motor	Viscera of thorax and abdomen
	GSS	Jugular ganglion	Auricular branch	Pinna of ear
Accessory (XI)	BM	Accessory nucleus	Side of neck	Trapezius and sternocleidomastoid muscles
Hypoglossal (XII)	SM	Hypoglossal nucleus	Side of tongue	Extrinsic and intrinsic tongue muscles

*BM = branchial motor.   GSS = general somatic sensory.   SM = somatic motor.   SSS = special somatic sensory.   VM = visceral motor.   VS = visceral sensory.

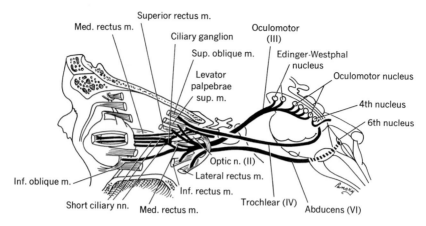

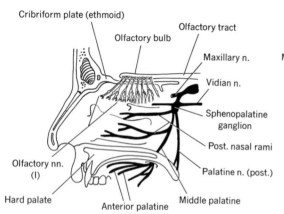

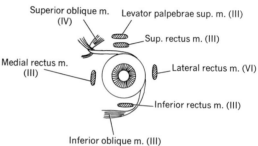

Figure 8-15. Cranial nerves I, II, III, IV, and VI. Distribution of the maxillary division of cranial nerve V to the nose and palate is also seen.

it joins to form the optic chiasma. Here the fibers that arise from the nasal side of the retina decussate with the corresponding fibers of the opposite side, whereas the fibers from the temporal side of the retina do not decussate but remain on the same side. The optic tracts arise on each angle of the chiasma and are continuations of the optic nerves. Each tract thus consists of fibers arising in the temporal half of the retina of the same side and the nasal half of the retina of the opposite side. The optic tracts then divide into lateral and medial roots. The lateral roots end in the gray matter of the lateral geniculate body and the superior colliculus. The medial root ends in the medial geniculate body and is not related to vision. From the lateral geniculate body fibers arise to enter the internal cap-

sule, where they are spoken of as the optic radiations and pass backward to terminate in the higher visual centers of the postcalcarine and calcarine occipital lobe cortex. Pressure about the optic nerve from the surrounding subarachnoid space may compress the central artery and vein, resulting in engorgement of the retinal veins and diminution of the artery, a condition spoken of as papilledema.

Cranial III. The oculomotor nerve (Figs. 8-9, 8-11, 8-14, 8-15) runs from the brain through the cavernous sinus, passes through the superior orbital fissure into the orbit, and divides into superior and inferior divisions. The superior supplies the superior rectus and levator palpebrae superioris muscle. The inferior supplies the inferior oblique and the medial and inferior recti. From the inferior division a short nerve passes to the ciliary ganglion, the ciliary muscle, and the sphincter muscle. Thus, this nerve supplies all the extrinsic muscles of the eye except for the external (lateral) rectus and superior oblique. A lesion of this nerve results in drooping of the upper lid (paralysis of the levator palpebrae superioris), dilated pupil (unopposed action of sympathetic fibers), loss of accommodation (paralysis of the ciliary muscle), and external strabismus (unopposed action of the external rectus and superior oblique muscles).

Cranial IV. The trochlear nerve (Figs. 8-9, 8-11, 8-14, 8-15) is the smallest of the cranial nerves and supplies the superior oblique muscle. Its pathway is similar to that of the third nerve. When the trochlear nerve is involved, the patient has difficulty in moving the eye outward and downward, by means of the "down-and-out" muscle. If there is an attempt to do so, the patient will see double.

Cranial V. The trigeminal nerve (Figs. 8-9, 8-11, 8-14, 8-16) has a wide distribution. It arises from the lateral side of the pons. It has a large sensory root, on which is found the semilunar (gasserian ganglion), and a motor root, which does not enter the ganglion but passes out of the skull with the mandibular division of cranial V. The nerve is divided into three major divisions: ophthalmic, maxillary, and mandibular. In general, the nerve provides the sensory supply to the face and anterior half of the scalp and sends motor branches to the four muscles of mastication (not the

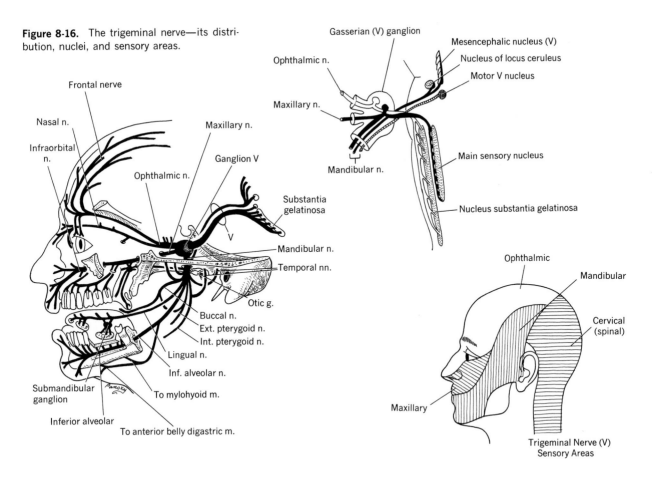

Figure 8-16. The trigeminal nerve—its distribution, nuclei, and sensory areas.

buccinator) and to the tensor veli palatini, tensor tympani, mylohyoid, and anterior belly of the digastric. The fifth nerve has five ganglia associated with it: the semilunar on the nerve trunk; the ciliary on the ophthalmic division; the sphenopalatine on the maxillary division; and the otic and submandibular, both on the mandibular division.

OPHTHALMIC DIVISION. This is the smallest division, is purely sensory, and after sending a small branch to the dura enters the orbit via the superior orbital fissure. Here it forms several branches: the lacrimal, which supplies the lacrimal gland, conjunctiva, and skin of the upper eyelid; the frontal, which subdivides into the supraorbital nerve, which reaches the scalp through the supraorbital foramen or notch to supply the scalp (the frontal nerve supplies the mucous membrane of the frontal sinus, skin of the upper eyelid, scalp, and forehead); the supratrochlear, which also goes to the scalp; and the nasociliary, which leaves the orbit via the anterior ethmoidal foramen and changes its name to the ethmoidal nerve, which then passes over the cribriform plate of the ethmoid to enter the nose through the nasal slit, where it supplies the nasal septum as well as the skin on the lower part of the nose. In the orbit the ophthalmic division gives off branches to the ciliary ganglion, nerves called the ciliary nerves to the eyeball and the infratrochlear nerve.

Ciliary Ganglion. The ciliary ganglion is found in the orbit near the optic nerve. It receives a sensory root from the nasociliary branch of the ophthalmic, a motor root from the oculomotor, and a sympathetic nerve from the plexus around the carotid artery. It gives off 12 to 14 short ciliary nerves to the ciliary muscles and iris.

MAXILLARY DIVISION. The maxillary division is purely sensory and leaves the cranial fossa via the foramen rotundum to enter the pterygopalatine fossa. From here it passes by way of the inferior orbital fissure to the inferior orbital groove and canal and enters the face at the infraorbital foramen as the infraorbital nerve. In its course, it gives a small meningeal twig to the dura, two branches to the sphenopalatine ganglion, and zygomatic branches to the orbit, which form the zygomaticotemporal and zygomaticofacial to the side of the face; posterior alveolar to the molar teeth; infraorbital to the three molars, canine teeth, and incisors; and facial branches to the lower eyelids (palpebral), side of the nose (nasal), and upper lip (labial).

Sphenopalatine Ganglion. The sphenopalatine ganglion has sensory roots that arise from the maxillary division of cranial V and motor fibers from sympathetic and nerve of the pterygoid canal (Vidian). It gives off orbital (secretomotor fibers to the lacrimal), pharyngeal, nasal, and palatine branches.

MANDIBULAR DIVISION. This consists of a sensory portion from the semilunar ganglion and a motor root. Both pass through the foramen ovale. Below the foramen, a twig to the meninges, the nervous spinosus, is given off, which passes back through the foramen spinosum. The nerve then gives off a branch to the medial pterygoid muscle and divides into two branches: the anterior (mainly motor) gives off deep temporal nerves, lateral pterygoid nerves, nerves to the masseter, and the buccal nerve; the posterior ends as the auriculotemporal, lingual, and inferior alveolar nerves to the lower teeth. In this posterior division the only motor twig is the mylohyoid branch of the inferior alveolar. Two ganglia are associated with this division: the otic and submandibular.

Otic Ganglion. The otic ganglion lies just below the foramen ovale and sends muscle branches to the tensor tympani and tensor palati.

Submandibular Ganglion. The submandibular ganglion lies on the hyoglossus muscle and joins the lingual nerve. Fibers are distributed to the submandibular and sublingual glands.

Cranial VI. The abducens nerve (Figs. 8-9, 8-11, 8-14, 8-15) supplies only one muscle, the lateral rectus of the eyeball. Involvement of the nerve paralyzes this muscle on the same side, and the now-unopposed medial rectus can displace the eye inward (internal strabismus). The lateral rectus abducts the eye.

Cranial VII. The facial nerve (Figs. 8-9, 8-11, 8-14, 8-17) is the motor nerve to the face and contains no

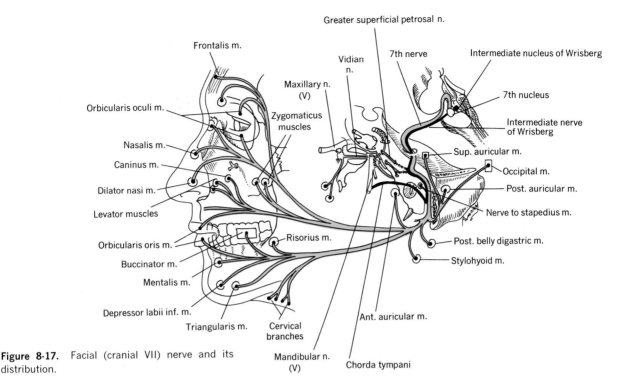

Figure 8-17. Facial (cranial VII) nerve and its distribution.

Labels in figure:
Greater superficial petrosal n.
Frontalis m.
Vidian n.
7th nerve
Intermediate nucleus of Wrisberg
Maxillary n. (V)
Orbicularis oculi m.
Zygomaticus muscles
7th nucleus
Intermediate nerve of Wrisberg
Nasalis m.
Sup. auricular m.
Caninus m.
Occipital m.
Dilator nasi m.
Post. auricular m.
Levator muscles
Nerve to stapedius m.
Orbicularis oris m.
Risorius m.
Buccinator m.
Post. belly digastric m.
Mentalis m.
Stylohyoid m.
Depressor labii inf. m.
Triangularis m.
Cervical branches
Ant. auricular m.
Mandibular n. (V)
Chorda tympani

cutaneous branches. It follows the eighth (auditory) nerve into the internal auditory meatus and passes through the temporal bone, where it gives off the superficial petrosal nerve, which supplies the mucous membrane of the soft palate and secretory fibers to the mucous glands; the nerve to the stapedius muscle; and the chorda tympani, which passes through the tympanic cavity (middle ear) to join the lingual nerve and supplies taste and sensation fibers to the anterior two thirds of the tongue and secretory fibers to the submandibular and sublingual glands. The facial nerve proper exits from the stylomastoid foramen, where it gives off the posterior auricular nerve to the ear muscles and the occipital muscle and branches to the stylohyoid and posterior belly of the digastric muscles. The nerve then continues on to terminal branches on the face, which form a nerve plexus, pes anserinus (goose's foot), that emerges at the anterior border of the parotid gland and radiates over the side of the face.

One sees the temporal branch (to the frontalis muscle and facial muscles above the zygoma), the zygomatic branch (to muscles about the zygoma), the buccal branch (to the buccinator and orbicularis oris muscles), the mandibular twig (to the lower lip and chin), and the cervical branch (to the platysma and depressors of the lower lip).

Intracranial lesions of the facial nerve characteristically involve only the lower half of the face (lesions may result from middle-ear diseases or fractures of the skull or face). Extracranial lesions result in facial paralysis in which the involved side of the face is expressionless and flat (Bell's palsy). The patient cannot whistle, blow out his cheeks, wrinkle his forehead, or show his teeth.

Cranial VIII. The acoustic or auditory nerve (Figs. 8-9, 8-11, 8-14, 8-18) consists of two parts: the cochlear (carrying auditory impulses) and the vestibular (dealing with equilibrium). The cochlear nerve is distributed

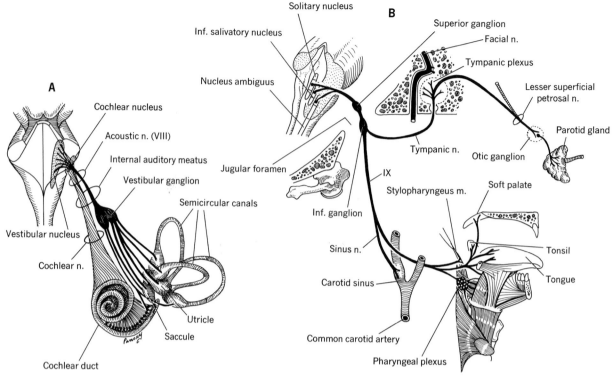

Figure 8-18. **A.** Course and distribution of the acoustic (VIII) nerve. **B.** Course and distribution of the glossopharyngeal (IX) nerve.

to the cochlear duct and spiral organ; the vestibular is distributed to the semicircular ducts, utricle, and saccule. Lesions of this nerve can produce deafness and a loss of equilibrium.

Cranial IX. The glossopharyngeal nerve (Figs. 8-9, 8-11, 8-14) leaves the skull through the jugular foramen, enclosed in a special dural compartment with the tenth and eleventh cranial nerves as well as the internal jugular vein. It supplies taste and sensations to the posterior one third of the tongue as well as the epiglottis, soft palate, tonsils, and pillars of the fauces. It supplies but one muscle, the stylopharyngeus. Lesions of the nerve create anesthesia of the posterior one third of the tongue as well as pharynx.

Cranial X. The vagus nerve (Figs. 8-9, 8-11, 8-14, 8-19) leaves the skull through the jugular foramen and

runs in the carotid sheath with the internal jugular vein and carotid artery. The right vagus gives off the recurrent laryngeal nerve as it crosses the right subclavian artery; the left recurrent laryngeal crosses under the aortic arch. Both run between trachea and esophagus to the laryngeal musculature. The right vagus continues into the thorax, where it forms the posterior pulmonary plexus as well as the esophageal plexus and continues below the diaphragm to supply the posterior surface of the stomach, sending branches to the celiac, splanchnic, and renal plexi. The left vagus helps supply the posterior pulmonary plexus and the esophageal plexus. This nerve then passes over the anterior surface of the stomach with branches to the liver. The most typical signs of a vagal lesion usually involve the laryngeal nerves and produce an immobile

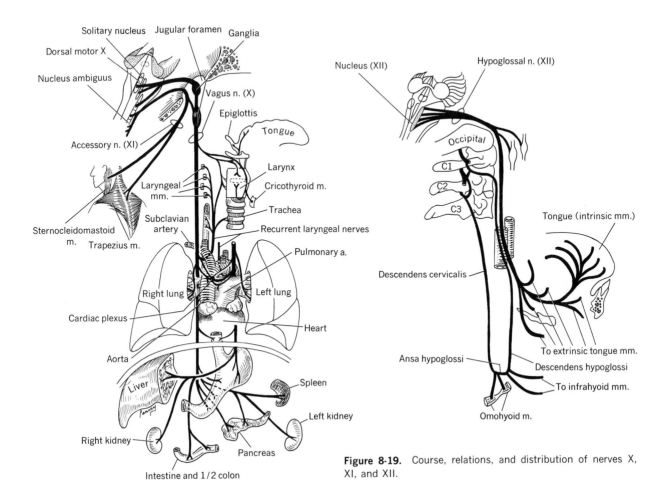

Figure 8-19. Course, relations, and distribution of nerves X, XI, and XII.

vocal cord and hoarseness. The vagus supplies the digestive system up to the middle of the transverse colon.

Cranial XI. The spinal accessory nerve (Figs. 8-9, 8-11, 8-14, 8-19) originates from the brain and partly from the upper spinal cord. The latter ascends in the spinal canal and enters the skull through the foramen magnum and joins the cranial part. Both leave the skull via the jugular foramen. The spinal accessory nerve supplies the sternocleidomastoid and trapezius muscles.

Cranial XII. The hypoglossal nerve (Figs. 8-9, 8-11, 8-14, 8-19) leaves the skull through the anterior condy-

lar canal. It leads to the tongue and supplies the extrinsic and intrinsic muscles of the tongue. When lesioned, the corresponding half of the tongue becomes atrophied, and the patient, when asked to protrude his tongue, finds it deviating toward the paralyzed side. The nerve is associated with cervical nerves C1 to C3 to form the so-called ansa hypoglossi, but the infrahyoid and suprahyoid muscles are innervated by the cervical nerves and not the hypoglossal.

Figures 8-9 and 8-20 show the brain stem, the cranial nerves, and the motor and sensory nuclei that they are associated with. Use these illustrations to supplement the previous discussion and for review.

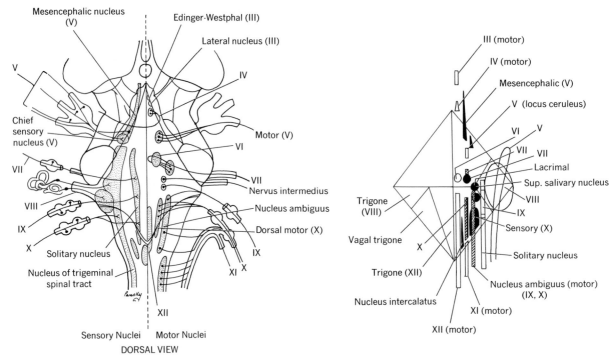

Figure 8-20. The cranial (brain stem) nuclei schematically shown in a dorsal view of the fourth ventricle.

The Spinal Cord and Associated Structures

Spinal or Vertebral Canal. The vertebral foramina make a single canal in the spinal column, whose anterior wall is closed by the bodies of the vertebrae and their disks and posterior and lateral walls consist of the bony arches (lamina). It is lined by periosteal and ligamentous structures. The canal is circular near the skull and becomes triangular in the lumbar region. It flattens in the sacrum but expands laterally. (See Chap. 5 for detailed discussion.)

Spinal Meninges. The meninges consist of three tubular fibrous membranes named, from outside to inside, the dura mater, arachnoid mater, and pia mater, all continuous with corresponding layers around the brain (Figs. 8-12, 8-13). The space between the dura and the walls of the vertebral canal is the epidural space, which is filled with semiliquid fat, areolar connective tissue, a venous network, and small arteries to the bones.

The cranial dura has two layers, but at the foramen magnum the outer layer blends or fuses with the linings of the spinal canal and the inner layer is the one that forms the covering for the cord and emerging spinal nerve roots. In the canal, the dura is a loose "envelope-like" covering attached firmly to the second and third cervical vertebrae but not closely associated with the periosteum elsewhere. It ends at the second sacral vertebrae (about five vertebrae lower than the spinal cord), with a prolongation of it investing the filum terminale of the cord, which runs to the back of the coccyx. The anterior and posterior nerve roots that "pierce" it actually carry tubular dural prolongations with them as far as the intervertebral canals, where they attach to the periosteum. In the area of S2 (where it ends), it forms a "cul-de-sac" that is pierced by the filum terminale, which also carries a prolongation of dura to its attachment to the coccyx (Fig. 8-13).

The spinal arachnoid is a fragile, thin, transparent

membrane. It, too, ends a little below S2 vertebra. It is connected posterior to the layer beneath it, the pia, by an incomplete posterior median septum in the cervical region. The arachnoid tends to follow dura, and the two move freely on one another in the capillary interval (subdural space) that exists between them. The subarachnoid space lies between arachnoid and pia and contains the cerebrospinal fluid. The space is traversed by delicate septa lined with flat arachnoid cells as well as by vessels going to the spinal cord (and brain). It communicates with its counterpart in the cranial space, and thus an increase in pressure as a result of a brain swelling or hemorrhage can be diagnosed by lumbar puncture (tapping of the spinal subarachnoid space and observing its pressure, contents, and appearance).

The pia mater is closely adhered to the spinal cord and dips into the anterior median fissure. It is a vascular, delicate membrane attached to the surfaces of the cord and carries blood vessels into its substance. At the lateral sides of the cord, the pia is attached to the dura by the thin, membranous denticulate (dentate) ligaments. These ligaments consist of 20 "toothlike" processes extending from pia to dura (pushing the arachnoid ahead of it). They serve to suspend the cord in the midline and separate the anterior and posterior nerve roots as they pass to and from the cord.

Cord Proper. The spinal cord (Figs. 8-21, 8-22) is

Figure 8-21. The spinal cord with its spinal nerves; general arrangement of gray and white matter and major cell columns at different levels.

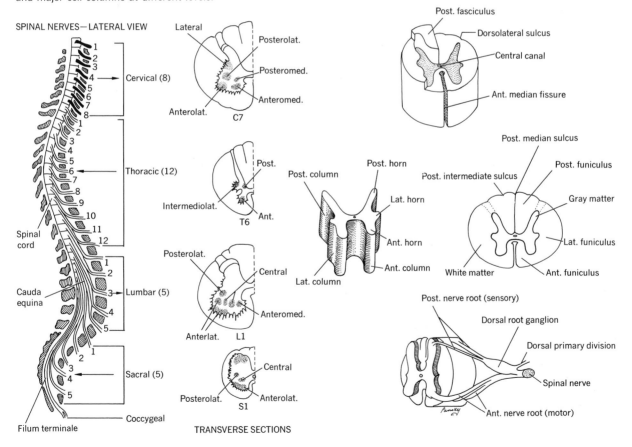

SPINAL NERVES—LATERAL VIEW

TRANSVERSE SECTIONS

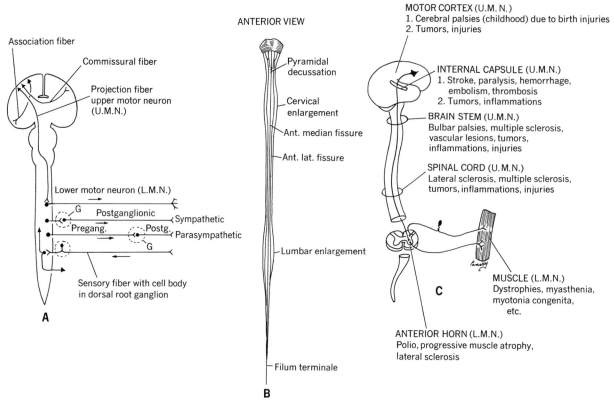

ANTERIOR VIEW

Association fiber

Commissural fiber

Projection fiber
upper motor neuron
(U.M.N.)

Pyramidal
decussation

Cervical
enlargement

Ant. median fissure

Ant. lat. fissure

Lower motor neuron (L.M.N.)

G   Postganglionic
Pregang.   Postg.
                          G

Sympathetic
Parasympathetic

Lumbar enlargement

Sensory fiber with cell body
in dorsal root ganglion

A

Filum terminale

B

MOTOR CORTEX (U.M. N.)
1. Cerebral palsies (childhood) due to birth injuries
2. Tumors, injuries

INTERNAL CAPSULE (U.M.N.)
1. Stroke, paralysis, hemorrhage,
    embolism, thrombosis
2. Tumors, inflammations

BRAIN STEM (U.M.N.)
Bulbar palsies, multiple sclerosis,
vascular lesions, tumors,
inflammations, injuries

SPINAL CORD (U.M.N.)
Lateral sclerosis, multiple sclerosis,
tumors, inflammations, injuries

MUSCLE (L.M.N.)
Dystrophies, myasthenia,
myotonia congenita,
etc.

C

ANTERIOR HORN (L.M.N.)
Polio, progressive muscle atrophy,
lateral sclerosis

Figure 8-22. **A.** Neuron types and connections in brain and spinal cord. *G* refers to ganglion. **B.** Spinal cord, anterior view. **C.** Upper motor neuron (*U.M.N.*) involvement and clinical relations; lower motor neuron (*L.M.N.*) involvement and clinical relations.

about 45 cm (18 in.) long and about 1.25 cm ($\frac{1}{2}$ in.) thick, and is generally a cylindric mass of nervous tissue lying in the upper two thirds of the vertebral canal in the adult. Until four months of fetal life the cord extends the full length of the vertebral canal; it extends as far as the sacrum by six months of fetal life and to the lumbar vertebrae at birth. In the adult it is at the upper border of the second lumbar vertebra. Since its diameter is less than that of the vertebral canal, the spinal column can bend and twist without placing great strain on the cord. It begins at the base of the skull (foramen magnum), where it is continuous with the medulla oblongata and ends in a tapered filament called the filum terminale (composed mostly of pia mater) to the back of the coccyx. The filum pierces

the dura and arachnoid and blends with the periosteum on the dorsum of the coccyx. (The subarachnoid and subdural spaces extend to the body of S2, but the cord proper ends at L2. Thus, between L2 and S2 there is no cord but only the filum terminale and roots of the lower spinal nerves, which are called the cauda equina since they resemble a horse's tail. A needle [in a spinal tap] that enters the subarachnoid space above L2 may damage the cord but distal to that would meet only terminal nerve.)

Opposite the fifth and sixth cervical vertebrae and again opposite the lower two thoracic vertebrae the cord has two enlargements, which are associated with the origin of two major plexi (branchial and lumbar) to the upper and lower extremities (Fig. 8-22).

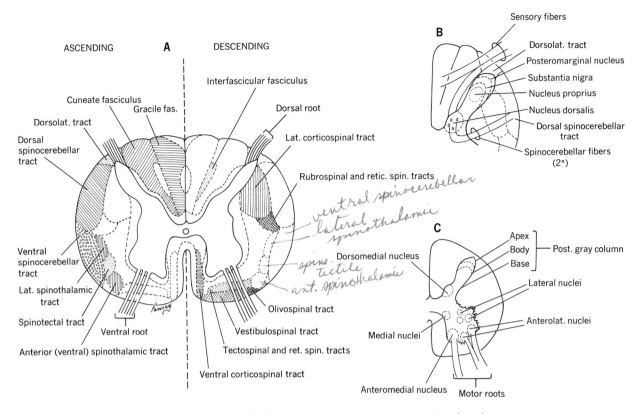

Figure 8-23. **A.** Diagram of the principal fasciculi of the spinal cord. **B.** Transverse section of cord showing dorsal gray column, its divisions, and major nuclei. **C.** Transverse section of cord showing major nuclear subdivisions of the gray matter.

In the cord, the gray matter is placed centrally and the white peripherally (Fig. 8-23). The gray is H shaped and contains anterior (ventral) and posterior (dorsal) limbs, spoken of as the anterior and posterior horns or gray columns. A groove (in front), the antero-median fissure, and a septum (behind), the postero-median septum, separate the right and left sides of the white matter. The H transverse limb connecting the right and left gray matter is the gray commissure and surrounds the central canal of the cord. One also sees, in the thoracic region, an additional lateral gray horn opposite the central canal (associated with the thoracic sympathetic system).

Spinal Nerves. Spinal nerves (Figs. 8-23, 8-24) are segmentally arranged and as indicated are attached to the cord by two roots: an anterior (motor or efferent) and a posterior (sensory or afferent). The posterior root has a ganglion on it. The anterior rami are larger and have a tendency to form the plexi, whereas the smaller posterior supply the muscles of the back and overlying skin.

From cells of the anterior horn, the fibers of anterior (efferent) nerve roots (motor) leave the cord at its anterolateral aspect. Opposite the posterior horn tip, the posterior (afferent) nerve roots, which are sensory, enter the spinal cord. This arrangement divides the surrounding white matter into three columns or white funiculi: an anterior white column, a lateral white column, and a posterior white column. The afferent and efferent neurons are not in physical continuity, but

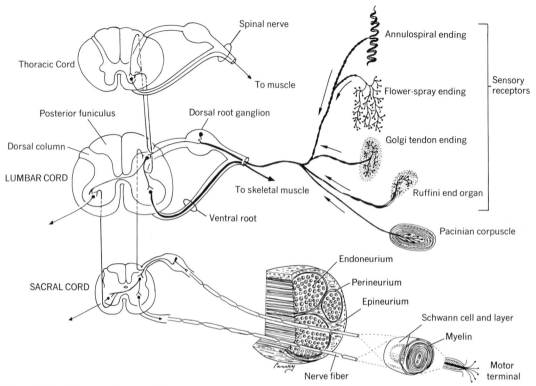

Figure 8-24. Diagram of some of the general proprioceptive (GP) reflex paths for afferent impulses to the central nervous system via spinal nerves.

the five terminals of the afferents merely make contact with the efferents. This transfer zone is called the synapse. In this way, the reflex arc is formed, composed of an afferent neuron with its receptor, a center where synapse occurs, and an efferent neuron with its effector (Fig. 8-24). A third neuron or association neuron is also seen, interposed between the afferent and efferent fibers, and increases the potential number of synapses possible. This creates a three-neuron reflex arc (see discussion of reflexes on p. 295).

Close to the vertebral column, the anterior rami of nerves C1 to C4 form the cervical plexus (Fig. 8-25); those of C5 to T1, the branchial plexus (Fig. 8-25); L1 to L4, the lumbar plexus (Figs. 8-26, 8-27); L4 to S4, the sacral plexus (Fig. 8-27); and S4 to Co1, the coccygeal plexus. On each posterior root, a ganglion

is located consisting of nerve cell bodies outside the CNS that give origin to central and peripheral fibers. They are all found (except the sacral and coccygeal) in the area of the intervertebral foramina. The sacral and coccygeal ganglia lie in the spinal canal. Near the foramina, each pair of nerve roots unites to form a spinal nerve (Fig. 8-24), which divides distally into anterior and posterior primary branches (Fig. 8-28). The spinal nerve (and its primary branches) are mixed nerves carrying both sensory and motor fibers. Distal to the division of the spinal nerve, a small recurrent branch to the meninges of the cord is given off (after receiving a branch from the sympathetic trunk). Each nerve root receives a covering of pia mater and arachnoid before it meets dura, and in the subarachnoid space is surrounded by cerebrospinal fluid. Beyond the

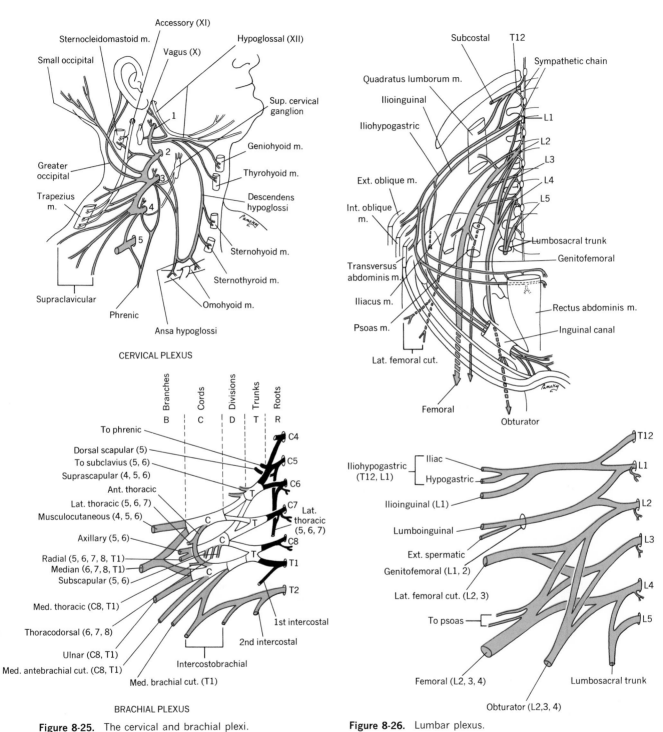

CERVICAL PLEXUS

Accessory (XI)
Sternocleidomastoid m.
Vagus (X)
Hypoglossal (XII)
Small occipital
Sup. cervical ganglion
Geniohyoid m.
Greater occipital
Thyrohyoid m.
Descendens hypoglossi
Trapezius m.
Sternohyoid m.
Supraclavicular
Sternothyroid m.
Phrenic
Omohyoid m.
Ansa hypoglossi

1
2
3
4
5

BRACHIAL PLEXUS

Branches (B)
Cords (C)
Divisions (D)
Trunks (T)
Roots (R)

To phrenic — C4
Dorsal scapular (5) — C5
To subclavius (5, 6)
Suprascapular (4, 5, 6) — C6
Ant. thoracic
Lat. thoracic (5, 6, 7) — C7
Musculocutaneous (4, 5, 6)
Lat. thoracic (5, 6, 7)
Axillary (5, 6) — C8
Radial (5, 6, 7, 8, T1)
Median (6, 7, 8, T1) — T1
Subscapular (5, 6)
Med. thoracic (C8, T1) — T2
Thoracodorsal (6, 7, 8)
1st intercostal
Ulnar (C8, T1)
2nd intercostal
Med. antebrachial cut. (C8, T1)
Intercostobrachial
Med. brachial cut. (T1)

Figure 8-25. The cervical and brachial plexi.

Subcostal — T12
Quadratus lumborum m.
Sympathetic chain
Ilioinguinal
Iliohypogastric — L1
Ext. oblique m. — L2
Int. oblique m. — L3
— L4
— L5
Transversus abdominis m.
Lumbosacral trunk
Genitofemoral
Iliacus m.
Rectus abdominis m.
Psoas m.
Inguinal canal
Lat. femoral cut.
Femoral
Obturator

Iliohypogastric (T12, L1) — Iliac / Hypogastric — T12
Ilioinguinal (L1) — L1
Lumboinguinal — L2
Ext. spermatic — L3
Genitofemoral (L1, 2) — L4
Lat. femoral cut. (L2, 3) — L5
To psoas
Femoral (L2, 3, 4)
Lumbosacral trunk
Obturator (L2, 3, 4)

Figure 8-26. Lumbar plexus.

Chap. 8 Neural Control Mechanisms **277**

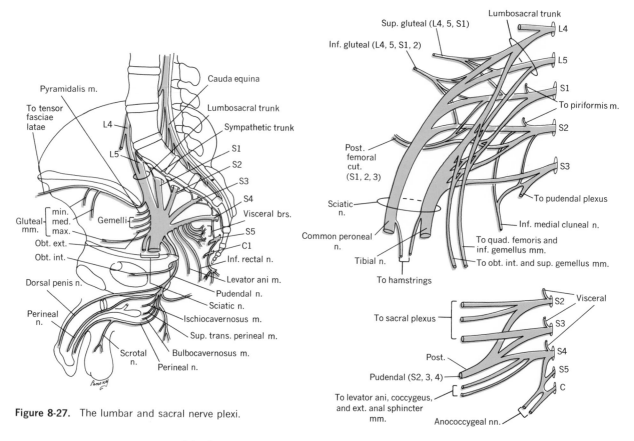

Figure 8-27. The lumbar and sacral nerve plexi.

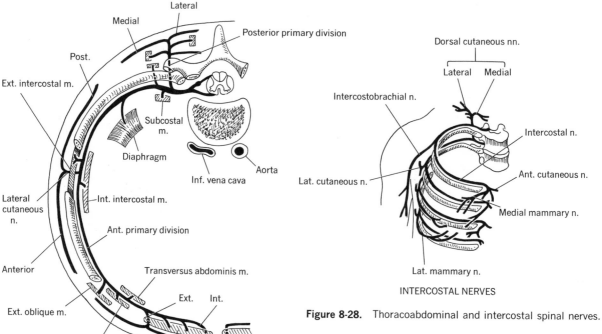

INTERCOSTAL NERVES

Figure 8-28. Thoracoabdominal and intercostal spinal nerves.

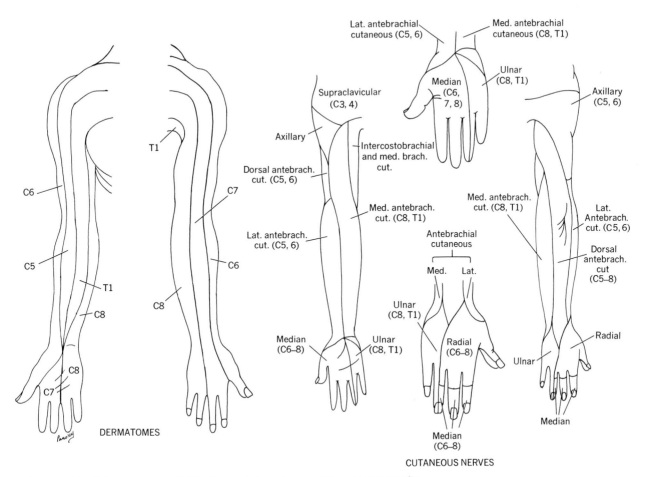

Figure 8-29. Cutaneous nerve distribution and dermatomes of the upper extremity.

space, they are covered by dura (which includes the sensory ganglion associated with the posterior [sensory] root).

The spinal nerves supply cutaneous and motor branches to the extremities and to the chest and abdominal walls. Individual nerve fibers, as they come off the spinal cord, supply small areas of the skin that can be mapped and are referred to as dermatome areas. Combined fibers that make up a spinal nerve supply a broader area and are referred to as cutaneous nerves (Figs. 8-29 and 8-30).

The motor and sensory nerves of the upper extremity

arise from the brachial plexus (Fig. 8-31). Those nerves supplying the lower extremity arise from the lumbar and sacral plexi (Fig. 8-32). See Table 8-2.

An intercostal nerve is the anterior nerve ramus (Fig. 8-28). There are 12 pairs of thoracic or intercostal nerves, 11 pairs of which are really intercostal and one pair of subcostal. The latter runs its course in the abdominal wall and not in an intercostal space. Each nerve sends a white ramus communicans to a corresponding sympathetic ganglion (carrying preganglionic sympathetic fibers to the ganglion) and receives a gray ramus communicans from it (carrying post-

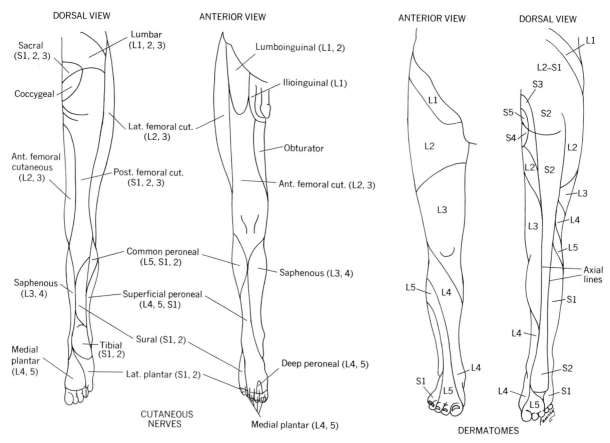

DORSAL VIEW   ANTERIOR VIEW   ANTERIOR VIEW   DORSAL VIEW

CUTANEOUS NERVES

DERMATOMES

Figure 8-30. Diagrams of cutaneous nerves of the lower extremity and dermatomes of the same areas.

ganglionic sympathetic fibers) (Fig. 8-33). A typical intercostal nerve supplies muscular and two cutaneous branches (a lateral cutaneous, to supply the side of the chest both anterior and posterior, and an anterior cutaneous, which is the terminal end of the nerve and supplies the skin over the front of the chest). The first joins the brachial plexus to the arm. The second intercostal is unique since its lateral branch does not divide but crosses the axilla as the intercostobrachial nerve to supply the skin of the posteromedial part of the arm down to the elbow. The seventh supplies the region of the epigastrium, the tenth the umbilical area, and the twelfth the area just above the pubis. Thus, nerves 2 to 6 are typical intercostals, whereas nerves 7 to 11 supply thorax and abdomen and are spoken of as thoracoabdominal nerves. The typical intercostal nerves supply the intercostal muscles and abdominal muscles and become cutaneous on the abdomen. The muscles supplied are external and internal intercostals, subcostals, transversus thoracis, levator costarum, external and internal oblique, transversus abdominis, pyramidalis, and, dorsally, the serratus posterior superior and inferior.

Nerve Tracts. The nerve fibers that make up the white matter of the cord consist mainly of two groups: ascending fibers arranged in tracts or fasciculi, which become larger as they pass upward, due to the addition of new fibers as you move up the cord; and descending

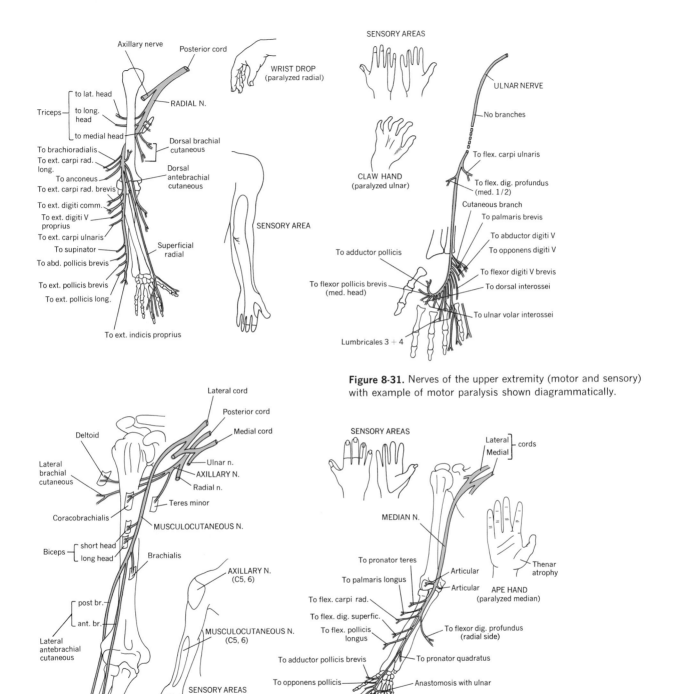

Figure 8-31. Nerves of the upper extremity (motor and sensory) with example of motor paralysis shown diagrammatically.

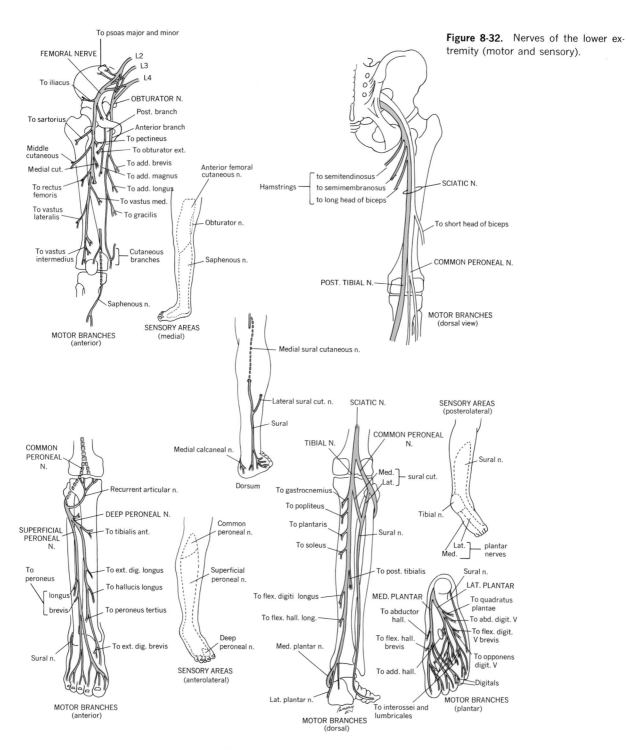

To psoas major and minor

FEMORAL NERVE

L2
L3
L4

To iliacus

OBTURATOR N.

Post. branch

To sartorius

Anterior branch

To pectineus

Middle cutaneous

To obturator ext.

Medial cut.

To add. brevis

To add. magnus

To rectus femoris

To add. longus

To vastus med.

To vastus lateralis

To gracilis

Anterior femoral cutaneous n.

To vastus intermedius

Cutaneous branches

Obturator n.

Saphenous n.

Saphenous n.

MOTOR BRANCHES
(anterior)

SENSORY AREAS
(medial)

Figure 8-32. Nerves of the lower extremity (motor and sensory).

Hamstrings
to semitendinosus
to semimembranosus
to long head of biceps

SCIATIC N.

To short head of biceps

COMMON PERONEAL N.

POST. TIBIAL N.

MOTOR BRANCHES
(dorsal view)

Medial sural cutaneous n.

Lateral sural cut. n.

Sural

SCIATIC N.

COMMON PERONEAL N.

TIBIAL N.

SENSORY AREAS
(posterolateral)

COMMON PERONEAL N.

Recurrent articular n.

Medial calcaneal n.

Dorsum

Med.
Lat.
sural cut.

Sural n.

DEEP PERONEAL N.

To gastrocnemius

Tibial n.

SUPERFICIAL PERONEAL N.

To tibialis ant.

Common peroneal n.

To popliteus

Sural n.

To plantaris

Lat.
Med.
plantar nerves

To peroneus
longus
brevis

To ext. dig. longus

To hallucis longus

Superficial peroneal n.

To soleus

To post. tibialis

Sural n.

To peroneus tertius

To flex. digiti longus

MED. PLANTAR

LAT. PLANTAR

To quadratus plantae

To ext. dig. brevis

To flex. hall. long.

To abductor hall.

To abd. digit. V

Sural n.

Deep peroneal n.

Med. plantar n.

To flex. hall. brevis

To flex. digit. V brevis

To add. hall.

To opponens digit. V

SENSORY AREAS
(anterolateral)

Lat. plantar n.

Digitals

MOTOR BRANCHES
(anterior)

To interossei and lumbricales

MOTOR BRANCHES
(plantar)

MOTOR BRANCHES
(dorsal)

Table 8-2. Summary of Muscle Innervation of Extremities

I. *Upper Extremity*
  A. Axillary nerve
    1. Shoulder girdle: teres minor, deltoid
  B. Suprascapular nerve
    1. Shoulder girdle: supraspinatus, infraspinatus
  C. Long thoracic nerve
    1. Shoulder girdle: serratus anterior
  D. Musculocutaneous nerve
    1. Arm: biceps brachii, coracobrachialis, brachialis
  E. Median nerve
    1. Forearm: pronator teres, flexor carpi radialis, palmaris longus, flexor digitorum superficialis, flexor digitorum profundus (lateral half), flexor pollicis longus, pronator quadratus
    2. Hand: abductor pollicis brevis, opponens pollicis, flexor pollicis brevis, lumbricales I and II
  F. Radial nerve
    1. Arm: triceps, anconeus, brachioradialis, extensor carpi radialis
    2. Forearm: extensor carpi radialis brevis, supinator, extensor digitorum communis, extensor digitorum V, extensor carpi ulnaris, abductor pollicis longus, extensor pollicis longus and brevis, extensor indicis proprius
  G. Ulnar nerve
    1. Forearm: flexor carpi ulnaris, flexor digitorum profundus (medial half)
    2. Hand: flexor digitorum V brevis, abductor digiti V, opponens digiti V, interossei, lumbricales III and IV, adductor pollicis.

II. *Lower Extremity*
  A. Superior gluteal nerve
    1. Buttock: gluteus medius, gluteus minimus, tensor fasciae latae
  B. Inferior gluteal nerve
    1. Buttock: gluteus maximus
  C. Femoral nerve
    1. Thigh: pectineus, sartorius, quadriceps femoris
  D. Obturator nerve
    1. Thigh: adductor longus, gracilis, adductor brevis, obturator externus, adductor magnus
  E. Sciatic nerve (tibial division)
    1. Thigh: semitendinosus, biceps (long head), semimembranosus
    2. Popliteal space (tibial nerve): gastrocnemius, plantaris, popliteus, soleus
    3. Leg: tibialis posterior, flexor digitorum longus, flexor hallucis longus
    4. Foot: abductor and adductor hallucis, abductor, opponens and flexor digiti minimi, dorsal interossei, lumbricales
  F. Sciatic nerve (peroneal division)
    1. Thigh: biceps (short head)
    2. Leg (deep peroneal nerve): tibialis anterior, extensor hallucis longus, extensor digitorum longus, peroneus tertius
    3. Leg (superficial peroneal nerve): peroneus longus and brevis
    4. Foot: extensor digitorum brevis

tracts or fasciculi, which become smaller as you follow them down since they are giving off fibers intrasegmentally, establishing local reflex pathways, and connecting gray matter at various levels (Fig. 8-23). In front of and behind the central gray matter, the anterior and posterior white commissures cross the midline to connect the two halves of the cord with each other, although not always at the same level, while some may even join one or another ascending tract after crossing to the opposite side.

ASCENDING OR SENSORY TRACTS. These tracts (Figs. 8-23, 8-24) conduct afferent impulses that may or may not be of a conscious nature. We speak of two groups: exteroceptive, receiving external stimuli from around us; and proprioceptive, receiving initial stimulation from our internal environment, such as joints and mus-

cles. Pain or heat sensations (on the basis of clinical evidence) ascend in the posterior white column associated with proprioceptive fibers. Crude tactile sensations appear to travel in the anterior spinothalamic tract.

Pain and thermal impressions (Fig. 8-23) are closely related to each other. The central processes enter the posterior horn of the gray matter and end by branching around cells in that area. They also connect by collaterals with higher and lower segments. From these cells, secondary fibers arise, cross the midline, and turn up in the lateral white column in the lateral spinothalamic tract. As this tract ascends, it receives additional fibers up to the level of the medulla, where it joins the anterior spinothalamic tract and the two together form the so-called spinal lemniscus.

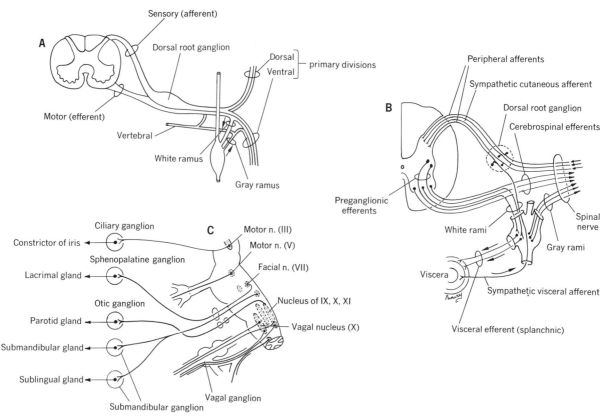

Figure 8-33. Autonomic nervous system. **A.** Autonomic connections to spinal nerves. **B.** Sympathetic ganglion and its connections. **C.** Centers of cranial parasympathetics.

Fibers carrying crude touch and pressure behave similarly, but their secondary neurons ascend in the posterior white column for a short distance before crossing the midline and turn up in the anterior (ventral) spinothalamic tract (found in the anterior white column).

The spinal lemniscus thus conveys all the thermal, painful, pressure, and crude tactile sensations from the spinal nerves of the opposite side of the body. In the pons (above the medulla), this lemniscus is associated with the so-called medial lemniscus, with which it ascends to end around cells of the thalamus. Here, a third set of fibers ascends through the posterior limb of the internal capsule to reach the cerebral cortex of the postcentral gyrus (sensory brain area).

The proprioceptive fibers (Figs. 8-23, 8-24) carry sensations of movement (passive or active), and their cell bodies are to be found in the spinal sensory ganglia. Their centrally passing fibers enter the cord and turn up in the posterior white column and are referred to as the fasciculus gracilis or Goll and the fasciculus cuneatus or Burdach. They are functionally the same except for location: the former is more medial and contains fibers from ganglia cells of the posterior roots of the sacral, lumbar, and lower thoracic nerves; the latter lies in the lateral part of the posterior white column and consists of fibers from the upper thoracic and cervical nerves. In the medulla oblongata, the fibers end by "arborizing" round cells in the gracile and cuneate nuclei. The second set of fibers from these nuclei become the so-called internal arcuate fibers and pass ventrally around the central gray matter, cross the midline (decussate) in the sensory decussation, and ascend. They pass through the pons and midbrain to end in the thalamus. A third neuron arises in the thalamus and ascends through the posterior limb of the internal capsule to the postcentral (sensory) gyrus. We are conscious of these sensations of movement.

A number of fibers that originate in the spinal cord pass to the cerebellum rather than the cerebrum. These activities are carried out subconsciously, and a constant flow of impulses from receptors in the trunk musculature reaches the cerebellum to keep the body in a proper state of tonus and adjustment (e.g., normal body posture). These fibers form the afferent pathway for the spinocerebellar tract (Fig. 8-23).

Some of the posterior nerve root fibers connect with the lower visual center in the superior colliculi. These connections are brought about by the spinotectal tract (Fig. 8-23). The fibers enter the cord and end in the posterior horn; then secondary neuron fibers cross the midline to form a small bundle that ascends medial to the anterior spinocerebellar tract and terminates in the superior colliculus, conveying somatic sensory impulses to reflex centers in the tectum.

Although it is too complex to discuss in detail, one should note and keep in mind that the thalamus with its many nuclei and interconnections and relay systems has functioned as a relay center and integrated sensory and motor activities to and from the cerebrum.

Here lies the basis of affective sensation (feeling of mental strain, well-being, malaise, pain, temperature, agreeableness, and tactile sensation) and discrimination (form, size, texture) eventually relayed to the cortex.

DESCENDING OR MOTOR TRACTS. These tracts are concerned with movement, both voluntary and visceral, and form the efferent paths for cerebral, equilibrium, visual, and other reflexes (Fig. 8-22).

The cerebrospinal or pyramidal tracts (Fig. 8-6) are involved with voluntary motor impulses from the cortex. The cells originate in the precentral (motor) gyrus and pass through the posterior limb of the internal capsule. Near its lower end, they form a compact bundle in the midbrain called the basis pedunculi. In the ventral part of the pons, the tract is broken up into many small bundles by the pontine nuclei and transverse pontine fibers. A compact projection tract is again reestablished in the upper medulla and lies on the anterior brain stem surface and is called the pyramid. In the lower medulla, the majority of the fibers decussate (cross) with those of the opposite side in the "decussation of the pyramid" and arrive at their final position in the middle of the lateral column of the spinal cord. The latter, the lateral cerebrospinal (crossed pyramidal) tract, decreases in size as it descends, since it gives off fibers that terminate by arborizing around the large anterior horn motor cells. From these motor cells, the anterior roots of the spinal nerves arise. The fibers that do not decussate in the medulla form the anterior cerebrospinal or direct pyramidal tract, which descends in the anterior white column near the anterior median fissure. Before they terminate, however, these fibers also cross the midline and end as crossed fibers about the anterior horn cells. As the fibers of the corticospinal tract descend, they also give off fibers to the nuclei of the cranial nerves (located in the brain stem) of both sides. Thus, cranial nerve innervation can be coordinated with peripheral motor nerve control.

The vestibulospinal tract (Fig. 8-23) is the efferent limb of the equilibratory reflexes and originates in the lateral vestibular nucleus in the pons of the medulla. It descends near the surface of the cord in the anterior white column and ends uncrossed around motor cells of the anterior horn.

The rubrospinal tract (Fig. 8-23) originates in the red nucleus of the midbrain. It decussates (crosses) with its fellow of the opposite side, descends through the dorsal pons and dorsolateral medulla, and reaches the lateral white column of the cord lying near the cerebrospinal tract. Its fibers end in association with motor cells of the anterior horn. It really is connected to the cerebellum and corpus striatum and is concerned predominantly with the maintenance of postural tone.

The tectospinal tract (Fig. 8-23) (along with the spinotectal tract) provides a pathway for visual reflexes and begins in the superior colliculus of the brain stem. It decussates (with its opposite fellow) in front of the

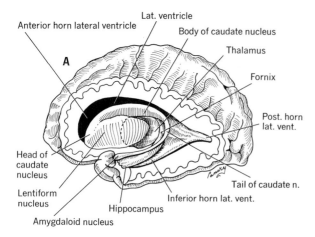

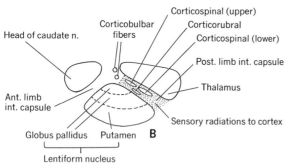

Figure 8-34. **A.** Lateral view of brain with insula removed to show the basal nuclei and their relations. **B.** Diagrammatic scheme of the internal capsule, its relations, and major components.

midbrain aqueduct and descends in the anterior white column of the cord to end like the rubrospinal tract.

Reflexes initiated in the auditory and visual systems apparently require cerebellar integration (Fig. 8-10). Voluntary movements initiated in the motor cortex of the cerebrum also require synergy, and, therefore, large pathways link the cerebrum and cerebellum (brachium pontis and conjunctivum). The cerebellum, in turn, influences voluntary muscle activity via the red nucleus and rubrospinal tract, the reticular formation, and reticulospinal tract, through the thalamus to the corpus striatum and other extrapyramidal pathways. Its action can be suppressing or facilitating.

Extrapyramidal control of muscle activity is a system that provides a "background" motor pattern (i.e.,

it sets up appropriate positions of body parts through the adjustment of large muscle groups), while the more precise and specific activities of individual muscles are added by the pyramidal system. The two balance each other. Here lie the gross movements of locomotion, expression, and posture adjustment along with semi-automatic movements, such as swinging one's arms while walking. The parts of the extrapyramidal system include the corpus striatum (caudate nucleus, putamen), globus pallidus, substantia nigra, subthalamus and nuclei, red nucleus, prerubral nucleus, and the many reticular nuclei (Fig. 8-34). Because of its many parts and connections it is a difficult system to evaluate and cannot be done here (for additional information, see textbooks of neuroanatomy and neurology).

Reticular formation or substance (Fig. 8-35) is the closely intermingled gray and white substance of the spinal cord, brain stem, and thalamus that contains groups of nerve cells with a variety of functions and interconnections. It involves most of the systems related to the extrapyramidal and thalamic interconnections—a sort of massive, complicated interrelay system.

The Autonomic Nervous System

One may also consider the nervous system as being both voluntary (concerned with consciousness, mental activities, and controlled, deliberate actions) and involuntary (independent, operating under unawareness, and not consciously controlled). The latter regulates our internal milieu, such as heartbeat, blood pressure, breathing, and digestion, yet also functions to provide for sudden changes outside of this normal, steady background. To do this it works in two parts: the sympathetic portion, which helps the body go into action (mass response) during emergencies by providing an increased rate of breathing and heartbeat and mobilizing glucose from glycogen to supply muscles with an added supply of oxygen and glucose; and the parasympathetic portion, which checks or counterbalances the sympathetic and tends to give more localized reactions. In reality the interaction between the sympathetic and parasympathetic systems determines how strongly smooth and cardiac muscles contract and how

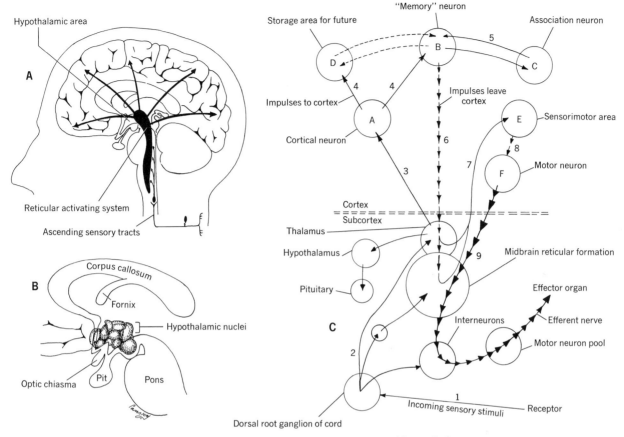

Figure 8-35. A. The reticular formation occupies an area in the brain stem and lower thalamus. Messages travel from receptor organs to sensory nerves, which synapse with nerve tracts in the spinal cord. The stimuli are conveyed along these tracts to the reticular formation, which acts as a central regulatory mechanism capable of accepting, sifting, integrating, modifying, coding, and patterning messages before they are sent on to the cortex. **B.** Hypothalamic nuclei. **C.** Integration by the reticular formation. Begin with the receptor and follow the numbers and arrows to see how the reticular formation is thought to relay messages.

much a gland secretes. Some organs, like sweat glands, receive nerves from only one system. Furthermore, either system may be excitatory: sympathetics cause vasodilation of the coronary arteries but vasoconstriction of those of the head. The parasympathetic system also helps to restore the body to a normal state after an emergency and prepares it for further efforts, starting up digestion after it has stopped and stimulating the mechanisms that control urination and defecation. The autonomic nerves are, for all practical purposes,

motor nerves. They differ from other peripheral nerves in that it is at least a two-chain system; the neuron that leaves the CNS (presynaptic or preganglionic) links with another neuron whose cells are located in ganglia (postsynaptic or postganglionic) rather than going directly to the organ supplied (Table 8-3).

The Sympathetic (Thoracolumbar) Division. This division (Figs. 8-33, 8-34, 8-35, 8-36) arises from preganglionic cell bodies found in the intermediolateral cell column of the 12 thoracic and upper three or four

Table 8-3. Effector Organs: Autonomic Nerve Supply

Organ	Sympathetic (Adrenergic)	Parasympathetic (Cholinergic)
Skin		
Muscles (pilomotor)	Contract	
Sweat glands	Slight secretion (local)	Generalized secretion
Lacrimal glands		Secretion
Salivary glands		
Parotid	No secretion	Profuse, watery secretion
Submandibular	Thick, viscous secretion	Profuse, watery secretion
Nasopharyngeal glands		Secretion
Autonomic ganglion cells		Stimulated
Adrenal medulla		Secretion of norepinephrine and epinephrine
Uterus	Variable	Variable (depends on stage of menstrual cycle and amount of circulation of estrogen and progesterone)
Digestive tract		
Stomach		
Motility and tone	Decreased	Increased
Sphincters	Contracted	Relaxed (usually)
Secretion	Inhibited	Increased
Intestine		
Motility and tone	Decreased	Increased
Sphincters	Contracted (usually)	Relaxed (usually)
Secretion		Increased
Gallbladder	Relaxed	Contracted
Urinary bladder		
Detrusor		Contracted
Trigone and internal sphincter		Relaxed
Ureter		
Tone and motility	Decreased	
Lung		
Bronchial muscle	Relaxed	Constricted
Bronchial glands		Stimulated
Heart		
Rate	Accelerated	Slowed
Rhythm	Ventricular extrasystoles, tachycardia, fibrillation	Bradycardia, AV block, vagal arrest decreased
Output	Increased	Decreased
Blood vessels		
Coronary	Dilated	
Skin and mucosa	Constricted	
Skeletal muscle	Constricted (epinephrine from adrenal medulla dilates blood vessels of liver and muscle, constricts outer vessels)	Dilated
Cerebral	Slight constriction	
Pulmonary	Constricted	
Abdominal viscera	Constricted (epinephrine from adrenal medulla dilates blood vessels of liver and muscle, constricts outer vessels)	
Eye		
Iris	Mydriasis	Miosis
Ciliary muscle	Relaxed for far vision	Accommodated for near vision

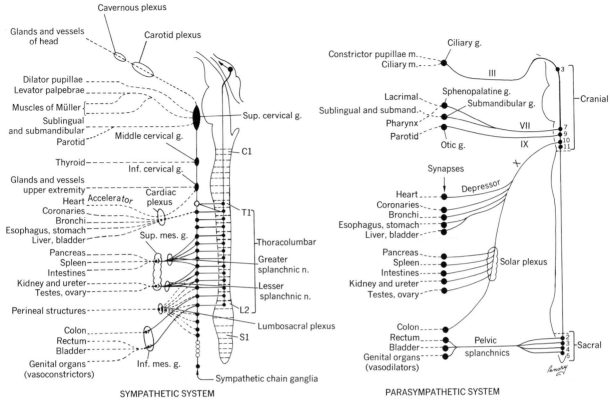

Figure 8-36. The efferent autonomic nervous system with sympathetic and parasympathetic divisions. The solid lines are preganglionic (presynaptic); the dotted lines are postganglionic (postsynaptic).

lumbar segments of the spinal cord. The preganglionic fibers are predominantly myelinated fibers, which pass through the ventral roots and form the white communicating rami of the thoracic and lumbar nerves to reach the ganglia of the sympathetic chain (trunk). The latter lie on the lateral sides of the bodies of the thoracic and lumbar vertebrae. The nerves synapse in the ganglia where they enter or pass up or down the sympathetic trunk to synapse at higher or lower levels. They may, however, pass through the trunk ganglia, to synapse on a collateral or intermediate ganglion. There is an extension of the sympathetic chain into the neck. Here, three ganglia form a chain lying behind the carotid sheath and on the prevertebral fascia. The ganglia are the superior (carotid), which is the

largest; the middle (thyroid); and the inferior (vertebral). They lie near the vessels after which they are named. The inferior ganglia often unite with the first thoracic to form the stellate ganglion.

From the sympathetic system come the following:

1. The unmyelinated postsynaptic fibers, as gray communicating rami, join all the spinal nerves and are distributed as vasomotor, pilomotor, and sweat gland nerves to somatic areas. Branches of the superior cervical ganglion enter into sympathetic plexi about the internal and external carotid arteries to be distributed to the head. The superior cardiac nerves arise from all three pairs of cervical ganglia and pass to the cardiac plexus to serve as "accelerator" fibers to the myocardium. Branches from the upper five thoracic

ganglia pass to the aorta (vasomotor) and posterior pulmonary plexus (dilator fibers to the bronchi).

2. The myelinated presynaptic fibers from the cord pass through the sympathetic chain as splanchnic nerves, arising from the lower seven thoracic ganglia. They pass to the celiac and superior mesenteric ganglia (collateral), where they synapse. Fibers from here pass to the abdominal viscera via the celiac plexus of nerves. The lumbar splanchnic convey fibers to the synaptic relays in the inferior mesenteric ganglia and small ganglia of the hypogastric plexus, through which fibers are sent to the lower abdominal and pelvic viscera.

The Parasympathetic (Craniosacral) Division. This division (Fig. 8-36) arises from preganglionic cell bodies in the gray matter of the brain stem and the middle three segments of the sacral cord. Its distribution is confined to visceral structures. Most of its preganglionic fibers run without interruption from their origin to the wall of the organ supplied. Nerves carrying preganglionic parasympathetic fibers to the head are III, VII, and IX. The vagus nerve (X) carries fibers to the thoracic and abdominal viscera via prevertebral plexi. The pelvic nerves (nervi erigentes) carry parasympathetic sacral fibers to most of the large intestine, pelvic visera, and genitalia by way of the hypogastric plexus. Autonomic supply to the head is unique. The sympathetics innervate the face and scalp via the superior cervical ganglion and fibers that follow the branches of the external carotid artery. The intrinsic muscles of the eye, salivary glands, and the mucous membrane of nose and pharynx have a dual supply mediated via four major cranial autonomic ganglia. The ganglia are (1) the ciliary ganglion, associated with nerve III; (2) the sphenopalatine ganglion, associated with nerves IX and VII; (3) the otic ganglion, related to nerve IX; and (4) the submandibular ganglion, associated with nerve VII via the chorda tympani.

Important Nerve Plexi. CARDIAC PLEXI (DEEP AND SUPERFICIAL). These are formed from cardiac sympathetic nerves and cardiac branches of the vagus and are distributed to the myocardium and walls of vessels leaving the heart.

RIGHT AND LEFT PULMONARY PLEXI. These are formed from both vagus and upper thoracic sympathetic nerves and supply vessels and bronchi of lung.

CELIAC (SOLAR) PLEXUS. This plexus is formed from vagal fibers via the esophageal plexus and sympathetics from nearby celiac ganglia and supplies most abdominal viscera. It has many subplexi, such as phrenic, hepatic, splenic, superior gastric, suprarenal, renal, spermatic (ovarian), superior and inferior mesenteric, and abdominal aortic.

HYPOGASTRIC PLEXUS. This plexus receives sympathetics from the aortic plexus and lumbar ganglia and parasympathetics from the pelvic nerves. It supplies pelvic viscera and genitalia via related subplexi.

Nervous Tissue Composition

The nervous system contains well over ten billion cells; however, although having varied and complex functions, the cells are essentially alike. Nervous tissue comprises about 2 percent of man's entire body weight. The brain accounts for about 90 percent of this, and the remaining 10 percent consists of the spinal cord and both spinal and cranial nerves. The most striking aspect of composition of nervous tissue is the very high lipid content, exceeding all tissues except adipose tissue itself. This consists of cholesterol, phosphatides, cerebrosides, and sulfolipids with little triglyceride fat. Nerve fibers, like muscle fibers, contain hundreds of fine filaments but, unlike the latter, are usually individual strands surrounded by fluid cytoplasm. Nervous tissue is specialized for storage, processing, and communication of information. The basic unit is the neuron, a nerve cell with its supporting specialized fibers.

Communication—The Messenger Carriers

Functional Description Based on Embryologic Development. The spinal cord contains a variety of nerve fibers. Somatic sensory (SA) fibers carry impulses from receptors in the skin and striated muscles; somatic motor (SE) fibers carry impulses to striated muscle; visceral sensory (VA) fibers carry impulses from receptors in smooth muscles; and visceral motor (VE) fibers carry impulses (autonomics) to smooth muscle, cardiac

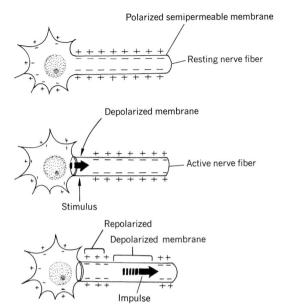

Polarized semipermeable membrane

Resting nerve fiber

Depolarized membrane

Active nerve fiber

Stimulus

Repolarized

Depolarized membrane

Impulse

Figure 8-37. Diagram to illustrate membrane theory of nerve transmission. A nerve fiber at rest resembles a charged condenser. It has an excess of negative ions on the inside and an equal excess of positive ions outside the membrane. When an impulse begins, the membrane of the axon loses its ability to keep the opposite electric charges separated. The inner wall of the membrane at the point of stimulation loses its negative charge temporarily (it is depolarized). Depolarization progresses along the fiber length, and thus the impulse is conveyed at about 130 yd per second as a wave of depolarization of the membrane. Polarization recurs after a short refractory period.

muscle, and glands. A simple reflex arc consists of a message from a receptor (pain, etc.) in the skin, which travels via an SA fiber to the gray matter of the spinal cord; this fiber synapses with a neuron that makes connections with the SE neuron, which sends an impulse via the SE fiber to a limb muscle, which then contracts. (See Chap. 4 for a detailed discussion of nervous tissue.)

Nerve Impulse. The nature of the message traveling along the nerve is reasonably clear (Fig. 8-37). The sensory nerve fibers are connected to specialized cells called receptors that detect a variety of stimuli; in the skin those that respond to heat and touch consist of a single cell. Special sense organs such as the eye and ear consist of thousands of receptors grouped together.

When a stimulus impinges on a receptor, an electrochemical alteration is set off. This stimulates the end of the sensory nerve fiber. A reaction called an impulse then travels along the fiber to the CNS at the speed of up to 130 yards per second. (Electrical impulse in wire travels at 327 million yards per second.)

The nature of the impulse is as follows: The cell has a high resting membrane potential that forms the bases of its excitability and conduction. The resting fiber has an unequal distribution of ions between the inside of the axon and the tissue fluid outside. In the resting state, the interior of the nerve, the axoplasm, has a negative electrical potential with respect to the surrounding liquid. This membrane potential arises from a biased distribution of ions. The membrane is permeable to potassium ions (K^{++}) found inside the fiber but less so to sodium ions (Na^+) found outside the membrane. Any Na^+ that enters is actively pumped out by some energy-consuming process that needs ATP as a source. The concentration of Na^+ is thus greater outside than that of the K^{++} inside, which produces a potential difference across the membrane of about 90 mv (resting potential), so that the axon is positive outside and negative inside. When the fiber is stimulated, a transient change in permeability of the membrane occurs and its membrane becomes permeable to Na^+, which diffuses in, making the inside positive with respect to the outside. This depolarization produces a reversal of potential from -70 mv to $+40$ mv, a change of 110 mv (action potential). The altered permeability persists for only a few milliseconds. The action potential causes ionic currents to flow forward and out of the adjoining inactive region, which in turn depolarizes. This spreads a wave of depolarization, which is propagated across the cell surface. The depolarized region now repolarizes (restores its original ionic balance to the resting state by pumping Na^+ out and K^{++} in, during which time [refractory period] no further impulse can pass). The rate of this wave of surface activity (nerve impulse) passes along the axon, depending on the size and character of the axon. For an axon of small diameter the impulse can be from 1 to 2 m per second, whereas in large-diameter axons it can be greater than 100 m per second. This also applies to myelinated nerves, but here the myelin confines depolarization, acting like an insulator to prevent escape of electricity; however, since the current flows across the cell membrane as well as along the nerve, the nodes allow ions to travel in and out at the nodes of Ranvier (the only exposed areas of the fiber) and reduce leakage of ionic currents. The impulse thus jumps from node to node, increasing velocity of conduction as compared to unmyelinated nerves.

A single nerve impulse lasts for approximately 0.001 second. As explained above, immediately after an impulse has traveled along a fiber section, a short resting period develops to enable ions to be actively pumped back to their original positions on either side of the cell membrane. This is necessary before a new stimulus can be transmitted.

The nature of the impulse is the same in every fiber. The impulse, furthermore, travels at full strength or not at all—i.e., one strength (all-or-none law). Thus, a stronger or weaker stimulus on a receptor does not initiate stronger impulses in its sensory fiber but produces impulses at variable frequency. The pressure of a shoe on a skin receptor of the foot may result in impulses at 1 per second, but if someone stomps on the foot the same receptor will produce the same impulse but at a faster rate, even to 1000 per second. However, a stimulus may be too weak to start an impulse but can lower the resistance of the fiber to weak stimuli that quickly follow after the first; therefore, weak but persistent stimuli can give similar results to strong but short.

According to the above-described ionic theory, excitation results from permeability change. It has been postulated that a compound called acetylcholine (ACh) is involved in development of permeability. Thus, electrical stimulation frees ACh from its storage reservoir (protein bound?). It then interacts with a receptor protein in the membrane, whose structure is altered, resulting in permeability change. The ACh is rapidly hydrolyzed by acetylcholinesterase to choline and acetate, which removes it. The receptor protein then reverts to its original configuration and decreased membrane permeability (Nachmansohn theory).

Synapses. When the impulse reaches the CNS, the inner end of the sensory fiber splits into many processes and these interlace with the processes of other neurons in the CNS at what is called synapses. These neurons in turn have processes that connect to still others. Thus, the original impulse is spread into a vast number of other nerve fibers in the CNS. There is never any structural joining of nerves, but a gap always exists between cells. When the impulses reach the end of the neuron, they cause the release of a transmitter substance (probably stored in the vesicles). This substance travels across the gap and depolarizes the membrane of the next neuron to start a new impulse. The transmitter is then destroyed by enzymes, with the entire process taking 0.001 second. Thus, an electrochemical impulse becomes a chemical reaction and back again to an electrochemical reaction in the new neuron. Nerve fibers also contain a unique structure not found in other cells. These are hollow, protein tubules visible only by electron microscopy. It is felt that these tubules may provide channels for transporting compounds necessary to the signaling system of the nerve ending. The function of the filaments is still unknown.

The synapse is a contact specialization peculiar to nervous tissue. Since the impulse travels normally in only one direction along a multineuronal path, the contacting parts of adjacent neurons that make up the synapse can be identified as presynaptic and postsynaptic components. A narrow intercellular space exists between parts of a synapse with no cytoplasmic continuity between cells here. The presynaptic element usually forms a small expansion in which one finds a cluster of mitochondria (suggesting localized metabolic activity). Close to the presynaptic membrane lie a group of small membrane-bound synaptic vesicles, each several hundred Angstroms in diameter. The intercellular space (150 to 200 Å) is the synaptic cleft (between presynaptic and postsynaptic membranes). Details of morphology vary somewhat at different sites, reflecting functional differences (see Chap. 4).

It is felt that the synaptic vesicles contain transmitter substance that is released into the cleft when a nerve impulse reaches the presynaptic terminal. The transmitter produces localized depolarization of the postsynaptic membrane that triggers off a nerve impulse. After its release, the transmitter substance is destroyed rapidly by enzyme action. Thus, the impulse passes from cell to cell in a chain of neurons through release of a chemical transmitter at synaptic contacts. The myoneural junction (motor end-plate) appears to operate similarly.

The ability to transmit in only one direction is due to the differences between the presynaptic and postsynaptic components. The slight delay in transmission of impulse across the synapse is accounted for by the time it takes for the discharge of the vesicles and diffusion of transmitter across the intercellular cleft. This intercellular portion of the pathway is also more accessible to drugs designed to interfere with nerve function than is the intracellular mechanism.

Nerve cells not only fire and release excitatory neurohumors but also use inhibitory signals to prevent firing. They release inhibitory neurohumors, which counteract the effect of stimulating humors. Thus, the nerve cell really exists in a "state of instability"; it can fire or remain silent. The final reaction of a cell depends on the total effect of signals reaching it. A nerve cell receives up to 25,000 signals from other cells and responds on an all-or-none basis, depending on the majority of stimulation effect. Considering that there are about ten billion cells in the nervous system, the chemical and electronic interplay and reaction are tremendous and very complex. Each nerve cell fires at a minimum rate of one time per second. The complexity is great; yet here, in the nervous system, there is a high degree of organization dating back to its many years in evolutionary development.

Various types of synapses are feasible: they occur between dendrites and the cell body, between dendrites and dendrites, and between axons and muscles (at the motor end-plate). Synapses allow nerve impulses to travel in one direction only and allow groups of neurons to act in a variety of patterns. The synapse, however, does not relay automatically. It can disregard an impulse below a certain frequency or relay a new set of impulses at a lower frequency. Furthermore,

synapses can receive signals from more than one type of receptor not to begin a fresh set of impulses in a new direction. This mixture of different factors at the synapse is the basis of all the computations made in the nervous system.

The chemical bridging the gap between two neurons differs according to the position of the synapse. The synapses of motor nerves use acetylcholine (ACh). The synapse, like the nerve fiber, also has a brief resting period between impulses. This is brought about by the enzyme cholinesterase, which destroys ACh as soon as it initiates the impulse, permitting another impulse to begin. Where the neurohumor is produced is uncertain, but it may be in the main body of the nerve cell and is somehow carried from here to the tip of the transmitting fiber. Rings of contraction have been noted down the surface of the cylindric fiber, and it may be that these substances are "squeezed" along down the nerve. The main cell body may package the neurohumors before sending them on. The packaging may be membrane fragments in the cells, cytoplasm, and the packages themselves, which are oval-shaped bodies containing ACh.

The ACh also transmits the impulse across the gap between the motor nerve and the muscle fiber (myoneural junction). In the autonomic nervous system, norepinephrine may start at the nerve–smooth muscle junction. This and other substances also activate synapses in various portions of the autonomic system. Several chemical transmitters have been found in the CNS, but how many there actually are and the role they play are unknown. Scientists do suspect that many nervous and mental disorders are caused by changes in makeup of these transmitters. It is also known that many drugs have their effect at the synapse and not on the nerve fiber. Curare, long used by South American Indians on their weapons, paralyzes by blocking all motor impulses at the muscle (except those to the heart), and the victim dies because he cannot breathe. Thus, curare has been used in surgery, along with an anesthetic or pain killer, to relax the body; however, the anesthesiologist must maintain a form of artificial respiration because the rib muscles and diaphragm are paralyzed.

Action at the synapse governs the behavior of the entire body. Every one of at least ten billion neurons in the brain communicates with at least 50 others—a mass of interwoven cell bodies and processes. Thus, at least 500 billion synapses are found in the brain and the different combination of impulses passing through this network is beyond conception. Yet, there is an orderly arrangement of neurons in the brain. Generally speaking, the cell bodies are concentrated in the outer surface of the brain as a layer of gray matter, approximately 3 mm thick. The extensions of these cell bodies intermingle below the surface area to function; the brain needs maximum interlocking and extension of neurons and achieves this in a solid mass and not on a spread-out surface.

General Function and Other Essential Areas. Reticular Formation. Millions of impulses bombard the CNS from the body's receptors continually. Imagine if all these impulses were at a conscious level—we would go mad. Fortunately a portion of the brain called the reticular formation (Fig. 8-35) stands almost like a barrier between the spinal cord (or medulla) and the cerebrum (the part of the brain that deals with all our conscious thoughts). The reticular formation holds back unimportant messages (not enough to engage our thoughts), and these messages are acted upon in an unconscious manner. The information may register consciously, but an unconscious act takes care of the problem. The reticular formation also promotes conditions leading to sleep. When tired, it holds back impulses so as not to stimulate conscious brain centers, and these centers inactivate and we fall asleep.

Thalamus and Hypothalamus. Lying near the reticular formation is the thalamus, an association center for impulses to and from the cerebrum and the location for sensations of pain and pleasure. Close to it is the hypothalamus, one of the major areas controlling the autonomic nervous system. It regulates heartbeat, rate of breathing, temperature, blood pressure, water metabolism, and hormone secretion. The hypothalamus does so most of the time at an unconscious level. If excessively stimulated (e.g., by a very hot room temperature), impulses pass to the conscious areas and a deliberate response, like opening a window, occurs.

However, like all areas of the brain, impulses travel in more than one direction. States of fear, anger, or excitement in the conscious centers communicate with the hypothalamus, resulting in sweating, increased hormone production, changes in pulse rate, or changes in breathing, to name a few. Prolonged hypothalamic stimulation by emotional states upsets bodily function and may lead to high blood pressure or duodenal ulcers. When the body responds to stresses in the mind, we speak of psychosomatic (Greek *psyche*, mind; *soma*, body) ailments. Time has shown us much in the way of related physical ills with an activated nervous system.

Medulla. A median section of the brain shows the medulla, an important sensory and motor area as well as an autonomic center that, in conjunction with the hypothalamus, controls the heart and breathing rates and digestion. The reticular formation runs from spinal cord through the medulla to the midbrain, and connects with all parts and regulates the information passing to the cerebrum. The midbrain is an association center for impulses from the ear, eye, and skin receptors.

Cerebral Hemispheres. The complexity of the brain is not the result of its enlargement as a whole, but its complexity is responsible for man's achievements. Areas of the human brain are comparable to similar parts in other animals, but it is the great development of the two cerebral hemispheres and its outer cortex that makes man's brain such a superior mechanism. The cortical layer is convoluted or folded to create gyri and sulci; these convolutions enable the layer of cell bodies to be six times larger than if it were smooth. The cerebrum is the seat of reasoning and memory.

All sensory impulses calling for conscious response go to the cerebral hemispheres, to definite sensory areas already mapped on the cortex. Closely associated areas are motor areas, sending messages along motor nerves to muscles and glands. There are areas of the cerebrum, however, that have as yet not been identified as serving a particular body part since no signals appear to go there or leave them. They are called association areas. Here the decisions taken are "dis-

Table 8-4. Major Reflexes

	Afferent (Sensory) Limb	Center	Efferent (Motor) Limb
A. Superficial:			
1. Corneal	(V)	Pons	(VII)
2. Nasal (sneeze)		Brain stem and upper cord	V, VII, IX, X, and spinal nerves of respiration
3. Pharyngeal and uvular (gag)	X	Medulla	X
4. Upper abdominal	T7–T10	T7–T10	T7–T10
5. Lower abdominal	T10–T12	T10–T12	T10–T12
6. Cremasteric	Femoral	L1	Genitofemoral
7. Plantar	Tibial	S1, S2	Tibial
8. Anal	Pudendal	S4, S5	Pudendal
B. Deep:			
1. Jaw	(V)	Pons	V
2. Biceps	Musculocutaneous	C5, C6	Musculocutaneous
3. Triceps	Radial	C6, C7	Radial
4. Wrist (flexion)	Median	C6–C8	Median
5. Wrist (extension)	Radial	C7–C8	Radial
6. Patellar	Femoral	L2–L4	Femoral
7. Achilles (soleus)	Tibial	S1, S2	Tibial
C. Visceral:			
1. Light	II	Midbrain	III
2. Accommodation	II	Occipital cortex	III
3. Oculocardiac	V	Medulla	X
4. Carotid sinus	IX	Medulla	X
5. Bladder and rectal	Pudendal	S2–S4	Pudendal and autonomics

cussed and weighed" and the overall situation considered in light of past experience. Thus, not only does the cortex receive information, plot it, store it, and route it but also association areas are concerned with past memory and hopes and plans, judgment, intelligence, and ability to learn.

CEREBELLUM. There are other areas of the brain, like the reticular formation, that receive and act on information without relaying it to the cerebrum. The cerebellum (Latin word meaning little brain) is another such region. Its convolutions resemble the cerebrum but on a smaller scale. The cerebellum receives a steady flow of impulses from receptors embedded in muscles and joints, from the eyes and parts of the ear—all involved in balance and posture. It registers our position (standing, sitting, leaning, etc.) as well as rate of movement and coordinates muscle movements.

The Reflex. Basic to all nervous system activity is the reflex, and much of man's behavior is built up of both simple and complicated reflexes (Table 8-4). In its most simple form it consists of a stimulus and a response. The pathway is referred to as the reflex arc and consists of a receptor, sensory fiber, one or more intermediate fibers in the CNS, a motor fiber, and an acceptor (muscle or gland).

All spinal reflexes have in common the fact that the response to a given stimulus is always the same and it happens quickly. Normally the brain is not involved and the reaction is not under willful control. Reflexes protect the body regardless of concern—heart, sweat, or withdrawing from pain or heat, etc. Although this pathway of reflexes is a simple pathway, it is difficult to find since the intermediate fibers between the sensory and motor fibers link up with other reflex arcs in the CNS and the major reflex tends to produce several secondary ones. If you withdraw your hand from pain (as from heat or cold), you also tend to grimace and cry out in pain or even move away.

Furthermore, it is possible for the brain to modify the response.

It has been known for a long time, dating back to the Russian scientist Ivan Pavlov, that many of the things we first do consciously may become reflex actions by repetition. These are called conditioned reflexes. If one associates a sound, for example, with dinner, eventually the sound itself may cause salivation; yet, if dinner does not come, the reflex may also disappear. Thus, the conditioned reflex is really an act of learning—the connection between sound and dinner. Large numbers of conditioned reflexes working together make up a habit, and these form a background to the higher mental activities of the cerebral cortex. The more we build complicated actions into habits, the more efficient we become, and in time, conscious efforts (like driving, playing an instrument, etc.) become habits. Yet man's consciousness plays such a great role in his life that, although we might sometimes believe that his behavior is no more than an amassing of conditioned reflexes, we feel he still greatly and consciously controls his major reactions.

Memory. The brain holds an enormous amount of information that is stored by an unknown complex process. Things remembered and all sensory information stream into the nervous system and represent intensities, colors, sounds, shapes, and so forth. Each is an electrical pulse lasting for only a fraction of a second, and yet these are somehow "frozen" in time and transformed into enduring records.

One theory suggests that memory is a result of structural changes in nerve cells caused by repeated passage of specific nerve impulses. It is felt these changes occur at the synapse or gap that the impulse bridges on its way from one cell to the next. Thus, this creates a closed loop of nerve cells and each loop represents a unit of memory.

Another theory, of a more chemical nature, suggests that memory is really the result of changes in nerve-cell proteins. Heightened activity that is related and nerve-cell proteins act as memory traces. Furthermore, some scientists feel that both DNA and RNA may play a role in our remembrance of things past. Somehow our memory files are also organized to let us deal with abstractions and create order from our experience, a cross-indexing of memory traces. In addition, not enough can be said of our "imagination." And yet, this complex nervous system of many functions and inter-relationships is all related to the behavior of living groups of cells and cell life!

Review Questions

1. What are the two major portions of the nervous system? Differentiate between them.
2. What are the three primary brain vesicles? What is the brain cavity called? What are its parts?
3. Name the lobes seen on a lateral view of the brain.
4. What major sulci (or fissures) and gyri are seen on its lateral surface?
5. On the brain's medial surface we note the corpus callosum. What is it? What are its named parts?
6. What is the fornix? To what function might you relate it?
7. What are the so-called commissures of the brain?
8. Review the structures seen on the ventral surface (base) of the brain and their general function.
9. What is the function of the medulla oblongata? The cerebellum? The cerebral peduncles?

10. Trace the blood supply of the brain beginning with the vertebral and carotid arteries. What is the circle of Willis?
11. What are emissary veins? What are the cerebral venous sinuses? Where are the latter found? Name the sinuses and general venous blood flow toward the jugular vein.
12. Describe the coverings of the brain and spinal cord and their relation to each other.
13. Trace the course of cerebrospinal fluid from its origin to its return to the venous system.
14. Name the cranial nerves, their general course and major function.
15. Describe a typical section through the spinal cord.
16. What are the major motor and sensory tracts of the spinal cord?
17. What is meant by a reflex arc? Describe a typical one.
18. What is the function of the so-called extrapyramidal system? Compare it with the pyramidal system. What is the reticular formation?
19. What is meant by the autonomic nervous system? What are its parts? What is the function, in general, of each of its divisions?
20. What is the nature of a nerve impulse? How does it come about?
21. What is a synapse? Describe a typical synapse and the relationship of structures there.
22. What is the significance of acetylcholine and cholinesterase in relation to question 21?
23. What function do reflexes generally serve? What is a conditioned reflex?
24. What might "memory" be?

References

Baker, P. F.: The Nerve Axon. *Sci. Am.*, **214**(3):74–82, 1966.

Barr, M. L.: *The Human Nervous System.* Harper & Row, Publishers, New York, 1972.

Butter, C. M.: *Neuropsychology: The Study of the Brain and Behavior.* Brooks/Cole Publishing Co., Monterey, Calif., 1968.

DeRobertis, E.: Ultrastructure and Cytochemistry of the Synaptic Region. *Science,* **156**:907, 1967.

Eyzaguirre, C.: *Physiology of the Nervous System.* Year Book Medical Publishers, Inc., Chicago, 1969.

House, E. L., and Pansky, B.: *A Functional Approach to Neuroanatomy,* 2nd ed. McGraw-Hill Book Co., New York, 1967.

Katz, B.: *Nerve, Muscle and Synapse.* McGraw-Hill Book Co., New York, 1966.

Miller, N. E.: Learning of Visceral and Glandular Responses. *Science,* 163:434, 1969.

Wooldridge, D. E.: *Mechanical Man: The Physical Basis of Intelligent Life.* McGraw-Hill Book Co., New York, 1968.

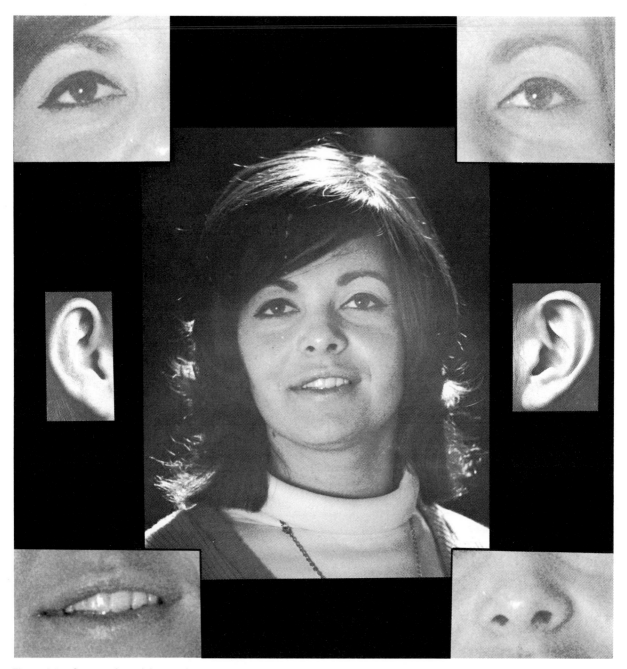

Figure 9-1. Organs of special sensation.

Perception of a Changing Environment: Special Senses

Man's entire world, inside and outside, continually reminds him of what is going on around him (Fig. 9-1). Some stimuli concern the position of the body's limbs, blood pressure, temperature, and other such changes. Other messages convey information from about the world outside.

The receptors are specialized cells whose major function is to react to change. They are connected to sensory nerve endings relaying impulses to the central nervous system. The somesthetic receptors are located all over the body and give us somatic (body wall or frame) exteroceptive sensations. Those that relay impulses from internal conditions are called proprioceptors (Latin *proprius*, one's own). Nerve fibers bringing proprioceptive sensations to the spinal cord are heavily myelinated, touch fibers are moderately myelinated, and pain fibers are unmyelinated. The work these receptors do passes unnoticed, but they maintain homeostasis and keep the body ready for action since they also relay impulses from muscles, tendons, and joints. Other receptors detect changes at the body surface, and of these we are more aware.

Specific receptors have been correlated with vision, hearing, taste, smell, change of head and body position, and sensations reaching the skin.

The Skin

It is the skin (Fig. 9-2) that makes direct contact with our environment. This is man's "covering" and greatly influences his interactions with others. The skin is the primary determinant of the cosmetic nature of man, and people spend significant parts of their lives changing the appearance of their "wrapping." The skin is multifunctional, resilient, and self-replacing and adjusts easily to environmental variations. It is a semipermeable layer dependent on its vascular, lymphatic, and nerve supply. Nevertheless, it carries on its own functions: production of keratin, sebum, sweat, and melanin; heat exchange; and protection. The skin also performs specialized functions involved in perception of sensation, wound repair, immunologic defense and inflammation, blood and lymph circulation, metabolism, and other activities.

The skin consists of an outer epidermis and underlying dermis, which is closely attached to the subcutaneous fatty connective tissue, and various specialized appendages, vessels, and nerves.

Epidermis

The epidermis (Fig. 9-2) is a stratified squamous epithelium derived embryologically from the ectoderm covering the embryo. It varies from 0.5 to 1.5 mm in thickness, normally becoming thicker on the palms and soles. Although it is usually considered as being layered, it is really a functional unit and should be thought of as such. The epidermal cells move from below upward to the cutaneous surface, where they are continuously shed as "cornified flakes," and thus the epidermis reconstitutes itself.

The undersurface of the epidermis consists of patterns of ridges and pockets that fit intimately over

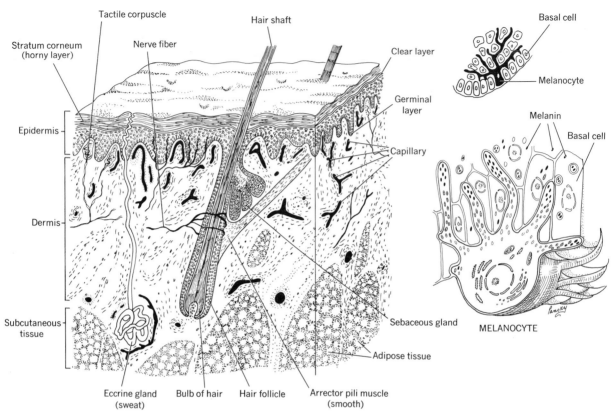

Figure 9-2. Zones of the skin are seen in a schematic section. The underlying dermis, the thickest part of the organ, is supported by a fat-rich subcutaneous layer. In the dermis are seen blood vessels, tactile and other nerves, smooth muscles, and a variety of special glands. Melanocytes, the pigment organs, produce the granules responsible for varying skin colors and lie at the base of the epidermis.

elevations of the dermis called dermal papillae. These papillae vary with the epidermal thickness and location, being most prominent in areas of the body under environmental stress.

The epidermis contains no blood vessels or lymph channels of its own and is dependent on the tissue fluids and vascular supply of the dermis for its existence.

Regions. The epidermis has been divided into four regions (from deep to superficial): basal cell layer, prickle cell layer, granular cell layer, and cornified layer, all representing stages in the keratinization process of the epidermal cell.

BASAL CELL LAYER. The basal cell layer is the deepest epidermal layer, lying next to the dermis, and its cells are columnar and line the dermoepidermal border. Repair following a wound results in marked mitotic activity in these cells, although normal mitosis occurs in the more superficial cells as well. In this basal layer one also sees melanocytes with small, dark nuclei and pale cytoplasm. They have tubelike processes that branch among adjacent cells and serve to spread melanin pigment to the adjacent epidermal cells.

PRICKLE CELL LAYER. The prickle cell layer (stratum spinosum) occupies most of the epidermis and is made up of polygonal cells that become flattened as they move toward the surface. The cells are joined together by protrusions that look like intercellular bridges (or prickles) but, when studied by electron microscopy, are referred to as desmosomes.

GRANULAR CELL LAYER. The granular cell layer (stratum granulosum) appears darker due to the presence of keratohyaline granules in the diamond-shaped cells. The chemical nature and function of the granules are uncertain.

CORNIFIED LAYER. The cornified layer (stratum corneum) is the end result of keratinization of the epidermal cell. It is a fibrous, tough layer of keratin that retains no cell structure. It varies in thickness and serves predominantly as a water-retaining barrier.

Dermis

The dermis is derived from mesenchyme, is the connective tissue layer closely associated with the overlying epidermis, and serves as a nutritional supply for the latter as well as exerting an influence on its growth and maintenance. It consists of an outer papillary layer and a looser, coarse-fibered, deeper reticular layer. It is composed of collagen fibers, elastic fibers, reticulum fibers, ground substance, and cellular elements. The collagen fibers make up most of the dermal substance. Reticulum, which is abundant during wound healing but scarce in normal dermis, is composed of mucopolysaccharides and other substances. It is amorphous, semiliquid, and nonvisible and changes with disease. The cellular elements of the dermis are fibroblasts, histiocytes, and mast cells.

Sweat Glands

Man has two kinds of sweat glands: eccrine and apocrine (Fig. 9-2). Both have ductal cells distinguishable from secretory cells (Fig. 4-1, p. 92).

Eccrine Glands. Eccrine glands are found practically everywhere skin is present. There are anywhere from two to five million sweat glands in the skin, being most numerous on the palms and soles. Only the lips and parts of the genitalia do not have them. They form before birth and become fully functional over a period of time from birth through the first year of life. Age apparently does not affect their number. They are under cholinergic sympathetic nerve control.

These glands extend to the epidermis from the dermis via a long coiled duct, which opens on the surface as an epidermal pore. The duct is surrounded by a ring of keratin and is easily blocked by inflammation or even hydration.

Apocrine Glands. Apocrine glands develop later and are larger than the eccrine type. They are recognizable at birth but do not become functional until near puberty. They are found in the axilla, nipples, outer ear, eyelids, and perianal and genital areas. They, too, are coiled glands and are associated with the hair follicles. They respond to adrenergic and emotional stimuli.

Hair

Hair follicles (Fig. 9-2) develop from both ectoderm and mesoderm and are seen histologically as early as the third intrauterine month, but emerge above the skin surface at about the fourth or fifth intrauterine month. The first hairs are fine and colorless and are called lanugo hairs and are subsequently lost. Vellus hairs are fine, nonpigmented hairs.

The hair shaft grows in a deep pocket of epidermal cells called the hair follicle. Each follicle is associated with a sebaceous gland (see below), and together they make up the pilosebaceous unit. Just beneath the sebaceous gland are one or more bundles of smooth-muscle fibers, the arrector pili muscle, which is innervated by adrenergic fibers. They pull the hair follicle, wrinkling the skin surface and creating "gooseflesh." (See any standard textbook of histology for details.)

All body hair (except on scalp and eyebrows) needs pituitary hormone stimulation for growth and maintenance. Scalp hair and eyebrow hair, however, are of greatest cosmetic concern. Hair follicle activity is cyclic, and each follicle acts independently. The cycle period varies with area and individual. Hair grows about 1 mm every three days, with the most active area being the temporal region. Hair loss in adults is about 70 to 100 hairs daily. Hair growth reflects one's state of health, and illness may cause cessation or slowing of growth. Hair plucking does not appear to cause permanent follicle damage.

Sebaceous Glands

Sebaceous glands (Fig. 9-2) develop from and are continuous with hair follicles, and, therefore, their secretion is released on the skin surface via the lumen of the hair follicle (regardless of the presence of a hair there). These glands secrete sebum, which is produced by degeneration of the central glandular cells in the gland with accumulation of neutral fat. Thus, they are holocrine glands (the secretion contains the cells of the gland). They are under hormonal control, but there is no apparent nervous control.

Nails

The adult nail (Fig. 9-3) consists of a plate of hard keratin situated on a nail bed over the distal phalanx of each digit. It appears to function mainly for protection and support. Nails arise from a thickening of the epidermis and a groove on its proximal aspect. One sees them by the fourth intrauterine month of life.

The nail fold is a semicircular groove at the nail base that houses the matrix or formative area for the hard keratin of the nail. This area is called the lunula. The free edge of the fold is continuous with the cuticle, an extension of the stratum corneum of the dorsum of the finger. Just beneath this is the eponychium, which is continuous with the root of the nail fold. The nail bed lies distal to the lunula and does not help in nail formation. The distal area of the finger has a rich local arteriolar blood supply.

Fingernails are replaced about every five and one-half months, while toenails take from one to one and one-half years. Individual variations, however, do occur and disease may alter the timetables.

Blood Supply of Skin

The epidermis has no blood vessels. The dermis, on the other hand, is interlaced with networks of arterioles rich in intercommunicating shunts between arterioles. Shunts also exist between venules as well as arterioles and venules. Capillary loops, of varying number, extend up into the papillae of the dermis to the dermo-epidermal junction. A special vascular shunt, the glomus body, occurs in the distal aspects of the fingers, toes, and palmar skin and is a neuromyoarteriolar structure concerned with temperature regulation.

Cutaneous Nerves

The skin has efferent sympathetic autonomic nerve fibers as well as sensory fibers (Fig. 9-4). The latter are myelinated except for the final free nerve endings, while the former are mostly unmyelinated. The autonomics supply blood vessels, the arrector pili smooth muscles, and the sweat glands. The existence of parasympathetic fibers to cutaneous structures is ques-

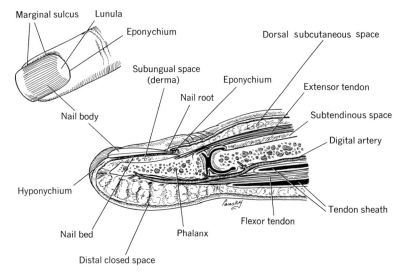

Figure 9-3. Median longitudinal section through the distal finger to show the relationships of the distal phalanx, nail, and surrounding structures.

Marginal sulcus Lunula

Eponychium

Dorsal subcutaneous space

Subungual space (derma)

Eponychium

Nail root

Extensor tendon

Subtendinous space

Nail body

Digital artery

Hyponychium

Tendon sheath

Nail bed

Phalanx

Flexor tendon

Distal closed space

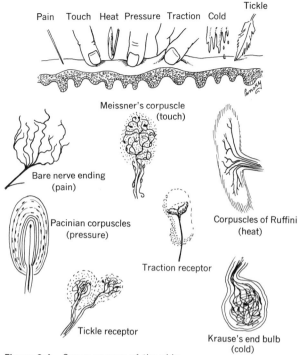

Figure 9-4. Sense organs of the skin.

nerve fibers. All stimuli—e.g., light, touch, pressure, and temperature—can initiate itching. Chemical agents responsible for itching include proteolytic enzymes, proteinases, and histamine.

Adaptation is one feature of sensory reception. Unimportant sensations become reduced in time, saving the conscious mind from bombardment with unimportant information (e.g., the feel of clothes against the skin disappears soon after dressing). In the brain, it is the reticular formation that helps suppress unwanted stimuli. However, the receptors also play their part in this mechanism.

The Visual System: The Eye

Many of the skin receptors give us general information regarding the environment at hand, but we must also be aware of it at a distance. Here the special sense organ of vision, the eye (Fig. 9-5), has taken over. It consists simply of a transparent front part and a lens that focuses images onto a light-sensitive layer, the retina, at the back.

The Eyelids (Palpebrae)

The eyelids (Figs. 9-5, 9-6) are two thin, movable folds. The upper lid is larger than the lower, is more movable, and has a muscle, the levator palpebrae superioris, that elevates the lid. Both lids are covered with skin superficially and mucous membrane or conjunctiva on the deep side. When the eye is open, the space between the lids is called the palpebral commissure or canthus. The two lids are separated medially by a triangular space, the lacus lacrimalis or "tear lake," which is bounded by a fold of conjunctiva called the plica semilunaris, a remnant of the so-called third eyelid. In the "lake" is a reddish elevation, the caruncle, made up of modified skin, hairs, sweat glands, and sebaceous glands. Near the medial angle of the eye, where the eyelids meet, the lashes stop and one sees rounded elevations, the lacrimal papillae, containing small openings, the punctum lacrimale, on the summit. These puncta connect to the lacrimal duct, which then passes to the lacrimal sac, which in turn,

tionable. There is also no evidence, as yet, to demonstrate specific innervation of sebaceous glands.

The afferent or sensory system is of spinal sensory nerve origin. The significant nerve endings or receptors consist of a well-developed network around the hair follicles, free nerve endings, Meissner corpuscles, mucocutaneous nerve end-organs, and Vater-Pacini corpuscles. All respond to touch and pressure. Pain is carried predominantly by the free-nerve endings. The bulbs of Krause record low temperature, and the organs of Ruffini react to high temperature.

Although many receptors have been identified in the skin, it is not always clear-cut as to which sensation corresponds to which receptor type, since, at times, any one receptor can register sensations of touch, pain, and temperature and it feels "hot." Thus, the mechanisms of skin sensation are not too well understood.

Itching (pruritus) is a sensation resulting from a disturbance in the regular pattern of activity arriving at the brain rather than from activation of known

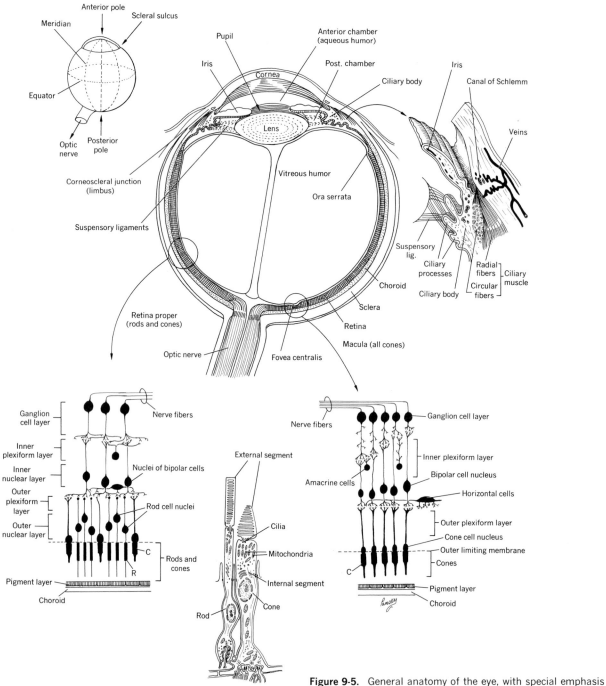

Figure 9-5. General anatomy of the eye, with special emphasis on the retina (histologically and by electron microscope).

Labels in figure:

Anterior pole
Meridian
Scleral sulcus
Pupil
Iris
Cornea
Anterior chamber (aqueous humor)
Post. chamber
Ciliary body
Iris
Canal of Schlemm
Veins
Equator
Lens
Optic nerve
Posterior pole
Corneoscleral junction (limbus)
Vitreous humor
Ora serrata
Suspensory ligaments
Suspensory lig.
Ciliary processes
Ciliary body
Radial fibers
Circular fibers
Ciliary muscle
Choroid
Sclera
Retina
Macula (all cones)
Retina proper (rods and cones)
Optic nerve
Fovea centralis

Ganglion cell layer
Nerve fibers
Inner plexiform layer
Inner nuclear layer
Nuclei of bipolar cells
Outer plexiform layer
Outer nuclear layer
Rod cell nuclei
C
R
Rods and cones
Pigment layer
Choroid

External segment
Cilia
Mitochondria
Internal segment
Cone
Rod

Electron Microscope View

Nerve fibers
Amacrine cells
Ganglion cell layer
Inner plexiform layer
Bipolar cell nucleus
Horizontal cells
Outer plexiform layer
Cone cell nucleus
Outer limiting membrane
Cones
C
Pigment layer
Choroid

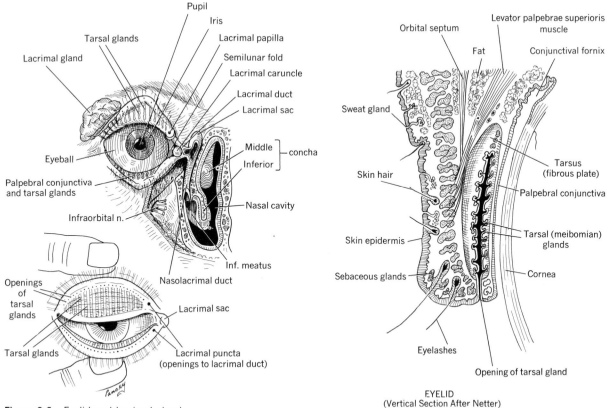

Figure 9-6. Eyelid and lacrimal gland.

Labels (left illustration, top):
Pupil
Iris
Tarsal glands
Lacrimal papilla
Lacrimal gland
Semilunar fold
Lacrimal caruncle
Lacrimal duct
Lacrimal sac
Middle — concha
Eyeball
Inferior
Palpebral conjunctiva and tarsal glands
Nasal cavity
Infraorbital n.
Inf. meatus
Nasolacrimal duct

Labels (left illustration, bottom):
Openings of tarsal glands
Lacrimal sac
Tarsal glands
Lacrimal puncta (openings to lacrimal duct)

Labels (right illustration):
Orbital septum
Levator palpebrae superioris muscle
Fat
Conjunctival fornix
Sweat gland
Skin hair
Tarsus (fibrous plate)
Palpebral conjunctiva
Skin epidermis
Tarsal (meibomian) glands
Sebaceous glands
Cornea
Eyelashes
Opening of tarsal gland

EYELID
(Vertical Section After Netter)

via the nasolacrimal duct, empties into the nasal cavity.

The eyelashes arise from the mucocutaneous junction of the lids and directly behind them are the openings of the tarsal (meibomian sebaceous) glands. The tarsal glands are located in the tarsal plate proper, and their secretion guarantees an airtight closure of the lid. An obstructed hair follicle (sty) will protrude on the front of the lid, but an obstructed gland or chalazion will protrude onto the globe of the eye as a cyst. The lids (Fig. 9-6) have six layers. From superficial to deep, they are: (1) skin, which is very thin with eyelashes and associated sebaceous glands (of Zeis) and sweat glands (of Moll); (2) loose subcutaneous tissue, with very little or no fat; (3) striped muscle, which includes palpebral fibers of the orbicularis oculi acting as an upper lid; (4) submuscular or areolar layer, in which lie the sensory nerves; (5) tarsal plates, which are two thin plates of dense connective tissue about 2.5 cm (1 in.) long and seen in each lid, contributing to the form and support of the lids (the tarsal glands are located in the plate proper); and (6) conjunctiva, which is the layer of mucous membrane that attaches the eyeball to the lid. The lines of reflection are termed superior and inferior fornices and over the bulb, the bulbar conjunctiva (cornea).

The arterial supply of the eye is via the ophthalmic artery (Fig. 9-7). The veins drain into ophthalmic veins (Fig. 9-7). The lymphatics pass to the preauricular glands (in front of the ear), the facial and submandibular group (Fig. 10-36, p. 375). The major motor nerve is the facial (VII) to the orbicularis oculi. The levator palpebrae is supplied by the oculomotor (III) (Fig. 8-15, p. 266).

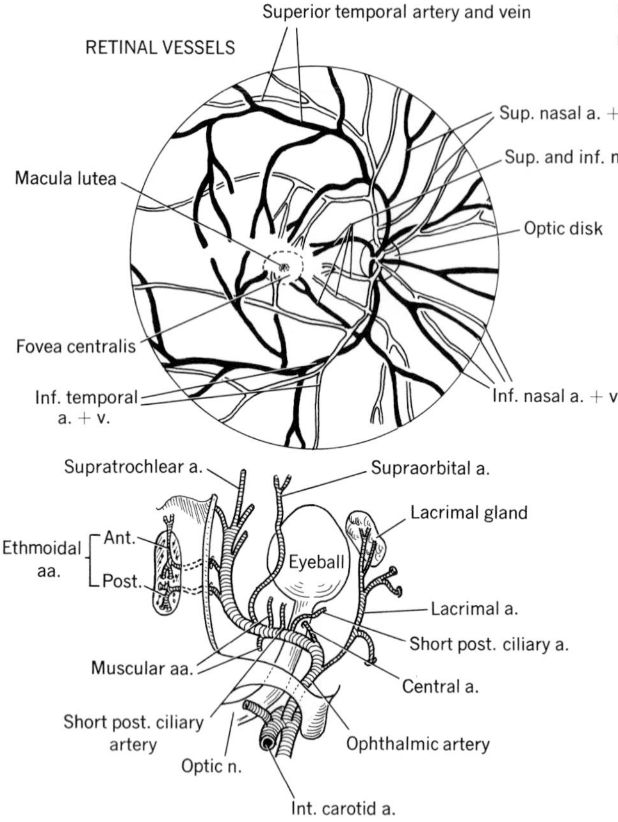

RETINAL VESSELS

Superior temporal artery and vein

Macula lutea

Sup. nasal a. + v.

Sup. and inf. macular a. + v.

Optic disk

Fovea centralis

Inf. temporal a. + v.

Inf. nasal a. + v.

Supratrochlear a.

Supraorbital a.

Lacrimal gland

Ethmoidal aa. [Ant. / Post.]

Eyeball

Lacrimal a.

Short post. ciliary a.

Muscular aa.

Central a.

Short post. ciliary artery

Ophthalmic artery

Optic n.

Int. carotid a.

Figure 9-7. Major vessels of the eye, including those of the retina (as seen by ophthalmoscope).

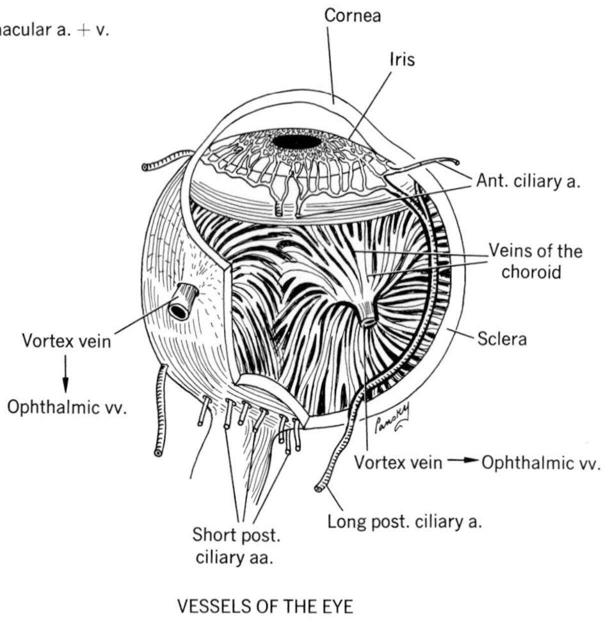

Cornea

Iris

Ant. ciliary a.

Veins of the choroid

Sclera

Vortex vein

Ophthalmic vv.

Vortex vein → Ophthalmic vv.

Long post. ciliary a.

Short post. ciliary aa.

VESSELS OF THE EYE

The lacrimal apparatus (Fig. 9-6) consists of the lacrimal glands, two lacrimal ducts, the lacrimal sac, and the nasolacrimal duct. The lacrimal gland is found in the superolateral angle of the orbit. It is about the size and shape of an almond. The gland pours tears into the upper fornix by means of 8 to 12 tiny ducts that travel over and lubricate the eyeball, pass into the lacrimal ducts via the puncta lacrimalia to the lacrimal sac, then go down the nasolacrimal duct into the nose.

The Orbit

The orbit (Fig. 9-8) is discussed in Chapter 5.

The Eyeball

The eyeball (Figs. 9-5, 9-6) sits in the anterior part of the orbit. It is about 2.5 cm (1 in.) in all diameters. Behind, it rests on a fibrous capsule that forms a socket

in which the "ball" moves freely. The eyeball consists of three coats that enclose three refractive media. The first coat is fibrous and consists of the sclera and cornea; the second is pigmented and contains the choroid, ciliary body, and iris; and the third is nervous and consists of the retina. The refractive media are the aqueous humor (clear, watery solution), vitreous humor (viscous, jellylike), and lens.

Outer Coat. THE SCLERA. Embryologically the sclera is an extension of dura and is the tough, fibrous outer capsule that covers the posterior five sixths of the eye and is continuous in front with the cornea. It thins out posteriorly where the optic nerve enters and becomes a "sievelike membrane," the lamina cribrosa. Beyond the margin of the cornea, the sclera is the "white" of the eye.

THE CORNEA. The cornea is the anterior transparent

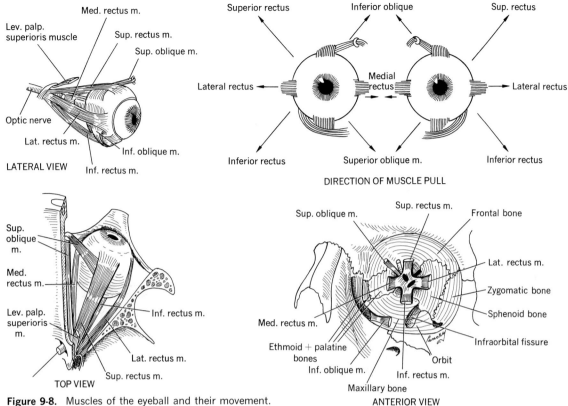

Figure 9-8. Muscles of the eyeball and their movement.

part of the outer coat. It has no blood vessels but receives nutrition from the lymph circulating in many lymphatic spaces. The superficial surface of the cornea is covered by a layer of stratified epithelium that is continuous with the conjunctiva. Posteriorly it is limited by a posterior elastic membrane (Descemet's), which is covered by a layer of mesothelium, which in turn is in contact with the aqueous humor. Its anterior basal membrane is thick and homogeneous and lies below the epithelium, sort of a modified basal lamina. At the peripheral margin, the fibers of this membrane divide into three groups: the innermost turn medially into the iris, the middle fibers give origin to the ciliary muscle, and the outermost fibers are continuous with the sclera. At the corneoscleral junction a circular venous space exists, the sinus venosus scleral (canal of Schlemm), which communicates with the

scleral veins and with the aqueous humor (the latter via the spaces of Fontana). The cornea is innervated by the ophthalmic division of the trigeminal (V) nerve by way of conjunctival and ciliary branches. These give warning of injury or a foreign body in the eye. If the nerve is damaged, the cornea becomes insensitive and may ulcerate and the eye may eventually be lost.

Middle Coat. THE CHOROID. Embryologically the choroid is a development of the pia-arachnoid layer and extends from the optic nerve to the ora serrata (a jagged edge where the true retina ends). It lies between sclera and retina. It is the so-called "nourishing" coat and consists predominantly of blood vessels. Its inner layer contains pigment cells. The vessels can be seen by ophthalmoscope and in the fundus produce the red background against which the retinal vessels appear. The veins lie external to the arteries and form four to five trunks, the venae vorticosae.

THE CILIARY BODY. The ciliary body consists of the ciliary muscle and ciliary processes. The muscle is made up of flat, nonstriated bundles of longitudinal and circular fibers, supplied by the oculomotor (III) nerve. The fibers run back from the corneoscleral junction to the choroid; when they contract, they pull the choroid forward, relaxing the suspensory ligament, which in turn allows the lens to become more convex. Thus, the ciliary muscle is the muscle of accommodation. The ciliary processes (about 70 in number) are structurally like the rest of the choroid, consisting essentially of blood vessels. They have much pigment and some glandular elements that may form the aqueous humor.

THE IRIS. The iris separates the anterior and posterior eye chambers, dividing the space between the cornea and lens into anterior and posterior chambers. It is like a diaphragm of a camera and has a central opening, the pupil. It regulates the amount of light reaching the retina. The pigment of the iris determines one's eye color. It is attached peripherally to the ciliary body but behind to the corneoscleral junction. It consists of a delicate stroma of connective tissue with blood vessels, nerves, pigment cells, and two groups of involuntary muscle fibers, a circular sphincter group (sphincter pupillae) that narrows the pupil, and a less defined dilator set (dilator pupillae). The sphincter is innervated by the oculomotor (III) nerve (parasympathetic) via the short ciliary nerves. The dilator pupillae is supplied by sympathetics. The vessels to the iris arise from the long and short anterior ciliary arteries.

Inner Coat or Retina. Embryologically the retina (Figs. 9-5, 9-7) is of neuroectodermal origin, arising from the forebrain. It resembles brain cortex and is the expanded termination of the optic nerve. The retina extends forward as far as the ciliary processes, where its visual portion ends in an irregular edge, the ora serrata. Beyond this it continues as the ciliary part of the retina, a thin layer that coats the deep side of the ciliary processes but contains no nerve fibers. The retina is normally transparent, and thus the subjacent choroid is visible as the red background of the eye.

The point of entrance of the optic nerve is called the optic disk (Fig. 9-7). The disk has a slight central depression that marks the point of diverging optic nerve fibers as they leave the eyeball. The disk is the physiologic "blind spot" of the retina since it contains no sensory receptors. Above and lateral to the disk is the yellow spot or macula lutea (Fig. 9-7), which is devoid of vessels and, in contrast to the disk, is the center of "most distinct vision." This has a central depression, the fovea centralis.

The central artery of the retina, a branch of the ophthalmic artery, enters the eyeball with the optic nerve and runs with it to the retina, where it enters the middle of the disk and then divides into superior and inferior branches, each in turn dividing into temporal and nasal branches (Fig. 9-7). The branches are end arteries and if "destroyed" result in blindness. They are seen with the ophthalmoscope. The veins of the eyeball follow the arteries and empty into the ophthalmic veins, which in turn pass back into the skull to form the cavernous sinuses. Pulsations are normal in these veins but not in the arteries of the eye.

The retina consists essentially of a pigmented epithelium and three layers of nervous tissue placed one on top of the other in synaptic series. They have been called the visual cells (rods and cones), bipolar cells, and ganglion cells (Fig. 9-5).

Eye Humors. AQUEOUS HUMOR. The aqueous humor is a clear fluid found in the area between the cornea (anteriorly) and lens (posteriorly). The iris divides this chamber into anterior and posterior, which communicate freely via the pupil. The aqueous humor is like the lymph since it has less albumin and does not clot unless altered pathologically. The humor is probably secreted by the ciliary body and passes through the pupil into the anterior chamber, from which it is drained away by the canal of Schlemm to the anterior ciliary veins.

VITREOUS HUMOR OR BODY. The vitreous humor (Fig. 9-5) is a soft, gelatinous substance that fills the space behind the lens. It supports the retina that surrounds it. It is enclosed in a delicate, transparent structure, the hyaloid membrane. A hyaloid canal runs from the optic disk through the vitreous to the lens capsule and is the remnant of a passage for the central artery of the retina that was seen in the fetus.

Lens. The lens (Fig. 9-5) is a biconvex, transparent, colorless structure found between the aqueous and vitreous humors. A capsule surrounds it and attaches it to the ciliary processes by the suspensory ligament (derived from the hyaloid membrane). Contraction of ciliary muscle draws the hyaloid membrane forward and relaxes the suspensory ligament, and the convexity of the anterior lens surface is increased. The ability of the lens (due to its elastic structure) to change its refractive power is called "the power of accommodation."

Orbital Fascia (Tenon's Capsule). Orbital fascia is a thickened connective tissue that suspends structures of the orbit and has prolongations that ensheathe almost all orbital structures. Its major parts are the fascia of the bulb (Tenon's capsule), the sheaths of the muscles, and check ligaments (that limit rotation of the eyeball).

Muscles of the Orbit. Muscles of the orbit (Fig. 9-8) include the four recti (superior, inferior, medial, and lateral), the two oblique (superior and inferior), and the levator palpebrae superioris. A common tendinous ring is found around the optic foramen for the origin of the muscles. The four recti insert into the sclera, each piercing the fascia of the bulb. The lateral rectus pulls the eye laterally (outward), and the medial does the reverse. The superior rectus, owing to its position, pulls upward and also creates medial rotation. The inferior rectus produces a combination of downward pull and medial rotation. The superior oblique passes through a fibrocartilaginous ring or pulley before passing back to the eyeball; therefore, when it contracts, it rotates the eyeball so as to make the pupil look down and out (laterally). Pure downward rotation occurs only when the superior oblique and inferior rectus act together. The inferior oblique makes the pupil look upward and out (laterally). To look straight upward, the inferior oblique and superior rectus must work together. The levator palpebrae superioris raises the upper eyelid (acts in opposition to the orbicularis oculi muscle).

Vessels and Nerves. OPHTHALMIC ARTERY. The ophthalmic artery (from the internal carotid) passes through the optic foramen (with the nerve) and ends by dividing into the supraorbital and supratrochlear arteries (Fig. 9-7). The branches of the ophthalmic are (1) the central artery of the retina, which runs in the substance of the nerve to the disk and divides into branches to supply the retina; (2) the ciliary arteries, which consist of a posterior group that ramifies in the choroid coat (two of them, the long posterior ciliary, run forward to anastomose with the anterior, which supplies the ciliary body and iris); (3) the supratrochlear and supraorbital, which leave the orbit to supply the superficial tissues of the forehead; and (4) muscular and palpebral branches and a branch to the lacrimal gland.

OPHTHALMIC VEINS. The ophthalmic veins are superior and inferior (Fig. 9-7). The superior follows the artery but passes through the superior orbital fissure to end in the cavernous sinus. The inferior divides as it leaves the orbit and joins the cavernous sinus through the superior fissure and the pterygoid plexus via the inferior orbital fissure (Fig. 10-29, p. 367).

ORBITAL NERVES. The orbital nerves (Fig. 8-15, p. 266) consist of:

1. The optic nerve, which passes through the optic foramen into the orbit and branches through the sclera and choroid coats as well as spreading out to form the inner layer of the retina. Its fibers, which are sensory, pass from the eye back to the brain.

2. The oculomotor (III) nerve, which supplies all the muscles of the orbit except the lateral rectus and superior oblique. Via the ciliary ganglion it supplies the sphincter muscle of the iris and ciliary muscle (of accommodation). If it is injured, one observes a ptosis or drooping of the upper lid due to paralysis of the levator palpebrae muscle, an external strabismus due to the unopposed action of the external rectus muscle and the inability of the eye to turn up or down, a dilatation of the pupil as a result of paralysis of the ciliary muscle.

3. The trochlear (IV) nerve, which supplies only the superior oblique muscle.

4. The abducens (VI) nerve, which supplies only the lateral rectus muscle.

Physiology of Sight

The anterior part of the eye bends or refracts light onto the retina (Fig. 9-9). The image must be carefully focused; however, unlike a camera where only a lens exists, the refractive part of the eye has not only a lens but also the cornea, the curved transparent portion lying in front of the lens proper. Here is where most of the light refraction actually occurs.

The cornea is protected from dust particles by continuous washings by antiseptic tears from the lacrimal gland. Simultaneously, the upper eyelid passes up and down over the eye like a windshield wiper. However, its greatest protection is the transparent membrane that lines the front of the cornea, the conjunctiva, which consists of self-repairing stratified epithelium.

The cornea is fixed and focuses images somewhere near the retina. Acting alone, the cornea would tend to focus images either in front of or behind the retina. It is the lens that makes the final adjustment and brings the image to focus on the retina. This it does by means of the ciliary muscles (circular and longitudinal), which are attached to the lens by suspensory ligaments. The muscles enable the lens to change its shape according to the distance of the image. The cornea, however, plays a larger role than the lens in focusing the image on the retina, because light rays are bent more in

Figure 9-9. **A.** Visual pathway. **B.** Reflex pathway for pupillary constriction when light hits the retina. **C.** Various eye-shape alterations and retinal images as related to vision.

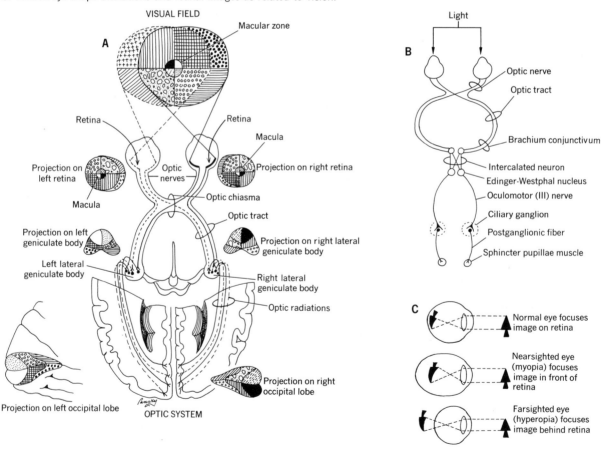

passing from air into the curvature of the cornea than in passing into and out of the lens. The light hits the cornea at different angles and is bent in different amounts as it is directed to the retina. Although the cornea performs the focusing, all adjustments for distance are made by the lens (accommodation). When viewing distant objects the ciliary muscle is relaxed and forms a circle of large diameter. This pulls the suspensory ligaments and flattens the lens, decreasing the refractive power so that objects are focused on the retina. Not only does the lens flatten, but also the pupil dilates, allowing more light to enter the eye. This dilatation occurs because the radial muscles of the iris contract and stretch the previously contracted circular muscles of the iris. When viewing near objects, the ciliary muscle contracts, making a circle of smaller diameter. This relaxes the ligaments, and the lens bulges because of its own elasticity and its refractive power increases. At the same time longitudinal muscles in the ciliary body contract to move the lens slightly forward. (Note that the eye muscles are exercised in viewing near objects rather than distant ones since it is at this time they contract!) With the bulging lens, the pupil contracts as more light enters, and this prevents overstimulation of the retina. Closure of the pupil also tends to increase visual acuity by focusing the light and preventing excess light from passing through the less optically efficient side of the lens.

Cells are added to the outer surface of the lens throughout life. The central cells age and die first, the lens becomes stiff, and accommodation is more difficult with age (requiring glasses). If the lens becomes opaque, we have the condition known as cataract and the defective lens must be removed surgically and compensating eyeglasses used.

The eye is not a faultless structure. Often the lens is distorted and cannot focus properly, and we get nearsightedness or myopia (Fig. 9-9). This can also occur if the eyeball is too long in relation to lens size. Near images are focused on the retina, but far objects are focused in front of it and one cannot see distant objects clearly. If the eye is too short for the lens, distant objects are focused on the retina but near objects are focused behind it. This condition is called farsighted or hyperopic vision (Fig. 9-9). In this case, near vision is poor.

Defects in the shape of the cornea lead to astigmatism, in consequence of which the rays from a luminous point are not focused at a single point in the retina but are spread out as a line in one or another direction. This can also be corrected by lenses.

The nerve cells can malfunction, and one of the results can be color blindness (an inherited condition). We normally see the entire range of colors (violet at one end of the visible spectrum to red at the other). It is felt that there is a different kind of receptor mechanism for each of the three primary colors of light. Light is the source of all colors, and it is the different wavelengths of light that stimulate the rods and cones of the eye. Sensation of any color can be gotten by the appropriate mixture of three lights: red, green, and blue. White light is a mixture of all wavelengths, and black is an absence of all light. Color vision appears to begin with activation of the pigments in the cone receptor cells, of which there apparently are three types. If any cone type is defective or absent, the shades of color they transmit will not be "seen" by the brain. Note, however, that what is a defect in humans may be normal for other creatures. In the human, red-green color blindness is most common and one cannot distinguish between the two colors. About 8.0 percent of men are color blind and only 0.5 percent of women. In the frog, the retina contains special neurons to distinguish between light and dark areas while other neurons respond to a sudden dimming of general light. Frogs' eyes are sensitive to motion, while stationary objects are dimly seen. On the other hand, dogs, horses, deer, cattle, and most mammals see no colors at all, and everything is seen in shades of gray. Insects have no rods or cones but have large compound eyes with hundreds of "transparent windows" and see things segmented and diffused; some insects do see color in the ultraviolet range not visible to humans. Only apes and higher monkeys have full color vision.

The eye is very delicate, and there is an optimum intensity of light needed for images to imprint on the retina. Excessive stimulation can damage or "burn" the light-sensitive surface. Thus, the light must be regulated, and this is done by the iris, located between the cornea and the lens. Its central opening, the pupil, dilates in dim light and narrows in bright light. Light thus passes through the cornea, the aqueous chamber (transparent fluid), the pupil, the lens, and the vitreous chamber (transparent fluid) to reach the retina, which is a light-sensitive surface consisting of more than 100 million receptors of rods and cones (Fig. 9-5).

The rods function when the intensity of light is low and give a picture of varying shades of gray. The rods are so sensitive that they can detect a candle flame at more than 16 km (10 miles), receiving less than 100 million-millionth of the original light. They increase in number as you move away from the retinal center. The rods, like other receptors, turn light (a stimulus) into a series of electrical impulses that travel along sensory nerves to the brain. How the rods perform is not certain. They react somewhat like the photographic film of a camera with its layer of silver salts that are light sensitive. When the salts absorb light energy, they break down, the breakdown being proportional to the strength of light. The retina has a chemical called rhodopsin (a photo pigment) that slowly breaks down when it absorbs low-intensity light. This occurs with a release of energy that sets off nerve impulses. Rhodopsin rebuilds itself and thus is ready to receive new stimuli.

The cones (Fig. 9-5) are more numerous at the center of the retina and are used in daylight vision and are responsible for perception of color. Most physiologists believe (although there is some argument) that three types of cones exist. The pigments that the cones contain absorb a variety of colors to different extents, and these pigments give rise to nerve impulses that are color dependent. The brain's visual cortex, in turn, interprets the impulses as color sensation.

The retina contains other types of nervous tissue. Both rods and cones are connected by two layers of nerve cells (ganglion and bipolar) that run over the surface of the retina so that light must pass through them to reach the rods and cones. The receptors and nerve cells are complexly joined. In some areas one receptor is joined to many nerve cells, and in other parts many receptors are joined to one nerve cell (as in the brain). Therefore, the nerve cells of the eye not only simply transmit light stimulation to the brain but also process the information before relaying it.

The nerve cells terminate in nerve fibers that form the optic nerve. Where the fibers emerge from the eyeball, the blind spot (optic disk) (Fig. 9-7), there are no receptors to respond to light. This offers no disadvantage, however; when light falls on the blind spot of one eye, it usually falls on a sensitive part of the retina of the other.

Most of the retina delivers vague signals to the brain except in a tiny area of the retina called the fovea centralis (Figs. 9-5, 9-7), which is located opposite about the middle of the lens. If we look directly at an object, it is focused on the fovea and the image of that object receives detailed interpretation while the surrounding area is blurred. Here there are no rods—only tightly packed cones. This area can record in such detail because here each cone is connected to its own individual fiber of the optic nerve whereas in other parts of the retina groups of up to 100 rods and cones feed impulses into a single fiber, producing only a coarse image. The fovea is further adapted to record fine detail because light can better reach its receptors here since the fovea is not covered by the "curtain" of blood vessels covering the rest of the retina. Particularly noteworthy are the retinal arteries and veins seen (through the ophthalmoscope) radiating from the area of the blind spot. There is also a thinner layer of nerve cells covering the receptors of the fovea. To add to its efficiency, a greater area of the cerebral cortex of the brain is devoted to receiving and analyzing messages from the fovea than from all of the remaining retina.

When looking at an object, the eye very quickly performs several functions. We notice things out of the corner of the eye even though this image may be vague. The eyes also converge so that the image falls on the fovea of both eyes and we do not see double. This covering is accomplished by the extrinsic muscles of the eyeball. When we are awake, the muscles are continuously positioning the eyeballs so that they converge when we look at objects; yet the retinal images in the two eyes differ just slightly since the eyes are

6.25 cm ($2\frac{1}{2}$ in.) apart and we see stereoscopically (in three dimensions), enabling us to judge distance (depth), size, and shape. The muscles, plus the lens and cornea to focus the image as well as the iris to adjust the light intensity, complete the picture.

At birth, the infant cannot converge the eyes. He must learn to do so; if he does not, he sees double and begins to squint. The double image is confusing, and the brain tends to suppress vision in the defective eye. However, since no anatomic or structural defect actually exists, reeducation can help this ineffective situation.

The visual perception system makes allowances for size and distance. The image of a near object occupies a certain retinal area; if it moves at a distance, the image appears smaller and yet we recognize the same shape of the object and do not take it for a "midget" of the first. All perception is automatically solved by the visual system. It is known that different cell groups in the visual cortex of the brain deal with a definite part of the retina and with images of a particular shape. Thus, the brain (cerebral cortex) has a very large area to interpret all this information since the cortex turns nerve impulses into the perception of shape and distance, depth, and color almost instantaneously.

The Vestibulocochlear System

Auditory Apparatus: Hearing

The auditory apparatus (Fig. 9-10) consists of an external ear, a middle ear (tympanum), and an inner ear. In the light, man relies on his eyes. In the darkness, his ears serve for both hearing and balance. The part of the ear we see is called the external or outer ear, which helps gather and guide sound.

External Ear. The external ear is made up of the auricle or pinna and the external auditory meatus. It collects and conveys sound waves to the tympanic membrane or eardrum.

AURICLE. The auricle (Fig. 9-10) has a cartilaginous framework to keep its shape, but its dependent lobule is made of fibrofatty tissue. It is supplied by several nerves and contains both intrinsic and extrinsic muscles

and ligaments. Its blood supply is via the external carotid and superficial temporal arteries, and its veins drain into the superficial temporal and external jugular veins. The lymphatics drain into nodes around the external ear and eventually into those deep in the neck.

EXTERNAL AUDITORY MEATUS. The external auditory meatus (Fig. 9-10) is a canal about 3.125 cm ($1\frac{1}{4}$ in.) long, the first half being cartilaginous and the rest osseous. It is not straight but, rather, S shaped. The skin of the cartilage is firmly bound to the bone cartilage and is supplied with hair follicles, sebaceous glands, and ceruminous glands. An infection gives rise to severe pain but little swelling. It is related to the cranial fossa above (suppuration can penetrate the bone and cause meningitis); to the parotid gland and jaw in front (gland abscess can reach it and a fall on the jaw can cause meatal bleeding); and to the mastoid behind (pus from the mastoid air cells can invade it).

Tympanic Membrane (Eardrum). The tympanic membrane (Fig. 9-10) separates the external ear and middle ear. It transmits the vibrations of sound waves along the auditory ossicles (attached to its medial or deep side) to the inner-ear labyrinth. It lies deep in the external meatus at about a 45° angle. The membrane bulges into the middle ear, and thus its deep part is concave and called the umbo. Through the otoscope, it appears pearly gray and consists of a flaccid part (Shrapnell's membrane) above and a dense part below. When it is observed through reflected light, a "cone of light" is seen running from the apex of the umbo, downward and forward to the periphery. The luminous triangle changes shape with disease. Shadows of some of the ear bones can also be seen through the membrane. It normally is taut, held by the tensor tympani muscle.

Middle Ear (Tympanic Cavity). The middle ear (Fig. 9-10) is an air space in the petrous part of the temporal bone, lined by mucous membrane that also covers the bony ossicles and extends into the eustachian tube and mastoid area. It contains the three auditory ossicles—the malleus, incus, and stapes—which transmit sound vibrations from the tympanic membrane to the internal ear. It is about 1.25 cm ($\frac{1}{2}$ in.) long, 1.25 cm ($\frac{1}{2}$ in.) high, and 0.41 cm ($\frac{1}{6}$ in.)

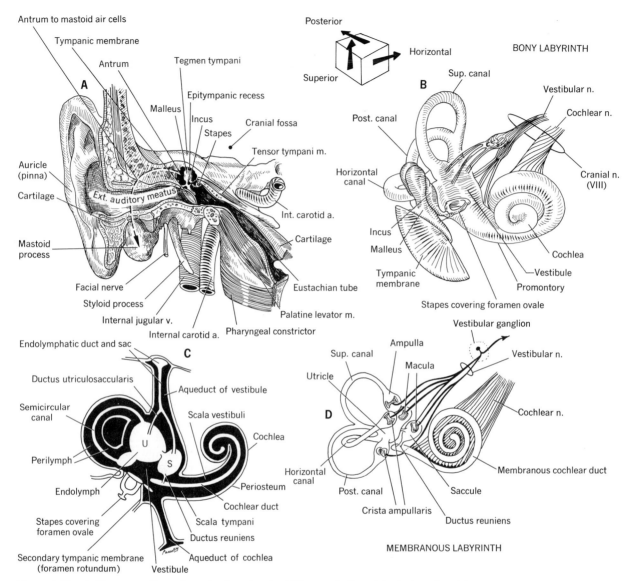

Figure 9-10. **A.** Diagrammatic section of external and middle ear. **B.** Bony labyrinth and ossicles. **C.** Endolymphatic and perilymphatic spaces and relationship. **D.** Membranous labyrinth and sensory receptors.

wide—very small! One describes it as having a roof, floor, and four walls (lateral, medial, anterior, and posterior).

Roof. The roof is a thin plate of bone (the tegmen tympani) separating the cavity from the cranial fossa and brain. A recess here is called the epitympanic recess or attic.

Floor. The floor is narrow, and a thin bone separates the cavity from the jugular fossa and the internal jugular vein.

LATERAL WALL. The lateral wall is formed mostly by the tympanic membrane. The chorda tympani nerve (a branch of cranial VII that goes to the tongue) passes between the malleolus and incus on this lateral wall to a small opening in front of the tympanic membrane.

MEDIAL WALL. The medial wall separates the middle ear and internal ear, and one sees the foramen ovale (fenestra vestibule), which leads into the vestibule of the inner ear and is occupied by the base of the stapes; the promontory, a rounded projection due to the first coil of the cochlea; the foramen rotundum (fenestra cochlea), seen behind the promontory and closed by a membrane (the secondary tympanic) which closes an opening leading to the cochlea; and the prominence of the facial canal, created by the facial nerve as it runs through the bone.

ANTERIOR WALL. The anterior wall opens directly into the eustachian tube, which in turn opens into the nasopharynx. The tube is an osseocartilaginous tube about 3.75 cm ($1\frac{1}{2}$ in.) long. The posterior one third is bony, and the anterior two thirds is cartilaginous. Through the tube air pressure is equalized on both sides of the eardrum. Microorganisms from the oronasal region often travel via the tube to the middle ear, causing infections. Above the tube is lodged the tensor tympani muscle, which swings across the middle ear to insert on the malleus, and next to it, running in bone, is the carotid artery.

POSTERIOR WALL. The posterior wall has a large opening, the aditus (doorway), leading into the tympanic antrum (hallway), which in turn passes into the mastoid air cells. The mastoid process, behind the ear, does not exist at birth but begins to develop at the end of the first year; as it grows, its bone is replaced gradually by air cells (the mastoid air cells) lined by membrane and filled with air. On the back wall one also sees an elevation, the pyramid, which contains the stapedius muscle that inserts on the stapes and stabilizes it.

Infection may result in erosion through the roof, causing meningitis; involvement of the floor encroaching on the vein, causing hemorrhage; extension medially to the facial nerve, causing facial paralysis; infection posteriorly into the mastoid causing mastoiditis; and involvement of the inner ear, causing deafness.

Auditory Ossicles. The auditory ossicles (Fig. 9-10) are a chain of three bones across the middle ear from the tympanic membrane to the wall of the inner ear. The bones form a system of levers that magnify pressure changes that occur on the eardrum and transmit them to the inner ear. The muscles of the middle ear (stapedius to the stapes and tensor tympani to the malleus) prevent the bones from transmitting excessive noise. The ossicles also form a mechanism that "slips" when too loud a noise hits the ear, and thus damage to the oval window is avoided. The bones are the malleus (hammer), the most lateral and attached to the tympanic membrane; the incus (anvil), the middle bone shaped like a two-fanged tooth; and the stapes (stirrup), which articulates via its foot plate with the foramen ovale that opens into the inner-ear vestibule. The bones are connected by joints lined by synovia and bound together and secured to the walls of the cavity by ligaments. To vibrate efficiently there must be equal atmospheric pressure on both sides of the eardrum. This is accomplished by the eustachian tube in the anterior wall of the middle ear. Thus, when we swallow, air passes in and out of the middle ear to equalize the pressure.

Internal Ear. The internal ear (Fig. 9-10) is located in the petrous part of the temporal bone and concerns itself with sound perception, orientation, and balance. It is made up of two labyrinths: a bony one, within which lies a membranous one. For the most part, the two are separated by a fluid, the perilymph.

BONY LABYRINTH. The bony labyrinth (Fig. 9-10) contains perilymph, is about 3 mm thick, and consists of the cochlea, which resembles a small "shell" making two and one-half turns. It has also been described as a "spiral staircase," turning around a central bony pillar, the modiolus. The base of the coil is medial, the apex lateral. A ledge of bone projects from the modiolus and divides the "shell cavity" of the cochlea into a scala vestibuli and a scala tympani. A duct, the aqueduct of the cochlea, leads from the scala tympani through the petrous bone to the cranial fossa and carries the perilymph of the canal into the subarachnoid space and communication with the cerebrospinal fluid. Posteriorly, the cochlea opens into a bony vestibule, which contains the saccule and utricle of the membranous labyrinth. The vestibule lies between the cochlea and three bony semicircular canals. The three bony semicircular canals (containing the membranous semicircular canals) open into the vestibule. They are named posterior (sagittal), superior (coronal), and lateral (horizontal). One end of the superior canal joins the upper end of the posterior to form a common channel, the common crus. Thus, there are but five apertures from the canals into the vestibule.

MEMBRANOUS LABYRINTH. The membranous labyrinth (Fig. 9-10) lies in the bony labyrinth and consists of "sacs" containing fluid called endolymph. It is a closed system and, unlike the bony perilymphatic space, it does not communicate with the subarachnoid space. It has the general shape and form of the bony labyrinth but is smaller and separated from the bone by the perilymph. In the vestibule, it differs by consisting of two membranous sacs, the saccule and utricle. Thus, it consists of the latter two structures, three semicircular canals, and a cochlear duct.

Cochlear Duct. The cochlear duct of the membranous labyrinth contains the spiral organ of Corti, the essential organ for hearing (Fig. 9-11). It lies lateral to the scala vestibuli, separated from it by the vestibular membrane. The elongated spiral ganglion is found in the bone that runs around the modiolus or core of the bony labyrinth. The peripheral branches from this

Figure 9-11. **A** and **B.** Diagrammatic scheme of the cristae of the ampullae of the semicircular canals and macula of the saccule and utricle. Both relate to vestibular function. **C.** A schematic representation of the histology of the cochlea, including the organ of Corti.

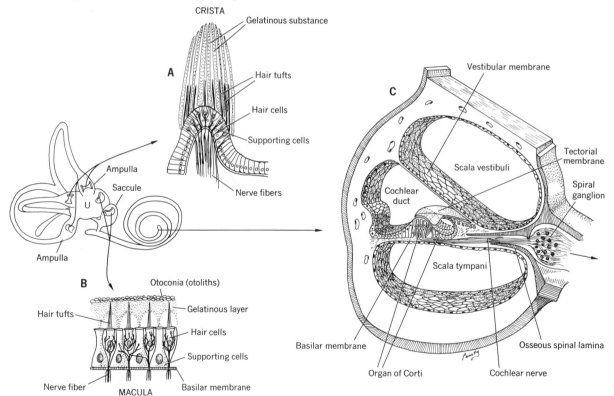

ganglion pass to the organ of Corti. The central branches leave the bone via foramina at the internal acoustic meatus and make up the cochlear part of the auditory (VIII) nerve. Posteriorly, the cochlear duct connects to the saccule by the ductus reuniens and terminates blindly.

Utricle and Saccule. The utricle and saccule (Fig. 9-10) are two membranous sacs seen in the vestibule and indirectly connected to each other. The utricle, the larger of the two, receives the openings of the membranous semicircular canals and a separate duct, the ductus utriculosaccularis, which leaves it to join a duct from the saccule, the endolymphatic duct. The junction of the two above-named ducts ends in a blind sac, the endolymphatic sac, which makes its way out of the petrous bone to lie on the posterior cranial cavity. The saccule is a round and small structure lying near the utricle.

Semicircular Canals. The membranous semicircular canals are similar to the bony canals in number, shape, and form. Each, however, has an expanded end, the ampulla. The ampullae of the semicircular canals, along with the utricle and saccule, receive the terminal branches of the vestibular nerve and contain special sense organs that register changes in position and motion.

Acoustic Nerve. As it leaves the inner ear the acoustic (VIII) nerve divides in the internal auditory meatus into two major branches: the cochlear and the vestibular (Fig. 9-10). The latter receives sensation from the utricle, saccule, and ampulla of the semicircular ducts. The cochlear portion receives sensation from the cochlea and the organ of Corti.

Physiology of Hearing

The function of the ear is to transmit to the brain an accurate pattern of the sound vibrations that come to it from the environment, their relative intensity, and the direction from which they come.

Characteristics of Sound. The sounds we hear are, in reality, vibrations of the air moving away from the source in all directions (like the ripples created in a pool of water by a thrown stone). These waves of sound have two characteristics determined by the source.

LENGTH OF WAVE. The first characteristic is frequency or wavelength. Since sound travels at a uniform speed in a given conducting medium, the closer the waves are together (shorter wavelength), the more frequently the waves will hit the eardrum. The number of vibrations per second is called the cycles per second (cps) or Hertz (Hz). The greater the frequency (shorter the wavelength), the higher is the pitch. Thus, 256 vibrations or cycles per second produce middle C, while the same note one octave higher has a vibration rate of 512 Hz.

The human ear, with much individual variation, can make out sounds over a fairly wide frequency range (from 16 to about 30,000 Hz). Perception of high frequencies is best in childhood, with a gradual decrease with age. A normal adult has difficulty with frequencies over 10,000 to 12,000 Hz. The ear is more sensitive to frequency changes than to loudness, and thus we distinguish about 370 changes in loudness but about 2000 changes in frequency. In fact, with age, sound by both air and bone conduction grows poorer as frequencies increase.

HEIGHT OF WAVE. The height of the wave or intensity of sound is expressed in decibels (dB). We can barely hear 0 dB intensity. With each increase of 10 dB there is a tenfold increase of sound intensity. Thus, a sound of 10 dB is ten times as intense as that barely audible; 20 dB is a sound one hundred times as intense; 60 dB is one million times as intense. At a distance of about 1.2 m (4 ft) an average whisper produces a level of 20 dB and ordinary conversation about 60 dB. In contrast, a riveter at 10.5 m (35 ft) away produces a sound of 100 dB intensity or ten billion times the intensity of barely audible sounds. Only when we approach 30 dB of hearing loss is impairment actually noticed, and the first thing noticed is the inability to hear a whisper or someone speaking at a distance.

Passage of Sound. The external ears (auricles) locate the direction of the source of sound, concentrate the sound waves, and conduct them to the external auditory meatus. The two ears provide us with stereophonic hearing so we can judge the direction of a sound. The ears locate the direction of the source of sound. When we listen to a sound from one side, that ear receives a wave at a moment of compression while the other receives it at a point of rarefaction. Thus, the waves are out of phase and the impulses in the auditory nerves are out of rhythm. The loudness, too, arrives at the ears unequally. The brain interprets the small differences coming to it from the two ears, and thus we know from what direction the sound came. If the sound is behind or in front of us, both ears receive waves in phase and this helps locate the origin of sound. This is due to differences in the phase of the vibration as it arrives at the two ears as well as to differences in intensity and quality (timbre), since in the far ear the sound must go around the corner. The external meatus shelters and protects the eardrum and maintains a constant passage of temperature and humidity. It also serves as a tubal resonator so that the vibrations of sound at the drum have a higher intensity of "pressure" than those at the auricle. The eardrum consists of circular and radial fibers that are kept tense by the tensor tympani muscle for better vibration reception (particularly of high frequency). At the drum, the pressure changes of sound waves are transformed into mechanical vibrations. Thus, the drum serves as a receptor of vibrations and as a barrier to protect the contents of the middle ear. It also provides an acoustic dead space, so that air vibrations in the middle ear will not break up and dissipate themselves due to the irregular shape of the middle ear walls and not exert pressure on the round window to compete with vibrations on the oval window.

Acting as a unit, the ossicles conduct vibrations from the drum to the oval window. These delicately suspended bones transmit vibrations without distortion and provide part of the increased power needed when going from air (a lighter conducting medium) to the inner ear perilymph (a heavier medium). Leverage does this, since the incus is shorter than the long process of the malleus and vibrations at the oval window are reduced in amplitude but increased in power by 2:1. The difference in size between the larger eardrum and the oval window increases the ratio to 10:1.

Vibrations into the oval window (by the footplate of the stapes) set up vibrations in the inner ear perilymph, which "bathes" the membranous labyrinth containing the organs of hearing and balance. The organ of Corti (hearing receptor), lying on the basilar membrane in the cochlear duct, is composed of a complex assortment of supporting cells interspersed between the hair cells, the sensory end-organs. The hair cells are arranged segmentally in well-defined rows, a single row of inner hair cells and several rows of outer hair cells. The total number of inner hair cells is about 3500 and outer hair cells about 12,000. Each hair cell ends on its free surface in a clump of hairs. Overhanging the hair cells is a gelatinous structure, the tectorial membrane, in which the hairs are embedded.

Vibrations of the stapes at the oval window are transmitted to the perilymph of the scala vestibuli of the cochlea. From the perilymph the vibrations are transmitted across the vestibular membrane to the endolymph of the cochlear duct and eventually through the basement membrane of the duct to the scala tympani on its other side, where they pass down to the round window. Vibrations of the round window coincide with those at the oval, a fraction of a second later and in opposite phase (when one is forced in, the other "gives" out).

Vibrations of the basilar membrane create a pull on the hair cells attached to the tectorial membrane. This action changes the energy, up to this point mechanical, into electrical impulses that stimulate the fibers of the eighth nerve to the brain and give rise to the action potentials needed for nerve transmission. The hair cells transmitting higher tones are located at the lower end of the cochlea, and those transmitting lower tones are near its apex.

Thus, each hair cell has one or more nerve fibers and each nerve fiber contacts one or more hair cells. The auditory nerve consists of 30,000 separate neurons. The nerve cells for these neurons are found in the spiral ganglia, from which axons pass via the cochlear part of the auditory (VIII) nerve to the dorsal and ventral cochlear nuclei of the pons. From these "way stations," fibers pass to various areas of the brain stem for synchronization of the stimuli and finally on to the cerebral cortex, where the tones of various pitch are perceived. The brain interprets low tones by the frequency of the nerve impulses in the nerve. How the brain interprets pitch, quality, and loudness is not fully understood.

To be deaf or even hard of hearing is very emotionally disabling. The baby born deaf learns to speak only with intensive training and even then usually remains six to eight years behind his fellow. Deafness later in life deprives one of a primary means of communication. Early diagnosis and help are truly essential.

Physiology of Balance and Equilibrium: The Vestibular System

The vestibular system contains mechanoreceptors that react to changes in position and movement of the head. The vestibular apparatus is found in the inner ear, bilaterally. It consists of the three membranous semicircular canals; the utricle with which they connect; and the saccule, which connects to the utricle as well as the membranous cochlear duct. The entire closed system of canals and sacs is filled with endolymph. The semicircular canals are arranged at right angles to each other and thus can respond to head movement in all directions. The actual sensory receptors are hair cells ensheathed by a gelatinous mass located in the ampullae or expanded ends of the canals and in the saccule and utricle. Afferent vestibular fibers lead to the brain from these receptors. When the head is moved, all parts (bony labyrinth, membranous labyrinth, and hair cells) turn with it. The endolymph fluid in the canal, because of inertia, tends to remain fixed. As the bodies of hair cells move with the skull, the hairs are pulled against the "relatively" stationary endolymph and are bent. Thus, the hair cells are stimulated. As the inertia is overcome, the hairs slowly return to their resting state and thus are only stimulated during changes in the rate of motion. If one moves at a constant speed, stimulation stops. Even when the head is motionless, the afferent nerve fibers are somehow activated at a relatively low resting frequency—how is unknown. Different movement occurs in different canals, on both one side and the other, due to their relative positions, and all is coordinated centrally. The semicircular canal receptors signal the rate of change of motion of the head, but the receptors in the saccule and utricle relay information regarding the position of the head relative to the direction of the forces of gravity. Here are found hair cells, with hairs that protrude into a gelatinous substance in which are embedded small calcium carbonate stones called otoliths (Fig. 9-11). When the head is bent, the gelatinous otolith material changes position, the forces against the hair cells bend the hairs, and the receptor cells are stimulated. Afferent fibers from both canals and sacs enter the brain stem as the vestibular nerve. The fibers end in the brain stem or cerebellum, and secondary fibers pass to the spinal cord interneurons, subcortical centers, and cerebral cortex.

Vestibular impulses control the eye muscles to the degree that, despite head movement, the eyes remain fixed on the same point. If the vestibular apparatus is destroyed in man, there is little disability as long as the visual system, joint position receptors (proprioceptors), and skin receptors are functioning. Such persons do, however, have trouble walking in the dark or up and down stairs since their vision is impaired. Thus, this system is essential where vision cannot be used or when swimming under water when the visual proprioceptive impulses are altered.

Taste and Smell

Receptors for the sensation of taste and smell in man (Figs. 9-12, 9-13) are not nearly as essential as those for vision or hearing. They serve to affect one's appetite and digestion but exert a relatively minor influence.

Taste

Taste is affected only by water or lipid-soluble chemicals. The special receptor organs are the taste buds, lying on the upper surface and edge of the tongue (Fig. 9-12) and on the pharynx, the epiglottis, and even the larynx. The circumvallate papillae of the tongue harbor the densest population of taste buds.

Taste sensations consist of sweet, sour, salty, and bitter, but specific taste bud receptors for each have

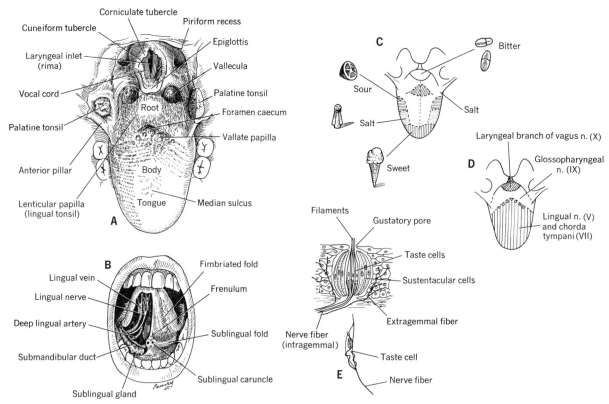

Figure 9-12. A. The tongue, as seen from above, with adjacent related structures. **B.** Undersurface of the tongue; right side shows vessels, nerves, and glands. **C.** Four basic taste sensations and their receptor locations on the tongue. **D.** Major nerve supply (sensory) to the tongue. **E.** Diagram of a taste bud and nerve fibers.

not been identified. A single receptor can apparently respond (in varying degrees) to any of the sensations. Apparently the chemical substance and the reactive site (bud) alter the receptor cell membrane, forming "pores" through which ions move to change the membrane potential of the cell, which then depolarizes. The latter affects the nerve fibers, which enter the bud from below and end on the receptor cells. One nerve fiber can innervate several cells, and one cell may be innervated by several neurons. Awareness of specific taste probably is related to the relative activity in a number of different neurons rather than in a specific one. Different afferent fibers show different firing patterns in response to various substances. The variation in sensitivity makes the firing pattern important. Temperature, texture, and odor all help when transmitted to the brain and interpreted together.

Most taste buds are innervated by the glossopharyngeal nerve at the posterior one third of the tongue. However, the facial (VII) nerve via the chorda tympani nerve innervates the anterior two thirds of the tongue. The vagus (X) receives branches from the epiglottis and larynx.

Smell

The olfactory receptors lie in the mucosa in the upper part of the nasal cavity, which is above the path of the major air currents (Fig. 9-13). The receptor cells

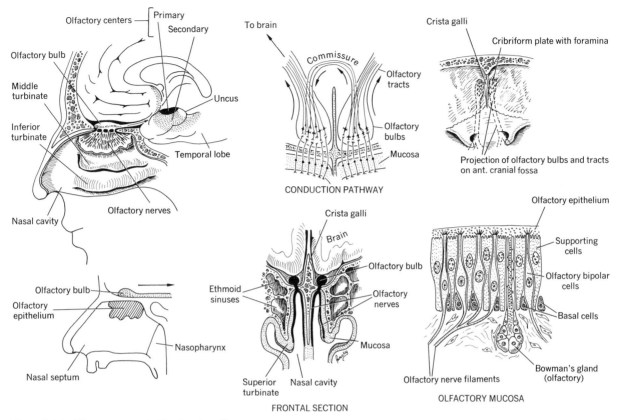

Figure 9-13. The anatomy of olfaction (smell).

are modified neurons having two processes: one passing to the brain to form the olfactory nerve (I), and the other having fine cilia that extend onto the surface of the nasal mucosa. Different odors (as with taste) do not have obvious different receptors, although differences may exist at the molecular levels.

The "odor" releases molecules that diffuse into the air and pass into the nose and dissolve in the layer of mucus covering the receptors. The membrane is depolarized and initiates an action potential in the afferent central nerve fiber. The physiologic basis for odor discrimination and differentiation is unknown. We may receive a blend of "primary odors" (like taste), but little information is available. As with taste, other stimuli enhance or depress the awareness of odors, and odors are affected by hunger, sex, smoke, and so forth.

Review Questions

1. What is meant by proprioception?
2. Describe the layers of the skin.
3. Differentiate between eccrine and apocrine sweat glands. How do they differ from sebaceous glands?

4. What is a hair follicle? Does hair growth generally reflect one's state of health?
5. When we speak of a nail, what are we referring to?
6. As you look at an eye, externally what can you see?
7. Describe the process of lacrimation—where found, how drained?
8. What are the layers of the eyeball? What is the function of each? Describe the subdivisions of each.
9. Are the aqueous and vitreous humors the same? Where are they found?
10. What is the arterial supply of the eye?
11. List the muscles of the eyeball and indicate their actions.
12. Follow a beam of light through the eye, naming what happens to it and the structures it passes through.
13. What is meant by accommodation? Nearsightedness? Farsightedness? Astigmatism? Color blindness?
14. Where are the rods and cones? What is their function?
15. Define the fovea centralis and the blind spot.
16. Trace a sound from the external ear until it is relayed via the cochlear nerve to the brain. List all structures crossed in this passage.
17. Describe the walls of the middle ear.
18. Differentiate between bony and membranous labyrinths. Where do you find endolymph and perilymph?
19. What are the parts of the membranous labyrinth?
20. Where is the organ of Corti (hearing) located? Approximately how does it function?
21. Where are the sensory receptors for both sound and balance located? What nerves are involved?
22. What are the two major characteristics of a sound wave? Relate the waves to function.
23. What is meant by:
    a. Hertz (Hz)
    b. Decibels of sound
24. Why is the vestibular system essential to man?
25. Are taste and smell as essential as the other modalities of sensory reception described in this chapter?
26. Where are the special taste receptors located? What are they?
27. Where are the receptors of smell found?

References

Alpern, M.; Lawrence, M.; and Wolsk, D.: *Sensory Processes.* Brooks/Cole Publishing Co., Monterey, Calif., 1967.

DeValois, R. L., and Jacobs, G. H.: Primate Color Vision. *Science,* **162:**533, 1968.

Dowhing, J. E.: Night Blindness. *Sci. Am.,* **215**(4):78–94, 1966.

Gregory, R. L.: *Eye and Brain: The Psychology of Seeing.* McGraw-Hill Book Co., New York, 1966.

Michael, C. R.: Retinal Processing of Visual Images. *Sci. Am.*, **220**(5):104–114, 1969.

Oakley, B., and Benjamin, R. M.: Neural Mechanisms of Taste. *Physiol. Rev.*, **46**:173, 1966.

Oster, G.: Auditory Beats in the Brain. *Sci. Am.*, **229**(4):94–102, 1973.

Rocke, I., and Harris, C. S.: Vision and Touch. *Sci. Am.*, **216**(5):96–104, 1967.

Stevens, S. S.; Warshofsky, F.; and the Editors of Life: *Sound and Hearing*. Time-Life Books, New York, 1965.

Internal Integration, Correlation, and Coordination

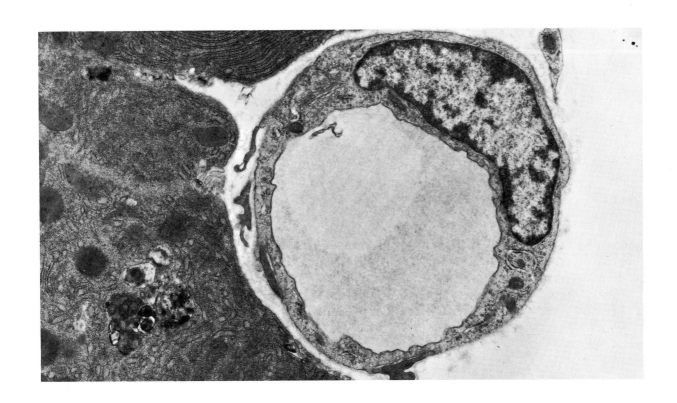

UNIT **IV**

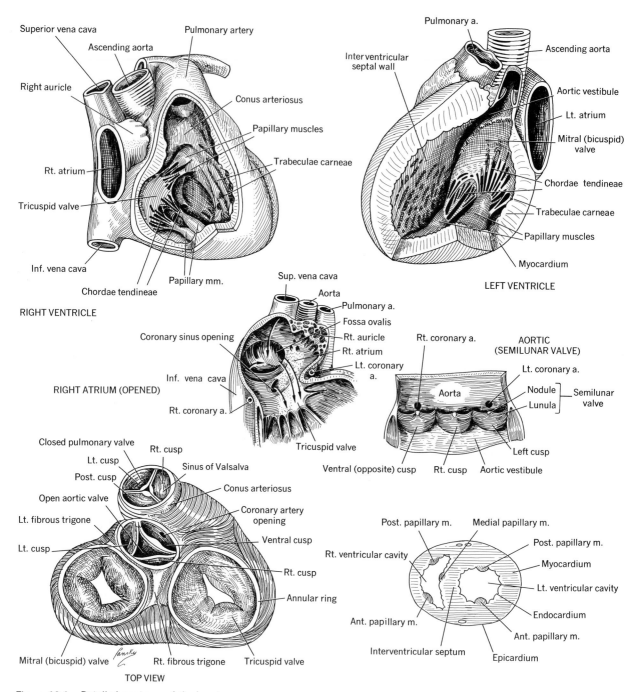

Figure 10-1. Detailed anatomy of the heart.

Circulation: The Cardiovascular System

The circulatory system and its related organs include (1) the cardiovascular apparatus (the heart and blood vessels); (2) the blood and lymph; (3) the lymphatic vessels and related structures; and (4) the blood-forming and blood-destroying tissues. Both the heart (Fig. 10-1) and its circuit are controlled by nerves and hormones.

The cardiovascular system is essentially a closed transport system for nutritive and waste materials, hormones, and heat energy. It is composed of two major subdivisions, a pulmonary circulation (carrying blood to and from the lungs) and a systemic circulation (carrying blood to all parts of the body and bringing it back to the heart). Each includes a pump (the heart) and a circuit of vascular tubes (the arteries, arterioles, capillaries, and veins). The arteries and veins are elastic tubes that connect, for the most part, with the lungs and tissues as well as the heart. The capillary bed of each circuit provides a specialized exchange surface (Table 10-1).

Anatomy of the Cardiovascular System

Thoracic (Chest) Cavity

The thoracic cavity contains the lungs and its coverings (the pleura) and a central area called the mediastinum, which holds the pericardium (fibrous sac that encloses the heart), the heart itself, the great vessels attached to the heart, and other structures.

Mediastinum. The mediastinum (Fig. 10-2) is bounded anteriorly by the sternum, posteriorly by the 12 thoracic vertebrae, superiorly by the thoracic inlet, inferiorly by the diaphragm, and on the sides by the pleurae (serous membranes that line the thoracic cage and cover the lungs). It is divided into a superior and inferior region by an imaginary line drawn from the sternal angle to the articular disk between the fourth and fifth thoracic vertebrae. The inferior region is further subdivided into three subdivisions: the part containing the heart is the middle mediastinum; that in front of it is the anterior mediastinum; and that behind it is the posterior mediastinum. (Details of the structures lying in each of these areas can be found in most anatomy texts.)

Pericardium. The pericardium (Figs. 10-3, 10-4) is a fibroserous sac found in the middle mediastinum and encloses the heart and roots of the great vessels leading to and away from it. Its outline generally corresponds

Table 10-1. Distribution of Blood Flow to Various Organs and Tissues of the Body at Rest and During Exercise (Blood Flow in Milliliters per Minute)*

	Rest	Exercise
Kidney	1100	600
Abdomen	1400	600
Muscle	1200	12,500
Skin	500	1,900
Brain	750	750
Heart	250	750
Other	600	400
Total	5800	17,500

* After studies by Chapman and Mitchell.

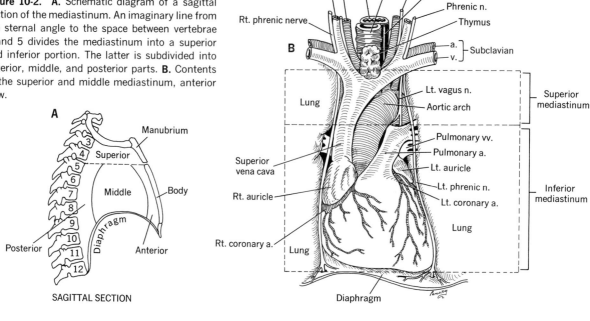

Figure 10-2. **A.** Schematic diagram of a sagittal section of the mediastinum. An imaginary line from the sternal angle to the space between vertebrae 4 and 5 divides the mediastinum into a superior and inferior portion. The latter is subdivided into anterior, middle, and posterior parts. **B.** Contents of the superior and middle mediastinum, anterior view.

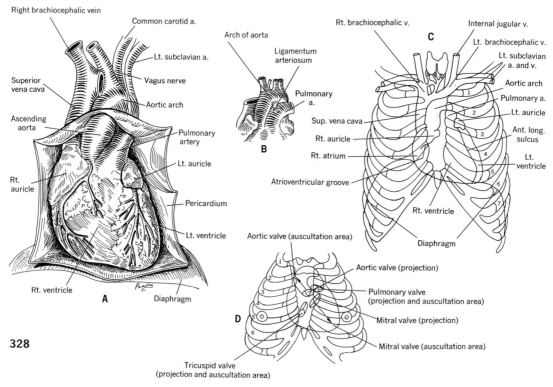

Figure 10-3. **A.** The pericardium (cut open to reveal structures), a fibroserous sac that encloses the heart and the roots of the great vessels. **B.** The ligamentum arteriosum is the "remains" of the ductus arteriosus, which connected the pulmonary and systemic circulation in the fetus. **C.** Topography of the heart and great vessels. **D.** Surface projection of the heart valves.

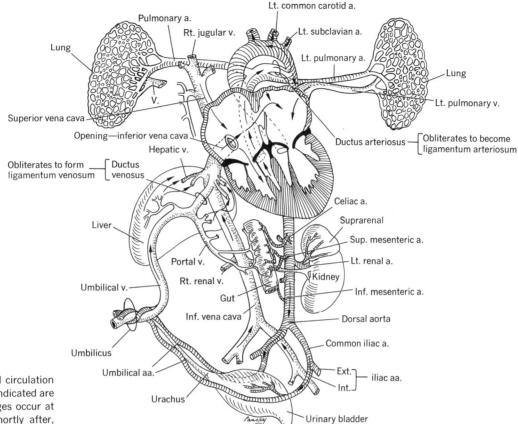

Figure 10-4. Fetal circulation before birth; also indicated are areas where changes occur at the time of, or shortly after, birth. (After studies by Patten.)

Labels in figure:
Pulmonary a. — Rt. jugular v. — Lt. common carotid a. — Lt. subclavian a. — Lt. pulmonary a. — Lung — Lung — Lt. pulmonary v. — V. — Superior vena cava — Opening—inferior vena cava — Hepatic v. — Ductus arteriosus — Obliterates to become ligamentum arteriosum — Obliterates to form ligamentum venosum — Ductus venosus — Liver — Celiac a. — Suprarenal — Sup. mesenteric a. — Portal v. — Lt. renal a. — Kidney — Rt. renal v. — Inf. mesenteric a. — Umbilical v. — Gut — Inf. vena cava — Dorsal aorta — Common iliac a. — Umbilicus — Ext. — iliac aa. — Int. — Umbilical aa. — Urachus — Urinary bladder

to that of the heart. It consists of three layers. The fibrous layer is entirely parietal and lines the sac, but the serous layer has both a visceral layer that covers the heart and a parietal layer that joins the fibrous one.

FIBROUS LAYER. The fibrous layer forms a saclike structure around the heart. It is fused with the outer surface of the parietal layer of the serous pericardium. It also fuses with the diaphragm inferiorly and blends with the outer coats of the great vessels as they enter and leave the heart. Two condensations, the upper and lower sternopericardial ligaments, attach it to the sternum anteriorly.

VISCERAL AND PARIETAL SEROUS LAYER. This layer (Figs. 10-3, 10-4) is a thin, glistening serous membrane.

The visceral layer or epicardium is thin and firmly adherent to the outer heart surface and cannot be easily separated. The visceral layer is reflected on itself where the vessels enter and leave the heart as the parietal layer of serous pericardium, which fuses with the fibrous pericardium. This reflection forms the upper portion of a cul-de-sac found behind the heart (between visceral and parietal layers) and called the oblique sinus of the pericardium.

The potential space between the visceral and parietal serous pericardium is known as the pericardial cavity and contains enough serous fluid to minimize friction between the surfaces. Another space, created by the reflecting visceral and parietal pericardium in the region of the vessels entering and leaving the heart, behind the aorta and pulmonary artery, is called the transverse sinus (Fig. 10-4).

Heart

With the pericardium opened, one sees the heart, a muscular organ that acts as a double pump with the right side handling deoxygenated blood and the left side oxygenated blood (Figs. 10-1, 10-3, 10-4). The human heart is divided longitudinally into right and left halves, each consisting of two chambers, an atrium and a ventricle. The cavities of the atria and ventricles on each side communicate with each other, but right and left do not.

Right Atrium. The right atrium (Fig. 10-1) begins at the opening of the superior and inferior venae cavae and receives deoxygenated blood from the veins of the body and heart, respectively. An indistinct depression, the sulcus terminalis, marks the junction between the embryonic sinus venosus and the right atrium and also marks a ridge seen on the inside of the right atrium called the crista terminalis. The embryonic sinus venosus becomes the part of the heart that receives the two veins. The posterior part of the atrium is generally smooth, but the front part is thrown into folds. These folds resemble the teeth of a comb and, therefore, are called the musculi pectinati. On the posterior wall of the atrium is the atrial septum (separating right and left atria). Near the center of this septum is a shallow, oval depression, the fossa ovalis, bounded by a ridge called the limbus of the fossa ovalis. This fossa was patent (open) in the fetus and allowed blood to pass from right to left atria and thus bypass the lungs. It normally closes at or shortly after birth. The inferior vena cava, as it enters the heart from below, has a fold called the eustachian valve. In the fetus this valve directs blood from the vein to the fossa ovalis and into the left atrium. An appendage of the atrium is the auricle, helping to increase the surface area of the atrium. The right atrioventricular opening or tricuspid orifice is found in the anterior wall of the right atrium and is large enough to admit three fingers. It opens into the right ventricle and is bounded by the tricuspid valve. Medial to this opening in the right atrium is that for the coronary sinus (carrying blood from the veins of the heart), which is also guarded by a valve, the coronary or thebesian valve.

Right Ventricle. The right ventricle (Fig. 10-1) receives blood from the right atrium and pumps it through the pulmonary artery to the lungs for oxygenation. It has a thick wall. Its uppermost part, the infundibulum, has smooth walls with no projecting muscle bundles and leads into the pulmonary artery. The rest of the ventricular inner surface is markedly irregular due to muscle columns called the trabeculae carneae. A number of conelike muscular projections, the papillary muscles, are attached to the ventricular wall and are connected to cusps of the right atrioventricular valve by a number of tendinous strands called the chordae tendineae. The right atrioventricular opening is about 2.5 cm (1 in.) in diameter and is surrounded by a fibrous ring. The valve guarding it has three cusps—anterior, posterior, and medial—and thus is called the tricuspid valve. The pulmonary orifice is also surrounded by a thin fibrous ring, to which are attached the bases of the three cusps of the semilunar, pulmonary valve.

Left Atrium. The left atrium (Fig. 10-1) has the openings of the four pulmonary veins from the lungs, unguarded by valves. Blood leaves the atrium to enter the left ventricle via the left atrioventricular bicuspid or mitral orifice, which is guarded by a valve of the same name. This valve is smaller than the right tricuspid. The interior of the chamber is generally smooth except in the auricular appendage portion, where musculi pectinati are also seen. The endocardium of this atrium appears pale due to its great thickness.

Left Ventricle. The cavity of left ventricle (Fig. 10-1) is longer and narrower than the right, but its walls are much thicker since it has to force blood to the head and body. Internally the trabeculae carneae muscles are finer and more numerous than are those seen on the right side. The papillary muscles, however, are less numerous and stronger, with chordae tendineae from each passing to both cusps of the mitral valve. The left atrioventricular orifice is guarded by the mitral valve with a large anterior and smaller posterior cusp. Blood leaves the left side of the heart via the aortic opening, which is surrounded by a fibrous ring to which the bases of the cusps of the aortic semilunar valve are attached. Like the pulmonary semilunar valve, this, too, has three cusps. Blood is pumped to the head and the body via the aortic artery.

General Topography of the Heart. A groove marks the separation between atria and ventricles: the atrioventricular groove or coronary sulcus, in which the right coronary artery lies (Fig. 10-3). Another groove divides the ventricles: the anterior longitudinal sulcus or interventricular groove. This contains the anterior descending branch of the left coronary artery. One can see all four chambers of the heart from the front. However, only a small part of the left atrium, the left auricle, is actually seen. The right border of the heart is convex and formed by the right atrium, whereas the left border is almost entirely left ventricle.

The walls of the heart are composed primarily of muscle (myocardium), which differs structurally from either skeletal or smooth muscle (see Chap. 4). The surface in contact with the blood in the chambers is lined by a thin layer of cells called the endocardium. The epicardium, or outside layer, is actually serous pericardium (see earlier discussion) (Fig. 10-1). The walls of the atria and ventricles are composed of layers of cardiac muscle, arranged in crisscross bands that are bound together and encircle the chambers. When the walls contract, they come together like a "squeezing fist" and exert pressure on the blood inside. Cardiac muscle differs from other muscle in that it consists of a continuous network of fibers, explaining how a contraction originating at one point spreads through the remainder of the heart. Waves of contraction and dilatation are rhythmically repeated at the rate of 70 times per minute or about 100,000 times per day, ejecting approximately 11,350 liters (3000 gal) of blood from the heart per day.

The performance of the heart as a pump depends largely on the contractile activity of the myocardium, and the latter reflects the integrated function of its individual contractile elements, the sarcomeres. (See Chap. 4 for a detailed discussion of the histology and electron microscopy of cardiac muscle.)

When myocardial muscle is significantly over-stretched, the fibrils begin to describe an irregular zigzag pattern. One cell can become longer than its individual components by allowing some parts to slide in one direction and other parts in another. This "slippage" is apparently important as a factor in chronic fatigue and dilatation.

In both acute and chronic dilatation there is an increase in sarcomere length, and in a chronically dilated heart the length is near peak, allowing for no reserve capacity. Cardiac function gets progressively worse, and the heart loses its ability to maintain a steady state.

With the help of an electron microscope, deterioration of myocardial fibers can be seen in many ways. The fibrils appear larger in diameter rather than increased in number. In cases of ischemia (deficiency of the blood) due to coronary artery involvement, the surface membranes are broken and disrupted and the T system becomes dilated. Finally, the mitochondria show vacuolization, and water droplets appear inside them as they lose their crystalline structure and become swollen and loosely packed, and dissolve—all of which are seen in hypoxia and severe congestive heart failure. Yet if blood flow is restored back to normal in 30 to 45 minutes, most of the changes are reversible and even the mitochondria regain their functional integrity rapidly.

Note that function deteriorates with time and age, and superimposed loads are even more of a burden; overloading finds little of the reserve that may be present in youth. Thus, failure is not a dilated heart, but the dilated heart is the compensation for the occurrence of myocardial failure, and as dilatation proceeds the function of the dilated heart fails.

Blood Vessels and Nerves

Pulmonary Artery. The pulmonary artery (Figs. 10-1, 10-3, 10-4, 10-5) passes up and back from the right ventricle and carries oxygenated blood to the lungs. The trunk of the artery is about 5 cm (2 in.) long and lies inside the fibrous pericardium. As it passes upward it is "embraced" by the left and right auricles and is in front of the aorta. It bifurcates below the arch of the aorta, like a letter T, into the right and

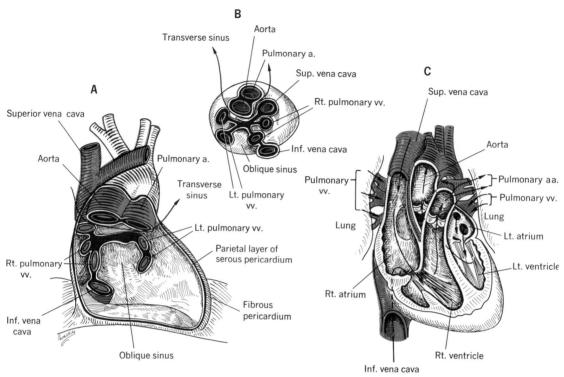

Figure 10-5. **A.** The transverse and oblique sinuses as seen from in front after the heart has been removed. **B.** The transverse and oblique sinuses as seen from on top and posterior. In front of the former is the arterial flow of the heart (aorta and pulmonary artery); behind the transverse sinus are the veins (venae cavae and pulmonary). **C.** Diagram of heart chambers, valves, and vessels. Arrows indicate the direction of blood flow. (Note that the traditional red for arteries and blue for veins have been used. Color is not related to oxygenation of blood.)

left pulmonary arteries. The right pulmonary artery is longer than the left and divides into three primary branches, one for each lobe of the right lung. The left pulmonary artery divides into two primary branches for the two lobes of the left lung.

Ligamentum Arteriosum. The ligamentum arteriosum (Fig. 10-3) is a fibrous band running from the bifurcation of the pulmonary trunk to the lower part of the aortic arch and is the remains of the ductus arteriosus of the fetus. The latter vessel, in the fetus, short-circuits the blood from the pulmonary artery into the aorta before it can get to the lungs, which are not functioning as yet (Fig. 10-4).

Aorta. The aorta (Figs. 10-1, 10-3, 10-5, 10-6) is the

great arterial trunk of the body and is found partly in the thorax and partly in the abdomen. It begins at the left ventricle, arches over and behind the root of the left lung, and descends in front of the thoracic vertebra through the diaphragm and enters the abdominal cavity. It has four parts: ascending aorta, arch of the aorta, descending or thoracic aorta, and abdominal aorta.

Coronary Arteries. The coronary arteries (Fig. 10-6) are the nutrient vessels of the heart and arise from dilatations at the root of the ascending aorta, which are called the aortic sinuses. There are two coronary arteries, a right and a left. The origins of both arteries are hidden from view by the right auricle and the

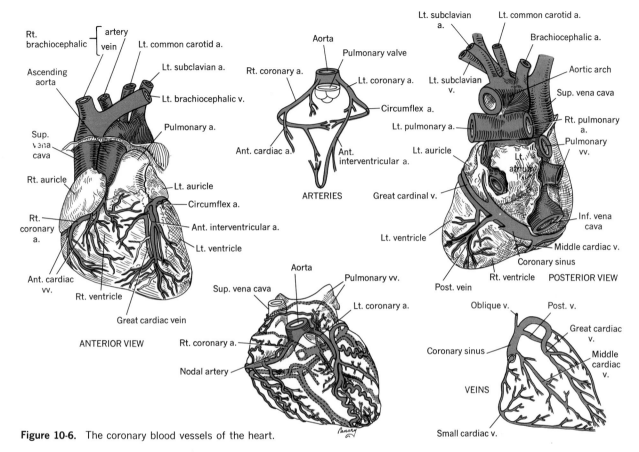

Figure 10-6. The coronary blood vessels of the heart.

Labels in figure:

ANTERIOR VIEW:
Rt. brachiocephalic [artery / vein]
Lt. common carotid a.
Ascending aorta
Lt. subclavian a.
Lt. brachiocephalic v.
Pulmonary a.
Sup. vena cava
Rt. auricle
Lt. auricle
Circumflex a.
Rt. coronary a.
Ant. interventricular a.
Lt. ventricle
Ant. cardiac vv.
Rt. ventricle
Great cardiac vein

ARTERIES:
Aorta
Pulmonary valve
Rt. coronary a.
Lt. coronary a.
Circumflex a.
Lt. pulmonary a.
Lt. auricle
Ant. cardiac a.
Ant. interventricular a.

POSTERIOR VIEW:
Lt. subclavian a.
Lt. common carotid a.
Brachiocephalic a.
Aortic arch
Sup. vena cava
Lt. subclavian v.
Rt. pulmonary a.
Pulmonary vv.
Lt. atrium
Great cardinal v.
Lt. ventricle
Inf. vena cava
Middle cardiac v.
Coronary sinus
Rt. ventricle
Post. vein

(lower center):
Aorta
Pulmonary vv.
Sup. vena cava
Lt. coronary a.
Rt. coronary a.
Nodal artery

VEINS:
Oblique v.
Post. v.
Great cardiac v.
Coronary sinus
Middle cardiac v.
Small cardiac v.

pulmonary artery. The vessels pass forward on either side of the pulmonary artery. The right coronary travels in the atrioventricular sulcus to the back of the heart to the posterior interventricular sulcus, there to give rise to the posterior descending interventricular artery. Along its course, it gives off the marginal branch along the heart's right margin. The left coronary reaches the anterior interventricular sulcus, into which it sends the anterior descending interventricular artery; however, the major trunk, the circumflex artery, continues around the left side of the heart to its posterior aspect.

Cardiac Veins. The cardiac veins (Fig. 10-6) follow the arteries in the sulci (usually lying superficial to the arteries). The companion of the left coronary artery is the great cardiac vein, which follows the interven-

tricular branch of the artery and the circumflex. The companion of the interventricular branch of the right coronary artery is the middle cardiac vein. Most of the venous blood of the heart terminates in the coronary sinus, which lies in the atrioventricular sulcus on the posterior aspect of the heart. It is about 3.75 cm ($1\frac{1}{2}$ in.) long and opens directly into the right atrium. Smaller veins also empty directly into the heart (thebesian veins).

Nerves. The nerves of the heart (Fig. 10-7) are derived from the vagus (X) and the cervical gangli-onated sympathetic chain, which contributes superior, middle, and inferior cardiac nerves. The nerves spread over the aortic arch and heart and are distributed with the branches of the coronary vessels via the superficial and deep cardiac nerve plexi.

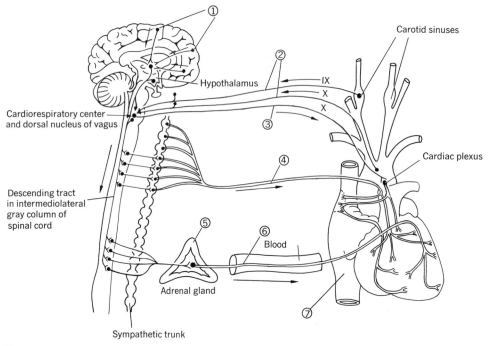

Figure 10-7. Neural and humeral regulation of cardiac function. **1.** Emotional stress stimulates sympathetic nerves by way of the hypothalamus. **2.** Afferent fibers from baroreceptors in carotid sinuses (via IX and X cranial nerves) and in aorta (via X) from afferent limb of reflex arcs to vagus (X) and sympathetic efferents. **3.** Vagal efferent cardiac fibers go to SA node and AV node. Stimulation causes release of acetylcholine, slowing heart rate and conduction. Inhibition of vagus causes acceleration of heart rate and conduction. **4.** Sympathetic efferent-fiber stimulation accelerates heart rate, increases force of contraction, and dilates coronary arteries by releasing norepinephrine at nerve endings, stimulating B-receptors. **5.** Output of catecholamines from suprarenal medulla promoted by sympathetic stimulation. **6.** Circulating catecholamines have same action as sympathetic efferent nerves on coronary arteries. **7.** Increased pH heightens catecholamine actions and lowers acetylcholine actions.

Heart Valve Sounds and Their Thoracic Projection

The audibility (ability to be heard) of the valves depends on (1) the depth of the valves from the chest surface, (2) the transmission quality of the tissue, and (3) the direction of the blood flow in the heart (Fig. 10-3). The tricuspid valve is heard best at the lower right part of the body of the sternum, where the right ventricle is nearest the body surface. The pulmonary valve is best heard at the third left costal cartilage near the sternum (opposite the actual position of the valve proper). The mitral valve is deeper than the valves of the right side and is actually located behind the left half of the sternum near the fourth costal cartilage; yet it is heard best at the cardiac apex in the fifth intercostal space (in the direction of blood flow). The aortic valve is heard best in the second right intercostal space at the sternal margin, although the valve itself is projected to where the third left rib meets the sternum.

Physiology of the Heart

Heartbeat Coordination

Contraction of heart muscle (like other muscle) is triggered by depolarization of the muscle membrane. If all the muscle fibers were to contract randomly, there would be a lack of coordination between pumping by each corresponding atrium and ventricle as well as a general lack of muscle coordination in the ventricles. The blood would move about in the ventricles instead of being ejected into the aorta and pulmonary arteries. Thus, the muscle mass of the ventricular "pump" must contract simultaneously for efficiency of action. This is made possible by (1) junctions or nexuses, which allow spread of an action potential from one muscle fiber to the next so that excitation in one fiber spreads throughout the heart; and (2) the presence of a specialized conducting system in the heart, to facilitate the rapid and coordinated spread of excitation.

Heartbeat Origin and Sequence of Excitation. Heart muscle is autorhythmic, or capable of spontaneous, rhythmic self-excitation (Fig. 10-8). Yet, even after it has been removed from the body and all nerves to it have been cut, certain cells have a faster rate of firing than others and, because of the junctions between cardiac cells, all the cells are excited at the rate set by the cell with the fastest autonomous rhythm. These special cells are the so-called pacers or pacemakers. The one with the fastest inherent rhythm consists of a small mass of specialized myocardial cells embedded in the right atrial wall near the entrance of the superior vena cava and is called the sinoatrial (SA) node. This is the normal pacemaker for the entire heart. In unusual circumstances, if another area of the heart becomes more excitable and develops a faster spontaneous rhythm than that of the SA node, then the new area determines the entire heart's rhythm.

The cells of the SA node make contact with the surrounding atrial muscle fibers as well as specialized internodal tracts. The wave of excitation spreads out in all directions from muscle cell to muscle cell across the right atrium and over the left atrium, and also via the tracts. Since it takes only about 0.1 second to reach the most distant part of the atria, both atria contract almost simultaneously. In the floor of the right atrium, the waves of excitation meet a second small mass of specialized cells, the atrioventricular (AV) node. This structure and the bundle of fibers from it (bundle of His) make up the only link between atria and ventricles, since the atria and ventricles are separated by the nonconducting connective tissue of the fibrous heart skeleton. Thus, excitation can travel from atria to ventricles only through the AV node and over the Purkinje fibers of the bundle of His. A malfunction of the AV node may completely dissociate atrial and ventricular contraction. Note, however, that at the AV node, the propagation of action potentials through the node is delayed by 0.1 second, allowing the atria to contract and empty into the ventricles before the latter start to contract.

After leaving the AV node, the impulse travels down on either side of the interventricular septum along specialized myocardial fibers (Purkinje), the right and left bundle branches of His, which finally spread through most of the right and left ventricular muscle. Due to the rapid conduction along the fibers, depolarization occurs in the right and left ventricular muscle cells almost simultaneously and a single, coordinated, simultaneous contraction takes place.

Heart pumping necessitates alternate periods of contraction and relaxation since prolonged contraction of cardiac muscle without relaxation would result in death. Thus, cardiac muscle has a long refractory period (rest period). In any excitable membrane, an action potential is followed by a period during which the membrane is insensitive to stimulation of any intensity (absolute refractory period). Following this delay is a second period during which the membrane can be depolarized but only by a more intense stimulus. The heart muscle rests almost ten times as long as skeletal muscle and cannot be excited to produce a summation.

Pathology. Disease commonly damages cardiac tissue and interferes with conduction through the AV node. Frequently, only a fraction of an atrial impulse is transmitted to the ventricles and the atrial rate of

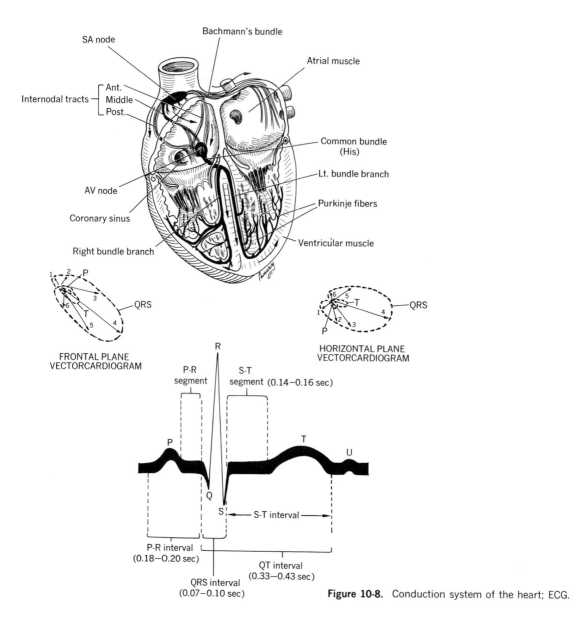

Figure 10-8. Conduction system of the heart; ECG.

Labels on heart diagram:

SA node

Bachmann's bundle

Atrial muscle

Internodal tracts — Ant. Middle Post.

Common bundle (His)

AV node

Lt. bundle branch

Coronary sinus

Purkinje fibers

Right bundle branch

Ventricular muscle

FRONTAL PLANE VECTORCARDIOGRAM

HORIZONTAL PLANE VECTORCARDIOGRAM

P-R segment

S-T segment (0.14–0.16 sec)

P-R interval (0.18–0.20 sec)

QRS interval (0.07–0.10 sec)

S-T interval

QT interval (0.33–0.43 sec)

80 beats per minute may only give us ventricular rates of 60 per minute (partial block). If there is complete block at the AV node, none of the atrial impulses get through. However, a portion of the ventricle just below the AV node usually begins to initiate excitation at its own spontaneous rate, which is quite slow (about 25 to 40 beats per minute). This is, however, out of synchrony with atrial contractions, which continue at a normal rate. Thus, the atria become ineffective as pumps since they contract against closed AV valves. Some people have transient, recurrent episodes of complete AV block and develop fainting spells due to

decreased cerebral blood flow, since the ventricles do not begin their own impulse generation immediately and cardiac pumping stops temporarily.

Partial AV block may be a normal adaptation. If, for some reason due to excess stimulation or ectopic foci or firing outside of the SA node, the atria beat at 300 beats per minute, ventricular rates at this level are inefficient since inadequate filling time develops. Here, the long refractory period of the AV node may prevent a number of impulses from reaching the ventricle and the latter can beat at a slower rate.

A prolonged or unusual conduction route in which the impulse always meets an area that is no longer refractory and thus keeps traveling around the heart is a so-called circus movement and may lead to continuous, disorganized contractions called fibrillation and even death.

Mechanical Events and the Cardiac Cycle

The heart, although housed in one organ, is actually two separate pumps: the right heart (right atrium and right ventricle) which pumps blood that has reached it from the rest of the body by way of the systemic veins, into the pulmonary artery, and to the lungs; and the left heart (left atrium and left ventricle), which pumps blood from the lungs via the pulmonary veins into the left atrium to the left ventricle, into the aorta, and out to the body. Normally the total volumes of blood pumped through the pulmonary and systemic circuits during a given period of time are equal; i.e., the right heart pumps the same amount of blood as the left heart.

The predominant feature of the cardiovascular system is the pumping of blood by the heart. In a resting, normal man, the amount of blood pumped simultaneously by each half of the heart is about 6 liters per minute. This may increase fivefold (30 liters per minute) under heavy work or exercise. The atria serve as a reservoir and can be considered as an extension of the venous system, whereas the ventricles provide the essential energy source for the pumping action. The pressure-actuated "flutter" valves at the entrance (right and left atrioventricular) and exit (pulmonary and aortic) of each ventricle ensure the flow of blood through the heart in one direction (unidirectional). Furthermore, fluid, including blood, flows from a re-

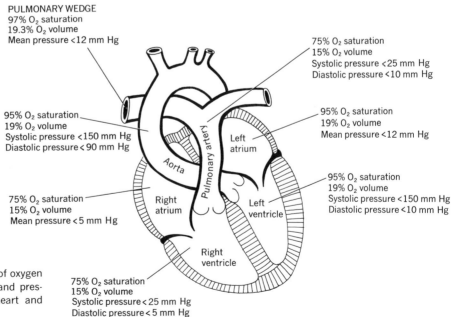

PULMONARY WEDGE
97% O$_2$ saturation
19.3% O$_2$ volume
Mean pressure <12 mm Hg

75% O$_2$ saturation
15% O$_2$ volume
Systolic pressure <25 mm Hg
Diastolic pressure <10 mm Hg

95% O$_2$ saturation
19% O$_2$ volume
Systolic pressure <150 mm Hg
Diastolic pressure <90 mm Hg

95% O$_2$ saturation
19% O$_2$ volume
Mean pressure <12 mm Hg

Aorta

Pulmonary artery

Left atrium

95% O$_2$ saturation
19% O$_2$ volume
Systolic pressure <150 mm Hg
Diastolic pressure <10 mm Hg

75% O$_2$ saturation
15% O$_2$ volume
Mean pressure <5 mm Hg

Right atrium

Left ventricle

Right ventricle

Figure 10-9. Values for percent of oxygen saturation, O$_2$ volume percent, and pressure in the chambers of the heart and great vessels.

75% O$_2$ saturation
15% O$_2$ volume
Systolic pressure <25 mm Hg
Diastolic pressure <5 mm Hg

gion of higher pressure to one of lower pressure (Fig. 10-9). Thus, each ventricle fills passively during its relaxation phase, or diastole, and empties actively during its contraction phase, or systole.

Cardiac Cycle

A single cycle is divided into two parts: ventricular contraction, or systole, and ventricular relaxation, or diastole (Fig. 10-10). The cycle takes about 0.8 second for a heart beating at 75 contractions per minute (cpm). Systole takes 0.28 second, while diastole takes 0.52 second. Note that diastole lasts almost twice as long as systole. When the heart rate increases, the period usually affected is the diastole filling phase.

The cardiac cycle takes place in the following sequence. Just before systole (late diastole), the intake AV valves are open because atrial pressure slightly exceeds ventricular pressure. The output aortic and pulmonary valves are closed because pressure in the vessels greatly exceeds ventricular pressure at this point. Thus, the ventricles receive blood from the atria throughout most of diastole and not only when the atria contract (in fact, about 80 percent of the ventricular filling takes place even before atrial contrac-

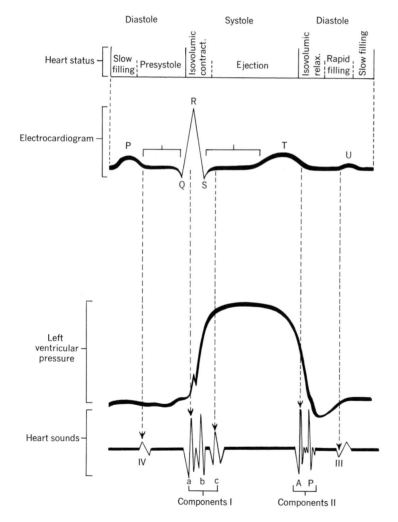

Figure 10-10. Cardiac cycle correlated with ECG, left ventricular pressure, and heart sounds.

tion). The vessel pressure, however, is slowly falling as blood moves away from the heart and down the arteries from a previous outflow. As the ventricle fills, its pressure gradually rises. Near the end of diastole, the SA node "fires," the atria depolarize (one sees the P wave on an electrocardiogram [ECG]), and the atria contract. With this contraction, an additional small volume of blood (end-diastolic volume) is added to the ventricle (about 20 percent).

As systole begins, the wave of depolarization passes through the ventricles (QRS complex on the ECG), and ventricular pressure begins to rise sharply. This immediately closes the AV valves to prevent backflow into the atria, producing the first heart sound, which signals the onset of ventricular systole.

Ventricular pressure continues to rise, taking about 0.05 second to reach the level of aortic pressure. During this period both intake and output valves are closed and no blood leaves or enters the ventricles so that the volume remains constant; this is isometric contraction. When ventricular pressure exceeds aortic and pulmonary pressure, the output valves open, blood begins to flow out of the ventricle into the vessels, and we have the ejection phase beginning—rapid at first but tapering off later. Vessel pressure now rises. The ventricles do not empty entirely but leave behind a small end-systolic volume.

When ejection begins, atrial pressure usually drops sharply since the atrial floor is pulled down by the contracting ventricles. However, it gradually rises as it receives blood from the veins. After a 0.23-second ejection, the myocardium begins to relax and ventricular diastole begins.

During early diastole, ventricular contraction ceases, the muscle relaxes, and ventricular pressure falls below aortic and pulmonary pressure and the semilunar valves close, producing the second heart sound, which signals the onset of diastole. After output valve closure, ventricular pressure continues to fall rapidly but needs about 0.08 second to reach atrial pressure level. During this 0.08-second interval, all valves are closed, ventricular volume remains constant, and we have isometric relaxation. When ventricular pressure falls below atrial pressure, the intake valves open, blood flows into the

ventricle, and the filling phase begins. Note that filling of the ventricle early in diastole ensures that filling of the ventricles will take place even during rapid heart rate when this period is shortened.

The right ventricular and pulmonary artery pressures are considerably lower in systole than are the left ventricular and aortic pressures since the pulmonary system is a low-pressure one. Yet the right ventricle ejects the same amount of blood as the left.

The pressure in each ventricle rises to a maximum during systole (systolic pressure) and falls to a minimum during diastole (diastolic pressure). These pressures are written as systolic/diastolic, and in an adult, normal man at rest, ventricular pressures are approximately $\frac{24}{2}$ mm Hg for the right and $\frac{124}{3}$ mm Hg for the left ventricle.

Heart Sounds. Two heart sounds (Fig. 10-10) are normally heard through the stethoscope on the chest wall. The first sound, a low-pitched lub, is associated with closure of the AV valves at the beginning of systole. The second sound is a high-pitched dub, which is associated with closure of the semilunar valves at the beginning of diastole. The sounds are normal and due to vibrations of valvular closure. Heart murmurs, on the other hand, can be a sign of heart disease and are due to turbulent flow that is produced by blood flowing through a narrowed valve or backward through a damaged leaky valve, or due to a hole between ventricles or atria in the septa separating them.

Pulmonary and Systemic Circuits

Ventricular Volumes and Flow Velocities. Both circuits consist of a distributing system of arteries, an exchange system of capillaries, and a collecting system of veins. As the arteries leave the heart, the vessels branch in such a way that the individual vessel diameter progressively decreases as we move away from the heart; yet the number of vessels increases so enormously that the total cross-sectional area of the capillary bed is over 600 times that of the aorta. As the vessels enter the venous system and return toward the heart, they get larger and their numbers decrease, as does the cross-sectional area of the vascular bed. Thus, the veins are of larger diameter than corresponding

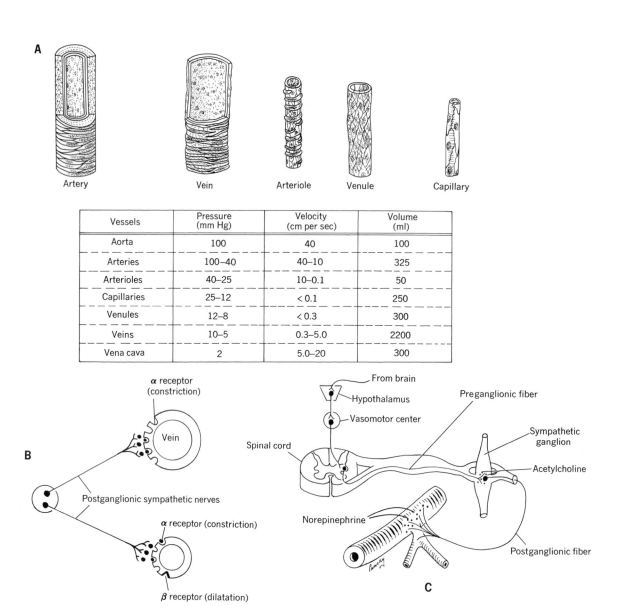

Vessels	Pressure (mm Hg)	Velocity (cm per sec)	Volume (ml)
Aorta	100	40	100
Arteries	100–40	40–10	325
Arterioles	40–25	10–0.1	50
Capillaries	25–12	< 0.1	250
Venules	12–8	< 0.3	300
Veins	10–5	0.3–5.0	2200
Vena cava	2	5.0–20	300

Figure 10-11. **A.** Hydrodynamic system is shown with blood passing in arteries at high pressure, maintained at high pressure by arteriolar resistance, and entering venules (after being in capillaries and permeating tissues) at low pressure. The average values for pressure, velocity, and volume for various parts of the vascular tree are noted in the table. **B.** Sympathetic nerves control the muscle of the blood vessel. Norepinephrine released at nerve endings affects the "alpha" receptors, resulting in constriction. If "beta" receptors are stimulated, dilation occurs. Arteries have both receptors, veins only the alpha. **C.** The stimulus for an arteriolar contraction begins in the brain; it then passes to the cord and along the nerve junction to the sympathetic ganglion, where it releases a chemical messenger, acetylcholine, that stimulates nerves; the signal is transmitted to the muscle wall of the vessel. Here another messenger, norepinephrine, causes the muscle to contract. Various antihypertensive drugs act at various sites along this pathway (e.g., oral diuretics act on the smooth muscle, methyldopa acts on the postganglionic fiber, and hydralazine or rauwolfia acts on the hypothalamus.)

arteries (Fig. 10-11). The total blood volume is normally distributed between the two major circuits in a ratio of about $3:1$ or $4:1$ (systemic to pulmonary). If the total blood volume is 6 liters, 5 liters occupies the systemic and 1 liter the pulmonary circuits. In each respective circuit, about 15 percent of the blood occupies the arterial portion, 5 percent is in the capillaries, and 80 percent is in the veins.

The volume of blood pumped by a ventricle per stroke is called its stroke volume and frequency (F). Since the latter is the same for both ventricles,

$$QR \text{ (liters/minute)} = F \text{ (frequency)} \times V_{SR} \text{ (or } V_{SL}\text{)}$$

The "steady state" for the cardiovascular system requires that $Q_R = Q_L$ (right output = left output). Thus, we usually speak of a single cardiac output (Q), which is the same volume output of either ventricle, previously indicated as being about 6 liters per minute. Since the frequency of both ventricles is also always identical, $V_{SR} = V_{SL}$, and we can also speak of a single stroke volume identical with that of either ventricle. Therefore, the normal adult man at rest is said to have a cardiac output of approximately 6 liters per minute with a frequency of 75 contractions per minute (cpm) and a stroke volume of about 0.08 liter (80 ml) output with every contraction or stroke. Transient variations may occur in which the blood volume in the pulmonary circuit must increase at the expense of the systemic ($Q_R > Q_L$) or vice versa. This must be brief, however, and the system does have a built-in mechanical self-regulation that limits the duration of such periods.

The maximum heart output occurs during heavy muscular exercise. Decompensation and failure are a result of maintaining a constant myocardial strength but a progressively decreasing muscle strength. The first seldom happens in man. The latter, however, develops slowly in chronic heart disease, and one finds a large heart performing a small amount of useful work.

The flow velocity or speed of flow varies in different parts of the system as a consequence of the differences in the cross-sectional area of the various parts of the vascular bed. This is an inverse relationship; ie., as the vascular bed increases, the flow velocity slows down (and vice versa). Thus, flow in the capillary bed is about $\frac{1}{1000}$ as great as in the aorta (42 cm per second versus 0.05 cm per second).

Circuit Pressures. In both pulmonary and systemic circuits pressures are highest where the pulmonary artery and aorta come off the heart and decrease as one passes out through the vessels. The drop in pressures represents energy lost in overcoming frictional resistance to blood flow. The resistance is greatest in the arterioles, where the pressure drop is greatest. A drop also occurs across the capillaries. The larger vessels offer little flow resistance (Fig. 10-11).

The pulmonary arterioles and capillaries offer less resistance than do the systemic, since less energy is needed to overcome frictional resistance in this part of the circuit. Thus, the mean pulmonary pressure is lower than the systemic pressure; it is about 14 mm Hg compared to 100 mm Hg.

Arterial pressures in the systemic and pulmonary circuits vary during the cardiac cycle, being maximal in systole and minimal in diastole. In a normal, resting man, pulmonary arterial systolic pressure averages 22 mm Hg and the diastolic is 8 mm Hg ($\frac{22}{8}$). Systemic arterial pressures average $\frac{120}{80}$ mm Hg. If the arteries were rigid tubes and the cardiac output did not change, arterial pressure would rise to a very high systolic peak and remain at zero throughout diastole. The volume flow rate through the arterioles would reflect ventricular volume output and would also be zero throughout diastole. Even if the arteries were elastic but flow resistance through the arterioles were negligible, no effective filtering action would occur. Thus, filtration (moving substances out of the vessels and into the cells and vice versa) requires a combination of arteriolar resistance (R) and vessel elasticity called compliance (C).

Systemic Circuit and Flow Distribution. The lungs are perfused by the total cardiac output (Q) from the right ventricle, but the many components of the systemic system receive only a portion of the total output, depending on their respective flow resistance (Fig. 10-11).

The resistance includes not only the arterioles but also the capillaries and veins, although discussions commonly refer to only arteriolar resistances (which do contribute the major share). The lowest resistances and highest blood flows occur in the splanchnic viscera (liver, spleen, gut) and kidneys and are intermediate in the brain and skeletal muscles. The highest resistance and lowest flow take place in the skin, the heart, and a "miscellaneous group."

If one considers organ weights, the kidney is in a class by itself while the heart ranks second instead of last. More significant functionally than the weight of an organ is its rate of metabolism or O_2 consumption. In this respect, the kidneys and skin have a relatively high blood flow in relation to their metabolism since the kidney performs a physical filtration process and the skin is concerned with heat dissipation. The value for the heart is quite low.

The cardiovascular system must be able to adjust blood flow rates to meet normal variations in tissue needs as well as to ensure that flows to the most critically vital organs, such as the brain and heart muscle, take first priority under emergency conditions. To manage this, the heart and its circuits are manipulated by neural and humoral factors. Feedback loops also exist to inform the "controller" of the adequacy of its operations.

Control of Heart Rate

The discharge of the SA node can take place spontaneously in the absence of any hormonal or nervous influences. However, it is under the continual influence of both. Many parasympathetic (vagus) and sympathetic fibers end on the SA node and other areas of the conducting system (Fig. 10-8). Stimulation of the parasympathetics to the SA node results in a slowing of the heart, and if strong enough even stops it. Stimulation of the sympathetics increases heart rate, and cutting the nerves slows the heart. In the resting state, the parasympathetic influence is apparently dominant.

Factors other than cardiac nerves can alter the heart rate. Epinephrine, a blood-borne sympathetic mediator hormone of the adrenal medulla, speeds the heart. Temperature, plasma electrolyte concentration, and

other hormones also affect the heart rate but are of lesser importance.

Electrocardiogram

The electrocardiogram (ECG) is a tool for evaluating the electrical phenomena in the heart (Figs. 10-8, 10-10). The action potentials of heart muscle, like batteries, cause current flow throughout the body. A minute fraction of these currents, which are intracellular action potentials at sites in the heart itself, can be detected at the body surface by placing two pickup electrodes (small metal plates) on the surface of the body (since the surrounding tissues conduct electricity) and connecting them to a voltmeter. With provision for graphic recording, this record is called an electrocardiogram (ECG)—a potential difference between two arbitrary points on the body surface plotted as a function of time during the cardiac cycle. Figure 10-10 illustrates a typical normal ECG recorded as the potential difference between the right and left wrists. The first wave, P, represents atrial depolarization.° The second complex, QRS, occurs about 0.1 to 0.2 second later and represents ventricular depolarization. The final wave, T, is ventricular repolarization. Because many myocardial defects change normal impulse propagation and thereby the shapes of the waves, the ECG is one tool used for diagnosing heart disease.

Observations between the voltages recorded from pairs of electrodes placed at standardized positions can be correlated with the spread of excitation through the cardiac tissue and can show abnormal electrical events occurring there. Vectorcardiography is concerned with the relationships between the different ECG tracings taken simultaneously at different sites on the same person and is based on the three-dimensional geometry of electrode placement and voltage distribution in the body as a "volume conductor." It emphasizes the spatial (geometric) features of the spread of excitement (Fig. 10-8).

° Depolarization is the breaking down of polarized, semipermeable membranes in the conduction of impulses. Polarization is the acquistion of electrical charges of opposite sign, as across semipermeable membranes.

The usual sequence of action potential development is that (1) the SA node fires first, (2) the atria depolarize and repolarize, (3) the AV node is triggered by the atrial depolarization, and (4) the Purkinje fibers conduct very rapidly to the ventricles, which depolarize and repolarize.

Leads. Standard positions have been designated for placement of the pair of pickup electrodes on the body surface—the ECG leads. Originally only three leads, the standard limb leads of Einthoven, were used: the right wrist, left wrist, and left ankle, with the right ankle serving as an electrical ground to minimize interference. They are also referred to as leads I, II, and III. Lead I is between the right and left wrists. An upright deflection in the ECG indicates that the left wrist is (+) with respect to the right. Lead II is between the right wrist and left leg, and again an upright deflection indicates that the left ankle is (+) to the right wrist. Lead III is between the left wrist and left ankle, with an upright indicating that the ankle is (+) to the wrist. The magnitude and polarity of the deflections depend on the relationship between the direction of the electrical axis of the heart and the direction of the line connecting the two measurement electrodes used to get the ECG deflection. Other leads are also utilized.

Pulse

In our physiologic system, we have elastic arteries distributed over a considerable linear distance (e.g., about 1.5 m [5 ft] from the aortic root to the dorsal pedis artery of the foot). This length is significantly large compared to the speed at which pressure changes are transmitted along the system. Thus, the arterial pressure varies not only with time but also with space, so that pressure recorded at the aortic arch is not the same as that recorded simultaneously at the radial (wrist) or femoral (thigh) arteries. Pressure changes originating at the aortic root are transmitted as a wave to the peripheral arterioles, with a finite velocity that can be measured. The arterial pulse then is a phenomenon dependent on the distribution and nature of the elastic arterial system and is very complex (as are all wave phenomena) (Fig. 10-11). With age, pulse wave

Table 10-2. Pulse Wave Velocity vs. Age

Age (Years)	Pulse Wave Velocity (Meters/Second)
27	8.8
51	13.0
76	13.7

velocity increases and is about 100 times faster than the linear velocity of the blood flow itself, which is about 10 cm per second (Table 10-2).

Circulation

The Blood

Every cell in the body is surrounded by tissue fluid (extracellular). Since the body's cells are so closely packed, the tissue fluid between them is a much smaller volume than the volume of the cells served. This might pose a problem in removing cellular wastes and supplying the cells with needed nutrients without depleting or polluting them, if it were not for the fact that the tissue fluid itself is in close contact with another fluid system capable of rapidly supplying and removing, namely, the blood. The route the blood follows is the network of blood vessels, and the force that drives it on and through the system is provided by the pumping action of the heart. Blood pressure at the arterial end of the capillary tends to force some blood out of the capillaries while osmotic pressure tends to suck the fluid back (Fig. 14-16, p. 497). Blood pressure (40 mm) is higher than osmotic pressure (25 mm), resulting in a net outward pressure of 15 mm; thus, a portion of the blood (lymph) is forced into the intercellular spaces, leaving the red blood corpuscles, white blood corpuscles, and most proteins behind. From the intercellular space, oxygen and nutrients diffuse into the cells. At the venous end of the capillary, the blood pressure falls (now 15 mm), while the osmotic pressure remains constant (25 mm), resulting in a net inward pressure of 10 mm; some lymph passes back into the capillaries, together with CO_2 and nitrogenous wastes that have diffused out of the cells.

More fluid leaves the capillaries than returns to them, but the excess in the intercellular spaces is drained into the lymphatic vessels. The latter, in turn, eventually drain into the venous system.

The blood participates in practically all body functions and is a very complex part of the transport system.

If you weigh 70 kg (150 lb), your body contains about 5.67 liters (1½ gal) of blood, of which 55 percent consists of an almost colorless watery fluid called plasma, in which are suspended specialized cells. If one takes blood out of the body and does not treat it, it gels or clots by itself. If one centrifuges the blood, the "gel" is removed and one is left with the serum. The latter differs from plasma in that protein and fibrinogen have been removed. The cells of blood consist of the red blood cells (erythrocytes), which give blood its color; the white blood cells (leukocytes); and the platelets, which assist in blood clotting (see Chap. 16). There are about 650 red blood cells to every white blood cell (these are discussed in detail in Chaps. 11 and 16).

Under normal circumstances the motion of the blood keeps the cells dispersed in the plasma; however, if the blood is allowed to stand or if a sample is taken and centrifuged, the cells "sink" to the bottom. By this means, the percentage of total blood volume made up by the cells, the hematocrit, is determined (Fig. 10-12). The normal hematocrit is about 45 percent. The total blood volume of an average man is about 8 percent of his total body weight. Thus, for a 70-kg man, the blood volume would be 5.6 to 6.0 kg × 1 liter (since 1 kg of blood occupies about 1 liter), or approximately 6 liters. Since the hematocrit is 45 percent, the total cell volume is equal to 6 × 0.45 = 2.70 liters. The plasma volume is 6.0 − 2.7, or 3.3 liters.

Plasma. Plasma is a clear, complex liquid consisting of many organic and inorganic substances dissolved in water. Whole plasma contains 7 to 8 percent protein, which is its most abundant solute by weight. The plasma proteins vary greatly in function and structure but are classified into two major groups (according to chemical and physical reactions): the albumins and the globulins. The former are three and four times more

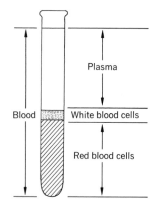

Figure 10-12. Whole blood after spinning down in centrifuge. Hematocrit can be read as: $\dfrac{\text{Red cells}}{\text{Blood}} \times 100 = \%$ HT.

abundant and usually of smaller molecular weight. The plasma proteins are synthesized by the liver, with the exception of the gamma globulins, which are formed in the spleen and lymph nodes. The plasma proteins serve many important functions but are not normally taken up by cells and used as metabolic fuel. In contrast, glucose uses the plasma as a vehicle of transport but functions in cells. The plasma proteins function in the plasma or, under certain conditions, in the interstitial fluid. Using free-boundary or gel electrophoresis at pH 8.6, five types of protein are found to migrate to the anode in normal human plasma. In order of decreased mobility, they are albumin, alpha (α) 1, alpha (α) 2, beta (β), and gamma (γ) globulin (Table 10-3).

Table 10-3. Blood Proteins via Electrophoresis

Protein*	Molecular Weight	Amount (%) In Human Plasma (% of Total Protein)
Albumin	69,000	55
α_1	—	5
α_2	—	9
β	93,000	13
γ	160,000	11
Fibrinogen	340,000	7

*Prothrombin and fibrinogen are two other proteins in suspension and help in the clotting of blood, which is essential to life and prevents one from bleeding to death.

Table 10-4. Essential Blood Electrolytes

Constituent	Grams/Liter	Millimoles/Liter
[Protein]	[70]	[2.5]
Sodium (Na^+)	3.39	144
Chloride (Cl^-)	3.55	100
Bicarbonate (HCO_3^-)	1.50	25
Potassium (K^+)	0.17	4.4
Calcium (Ca^{++})	0.10	2.5
Phosphate (HPO_4^{--} or $H_2PO_4^-$)	0.10	1.0
Magnesium (Mg^{++})	0.04	1.5

Under normal conditions the blood proteins never leave the vessels and remain as permanent ingredients of the plasma. Other substances in the blood are only passengers and are continually entering and leaving the vessels and cells of the body. These include gases, such as oxygen and carbon dioxide; mineral salts, for building bones; nitrogenous substances, from which proteins are built; waste products of protein breakdown; sugar, in the form of glucose; hormones; and enzymes. Of the large number of mineral electrolytes, some of the essential ones are shown in Table 10-4.

Note that the protein molecules are so large that, in comparison to sodium, for example, a very small number of them greatly outweighs a much larger number of sodium ions. Nevertheless, the osmolarity (water concentration) of a solution depends on the number, not the weight, of the solute particles; thus, sodium is the most important determinant of total plasma osmolarity. In fact, all these constituents play a role in the blood's osmotic pressure and the exchange of materials with the tissue fluids.

Blood Cells. All blood cells originate from undifferentiated mesenchymal cells. From these "stem" cells, clones of cells differentiate and ultimately appear in the circulating blood as red blood cells, various types of white blood cells, and platelets. The earliest cells of each cell line have similar morphologic characteristics and cannot be differentiated from each other by appearance. They are given names such as myeloblast, lymphoblast, or rubroblast, depending on the tissue in which they are found, the cells with which they are associated, and the definitive cell they finally become.

Immature cells as a class are generally large and become progressively smaller as they mature. The nuclei of young cells of a maturation sequence are large and relatively large in relation to the cytoplasm. With aging, the absolute and relative size of the nucleus decreases and in the red-cell series the small and degenerated nuclei in older cells are actually lost.

The cytoplasm of the primitive cell stains predominantly blue and contains much RNA (affinity for basic blue dye). As the cytoplasmic structures and secretory products are produced, the cytoplasm tends to be more red since nuclear chromatin strands of immature cells have DNA, which has an affinity for acidophilic red dye. Thus, the most reliable criterion for the age of the cell is not its size or color but the structure of its nuclear chromatin. One of the signs of immaturity in blood cells, on the other hand, is the presence of nucleoli in the nucleus. These are signs of metabolic activity and growth.

The shape of the cell is influenced by the mechanical trauma it is subjected to, the pressure of surrounding cells, and its ameboid activity. Primitive cells are fixed by cytoplasmic extensions in the ground substance, but when torn away (bone marrow aspiration) the edges are frayed and jagged. Mature and free cells in the circulation have smooth margins. Fixed tissue and early cells that move slowly have oval or round nuclei only slightly indented. Active, motile, and mature cells (monocytes and granulocytes) have indented, kidney-shaped, lobulated, or segmented nuclei. The nuclei, however, of red cells and plasma cells are always round.

Immature cells that are metabolically active have a "light zone" near their nucleus, the Golgi area, which contains smooth endoplasmic reticulum and centrioles as well as mitochondrial aggregations.

Granules are not present in stem cells (blasts). As the cells mature, granules become dark and blue; with maturity, they become lighter and more red. Manufactured products in the cells, such as globulin in plasmocytes or hemoglobin in red cells, are signs of maturation.

Phagocytosis is a manifestation of functional activity characteristic of differentiated cells. The cytoplasmic, granular, and nuclear characteristics are usually well synchronized in normal cells, but the steps may be out of step in pathologic conditions. (See Chap. 11 for a detailed discussion of red blood cells and Chap. 16 for information concerning white blood cells.)

Blood Vessels

Despite the importance of the heart, the blood vessels (Fig. 10-11) play critical roles in the circulation of blood. They are not merely inert "plumbing" but have characteristic function and structure.

The arteries (Greek *arteria*, containing air) carry blood away from the heart; the veins (from the Latin *vena*) return blood to it. The arteries have thick walls containing much elastic tissue. They have smooth muscle as well; however, because there is little alteration in the state of the muscle activity, the arteries have been considered to be elastic "tubes." Since the arteries have large radii, they offer low resistance. They also serve as a pressure reservoir, driving blood through the tissues.

The contraction of the ventricles ejects blood into both systemic and pulmonary arteries during systole. If an equal amount of blood were to come out of the arteries into the arterioles, the total blood volume in the arterioles would remain constant and the pressure not changed. However, only one third of the stroke volume actually passes out of the arteries in systole. The rest distends the arteries and raises the arterial pressure. Thus, when the ventricular contraction ends, the stretched vessel wall recoils passively and the arterial pressure continues to drive the blood through the arterioles. As blood finally leaves the arteries, the pressure falls slowly. The next ventricular contraction, however, takes place while some blood is still in the arteries, to keep them partially stretched so that the pressure never falls to zero. The maximum pressure is reached at ventricular ejection (systolic pressure). The minimum pressure occurs before ventricular contraction (diastolic pressure). Thus, systolic/diastolic = 125/75 mm Hg. The pulse felt in the artery is due to the difference between the pressures of $125 - 75 = 50 = $ pulse pressure. The latter is altered by decreased heart rate, decreased arterial distens-

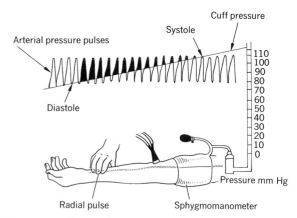

Figure 10-13. Taking blood pressure. When pressure is increased over arterial blood pressure, arteries under the cuff are occluded and no pulse is felt at the wrist. With release of pressure, systolic pressure exceeds cuff pressure and blood enters the arteries below the cuff and pulse is felt. When the pulse is heard or felt, the pressure is at systolic. As the cuff pressure decreases further, the sound increases and is suddenly muffled at diastolic pressure when the arteries are open for the entire pulse wave.

ibility (arteriosclerosis), or increased stroke volume due to arterial stretching.

Arterial pressure changes constantly. Only complex methods can measure true mean arterial pressure, but it can be calculated as follows: diastolic pressure $(75) + \frac{1}{3}$ of pulse pressure $(50) = 91$ mm Hg. Both systolic and diastolic blood pressure are measured with a sphygmomanometer (Fig. 10-13).

The large arteries nearest the heart divide into smaller divisions, which in turn divide into still smaller ones, and so on. The smallest arteries are known as arterioles. These lead into vessels of very narrow caliber called capillaries (Latin *capillus*, a hair). The capillaries branch and form a network throughout the area supplied by the artery. After passing through this network, the blood collects in the venules, which unite to form small veins. These, in turn, combine to form larger vessels, until eventually the great veins open into the heart. The veins are equipped, at intervals, with valves that prevent reverse blood flow.

Arteries. Arteries (Fig. 10-14) are constructed to continuously distribute blood from the heart to the tissues and are under the "steady" strain of blood

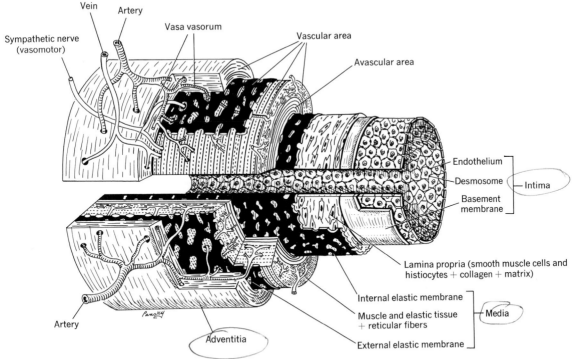

Labels on figure:
Vein
Artery
Sympathetic nerve (vasomotor)
Vasa vasorum
Vascular area
Avascular area
Endothelium
Desmosome — Intima
Basement membrane
Lamina propria (smooth muscle cells and histiocytes + collagen + matrix)
Internal elastic membrane
Muscle and elastic tissue + reticular fibers — Media
External elastic membrane
Artery
Adventitia
Pansky

Figure 10-14. Structure of a blood vessel. (After studies by Nelter and Garven.)

pressure. They perform as "active organs" and not just "tubes" that propel blood passively. Work energy is obtained from the vessel wall itself, which contains cells with organelles for energy production as well as synthesis of lipids, carbohydrates, proteins, and muco-polysaccharides.

Under the electron microscope one sees that the three arterial layers (intima, media, and adventitia) differ from artery to artery and in various segments of the same vessel, depending on the function of any given segment. One notes three types of arteries: (1) large or elastic, (2) muscular or medium sized, and (3) small arteries and arterioles. Atherosclerosis affects vessels in only the first two categories.

LARGE, ELASTIC ARTERIES. The aorta and pulmonary vessels are vessels that fall into this group. They serve to keep the vascular stream in a constant flow by maintaining pressure in the arterial system. This is possible due to the elastic tissue in the vessel media,

which maintains a degree of tension that opposes the expanding action of the blood pressure, cushions the force of the heart pump, and absorbs the impact of blood delivered into the vessel lumen.

Intima. The intima (inner vessel lining), is thin at birth and consists of a layer of endothelial cells separated from an internal elastic lamina by a narrow layer of connective tissue. With age, it thickens and, later in life may equal or even exceed the size of the media.

Media. This consists of concentric layers of smooth muscle cells that are interposed between thick, fenestrated plates of elastic lamellae arranged in a parallel or slightly spiral manner. All are embedded in a ground substance rich in acid mucopolysaccharide. The outer one half to two thirds of the media is invaded by penetrating branches of blood vessels, the vasa vasorum.

Adventitia. The adventitia (outer layer) is usually thin but varies from vessel to vessel and is made up of irregularly arranged, mature collagen bundles, scattered fibroblasts, and a few elastic fibers as well as the vessels from which the vasa vasorum arise.

MEDIUM-SIZED OR MUSCULAR ARTERIES. These arteries distribute blood from the elastic arteries to the organs and tissues (i.e., coronary arteries, facial, etc.). The changes of blood volume depend on the ability of these vessels to alter their luminal size (regulated by nervous control of the smooth muscle of the media).

Intima. The intima is usually thinner than in the elastic vessel. A variation is the coronary (heart) arteries, where the intima is thicker than the media. However, the intimal layers in other arteries, except for thickness, are like those seen in the elastic vessels.

Media. The media consists of circularly disposed smooth muscle with only a small number of fine collagen and elastic fibers interposed between them. A small amount of ground substance acts as a lubricant for embedded elements. No vasa vasorum are seen in the media here.

Adventitia. The adventitia varies in thickness from one half to two thirds of the media and resembles that layer in the elastic arteries but there are more elastic fibers.

SMALL ARTERIES AND ARTERIOLES. Small arteries and arterioles (Figs. 10-15, 10-16), particularly the latter, are responsible for control of blood-flow distribution. Their walls contain smooth-muscle fibers with little to no elastic tissue.

Each organ receives only a portion of the heart output. The digestive tract (and liver), kidneys, and brain receive the most blood. Where the flow to the brain appears constant, the muscle tissue supply varies

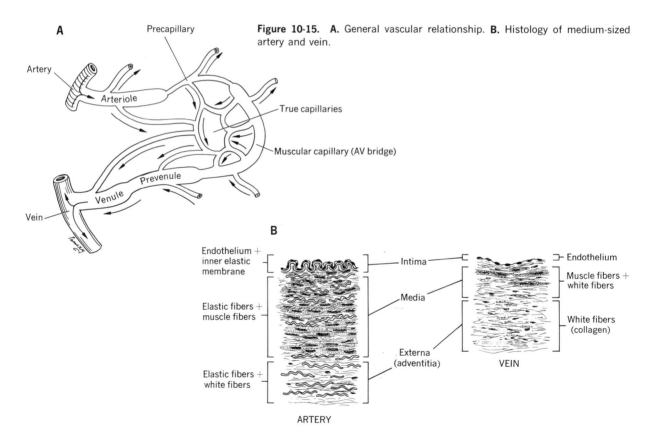

Figure 10-15. **A.** General vascular relationship. **B.** Histology of medium-sized artery and vein.

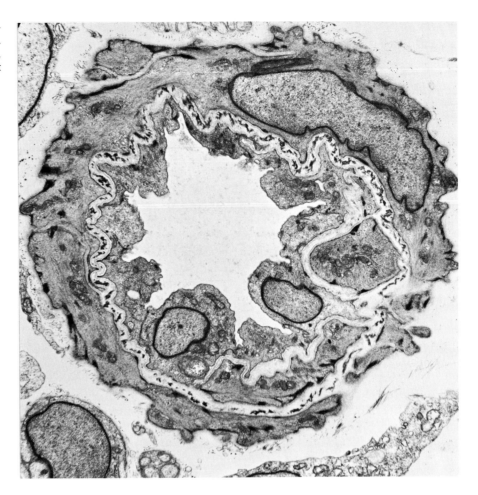

Figure 10-16. Electron micrograph of an arteriole. ×7333. (Courtesy of Dr. G. Colin Budd, Medical College of Ohio at Toledo.)

with muscle activity. Thus, with exercise, blood to muscle and skin increases, kidney flow decreases, but brain flow is unchanged. Cardiac output is distributed to the organs and tissues according to the needs and functions at any given time.

The major arteries are so large that they contribute little resistance to flow. The smaller terminal ones offer some resistance, but slight. The arterioles, on the other hand, are narrow enough to offer considerable resistance. Their size is under precise physiologic control and, since their walls have little elastic tissue but much smooth muscle, which can relax or contract, blood-flow distribution depends on the degree of arteriolar smooth-muscle constriction in each organ and tissue.

Increased blood flow due to increased activity (active hyperemia) is the direct result of arteriolar dilation in the active organ and does not depend on the presence of nerves or hormones but is locally mediated. It may be due to a locally decreased oxygen concentration as a result of an increased use of oxygen by more active cells, increased local metabolites such as CO_2 or hydrogen ions, and so forth. All these chemical changes apparently act directly on the arteriolar smooth muscle, causing it to relax (dilate). This active hyperemia is most common in skeletal muscle, heart muscle, and the gastrointestinal tract. Injury to tissue causes local release of histamine, which causes arteriolar smooth muscle to relax locally.

Most arterioles are supplied by many sympathetic, postganglionic nerve fibers, which usually release norepinephrine, which acts on the vessel smooth muscle to cause vasoconstriction. Only the arterioles of the heart and brain are not significantly influenced by these constrictor fibers. Dilation is achieved by decreasing the rate of sympathetic activity below basal level since the nerve fibers are continually discharging at some "finite rate." Thus, control of the sympathetic fibers can accomplish either dilation or constriction. These nerves are involved with reflexes that maintain an adequate blood supply to the organs rather than local coordination of metabolic needs and blood flow. It should be noted that there exists a group of sympathetic nerves (not parasympathetic) that cause arteriolar dilation and increased flow. These so-called vasodilator fibers respond apparently only to stress or exercise.

Except for the blood vessels of the genital tract, there are no known parasympathetic nerves to arterioles.

Capillaries. The exchange of nutrients and metabolic end products takes place at the capillaries (Fig. 10-11). The rest of the vascular tree sees to it that there is an adequate blood flow through the capillaries. No cell of the body is more than $0.125 \mu m$ (0.005 in.) from a capillary; thus, diffusion distances are small and exchange efficient. There are about 60,000 miles of capillaries (laid end to end) in an adult, with each individual one being about 1.0 mm long. All filled with blood, the capillary network would hold about 7000 ml of blood but it usually contains only about 5 percent of the total circulating blood. Thus, most of the capillaries, at any moment in time, are closed.

The capillary is a thin-walled, single-celled "chamber" of epithelial cells called endothelium. It has no elastic tissue, connective tissue, or smooth muscle to interfere with the passage of water or solutes. There are no pulsations here, and blood passes through as a result of pressure created by the pumping heart. The thin capillary membrane acts like it has small pores or pinocytic vacuoles through which water and solute particles (smaller than proteins) pass easily. The permeability varies from area to area (highest in the liver,

lowest in the brain), but nowhere can red blood cells permeate normally.

Blood enters the capillaries from the arterioles. There are two types of capillaries: the arteriovenous (AV) and the true capillaries. The former connect the arterioles and venules directly, whereas the latter are those across which exchange of materials from the cell occurs. Where the true capillary begins, one does find a ring of smooth muscle, the precapillary sphincter, which opens and closes to permit intermittent flow through any given capillary. The sphincters function with the arteriolar smooth muscle to regulate total flow through the tissue capillaries as well as the number of capillaries functioning at any one time.

The capillary, although it offers resistance to flow, does afford such a huge total number and cross-sectional area that the total resistance of all is less than that of the arterioles and, since there is no smooth muscle, the radius is not under active control. Blood flows through the capillaries so slowly (0.07 cm per second) that there is time for exchange between blood and tissue.

CAPILLARY DIFFUSION. Tissue cells do not exchange materials directly with blood. The interstitial fluid acts as an intermediary. Nutrients diffuse across the capillary wall into the interstitial fluid and then enter the cell. The tissue fluid is called lymph and has almost the same composition as plasma except that the larger particles such as blood proteins or red cells cannot filter through and are not normally found in the lymph. Thus, two membrane transfers are involved: across the capillary wall and across the cell membrane. Transcapillary diffusion takes place primarily in one direction because of diffusing gradients between the blood and interstitial fluid. (There is no active transport of solute across the capillary wall; it is by simple passive diffusion.) Fat-soluble substances penetrate cell membranes easily and may pass directly through the endothelial capillary cells; others may pass through or between endothelial cells. The capillary wall is highly permeable to water and all the solutes of the plasma, with the exception of the plasma proteins, and behaves like a highly porous filter through which protein-free plasma moves by bulk flow, ultrafiltration under a hydrostatic pressure gradient due to a difference between pressure inside and outside the capillary.

All the plasma from the capillaries does not filter out into the interstitial space, because of the principle of osmosis. Both plasma and interstitium contain large amounts of low-molecular solutes (e.g., sodium, chloride, glucose, etc.). They all have similar concentrations in both areas since the capillary wall is very permeable to all. In contrast, plasma proteins diffuse across the wall very slightly and there is a low interstitial-fluid level. Thus, the difference in protein concentration between the inside and outside of the capillary means that the water concentration of the plasma is lower than the interstitial fluid, creating an osmotic flow of water from the interstitium into the capillary. Along with water are carried the dissolved low-molecular solutes. It is osmotic flow that maintains the concentrations of the solutes in plasma and interstitial fluid. Thus, hydrostatic pressure favors movement out of the capillary whereas water-concentration differences result in protein-concentration differences and favor osmotic movement back into the capillary.

Hydrostatic pressure at the beginning of the capillary is about 40 mm Hg. This decreases (due to resistance to flow) to 15 mm Hg at the end of the capillary. The interstitial pressure is near 0 mm Hg. The osmotic flow pressure from interstitium to capillary is about 25 mm Hg. Thus, at the first part of the capillary hydrostatic pressure is greater than osmotic pressure and fluid passes out. In the last portion, the reverse is true and fluid passes back into the capillary (absorption). Thus, over all, they cancel each other out and there is little net loss or gain of fluid.

In disease (e.g., liver disease), however, protein synthesis decreases, plasma protein concentration is lowered, plasma water concentration increases, net filtration into the interstitial space results, and fluid accumulates in the interstitial space (edema). It can be seen that it is the capillary filtration (absorption equilibrium) that determines the distribution of the extracellular (interstitial) fluid volume between the interstitial and vascular compartments.

ATHEROSCLEROSIS. Atherosclerosis (hardening of the arteries) is the most common arterial disorder. It is the major cause of morbidity and mortality in many nations. It is a disorder of the vascular wall, specifically the intima of large (elastic) and medium-size (muscular) arteries. It is characterized by focal intimal lesions. Calcification, ulceration, thrombosis, and hemorrhage are secondary changes. Atherosclerosis is but one form of arteriosclerosis, which is a broader term for conditions of the other vascular layers and for variable etiologies and pathogenesis (see p. 377). Its prevention is hampered by little precise knowledge of its cause as well as its progress.

NERVE SUPPLY, CIRCULATION, AND VESSEL NUTRITION. The regulation of blood flow (according to need in different areas) is regulated by both nervous and humoral factors (Fig. 10-11). The nervous control is mediated by a reflex mechanism, whose efferent fibers are responsible for the release of norepinephrine at their terminations in the arterial and capillary wall. The humoral control is mediated by epinephrine, which enters the bloodstream after being secreted by the adrenal medulla. Both cause contraction of smooth muscle.

The efferent nonmyelinated motor fibers extend into the medial smooth muscle of the blood vessels and are called vasomotor nerves. They are part of the autonomic system and are distributed as sympathetic nerves. When they are stimulated, there is a decrease in the size (narrowing) of the muscular arteries, especially arterioles (vasoconstriction). This is followed by a rise in blood pressure proximal to the constriction and a decrease in blood supply distal to it. This permits blood to be diverted elsewhere, say to the intestine during digestion or to the skin for cooling. The sympathetic outflow from the spinal cord is regulated by impulses from still higher centers or from incoming sensory fibers. Vasomotor fibers from parasympathetic nerves are found in only a few areas, such as in the pelvic splanchnic nerves and some cranial nerves. Their function is uncertain.

Vessels in which the smooth muscle is circularly arranged can dilate only by muscle relaxation, and their dilation does not depend on impulses from either sympathetic or parasympathetic fibers. Vessels in which the muscle is longitudinal or spiral increase in diameter with contraction but also shorten at the same time.

The nervous control of arterial pressure levels is maintained by the innervation of the arterioles and by special structures located at certain sites in large vessels.

Pressor receptors in the walls of the atria, in the aortic arch, in the carotid sinuses (located in the internal carotid artery), and in the large veins near the heart are part of a reflex mechanism involving blood pressure. The afferent fibers of these receptors pass centrally via the vagus nerves (glossopharyngeal from the carotid sinuses) to vasomotor centers in the brain stem. The former are stimulated by an increase in pressure in the heart and vessels (stretching stimuli), and the reflex effects are a lowering of blood pressure and a decrease in heart rate.

Receptors sensitive to the oxygen and carbon dioxide content of the blood are found in the carotid bodies (small condensations of tissues in the wall of the internal carotid artery, consisting of epithelial-like cells rich in nerve endings). Afferent nerves from here join the glossopharyngeal nerves. Stimulation of these receptors by either an increase in CO_2 or a decrease in O_2 concentration leads to an increase in rate and depth of respiration and thus indirectly affects circulation. Similar structures are seen in the aortic arch and origin of the pulmonary and subclavian arteries. Note that some afferent fibers from blood vessels involve pain and that the puncture of an artery may be painful.

The normal arterial intima is not vascularized and, with the exception of the outer part of the elastic arteries, neither is the media. Pressure in the lumen of arteries is always high, and capillaries from the vasa vasorum (with blood under low pressure) would only collapse if present in the media and intima. High intraluminal pressure actually forces the entry of nutrients from the blood into the intima via pressure and diffusion gradients across the vessel wall. Thus, plasma proteins enter the intima via the lumen, and the ground substance plays an essential role in this process of diffusion and filtration, acting like a sponge. A fine balance normally exists between blood pressure and the amount and rate of entry and clearing of substances

derived from the blood, local O_2 tension, and wall thickness.

Veins. The veins (Fig. 10-11) are more passive channels than the arteries, have valves, and carry blood to the heart. They are capable of distending more than the arteries and at any given time may contain up to 70 percent of the total body blood. They are "capacity vessels" as well as low-resistance "canals." Large amounts of blood may accumulate in the lower parts of the body in exercise, blood loss, or heart failure, and the veins must function actively to maintain venous pressure and redistribute the body blood.

The veins have a small amount of smooth muscle and elastic tissue in their walls compared to the arteries and do not pulsate. Blood in the veins in many cases travels against gravity. Many of the larger veins have valves that direct the blood toward the heart and prevent backflow. The veins transport at a slower speed than arteries, but this is compensated for by their relative large caliber. Body muscle (striated) action helps return flow by the veins by compressing the veins between muscles, the so-called muscle pump. The return flow is further assisted by the muscle walls of the abdomen exerting pressure on the large abdominal veins. Blood is also helped to return by the fact that there is a slight negative pressure in the thorax, the so-called thoracic pump.

It has long been known that changes in tone of the arterial vessels occur in response to a variety of stimuli and exert an immediate influence on blood pressure. The veins, too, constrict actively, not only to preserve blood pressure but also to shift blood from the periphery to the central circulation, as needed. Thus, the veins serve as a "reservoir" of the circulatory system.

The veins have thinner walls than the arteries and are more distensible and can accommodate large volumes of blood with little increase of internal pressure. Therefore, you find 50 percent of the total blood volume at any one time, in the veins at about 10 mm Hg but only 15 percent in the arteries at 100 mm Hg. This pressure-volume relation permits the veins to act as a blood reservoir.

Most of the pressure imparted to the blood by the heart is dissipated in the arterioles and capillaries so that the pressure in the small venules is only about 15 mm Hg. This low resistance in the venules (due to the large diameter of the vessels) enables them to function so efficiently that the total pressure drop from venule to right atrium is only 10 mm Hg and the right atrial pressure is about 0 to 5 mm Hg.

The walls of the veins contain smooth muscle richly innervated by sympathetic vasoconstrictor nerves. Stimulation increases the stiffness of the wall (vasoconstriction), making it less distensible and raising the venous pressure. This drives more blood out of the veins into the right heart. Thus, when total blood volume decreases, pressure in the circulation is reduced, including the veins, and venous return to the right heart decreases and heart output decreases. Increased sympathetic discharge to venous smooth muscle causes contraction, raising venous pressure to normal, and restores venous return and heart output.

During skeletal muscle contraction, the veins in the muscle are partially compressed, their diameters are reduced, and venous capacity is decreased. This also happens during inspiration when the diaphragm descends and pushes the abdominal organs, increasing abdominal pressure, which compresses the large abdominal veins. This facilitates venous flow only toward the heart (due to venous valves that prevent backflow).

A concept of nerve function states that small protein receptor complexes on the walls of the smooth muscle cells determine the response the cell will make when the receptor is stimulated by a suitable substance. Stimulation of an alpha (α) receptor results in contraction (blood vessel constriction). Stimulation of a β receptor results in relaxation (dilatation of the vessel). Arteries have both α and β receptors so that specific stimuli may result in either constriction or dilatation. Norepinephrine predominantly affects the α receptors, and the net effect of norepinephrine on the arteries is constriction. In veins, however, only α receptors are found, so that nerve stimulation results only in constriction.

The nervous system causes the heart to increase and decrease its pumping action to maintain the flow of blood needed by the body. The arterial side of the system tends to respond to the need for flow to specialized tissues by maintaining arterial pressure (the brain circulation remains open while others of less essential need may close down under stress). On the venous side, nervous impulses act on the veins to maintain sufficient pressure and volume of blood in the central veins to form an adequate reservoir of blood for the heart to pump back into the lungs and into the arterial system.

The distensibility of any hollow organ is described by the way its volume varies with the pressure of the fluid in it. The volume in veins is recorded by measuring, with a plethysmograph, changes in the size of a patient's arm or leg. This is possible because so much of the blood in an extremity is contained in the veins and because the capillaries and arteries change their volume relatively little. Thus, fluctuations in volume of an extremity can be ascribed almost completely to changes in vein volume.

Temperature changes also prompt a venous response since an important function of the circulation is to help maintain a constant body temperature. At room temperature (83° F) arterioles dilate but veins do not and so the blood velocity in the veins increases. At 68° F, the arterioles constrict and reduce blood flow, thereby restricting venous flow. Here the veins constrict and restore the velocity to normal.

A major physiologic problem for man is to maintain blood flow to all parts of the body in the erect position. When one stands, the hydrostatic pressure in the leg veins approaches 100 mm Hg and a large volume of blood tends to settle in the distensible vessels of the lower legs. This blood pooling is counteracted by (1) the simple flap valves in the veins that prevent the return flow of blood moving up, thus holding the blood in a series of short, low-pressure columns between valves; and (2) a generalized venous constriction not only in the legs but everywhere in the body. These responses help maintain the pressure in the venous system, particularly near the heart. If this pressure is too low, there is inadequate flow into the heart and inadequate cardiac output and reduced flow to the brain, which may cause fainting. Abnormally great increases in venous resistance (due to a clot or compressing tumor) can impede blood flow, and blood accumulates behind the lesion. This results in increased pressure in the small veins and capillaries being drained by the blocked vessel, the capillary filtration increases, and the tissues become edematous (swollen with fluid).

Exercise requires additional blood supply to the tissue. The heart rate increases to meet the need, but the heart may be limited in its ability to respond if not enough blood is returned to it by the veins. The veins constrict in exercise to shift blood toward the heart to facilitate increased cardiac output. If the heart is subjected to a handicap that impairs its pumping ability, such as coronary disease, high blood pressure, or valve damage, then cardiac failure may result with symptoms due to insufficient blood being pumped. The normal heart can meet its total blood flow needs even during mild exercise without venous constriction. When a heart is handicapped, however, the veins are constricted even at rest and this chronic state is associated with a greater-than-normal total blood volume characteristic of heart failure, which results in high

Figure 10-17. A. Phlebothrombosis is due to a clot and is unassociated with inflammation of the vein wall. Thrombophlebitis, on the other hand, is an inflammation of the vein wall and precedes thrombus formation. **B.** Thrombophlebitis can be caused by pressure, infection, injection, or contusion. **C.** Phlebothrombosis can be the result of prolonged immobilization (sitting too long), which results in a steady venous compression and stasis. **D.** When the valve weakens, the vein wall dilates and you get secondary incompetence. Increased venous pressure leads to further vessel dilatation with further secondary incompetence and varicose veins. **E.** Varicose veins, which can lead to thrombophlebitis.

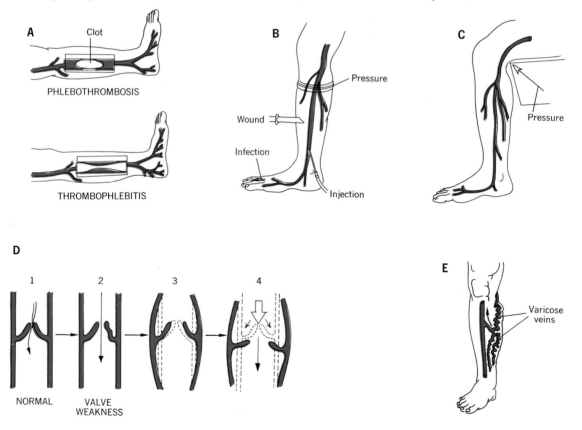

pressure in veins and capillaries. This forces large quantities of fluid through the thin capillary walls into the tissues (edema). Of interest is the fact that what the tissues need in failure and exercise is oxygen. Similarly, when a man is exposed to an altitude of more than 0.3 km (1000 ft), the responses of the veins simulate heart failure in many respects.

When a person has high blood pressure not related to kidney disease or a known cause, he is said to have essential hypertension. The cause is unknown, but the mechanical problem is an arteriole constriction. The veins, however, are not usually constricted.

When blood flows too slowly, it tends to clot (Fig. 10-17). Clots forming in veins (thrombophlebitis) can break off (emboli) and travel to the lungs, causing pulmonary emboli. They may remain stationary and obstruct blood flow and cause serious edema. Thrombophlebitis may damage the valves in a vein. This is one cause of varicose veins, where lack of valvular function causes blood to pool in the leg veins, which then become chronically dilated. Varicosities can also arise from valve injury, congenital absence of valves, or even an inherited venous distensibility that makes for lack of tone in the leg veins (Fig. 10-17).

Pregnant women have a predisposition to thrombophlebitis and varicose veins. Pulmonary emboli from thrombophlebitis have also been reported as a result of oral contraceptive drugs. Varicose veins usually involve superficial vessels that are not essential to the circulation. They can be tied off and removed by surgery. When a number of deep veins are involved, this is not possible. Furthermore, a constant high pressure of pooled blood with little valve action can cause marked edema, which results in tissue breakdown and ulcers.

Gross Morphology: The Vessels and Their Distribution

Pulmonary Circulation

The pulmonary circulation carries the blood from the heart to the lungs via the pulmonary arteries and back to the heart via the pulmonary veins. The pulmonary artery takes origin from the right ventricle and divides into right and left branches that enter the roots of the lungs. In the lung, each trunk breaks up into small branches that follow the subdivision of the bronchi fairly close. The terminal arterioles open into a capillary network that courses over the alveolar sacs. The pulmonary veins enter the left atrium via four vessels, two from each lung. Their smallest tributaries arise as venules from the capillary network.

Systemic Arteries

The aorta leaves the left ventricle of the heart. After distribution to all parts of the body, blood returns to the right atrium through the two veins, the superior and inferior venae cavae.

Aorta. The aorta (Fig. 10-18) is the largest artery in the body. It arises (ascending portion) from the left ventricle and arches (aortic arch) to the left and descends (descending or thoracic aorta) on the posterior thoracic wall. It then passes behind the diaphragm as the abdominal aorta to the fourth lumbar vertebra, where it ends by dividing into right and left common iliac arteries. The various parts of the aorta obviously have many important anatomic as well as clinical relations (see any standard anatomy text for details).

The major branches of the aorta (Fig. 10-18) are as follows:

1. From the ascending aorta: the coronary arteries, right and left, which provide blood to sustain the heart muscle. A branch of the right supplies the SA node of the conduction system.
2. From the aortic arch: the brachiocephalic, which divides into the right common carotid and right subclavian arteries and the left common carotid and left subclavian arteries.
3. From the descending thoracic aorta:
   a. Parietal branches: intercostal, subcostal, superior phrenic.
   b. Visceral branches: pericardial, bronchial, esophageal, and mediastinal.

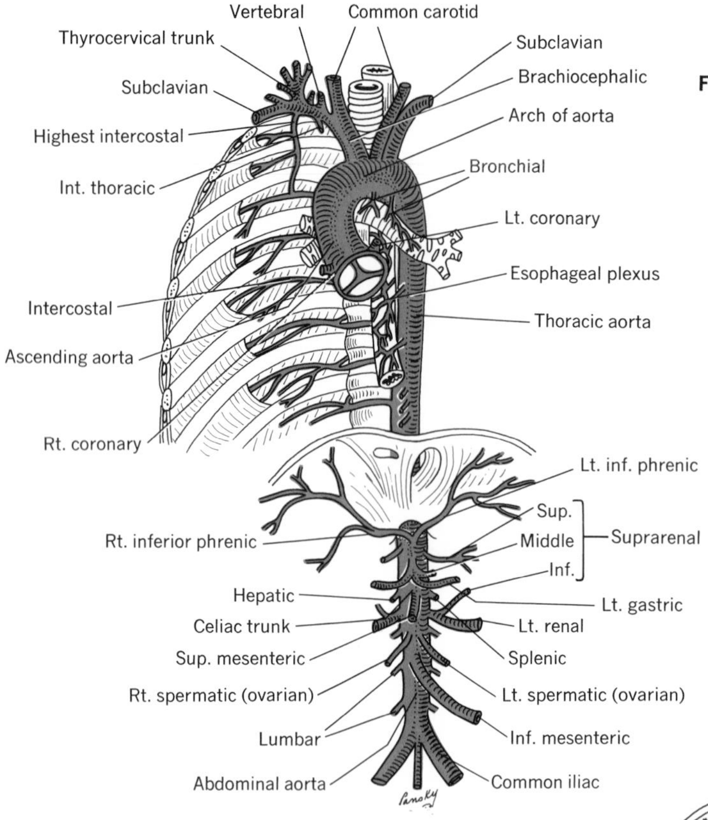

Figure 10-18. Branches of the aorta.

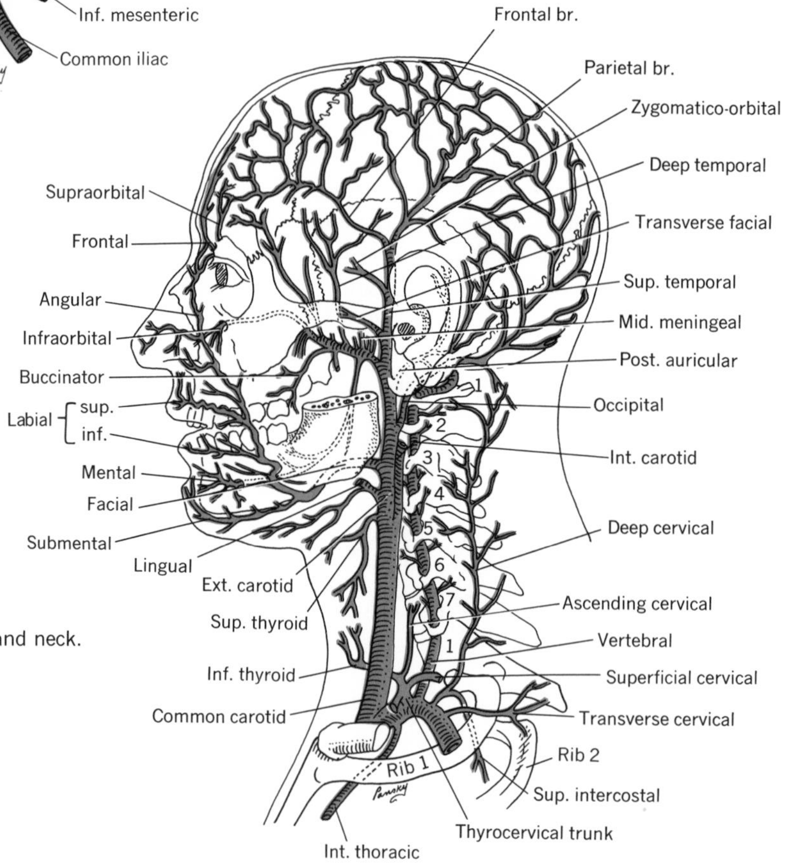

Vertebral
Thyrocervical trunk
Common carotid
Subclavian
Subclavian
Brachiocephalic
Highest intercostal
Arch of aorta
Int. thoracic
Bronchial
Lt. coronary
Esophageal plexus
Intercostal
Thoracic aorta
Ascending aorta
Rt. coronary
Lt. inf. phrenic
Sup.
Rt. inferior phrenic
Middle — Suprarenal
Inf.
Hepatic
Lt. gastric
Celiac trunk
Lt. renal
Sup. mesenteric
Splenic
Rt. spermatic (ovarian)
Lt. spermatic (ovarian)
Lumbar
Inf. mesenteric
Abdominal aorta
Common iliac

Figure 10-19. Arteries of the head and neck.

Frontal br.
Parietal br.
Zygomatico-orbital
Deep temporal
Supraorbital
Transverse facial
Frontal
Sup. temporal
Angular
Mid. meningeal
Infraorbital
Post. auricular
Buccinator
Occipital
Labial { sup.
inf.
Int. carotid
Mental
Facial
Deep cervical
Submental
Lingual
Ext. carotid
Ascending cervical
Sup. thyroid
Vertebral
Inf. thyroid
Superficial cervical
Common carotid
Transverse cervical
Rib 1
Rib 2
Sup. intercostal
Thyrocervical trunk
Int. thoracic

356

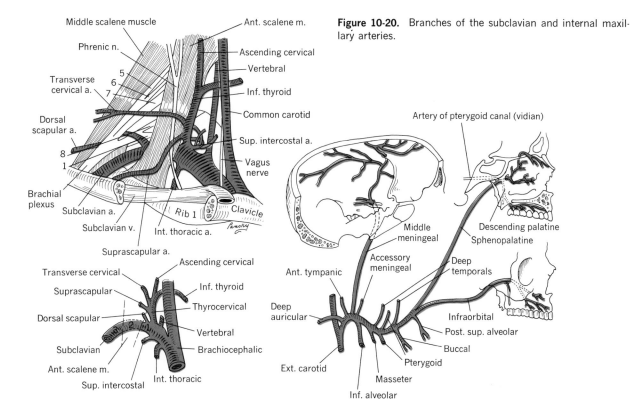

Middle scalene muscle
Phrenic n.
Transverse cervical a.
5
6
7
Dorsal scapular a.
8
1
Brachial plexus
Subclavian a.
Subclavian v.
Int. thoracic a.
Suprascapular a.
Rib 1
Clavicle
Ant. scalene m.
Ascending cervical
Vertebral
Inf. thyroid
Common carotid
Sup. intercostal a.
Vagus nerve

Transverse cervical
Suprascapular
Dorsal scapular
Subclavian
Ant. scalene m.
Sup. intercostal
Int. thoracic
Ascending cervical
Inf. thyroid
Thyrocervical
Vertebral
Brachiocephalic

Artery of pterygoid canal (vidian)
Middle meningeal
Accessory meningeal
Ant. tympanic
Deep auricular
Ext. carotid
Inf. alveolar
Masseter
Pterygoid
Buccal
Post. sup. alveolar
Infraorbital
Deep temporals
Descending palatine
Sphenopalatine

Figure 10-20. Branches of the subclavian and internal maxillary arteries.

4. From the abdominal aorta:
   a. Parietal branches: inferior phrenics, lumbars, middle sacral (unpaired).
   b. Visceral branches: celiac (unpaired), superior mesenteric (unpaired), inferior mesenteric (unpaired), middle suprarenals, renals, testicular, and ovarian.
5. Terminal branches: common iliacs.

Arteries of the Head and Neck. The *two common carotid* arteries pass up into the neck and end opposite the thyroid cartilages of the larynx by dividing into two branches, the external and internal carotid arteries (Fig. 10-19).

EXTERNAL CAROTID ARTERY. This artery (Figs. 10-19, 10-20) supplies almost everything in the neck and head regions except the contents of the cranial and orbital cavities. However, since it arises high on

the neck, it is the subclavian arteries that tend to supply the lower neck as they cross the root of the neck on their way to the upper extremities. The external carotid terminates near the mandibular joint by dividing into two terminal branches, the superficial temporal and internal maxillary arteries. Its major branches are (Fig. 10-19):

1. Superior thyroid: to the thyroid gland and structure of the larnyx via the superior laryngeal artery.
2. Lingual: which supplies the tongue.
3. Facial: which crosses the lower border of the mandible (can be felt pulsating here) and enters the face to supply it and the muscles of facial expression, through which it passes.
4. Occipital: to the occipital region.
5. Posterior auricular: to structures of the scalp behind the ear.

6. Ascending pharyngeal: which runs up the pharyngeal wall and supplies it.
7. Superficial temporal: which supplies the side of the scalp.
8. Internal maxillary (Fig. 10-20): which is the largest of the terminal branches of the carotid and passes deep to the neck of the mandible. It supplies the outer eardrum and external acoustic meatus; the teeth, via the inferior and superior dental (alveolar) arteries; the meninges (brain coverings), by way of the middle meningeal artery; the muscles of mastication, by muscular branches; and the mucous membrane of the nose, pharynx, and palate.

INTERNAL CAROTID ARTERY. The internal carotid artery (Fig. 10-21) has no branches in the neck. It runs in the carotid sheath and enters the middle cranial cavity, where it makes a "hairpin" turn to divide into three terminal branches: the ophthalmic artery, which follows the optic nerve to the orbit and eye; and the anterior and middle cerebral arteries, which go to the brain (see Chapter 8).

SUBCLAVIAN ARTERY. The subclavian artery (Figs. 10-20, 10-22) is the vessel of the upper limbs. Each vessel arches in the root of the neck and then passes into the armpit, where it changes its name to axillary. It begins and ends behind the clavicle. Three of its

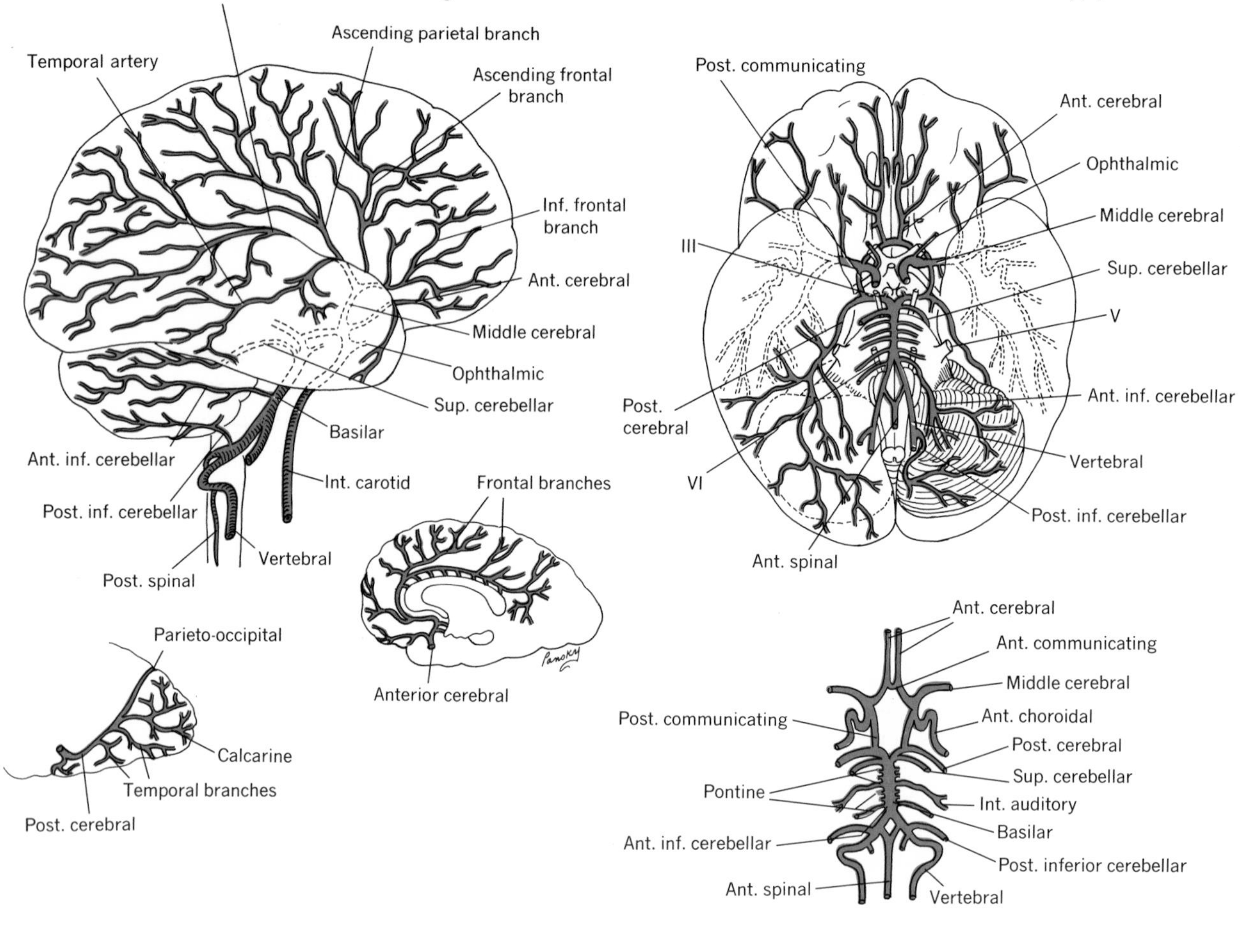

Figure 10-21. Branches of the internal carotid and arterial blood supply to the brain.

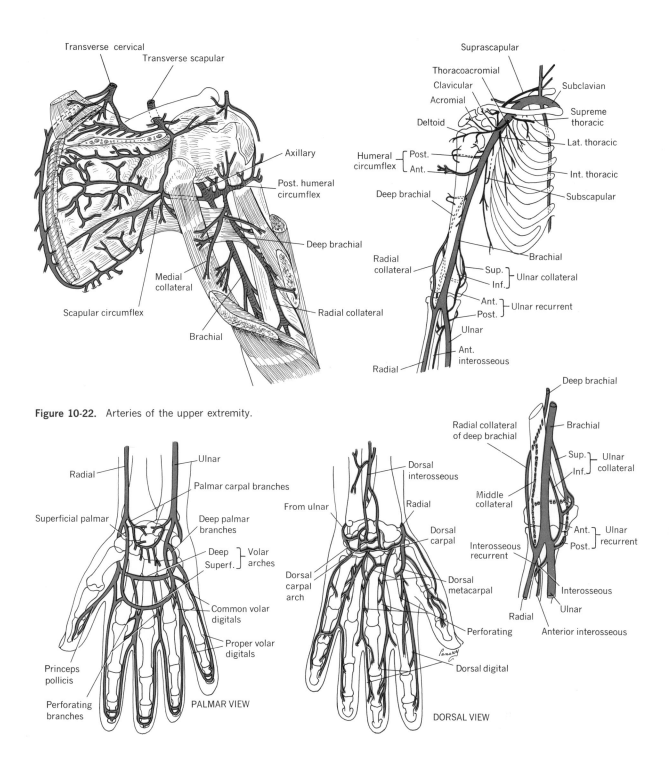

Figure 10-22. Arteries of the upper extremity.

Labels in figure:

Transverse cervical
Transverse scapular
Axillary
Post. humeral circumflex
Deep brachial
Medial collateral
Scapular circumflex
Radial collateral
Brachial

Suprascapular
Thoracoacromial
Clavicular
Acromial
Deltoid
Humeral circumflex { Post. Ant.
Deep brachial
Radial collateral
Subclavian
Supreme thoracic
Lat. thoracic
Int. thoracic
Subscapular
Brachial
Sup. } Ulnar collateral
Inf.
Ant. } Ulnar recurrent
Post.
Ulnar
Ant. interosseous
Radial

Deep brachial
Radial collateral of deep brachial
Brachial
Sup. } Ulnar collateral
Inf.
Middle collateral
Interosseous recurrent
Ant. } Ulnar recurrent
Post.
Interosseous
Ulnar
Radial
Anterior interosseous

Radial
Ulnar
Palmar carpal branches
Superficial palmar
Deep palmar branches
Deep } Volar arches
Superf.
Common volar digitals
Proper volar digitals
Princeps pollicis
Perforating branches
PALMAR VIEW

From ulnar
Dorsal interosseous
Radial
Dorsal carpal
Dorsal carpal arch
Dorsal metacarpal
Perforating
Dorsal digital
DORSAL VIEW

four branches are found near its arch, and only one, the internal thoracic, arises from its concavity.

Internal Thoracic Artery. The internal thoracic (Fig. 10-22) descends vertically into the thorax behind and near the sternum. As it crosses the anterior ends of the upper six intercostal spaces it gives off small anterior intercostal arteries, which anastomose with terminal branches of the posterior (aortic) intercostals. The artery divides at the lower sternum into the musculophrenic artery, supplying the diaphragm and lower intercostal spaces, and the superior epigastric artery, which enters the rectus sheath.

Vertebral Artery. The vertebral artery (Figs. 10-20, 10-21) ascends in the neck through the foramina of the cervical transverse processes and enters the skull via the foramen magnum. It then lies on the front of the brain stem, where it meets its opposite fellow to form a single, midline trunk, the basilar artery. The vertebral arteries supply branches to the upper spinal cord (aortic intercostals supply the lower) as well as the brain stem, cerebellum, and posterior cerebrum. The basilar and its branches along with the internal carotid and its branches form a circle of arteries on the base of the brain, the circle of Willis.

Thyrocervical Trunk. The thyrocervical trunk (Fig. 10-20) is a short stem, dividing into (1) the inferior thyroid, to the lower half of the gland; (2) the transverse cervical, which runs around the root of the neck to the vertebral border of the upper end of scapula; (3) the suprascapular, which aids in supplying the muscles around the scapula; and (4) the costocervical trunk, which supplies the first and second intercostal spaces. The trunk also supplies a deep cervical artery to some muscles at the back of the neck.

Arteries of the Upper Extremity. AXILLARY ARTERY. The axillary artery (Fig. 10-22) begins as a continuation of the subclavian artery. It courses through the axilla surrounded by the nerves of the brachial plexus and two or three accompanying veins (venae comitantes). It and its companion structures are wrapped together in a sheath of fibrous tissue, the axillary sheath. The axillary artery provides a branch for each wall of the axilla. The branches anastomose freely with each other and with scapular branches of

the subclavian to create collateral circulation around the shoulder. The branches are (1) the thoracoacromial, to the anterior axillary wall; (2) the lateral thoracic, which descends on chest and supplies the medial axillary wall with a large branch to the female breast; (3) the subscapular, the largest branch, which runs along the scapula and supplies the posterior wall; and (4) the anterior and posterior humeral circumflex, which circle the neck of the humerus and supply the shoulder joint and shoulder muscles.

The axillary artery, as it enters the "free limb" below the pectoralis minor muscle, changes its name and becomes the brachial artery.

BRACHIAL ARTERY. The brachial artery lies in close association with the three principal nerves of the limb. Throughout its course, it is accompanied by the median nerve. Immediately distal to the elbow joint, it bifurcates into its two terminal branches, the radial and ulnar arteries. Branches of the brachial are (1) the deep (profunda) brachii, the largest collateral branch, which runs with the radial nerve and supplies the triceps muscle and branches to the extensor muscles of the forearm; (2) the ulnar collateral, which runs with the ulnar nerve and contributes to the supply of the flexor muscles of the forearm; and (3) the inferior ulnar collateral, which arises just above the elbow and also supplies the flexor muscles.

Collateral anastomoses about the elbow joint are rich and important and take place between branches of brachial, radial, and ulnar arteries (Fig. 10-22).

RADIAL ARTERY. The radial artery runs down the lateral (radial) side of the forearm and supplies muscles and tissues along its course. Near its origin it gives off the radial recurrent artery to anastomose with the deep brachial. It reaches the wrist at the base of the thumb (pulsations can be felt here). Turning to the back of the wrist, it supplies some branches to the back of the hand, but its major supply plunges it deep through the muscles that fill the space between the thumb and index finger into the palm. Here it supplies the thumb and index finger and helps to form the deep palmar arch (Fig. 10-22).

ULNAR ARTERY. The ulnar artery runs along the medial side of the forearm and enters the hand. It is mainly concerned in the formation of the superficial palmar arch. Near its origin it gives off the common interosseous artery, which divides into anterior and posterior interosseous arteries running on either side of the interosseous membrane. The anterior supplies the deep flexor muscles, the posterior the forearm extensors.

SUPERFICIAL PALMAR ARCH. This arch runs across the palm about at the level of the distal palmar crease or outstretched thumb and comes from the ulnar and radial arteries.

DEEP PALMAR ARCH. This arch is proximal to the superficial but lies deeper at the base of the metacarpal bones and arises from the radial and a branch of the ulnar.

Both superficial and deep arches provide palmar branches that run forward and unite at the finger clefts.

Each common stem then divides to give off digital arteries to the sides of the fingers, ending in a network of small vessels at the fingertips (Fig. 10-22).

Arteries of the Thorax and Abdomen. THORACIC AORTA. Eleven pairs of arteries (Figs. 10-18, 10-23) come off the descending thoracic aorta, of which the first nine are posterior intercostal arteries for intercostal spaces 3 to 11. The tenth pair are the subcostal arteries since they lie below the last rib. The eleventh pair are the superior phrenic arteries that supply the thoracic side of the diaphragm. Other small vessels, for adjacent viscera, also arise from the thoracic aorta, namely, bronchial arteries for the bronchial tree, esophageal, and pericardial.

ABDOMINAL AORTA. The abdominal aorta (direct continuation of thoracic aorta) (Figs. 10-18, 10-24, 10-25) has both paired and unpaired branches.

Figure 10-23. Arteries of the thoracic and abdominal walls.

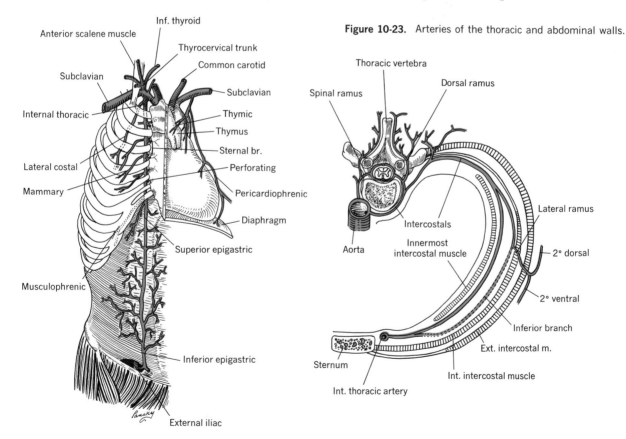

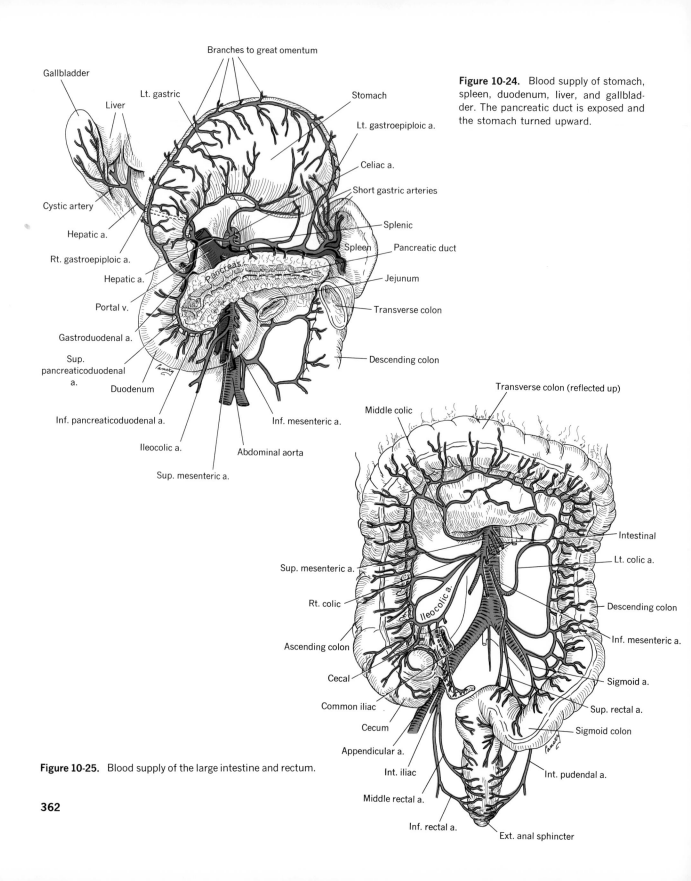

Gallblodder

Branches to great omentum

Lt. gastric

Stomach

Liver

Lt. gastroepiploic a.

Celiac a.

Short gastric arteries

Splenic

Cystic artery

Spleen

Pancreatic duct

Hepatic a.

Rt. gastroepiploic a.

Jejunum

Hepatic a.

Transverse colon

Portal v.

Gastroduodenal a.

Descending colon

Sup.
pancreaticoduodenal
a.

Duodenum

Inf. pancreaticoduodenal a.

Inf. mesenteric a.

Ileocolic a.

Abdominal aorta

Sup. mesenteric a.

Figure 10-24. Blood supply of stomach, spleen, duodenum, liver, and gallbladder. The pancreatic duct is exposed and the stomach turned upward.

Pancreas

Transverse colon (reflected up)

Middle colic

Intestinal

Lt. colic a.

Sup. mesenteric a.

Descending colon

Rt. colic

Inf. mesenteric a.

Ileocolic a.

Ascending colon

Sigmoid a.

Cecal

Sup. rectal a.

Common iliac

Sigmoid colon

Cecum

Appendicular a.

Int. iliac

Int. pudendal a.

Middle rectal a.

Inf. rectal a.

Ext. anal sphincter

Figure 10-25. Blood supply of the large intestine and rectum.

Paired Branches to the Abdominal Walls. Five pairs arise from the back of the aorta: the inferior phrenic arteries to the abdominal surface (inferior) of the diaphragm and four pairs of lumbar arteries supplying the posterior abdominal wall. All posterior intercostals as well as lumbars send branches into the intervertebral foramina to supply the spinal cord, its nerve roots, and the spinal meninges.

Paired Branches to Three Paired Glands. These arise from the sides of the aorta: renal arteries to the kidneys; suprarenal arteries to the suprarenal glands (these glands also receive branches from the inferior phrenic and renal vessels); and testicular (internal spermatic) or ovarian arteries to the testis or ovary and uterine tube.

Unpaired Arteries. Unpaired arteries are disposed to the digestive tract and accessory organs. There are three that arise from the front of the aorta.

The celiac artery (Fig. 10-24) is a very short trunk with three branches: (1) the left gastric supplies the lower end of the esophagus and part of the stomach (along the lesser curvature); (2) the splenic, its largest branch, runs along the upper border of the pancreas, giving off (a) pancreatic branches, (b) short gastrics to the fundus of the stomach, and (c), before entering the splenic hilum, a large branch, the left gastroepiploic, which runs along the greater curvature of the stomach to supply it and the greater omentum; and (3) the hepatic artery runs in the free edge of the lesser omentum to the liver. The hepatic artery gives off a cystic branch to the gallbladder; a small right gastric branch to the pylorus, which continues in the lesser curvature to anastomose with the left gastric; and a large gastroduodenal artery, which passes behind the duodenum and gives off two branches, the right gastroepiploic (which anastomoses with the left) and the superior pancreaticoduodenal (for the duodenum and head of the pancreas). The hepatic artery terminates into the right and left hepatic arteries.

The superior mesenteric artery (Fig. 10-25) arises below the celiac, runs behind the neck of the pancreas, and crosses the head of the pancreas and horizontal part of the duodenum to enter the root of the mesentery. It supplies all of the small intestine (except for part of the duodenum) and the right half of the large intestine. Thus, it has intestinal branches (10 to 15) that form anastomotic loops and supply about 6 m (20 ft) of intestine; an ileocolic artery to the terminal ileum, cecum, appendix, and beginning of the ascending colon; and the right and middle colic arteries to the ascending colon, hepatic flexure, and transverse colon.

The inferior mesenteric artery (Fig. 10-25) is the smallest of the digestive tract branches and runs down and to the left. It terminates as the superior rectal artery and supplies the gut as low as the anal canal. In the abdomen it gives off the superior and inferior left colic branches to the splenic flexure and descending and sigmoid (pelvic) colon.

COMMON ILIAC ARTERIES. These arteries (Figs. 10-25, 10-26, 10-27) have no collateral branches. They divide into an internal iliac to the pelvis and perineum, and an external iliac artery, the vessel of the lower extremity.

INTERNAL ILIAC (HYPOGASTRIC) ARTERIES. The internal iliac arteries (Fig. 10-26) are usually described as having an anterior and a posterior division, but regional distribution is easily seen. The branches and regions supplied are:

Gluteal Region. The gluteal region receives two branches that arrive there through the greater sciatic foramen: the superior and inferior gluteal arteries.

Perineal Region. This is below the level of the pelvic diaphragm. It is supplied by the internal pudendal artery and includes the external genitalia and anal canal.

Wall of the False Pelvis. This is supplied by the iliolumbar artery.

Wall of the True Pelvis. This is supplied by the lateral sacral and the obturator, which after supplying the inside of the lateral wall leaves the pelvis to supply the medial thigh.

Vessels to the Pelvic Organs. These vessels (Fig. 10-26) are the superior and inferior vesical arteries to the bladder, prostate gland, seminal vesicles, and vas deferens; the uterine to the uterus and vagina; and the middle rectal to the rectum.

External Iliac Arteries. These arteries (Figs. 10-26, 10-27) course along the pelvic brim. Before they be-

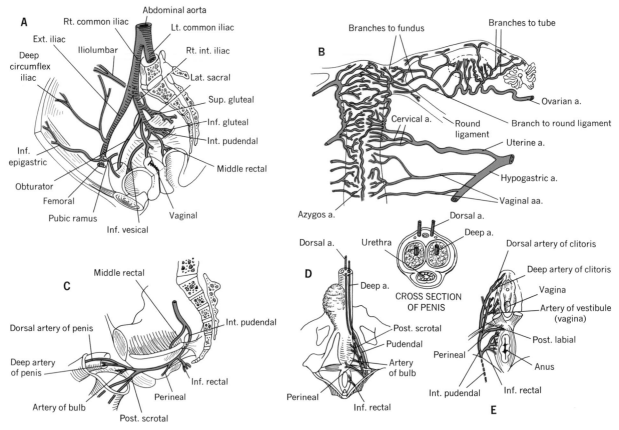

Figure 10-26. **A.** The internal iliac and its branches. **B.** Arteries of the internal organs of generation of the female, posterior view. **C.** Internal pudendal artery and its branches. **D.** Arteries of the male external genitalia. **E.** Arteries of the female external genitalia.

come femoral arteries, they give off the deep circumflex iliac, to the lower and lateral half of the abdominal wall; and the inferior epigastric, which enters the lower part of the rectus sheath and eventually anastomoses with the superior epigastric.

Arteries of the Lower Extremity. FEMORAL ARTERY. The femoral artery (Fig. 10-27) supplies the lower limb. It descends vertically until just above the knee, where it passes backward through an opening in the adductor magnus muscle (adductor hiatus) to the back of the knee and becomes the popliteal artery. As it passes into the thigh it gives off small branches to the lower abdominal wall and to the scrotum or labium

majus. In the thigh it gives off three major branches (supplying adjacent muscles): (1) the medial femoral circumflex, which passes backward and anastomoses with the inferior gluteal to supply the hip joint and upper back of thigh as well as head of the femur; (2) the lateral femoral circumflex, which is the great muscular branch to the quadriceps femoris and has branches that ascend and descend the whole length of the lateral thigh from hip to knee; and (3) the deep (profunda) femoral, which provides blood to the back of the thigh via a series of three or four perforating branches.

POPLITEAL ARTERY. The popliteal artery (Fig.

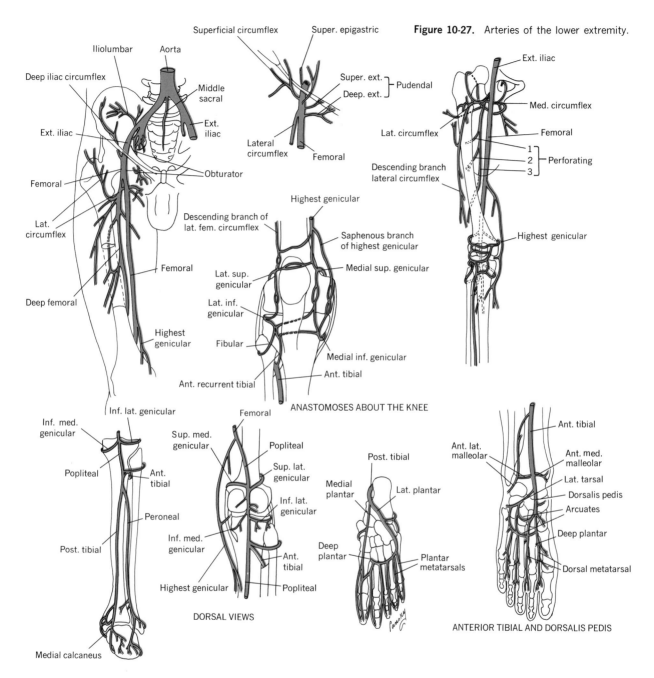

Superficial circumflex — Super. epigastric

Figure 10-27. Arteries of the lower extremity.

Iliolumbar — Aorta

Deep iliac circumflex

Ext. iliac

Middle sacral

Ext. iliac

Super. ext. ⎫ Pudendal
Deep. ext. ⎭

Femoral

Lat. circumflex

Obturator

Lateral circumflex — Femoral

Deep femoral

Femoral

Highest genicular

Ext. iliac

Med. circumflex

Lat. circumflex — Femoral

1
2 ⎫ Perforating
3 ⎭

Descending branch lateral circumflex

Highest genicular

Highest genicular

Descending branch of lat. fem. circumflex

Saphenous branch of highest genicular

Lat. sup. genicular

Medial sup. genicular

Lat. inf. genicular

Fibular

Medial inf. genicular

Ant. recurrent tibial

Ant. tibial

ANASTOMOSES ABOUT THE KNEE

Inf. med. genicular

Inf. lat. genicular

Femoral

Ant. lat. malleolar

Ant. tibial

Popliteal

Ant. med. malleolar

Sup. med. genicular

Popliteal

Post. tibial

Sup. lat. genicular

Medial plantar

Lat. plantar

Lat. tarsal

Ant. tibial

Inf. lat. genicular

Dorsalis pedis

Peroneal

Inf. med. genicular

Deep plantar

Arcuates

Post. tibial

Ant. tibial

Deep plantar

Plantar metatarsals

Dorsal metatarsal

Highest genicular

Popliteal

Medial calcaneus

DORSAL VIEWS

ANTERIOR TIBIAL AND DORSALIS PEDIS

10-27) is the artery of the back of the knee and is deep, lying against the back of the knee joint. It ends near the top of the tibia by dividing into anterior and poste-rior tibial arteries. The popliteal gives off a series of four genicular arteries that anastomose about the knee joint and supply the joint and adjacent muscles.

ARTERIES OF THE LEG AND FOOT. *Anterior Tibial.* The anterior tibial artery (Fig. 10-27) passes above the interosseous membrane and runs down the front of the leg, giving branches to muscles and crossing the front of the ankle joint in the midline. It enters the dorsum of the foot as the dorsalis pedis artery, supplying that region, and terminates by piercing the space between the first and second toes to anastomose with the plantar arteries on the sole of the foot.

Posterior Tibial. The posterior tibial artery (Fig. 10-27) runs down the back of the leg, giving off muscle branches. It provides a large peroneal artery to the lateral side of the leg. Reaching the back of the ankle joint, the tibial artery enters the bottom of the foot behind the medial malleolus and divides into medial and lateral plantar vessels, which supply the structures in the sole of the foot, with the medial going to muscles of the great toe and the lateral to just about everything else. The lateral and dorsalis pedis (see above) form a single plantar arch, off of which digital vessels arise.

Systemic Veins

The blood from the systemic group of vessels, via tributaries rather than branches, returns to the heart by three major channels: (1) that from cardiac veins of the heart returns through the coronary sinus; (2) that from the head, neck, upper extremities, and portions of the trunk returns through the superior vena cava; and (3) that from the rest of the trunk and lower extremities returns through the inferior vena cava. The portal system, draining the digestive tract, will be considered separately.

Coronary Sinus. The coronary sinus (Fig. 10-6) receives most of the cardiac veins of the heart, lies on the posterior aspect of the heart, and terminates by emptying into the right atrium.

Veins of the Head and Neck. DEEP VEINS. *Cranial Venous Sinuses.* These sinuses (Fig. 8-12, p. 260) course between the layers of the dura and drain blood from the cranial cavity. One of the more important ones includes the superior sagittal on the upper edge of the falx cerebri, which receives the many superior cerebral veins from the surface of the hemispheres and

ends by turning (usually to the right) into the right transverse sinus (in the tentorium cerebelli). The latter makes a bend as the sigmoid sinus and leaves the skull via the jugular foramen, to end in the internal jugular vein. The left transverse sinus follows a similar path, but usually begins by receiving the straight sinus, which is midline and drains from the internal great cerebral vein of Galen (from the interior of the brain). The cavernous sinuses lie on either side of the sphenoid body. Venous channels unite it, front and back, to its fellow of the opposite side. It receives the ophthalmic veins anteriorly from the orbital cavity and drains posteriorly into the inferior petrosal sinus, which joins the internal jugular vein, and the superior petrosal sinus, which joins the sigmoid sinus.

Emissary Veins. These veins (Fig. 10-28) connect the above-listed sinuses to veins on the outside of the skull by pathways through the skull.

Internal Jugular Veins. These veins (Figs. 10-28, 10-29) are the chief veins draining the head and neck. They descend in the neck with the internal carotid artery and vagus nerve in a fascial sheath, the carotid sheath, on the side of the pharynx and esophagus. In the neck, these veins receive tributaries corresponding to branches of the external carotid artery. At the root of the neck, the internal jugular meets the subclavian vein to form the right or left brachiocephalic vein.

SUPERFICIAL VEINS. The external jugular (Figs. 10-28, 10-29) descends in the neck after it is formed by the union of a vein from the posterior part of the face, the posterior facial, with one from the scalp above and behind the ear, the posterior auricular. In the root of the neck, the jugular receives the transverse cervical, suprascapular, and anterior jugular veins from the front of the neck. The external jugular then ends in the subclavian vein. This jugular, being superficial, is not accompanied by a corresponding artery.

Veins of the Upper Extremity. SUPERFICIAL VEINS. The superficial veins (Fig. 10-30) begin as a dorsal venous arch on the back of the hand. Near the lateral end of the arch, the cephalic vein originates and runs up the lateral side of the front of the extremity to reach the deltopectoral groove. It then passes below the clavicle and ends in the axillary vein. At the medial

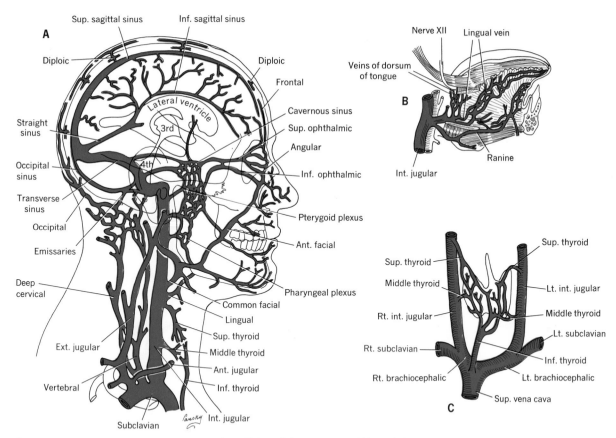

Figure 10-28. **A.** Veins of the head and neck. **B.** Veins of the tongue. **C.** Veins of the thyroid gland.

Figure 10-29. **A.** Superficial and deep veins of the face. **B.** Facial and ophthalmic veins.

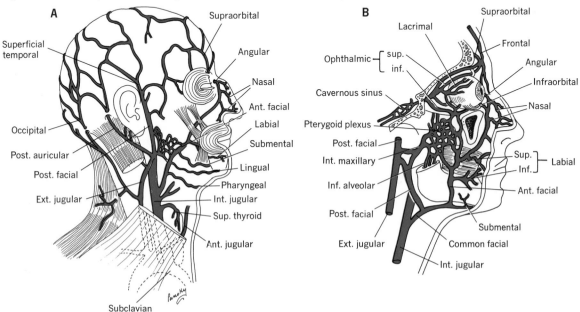

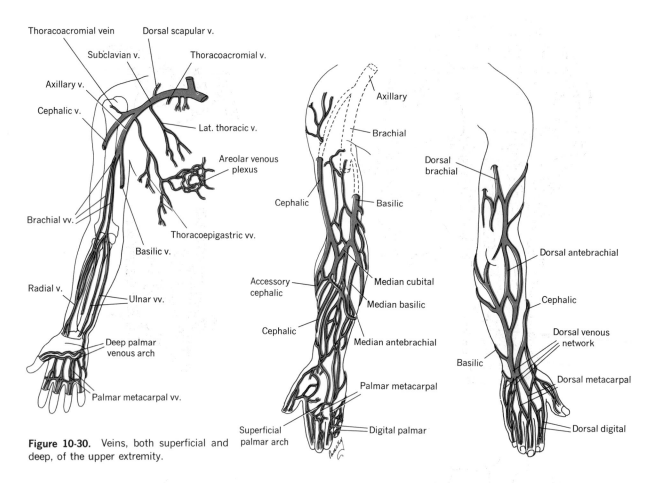

Figure 10-30. Veins, both superficial and deep, of the upper extremity.

side of the dorsal venous arch, the basilic vein begins. This vessel runs up the medial side of the forearm to the elbow. In front of the elbow joint it receives a communication from the cephalic vein, the antecubital vein. The basilic vein in the middle of the arm turns deep, to run with the brachial artery and become the major tributary of the axillary vein.

DEEP VEINS. The deep veins (Fig. 10-30) follow the arteries, have similar names, and require no detailed description. Note that the axillary veins become the subclavian.

Veins of the Thorax. SUPERIOR VENA CAVA AND TRIBUTARIES. Behind the sternoclavicular joint (of each side) the subclavian veins from the upper extremity (which have already received the external jugular

veins) meet the internal jugulars to form the brachio-cephalic veins (Fig. 10-31). Although there is only one brachiocephalic artery (right), there are two veins. The left brachiocephalic vein crosses behind the sternum and joins the right one along the right border of the sternum to become the superior vena cava. The latter descends and enters the right atrium. Halfway along its course it receives the azygos vein for drainage from the thoracic and abdominal cavities.

AZYGOS VEIN AND TRIBUTARIES. The azygos (Greek word meaning unpaired) vein is a long, slender vessel that ascends vertically on the posterior thoracic wall to the right of the descending thoracic aorta (Fig. 10-31). It begins in the abdomen, often by union of lumbar veins.

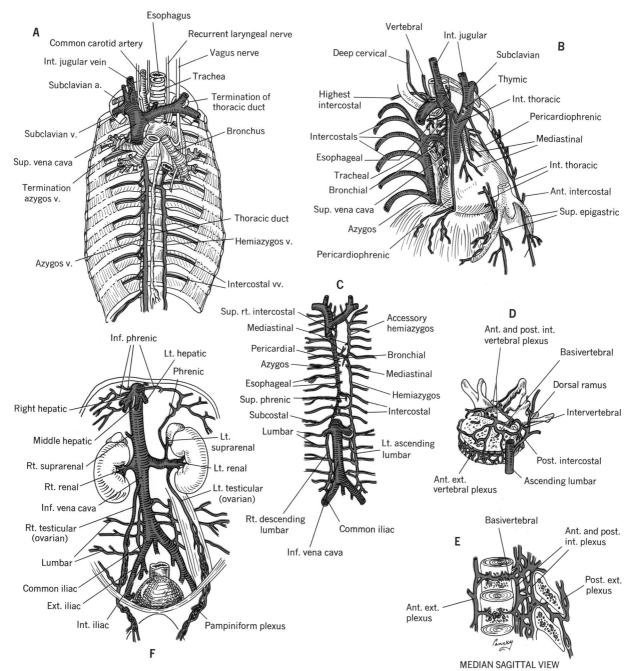

Figure 10-31. **A.** Veins of the posterior thoracic wall. **B.** Veins of the thoracic cavity. **C.** Veins of the posterior abdominal and thoracic wall. **D** and **E.** Vertebral veins. **F.** Veins of the abdominal cavity (portal system is not seen).

The azygos vein receives intercostal veins from both sides of the thoracic walls (companions of the posterior intercostal arteries). The azygos vein finally comes forward by arching over the root of the right lung to join the back of the superior vena cava.

Since the vein ascends no higher than the lung root, the veins draining the upper intercostal spaces on the right side unite to form the right superior intercostal vein, which descends to join the azygos before it enters the vena cava. The upper intercostal spaces on the left side are drained by the left superior intercostal vein, which, however, drains into the left brachiocephalic vein since it has no direct access to the azygos. The left lower and middle portions of the chest drain into the hemiazygos and accessory hemiazygos, which in turn cross the middle to join the azygos (Fig. 10-31).

Veins of the Lower Extremity. SUPERFICIAL VEINS. These veins (Fig. 10-32) lie directly beneath the skin and consist of two major cutaneous veins. Like the upper extremity, they begin at the extremities of a dorsal venous arch that lies on the dorsum (top) of the foot just proximal to the origin of the toes. From the lateral end of the arch, the short saphenous takes origin and runs along the lateral side of the foot and up the back of the leg to the knee, where it passes deep to join the popliteal. From the medial end of the arch arises the important long saphenous vein, which runs up the medial leg and thigh to terminate in the femoral at the upper thigh. The long saphenous vein is important because of the frequency with which it becomes

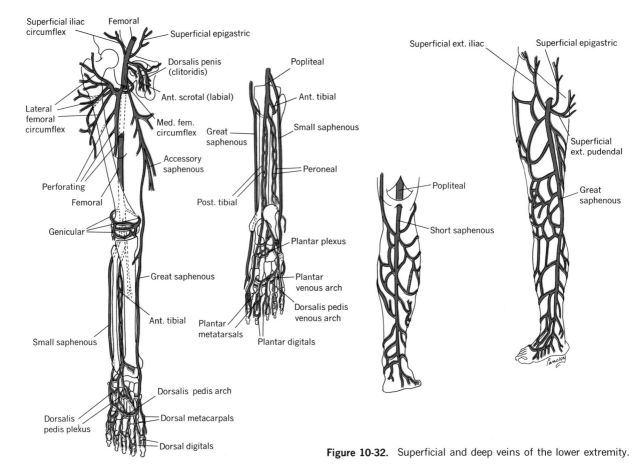

Figure 10-32. Superficial and deep veins of the lower extremity.

Figure 10-33. Veins of the pelvis and external genitalia.

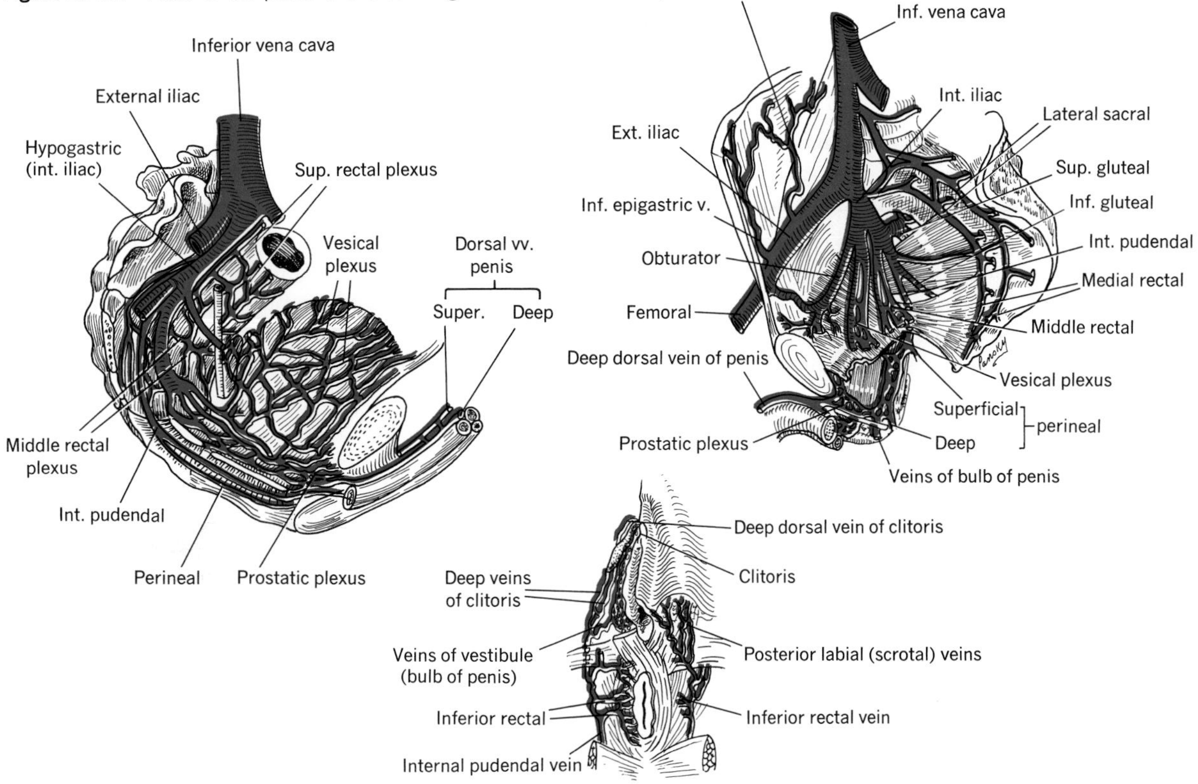

Inferior vena cava

External iliac

Hypogastric (int. iliac)

Sup. rectal plexus

Vesical plexus

Dorsal vv. penis

Super.　Deep

Middle rectal plexus

Int. pudendal

Perineal　Prostatic plexus

Deep iliac circumflex v.

Inf. vena cava

Int. iliac

Lateral sacral

Ext. iliac

Sup. gluteal

Inf. epigastric v.

Inf. gluteal

Obturator

Int. pudendal

Medial rectal

Femoral

Middle rectal

Deep dorsal vein of penis

Vesical plexus

Prostatic plexus

Superficial

Deep

perineal

Veins of bulb of penis

Deep dorsal vein of clitoris

Deep veins of clitoris

Clitoris

Veins of vestibule (bulb of penis)

Posterior labial (scrotal) veins

Inferior rectal

Inferior rectal vein

Internal pudendal vein

varicosed. The two saphenous veins usually communicate with each other.

DEEP VEINS. These veins (Fig. 10-32) correspond to the arteries.

Veins of the Pelvis and Abdomen. The course and tributaries of the external, internal, and common iliac veins correspond to those of their companion arteries (Fig. 10-33).

INFERIOR VENA CAVA. The inferior vena cava (Figs. 10-31, 10-33) is the largest blood vessel in the body (about 3.125 cm [1¼ in.] thick). It is longer than the superior vena cava. It begins at the level of the fifth lumbar vertebra by the union of the two common iliac veins and runs up the right side of the lumbar vertebrae and behind the peritoneum. Its upper portion lies buried in the back of the liver. It pierces the diaphragm and pericardium to enter the right atrium of the heart.

In the abdomen, the inferior vena cava receives tributaries that correspond to the lumbar arteries, renal arteries, suprarenal arteries, and testicular or ovarian arteries. It also receives the hepatic veins.

Of great importance is the fact that it does not receive veins from the digestive tract, pancreas, or spleen. The veins draining these structures constitute the portal system.

(All systemic veins, except the superior and inferior venae cavae, contain a series of bicuspid valves that permit blood to flow toward the heart but not in the opposite direction. The valves are thin, semilunar, pocket-forming flaps of endothelium [intima] that can approximate each other to prevent backflow. The pulmonary and portal systems of veins do not have valves.)

Portal System. The mucous membrane of the small intestine contains villi that project from its surface into the gut lumen and give it a "velvety" appearance. The villi are fingerlike processes that contain microscopic minute arterioles, venules, and lymph vessels with associated capillaries. The arteriole nourishes the cells of the villus, while the venule receives the products of digestion that were absorbed by the blood capillaries. The lymph vessels receive emulsified fat absorbed by the lymph capillaries. The venules, with digestive products, unite with one another to form veins, which become the tributaries of the superior mesenteric vein.

Venous blood from the spleen and pancreas via the splenic vein joins the venous blood from the superior mesenteric, behind the pancreatic neck, to form the portal vein (Fig. 10-34). The inferior mesenteric vein joins the splenic, the superior mesenteric, or the angle of union between them. Veins corresponding to the gastric and gastroepiploic arteries drain the stomach and enter either the portal vein or the superior mesenteric. No vein exists comparable to the celiac artery.

The portal vein runs with the hepatic arteries and bile ducts in the lesser omentum (portal triad). Thus, the liver receives blood from the digestive tract, pan-creas, and spleen. The terminal branches of the portal vein lie between the liver lobules and are called inter-lobular veins. From these, the blood enters a lobule at its periphery and runs in spaces (sinusoids) between plates of liver cells to the center of the lobule. The liver cells absorb the glucose from the digestive tract, alter it to glycogen, and store it.

The center of the lobule contains a venule, which leaves the lobule to unite with other venules to form a sublobular vein. The latter unite to form a tributary of a hepatic vein. The hepatic veins finally empty into the inferior vena cava and, indeed, are the last veins received by the vena cava.

Lymphatic System

The lymphatic system (Fig. 10-35) consists of (1) a large number of lymphatic capillary vessels, distrib-uted widely through the tissues; (2) a system of collect-ing ducts, which open into the bloodstream; and (3) a number of lymph nodes or glands, interspersed along the lymph channels. The lymphatics constitute a

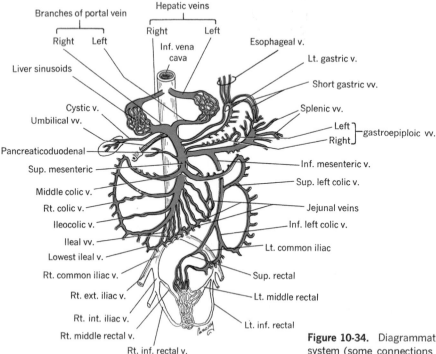

Figure 10-34. Diagrammatic presentation of the portal venous system (some connections to the systemic system are noted).

Figure 10-35. Lymphatic drainage of the body. Generally speaking the right lymphatic duct drains the heart, lungs, part of the diaphragm, the right upper part of the body, and the right side of the head and neck and empties into the right subclavian vein. The thoracic duct drains the rest of the body.

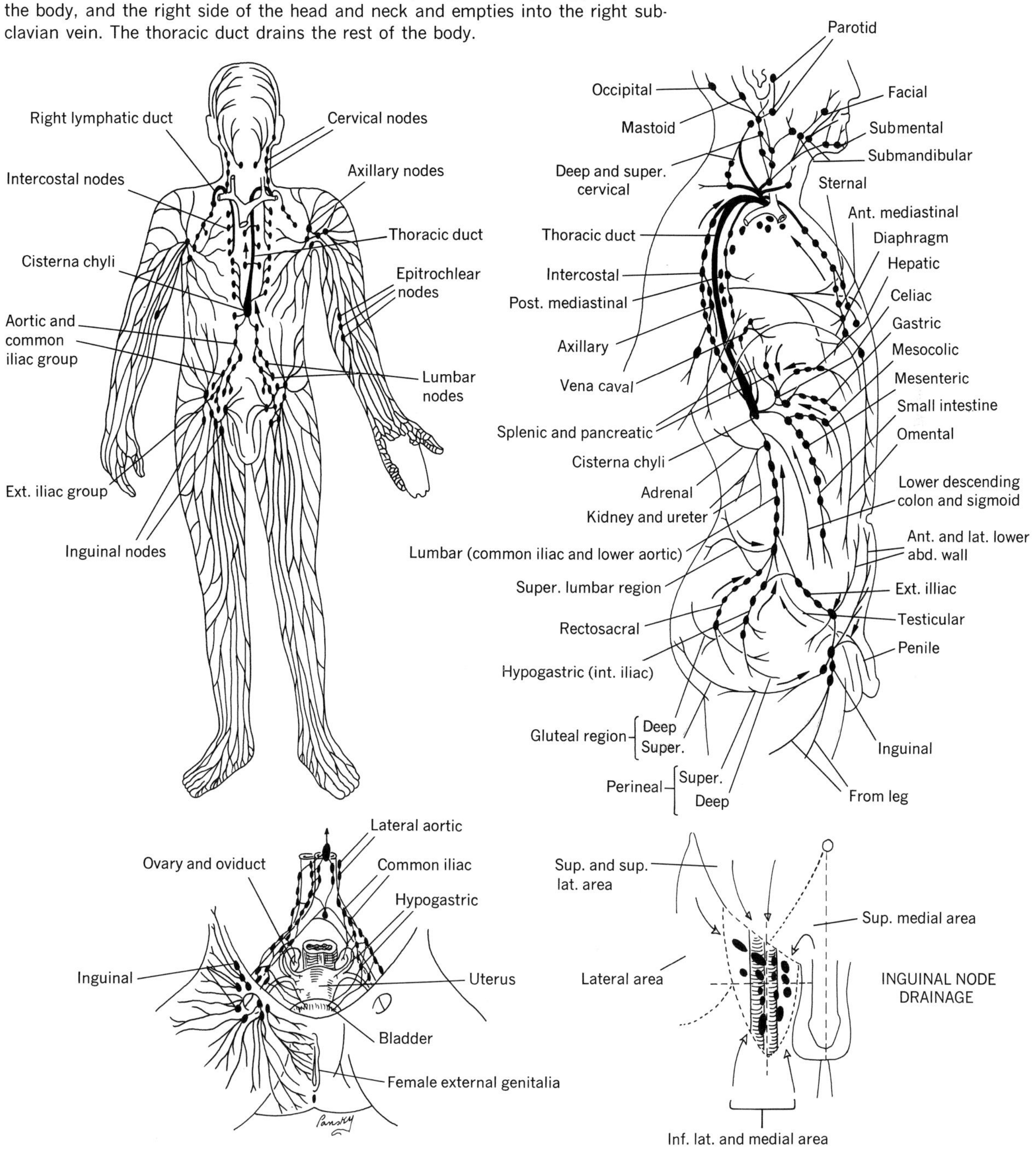

Right lymphatic duct

Intercostal nodes

Cisterna chyli

Aortic and common iliac group

Ext. iliac group

Inguinal nodes

Cervical nodes

Axillary nodes

Thoracic duct

Epitrochlear nodes

Lumbar nodes

Parotid

Occipital

Mastoid

Deep and super. cervical

Thoracic duct

Intercostal

Post. mediastinal

Axillary

Vena caval

Splenic and pancreatic

Cisterna chyli

Adrenal

Kidney and ureter

Lumbar (common iliac and lower aortic)

Super. lumbar region

Rectosacral

Hypogastric (int. iliac)

Gluteal region { Deep Super.

Perineal { Super. Deep

Facial

Submental

Submandibular

Sternal

Ant. mediastinal

Diaphragm

Hepatic

Celiac

Gastric

Mesocolic

Mesenteric

Small intestine

Omental

Lower descending colon and sigmoid

Ant. and lat. lower abd. wall

Ext. illiac

Testicular

Penile

Inguinal

From leg

Lateral aortic

Ovary and oviduct

Common iliac

Hypogastric

Inguinal

Uterus

Bladder

Female external genitalia

Sup. and sup. lat. area

Lateral area

Sup. medial area

INGUINAL NODE DRAINAGE

Inf. lat. and medial area

Chap. 10 Circulation: The Cardiovascular System **373**

"one-way" passage from the interstitial fluid to the blood. The system differs from the vascular system in that it does not form a complete "tubular" circuit because the lymphatic capillaries begin as blind-end lymph capillaries.

The fluid in the lymphatic vessel is called lymph (Latin *lympha*, pure spring water), which is a colorless fluid derived from tissue juices by the lymph capillaries. The capillaries are apparently permeable to practically all interstitial-fluid constituents, including protein, which either diffuse or filter into them. Lymph in the lacteals of the intestinal villi, which is called chyle, contains a considerable amount of fat and appears milky.

Lymph flow depends primarily on forces external to the vessels, such as the "pumping action" of the muscles through which the lymphatics flow and the effects of respiration on the thoracic cage pressures. External pressure, due to the presence of valves, allows only unidirectional flow. Lymphatics do not pulsate nor can they depend on the heart to propel the fluid in them.

Lymph Capillaries and Vessels. The lymph capillaries and vessels (Fig. 10-36) are structurally similar to blood capillaries but are larger in diameter and also have many blind ends (lacteals of the intestine). They unite to form larger vessels, which are similar to small veins and, like them, are provided with valves and offer an alternative pathway for the return of tissue fluid to the bloodstream. They are small, delicate, and thin walled, and they tend to run side by side as "leashes of vessels" rather than to unite to form large trunks as do blood vessels. The flow of lymph is sluggish but increases with activity, which seems to "massage" the vessel walls. They are predominantly associated with skin, glands, and membranes (mucous, serous, synovial, etc.). They tend not to be present if blood vessels are also absent and are not found in the central nervous system, skeletal muscles, bone marrow, splenic pulp, or avascular structures such as hyaline cartilage, nails, hair, and the eye. Along their courses are found lymph glands (nodes) through which the vessels pass.

In a normal person the fluid filtered out of the capillaries each day slightly exceeds that which is reabsorbed. Thus, a malfunction leads to increased interstitial fluid (edema). Since there is a steady small loss of protein from the blood into the interstitial fluid, a breakdown in this removal is probably the major cause of marked edema in patients with lymphatic malfunction.

Lymph Glands (Nodes) and Major Drainage. The lymph glands (Figs. 10-35, 10-36) are flattened, ovoid, or kidney-shaped bodies located at intervals along the lymph vessels. They are particularly numerous at the roots of the limbs, armpit (axilla), and groin (inguinal); at the side of the neck (cervical); and along the course of the abdominal aorta and its branches.

They range from 1 to 25 mm (pinhead size to that of a large bean). Each gland has several lymph vessels entering it and a smaller number leaving it. The entering (afferent) vessels are found scattered over its outer surface, while those leaving it (the efferent vessels) pass out from an indented area called the hilum. Microscopically the glands are composed largely of lymphoid tissue, which consists of lymphocytes of different sizes supported by reticular cells and fibers and by collagenous, elastic, and smooth-muscle fibers. Each gland consists of two easily differentiated parts, the outer (superficial) cortex and a deeply (inner) placed medulla. In the cortex are many lymphoid follicles, while in the medulla the lymphoid tissue is of the diffuse type. The organ, as a whole, is permeated by numerous large channels or sinuses that transmit the fluid lymph through the substance of the gland from afferent to efferent vessels. In its passage through the node, the lymph loses little of its volume but carries off the lymphocytes produced in the node (just as the blood carries off those produced in the spleen).

After passing through at least one, and often more than one, node, the lymph is finally emptied into the venous bloodstream at the root of the neck. The principal return is into the left or right brachiocephalic veins or into one of the two veins that unite to form it (left subclavian or left internal jugular) or between them.

The lymph vessels from the intestinal villi empty by way of intestinal trunks into a reservoir, the cisterna

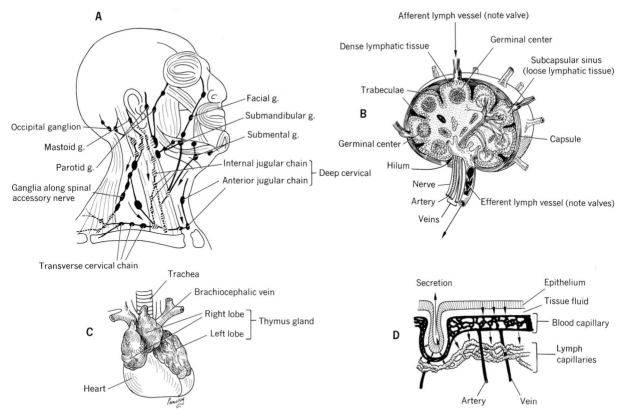

A

Facial g.
Submandibular g.
Submental g.
Internal jugular chain ⎤
Anterior jugular chain ⎦ Deep cervical

Occipital ganglion
Mastoid g.
Parotid g.
Ganglia along spinal accessory nerve

Transverse cervical chain

Trachea
Brachiocephalic vein
Right lobe ⎤
Left lobe ⎦ Thymus gland

C

Heart

Afferent lymph vessel (note valve)
Germinal center
Dense lymphatic tissue
Subcapsular sinus (loose lymphatic tissue)
Trabeculae
B
Germinal center
Capsule
Hilum
Nerve
Artery
Efferent lymph vessel (note valves)
Veins

Secretion
Epithelium
Tissue fluid
Blood capillary
D
Lymph capillaries
Artery
Vein

Figure 10-36. A. Lymph drainage of head and neck (g = ganglia). **B.** Diagram of small lymph node, illustrating cellular detail as well as blood vessels and nerves. (After studies by Garven.) **C.** Thymus gland (gross). **D.** Illustration (schematic) of relationship of epithelium, blood capillaries, lymph capillaries, and tissue fluid. (After studies by Garven.)

chyli. It lies to the right of the abdominal aorta just below the diaphragm. The cisterna also receives the right and left lumbars, which bring lymph from the lower limbs, pelvis and pelvic viscera, kidneys, adrenals, deep lymphatics of the abdominal walls, and the intestinal trunks, which carry lymph from the stomach, spleen, liver, and intestines. From its upper end proceeds the thoracic duct, a pale, thin-walled structure of about 45 cm (18 in.), which enters the thorax in the posterior mediastinum. It continues up on the left side of the esophagus into the neck, where it arches behind the carotid sheath to join (as a single vessel) the junction of the left internal jugular and subclavian veins. This entrance is guarded by valves to prevent regurgi-

tation of blood. The duct returns lymph from both lower limbs, abdomen (except upper surface of the right lobe of the liver), left side of thorax, left side of head and neck, and left upper limb.

The right lymphatic duct is short (about 2.5 cm [1 in.] long) and is formed by the union of the right jugular, subclavian, and bronchomediastinal lymph trunks. It opens into the junction of the right internal jugular and right subclavian veins. It returns lymph from the upper surface of the right lobe of the liver, right lung and pleura, right side of the heart, and right upper limb.

Distribution of Major Lymph Glands. In the head and neck there are seven groups of lymph glands that are recognizable (Fig. 10-35): (1) submental, beneath the chin; (2) submandibular, under the cover of the jaw angle; (3) preauricular, in front of the ear; (4) postauricular, behind the ear near the mastoid process; (5) occipital, in the occipital region; (6) superficial cervical, in the neck over the sternocleidomastoid muscle; and (7) deep cervical, deep in the neck in relation to the carotid artery and internal jugular vein.

In the arm are two chief groups: (1) the superficial cubital, near the medial epicondyle of the humerus, which drains the hand and forearm; and (2) the axillary nodes, alongside the axillary vessels to receive lymph from the arm, breast, and lateral chest wall (Figs. 10-35, 10-37).

In the leg the inguinal nodes lie in the groin to drain the leg, external genitalia, and superficial perineal region.

In the thoracic and abdominal cavity are numerous groups: (1) tracheobronchial; which drain the lungs, heart, and mediastinum; (2) gastric, along the lesser curvature of stomach; (3) mesenteric, between layers of mesentery and which drain the intestine; and (4) numerous glands alongside the aorta, common iliacs, and external and internal iliacs, which drain the lower extremity and pelvic viscera.

Figure 10-37. **A.** Scheme of body lymphatic drainage. **B.** Lymph drainage of the breast (cutaneous and glandular lymphatics). **C.** Diagram of lymph node (ganglion) groups in axillary region.

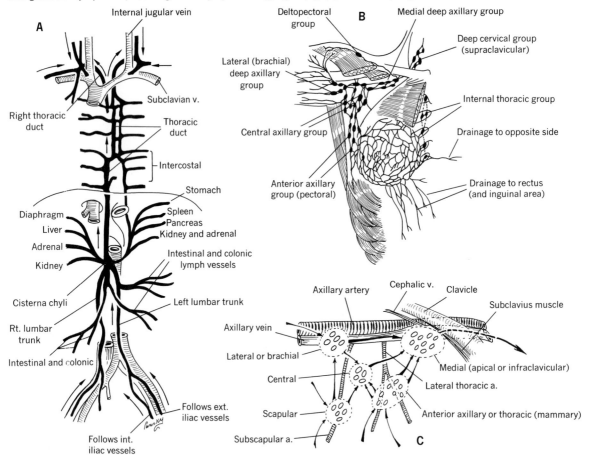

Function. The production of lymphocytes is the major function of lymphatic tissues and organs. These tissues may also play a part in the storage of vitamins and in the metabolism and transport of fat and protein, but these functions have not been proven. The lymphatic tissue also appears to have a minor role in red-cell destruction. Lymphocytes have important roles in the development of antibodies and immune reactions. They are capable of movement but are not normally phagocytic. They carry some viruses and may contain proteolytic enzymes.

The fact that lymphatic tissues serve as "filters" has given way to the theory (barrier theory) that these tissues play a role in the body's defense mechanisms. Inanimate materials, like carbon, are held up in this tissue. Bacteria, cancer cells, viruses, and red blood cells are also retained to a variable degree. Thus, the lymphatic tissues are barriers only to a degree; however, one must remember that their efferent vessels may also facilitate the spread of infections and malignant growths to other organs and tissues.

Diseases and Abnormal Conditions

Hypertension

Hypertension (high blood pressure) is a chronically increased arterial pressure, with the dividing line between normal and hypertension being $^{140}/_{90}$ mm Hg. The diastolic pressure is the most important index of hypertension. About six million people suffer from high blood pressure, and it is one of the major causes of illness and death in the United States.

Congestive Heart Failure

Congestive heart failure is a weakening of the heart, resulting in impaired performance with exercise, shortness of breath, and early fatigue. The basic defect is decreased contractility of the cardiac muscle. The exact mechanism on a molecular basis is unknown. One consequence of the decreased heart output is reduced kidney flow and abnormal sodium and water excretion.

The retained fluid creates extracellular volume expansion, increased venous pressure, and venous return. Edema is a result of increased capillary filtration. With the additional factor of gravity, the edema often is seen in the legs and feet but the same engorgement is occurring in other organs and can create malfunction.

Infarction

Infarction is death of a portion of the myocardium due to insufficient coronary blood flow. Its cause is manifold.

Arteriosclerosis (Atherosclerosis)

Arteriosclerosis (hardening of the arteries) is characterized by a thickening of the arterial wall with connective tissue and cholesterol deposits. The cause is unknown; yet smoking, obesity, high-fat diets, nervous tension, and other factors predispose to this aging condition. Cholesterol may be involved since it is the precursor of bile acids and hormones, but the evidence is not absolute. The incidence of this disease in the United States is high, and there are about 500,000 deaths a year. If arteriosclerosis reaches a point of vessel occlusion, inadequate blood flow results at the time of tension or exercise. The associated pain is called angina pectoris. The cause of death in sclerosis or infarction is either severe hypotension or disordered heart rhythm due to a damaged cardiac conducting system.

Arteriosclerosis can involve most arteries of the body and not only the coronaries. Cerebral occlusion (stroke) is common in old age and accounts for about 20,000 deaths per year.

Hemorrhage

Hemorrhage (bleeding) can cause a drop in blood volume and, therefore, a drop in blood pressure (hypotension). It can be dangerous in that it can reduce blood flow to the brain and cardiac muscle, leading to numerous problems, shock, and even death.

Abnormal Conditions

The autonomic system of controlling blood pressure is quite adequate for normal activities, e.g., walking, working, and running. However, modern times have made man's body liable to undue stresses. In an aviator making a high-speed turn, the autonomic mechanisms of blood flow are disturbed by centrifugal force, which can result in a failure of blood to return to the heart, thereby leading to a reduced arterial supply to the brain. The result is a brief loss of vision due to a shortage of blood in the retina and possibly loss of consciousness—a "blackout." This may be harmless if only short lived. A pressure suit to prevent blood from stagnating in the lower limbs and abdomen helps to reduce the risk of blackouts.

Space travel presents other obstacles. There is tremendous acceleration at "lift-off" to the stage where the astronaut is ready to go into orbit, imposing a "load" on the body up to nine times the force of gravity. This intensifies peripheral blood stagnation by a factor of nine. A pressure suit, as well as special posture, restricts the flow of blood in these areas to prevent blackouts. Weightlessness is another hazard since the autonomic control of blood flow, especially to the brain, normally operates in a gravitational field. With reduced gravitational pull, the system is varied. The heart rate in any event tends to be slower than normal because of cramped space conditions that prevent the normal exercise and movement to enable body use of the "muscle pump." Deceleration on the return to earth is another major stress, and the acclimation once back to earth may itself take many days for readjustment.

Review Questions

1. What does the circulatory system and its related organs include?
2. How do the pulmonary and systemic circulations differ?
3. What is meant by the mediastinum? Describe its parts.
4. Define the term *pericardium*. Relate it to the heart.
5. Describe the four chambers of the heart. Compare them.
6. Describe a semilunar and atrioventricular valve. How do they differ?
7. Trace the flow of blood to, through, and away from the heart.
8. Which vessels lead to and away from the heart?
9. How is the heart itself supplied by blood? Which vessels drain it?
10. Discuss the conduction system of the heart. Trace an impulse through the heart.
11. What is meant by cardiac cycle? Describe this process. What are systole and diastole?
12. Why do we have heart sounds?
13. What is the cardiac output per minute? How is it calculated?
14. What are the two major factors in blood filtration? Where is the major resistance to blood flow?
15. Correlate the electrocardiogram with the cardiac cycle. What is an electrocardiogram? How is it measured?
16. What makes up our blood? Describe its essential constituents.
17. What various types of vessels do we have in the body? How would you tell them apart?
18. Why does capillary diffusion take place?
19. Is there nervous control of blood vessels? Of the heart?
20. Differentiate between:
    a. Thrombus   b. Embolus   c. Varicose veins

21. Be able to trace the following circulations:
    a. Pulmonary
    b. Common carotids to head and neck
    c. Subclavian to upper extremity
    d. Aorta, thoracic and abdominal, and its branches
    e. Blood supply of chest wall
    f. Blood supply of abdominal organs
    g. Internal iliacs to pelvic structures
    h. External iliacs to lower extremities
22. Be able to retrace the venous blood back to the heart. Note the variation in drainage from the head and cranial sinuses. What are emissary veins?
23. Trace an infection, via the lymph drainage from:
    a. The foot
    b. The hand
    c. The abdominal organs
    d. The chest
    e. The head and neck
24. What makes up the portal system? What does it drain?
25. What is the cisterna chyli? What is lymph?

References

Adolph, E. F.: The Heart's Pacemaker. *Sci. Am.,* **216**(3):32–37, 1967.

Back, N.: Fibrinolysin System and Vasoactive Kinins. *Fed. Proc.,* **25**:77–83, 1966.

Berne, R. M., and Levy, M. N.: *Cardiovascular Physiology,* 2nd ed. C. V. Mosby Co., St. Louis, 1972.

Burton, A. C.: *Physiology and Biophysics of the Circulation,* 2nd ed. Year Book Medical Publishers, Inc., Chicago, 1972.

Friedberg, C. K.: *Diseases of the Heart,* 3rd ed. W. B. Saunders Co., Philadelphia, 1966.

Ham, A. W.: *Histology,* 6th ed. J. B. Lippincott Co., Philadelphia, 1969.

Henry, J. P., and Meehan, J. P.: *The Circulation.* Year Book Medical Publishers, Inc., Chicago, 1971.

Pickering, G. W.: *High Blood Pressure,* 2nd ed. Grune & Stratton, Inc., New York, 1968.

Ratnoff, O. D.: The Biology and Pathology of the Initial Stages of Blood Coagulation. *Prog. Hematol.,* **5**:204–45, 1966.

Rushmer, R. F.: *Cardiovascular Dynamics,* 3rd ed. W. B. Saunders Co., Philadelphia, 1970.

————: *The Structure and Function of the Cardiovascular System.* W. B. Saunders Co., Philadelphia, 1972.

Soloff, L. A.: Coronary Artery Disease and the Concept of Cardiac Failure. *Am. J. Cardiol.,* **22**:43–48, 1968.

Spain, D. M.: Atherosclerosis. *Sci. Am.,* **215**(2):48–56, 1966.

Wood, J. E.: The Venous System. *Sci. Am.,* **218**(1):86–96, 1968.

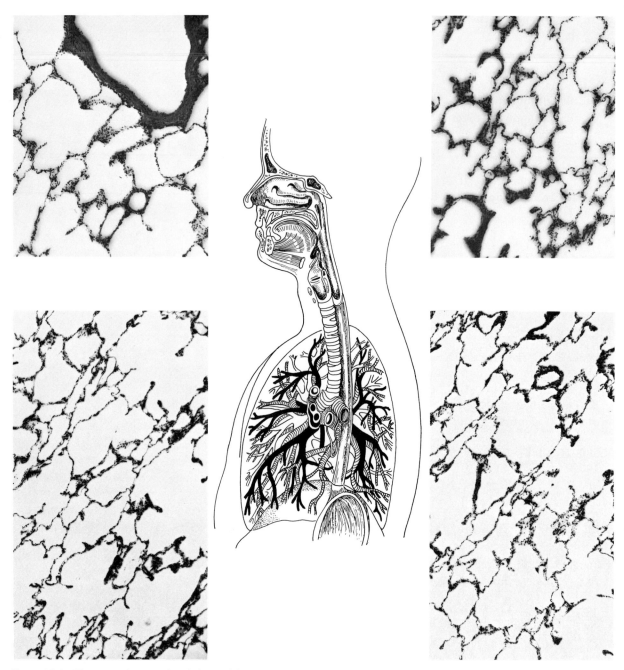

Figure 11-1. The general anatomic relations of the respiratory system, including the right lung, are seen in the center drawing. The four photographs are of human lung, showing the alveolar spaces and connecting channels (the actual air-exchange regions). ×200. (Micrographs courtesy of Dr. G. Colin Budd, Medical College of Ohio at Toledo.)

Transfer of Gases: The Respiratory System

The oxidation or burning up of food goes on perpetually in the body. The product of this chemical reaction is energy. Energy is required for muscle action and active transport of substances across cell membranes. Oxygen, the gas, is as important as food since the tissues require both continuously. The breakdown of "food," or the major end product of these oxidations, releases a gas, carbon dioxide (CO_2), that must be removed. A unicellular organism can exchange oxygen and carbon dioxide directly with the external environment. In a complex organism like man, however, only a small fraction of the total cells is in direct contact with the external environment (skin, gastrointestinal lining, respiratory tract lining, and so forth). Thus, specialized systems have been developed to supply oxygen and remove carbon dioxide. In man, this is the respiratory system, in which the organs of gas exchange are the lungs.

The lung is the organ of pulmonary circulation and through it passes, in a unit of time, the same amount of blood as through the organs of the rest of the body. The circulation of blood in the lung is only fully established at birth and is associated with changes in the heart and blood vessels. The pulmonary artery contains venous blood (reduced oxygen), while the pulmonary veins contain arterial blood (almost completely oxygenated). The lung is supplied not only with venous blood at lower pressure in larger amounts by the pulmonary artery but also with arterial blood (bronchial) at higher pressure in smaller amounts. The two systems have connections and play a role in disease. During life the lung undergoes a regular rhythmic and exten-sive change in volume associated with its large amount of elastic fibers. These fibers often alter with age, and this accounts for the increasing incidence of respiratory disease in older age groups. The lung contains many air-filled, thin-walled spaces providing a large surface area for gas exchange per unit of time (about 700 million minute sacs or alveoli). The lung, too, has the provision of a wide margin of safety (reserve) and, therefore, only about 1/20 of the possible oxygen intake is used in normal quiet breathing.

Specialized blood components are present to allow for transportation of large amounts of gases between the lungs and cells. In man (at rest) the cells use up about 200 cc of oxygen per minute, and under stress this may increase almost 30-fold. Equivalent amounts of CO_2 are eliminated. Thus, coordination must exist between breathing and metabolic needs. The movement of gases into the body and out of it is known as respiration. Respiration takes place in two different areas: (1) the gas is transported in the blood, with gas exchange taking place between the blood and body tissues (internal respiration) via a complex enzymatic process in the cell; and (2) the exchange of gas between the blood and the external environment (external respiration) consists of a movement of volumes of air in and out of the lungs (breathing), a movement of molecules across a membrane (oxygen passes in and CO_2 out), and a movement of blood (with the necessary oxygen-carrying pigment called hemoglobin) to and from the surface where the gaseous exchange takes place (the pulmonary circulation). The term *respiration* in common usage usually connotes the external type, and comprises the lungs, the many passageways leading to them, and the chest structures around them that are needed for movement of air in and out of the lungs.

The circulatory and respiratory systems function as a unit to provide the tissues with sufficient oxygen at adequate tensions to meet metabolic needs. The function of respiration is to operate on alveolar air (in the lungs) and to maintain the pO_2 (partial pressure of oxygen) at a normal level. It does so by "ventilating" the alveoli with "fresh air" from the external environment. Thus, arterial pO_2 is a criterion of adequate respiration. Similarly, the circulation operates on tissue fluid to maintain its pO_2 at a normal level by perfusing the tissue fluid with an adequate arterial blood supply. Therefore, the criterion of adequate circulatory performance is the level of tissue fluid or venous pO_2.

If respiration fails to perform its function adequately and alveolar and arterial pO_2 fall, we have arterial anoxemia, due either to the abnormality of the composition of the air being supplied or to a rate of ventilation that is too low relative to the metabolic rate. If the circulatory system fails to perform, tissue and mixed venous pO_2 fall and we have venous anoxemia. This can occur as a result of the abnormal composition of arterial blood or because the rate of blood flow is too small relative to metabolic rate. Arterial blood may be abnormal with regard to its O_2 content due to low O_2 saturation or low oxygen capacity or both. Thus, venous anoxemia is caused by circulatory inadequacy and arterial anoxemia by respiratory inadequacy, and they are dependent on each other since, if there is arterial anoxemia due to respiratory failure, there will also be a venous anoxemia unless the system compensates for the respiratory defect.

Regardless of the great flexibility, there are restrictions on respiration that determine the value of the arterial (alveolar) pO_2 and these are ventilation, metabolic rate, respiratory quotient, and the composition of inspired air. There are also restrictions on circulation and mixed venous pO_2 and these are cardiac output, metabolic rate, and arterial blood compensation.

Respiratory Tract Anatomy

The respiratory system is divided morphologically into the following components:

1. A conducting system, extending from the nose to the terminal fine bronchioles via the pharynx, larynx, trachea, and bronchi (Fig. 11-1). It delivers air in a warmed, cleansed state to the respiratory units.

2. The respiratory units, which are composed of a respiratory bronchiole and its divisions—the alveolar ducts, alveolar sacs, and pulmonary alveoli. Here is where the gas exchange actually occurs (across the alveolar walls and their capillary beds).

3. The large incoming blood vessels, the pulmonary arteries (and their branches), which are associated with the air-conducting system. Each part of the bronchial tree is associated with a branch of the pulmonary artery and, in the bronchial wall, a branch of the bronchial artery. The capillary bed lies in the alveolar wall. The pulmonary veins carry oxygenated blood back to the heart.

Nose

External Nose. The external nose (Fig. 11-2) forms a triangular pyramid, with its upper angle (near the forehead) called the root and its free angle the tip or apex. The lateral walls expand to form the alae (wings), which are supplied with sweat glands and sebaceous glands. The base has two elliptic openings, the nares. Small hairs, the vibrissae, are seen along the margins of the nares to help protect against foreign substances carried in during respiration. The dorsum of the nose is formed by the union of the two lateral sides in the midline, and the bridge is the upper part of the dorsum. The skin is thin over the root and dorsum; however, it is thick over the alae and adherent there, with a rich blood supply making it suitable for plastic surgery and healing. The nose is made up of five cartilages, which add to its shape and support. The external nose is supplied by the nasal branch of the first and second divisions of the trigeminal (V) nerve. The fact that one of the nasal nerves is a branch of the ophthalmic trunk, which has connections with the eye, ac-

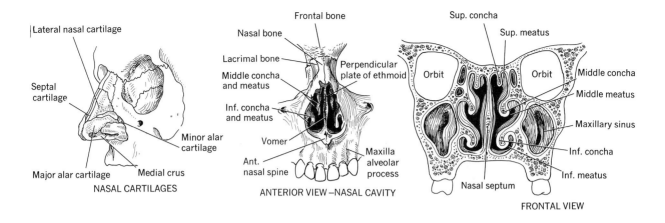

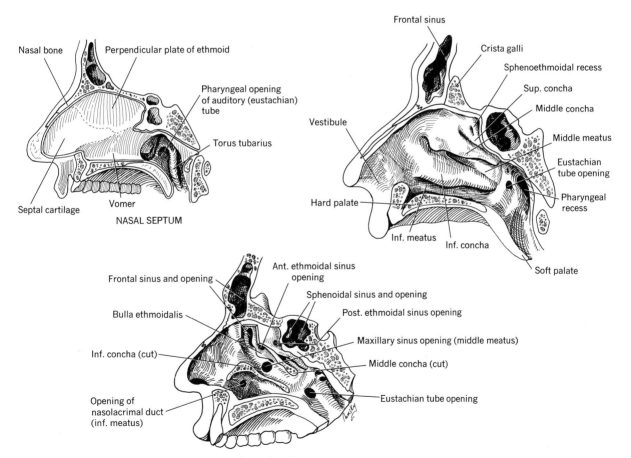

Figure 11-2. Anatomy of the external nose and nasal cavity.

counts for lacrimation (tearing) with painful infections around the nostrils. The arteries of the nose are branches from the facial and ophthalmic arteries, with the veins following a similar path. Those following the ophthalmic artery end in the cavernous venous sinus of the brain, and thus an infection of the nose may travel into the intracranial circulation and result in septic emboli to the brain and meningitis.

Internal Nose. In the internal nose (Fig. 11-2) are two nasal cavities adapted for detecting odors, warming air, and filtering air on its way to the lungs. These cavities are separated by a nasal septum, and they communicate anteriorly to the outside via the anterior nares and posteriorly with the nasopharynx via the posterior nares or choanae. The vestibule is the doorway or aperture (slightly dilated) into the cavities. Large hairs with sebaceous glands and sweat glands are found here.

The roof of each nasal cavity is thin and delicate and perforated by olfactory nerves and blood vessels and lies just below the cranial cavity. A fracture here can lead to meningitis. The floor is formed by the hard and soft palate separating the cavity from the mouth. The medial or septal wall is usually deflected from the midline (deviated septum) and made up of the vomer bone, perpendicular plate of the ethmoid bone, and a septal cartilage. The lateral wall has a very irregular surface due to three elevations, the superior, middle, and inferior conchae or turbinates. Below and lateral to each concha are found the corresponding nasal passages or meatuses, and above the superior is the sphenoethmoidal recess into which the sphenoidal sinus opens. The superior and middle conchae arise from the ethmoid bone; the inferior is an independent bone. The conchae contain air cells lined with mucous membrane.

The anterior and middle ethmoid cells, the frontal and maxillary sinuses, and the infundibulum (a short passage by means of which the frontal sinus enters the middle meatus) all open into the middle meatus. The inferior meatus receives the nasolacrimal duct emptying secretions from the medial side of the eyes.

The mucous membrane of the nose is continuous with all the cavities communicating with it. The upper or olfactory part is thin and less resistant, covers the cribriform plate of the ethmoid, and contains the olfactory nerve endings. The respiratory part is thicker, more vascular (particularly thick over the middle and inferior conchae), making the cavities and apertures of the nose smaller in the living human than in the skeleton. The mucous membrane is lined by columnar epithelium that is ciliated in the respiratory part and nonciliated in the olfactory portion. The cilia in the anterior one third are not active, while the rest create a current of mucus and fluid moving toward the nasopharynx. The respiratory part of the mucous membrane enables the air to be warmed and made moist as it passes through the nose. The membrane is supplied by many glands, conspicuous over the lower and back portions of the outer wall and over the septum. They may hypertrophy and are capable of copious watery secretion. Inflammation is common here.

The blood supply of the mucous membrane is chiefly via the terminal branches of the internal maxillary artery (its sphenopalatine branch) and by the anterior and posterior ethmoidal branches of the ophthalmic artery (Fig. 10-20, p. 357). The veins follow the arteries and form a network beneath the membrane. The lymph drainage is into the deep cervical glands in the neck. The nerves are derived from (1) the olfactory nerves to the upper one third of the cavities and (2) the sensory nerves from the ophthalmic and maxillary divisions of the trigeminal (V) nerve (Fig. 8-16, p. 267).

Paranasal Sinuses

The paranasal sinuses (Figs. 11-2, 11-3) are irregular air spaces originating from buds of mucous membrane that arise from the nasal cavities and grow into the diploic layer of some bones. Each is named from the bone it occupies, communicates with the nasal cavity with which the mucous membrane is continuous, and is filled with air. The narrow channel communications sometimes occlude due to congestion of the membrane. Like the lining of the nose, the membrane of the sinuses is covered with ciliated epithelium. The sinuses are as follows: (1) the maxillary, which is the largest and the first to appear embryologically and which opens into the middle meatus; (2) the frontal (respon-

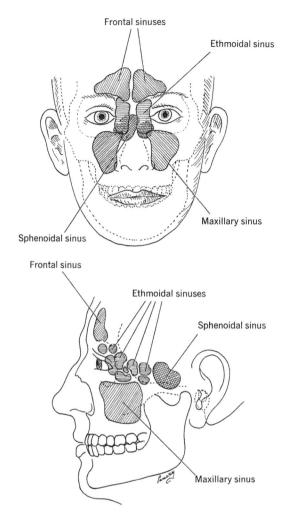

Figure 11-3. Accessory sinuses of the nose (paranasal sinuses). (After studies by Jones and Shepard.)

dle open into the middle meatus, the posterior into the superior meatus.

Nasopharynx

The nasopharynx (Fig. 11-2) is found behind the nasal cavities, above and behind the soft palate. It leads the air from the back of the nasal openings to the oropharynx and larynx. It belongs to the respiratory tract and not the digestive tract. The walls of the nasopharynx, except for the soft palate, are immovable and, therefore, its cavity remains patent and its form never alters. In its posterior wall (especially in children) is a mass of lymphoid tissue called the pharyngeal tonsils or adenoids, which may fill the nasopharynx and hinder or even completely block nasal breathing. The uvula and soft palate, which move upward and backward in its floor, prevent food and liquid from regurgitating into the nose. In the lateral walls of the nasopharynx, just behind the inferior concha, are the openings of the eustachian tubes, which connect this region with the middle ear. The tonsils lie in a pharyngeal recess near the tube opening. Thus, when the adenoids occlude the orifice of the tube, the air in the middle ear is gradually absorbed and deafness may result. The middle ear can, however, be inflated via the tube by means of a eustachian catheter.

Oropharynx

The oropharynx is partly respiratory and partly digestive. It connects the nasopharynx and mouth with the regions below, namely, the laryngeal pharynx for food to the esophagus and the larynx for air (see Chap. 12).

Larynx

The larynx (Fig. 11-4) is the specialized upper portion of the "windpipe," extending from the epiglottis to the trachea and forming the organ of the voice as well as respiratory passage. It lies in the front and upper neck, below the tongue and hyoid bone and between the great neck vessels. It opens below into the trachea and above into the pharynx and consists of cartilages, ligaments, and muscles lined by a thickened mucous membrane. It is embraced by the thyroid

sible for the prominence of the forehead above the eyebrow, where it lies in the bone), which also empties into the middle meatus (it is not present at birth and not recognizable until age two to seven years); (3) the sphenoidal, which is found in the body of the sphenoid bone and opens into the sphenoethmoidal recess; and (4) the ethmoid sinuses or cells, which consist of eight to ten thin-walled, intercommunicating cavities in the ethmoid bony labyrinth and which are divided into anterior, middle, and posterior. The anterior and mid-

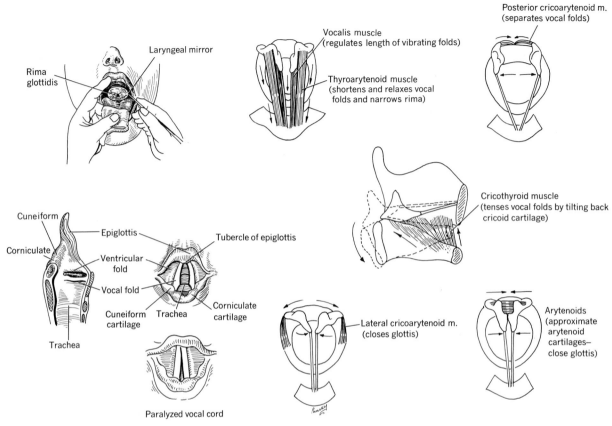

Figure 11-4. The rima glottidis and the intrinsic laryngeal muscles that control its opening and closing.

gland. The larynx is usually smaller in the female than in the male.

Cartilages. The cartilages of the larynx are nine in number (Fig. 11-5).

EPIGLOTTIS. The epiglottis, a leaflike structure of yellow fibrocartilage covering the superior laryngeal aperture, is found behind the tongue, hyoid bone, and thyroid cartilage and is attached to the latter by the thyroepiglottic ligaments. It is connected to the tongue by the median, right, and left glossoepiglottic folds, which create two depressions between them and behind the tongue, the valleculae.

THYROID CARTILAGE. The thyroid ("shieldlike") cartilage is hyaline cartilage, is the largest of the laryngeal cartilages, and is formed by a pair of laminae that fuse in front but remain separated behind. The fusion has been called the laryngeal prominence or Adam's apple. The laminae are separated above by the thyroid notch, which can be felt subcutaneously. The free posterior borders form the so-called superior and inferior horns. The former attaches to the hyoid bone above by the lateral thyrohyoid ligament, while the inferior horn articulates with the cricoid cartilage. One also sees an oblique line on the outer, lateral surface of the laminae, which gives insertion to the sternothyroid and thyrohyoid muscles of the neck. The hyoid bone and thyroid cartilage are also joined by the thyrohyoid membrane.

CRICOID CARTILAGE. The cricoid cartilage is also hyaline cartilage and is shaped like a "signet ring" with the wide portion posteriorly. It marks the beginning of the trachea and the level of origin of the esophagus. It consists of an anterior arch and a posterior lamina. The arch is attached to the thyroid cartilage by the cricothyroid membrane and to the trachea via the cricotracheal ligaments. Both the thyroid and the cricoid cartilages tend to calcify early in life and may ossify thereafter.

ARYTENOID CARTILAGES. The two arytenoid cartilages look like triangular pyramids whose bases articu-

Figure 11-5. Cartilages of the larynx.

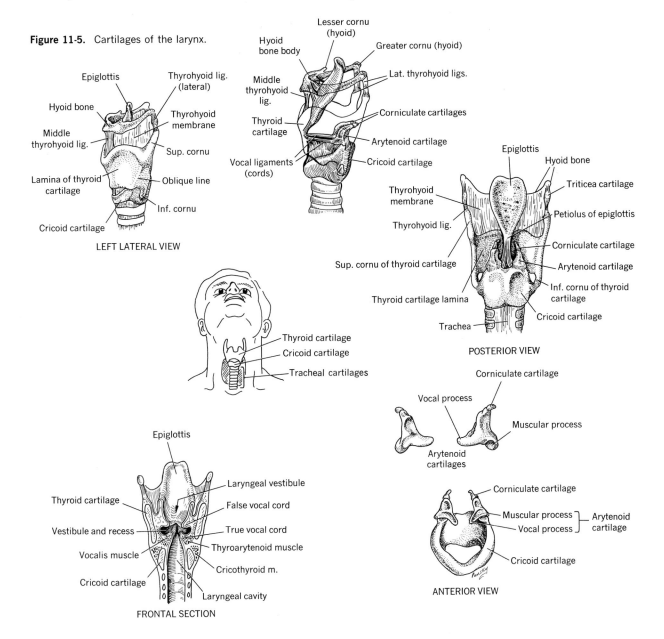

LEFT LATERAL VIEW

POSTERIOR VIEW

FRONTAL SECTION

ANTERIOR VIEW

late with the upper part of the cricoid cartilage. The posterolateral angle of the base is the muscular process, which gives attachment to the cricoarytenoid muscles. The anterior angle of the base is called the vocal process and gives attachment to the vocal cords.

CORNICULATE CARTILAGES. The two corniculate cartilages are small and conical, attach to the apices of the arytenoid cartilages, and give attachment to the aryepiglottic folds.

CUNEIFORM CARTILAGES. The two cuneiform cartilages are a pair of rod-shaped cartilages found in the aryepiglottic folds in front of the corniculates.

Membranes and Ligaments. Connecting the ensemble together are a series of membranes and ligaments. There are thyrohyoid and cricothyroid membranes, mentioned above. Of importance is the fact that the lateral portion of the cricothyroid membrane or conus elasticus is free and unattached above; however, it is fixed anteriorly to the thyroid cartilage and posteriorly to the vocal processes of the arytenoids. The free border is covered by mucous membrane (nonkeratinized stratified squamous) and forms the true vocal cords (Figs. 11-4, 11-6), which appear as short, straight folds characterized by the pallor of the covering mucous membrane produced by the lack of loose submucous tissue as well as blood vessels. There is also a ventricular band covered by membrane that extends from the angle between the thyroid laminae and runs back to the lateral surface of the arytenoids above the vocal cords, thereby forming the false vocal cords.

Laryngeal Cavity. The laryngeal cavity (Fig. 11-4) is divided into three areas: vestibule, glottis, and infraglottis.

VESTIBULE. There is a vestibule or supraglottic por-

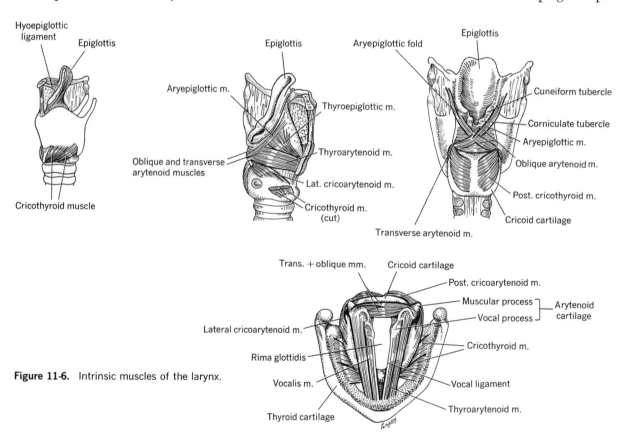

Figure 11-6. Intrinsic muscles of the larynx.

tion whose inlet is bounded by the epiglottis, aryepiglottic folds (between the epiglottis and arytenoid cartilages), and interarytenoid fold.

GLOTTIS. The middle compartment or glottis is the narrowest part of the laryngeal interior. It has a small sinus or ventricular recess between the true and false vocal cords.

The rima glottidis is the fissure (slit) between the true vocal cords and arytenoid cartilages. It is triangular in shape at rest, linear during phonation, and lozenge shaped in respiration (Fig. 11-4). The glottis is closed after inspiration, which helps fix the diaphragm in efforts of urination, vomiting, defecation, and parturition.

INFRAGLOTTIC PORTION. The infraglottic portion (lowest part) extends from the vocal cords to the first tracheal cartilage.

Laryngeal Muscles. The laryngeal muscles are extrinsic and intrinsic (Figs. 11-4, 11-6). They serve to open the glottis and permit breathing, close the glottis and vestibule during swallowing, and regulate the tension of the vocal cords. The first two functions are automatic and controlled by the medulla; the third is voluntary and controlled by the cerebral cortex.

EXTRINSIC MUSCLES. Extrinsic muscles (Figs. 11-4, 11-6) act on the "voice box" as a unit. They are the omohyoid, sternohyoid, sternothyroid, thyrohyoid, and some suprahyoid muscles (stylopharyngeus, palatopharyngeus, and the middle and inferior constrictors of the pharynx).

INTRINSIC MUSCLES. Intrinsic muscles (Figs. 11-4, 11-6) act to modify the size of the laryngeal opening (rima glottidis) as well as to control tension of the vocal ligaments. The major muscles are listed below.

Cricothyroid. The cricothyroid is the only muscle lying on the exterior of the larynx. It is the chief tensor of the vocal ligaments by pulling the arch of the cricoid up around its articulation with the inferior horn of the thyroid cartilage, thus forcing the arytenoids back and stretching the cords.

Posterior Cricoarytenoid. The posterior cricoarytenoid is undoubtedly the most important of the muscles, since it is the only abductor (muscle that separates the cords and widens the opening rima glottidis). Bilateral

paralysis of abduction muscles results in suffocation.

The posterior cricoarytenoid pulls back on the muscular process and thus separates the cords (all the other muscles listed below close the larynx by a "sphincteric" adduction action).

Transverse Arytenoid. The transverse arytenoid, the only unpaired muscle of the larynx, is thin and flat and passes across the backs of the arytenoids, which together help close the laryngeal inlet during swallowing.

Oblique Arytenoids. The oblique arytenoids are weak slips lying on the back of the transverse muscle. Each oblique muscle continues into the aryepiglottic fold as the aryepiglotticus. It draws the arytenoids together and shortens the fold approximating the epiglottis and arytenoids. It acts as a sphincter of the laryngeal inlet during swallowing.

Lateral Cricoarytenoids. The lateral cricoarytenoids are adductors of the cord and reduce the width of the rima glottidis.

Thyroarytenoids. The thyroarytenoids are upward continuations of the lateral cricoarytenoid that pull the arytenoid cartilage forward, slackening the cords (the upper fibers run into the aryepiglottic folds as the thyroepiglottic muscles).

Vocalis. Some of the deepest fibers of the thyroarytenoid (but still a part of it) form a muscle bundle, the vocalis muscle, that draws the vocal process forward, relaxing the vocal ligaments.

Vascular Supply. The vascular supply (Fig. 10-19, p. 356) is derived from the superior laryngeal branch of the superior thyroid artery and inferior laryngeal branch of the inferior thyroid artery.

Lymphatics. The lymphatics from the larynx end in glands in the neck and around the larynx and trachea.

Nerve Supply. The nerve supply is by the vagus nerve via its two branches, the superior and inferior (recurrent) laryngeal. Cutting the superior results in a loss of sensation to the laryngeal mucous membrane, making it hard for the patient to perceive a foreign body in the larynx. The cricothyroid muscle is also weakened or paralyzed, producing a husky voice. The inferior nerve supplies all the remaining intrinsic muscles and mucous membrane.

Thoracic Cavity (Pleural Cavities) and Pleurae

Thoracic Cavity

The lungs lie inside an "atmospheric pump" called the thorax or chest (Figs. 11-7, 11-8), which is walled in and protected by the 12 pairs of ribs and the muscles between them (the intercostals), the backbone (thoracic vertebrae), the breastbone (sternum), and the

Figure 11-7. Schematic presentation of the topography (surface projection) of the lungs and pleura.

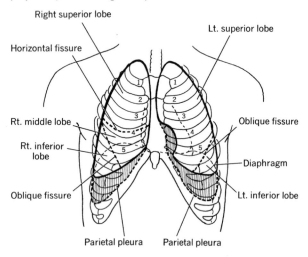

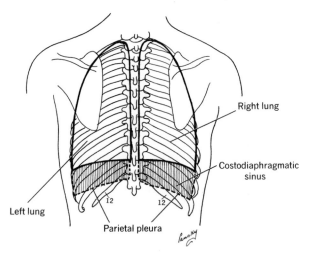

muscles that cover these bones. At its top the thorax is continuous with the base of the neck; the floor of the thorax consists of a musculofibrous sheet called the diaphragm, which separates it from the abdominal cavity. The two lungs almost fill the thoracic cavity, but space is provided in the center for the heart, large vessels leading to and from it, esophagus, thoracic duct, trachea, thymus, and nerves.

The thoracic cavity can be divided into right and left pleural cavities and an area between called the mediastinum (see Chap. 10). The lung in its development invaginates the pleural cavity completely, so that only a potential space remains. Therefore, pleura is divided into a visceral (lung) layer and a parietal (wall) layer (Figs. 11-7, 11-8). The little space between the visceral pleura and the parietal pleura is called the pleural cavity (or space) and is filled with fluid that lubricates the pulsating lungs when they make contact with the thoracic wall.

Pleurae

Visceral Pleura. The visceral pleura invests the lung, dipping into all of its fissures, and adheres firmly to the lung tissue and cannot be normally stripped off.

Parietal Pleura. The parietal pleura is divided into four parts.

COSTAL PLEURA. The costal pleura lines the ribs and cartilages, the sides of vertebral bodies, and the back of the sternum and is the thickest part. It is separated from the thoracic wall by a thick layer of endothoracic fascia.

CERVICAL PLEURA. The costal pleura extends to cover the apex of the lung, as cervical pleura or cupola, which extends into the root of the neck behind the sternomastoid muscle as high as the first rib.

DIAPHRAGMATIC PLEURA; MEDIASTINAL PLEURA. The diaphragmatic pleura is thin, adheres to the diaphragm, and is continuous with the costal pleura laterally and the mediastinal pleura medially. It meets the costal pleura behind the sternum (sternal reflection) and in front of the vertebra (vertebral reflection). Where the visceral pleura becomes continuous with the mediastinal layer of parietal, its upper portion contains those structures making up the root of the

Figure 11-8. A. Expansion of lung (with pleura) during inspiration; note descent of diaphragm and enlarging of chest wall. **B.** In this condition the lungs fail to expand. **C.** Collapse is due to external pressures (effusion, pneumothorax, hemothorax, etc.).

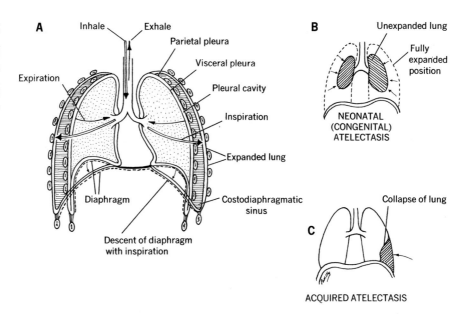

A

Inhale — Exhale
Parietal pleura
Visceral pleura
Pleural cavity
Inspiration
Expanded lung
Costodiaphragmatic sinus

Expiration
Diaphragm
Descent of diaphragm with inspiration

B

Unexpanded lung
Fully expanded position

NEONATAL (CONGENITAL) ATELECTASIS

C

Collapse of lung

ACQUIRED ATELECTASIS

Figure 11-9. Lateral and mediastinal (medial) surfaces of the right and left lungs.

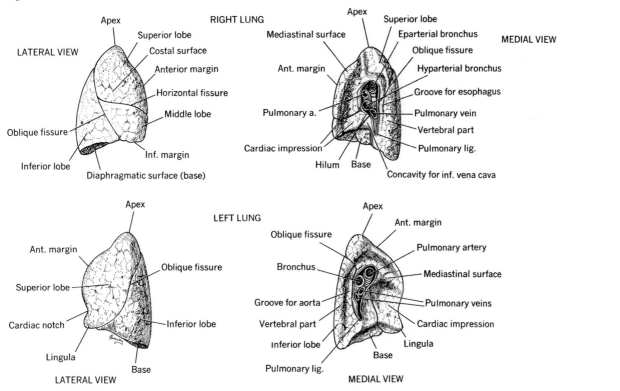

RIGHT LUNG

LATERAL VIEW

Apex
Superior lobe
Costal surface
Anterior margin
Horizontal fissure
Middle lobe
Oblique fissure
Inf. margin
Inferior lobe
Diaphragmatic surface (base)

MEDIAL VIEW

Apex
Superior lobe
Mediastinal surface
Eparterial bronchus
Oblique fissure
Ant. margin
Hyparterial bronchus
Groove for esophagus
Pulmonary a.
Pulmonary vein
Vertebral part
Cardiac impression
Pulmonary lig.
Hilum   Base
Concavity for inf. vena cava

LEFT LUNG

LATERAL VIEW

Apex
Ant. margin
Oblique fissure
Superior lobe
Cardiac notch
Inferior lobe
Lingula
Base

MEDIAL VIEW

Apex
Ant. margin
Oblique fissure
Pulmonary artery
Bronchus
Mediastinal surface
Groove for aorta
Pulmonary veins
Vertebral part
Cardiac impression
Inferior lobe
Lingula
Base
Pulmonary lig.

lung (bronchus plus pulmonary vessels). Its lower portion is empty and forms the so-called pulmonary ligament (Fig. 11-9), a "sleeve" of pleura that narrows below the root structures.

Only on deep inspiration do the lungs and parietal pleurae come in contact with each other. In normal breathing the lungs are not completely expanded and the edges of the pleurae fall together, thereby preventing the formation of a cavity. During quiet breathing the costal and diaphragmatic pleurae remain in opposition below the lower lung border. The space thus formed is known as the costodiaphragmatic recess. (For surface markings of the lungs and pleurae, see Fig. 11-7.)

The Lungs (Pulmones)

Embryology

The pulmonary system arises as an outgrowth of the "digestive tract," beginning as a small, simple bud from the ventral midline side of the pharynx. The bud increases in length and grows down into the chest to form the trachea, which then divides into a right and left primary bronchus. The ends of these bronchi grow and divide to form a complex bronchial tree with terminal alveoli. Only at birth, however, is the structure distended by air. In this development, the trachea retains its original ciliated epithelium. (The esophagus, on the other hand, loses its cilia and develops a stratified squamous lining.) The smooth muscle of the bronchi develops, as does that of the primitive gut, and retains similar nervous connections throughout life and thus similar reactions to epinephrine and acetylcholine. The development of cartilage to make the walls relatively incollapsible is new, as is the large number of elastic fibers, which is correlated with extensive changes in lung volume.

Anomalies of the respiratory tract occur less frequently than do malformations in the other major organ systems, and for the most part variations in development are compatible with survival. Tracheoesophageal fistula (a connection between the trachea and esophagus) is the most frequent major malformation of the respiratory system and creates a hazard of pneumonia in early neonatal life. Initially the pleural cavity is in direct communication with the abdominal cavity (together constituting the celom), but they are later separated by the transverse partition of the diaphragm.

Anatomy of the Trachea and Lungs Proper

Trachea and Extrapulmonary Bronchi. The trachea (Fig. 11-10) begins where the larynx ends at the lower border of the cricoid cartilage. It is 10 to 12.5 cm (4 to 5 in.) long, half in the neck and half in the thorax. About 15 to 20 horseshoe-shaped rings of hyaline cartilage keep the lumen of the trachea open and support it except posteriorly, where it is flattened and closed by the trachealis muscle, which is supplied by the recurrent laryngeal (X) nerve. It divides into a right and left bronchus.

The bronchi diverge from each other on their way to the lung roots. They have firm, elastic, fibrocartilagenous plates. The right bronchus is about 2.5 cm (1 in.) long and is shorter, wider, and more vertical than the left and thus foreign bodies from the trachea usually pass to the right. The right bronchus gives off an eparterial branch, which passes to the upper lobe of the right lung and continues as the hyparterial branch, which in turn divides to supply the other two lobes. The left bronchus travels farther than the right and thus is twice as long and less vertical.

The carina (Fig. 11-10) is a sagittal spur located where the trachea divides into the primary bronchi. It is seen through a bronchoscope and is used as a guide to the bronchi.

Lungs. The lungs (Fig. 11-9) are paired, comparatively light, and conical in shape, and each has an apex, a base, two surfaces, three borders (anterior, inferior, posterior), and a root. They lie in the pleural cavities but do not fill all the available space during normal breathing. Healthy normal lungs (with an average amount of blood) weigh about 620 gm for the right and 570 gm for the left. The lungs lie free in the pleural cavities, attached at only two points: (1) at their roots and (2) at the pulmonary ligaments. The two lungs are not symmetric in shape due to the higher

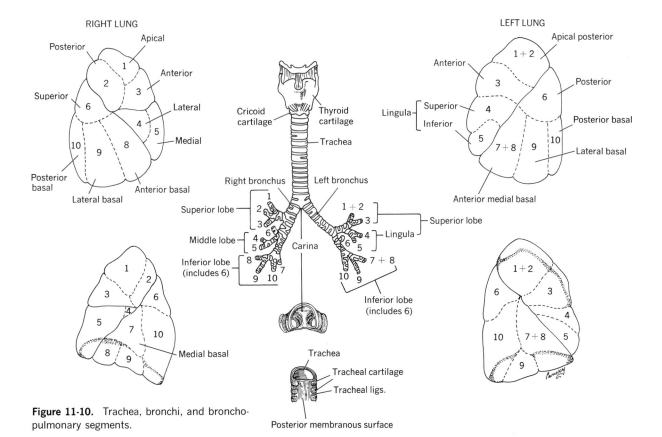

Figure 11-10. Trachea, bronchi, and broncho-pulmonary segments.

level of the diaphragm on the right, the direction of the heart to the left of the midline, and the impressions of surrounding structures on their surface. At birth the lungs are pinkish-white, but they turn dark gray with age, and as time passes the "mottling" becomes almost black due to carbonaceous deposits.

APEX. The apex is rounded and extends for about 3.75 cm (1½ in.) into the root of the neck and above the first rib because the thoracic inlet is oblique. A layer of fascia, Sibon's fascia, protects the apex and the pleura, which may be torn during any severe neck trauma or fracture of the clavicle.

BASE. The base (diaphragmatic surface) is concave and is related to the diaphragm, which separates the right lung from the liver and the left lung from the stomach, spleen, and liver. Due to the liver, the right lung is shorter.

SURFACES. There are two surfaces, a costal and a medial. The former is smooth and convex and is related to the inner surface of the ribs, the costal cartilages, the intercostal spaces, and, slightly, the sternum. The medial surface is the part related to the mediastinum and vertebral column.

Right Side. On the right side, at the hilum (where the vessels and bronchi pass to the lung [Fig. 11-9]), one sees an eparterial and hyparterial bronchus, a right pulmonary artery, and an upper and lower pulmonary vein with numerous lymph glands and nerves. Impressions of related structures are also seen.

Left Side. On the left side, the hilum (Fig. 11-9) reveals a single left bronchus, left pulmonary artery, upper and lower pulmonary veins, lymph glands, nerves, and related structure impressions.

BORDERS. Three borders are seen: (1) anterior, (2)

inferior, and (3) posterior. The anterior border is thin and sharp, being squeezed between the sternum and pericardium. It is straight on the right side but has a deep notch at the fourth and fifth intercostal spaces on the left, the cardiac notch. Here the lung falls short of the sternum by about 2.5 cm (1 in.) and the pericardium is thus separated from the sternum by pleura alone (not lung), creating an "area of superficial cardiac dullness." The inferior border separates the base of the lung from the costal and medial surface. The posterior border is rounded and indistinct.

FISSURES, LOBES, AND BRONCHOPULMONARY SEGMENTS. Each lung has a complete oblique fissure. The right lung has a second fissure, the transverse fissure, which runs horizontally to meet the oblique in the midaxillary line. The lobes (three of right lung, two of left lung) are subdivided into smaller units called bronchopulmonary segments, defined as the area of distribution of a bronchus. On inspiration (breathing in), air passes along the trachea and the 18 to 20 subdivisions of bronchi and bronchioles, which are the conducting airways. Peripherally, in the lung, there are two or three divisions of respiratory bronchioles and finally alveolar ducts, which bear the alveoli. The 18 bronchopulmonary segments are constant in position

and size, but the divisions of the airways may be variable (Figs. 11-9, 11-10).

ROOT. The root of the lung (Fig. 11-9) is a short, broad pedicle with three essential structures: pulmonary artery, pulmonary veins, and bronchi, with associated bronchial vessels, lymph glands, and nerves. The pulmonary ligament may be considered with the root. The bronchi usually lie posterior to the vessels and the veins below the arteries, but variations are common.

Pulmonary Veins. The pulmonary veins (Fig. 10-5, p. 332; Fig. 11-11) are usually two on each side and return oxygenated blood to the left atrium of the heart. They drain toward the hilum within the connective tissue septa between lobes.

Pulmonary Artery. The pulmonary artery (Fig. 10-5, p. 332; Fig. 11-11) divides to send a right branch to the right lung and a left branch to the left lung. In the lung hilum, the right divides into two branches and the left remains single. These vessels end in a capillary network over the alveolar tissue, carrying "impure" venous blood from the right heart ventricle to the lung. The pulmonary arteries follow the path of the airways. Thus, a thrombus of the right side of the heart (of venous origin) can release an embolus that can block a branch of the pulmonary artery and inter-

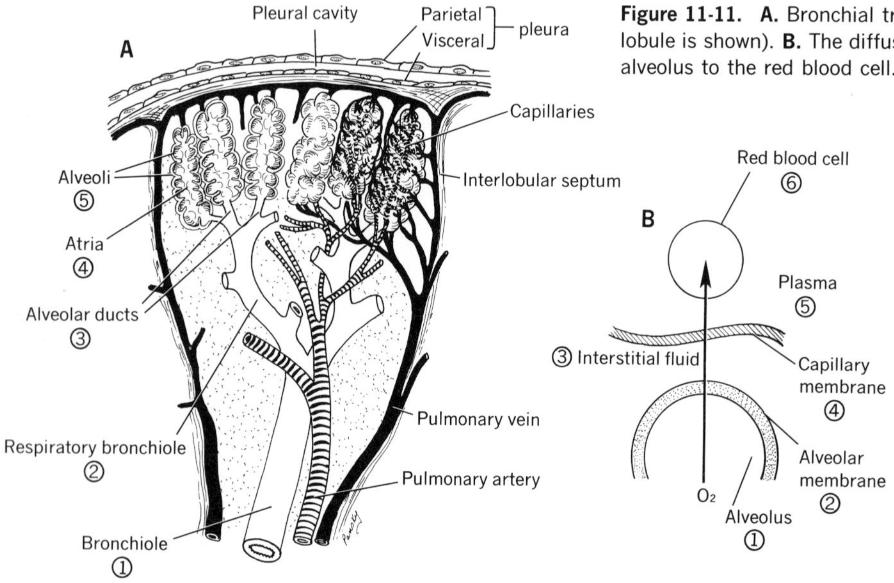

Figure 11-11. A. Bronchial tree and its vascularity (pulmonary lobule is shown). **B.** The diffusion pathway in the lung from the alveolus to the red blood cell.

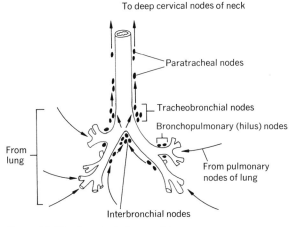

Figure 11-12. Lymph drainage of lung.

To deep cervical nodes of neck

Paratracheal nodes

Tracheobronchial nodes

Bronchopulmonary (hilus) nodes

From lung

From pulmonary nodes of lung

Interbronchial nodes

Figure 11-13. Regulation of breathing by lower brain plus receptors and reflex centers.

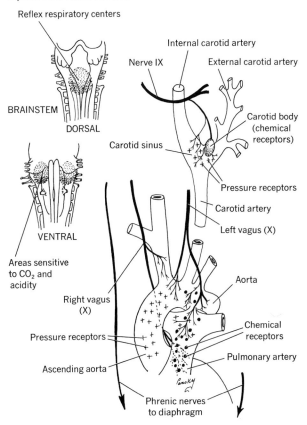

Reflex respiratory centers

Internal carotid artery

Nerve IX

External carotid artery

BRAINSTEM

DORSAL

Carotid body (chemical receptors)

Carotid sinus

Pressure receptors

Carotid artery

VENTRAL

Left vagus (X)

Areas sensitive to CO_2 and acidity

Aorta

Right vagus (X)

Chemical receptors

Pressure receptors

Pulmonary artery

Ascending aorta

Phrenic nerves to diaphragm

fere with lung function. This results in a pulmonary infarct and may lead to degeneration or even a pulmonary abscess.

Bronchial Arteries. The bronchial arteries (Fig. 10-18, p. 356) lie in the interlobular connective tissue septa and do not follow the airways. They supply the lung stroma and arise from the aorta or intercostal arteries. They are accompanied by bronchial veins. The bronchial arteries terminate as capillaries on the alveoli of the respiratory bronchioles and anastomose here and in the pleura with capillaries carrying blood from the pulmonary veins. No vessels of bronchial arterial origin (arteriolar or capillary) pass further down to the alveolar ducts or alveoli.

Lymph Glands. There is both a periarterial and a perivenous lymphatic system that is especially abundant in the connective tissue around the bronchi and arteries (Fig. 11-12). In general, the lymphatic system of the lungs follows the veins.

Nerves. The lung receives parasympathetic fibers from the vagus nerve and sympathetic fibers from the second to the fourth thoracic segments of the spinal cord. Afferent fibers also pass from the lung to the brain stem (Fig. 11-13).

At the root of the lung, these fibers form the anterior and posterior pulmonary nerve plexi. Extensions from these pass into the lung to follow the bronchial tree, pulmonary artery, and pulmonary vein. Nerve plexi are also present in the pleura and may initiate reflex coughing.

General Cellular Structure

The bronchi are lined by pseudostratified, ciliated (short microvilli lie between the bases of the cilia), columnar epithelium with many goblet cells. The cilia act with a quick, lashlike upward movement (toward the pharynx) and a long, slower downward recovery. Waves of motion pass at about 10 per second. The cilia increase their activity with a rise in temperature and are insensitive to a lack of oxygen, even over a long time; increased CO_2 tension or anesthetics decrease their activity markedly. The cilia act on the "covering" of mucus that lines the surface and, therefore, debris is moved upward toward the pharynx and out of the lung. The walls, in addition to smooth muscle, contain plates of cartilage and submucosal glands.

Bronchiolar epithelium has fewer cilia, and goblet cells are rare or absent. The walls of the bronchioles consist almost entirely of smooth muscle. Respiratory bronchioles are short and lined by an increasing number of alveoli as you pass farther distally. One can see neither epithelium nor muscle at this level under ordinary microscopy. However, with the electron microscope the alveoli, alveolar ducts, and respiratory bronchioles are seen to be covered by attenuated "squamous" epithelial cells (type I), whose cytoplasm is about 200 Å thick, and in areas only a basement membrane separates them from adjacent capillary endothelium. There are no cartilages or glands at this level; yet smooth muscle and elastic fibers are present.

The alveolar ducts are relatively long ducts arising from the respiratory bronchioles and ending in rounded dilatations, the atria. They have many alveoli. One still sees fine elastic fibers and a few smooth muscle fibers. Beyond this point, no muscle exists. The air sacs are dilated spaces on the ends of the alveolar ducts or atria; they have delicate walls encircled by elastic fibers and thus can collapse easily and quickly. In the corners of the alveoli one may see a cell that is cuboidal in shape (the great alveolar cell, type II). This cell contains laminated ultrastructural bodies associated with surfactant, an antiatelectasis factor that is responsible for the pressure/volume characteristics of the lung. The alveolar and bronchiolar interstitial tissue normally contains a variety of monocytes, fibroblasts, and histiocytes. Intra-alveolar macrophages exist and are "scavengers" of the lung, acting in cases of injury and irritation.

Functional Considerations

The lungs are not simply two hollow bags filled with air and fed by many blood vessels. In the thoracic space, the lung provides about 600 million round, blind pockets called alveoli, which look like tiny bubbles of air and give the lungs a foamy appearance and across which O_2 and CO_2 are exchanged between inspired air and the pulmonary circulation. One takes about 15,000 breaths per day, which diffuses over 1 sq m per kilogram of weight or over an area greater than that of a basketball court.

Blood transports oxygen and carbon dioxide throughout the body. Oxygen must pass quickly through the lung surface into the bloodstream. It is not sufficient to just inhale oxygen into the lungs. Also true is the fact that CO_2 must travel quickly from the blood and out through the lung cavities. Thus, the lungs need a large surface area and an extensive blood supply. The major gas exchange occurs at the alveoli (Fig. 11-11). As mentioned above, large numbers of alveoli are grouped together to form sacs, which, in turn, lead through alveolar ducts into larger tubes called bronchioles. Thousands of bronchioles join repeatedly to form larger tubes called bronchi, which branch off from each lung. The two bronchi join to form the windpipe or trachea, which leads to the larynx and then to the pharynx (throat). The pharynx, in turn, opens into the nose and mouth, thereby connecting the lungs to the outer air. The pulmonary arteries carry deoxygenated blood from the right side of the heart to the lungs. They divide into a network of capillaries that surround each alveolus (Fig. 11-11). Capillaries merge into the pulmonary veins, which return oxygenated blood to the left side of the heart.

The alveoli are where O_2 and CO_2 are exchanged, and where gases pass through the alveolar and capillary walls by diffusion. This passive process is very efficient because the wall of an alveolus is so thin that gases have a very short distance of only about $1.0 \mu m$ to travel between alveolus and the blood. The endothelial lining of the blood capillary lies snugly against the epithelial alveolar lining. Thus, to reach the blood, O_2 passes (1) through very thin ($0.2 \mu m$), nucleated cells, (2) then through their basement membrane, and finally (3) through the endothelium of the capillaries (Fig. 11-11). Many white cells line the alveoli. The exact structure of the alveolar lining is still uncertain. Furthermore, each alveolus is surrounded by such a very dense net of blood capillaries that there is a large amount of blood available to collect and deliver the gas and this can be a rapid exchange. The many blood vessels constitute a large portion of the total lung substance. Between the air tubes and blood vessels of the lungs are large quantities of elastic connective tissue that play an important role in breathing. Since the lungs lack muscle, they are passive "elastic containers" with no ability to increase their volume. Lung expansion is accomplished by the action of the diaphragm and the muscles that move the ribs. The alveolar lining is also moist, so that the gases can first dissolve in order for diffusion to occur. The reason there is no serious water loss in the air breathed out is that mucus covers all the respiratory surfaces and holds the water and checks evaporation.

The passages and channels (the conducting system) that connect the outside with the lungs do more than convey air. The ducts vary in structure and serve several functions. They air-condition the system, starting in the nose, where hairs filter larger particles from the air. The air is also warmed and moistened as it is purified, since it passes between the bony nasal turbinates (meatuses) lined with mucosa, which contain air cells richly supplied with warm blood. Moistening of the air also takes place by contact with the mucus surface of all the passages that it passes as it makes its way to the lungs.

The nasal passages lead by way of the pharynx into the trachea. The trachea and bronchi are lined with cilia and mucus. Small dust particles that have gotten by the nasal barrier become trapped in the mucus, which lines the passages as far down as the bronchioles, and by ciliary movement these particles are "swept" up into the throat and there sneezed and coughed away or swallowed and eliminated in the feces. This mechanism is important in the body's total defenses against bacterial infection since bacteria enter the body on dust particles. A major cause of lung infection is undoubtedly due to paralysis of ciliary action by noxious agents. Phagocytic cells, which are present in great numbers in the respiratory tract lining, also are protective. The air finally reaching the respiratory surface is moist, warm, and nearly dust-free.

The walls of the upper respiratory tract contain smooth muscle highly innervated and sensitive to circulating hormones such as epinephrine. Contraction or relaxation, particularly in the bronchioles, alters resistance to the flow of air and plays a role in the conducting mechanism.

The act of breathing is deceptively simple, and hence its mechanism is often misunderstood. We often speak of "filling our lungs with air," disregarding chest expansion as a secondary effect. The process is really the opposite. The thoracic cavity (chest) is an airtight box and its volume can be increased in two ways: (1) by flattening the dome-shaped diaphragm and (2) by a movement of the ribs upward and outward. The lungs usually occupy most of the available space in the thoracic cage and are permanently connected to the outside air. The pressure of the atmosphere (approximately 15 lb/sq in. at sea level) not only presses over the entire body surface but also is present on the inside of the lungs. The muscles of the diaphragm and ribs are strong enough to expand the "box" against normal atmospheric pressure but can do so only because the air pressure exists on the inside of the lungs as well. Hence, as the thorax expands, air flows into the lungs and the extra air expands the lungs to fill the extra space.

Respiration involves the exchange of air between the external environment (atmosphere) and the alveoli and includes the movement of air in and out of the lungs and its distribution in the lungs. New air must be delivered constantly to the alveoli and distributed. This is referred to as the bulk flow of ventilation. There must be exchange of O_2 and CO_2 between alveolar air and lung capillaries by diffusion; consequently, the volume and distribution of pulmonary blood are important. Furthermore, O_2 and CO_2 must be transported by the blood, and there must be an exchange of gases between the blood and body tissues by diffusion as blood traverses the tissue capillaries (Figs. 11-14, 11-15).

Breathing in (inspiration) is an active process. The muscles of the diaphragm and ribs contract, expanding the rib cage and pulling the diaphragm down. The ribs are linked to the sternum (breastbone) and the vertebral column by cartilage attachments (and joints),

Figure 11-14. A. Mechanism of O_2 transfer from alveoli to tissues. Alveolar O_2 pressure is higher than that of blood arriving at lung because the blood has lost its O_2 to the tissues and thus O_2 diffuses into the lung capillaries. At the tissues the blood O_2 pressure is higher so that O_2 diffuses from the blood into the tissues. **B.** Mechanisms of CO_2 transfer from tissues to alveoli. When glucose is burned in cell respiration, CO_2 is produced. Since CO_2 pressure is higher in the tissues, it diffuses into the blood. Similarly at the lungs, CO_2 passes into the lung alveoli to be removed.

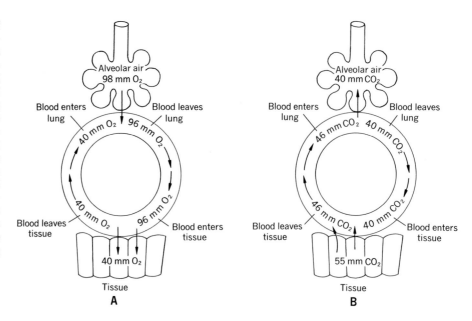

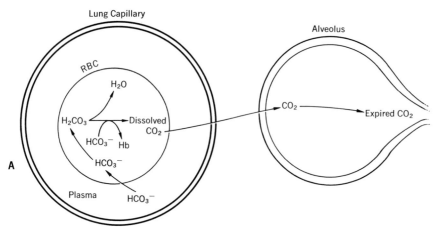

Figure 11-15. A. CO_2 movements and reaction as blood flows through lung capillaries. All movements are by diffusion. H_2CO_3 breakdown catalyzed by carbonic anhydrase. **B.** In the tissues all movements also take place by diffusion. Most of the CO_2 is converted to HCO_3^- in the red blood cell. **C.** The oxygen pathway—lung to cell.

B	Cells	Interstitial Fluid	Plasma	RBC		Plasma
	CO_2 produced $\rightarrow$	Dissolved CO_2 $\rightarrow$	Dissolved CO_2 $\rightarrow$	Dissolved CO_2 $\rightarrow$ HOH / Hb → H_2CO_3 / $HbCO_2$ $\xrightarrow{Enzyme}$ $HCO_3^- + H^+$	Most diffuses out $\rightarrow$	HCO_3^-

C	Alveoli	Plasma	RBC	Tissue Capillary Plasma	Interstitial Tissue	Cells
Inspired air $\rightarrow$	O_2 $\rightarrow$	Dissolved O_2 $\rightarrow$	+ Hb = HbO_2 $\rightarrow$	Dissolved O_2 $\rightarrow$	Dissolved O_2 $\rightarrow$	Dissolved $O_2 \rightarrow O_2$ Consumption

which are pliable and act as hinges. Some of the rib muscles are attached to the vertebrae, and when they contract the ribs are pulled up and out (inspiration). The lungs then expand passively as air flows into them.

Breathing out (expiration) is a "let-go" process. The ribs drop downward and inward to their original position and the diaphragm moves upward. As the thoracic cavity diminishes, air is forced out of the lungs. Here the lung also gives some assistance, since each alveolus is surrounded by elastic tissue that was stretched during inspiration and the elastic recoil now helps expel the air.

The outside air comes into contact with the alveoli, and one must examine how the gases move in relation to the bloodstream. The exchange of O_2 and CO_2 in the alveoli and at the individual cells takes place simultaneously. Considering each gas makes the concepts easier to comprehend.

At sea level, atmospheric pressure is 760 mm Hg and remains constant. Since intra-alveolar air pressure is less than atmospheric, air flows into the alveoli. The air filling the alveolus contains a sufficient amount of oxygen. Oxygen is present in the alveolar air under pressure of about 98 mm Hg. Alveolar oxygen pressure is always higher than that of the blood just arriving at the lung from the body, because the blood has already lost oxygen to the tissues. Oxygen, therefore, diffuses into the lung capillaries, since there is a difference in concentration between the alveolar O_2 and the blood O_2 of about 40 mm Hg (Figs. 11-14, 11-15).

At the tissue capillaries, the gas gradient in the blood is reversed. The O_2 pressure in the red blood cells (96 mm Hg) is higher than that in the tissue cells outside the capillaries (40 mm Hg). There is a natural tendency for O_2 to leave the hemoglobin and diffuse into the tissue cells (Table 11-1). The degree to which hemoglobin binds with O_2 depends on the amount of CO_2 present. A rise in blood CO_2 reduces the affinity of hemoglobin for O_2, and O_2 release from oxyhemoglobin is stimulated (Figs. 11-14, 11-15).

The saturation of hemoglobin with oxygen varies with the oxygen pressure (Fig. 11-16). An alveolar pressure of 100 mm Hg produces a saturation of 98 percent. On arrival at the tissues, where oxygen pres-

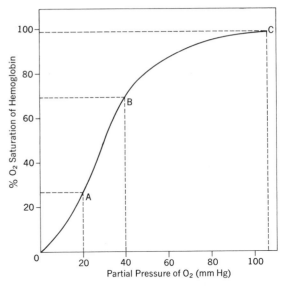

Figure 11-16. Saturation of hemoglobin with O_2 varies with O_2 pressure. An alveolar pressure of 100 mm Hg produces a saturation of about 98 percent **(C)**. At the tissues, where O_2 pressure is about 40 mm Hg, hemoglobin gives up O_2 so that its saturation is reduced to about 70 percent **(B)**. When tissues are very active, their O_2 pressure may be lower **(A)**. Hemoglobin never parts with all its O_2 since the pressure never reduces that low.

sure is about 40 mm Hg, hemoglobin gives up O_2 so that its saturation is reduced to about 70 percent (Table 11-1). When tissues are very active, their oxygen

Table 11-1. O_2 Transport and Removal of CO_2

O_2 Transport	O_2 Tension (mm Hg)	
Inhaled air	158	
Alveolar air	100	$Hb + O_2 = HbO_2$ (in lungs)
Arterial blood	90	
Capillary	40	$HbO_2 = Hb + O_2$ (at cells)
Interstitial fluid	30	
Cell interiors	10	

CO_2 Removal	CO_2 Tension (mm Hg)
Tissues	50
Venous blood	46
Alveolar air	40
Exhaled air	32
Atmosphere	0.3

pressure may be as low as 20 mm Hg and more O_2 is released, leaving the hemoglobin still 25 percent saturated. Note that the pressure in the tissues never falls low enough to make the hemoglobin part with all its O_2.

Hemoglobin and Oxygen Transport

Oxygen is present in the blood in two forms: (1) physically dissolved in the blood "water" and (2) chemically bound to hemoglobin molecules. Only 3 cc of oxygen can be dissolved in 1 liter of blood at normal alveolar and arterial pO_2 of 100 mm Hg, because O_2 is relatively insoluble in water. In contrast, 197 cc of O_2 per liter of blood (about 98 percent) is chemically bound to hemoglobin and carried in the red blood cells.

Hemoglobin is the major agent and is indispensable for O_2 transport by the bloodstream. It is found in the red blood cells, accounting for a major part of their mass. This protein is unique in its capacity to bind molecular oxygen in a readily reversible attachment. Mammalian hemoglobin is a conjugated protein of molecular weight 67,000.

The hemoglobin molecule is composed of two alpha and two beta polypeptide chains, four protoporphyrin molecules, and four atoms of iron. The polypeptide part of the molecule, exclusive of the heme groups, is called globin. At natural pH, globin combines with ferroprotoporphyrin to form hemoglobin or with ferriprotoporphyrin to yield methemoglobin. The latter lacks the ability to combine reversibly with O_2 and holds on to O_2 tenaciously. The function of hemoglobin is to shuttle oxygen from lung to tissue and carbon dioxide from tissue to lung on the return trip. This ability of the blood to take up and release gas depends on the presence of iron in the heme portion of the assembled hemoglobin molecule. Without adequate iron, an important link in the tissues' oxygen supply is weakened and blood and tissue hypoxia result. Unique to hemoglobin is its ability to form a stable O_2 complex while the iron atom of heme is in a ferrous state.

Normally man has 40 to 50 mg of iron per kilogram of body weight. Seventy per cent is functional or essential, and 30 percent is stored or nonessential iron. Women, with lower hemoglobin levels and less iron storage, have 35 to 50 mg of iron per kilogram of body weight. Most of the functional iron is in the red blood cell (Fig. 11-17).

Iron loss from the body is very small since it amounts

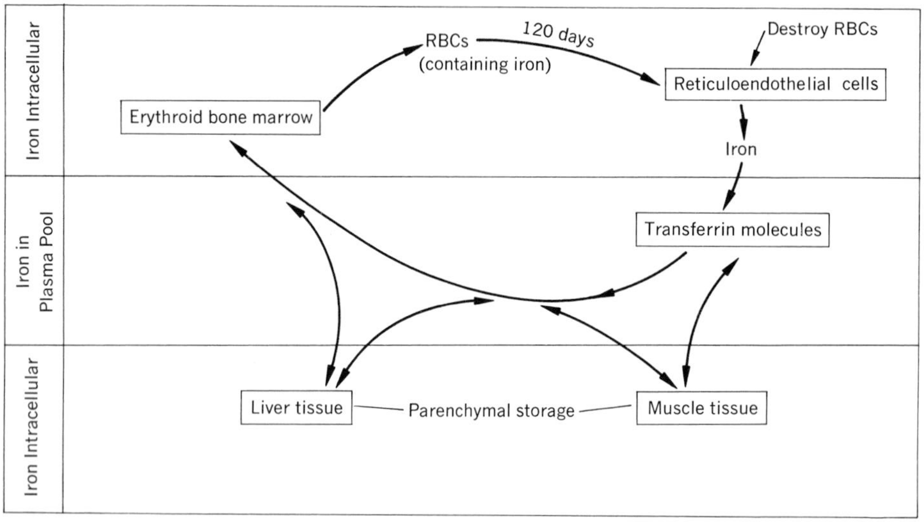
Iron Intracellular
Iron in Plasma Pool
Iron Intracellular
RBCs
(containing iron)
120 days
Destroy RBCs
Reticuloendothelial cells
Erythroid bone marrow
Iron
Transferrin molecules
Liver tissue
Parenchymal storage
Muscle tissue

Figure 11-17. Iron cycle in the body. Erythrocytes are produced predominantly in the bone marrow. After about 120 days they are destroyed by the reticuloendothelial cells and virtually all the iron is released and taken up by transferrin molecules. These return most of the iron back to the erythroid marrow for production of new red cells. However, depending on need, some iron is shuttled to and from parenchymal storage areas. Storage is limited and additional iron from food is needed each day.

to about 10 to 15 percent a year in the male or less than 1 mg per day. Three fourths is lost through the gastrointestinal tract, some from exfoliation of intestinal epithelial cells, some from bleeding, and some from bile excretion. Normal exfoliation of cells of the skin causes loss of about one fourth of our body iron. When a deficiency occurs, it is usually due to bleeding. In women, the loss of menstrual blood represents a loss of about 0.5 mg per day. Pregnancy puts another demand on iron need since 270 mg is transferred to the fetus and another 240 mg is lost at delivery. Thus, iron supplements are greatly needed.

Food-iron absorption in man usually ranges from 5 to 15 percent of that available from intake (Fig. 11-18). The degree of gastric acidity influences solubility and availability of ferric iron. Other substances, too, influ-

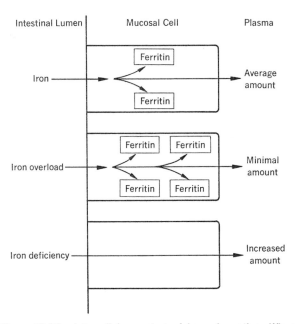

Figure 11-18. Intracellular control of iron absorption. When body stores are normal, soluble iron enters duodenal mucosal cells and a portion passes into the plasma. The unneeded iron combines with apoferritin to form ferritin, which remains in the cell until it is exfoliated and excreted in the feces. In iron "overload" most entering iron is trapped in the mucosa as ferritin molecules, then exfoliated and excreted. In iron deficiency intracellular control alters and iron is permitted to pass through into the plasma and reach the marrow.

ence iron absorption. Dietary iron, occurring in various states, is partially converted to iron salts by the gastric hydrochloric acid. The control of iron balance takes place in the intestinal epithelium, where iron from ingested food is absorbed. The intestine absorbs only a fraction of the iron ingested, depending on the body's need and iron balance (in reduced form) and mediated by the very iron content of the intestinal epithelium. When stores are sufficient, the epithelial iron store is increased and this prevents iron absorption. When there is hemorrhage, resulting in a drop in iron reserves, iron stored in the epithelium enters the blood from the ferritin of mucosal cells and is transported to the marrow (storage reservoir) as a complex with the plasma protein transferritin, resulting in a lowered epithelial iron, and more is absorbed.

About 70 percent of the total body iron is in hemoglobin, and with an iron deficiency most people look pale and even develop "air hunger," taking exaggerated deep breaths in an effort to get O_2 into the cells. This "air hunger" is also seen at high altitudes, where the air pressure is not strong enough to saturate hemoglobin with oxygen. One can acclimate to this "rarefied" air in a day or two. At altitudes of over 3 km (9800 ft), exertion can result in fainting. If one does live at higher altitudes, the body compensates for the O_2 deficiency by breathing deeper, with the development of an expanded chest and lung capacity; by increasing the number of red blood cells from five to seven million, thus increasing the total carrying capacity and the total amount of blood hemoglobin; and by increasing the total blood volume (from 5.67 to 7.57 liters [$1\frac{1}{2}$ to 2 gal]). Iron that is not in hemoglobin is stored in the spleen, bone marrow, and liver, existing as a complex with the protein ferritin. A second storage protein is hemosiderin, which occurs with ferritin. As the red blood cell is destroyed, the iron released is returned to these iron depots (Fig. 11-17).

The combination of O_2 with hemoglobin is very rapid (complete in 0.1 second under physiologic conditions). The degree of oxygenation of hemoglobin is a function of the partial pressure of O_2, increasing with O_2 tension. A maximum of 402 molecules of oxygen can be bound by one molecule of hemoglobin. The affinity of hemoglobin for O_2 also depends on pH. If one were to change the pH (acidity) or hydrogen ion concentration, one would see that, with an increased number of hydrogen ions (when the acidity is high), hemoglobin has less affinity for oxygen. Since the hydrogen ion concentration in the tissue capillaries is greater than in the arterial blood, blood flow through the tissue capillaries is exposed to this elevated concentration and loses even more O_2 than if only the pO_2 is decreased. Conversely, the hydrogen ion concentration is lower in the lung capillaries than in the systemic venous blood, so that hemoglobin picks up more O_2 in the lungs than if only the pO_2 were involved. Also, the more active a tissue is, the greater is its hydrogen ion concentration, and hemoglobin releases even more O_2 as it passes through these tissues.

The reduction in affinity of hemoglobin for O_2 in the presence of CO_2 (as seen in the tissues) is also largely due to a drop in pH. Furthermore, the formation of carbaminohemoglobin, which also has less affinity for O_2 than hemoglobin itself, is significant. The extent to which the hemoglobin binds O_2 also depends on temperature. Actively metabolizing tissues (exercising muscles) have an elevated temperature that facilitates the release of O_2 from hemoglobin as the blood flows through the capillaries.

Carbon monoxide (CO) can replace O_2 in its complex with hemoglobin. Its binding is so strong that small amounts in the atmosphere can block O_2 transport. Carbon monoxide is odorless and combines with hemoglobin 300 times more readily than does oxygen. A concentration of only 0.5 percent carbon monoxide in the air breathed can monopolize the blood's hemoglobin and result in death. This gas is found in exhaust fumes when an engine is cold and fuel combustion is incomplete, in coal gas, and in furnaces or stoves burning with poor ventilation, just to mention a few examples.

Normal adult human hemoglobin (HbA) is made up of two equivalent half molecules, each containing an alpha chain and a distinct (but similar) beta chain. In fetal hemoglobin (HbF), the two beta chains are replaced by two gamma chains. A number of hereditary diseases are known that are characterized by changes in the amino acid sequences of the chains. One of these is sickle cell hemoglobinemia (HbS). In each case, the variant protein arises from a point of mutation in the chain responsible for the alpha and beta polypeptides. This is mendelian inherited.

Myoglobin, an iron-containing protein with a molecular weight of 17,000, is found in skeletal muscle cells of all vertebrates and invertebrates. It resembles hemoglobin but has a greater affinity for O_2 than hemoglobin and tends to hold its O_2. Thus, if exercise is mild, the pO_2 of the cells is at 20 mm Hg and myoglobin holds its O_2. During severe exercise, the intracellular pO_2 may reach 0 mm Hg and myoglobin gives up its O_2. Thus, myoglobin serves as a cellular storage or oxygen-retaining protein in resting muscle to be used during severe exercise. This may account for the fact that intermittent muscle contraction over a longer time is less tiring than shorter periods of prolonged exercise of less magnitude.

Red Blood Cells (Erythrocytes)

Number and Structure

Since each milliliter of blood contains about five million red blood cells (Fig. 11-19) and since there are about 5 liters of blood totally, the total number of red blood cells is about 25 trillion (containing about 950 gm of hemoglobin). The red blood cell looks like a biconcave disk, thicker at the edges than in the middle (Figs. 11-19, 11-20). It is about 2 μm thick at the margin and 1 μm at the center; the average diameter is about 7 to 8 μm. The shape is adaptive for rapid diffusion of O_2 and CO_2 since it allows for a larger surface-to-volume ratio than other body cells. A red blood cell consists of an external sac or stroma that surrounds and contains the hemoglobin. A semipermeable membrane composed of protein and lipid is present at the surface. In contrast to plasma, the principal internal cation is potassium (K^+). No nucleus is present. Various chemical and physical treatments (exposure to solvent of low salt or snake venom) cause rupture of the stroma and hemoglobin release, and this is referred to as hemolysis (Fig. 1-25, p. 33).

Formation, Development, Destruction, and Renewal

Stages of Development. Red blood cells are formed in the red bone marrow (breastbone or backbone, ribs, and long bones of the limbs). There are several stages of development: megaloblast to early erythroblast, to intermediate erythroblast, to late erythroblast, to reticulocyte, to erythrocyte (Fig. 16-1, p. 548). All the cells up to the reticulocyte are nucleated (Fig. 11-21).

The number of red blood cells in circulation is generally constant, with destruction and formation being at equilibrium. The greatest portion of breakdown occurs in the spleen and less in the bone marrow.

Hemoglobin and Oxyhemoglobin. The red blood cell contains the iron-containing protein hemoglobin, which makes up about one third of the total cell weight and combines with O_2 in the lung under rela-

Figure 11-19. Development of the red blood cell. The red blood cell is round and biconcave and has lost its nucleus.

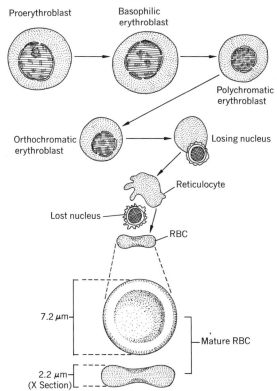

Proerythroblast

Basophilic erythroblast

Polychromatic erythroblast

Orthochromatic erythroblast

Losing nucleus

Reticulocyte

Lost nucleus

RBC

7.2 μm

Mature RBC

2.2 μm (X Section)

tively high pressure yet easily separates from O_2 when surrounded by tissues where the pO_2 is low. Oxygen and hemoglobin form oxyhemoglobin. The enzyme carbonic anhydrase is also present in the red blood cell and facilitates transportation of CO_2.

Destruction and Renewal. The volume and hemoglobin content of red blood cells are not fixed but subject to physiologic control. Red blood cells cannot reproduce themselves or maintain their normal structure for too long since they are incomplete cells having no nuclei. As enzymes in them deteriorate, the cells age and die. The average life of a red blood cell is about 120 days. Thus, almost 1 percent of all the erythrocytes in the body are destroyed per day. The destruction and renewal of red blood cells take place at a rate of about 210 thousand million per day or 2.5 million cells per second. Their destruction takes place in the liver, spleen, bone marrow, and lymph nodes by phagocytes or macrophages (Greek *macros*, big; *phagen*, to eat), which ingest and destroy them by breaking down their complex molecules into component parts: (1) the iron-containing heme and (2) the protein globin made up of amino acids that can be used again in the development of new cell proteins. The hemoglobin molecule is broken down (degradation of heme groups) and converted to a yellow molecule called bilirubin, which is released into the blood. Bilirubin is picked up by the liver, conjugated with glucuronic acid, and added to the bile. Thus, bilirubin and glucuronic acid are the chief bile pigments. From the gallbladder they enter the intestine and undergo conversions, and much is excreted by the intestine. If the liver is damaged (or overloaded due to an abnormally high rate of red-cell destruction), the bilirubin accumulates in the blood and gives the skin a yellow color, which is called jaundice. Most of the iron returns to the bone marrow and is used again for making hemoglobin (Fig. 11-17).

Required Nutrients. The formation and development of red blood cells require nutrients and structural materials: lipids, amino acids, and carbohydrates as well as growth factors (vitamin B_{12} and folic acid). There must also be materials to make hemoglobin, namely, amino acids, iron, and other organic mole-

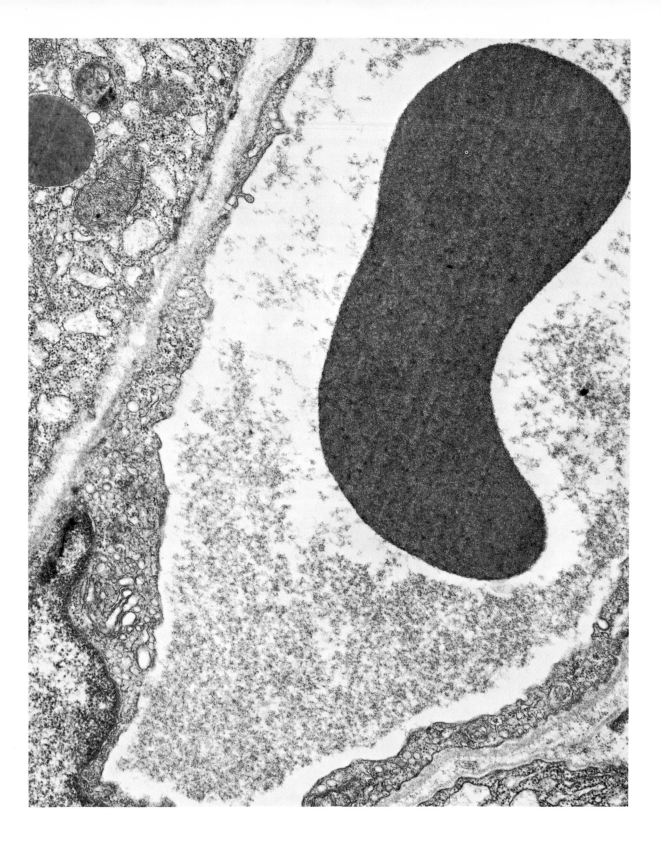

Figure 11-20. [OPPOSITE] Electron micrograph of a red blood cell in a vessel. ×25,000. (Courtesy of Dr. Dale E. Bockman, Medical College of Ohio at Toledo.)

Figure 11-21. Electron micrograph of a nucleated red blood cell seen in bone marrow. ×22,200. (Courtesy of Dr. Dale E. Bockman, Medical College of Ohio at Toledo.)

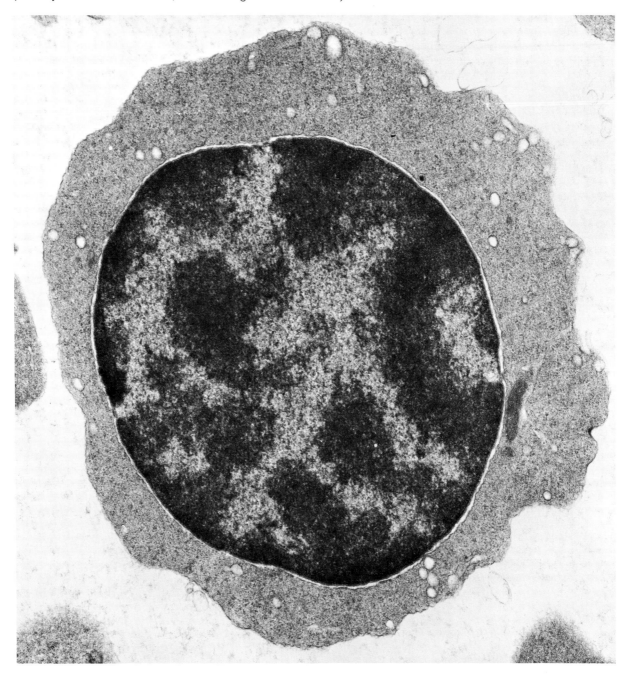

cules. A lack of these results in reduced red-cell circulation, or anemia. Most often lacking are vitamin B_{12} and iron.

VITAMIN B_{12}. Normal formation and maturation of red blood cells require a minute amount of the cobalt-containing vitamin B_{12}. This vitamin is not synthesized in the body and needs to be ingested in the diet. Absorption of B_{12} requires a still-unidentified substance normally secreted by the epithelial lining of the stomach. Its absence prevents absorption, causes deficient red-cell production, and results in pernicious anemia.

Endocrine Control. Control of red-cell production is exerted by a hormone called erythropoietin (the site of production is uncertain but it may be the kidney). Normally, a small amount is circulating and stimulates the marrow to produce red blood cells at a basal rate.

Carbon Dioxide Transport

The quantity of carbon dioxide that can dissolve in blood at physiologic partial pressures (pCO_2) is small. Certainly it is much smaller than the large volume of CO_2 that must be regularly transported from tissues to lungs (Fig. 11-22). However, the addition of CO_2 to a liquid (blood) results in bicarbonate (HCO_3^-) and hydrogen ions (H^+). It also can combine with hemoglobin to form carbaminohemoglobin. Thus, when arterial blood flows through tissue capillaries, oxyhemoglobin gives up O_2 to the tissues and CO_2 diffuses into the blood, where (1) a small fraction (8 percent) remains physically dissolved in the red blood cells and plasma, (2) 67 percent is converted to bicarbonate and hydrogen ions, primarily in the red blood cells since they contain large amounts of the enzyme carbonic anhydrase (plasma does not), and (3) 25 percent reacts directly with hemoglobin to form carbaminohemoglobin (Fig. 11-23).

The CO_2 story begins at the cells, where there is more CO_2 than in the nearby blood capillaries. The pCO_2 in the cells is approximately 55 mm Hg, whereas it is only -46 mm Hg in the venous capillaries and 40 mm Hg in the arterial capillaries. Thus, CO_2 diffuses from the cells into the blood. Simple as this may sound, the transport of CO_2 in the blood is not as simple as it was with O_2. For one thing, CO_2 is poisonous and unlike O_2 combines chemically with water. How does one then transport CO_2 from cells

Figure 11-22. Chemical reactions involved in CO_2 carriage by blood. As noted above, 8 percent is carried as dissolved CO_2 by both plasma and red blood cell; 27 percent is carried by formation of carbamino compounds (e.g., carbamino hemoglobin); 65 percent is carried by formation of bicarbonate, which takes place in the red blood cell. The CO_2 entering the cell is hydrolyzed to carbonic acid (H_2CO_3), which is neutralized by hemoglobin, with the bicarbonate (HCO_3^-) ion diffusing back into the plasma. At the same time chloride passes into the red blood cell to maintain electrical neutrality.

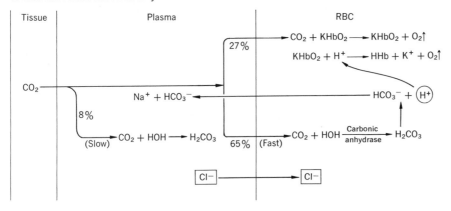

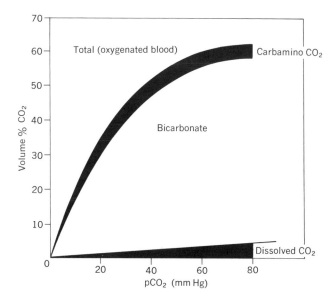

Figure 11-23. The CO_2-blood dissociation curve, illustrating the amount of CO_2 present in the various forms in which it is carried by the blood.

to lung without causing bodily damage along the way? Carbon dioxide dissolves readily in the watery plasma of blood to form carbonic acid (H_2CO_3). Yet cells cannot survive unless the acidity is kept within a very narrow margin. Therefore, the H_2CO_3 cannot remain in the blood for a long time. Most of it quickly turns into bicarbonate (HCO_3^-) inside the red blood cells. This chemical substance is harmless and is transported away in the plasma. A small amount of CO_2 leaving the tissues is carried away in simple solution, and the rest combines with the plasma proteins and the protein of hemoglobin itself.

CO_2 thus reaches the lung, predominantly in the form of HCO_3^- (bicarbonate), where the HCO_3^- breaks down to form CO_2 with a pressure of 46 mm Hg, which diffuses into the alveoli where the concentration of this gas is very low (alveolar air contains CO_2 at a pressure of 40 mm Hg).

The Process of Breathing; Lung Function

The need for O_2 depends on how active the body is at any specific time. Muscle requires over 40 times

more oxygen when active than when at rest. The oxygen supply can be varied only by altering the rate and depth of breathing. One normally breathes at a rate of 14 to 20 times per minute, and about 500 cc of air pass in and out of the lungs with each single breath— the tidal volume (in and out like the tides). The volume of air that can be inspired over and above the resting tidal volume is called inspiratory reserve and amounts to about 2500 to 3000 cc of air. During severe exercise each breath, on the other hand, may carry as much as eight times normal or 4000 cc. This is referred to as vital capacity, the maximum volume that can be exchanged at any one time, and consists of normal tidal plus inspiratory and expiratory reserve volumes. When we breathe out (expiration), the thorax does not completely collapse and some air remains in the lung and is not squeezed out. The amount remaining is called the functional residual capacity and may amount to about 3000 cc. There is also a respiratory reserve of about 1000 cc, a volume that can be forcibly exhaled using expiratory muscles. Note, therefore, that although we say we are breathing "fresh air," the air taken in is always mixed with "stale air" already residing or "residual" in the lungs (Fig. 11-24).

Breathing and breathing movements are subconscious movements under normal conditions; yet during exercise we certainly notice our increase in breathing. The total pulmonary ventilation per minute is determined by the tidal volume times the respiratory rate (breaths per minute). At rest this is about 500 cc of an "in and out" of the lungs with each breath and 10 breaths per minute, resulting in a total minute ventilatory volume of 500 cc $\times$ 10 = 5000 cc of air per minute.

It has been noted that during inspiration active muscle contraction provides the energy needed to expand the thorax and lungs. The very stretchability of these structures determines to a degree the work needed to perform ventilation. The easier they stretch, the less energy is needed for expansion. Thus, much work goes into stretching the elastic tissue of the lung but even more goes into stretching a different kind of "tissue"—water itself. The air in the alveolus is separated from the alveolar membranes by a thin layer of fluid. Therefore, the alveolus is like an air-filled bubble lined with water, and at the surface the forces between water molecules cause them to squeeze in the air (surface tension) and resist stretching. In inspiration energy is thus needed to expand the lungs against this tension. To reduce tension the alveolar cells produce a phospholipid material that interspereses with the water molecules, reduces their cohesive force, and lowers surface tension. This material is called pulmonary surfactant.

In general, breathing even during exercise uses up only about 3 percent of the total energy expenditure. Only in disease due to structural change of the lung or thorax, loss of surfactant, or increased airway resistance does breathing become an exhausting phenomenon.

The brain somehow must respond to the amount of O_2 in the bloodstream. However, the brain is not usually directly affected or influenced by slight O_2 variations. What occurs is that the respiratory center in the brain is indirectly affected by the amount of CO_2 in the blood passing through it. After the concentration of CO_2 rises, the brain causes an increase in the rate of breathing. Epinephrine also influences the rate of breathing by altering the tone of the smooth muscle in the smaller bronchioles in order to regulate the amount of air reaching the alveoli.

The breathing rate is geared to the needs of the body's cells for O_2 and is therefore affected by the blood pressure, amount of O_2 and CO_2 in the blood, and body temperature. At rest, during each minute, body cells consume about 200 cc of O_2 and produce the same amount of CO_2. The ratio of CO_2 produced to O_2 consumed is the respiratory quotient (RQ). At rest, the total pulmonary ventilation equals about 5 liters of air per minute. Since only about 20 percent of the air is O_2 (the rest being nitrogen), the total O_2 input is 20 percent of 5 liters = 1 liter of O_2 per minute. Of this, about 200 cc crosses the alveoli into the pulmonary capillaries, and the rest (800 cc) is exhaled. This 200 cc of O_2 is carried by 5 liters of blood (cardiac output per minute). The blood is then pumped by the left ventricle through the tissue capillaries, and about 200 cc of O_2 leaves the blood and is utilized by the cells. Note that the quantities of O_2 added to blood in the lungs and removed in the tissues are identical. The story is just reversed for CO_2. The force that induces the net movement of the gas molecules across the alveolar, capillary, and cell membranes is passive diffusion. There is no active membrane transport for O_2 or CO_2. These alterations or variations are interpreted by the respiratory center in the brain. Momentary interruptions to regular breathing are also under conscious control (as in talking and laughing), and the process of swallowing also inhibits breathing momentarily.

Many daily factors influence and result in changes in our normal breathing patterns. Our emotions affect the brain's respiratory center. Laughter consists of repeated staccato expirations alternating with inspira-

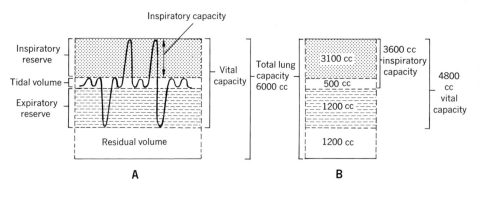

Figure 11-24. **A.** Components of lung volume. **B.** Normal values for lung volumes (measured at body temperature and pressure, saturated with water vapor).

tions. Crying and sobbing involve similar movements. Anxiety and fear result in rapid breathing, while mental concentration suppresses breathing movements. Even our yawn is a large inspiration with the mouth open, often as a result of mental fatigue. Pain stimuli anywhere in the body can produce reflex stimulation of the respiratory centers.

The voice is closely associated with breathing. Sound is produced in the larynx (voice box) situated at the top of the trachea or windpipe (already noted as one of the passages from nose to lungs). When breathing quietly, air moves past the two vocal cords in the larynx without making a sound. However, as soon as we wish to produce a sound, the cords move into new positions and, as air is forced past them, the cords vibrate. The faster the air passes through the larynx, the louder is the sound produced by the vibrating cords. Furthermore, the pitch and quality of the sound are varied by contracting muscles that pull on the vocal cords, altering their length and the distance between them. The loudness and quality of sound are modified by the resonance of the mouth, nose, head bones, throat, and chest. Language is uniquely human although animals do make sounds and partake in a form of speech. Air passing over the cords produces only a continuous "singing" tone, which is then modulated, diverted, and broken up by the epiglottis, teeth, tongue, and lips in an infinite number of variable sounds; and man, by modulating and ordering these sounds, builds words and sentences and creates a language.

Control of Respiration

Neural Control

The inspiratory muscles normally contract rhythmically. Since the diaphragm and intercostal muscles are skeletal muscles, they are not capable of self-excitation like the heart, but require nerve stimulation. Thus, breathing depends on the respiratory muscles being excited by the phrenic nerves (to the diaphragm) and the intercostal nerves (to the intercostal muscles). If the nerves are destroyed or paralyzed, the muscles

cease to function and death follows unless some form of artificial respiration is used.

The respiratory cycle is controlled primarily by neurons with cell bodies in the medulla of the brain stem. Inspiration is initiated by the spontaneous discharge of these inspiratory neurons. The neurons stop "firing" as a result of (1) their own self-limitation and (2) inhibitory impulses from medullary expiratory neurons and pulmonary stretch receptors acting via higher brain centers. A marked drop in the firing of the inspiratory motor units indicates a "turn-off" of the inspiratory center, and passive expiration begins. In addition, active expiratory movements are synchronized with this passive component of expiration as a result of reciprocal connections between the inspiratory and expiratory centers in the medulla. Thus, when the inspiratory neurons fire, impulses travel to the expiratory neurons via inhibitory interneurons and restrain them. Conversely as the inspiratory neuron impulse decreases, expiratory neurons are released from inhibition and begin to fire. Also, via inhibitory interneurons, they help terminate the discharge of the inspiratory neurons. Active expiration by way of descending pathways from the expiratory neurons stimulates the nerves to expiratory muscles.

Chemical Control

The depth and force of breathing alter and determine the volume of air in the alveoli. We have mentioned the motor pathways that control respiration, but what of the afferent or sensory input? Certainly the most obvious regulators are the plasma levels of O_2 and CO_2.

The receptors stimulated by a low concentration of dissolved oxygen and not total blood oxygen are located at the bifurcation of the common carotid arteries and in the arch of the aorta, near but distinct from the so-called baroreceptors. These receptors are the carotid and aortic bodies, which are made up of epithelial-like cells in close contact with the arterial blood. Afferent nerves located on or between these cells run to the medulla, where they synapse ultimately with the neurons of the medullary centers. Thus, a low pO_2 increases the rate at which the receptors "fire" and results in stimulation of the inspiratory neurons of the medulla.

The pCO_2 is the major determinant of respiratory center activity. An increase in pCO_2 stimulates ventilation and promotes CO_2 excretion.° A decrease in pCO_2 allows metabolically produced CO_2 to accumulate and return the pCO_2 to normal.

Other Controls

Temperature. Increased temperature (often as a result of increased activity) stimulates alveolar ventilation, probably due to a direct effect of increased temperature on the respiratory-center neurons and stimulation from hypothalamic thermoreceptors.

Cortical Control. There are fibers from the cortex to the hypothalamus and respiratory centers.

Epinephrine. Secretion of epinephrine increases with exercise and induces increased ventilation.

Muscle and Joint Reflexes. Afferent pathways from both areas may stimulate the medullary centers during exercise.

Others. Other factors regulating respiration include pain, emotion, controlled breathing, and so forth.

Pathology and Aging

The lungs are rich in elastic connective tissue fibers, and with age, rather unexpectedly, there is an increase in the amount of this tissue. There is apparently no age-related pigmentation as seen in arteries in other areas, and there are no spectacular calcific changes as are so often present in arteries of old age. Diffuse lymphoid tissue along the branches of the respiratory tree and small blood vessels is increased.

Some common disorders of the respiratory system are listed below.

° It appears that the effects of CO_2 on ventilation are due not to CO_2 itself but to the associated changes in H^+ concentration. The critical H^+ concentration seems to be brain extracellular fluid rather than arterial blood concentration. Thus, an increased H^+ concentration increases the rate of discharge of the inspiratory neurons by acting directly on them or nearby chemosensitive cells that synapse with them, and a decrease in H^+ concentration inhibits their discharge.

1. Atelectasis is collapse of the lung tissue due to extrinsic pressure or intrinsic changes in the pulmonary airways. It appears to be related to local ischemia that impairs the synthesis of surfactant, and edema washes this surface-tension–lowering compound from the alveolar lining. Hyaline membrane disease in the newborn is an example of this deficiency.

2. Bronchitis is characterized by the production of copious sputum and is often related to respiratory infection. Chronic bronchitis can be accompanied by right heart failure. It is basically an inflammation.

3. Emphysema means "inflation" and is defined as an increase, beyond normal, in the size of air spaces distal to the terminal nonrespiratory bronchioles due to either dilatation or destruction of their walls. It results in impaired pulmonary mixing and diffusion of gases and increases the total lung capacity and residual volume. Its etiology is unkown.

4. Bronchial asthma is characterized by intermittent rather than persistent airway obstruction due to bronchiolar constriction and spasm, most often caused by an allergic response.

5. Pneumonia is an acute inflammatory response usually due to airborne organisms, either viral or bacterial, that have access to the lung because of the large volume of air exchanged.

6. Bronchiectasis is a condition in which the bronchi are dilated and distorted and is more often than not bacterial in origin.

7. Tuberculosis is now described as a delayed-type hypersensitivity to the tubercle bacillus and is the classic immunologic lung disorder.

8. The pulmonary circulation may serve as a filter for dislodged venous thrombi, and emboli. If large enough to obstruct a main pulmonary artery, a thrombus may even cause sudden death.

9. Pulmonary edema results in intra-alveolar fluid accumulation, involving an imbalance of intravascular pressure, plasma osmotic forces, and altered capillary permeability. The lungs become heavy and congested, and this is associated with heart disease.

10. There are many other lung conditions, too many to mention let alone describe, such as sclerosis, tumors, abscesses, cysts, and others.

Review Questions

1. Trace a breath of air from the nose to the lungs.
2. Describe the inside of a nasal cavity.
3. What are the turbinates? The meatuses?
4. List the paranasal sinuses and indicate where they empty into the nasal cavity.
5. Where is the larynx located? What are its cartilages? How are they connected?
6. What is the rima glottidis? Describe the muscles that open and close it.
7. What is the major nerve supply of the larynx?
8. Describe the pleura and its parts.
9. Where are the lungs found? How many lobes does each lung have?
10. What do we see in the root of a lung?
11. What is the difference between the function of the pulmonary and bronchial arteries?
12. What is meant by bronchopulmonary segments?
13. Trace air (or CO_2) through the lung's channels to the point of exchange.
14. Describe the "act of breathing."
15. Differentiate between inspiration and expiration.
16. Why is there an exchange of O_2 and CO_2 at the lungs as well as at the tissues?
17. How are oxygen and CO_2 transported in the blood?
18. What function does iron play in O_2 transport?
19. Describe a red blood cell. What are needed for its development and formation?
20. What is anemia?
21. Define: **a.** Tidal volume. **b.** Inspiratory reserve. **c.** Vital capacity. **d.** Residual capacity. **e.** Respiratory quotient.
22. What is surfactant?
23. What factors influence breathing and the respiratory center?
24. What controls respiration? Neural factors? Chemical control? Others?
25. Define: **a.** Atelectasis. **b.** Bronchitis. **c.** Bronchial asthma. **d.** Bronchiectasis. **e.** Emphysema. **f.** Pulmonary edema.

References

Bates, D. V.: Chronic Bronchitis and Emphysema. *N. Engl. J. Med.,* **278:**546–51, 600–605, 1968.

Comroe, J. H., Jr.: The Lung. *Sci. Am.,* **214**(2):56–68, 1966.

Davenport, H. W.: *The ABC of Acid-Base Chemistry,* 5th ed. University of Chicago Press, Chicago, 1969.

Finch, C. A. (ed.): *Red Cell Manual,* University of Washington, Seattle, 1969.

Harris, J. W., and Kellermeyer, R. W.: *The Red Cell.* Harvard University Press, Cambridge, Mass., 1970.

Krantz, S. B., and Jacobson, L. O. (eds.): *Erythropoietin and the Regulation of Erythropoiesis.* University of Chicago Press, Chicago, 1970.

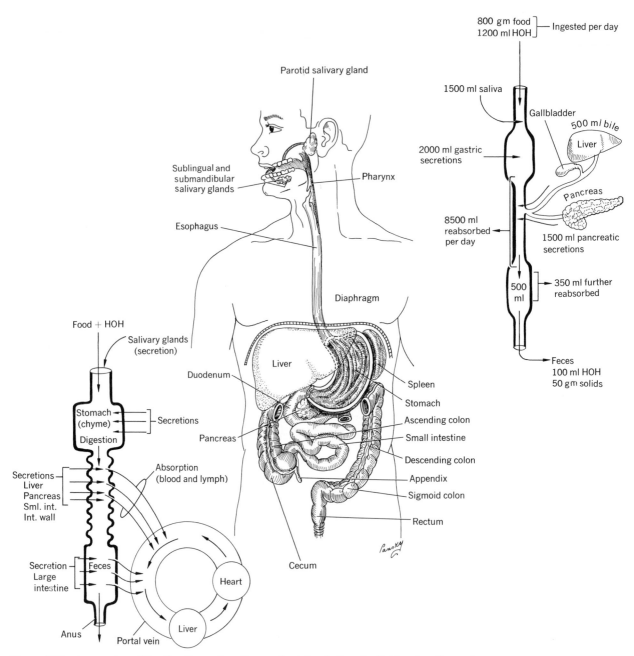

Figure 12-1. An overview of the digestive system. Summarized, too, is its potential for digestion and reabsorption (seen in the smaller figures).

Digestion and Food Absorption: The Digestive System

In early embryonic development, the primitive gut (archenteron) is a simple straight tube with a small surface area lined by a single layer of epithelial cells (the undifferentiated endoderm). The mouth end (stomodeum) is formed by an ectodermal invagination, as is the hind end (proctodeum). This early gut gradually differentiates to form all the structures from the pharynx to the rectum (Fig. 12-1).

Early in development, this tube of simple cuboidal or columnar endodermal epithelium is capable of secretion of digestive juices containing enzymes and absorption. Throughout life, this epithelial layer is the most important part of the gut wall and retains these two essential functions. The following changes take place in development: (1) There is growth in length with intestinal coiling. (2) There is a marked increase in surface area as a result of (a) extensive proliferation of epithelial cells, creating pockets (diverticula) to form glands such as the liver and pancreas, which retain their connections with the gut or crypts with a predominant secretory function, and (b) long finger-like processes or villi, which extend into the gut lumen and serve an absorptive function. (3) There are foldings of the entire inner mucosal layer, some permanent, such as the spiral folds (plicae circulares) of the lower duodenum and jejunum, and others not permanent, such as the stomach rugae.

The original endodermal cells are undifferentiated; however, with growth, highly differentiated cells develop. Some of the cells that secrete into the lumen are of exocrine function: (1) acid-secreting cells (the oxyntic cells of the glands of the stomach) that liberate hydrochloric acid, which is bactericidal and digests protein; (2) mucus-secreting cells (columnar epithelium of the stomach, goblet cells, and so forth), which are lubricating; and (3) enzyme-secreting cells. There are also endocrine cells, which are ductless and whose secretion is absorbed into the bloodstream. Note, too, that part of the function of the lining epithelial cells is also excretion into the lumen.

With development, other tissues are added to the epithelial cells: connective tissue for strength, muscle tissue for mobility, nervous tissue to coordinate secretion and mobility, lymphatic tissue to aid in the antibacterial battle, and blood and lymph vessels to supply the tissue and remove absorbed materials.

Histology of the Wall of the Digestive System

There are four major "coats" or layers that form the wall of the tract, from the luminal surface outward: mucosa (mucous layer), submucosa (submucous layer), muscularis externa (muscle layer), and serosa (serous layer) (Fig. 12-2). The first is most characteristic and said to be most important.

Mucosa
The mucosa consists of the surface epithelium and associated glands; the lamina propria (subepithelial) of loose connective tissue, in which the glands lie and which is well endowed with capillaries, small blood

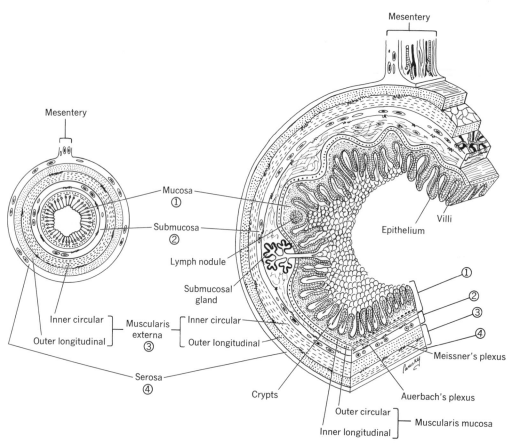

Figure 12-2. Drawings showing the general plan of the gastrointestinal tract. Note arrangement in four basic layers.

vessels, and lymphatic vessels; and a thin layer of muscle called the muscularis mucosae (mucosal muscle). The mucosa has a secretory and absorptive function and is the first line of defense against infection. It can increase and decrease its folds (via its muscle) and continually alters the contact surface of the bowel with relationship to its contents.

Submucosa

The submucosa is a layer of denser connective tissue containing larger blood vessels, lymphatic vessels, and the nerve plexus of Meissner (fine ganglionated nerve plexus). Glands are seen in this layer only in the esoph-

agus and the first portion of the duodenum. This layer is a "strengthening" one due to its connective tissue. It "supplies," as a result of large vessels, and coordinates by its nerve plexi (motor and sensory).

Muscularis Externa

The muscularis externa consists of muscle fibers arranged into inner circular and outer longitudinal bands. Between the two is the ganglionated nerve plexus of Auerbach (myenteric). Large blood and lymphatic vessels are also seen here. This layer regulates the diameter of the digestive tract and local, as well as general, movements of the wall.

Serosa

Serosa is present where there is mesentery to suspend and wrap around the tract. Where no mesentery exists, it is absent and is replaced by adventitial connective tissue. Serosa is a layer of peritoneal mesothelium with connective tissue, in which are found blood and lymph vessels and nerve branches extending out from the mesentery.

Note that in the mucosa and submucosa are found solitary lymph follicles, which are scattered irregularly along the gut but are most common in the small and large intestine. Aggregations of lymph follicles in groups are seen in the tonsillar ring around the oropharynx, in the ileum (Peyer's patches), and in the vermiform appendix. All are related to the defense mechanism.

Regulation of the Digestive System

Nerves and hormones regulate glandular (exocrine and endocrine) secretion and smooth muscle contraction in the system. Included as exocrine glands found associated with the system are the salivary glands, the pancreas, and the liver. Intrinsic glands are mucosal and submucosal in the gastrointestinal tract wall. The exocrine glands secrete via ducts that empty into the intestinal lumen, while the endocrine glands in the mucosa secrete into the bloodstream and are carried to the heart and returned to the tract via the arterial system. The endocrine secretions can thus act on glands or smooth muscle at sites far removed from their source of origin.

Located in the walls of the gastrointestinal tract from the esophagus to the anus are two nerve plexuses, the myenteric plexus (intermuscular between the circular and longitudinal) and the submucous plexus. Many synapses are formed, even between the plexuses, so that neural activity is integrated between the two. The axons branch profusely, and stimulation can involve the entire gastrointestinal tract.

Sympathetic and parasympathetic nerve fibers enter the tract and synapse with neurons in the internal nerve plexuses and exert their effects on glands and muscles of the tract through the internal plexuses. The sympathetic fibers may bypass the plexuses, however, and end in close approximation to gland cells and smooth muscle of vessels. The major autonomic nerve supplying the tract is the vagus nerve (parasympathetic), which sends nerves to the entire tract down to about the middle of the transverse colon. It contains both motor (efferent) and sensory (afferent) fibers. The afferent supply is complicated by the fact that the walls of the tract contain many sensory receptors that are part of the internal neuron plexuses. Thus, sensory information passes to these plexuses as well as to the central nervous system via the vagus nerve. The former involve short reflex circuits and give the tract a degree of local self-regulation. Complex behavioral influences, such as sight, emotion, and smell, are under central nervous system control.

Anatomy of the Digestive System

The gastrointestinal or digestive tract consists of the mouth, oropharynx, esophagus, stomach, and small and large intestine. It is supplemented by the salivary glands and portions of the liver and pancreas. It transports food and water from the external to the internal environment, where they are eventually carried to the cells via the vascular system. The tract is of variable diameter and about 4.5 m (15 ft) in length from mouth to anus.

The Mouth (and Surrounding Area)

Lips. The lips (Fig. 12-3) are fleshy folds that circumscribe the mouth and close the buccal cavity. They unite at the sides to form the commissures. The lips consist of five layers: skin, thick and adherent to tissue below; superficial fascia, which is loosely arranged and therefore causes edema on bruising; orbicularis oris

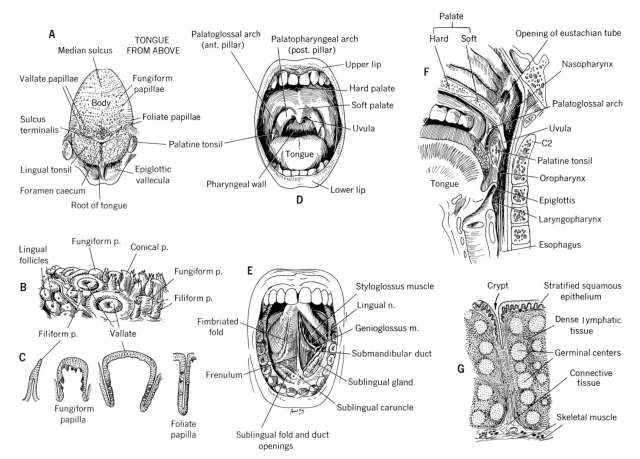

Figure 12-3. The mouth. **A.** The tongue, seen from above. **B** and **C.** The lingual papillae. **D.** Interior of the mouth, anterior view. **E.** Undersurface of the tongue, partially dissected to show deep structures. **F.** Interior of the pharynx, side view. **G.** Palatine tonsil, histologic representation.

muscle, which acts as a sphincter; submucous tissue, with its vessels and mucous labial glands as well as an arterial circle that provides a rich blood supply for the lips; and the mucous membrane, which is the innermost layer and covered by stratified epithelium.

Mouth Cavity. The mouth cavity (Fig. 12-3) is divided into the vestibule (between gums and cheeks) and the mouth proper (behind and in the arch of the teeth). The latter is bounded in front and sides by the gums and teeth, above by the hard and soft palates, and below by the tongue and sublingual area and opens posteriorly into the oral pharynx.

FLOOR. The floor of the mouth (Figs. 12-3, 12-4) consists of two deep grooves between the mandible and root of tongue. The anterior two thirds of the tongue rises from the floor, and the lingual frenulum is the midline fold from the bottom of the tongue to the floor. Two small papillae (mounds) are seen on either side of the frenulum, at the tips of which are located the openings of the submandibular glands. A fold of membrane laterally covers the sublingual glands, the smallest of the salivary glands, whose ducts open separately into the floor of the mouth.

VESTIBULE. The vestibule communicates with the

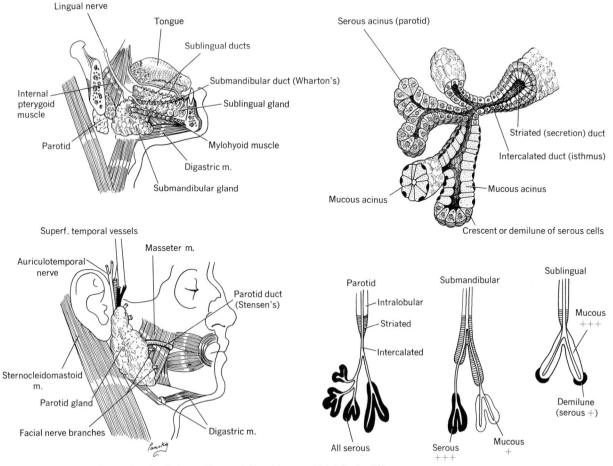

Figure 12-4. The salivary glands, their position, relationships, and histologic differences.

mouth behind the last molars. The parotid duct opens here opposite the second upper molar and is easily seen as a small papilla. The cheek resembles the lips in structure. In its subcutaneous fat is the suctorial fat pad (gives roundness to the baby's cheeks and is useful in sucking).

GUMS. The gums (gingivae) consist of dense, vascular fibrous tissue covered by mucous membrane and attached to the alveolar margins of the jaw and are continuous with the mucosa of the vestibule, palate, and floor of mouth. The gums are vascular but not very sensitive. A portion of the gingiva projects into each interdental space and surrounds the neck of the tooth.

TEETH. The teeth (Fig. 12-5) appear usually at six months of life, but the early formation of both temporary and permanent teeth is present at birth. Teeth develop from "skin," the dentin being derived from dermis and the covering enamel from epidermis.

The first dentition consists of 20 temporary (deciduous or milk) teeth that appear between the sixth and twenty-fourth months (lowers before uppers). There are four incisors, two canines, and four molars in each jaw with no premolars. Approximately four years later, at the age of six, the permanent (second dentition) teeth begin to erupt and do so until about age 25. In the permanent set, the six-year molars come first, the

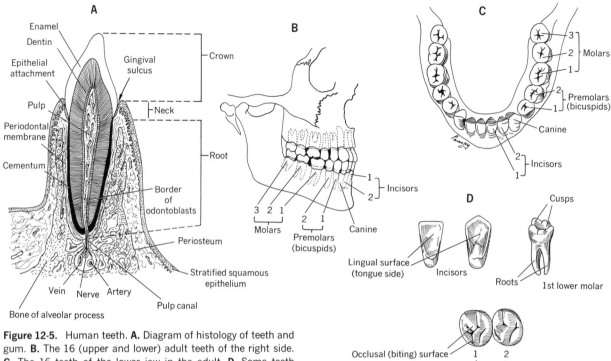

Figure 12-5. Human teeth. **A.** Diagram of histology of teeth and gum. **B.** The 16 (upper and lower) adult teeth of the right side. **C.** The 16 teeth of the lower jaw in the adult. **D.** Some teeth surfaces are shown.

12-year molars second, and the 18-year molars (wisdom teeth) third. The latter may erupt late in life or not at all.

There are 32 permanent teeth, 16 per jaw. These consist of (1) the incisors or cutting teeth, whose crowns are chisel shaped and whose roots are single; (2) the canine teeth, which also have pointed crowns with single, long roots (the upper canines are larger than the lower and are often referred to as the "eye" teeth); (3) the bicuspids or premolars, which have two cusps, each with a single root; and (4) the molars, distinguished as the first, second, and third and called the "grinding" teeth. The uppers have four cusps and the lowers five, and all are characterized by having large crowns. Each has either two or three roots.

Each tooth has a crown, a neck encircled by gum, and a root embedded in the jaw (Fig. 12-5). At the

apex of each root is the apical foramen, a pinpoint opening running through a widening canal to the pulp cavity, which is filled with tooth pulp, vascular connective tissue, and many vessels and nerves. The latter enter and leave via the foramen. Each tooth consists of enamel (the insensitive covering of the crown), dentin (very sensitive yellow foundation of the tooth), cement (bony covering for the root and neck), pulp (fibrous material containing the vessels and nerves), and the periodontal membrane (continuous with the lamina propria of the gum and attached to the cement and alveolar wall). The teeth are rooted in the alveolar cavities (sockets).

The upper teeth are supplied by branches of the maxillary and infraorbital nerves (V), and their lymph drainage ends in glands on the surface of the parotid glands and below the mandible. The lower teeth are supplied by the mandibular nerve (V), and their lymph vessels end in glands below the mandible and in the neck.

Pyorrhea (alveolaris) is a purulent inflammation of the dental periosteum. It involves the gingivae and alveolar walls and often requires tooth removal.

HARD AND SOFT PALATES. The hard and soft palates (Fig. 12-3) form the roof of the buccal cavity and the floor of the nasal cavities. The hard palate, which consists of maxilla and palatine bone covered by mucous membrane tightly bound to the underlying periosteum, forms a partition between the mouth and nasal cavities. The membrane is thin in the midline but thick at the sides due to the many glands found laterally. One sees a median raphe and transverse ridges called palatine rugae. The soft palate is a fleshy "curtain" that is attached to the back edge of the hard palate and consists of connective tissue (palatal aponeurosis), muscles, blood vessels, nerves, and glands. It shuts off the mouth from the pharynx during nasal breathing. In swallowing or mouth breathing, it is raised to a horizontal position to close the oral part of the pharynx from the nasopharynx, to prevent food from entering the nasal cavities. Its posterior edge is free, and there is a downward projection in the midline called the uvula. The soft palate has five muscles: the palatoglossus and palatopharyngeus descend from it to the tongue and pharynx; the levator palati and tensor palati descend to the palate from the base of the skull; and the musculus uvulae lies in its substance. The levator and tensor function as named. The palatoglossi aid in elevating the dorsum of the tongue and in closing the oropharyngeal opening during early swallowing. The palatopharyngeal muscles aid in elevation of the pharynx and thus shorten the pharynx in swallowing. The uvulae muscle shortens and raises the uvula. The levator palati, musculus uvulae, and palatopharyngeus are innervated by branches from the pharyngeal plexus (like pharyngeal muscles); the palatoglossus, by the pharyngeal plexus and the tensor palatini, by a branch of the fifth cranial nerve. The sensory nerves of the mucous membrane are derived from palatine, nasopalatine, and glossopharyngeal nerves.

The blood supply of the palate is by way of the ascending pharyngeal, maxillary, and facial arteries.

TONGUE. The tongue (Fig. 12-3) is a solid mass of muscle covered with mucous membrane and attached to the floor of the mouth, hyoid bone, and mandible. It consists of two parts that differ structurally, functionally, in nerve supply, and in appearance. The anterior two thirds is called the body (oral part), and the posterior one third is the root (pharyngeal part). An inverted V-shaped groove, the terminal sulcus, separates the two dorsally. A depression, the foramen cecum, is seen at the apex of the sulcus.

The dorsal surface of the body has a rough mucous membrane with numerous papillae, whereas the root surface has masses of lymphoid nodules, the lingual tonsils. The latter form the lowest part of a ring of lymphatic tissue around the pharynx, which includes the palatine tonsil and pharyngeal tonsil or adenoids and is called the ring of Waldeyer.

The three types of papillae seen on the anterior two thirds of the tongue (Fig. 12-3) are vallate, fungiform, and filiform. The vallate papillae (10 to 12 cylinders surrounded by trenches) lie just in front of the sulcus terminalis and run parallel to it. Serous glands open into a trench that surrounds the papillae, and taste buds lie here. The fungiform papillae (rounded head and narrow base) are small, look like red spots near the top and margin of the tongue, and also have taste buds. The filiform papillae are the smallest and most numerous papillae and give the tongue its "velvety" look. They, too, are found in rows parallel to the sulcus terminalis. They contain touch corpuscles but no taste buds.

The tongue's inferior surface has no papillae, is smooth, and is connected to the floor by the frenulum. The muscles of the tongue are right and left and are separated by a median fibrous septum, each half having intrinsic and extrinsic muscles (Fig. 7-6, p. 198). The latter alter the shape of the tongue and move it, while the former only alter its shape. The four extrinsic muscles are the genioglossus (protrudes tongue), styloglossus (retracts it), hyoglossus (depresses it), and glossopalatine (elevates root). The four intrinsic muscles are the superior and inferior longitudinal, transversus, and verticalis.

The motor nerve to all of the muscles of the tongue is the hypoglossal (XII). The glossopharyngeal (IX) nerve supplies taste and sensation to the posterior one third. The lingual (V) nerve supplies sensory fibers to the anterior two thirds, while the chorda tympani (VII), incorporated in the lingual, supplies taste to the anterior two thirds.

The lingual artery, a branch of the external carotid, supplies the tongue. Four veins pass back to form the lingual vein. The lymph drainage consists of apical (from tip) to nodes under and near the mandible, marginal from the sides of tongue to nodes under the mandible and deep cervical glands, and central from either side of the midline and posterior, from the back, both draining into deep neck lymph nodes (Fig. 10-36, p. 375).

Salivary Glands. The glands that pour their secretions or saliva into the mouth are called the salivary glands (Fig. 12-4). In man there are usually three large paired glands, which in order of size are the parotid, submandibular, and sublingual. The glands are often classified by their secretory cell type as pure mucous (sublingual), pure serous (parotid), or mixed mucous and serous (submandibular). The serous cells form the salivary enzyme ptyalin, and the secretion is clear and watery. The mucous cells secrete a viscous, slimy fluid that contains mucin. The arteries that supply the glands follow the interlobular connective tissue and give off branches that follow the ducts to the lobules to form a rich capillary plexus between the acini. This is drained by venules and veins. Lymphatic capillaries are not numerous. Sensory nerve fibers from the trigeminal nerve pass into the paired salivary glands and form endings there. Motor fibers from both the sympathetic and parasympathetic systems enter the glands to form rich plexi of fine fibrils. Stimulation of sympathetic fibers results in a scanty flow of secretion, rich in enzyme. Vasoconstriction also occurs. The parasympathetic fibers to the parotid arrive with the glossopharyngeal (IX) fibers associated with the otic ganglion. The parasympathetic fibers to the submandibular and sublingual glands pass in the chorda tympani (VII) to join the lingual branch of the trigeminal nerve, and they synapse in the cells of the submandibular ganglion, from which fibers pass into the plexus in the glands. Stimulation of these fibers causes a copious secretion of a watery consistency with a high content of salts. Vasodilation also occurs.

PAROTID GLAND. The parotid or Stensen's duct begins at the anterior part of the gland, which lies in front of the ear, runs across the upper part of the cheek, and opens on a papilla in the mouth vestibule, opposite the upper second molar tooth.

SUBLINGUAL GLAND. The sublingual gland is the smallest of the salivary glands and lies beneath the mucous membrane of the floor of the mouth at the side of the frenulum of the tongue. It is narrow, flattened, and almond shaped and weighs about 2 gm. Its excretory ducts are from 8 to 20 in number and open separately into the mouth on the elevated crest of mucous membrane on either side of the frenulum (Fig. 12-3).

SUBMANDIBULAR GLAND. The submandibular gland is irregularly shaped and the size of a walnut. It lies under the mandible on the mylohyoid muscle and has a deep process that folds around behind the same muscle. From this deep process, a submandibular duct (Wharton's duct) runs forward and opens, by a narrow orifice, on the summit of a small papilla that is at the side of the frenulum linguae at the base of the tongue.

The Pharynx

Tonsil. The name *tonsil* usually refers to the faucial or palatine tonsils (Fig. 12-3). The tonsillar area is located in the anterolateral pharynx as two masses of lymphoid tissue, found in the tonsillar fossa between the palatoglossal and palatopharyngeal arches just above the back part of the tongue and below the soft palate. Its surface can be seen through the mouth with the tongue depressed. One notes 12 to 30 rounded or slitlike openings on its surface, the tonsillar crypts. The surface is covered with mucous membrane (squamous epithelium), which lines the crypts. Its outer surface is covered by fascia, called the tonsillar capsule. The blood supply of the tonsil is very profuse, its main vessel being the tonsillar artery (branch of the external maxillary). Other small vessels also help supply it, and all are freely anastomosing. The veins form a plexus and end in the pharyngeal plexus, which empties into the internal jugular vein. The lymphatics end in the superior deep cervical chain. The nerve supply is derived from the glossopharyngeal via the pharyngeal plexus.

Pharynx Proper. The pharynx proper (Fig. 7-9, p. 202) is a large vestibule common to the respiratory and digestive systems. It is a musculomembranous tube extending from the base of the skull to the level of the cricoid cartilage, where it continues as the esophagus. It is about 12.5 cm (5 in.) long and is wider from side to side than from front to back. It communicates with the nose, mouth, and larynx, and thus we speak of a nasopharynx (see Chap. 11), oropharynx, and laryngopharynx.

The pharyngeal wall has four distinct layers. From without in they are as follows:

1. Buccopharyngeal fascia, a layer of fibrous tissue covering the pharyngeal muscles.

2. The muscular layer, which is made up of five paired voluntary muscles: the superior, middle, and inferior constrictors, forming an outer circular layer; and the stylopharyngeal and palatopharyngeal muscles, forming an inner longitudinal layer. The constrictors are arranged like three flowerpots sitting in each other from above down, and each is attached by its front end to the side walls of the nasal, oral, and laryngeal cavities and meet posteriorly in a median raphe. All are supplied by the pharyngeal plexus (the inferior also gets branches from the vagus).

3. The fibrous coat (pharyngobasilar fascia) or submucous coat, which is most distinct near the skull base and is the major attachment of the pharynx to the skull.

4. The mucous coat, consisting of columnar ciliated epithelium in the nasopharynx and stratified squamous in the lower part.

OROPHARYNX. The oropharynx (Fig. 12-3) is partly respiratory and partly alimentary. It is the posterior extension of the oral (mouth) cavity, lying behind the mouth and tongue. It is lined by pharyngeal muscles and the membrane over it.

LARYNGOPHARYNX. The laryngopharynx (Fig. 12-3) is the longest division and lies behind the larynx. Its upper part is both digestive and respiratory; its lower part is strictly digestive. The mucous membrane is supplied by the internal laryngeal nerve (vagus, X). If a crumb lodges here, uncontrollable coughing develops.

The Esophagus (Gullet)

Embryology. Initially the stomach is separated from the pharynx by a constriction (the future esophagus). As the lungs develop, the constriction lengthens; however, prior to this elongation the trachea and esophagus form a single tube. This tube is then divided by the growth of two lateral septa that fuse, forming the trachea anteriorly and esophagus posteriorly. Thus, esophagotracheal communications for fistulae can occur in development if the septa do not fuse.

Adult Esophagus. The adult esophagus (Fig. 12-6) is a muscular tube, essentially a passage conveying the bolus of solid food or swallowed liquid from the pharynx to the stomach. It is lined by mucous membranes and runs from the end of the pharynx through the lower neck, through the superior and posterior mediastinum (chest), and through the diaphragm to enter the stomach. The last portion is about 2.5 cm (1 in.) long and may act as a sphincter. From teeth to stomach it is about 40 cm (16 in.). It is the most heavily muscularized part of the digestive tract and, except during the passage of food, it is closed by longitudinal folds of mucosa. Three permanent constrictions are seen in it: at its beginning, behind the bifurcation of the trachea, and as it passes through the diaphragm (Fig. 12-6).

VESSELS AND NERVES. The arteries of the esophagus (Fig. 10-24, p. 362) are derived from the esophageal branches of the inferior thyroid (cervical part), right intercostal and bronchial arteries from the thoracic aorta (thoracic part), and left gastric and left inferior phrenic from the abdominal aorta. Venous return is via esophageal venous plexi to the inferior thyroid veins, to the azygos and hemiazygos veins, and to coronary and short gastric veins of the stomach. Lymph drainage of the esophagus is into the numerous nodes that lie along its course. The nerves are derived

Figure 12-6. The esophagus. The numbers **(1-3)** in the left-hand drawing represent the areas of esophageal constriction: at its origin, at its termination, and where it crosses behind the arch of the aorta.

Figure 12-7. The stomach and its mucosa. Epithelial cells that secrete mucus cover the stomach surface and line the gastric pits. The parietal, or oxyntic, cells produce hydrochloric acid (HCl), and the chief, or zymogenic, cells produce pepsinogen. The parietal and chief cells lie in deep tubules, and their secretions reach the surface via gastric pits.

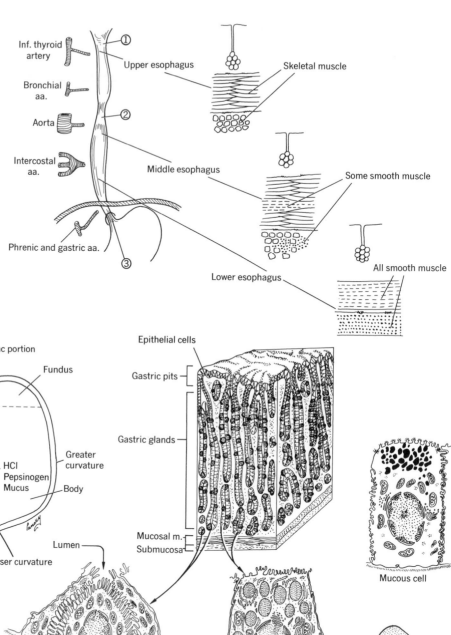

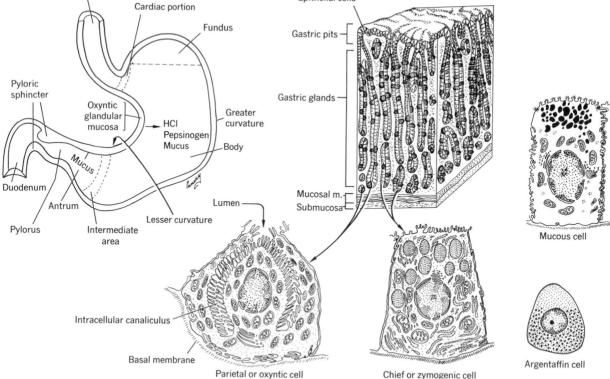

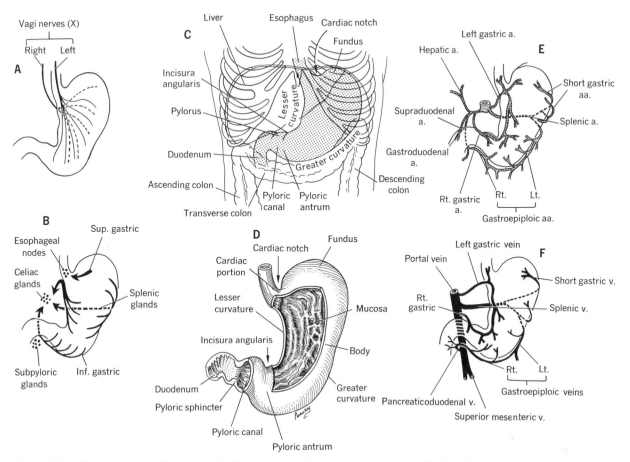

Figure 12-8. The stomach. **A.** Nerve supply. The parasympathetics are shown; sympathetics also supply the region. **B.** Lymphatics of the stomach. **C.** Topography of the adult stomach. **D.** Detailed anatomy of the stomach. **E.** Arterial supply of the stomach. **F.** Venous drainage of the stomach.

from sympathetic and parasympathetic systems via the cervical and thoracic chains and the vagus nerve, respectively.

The Stomach

Embryology. The stomach develops from the foregut. At the fourth week, it is in the neck. During weeks six and seven, the growth of the lung buds causes an elongation of the esophagus and a migration of the stomach backward so that it is now found in the lower thorax. Embryologically its long axis is in the median plane with its greater curvature directed backward, but during rotational development it comes to lie almost transversely with its greater curvature facing down and to the left and lesser curvature up and to the right. This creates a lesser peritoneal cavity (omental bursa) behind the organ and a greater cavity anterior to it. The opening from the greater cavity into the omental bursa is called the epiploic foramen (of Winslow).

Adult Stomach. The adult organ (Figs. 12-7, 12-8) is the most dilated part of the digestive tract and is about 25 cm (10 in.) long, 12.5 cm (5 in.) wide, and normally holds about 1 liter of "fluid." It is highly dilatable but can also shrink to a "tubular form" when empty. It is somewhat fixed by the esophagus at the diaphragm and with its connection to the duodenum. Its position and shape, however, are not fixed and may vary considerably, depending on body posture, gastric contents, stages of digestion, degree of abdominal and gastric muscle contraction, pressure of surrounding viscera, respiration, and gastric tonus. The organ resembles a "horn of plenty" and lies in the upper left part of the abdomen. It has two openings, two curvatures, two surfaces, and two incisurae (notches).

The two openings are the cardiac junction between esophagus and stomach (cardia) and the pyloric orifice where the stomach enters the duodenum. The two curvatures are the lesser on the right and the greater on the left. The former affords attachment for the lesser omentum (gastrohepatic part connecting stomach to liver) and contains the arterial circle of right and left gastric arteries. The greater curvature gives attachment to the greater omentum. The right and left gastroepiploic vessels create an arterial circle here

between peritoneal layers. The two surfaces are anterior and posterior, and the two incisures or notches are the cardiac notch on the left border where the esophagus meets the stomach and the angular notch in the lesser curvature (Figs. 12-7, 12-8).

The stomach is subdivided into a fundus (above and to the left of the cardiac notch); a body (between cardiac notch and angular notch), which is the area that secretes pepsin and hydrochloric acid; and a pyloric part (the rest), the mucus-secreting area of the gastric mucosa, which is subdivided into a dilated pyloric antrum proximally and a tubular pyloric canal (distally). The pylorus is considered to be the last 1.25 cm (1/2 in.) of the stomach and has thicker walls than the rest of the stomach due to an increase in circular muscle fibers that close and relax the pyloric orifice and are referred to as the pyloric sphincter. The pylorus is normally closed but when open admits a fingertip.

As indicated above, two layers of peritoneum envelop the stomach. They meet at the lesser curvature to form the lesser omentum, which attaches to the liver and diaphragm (Fig. 12-9). It has two parts: the gastro-

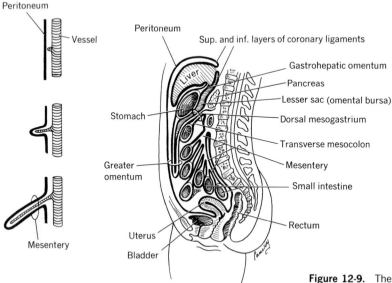

Figure 12-9. The peritoneum and its reflections. A mesentery is a double layer of peritoneum.

hepatic part (avascular, thin, with no important structures except near the stomach) and a duodenohepatic part (which is thick and contains the common bile duct, hepatic artery, and portal vein). The two layers of peritoneum also meet at the greater curvature to form a great fold. Different parts of the latter are named according to attachment: gastrophrenic ligament (to diaphragm), gastrosplenic ligament (to spleen), and greater omentum (to transverse colon and over it as an apron). That part of the omentum between the stomach and colon is called the gastrocolic ligament.

Histology. The gastric mucosa is grossly thrown into "plastic" folds and ridges. Under a light microscope, the surface is seen to be covered by a single layer of cells, an epithelium consisting of mucus-producing columnar cells (Fig. 12-7). If one uses a scanning electron microscope with magnification about $500\times$, the surface appears slippery, slimy, and smooth because of adherent mucus; however, the surface is also dimpled by many "holes," which are the openings of pits (gastric pits) that lead, in turn, to the gastric glands from which the specific gastric secretions emerge. Parietal (oxyntic), chief (peptic), and mucus cells (neck cells) line the canal and are found at the center of each gland. Other cells are also seen: the argentaffin ("silver-loving") or enterochromaffin cells, which take up silver stain and are known to make serotonin, an agent that influences body functions (including blood vessel contraction). The reason for their being in the stomach is unknown. Thus, all these various cells cluster about a central canal and constitute a gastric gland. These glands are very dense in the middle two thirds of the stomach, but at either end the chief and parietal cells are less prevalent. The stomach kneads (like a mill), stews (breaks down materials with acids), holds food (trough), and produces hydrochloric acid.

The muscularis mucosae and its interglandular strands are present as a complex layer, which forms the boundary of the mucosa on its submucosal side. The lamina propria fills the space between the surface epithelium, the glands, and the muscularis mucosae. The submucosa shows no variations. The muscularis externa has bundles of smooth muscle irregularly arranged and may even consist of three layers in the body of the stomach instead of the usual two. The stomach is covered by peritoneum and thus has a serosal layer.

Vessels and Nerves. ARTERIAL SUPPLY. The arteries (Fig. 10-24, p. 362; Fig. 12-8) are derived from the celiac axis. The lesser curvature is supplied by the left gastric artery, which anastomoses with the right gastric artery (from the hepatic). The greater curvature receives blood from short gastrics (four or five) from the splenic; the left gastroepiploic, also from the terminal end of the splenic; and the gastroduodenal, which arises from the hepatic and ends at the lower border of the duodenum by dividing into a superior pancreaticoduodenal and a right gastroepiploic. The latter runs in the layers of the greater omentum, supplying the stomach and anastomosing with the left vessel of the same name.

VENOUS DRAINAGE. This corresponds to the arteries and usually terminates in the portal vein. Two loops are formed: on the lesser curvature, the left gastric or coronary and right gastric or coronary end in the portal vein; on the greater curvature are the right and left gastroepiploics, which end in the splenic vein. The left gastric anastomoses with the lower esophageal vein, and thus drainage may pass here to the azygos system and then into the caval venous system (a portacaval anastomosis). Varicosities (dilated veins) may occur at this location in liver disease, and rupture can lead to severe hemmorrhage and death.

LYMPH DRAINAGE. Lymph drainage (Fig. 10-35, p. 373; Fig. 12-8) follows the three branches of the celiac axis—namely, hepatic, gastric, and splenic. The celiac or preaortic glands empty into the cisterna chyli, which in turn passes to the thoracic duct.

NERVES. Both sympathetic and parasympathetic nerves innervate the stomach. The latter is via the vagus (Fig. 12-8). The left vagus supplies the liver and anterior stomach wall, the right vagus passes to the celiac plexus, and the rest goes to the posterior stomach wall. The sympathetic supply is from the celiac (solar) plexus, whose fibers originate from the greater and lesser splanchnic nerves. The greater arises from the sympathetic chain between the fifth and tenth thoracic ganglia. The lesser arises from the chain in the area of the ninth and tenth thoracic ganglia. The nerves from the plexus follow the branches of the celiac artery. In the gastric wall, a plexus is found between muscle layers (myenteric plexus) and in the submucosa (submucous plexus) (Fig. 12-2). These contain both sympathetic and parasympathetic fibers and supply the muscles and glands of the stomach (see Chap. 8).

The Pyloroduodenal Junction

The change from pyloric structure to duodenal is abrupt (Fig. 12-8). The surface epithelium changes from the mucus-secreting columnar to the striated-border columnar type, and goblet cells appear (not seen in stomach). Villi appear, mucus thickens, and the

Figure 12-10. The small and large intestines. The upper right-hand illustration is a diagrammatic representation of the various layers of tissue in the intestinal tract as you pass from duodenum to rectum.

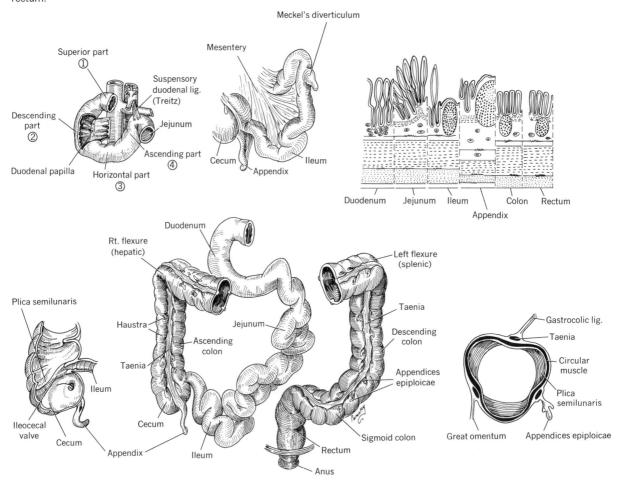

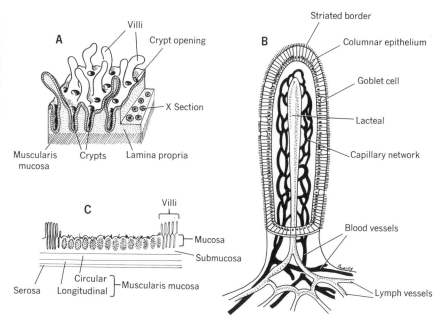

Figure 12-11. Intestinal villus. **A.** Lining of the small intestine; villi with cores of lamina propria that extend into the lumen. Note that the crypts (of Lieberkühn) are glands that dip down into the lamina propria. **B.** An enlarged view of a typical villus. **C.** Diagram of lymph follicle aggregates (Peyer's patch).

branched pyloric glands are replaced by the unbranched crypts of Lieberkühn. One also sees the glands of Brunner. There is a thickening of the muscularis externa, and much lymphatic tissue is seen.

The Small Intestine

The small intestine (Fig. 12-10) is traditionally divided into duodenum, jejunum, and ileum. It is better divided into (1) that part characterized by glandular acini in the submucosa (the duodenal glands of Brunner), which occurs in the duodenum above where the bile and pancreatic ducts enter; (2) that part containing permanent circular mucosal folds, the valvulae conniventes, normally comprising the entire lower part of the duodenum, jejunum, and upper ileum; and (3) that part characterized by large linear aggregations of lymph follicles or typical Peyer's patches, as seen in the lower ileum.

All parts of the small intestine show: (1) villi, which are long, fingerlike processes of the mucosa (Figs. 12-10, 12-11); (2) test-tube-like, straight glands or crypts of Lieberkühn; (3) an entire surface of villi and crypts, covered by the specialized absorbing epithelium, the simple, striated-border columnar type; (4) single mucus-secreting goblet cells; (5) a regular arrangement of the muscularis externa (inner circular, outer longitudinal); and (6) a regular arrangement of the intrinsic nerve plexuses, the submucosal plexus of Meissner and the intramuscular plexus of Auerbach.

Duodenum. EMBRYOLOGY. The duodenum, like the stomach, is embryologically placed with its long axis in the median plane. With rotation, the first portion of the duodenum comes to lie to the right, the second portion becomes related to the right kidney, and the third portion crosses the abdomen so that the duodenojejunal junction lies to the left.

ADULT DUODENUM. The adult duodenum (Fig. 12-10) is about 25 cm (10 in.) long and is the shortest, thickest, and most fixed portion of the small bowel. It extends from the pylorus to the duodenojejunal junction, forming a C-shaped curve with the curve occupied by the head of the pancreas. It is covered by peritoneum except where crossed by the transverse colon. Its posterior surface has no peritoneum. In general, it lies on the posterior abdominal wall above the umbilicus and is divided into four parts: (1) superior part (5 cm [2 in.]), which begins at the pylorus and ends at the neck of the gallbladder (the term *duodenal bulb* refers to the first 2.5 cm); (2) descending part (7.5 cm [3 in.]), which passes vertically; (3) horizontal part (10 cm [4 in.]); and (4) ascending part (2.5 cm [1 in.]). The common bile duct and main pancreatic ducts open into the second part. The last portion terminates in the duodenojejunal flexure and is fixed by the suspensory ligament (muscle) of Treitz, a band of fibrous, muscular tissue running from the flexure to the back of the abdominal wall.

VESSELS, NERVES, AND LYMPHATICS. Many vascular patterns are present, and anomalies are frequent. Since the duodenum is closely related to the head of the pancreas, their blood supply overlaps (Fig. 10-24, p. 362). The usual description is related here. The gastroduodenal artery (branch of the hepatic artery) terminates near the lower border of the first part of the duodenum by dividing into the right gastroepiploic and anterior superior pancreaticoduodenal arteries. The gastroduodenal gives off a posterior superior pancreaticoduodenal artery to the superior margin of the duodenum. The first part of the duodenum receives two vessels, the supraduodenal to its superior wall and anterior surface and the retroduodenal to the lower two thirds of the posterior wall. The rest of the first part is supplied by branches from the right gastroepiploic and superior pancreaticoduodenal arteries. The superior anterior and posterior pancreaticoduodenal arteries anastomose with corresponding anterior and posterior inferior pancreaticoduodenal arteries, which arise as a common trunk from the superior mesenteric. Thus, two arcades are created, one on the anterior and the other on the posterior head of the pancreas. The duodenum receives anterior and posterior sets of vasa recta from the two arcades.

The nerves to the duodenum take origin from the celiac and superior mesenteric plexuses and follow the arteries (Fig. 8-36, p. 289).

The duodenal lymphatics are related to those of the pancreas. There are anterior and posterior sets of glands that drain into the superior pancreatic and pancreaticoduodenal glands that lie between duodenum and pancreas. The efferent vessels from these glands pass either upward to the hepatic lymph glands or downward to the preaortic lymph glands around the superior mesenteric artery.

Jejunum and Ileum. The jejunum and ileum (Fig. 12-10) together measure about 6 m (20 ft). They make up the second and third portions of the small intestine, with the jejunum making up two fifths of the length and ileum three fifths. No sharp, definitive line exists between them; however, they differ from the duodenum in not being fixed, but are covered by peritoneum and are suspended from the back abdominal wall by a peritoneal fold, the mesentery. The size of the intestinal lumen diminishes from above downward, being narrowest at its termination. To accommodate the intestinal length, the mesentery is thrown into many convolutions. The mesentery itself has a broad attachment along the back abdominal wall.

Since the coils of small intestine are movable, their position in the abdomen is variable and parts even extend down into the pelvis. From the pylorus to the ileocecal valve, the small intestine has a complete external muscle coat (longitudinal fibers) and an internal coat (circular fibers). The mucous membrane, however, differs in different parts of the gut. In the first portion of the duodenum, the inner lining is smooth but the branched, tubular, duodenal glands are numerous near the pylorus. The plicae circulares are true folds of mucous membrane set at right angles to the long axis of the gut. The folds begin in the second part of the duodenum and become closely packed below the duodenal papilla. They are also close in the upper jejunum but diminish near the lower end. In the ileum they become widely separated and in its terminal end disappear completely. The plicae increase the absorptive surface area without increasing the length of the gut. Small collections of lymphoid tissue create elevations in the mucous membrane and are referred to as solitary lymph nodules. Larger collections of lymphoid tissue are the "aggregated" lymph nodules or Peyer's patches (most numerous near the terminal end of the ileum), which form elongated, oval areas on the anti-mesenteric border of the intestine. They are absent in the upper two thirds of the jejunum.

In general, one can differentiate between jejunum and ileum by the fact that the upper or proximal intestine is thicker and its diameter diminishes as you pass distally. Also, the blood vessels of the small intestine are larger proximally; there are only primary vascular loops associated with the vasa recta (about 3 to 5 cm [$1\frac{1}{5}$ to 2 in.] long) proximally, but distally secondary vascular loops or arcades appear, producing "smaller windows or lunettes" as well as shorter vasa recta (about 1 cm); as you pass distally, the amount of fat increases, which obscures the lunettes and even the vessels.

ARTERIAL SUPPLY. The arteries (Fig. 10-24, p. 362) that supply the jejunoileum arise from the left side of the superior mesenteric artery. The mesenteric artery terminates by anastomosing with one of its own branches, the ileocolic artery.

The jejunal and ileal arteries make up about 12 or more branches that spread out from the left side of the superior mesenteric artery and pass between layers of mesentery to reach the jejunoileum. They unite to form loops (or arches) from which the straight terminal branches (vasa recta) pass alternately to opposite sides of the jejunum and ileum. In the wall of the intestine, the vessels run parallel with the circular muscle coat, transversing serous, muscle, and submucous layers successively. The vasa recta do not anastomose themselves but pass to the submucous plexus, where their ramifications anastomose freely.

VENOUS DRAINAGE. The superior mesenteric vein (Fig. 10-31, p. 369) returns blood from the intestines as well as the ascending and transverse colon. Behind the neck of the pancreas, it unites with the splenic vein to form the portal vein.

LYMPH DRAINAGE. The lymphatics (Fig. 10-35, p. 373) of the jejunoileum drain into the superior mesenteric glands that lie in the mesentery near the arterial arches.

NERVES. The nerve supply (Fig. 8-36, p. 289) is derived from the celiac plexus of the sympathetic system and via the vagus. Referred pain from this area is felt in regions supplied by the ninth to eleventh thoracic nerves (related to the splanchnic origin) and thus can be felt around the umbilical region, lumbar region, and back.

MECKEL'S DIVERTICULUM. Meckel's diverticulum (Fig. 12-10) is a blind outpouching, at a right angle near the terminal part of the ileum opposite its mesenteric attachment. It therefore occurs near the terminal 60 cm (2 ft) of the ileum, is about 5 cm (2 in.) long, is seen in 2 percent of people, and occurs more frequently in males. Embryologically, it is a portion of the omphalomesenteric duct, connecting intestine to the yolk sac. It normally occludes and disappears completely. If it remains patent, a congenital fecal fistula exists between intestine and umbilicus.

ILEOCECAL VALVE. The ileocecal valve (Fig. 12-10) guards the end-to-end opening of the small intestine into the large intestine. The actual opening is only a transverse slit formed by the upper and lower lips, just on the edge, of the fold between the cecum and the first part of the ascending colon, and it lies just above the opening of the appendix into the cecum.

The Large Intestine (Colon)

Embryology. The cecum, ascending colon, and about one half of the transverse colon are derived from the midgut, whereas the rest of the colon is derived from hindgut. The cecum is fixed to the right side near the crest of the ileum, and at this stage the colon runs obliquely upward to the left of the stomach, where it curves as the splenic flexure to form the descending colon. As the liver increases in size, the hepatic flexure appears in the oblique part of this proximal colon and demarcates the transverse colon. Posterior fixations of peritoneum occur, so that the ascending mesocolon and colon fuse to the body wall on the right and the descending colon fuses on the left. The transverse colon and mesocolon do not fuse but hang suspended from the back abdominal wall. The redundant sigmoid colon also does not fuse, and thus we have a mesosigmoid. The rectum is the only part of the entire tract that maintains its primitive sagittal position and has no mesentery.

Adult Large Intestine. The following differences are found between large and small intestine:

1. The large bowel is sacculated; the small is smooth.

2. The large intestine has taeniae coli, which are the remains of the incomplete outer longitudinal muscle of the bowel arranged in three longitudinal bands that are shorter than the gut itself, causing a puckering or sacculation; the small bowel has none since its outer muscle coat is complete.

3. The large bowel has appendices epiploicae; the small bowel does not. These are little fatty "tags" projecting from the serous coat of the bowel.

4. The large bowel has a greater diameter than the small one. The size diminishes from the cecum distally.

5. Internally, the colon has no aggregated lymph nodules, villi, or circular folds, which are seen in the small bowel. The mucous membrane of the colon is thrown into folds opposite the constrictions between sacculations, but these are not permanent (as in the intestines) and can be smoothed out by cutting the taeniae.

The large bowel (Fig. 12-10) begins as a blind head, the cecum, in the right iliac fossa, to which is attached the vermiform appendix. The cecum continues upward into the ascending colon on the right half of the posterior wall of the abdomen. At the liver, the ascending colon bends sharply to the left to form the right colic or hepatic flexure and continues as the transverse colon. The latter crosses from right to left and bends near the left kidney and spleen as the left colic or splenic flexure to pass into the descending colon. The latter continues down the left side of the posterior abdominal wall to the crest of the ilium, turns medially to become the pelvic or sigmoid colon, which ends at the center of the midsacrum, where it becomes the rectum. Thus, the large bowel is about 1.5 to 1.8 m (5 to 6 ft) long and forms a three-sided "frame" around the small bowel, leaving the inferior area open to the pelvis. The bowel is called large because of its great ability to distend.

Between the ileum and cecum is the ileocecal valve, which is a shelflike projection of mucous membrane where the wall of the ileum has become invaginated into the cecal lumen. The cecum itself is the widest part of the colon and also its thinnest, and thus all spontaneous perforations of colon usually occur here. The taeniae coli are anterior, posterior, and medial, and all converge on the base of the vermiform appendix, for which they supply a complete longitudinal muscle coat.

The position of the appendix is variable and may be difficult to locate surgically (Fig. 12-10). It is usually 7.5 to 10 cm (3 to 4 in.) long but varies from 2.5 to 22.5 cm (1 to 9 in.), is entirely covered by peritoneum, and has a mesentery (mesoappendix). It is larger in children than in later life. The appendicular artery from the ileocolic supplies the appendix. A large amount of lymphoid tissue is found here and it is often called the "abdominal tonsil," but this tissue decreases with age. When inflamed and in normal position, it may be located by a point of tenderness, McBurney's point, at the junction of the middle and outer thirds of a line drawn between the umbilicus and right anterior superior iliac spine.

The Rectum

The rectum (Fig. 12-12) begins where the colon loses its mesentery, at about the third sacral vertebra. It is about 12.5 cm (5 in.) long and lies in the posterior part of the pelvis, following the curve of the sacrum and coccyx; it ends about 2.5 cm (1 in.) in front of the tip of the coccyx by bending down and back into the anal canal. The rectum is not sacculated and has no taeniae or appendices epiploicae. Its upper one third is covered anteriorly and laterally by peritoneum, its middle one third is covered only anteriorly, and its lowest third has no peritoneum. it has a series of flexures that, internally, correspond to three rectal valves (of Houston), which are crescentric, horizontal folds arising transversely from the lateral sides of the bowel and which, except for the outer longitudinal muscle, contain all layers of the rectal coat.

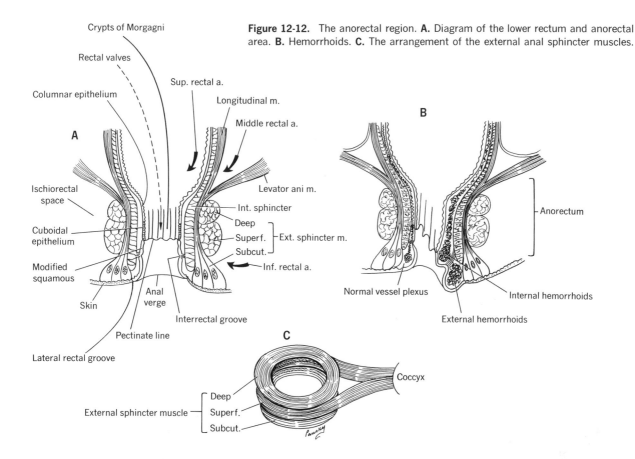

Crypts of Morgagni

Rectal valves

Columnar epithelium

Sup. rectal a.

Longitudinal m.

Middle rectal a.

A

Levator ani m.

Int. sphincter

Deep
Superf. ⎱Ext. sphincter m.
Subcut. ⎰

Ischiorectal space

Cuboidal epithelium

Inf. rectal a.

Modified squamous

Anal verge

Skin

Interrectal groove

Pectinate line

Lateral rectal groove

B

Anorectum

Normal vessel plexus

Internal hemorrhoids

External hemorrhoids

C

Coccyx

External sphincter muscle ⎱ Deep
Superf.
Subcut. ⎰

The Anal Canal

The anal canal (Fig. 12-12) is the terminal part of the large intestine and is about 3.75 cm (1½ in.) long, opening on the exterior as the anus. Anteriorly in the male it is related to the bulb of the penis and in the female to the lower vagina. Posteriorly it is related to the coccyx.

Anal Sphincters. The anal sphincters are two in number, the external and the internal (Fig. 12-12). The internal sphincter is the lower, thickened portion of the circular muscle of the bowel; it is about 2.5 cm (1 in.) long, can be felt digitally, and closes the upper two thirds of the canal. The external sphincter is supplied by the perineal branch of the fourth sacral nerve and inferior rectal nerves.

Landmarks. If one views the anal canal through an anoscope, there are four landmarks: the anocutaneous line, Hilton's white line, the pectinate line, and the anorectal line.

ANOCUTANEOUS LINE. The anocutaneous line or anal verge is thrown into folds by the "corrugator" muscle of the anal skin.

HILTON'S WHITE LINE. The Hilton's or intersphincteric white line marks the interval between the internal and external sphincters and feels like a depression. It lies halfway between the anal verge and the more superior pectinate line. Above it is an area about 0.8 cm (⅓ in.) wide called the "pecten" because arising from its upper edge is a serrated margin like the teeth of a comb.

PECTINATE LINE. The pectinate line or dentate line is the upper border of the pecten. Here we see the anal papillae continuous above with the columns of Morgagni. The bases of the papillae are connected by irregular folds called anal valves. Between the vertical columns are the so-called crypts or pockets of Morgagni.

The pecten has structural as well as clinical significance: (1) Stratified squamous epithelium is found below it and columnar epithelium above; the line itself is said to be covered by transitional epithelium. (2) At this area the external sphincter replaces the internal. (3) The sympathetic and cerebrospinal nerves meet here. The skin distal to the pecten is supplied by the inferior rectal nerve, which carries pain fibers; however, the mucous membrane proximal is supplied by sympathetic nerves, which have no pain fibers. (4) Internal hemorrhoids occur above the pecten, external hemorrhoids below it. (5) The area divides the lymph drainage. Above the line lymph drains into the pelvic lymph glands but below it into the superficial inguinal glands. (6) The pecten marks the dividing line between superior and inferior rectal vessels. (7) Focal infections take place here because of the crypts of Morgagni. (8) Developmental defects often occur here.

ANORECTAL LINE. The anorectal line is not the anocutaneous line. It lies about 1.5 cm ($\frac{3}{5}$ in.) above the pectinate line.

Vessels and Nerves of the Colon, Rectum, and Anal Canal. ARTERIAL SUPPLY. The colon receives its blood supply from the superior and inferior mesenteric arteries. The superior supplies that portion of the intestinal tract from the second part of the duodenum to the right half of the transverse colon. The artery is accompanied by a vein. Branches of the superior mesenteric artery are the jejunoileal, the middle colic to the transverse colon, the right colic to the ascending colon, and the ileocolic to the cecum, appendix, and distal ileum. The inferior mesenteric artery supplies the descending and sigmoid colons as well as the proximal rectum via the superior rectal artery. It gives off a left colic artery, sigmoid artery, and superior rectal artery. There is generally good anastomosis between all vessels except between the sigmoid and rectal branches.

The vessels of the colon anastomose with each other to form the so-called marginal artery (of Drummond), which extends from the ileum to the pelvic colon.

The arteries of the rectum and anal canal are the superior rectal (from the inferior mesenteric), the middle rectal (from the internal iliac), and the inferior rectal (from the internal pudendal). These all form rich anastomoses around the anorectal region.

VENOUS DRAINAGE. The rectal veins are arranged like the arteries but form a rectal (hemorrhoidal) plexus in the thickness of the bowel. It is best developed in the anal region, occupying the columns of Morgagni in the anal canal. In general, the greater part of the blood drains to the superior rectal vein and via the inferior mesenteric to the portal circulation. The middle and inferior rectal veins drain into the iliac vein and systemic circulation. Structural factors are essential to the development of varicosities of these veins, which are known as hemorrhoids. These factors are absence of valves and oblique passage of veins through the muscle wall.

Portal Vein. The portal vein (Fig. 10-34, p. 372) is formed behind the neck of the pancreas by the union of the superior and inferior mesenteric veins and the splenic vein. It enters the lesser omentum and divides into right and left branches, receiving, in its course, the left and right gastric veins from the lesser curvature of the stomach and the pancreaticoduodenal vein. The portal vein delivers to the liver blood that has circulated through the spleen, pancreas, and entire length of the digestive tract from the lower end of the esophagus to the upper end of the anal canal. The hepatic veins carry blood from the liver to the inferior vena cava. If the liver becomes diseased, the portal vein may obstruct. Communications exist between the portal circulation and the systemic circulation—the accessory portal system. These communications are found at (1) the lower end of the esophagus (esophageal of left gastric with esophageal of azygos veins); (2) in the bare area of the liver (where the diaphragmatic veins unite with the hepatic veins); (3) at the lower end of the rectum (superior rectal vein anastomoses with the middle and inferior rectal); (4) around the umbilicus (veins of the liver communicate with the epigastric veins around the umbilicus—an enlargement of veins, here spoken of as caput medusae); and (5) posterior ascending and descending colon (colic veins anastomose with lumbar veins).

LYMPH DRAINAGE. The lymphatics follow the blood vessels centrally to terminate in nodes along the aorta.

NERVES. The nerve supply to the large bowel is derived from the autonomic nervous system, except for the lower end of the anal canal, which is supplied by the inferior rectal nerve. The colon thus has both sympathetic and parasympathetic fibers (Fig. 8-36, p. 289).

Sympathetic Fibers. The sympathetic fibers are derived from the lower thoracic and upper lumbar segments of the spinal cord. These fibers then proceed to the celiac plexus via the lesser splanchnic nerves. From here they follow the superior mesenteric plexus, along the same artery, to supply the cecum, appendix, ascending colon, and transverse colon. The lumbar sympathetic nerves leave the sympathetic chain via the lumbar splanchnic nerves and follow the inferior mesenteric artery to the descending colon, sigmoid colon, and upper rectum. Intermesenteric nerves pass down as the hypogastric plexus, which divides into right and left pelvic plexuses that supply the bladder, prostate, and pelvic organs and form the so-called rectal plexus.

Parasympathetic Fibers. The parasympathetic fibers arise from the vagus and pelvic nerves. The nervi erigentes are the sacral autonomic nerves from the second to the fourth sacral nerves. They join the pelvic plexuses and are distributed to the bowel with the sympathetic fibers.

Pudendal Nerve. The pudendal nerve originates from the second to the fourth sacral nerves; follows the pudendal artery into the perineum, giving off the inferior rectal nerve to the external sphincter and lower cutaneous part of anal canal; and continues on to supply the external genitalia.

Physiology of Digestion

Nutrition (Food)

In the human body, the billions of cells are isolated from direct contact with the external environment. Thus, a problem exists in getting food to those cells. Obtaining oxygen is relatively simple since we live in a gaseous environment that includes oxygen, but food supplies must somehow enter the bloodstream from the outside world in order to reach the cells.

The food we eat is variable and consists of complex, large molecules, all built up from small, simple molecules. However, the protein substances of animals and plants are not always the same as human proteins, and the nonprotein part of foods contains many substances that man cannot utilize. In plants, the cell walls are made of cellulose, which plays no role in nourishing man since he lacks the special intestinal bacteria to convert cellulose into useful sugar. Thus, some food exists in a form acceptable for absorption into the individual human cells and much of what we eat is discarded since it is not usable.

The guiding principle of cell nutrition is that all foods must come to the cell as small, simple molecules since nutrients can reach the cell only in solution by way of the bloodstream and tissue fluids. As a rule, it is only the small molecules of any compound that are water soluble. The cell must retain its large protein molecules and still admit the smaller ones with nutrient, and from these form the "building blocks" into large proteins. Once created, the proteins cannot escape without breaking down into amino acids, from which they have been made.

Cell Needs

One of the cell's most essential needs is water, which may be classified as a "food." In fact, 50 percent of the cell's weight is water, and few reactions take place without it. The cell requires a watery medium in order to accept the substances brought to it in solution. Other substances also not usually thought of as foods are minerals (salts) and vitamins, which exist in very small amounts in most things we eat but fall under the classification of food since the cell cannot make most of them from other raw materials. Vitamins and minerals are necessary to almost all of the body's reactions.

The next essential need is an energy source. Just moving the body around requires a good deal of energy, as do the internal movements of structures such as the digestive organs or heart. Energy is consumed in the very process of active transportation across the membranes of all cells and goes on at all times. Growth, involving the synthesis of complex compounds from simple substances, uses up energy, and the reactions that build proteins from amino acids also absorb energy. The basic source of cell energy is carbon. Carbon and oxygen form heat and carbon dioxide. Yet pure carbon is insoluble in water and burns at a very high temperature and, therefore, for the cells' needs, it must be supplied in a very soluble form that can be burned at a controlled rate and at a low temperature. This form of carbon comes to the cell as glucose (sugar) and is broken down in the cell by the mitochondria. Thus, with little waste of heat, the cell acquires its energy in a usable form.

Heat and energy are closely related, and foods are rated in terms of the amount of heat energy that can be extracted from them. We deal in units called calories, one of which is defined as the amount of heat needed to raise the temperature of 1000 gm of water by 1° C. Thus, foods are classified according to their caloric value, which is the amount of heat, in calories, that a given weight of a compound would yield if it were completely oxidized (burned). Yet the energy released by oxidation does not always appear as heat. The release of muscle energy takes two forms: motion energy and heat energy.

A large proportion of food energy is used in staying alive, keeping warm, and performing vital chemical processes. Energy is consumed in driving the muscles of digestion and respiration and the heart. All other activity, of any kind, such as walking, talking, and even thinking, requires energy.

How are the cell's needs met by the foods we eat? Mother's milk is a guide to the human "perfect" diet; it is a complete food and sustains growth and activity

Table 12-1. Comparison of Human and Cow's Milk*

	Human Milk	Cow's Milk
Water	87%	87%
Protein	1.5%	3.5%
Fat	3.5%	3.5%
Sugar	7%	4.5%
Minerals	0.25%	0.75%
Reaction	Alkaline or amphoteric	Acid or amphoteric
Curd	Soft, flocculent	Hard, large
Digestion	More rapid	Less rapid
Calories	20/oz	20/oz

*Breast milk is sterile, cheaper, and available at proper temperature and has a greater amount of vitamin C.

for the early months of a child's life (Table 12-1). If one basic nutrient is absent from the daily diet, one may spend a lifetime in poor health and even die prematurely. In fact, two out of three people living today are either undernourished or have a vitamin deficiency or both. Protein starvation is the commonest form of malnutrition since protein foods are expensive and only affluent societies can afford a steady supply. A chronic deficiency of protein, minerals, and vitamins undermines the physical and mental health of two thirds of the world's population.

Although human milk is an ideal balanced diet, all the substances in milk cannot be directly absorbed. Apart from water, minerals, and vitamins, most everything else needs treatment before the cells can utilize them. This is particularly true of vegetable and animal dishes in the adult diet. There must be a breakdown of complex molecules into simpler substances, and anything that cannot be absorbed by the cells must be rejected.

Raw food has one advantage over cooked food: more vitamins and salts are retained. Yet cooking has three advantages: (1) man is able to make food more tender, and it breaks down more easily for processing in the body; (2) it is possible to process foods otherwise not capable of being absorbed by the body; and (3) one can mix and heat a variety of raw materials to make eating more palatable.

Cooking makes the molecules of food more vulnerable to attack by the chemical agents of the body. Oats,

Table 12-2. Summary of Digestion

Type of Digestion	Location of Digestion	Digestive Juice and Enzymes	Compounds Acted on	Products of Digestion
Salivary	Mouth, esophagus, and stomach	*Saliva*		
		Salivary amylase	Starch	Dextrins
Gastric	Stomach	*Gastric juice*		
		Pepsin	Proteins	Simple proteins and polypeptides
		(Rennin)	Milk Protein	Curdles protein
		Lipase	Fats	Fatty acids and glycerol
		(HCl)	Pepsinogen	Pepsin
Intestinal	Small intestine	*Pancreatic juice*		
		Trypsin	Proteins and polypeptides	Simpler peptides
		Chymotrypsin	Proteins and polypeptides	Simpler peptides
		Carboxy peptidase A and B	Dipeptides and higher polypeptides	Amino acids
		Nucleases	Nucleic acid	Nucleotides
		Pancreatic amylase	Starch (dextrins)	Dextrins
		Lactase	Lactose	Galactose and glucose
		Sucrase	Sucrose	Fructose and glucose
		Steapsin	Lipids	Diglycerides, monoglycerides, and fatty acid salts
		Intestinal juice		
		Oligo-1,6-glycosidase	Dextrins	Maltose
		Maltase	Maltose	Glucose
		Sucrase	Sucrose	Fructose and glucose
		Lactase	Lactose	Galactose and glucose
		Enterokinase	Trypsinogen	Trypsin
		Amino peptidases	Peptides	Amino acids
		Dipeptidases	Dipeptides	Amino acids
		Lipase	Lipids	Diglycerides, monoglycerides, and fatty acid salts
		Amylase	Dextrins	Maltose
		Nucleotidases	Nucleotides	Nucleosides
		Nucleosidases	Nucleosides	Purines, pyrimidines, ribose
		Phosphatases	Phosphate esters	Phosphoric and other compounds

a raw grain, cannot be "pulverized" by human teeth or jaws and cannot be eaten unless boiled. Boiling makes the starch granules inside the cells swell and burst their indigestible cellulose cell walls, which is true of many vegetables. In meat and fish, heat turns the indigestible collagen of flesh into digestible gelatin. The elastin and proteins shrink so that the muscle fibers of the meat separate, making chewing easier. Yet the nutritional value of the fat and protein is not altered. What is lost are the vitamins.

Steps in the Digestive Process

Food is processed in the gastrointestinal tract (also called the alimentary canal or digestive tract), whose interior is really external to the body cells and, therefore, food will remain outside the cells until it is transported across the canal's wall into the blood transportation system. Mechanical breakdown is completed in the first parts of the canal, but the major process in the canal is a unique chemical treatment referred to as the digestive process (Table 12-2). The gastrointestinal tract is not primarily an excretory system, since the feces eliminated consist mostly of bacteria and substances not digested and absorbed. At times, however, volumes of salt and water are excreted and may endanger the entire homeostasis of the body.

Food breakdown in the digestive tract is a form of chemical reaction called hydrolysis (Greek *hydro,* water; *lysis,* loosening); therefore, water is absolutely essential for the process. Most reactions are speeded up by substances called catalysts, which are not changed by the chemical reactions that they take part in. The catalysts produced by living organisms are called enzymes. An enzyme promotes reactions at body temperature and at regular atmospheric pressure. The body's enzymes do, however, deteriorate in time and must be replaced (see Chap. 1).

Enzymes are made up of large protein molecules; if the protein is destroyed, their activity terminates. They are affected by temperature and degree of acidity. Different enzymes in the digestive tract operate best at different degrees of acidity, as is found in the digestive system. Furthermore, enzymes are most efficient at 37° C (98.6° F), and the body temperature maintains this level. Above 40° C, enzyme activity decreases since the enzymes are destroyed. Enzymes consist of more than proteins, and the need in our diet for vitamins and minerals is apparent, for these form an integral part of certain enzymes (neither enzymes nor vitamins can work on their own). Much is known about enzymes and the reactions they catalyze; yet not all is known as to what happens to the substrate, the substance they catalyze. Each enzyme has a unique surface with a particular area where the substrate molecule joins it.

Step I: Mastication and Salivary Digestion. Digestion begins in the mouth, where large particles are taken in and consist of high-molecular-weight materials that cannot cross cell membranes as they are and must be broken down into smaller molecules. A partial breakdown of starch (polysaccharide) by the salivary enzyme amylase begins in the mouth. The enzyme has also been called ptyalin (Greek *ptyein,* to spit). It continues its activity in the stomach until inhibited by

hydrochloric acid. The ptyalin is carried to the mouth in the watery saliva from the salivary glands. Saliva, like blood, acts as a transport fluid and also dissolves some of the food molecules so they reach the chemoreceptors and give us the sensation of taste. Saliva provides the water that enzymes need to function; it moistens and softens the food, which enables the tongue to mold and form it into a bolus (Greek word for ball) so that it can be swallowed; and supplies the mucus. The major proteins of saliva are the mucins, which contain small amounts of carbohydrate attached to amino acids. When mixed with water, the mucins form a very viscous solution, the mucus, which lubricates the food, thereby making it possible for the food to pass smoothly and easily down the digestive tract. However, before ptyalin can function properly, the surface area of the food must be increased by the cutting and grinding action of the teeth, a process called mastication. Chewing thus becomes an essential part of the entire process. The teeth bite off and grind the food into pieces capable of being swallowed. Chewing is a combination of voluntary and reflex action of the skeletal muscles of the mouth and jaw. Prolonged chewing does not apparently alter the rate of digestion and absorption. Note, however, that swallowing too large a piece may result in choking, as the pieces may block the trachea, or "windpipe."

The secretion of saliva is triggered by a variety of stimuli: the thought or sight of food makes the mouth "water" when we are hungry; the smell of food is a stimulation; and the strongest stimulus of all is the sense of taste when food reaches the surface of the tongue. Here are located four different types of receptors, which register the basic tastes of sweet, sour, salt, and bitter (Fig. 12-3). Surprisingly enough, the nose, too, has similar receptors. All the receptors transmit stimuli to the brain, where the messages are sorted and interpreted so we can consciously recognize the different odors and tastes. The sensations of taste and smell are closely linked. Often we think we are tasting when we are actually responding to the odor entering the back of the nose from the mouth. Note, further, that the molecules that make up foods cannot register as either sensation until they are in solution. The secretion of saliva is controlled by nerves and not hormones. The glands are innervated by both sympathetic and parasympathetic fibers and both types stimulate secretion, with the parasympathetic stimulus creating a greater fluid volume. Little saliva is secreted during sleep, and at other times a base level of about 0.5 ml per minute keeps the mouth moist. The major stimulus to secretion is the actual presence of food (acid solutions being the most potent) in the mouth, and this is mediated by nerve fibers from chemoreceptors and pressure receptors in the walls of the mouth and tongue. The medulla contains the integrating center controlling secretion. Between 1 and 2 liters of saliva are secreted per day, and most is swallowed. The proteins break down into amino acids in the stomach and intestine, and the amino acids, in turn, are reabsorbed—along with water and minerals—into the circulation across the digestive tract wall.

Step II: Swallowing. The second step is swallowing (a reflex action). This begins with the tongue pressing against the hard palate, which forces the bolus into the pharynx (or throat). Food entering the oral pharynx involves a risk uncommon to the rest of the tract, for the food may get into the trachea as it crosses it or even into the nasal cavities, since food and air passages form a common crossroad in the pharynx (Fig. 7-9, p. 202).

Several mechanisms prevent this. When the tongue presses back and up against the roof of the mouth, the bolus of food is forced to the back of the pharynx. The bolus stimulates pressure receptors at the back of the throat and tongue, which send afferent impulses to the swallowing center in the medulla, which in turn coordinates the skeletal muscles of the pharynx, larynx, and upper esophagus as well as the smooth muscle of the lower esophagus and inhibits or slows down breathing. The rear entry to the nose is closed by the rising of the soft palate. The bolus also stimulates other nerve endings so that the pharynx squeezes it into the esophagus. At the same time the larynx is carried up, and the epiglottis, a cartilage attached to both the tongue and larynx, shuts off the opening to the trachea (windpipe) so food cannot enter. Although swallowing begins as a voluntary movement, choking is prevented by this chain of involuntary reflex events. Once swallowing is initiated, it cannot be stopped voluntarily, even though it involves skeletal muscles. As noted, the upper one third of the esophagus has skeletal muscle, the lower two thirds smooth. The esophageal opening is normally closed by the passive elastic tension of the walls. The swallowing center initiates esophageal muscle contraction, and food passes into the esophagus. After the bolus passes, the sphincterlike muscles relax, the opening closes, the glottis opens, and breathing continues. This pharyngeal phase takes about one second.

Step III: Peristalsis. For food to be treated in a variety of ways in the digestive tract, it must pass from one end of the tract to the other. There is a special movement of the smooth muscle of the tract called peristalsis or peristaltic waves, which begins in the esophagus where the circular muscle relaxes in front of the food and contracts behind it, forcing the food forward and into the stomach. The above action is coupled with a pendulum movement that thoroughly mixes the contents. The contraction and relaxation of the esophageal wall last about three to nine seconds, moving toward the stomach at a rate of about 2 to 4 cm ($\frac{4}{5}$ to $1\frac{3}{5}$ in.) per second. After a meal, stomach distention stimulates peristalsis in the colon (gastrocolic reflex), which in turn stimulates defecation. Every act of swallowing is followed by a contraction and relaxation wave carrying food to the stomach. The progression of the wave is controlled by autonomic nerves coordinated in the medullary swallowing center. Note that in the rest of the tract this control is predominantly through the internal nerve plexuses of Auerbach and Meissner, which are found in the gut wall. Swallowing can take place even when standing on one's head since it involves the peristaltic waves and not primarily gravity.

In the region of the last 4 cm (1 3/5 in.) of the esophagus there is a so-called gastroesophageal sphincter. Although it does not appear grossly different from the rest of the esophagus, this area remains tonically contracted when at rest. As the peristaltic wave begins in the esophagus, the sphincter relaxes, allowing the bolus into the stomach, and then contracts again. The barrier here is helped by the fact that the last portion of the esophagus lies in the abdominal cavity and is subject to abdominal pressures. During pregnancy, as the fetus grows, the terminal part of the esophagus is displaced upward into the thoracic cavity and there is a tendency for increased pressures in the abdominal cavity to force stomach contents into the esophagus. The hydrochloric acid from the stomach irritates the esophageal walls, causing contractile spasms of smooth muscle, and is felt as pain and generally referred to as heartburn since the feeling appears located in the heart area. It subsides late in pregnancy as the uterus descends before delivery. A newborn child has his esophagus entirely in the thorax and thus has a great tendency to regurgitate. Liquids and air are also swallowed during eating, but most of the air travels no further than the esophagus (eventually expelled by belching) while the rest reaches the stomach or even the intestine.

Step Four: Gastric Digestion and Absorption. The stomach is the expanded and highly muscular part of the tract, designed to accommodate the series of boluses of food (Fig. 12-13). The stomach is a preparatory organ for digestion and absorption that take place in the intestine. It stores a meal and releases it slowly at a rate attuned to that at which the small intestine can digest and absorb the material. If the meal is too concentrated, it is retained for dilution by gastric secretion; if too coarse, it is homogenized and chemically prepared for the small gut. The protein constituents are partially digested for optimal digestion in the small intestine.

SECRETION OF DIGESTIVE JUICES. Even before the food reaches the stomach, digestive juices have already been secreted in the stomach. Once in the stomach, powerful muscular action churns the food and mixes it with the gastric juices for further digestion. A pasty substance called chyme results.

Hydrochloric Acid. The stomach's most essential function is to regulate the rate at which ingested mate-rials enter the small intestine, where most digestion and absorption take place. The stomach secretes hydrochloric acid (HCl) and several digestive enzymes, which, along with salivary amylase, partially digest the food prior to its release into the intestine (Fig. 12-13).

The delicate, glandular mucosal lining of the healthy human stomach contains about 1 to 1.5 billion parietal cells, which either can be dormant or, in response to eating or emotional tension, can manufacture a cup (200 ml) of 0.17 N HCl per hour. As seen by electron microscopy, each parietal cell is lined by a myriad of tiny projections or microvilli, and from these (presumably) the HCl drips, later to be discharged into the main channel of the gastric gland that leads to the stomach lumen.

The exact source of the H^+ of the HCl secreted is not certain, but it may be generated by splitting of water into H^+ and OH^-. The freeing of the H^+ by the parietal cell leaves behind the waste product OH^-, which has strong basic properties and, if left to accumulate, could chemically destroy stomach tissue just like acid. However, buffering mechanisms exist to prevent this. In any event, regardless of source, the production of HCl proceeds in a series of intermediate steps in which CO_2 and H_2CO_3 take part. It also requires energy. About 324 calories are needed to produce 30 mEq of H^+ ions per hour in a healthy stimulated stomach.

Since positive and negative ions are always in balance, the positively charged hydrogen ions are accompanied by negatively charged ions, chlorine, thus accounting for HCl. This ion is provided by simple NaCl, found in the body's fluids. Sodium (the extra positively charged ion) is left behind and is used in the buffering process that keeps the OH^- from injuring the tissues.

Pepsin and Other Enzymes. Not only is HCl secreted but also gastric mucosa makes enzymes that operate in an acid medium and have the ability to digest proteins. Pepsin is the most important of these and is produced in the chief cell. Pepsin works well in the acid milieu of the stomach and can break up a protein like egg albumin (made up of about 500 amino acids) into numerous fragments or peptides, each containing

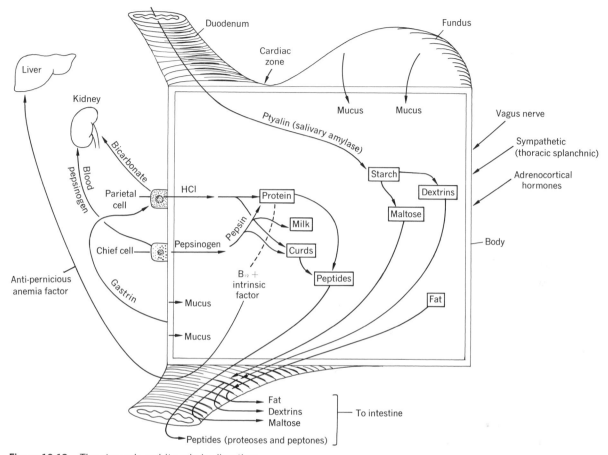

Figure 12-13. The stomach and its role in digestion.

from 2 to 50 amino acids per peptide. Gastric digestion, however, is only partial because 90 percent of the reduction of protein to absorbable peptides takes place in the small intestine. If the stomach contents are neutralized and cannot produce HCl or are inactivated by an antacid, pepsin becomes inactive and little protein digestion takes place. It is the combination of HCl and pepsin that is the "power" in breaking up proteins. If that "power" becomes undisciplined and attacks the lining of the stomach or intestine, ulcers occur.

The stomach lining also produces proteolytic enzymes when the gastric contents are near neutral.

There is also a gastric lipase that splits fats. No enzyme is produced that attacks carbohydrates, and yet some starch digestion takes place here if the starches are protected from the acid and exposed to swallowed salivary or regurgitated duodenal (pancreatic) amylase.

Mucus. The third most important product of the gastric mucosa (in addition to HCl and pepsin) is mucus, which is secreted by the mucus cell. Mucus is slippery, slimy, colorless, and transparent and consists of a complex mixture of proteins and sugars. It is credited with adhering to the gastric mucosa and giving it a protective coat. It also lubricates the mucosal surface and protects it from contact with ingestive substances (that might scrape it) and from endogenous and exogenous chemicals (that could irritate it).

Water and Electrolytes. Although gastric juice is known for its HCl, pepsin, and mucus, its most abundant constituent is water. In addition, it contains electrolytes. The total gastric secretion, after an average meal, may amount to about 200 to 400 ml and is a mixture of nonspecific and specific secretory mechanisms. The concentration of electrolytes usually reflects that seen in the blood, with the exception of H⁺ (200,000 to 1,000,000 times that in the blood).

Finally, the stomach contents may be influenced by swallowed material and often by regurgitation from the duodenum.

STOMACH ABSORPTION. The surface columnar epithelium of the stomach allows movement of aqueous solutions in a restricted fashion. Body water and contained dissolved salts (chiefly chlorides and carbonates of sodium, potassium, and calcium) flow from tissues into the stomach lumen but not in reverse. Also, the stomach's own secretion of HCl is normally not permeable, and H⁺ cannot return via this route into the body. Larger water-soluble molecules such as glucose are completely blocked and do not pass in either direction.

Partially or totally fat-soluble materials, particularly of small molecular size, can diffuse from the lumen into the tissues of the stomach. The outer membranes of the columnar cells are themselves partly fat and allow this passage. These substances are often drugs, a common one being ethyl alcohol. About 20 percent of the alcohol one drinks enters the body via the stomach without entering the small intestine first. A number of drugs are either water or lipid soluble, depending on the acidity of their environment. Aspirin in a medium of pH 3.5 or greater is chiefly water soluble, ionized, and incapable of entering the gastric mucosa. In a more acid medium (pH of 2 or less), aspirin is mainly fat soluble and nonionized and can diffuse across the membranes of the gastric epithelium. Thus, the stomach is active, to a degree, in the absorption of drugs, but in regard to ordinary nutrients its absorptive function is negligible.

STORAGE AND MIXING. The empty stomach has a volume of about 50 ml. Filled with food it expands and the folds and ridges seen in the empty state get smaller. Emptying the stomach depends predominantly on the contractile activity of its smooth muscle layers.

The stomach permits one to eat briefly and intermittently. Digestion and absorption take place slowly as the stomach gradually discharges its contents into the small intestine. We complete a meal in about one-half hour, but the meal is discharged into the intestine over three to five hours. In general, as the stomach fills with the meal, it is puddled in the relatively inert upper half while the meal in the lower half is agitated and compressed by a series of muscular constriction rings that keep moving down from the midstomach to its most aboral end. Here, too, peristalsis pushes the contents into the duodenum. Peristaltic activity is weak during the first one-half hour after a meal but then increases in intensity and frequency. Because of the size of the stomach and the frequency of contraction, two or even three waves may be proceeding over its surface simultaneously. At other times a part of the meal in the lowest part of the stomach does not advance but squirts back through the advancing peristaltic waves toward the midstomach. The upper half of the stomach, like a reservoir, presses out its contents a little at a time as the lower or pyloric end is prepared to receive it. The stomach wall, when contracted, almost obliterates the gastric cavity; yet with food it adjusts rapidly and the cavity enlarges. Volumes up to 1 liter are easily taken and adjusted to. Volumes beyond that lead to discomfort.

At either end of the stomach, contents are prevented from regurgitating by means of sphincters: at the esophagus, which opens when we swallow or vomit or burp; and at the pylorus, which controls intestinal contents from regurgitating and also controls stomach emptying. The sphincters are complex mechanisms influenced by muscle function, anatomic arrangements, and differential pressures.

SERVOMECHANISMS. Feedbacks from the central nervous system, bloodstream, and local circuits regulate the action of the stomach. The brain inhibits stomach muscular action in some cases and via the vagus nerve stimulates appetite secretions. However, in addition to "psychic pathways" mechanisms exist to cause the stomach to react to food in its lumen and start digestion. Two major mechanisms, "thinking" of food and "enjoying" its taste, stimulate the vagus (mediated psychic secretion). Food in the stomach stimulates the release of gastrin (from the mucosa of the gastric antrum), which stimulates acid secretion. Once the gastric contents turn strongly acid because of secreted HCl, gastrin ceases to be secreted. Thus, gastrin is a perfect servomechanism—a regulator and a feedback control.

Food substances well recognized as potent stimulators of gastric secretion are alcohol and caffeine.

THE BLOOD AND THE STOMACH. The normal stomach produces and secretes intrinsic factor (IF), probably by the parietal cell. Its exact nature is uncertain, but it appears to be a relatively large molecule with a molecular weight of about 60,000. It is not HCl. Extrinsic factor (EF) has been identified as vitamin B_{12}. In the gastric lumen, IF latches onto EF and the complex passes through the small intestine to the terminal ileum, where it is actively absorbed by a special mechanism that mostly ignores free vitamin B_{12}. If there is no IF (if the stomach does not produce HCl and has no parietal cells due to disease or the stomach is surgically removed), then pernicious anemia follows and the patient must be given vitamin B_{12} by injection to survive.

GASTRIC EMPTYING (CONTROL). The amount of material passing into the duodenum depends on the strength of antral muscle contraction. The stomach empties at a rate proportional to the volume in it at any given time. This effect is mediated by the internal nerve plexus or an effect of stretching the smooth muscle. Excessive stretching, however, actually inhibits motility.

Gastric volume is not the major factor controlling emptying; rather, it is the chemical composition and amount of chyme in the duodenum. When the latter contains fat, acid, or hypertonic solutions or is distended, gastric motility is reflexly inhibited. These reflexes caused by nerve fibers are enterogastric reflexes, and the hormones are called enterogastrones (starting in the intestine and acting on the stomach). Two hormones that are known are secretin and pancreozymin (Fig. 12-14). Fat is apparently the most potent stimulus for inhibition of gastric motility, requiring more time for digestion and absorption in the intestine. Hydrochloric acid is neutralized in the duodenum by sodium bicarbonate secreted by the pancreas.

Vomiting. Vomiting is the reverse expulsion of stomach and upper intestinal tract material through the mouth and is a complex reflex coordinated by the vomiting center in the medulla. It is preceded by salivation, sweating, nausea, and increased heart rate. It commences with a deep inspiration, closure of the glottis, and elevation of the soft palate; then there is a contraction of the thoracic and abdominal muscles, raising the intra-abdominal pressure. The gastroesophageal sphincter relaxes, and the stomach contents are forced into the esophagus and into the mouth.

Step V: Intestinal Digestion and Absorption. To this point, the digestive effort appears only half-hearted, with the digestion of proteins and starches having proceeded only part way. By both mechanical and chemical action, the food is now in a state that it can be completely broken down. It is passed through the pyloric valve to the first 22.5 to 25 cm (9 to 10 in.) of the small intestine, known as the duodenum, where the food is either partially or completely digested. From here, it proceeds into the jejunum and ileum, where the remaining fats and proteins are digested and the digestion of certain carbohydrates actually begins. Thus, the small intestine completes the digestive process (Fig. 12-14).

Three digestive juices are secreted into the intestinal lumen and onto the food. As all are alkaline, they tend to neutralize the output of the stomach's acid. Two secretions empty into the duodenum through ducts: the bile from the liver via the hepatic and common bile ducts; and the pancreatic juice, which contains many enzymes, from the pancreas. The third secretion, the succus entericus, is made by the wall of the small

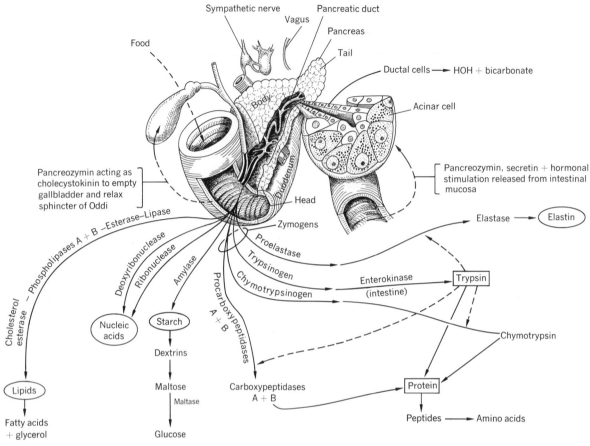

Figure 12-14. Scheme of the physiologic secretion of pancreatic enzymes. The pancreatic secretion is greatly augmented by food via the basal hormonal, neurogenic, and vascular mechanisms. Estimates of an average 24-hour volume vary from 1 to 2 liters.

intestine itself. It is a watery, mucous secretion with enzymes that act on all solid foods (protein, fats, and carbohydrates). It may also protect the duodenal wall from digestion by its own enzymes; in fact, this may be true of mucus in the entire digestive tract.

THE PANCREAS AND ITS EXOCRINE SECRETIONS. *Embryology.* The adult pancreas (Fig. 12-14) develops from two primitive pancreatic buds called the dorsal pancreatic and ventral pancreatic buds. The dorsal arises from the dorsal border of the duodenum, and its duct system opens directly into the duodenum. The ventral bud arises from the ventral border of the duodenum and originates with the primitive bile duct system; therefore, its duct system communicates with the gallbladder and liver. In development, the ventral bud rotates backward behind the duodenum and the two buds fuse. The ventral bud gives rise to a part of the head and uncinate process of the pancreas, while the dorsal bud becomes the rest of the head, the neck, the body, and the tail of the pancreas. In the adult the pancreas is retroperitoneal except for its tail. When the buds fuse, communication is created between

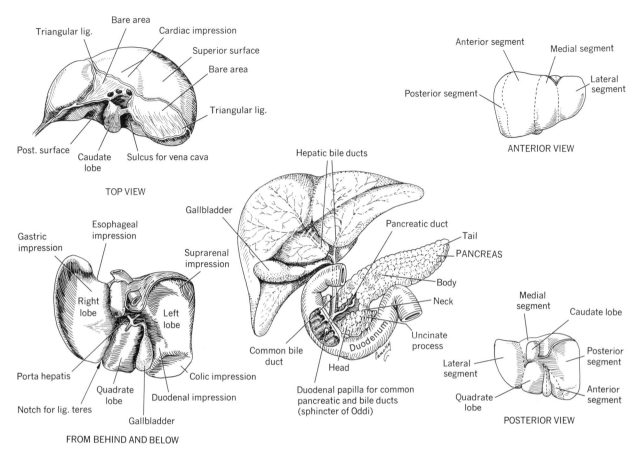

Figure 12-15. The anatomic relations and morphology of the liver and biliary tract.

their ducts. The duct of the dorsal bud becomes the accessory duct of Santorini, while that of the ventral bud becomes the major duct of Wirsung.

Adult Pancreas. The adult pancreas (Figs. 12-14, 12-15) looks like a pistol. Its length varies from 10 to 15 cm (4 to 6 in.), and it is divided indistinctly into a head, the neck, the body, and the tail. The uncinate process is a downward projection from the lower part of the head.

VESSELS. Arterial arches (Fig. 10-24, p. 362) are formed in front and behind the head of the pancreas by the superior pancreaticoduodenal (from the gastroduodenal) and the inferior pancreaticoduodenal (from the superior mesenteric). By means of these arches,

the head of the gland is supplied (as is the duodenum). The body and tail are supplied by the splenic artery.

Pancreatic veins terminate in the splenic vein, which joins the superior mesenteric to form the portal vein.

Lymphatics of the gland drain either directly or indirectly into the celiac glands around the celiac artery.

DUCTS. The pancreatic duct of Wirsung forms the major excretory channel of the gland and usually opens independently into the ampulla of Vater. It may also join the common bile duct higher up, and both enter the ampulla by a common channel. The accessory duct remains patent in most instances and opens into the duodenum above the ampulla or may join the duct of Wirsung.

Normal Physiology and Biochemistry of the Exocrine Pancreas. The normal human exocrine pancreas is a "synthetic factory," secreting more protein per gram of tissue than any other body organ. It secretes about 2500 ml per day of pancreatic juice, which contains about 6 to 12 gm of digestive enzymes. The endocrine cells secrete the hormones insulin and glucagon into the blood (neither is involved in the gastrointestinal function and is described elsewhere).

The exocrine portion secretes two solutions that take part in the digestive process: one contains a high concentration of sodium bicarbonate, secreted by the cells lining the early portions of the ducts leading from the acinar cells; the other contains a large number of digestive enzymes, released from the acinar cells at the base of the exocrine glands. Both are secreted into ducts and eventually into the duodenum.

BICARBONATE SECRETION. This is an active process, and during the day a total of 1500 to 2000 ml of solution is emptied into the duodenum. On reaching the duodenum, the bicarbonate ions combine with the hydrogen ions coming from the stomach to form carbonic acid, which splits into CO_2 and HOH. Thus, the bicarbonate neutralizes the acid coming from the stomach. If one ingests a bicarbonate for an upset stomach, this neutralizes stomach acidity (not intestinal acidity). The net increase in bicarbonate ions (and thus net alkalinization in the blood) is restored by elimination of these ions by the kidneys.

ENZYME SECRETION. The pancreas secretes enzymes that digest protein, fats, and carbohydrates (Fig. 12-14).

The enzymes that digest protein (proteolytic enzymes) are secreted as inactive precursors (zymogens), such as trypsinogen, chymotrypsinogen, proelastase, and procarboxy peptidases A and B. (Proteins are chains of amino acids, and they digest into polysaccharides and then amino acids.)

When trypsinogen enters the duodenum (via the pancreatic duct), it is activated by enterokinase (an enzyme derived from the duodenal mucosa), which cleaves a lysine-isoleucine bond in the trypsinogen to form the active enzymine trypsin.

Trypsin is the "key" to the entire activation process of the proteolytic zymogens since it activates other trypsinogen molecules to form free trypsin, and these trypsin molecules then activate the other zymogens to produce chymotrypsin, elastase, and carboxy peptidases A and B. These enzymes are so potent and numerous that they can split the dietary proteins ingested in a meal into dipeptides and amino acids by the time they reach the jejunum.

Pancreatic juice also contains ribonuclease and deoxyribonuclease, to break down the nucleic acids (RNA and DNA into free mononucleotides) present in food; an alpha-amylase, which splits polysaccharides (starch, glycogen) into hundreds of simple sugar units, which in turn are broken down into disaccharides having two simple sugar units and then monosaccharides such as glucose; and four fat-splitting enzymes (lipolytic enzymes), which are lipase (splits lipids into free fatty acids and 2-monoglycerides), phospholipases A and B, and cholesterol esterase. Pancreatic juice also contains a nonspecific esterase that splits water-soluble esters of fatty acids.

Conjugated bile salts are also very important for the digestion of lipids in the small intestine since they emulsify triglycerides and make the lipids more susceptible to lipase digestion. The bile salts directly activate the phospholipases, cholesterol esterases, and nonspecific esterase.

HORMONE CONTROL. The secretion of pancreatic enzymes is controlled by both hormonal and vagal stimulation. When acid and gastric contents enter the duodenum, the hormone pancreozymin is released, circulates in the blood, and stimulates the exocrine acinar cells of the pancreas to secrete the zymogen granules that contain the digestive enzymes. Thus, the hormone acts like cholecystokinin (may be the same as pancreozymin), another enzyme that causes emptying of the gallbladder and relaxation of the sphincter of Oddi to add bile salts to the duodenal juice.

When gastric juice enters the duodenum, another

hormone, secretin, is released and stimulates the ductal cells of the pancreas to produce water, bicarbonate, and other electrolytes.

Stimulation of the vagus nerve either centrally or by gastric distention also releases pancreatic enzymes. Gastrin (secretion of the stomach antrum) not only stimulates the parietal cells of the stomach to produce acid but also acts directly on the exocrine acinar cells of the pancreas to release enzymes into the pancreatic juice.

When all is functioning normally, the food ingested is split into amino acids and dipeptides, disaccharides, and fatty acids to be absorbed by the mucosa of the small intestine. Note that the very powerful enzymes acting in the digestive tract lumen could be dangerous if activated in the pancreas itself, since they would cause autodigestion of the organ. To prevent this, the proteolytic enzymes are produced as inactive zymogens, which must be activated by trypsin. Second, the zymogens are not synthesized and released into the cell cytoplasm but are wrapped in lipoprotein membranes and condensed into zymogen granules. Third, the pancreas produces trypsin inhibitors, one present in the gland and the other secreted into the pancreatic juice. These proteins bind to any trypsin inadvertently produced and inactivate it so digestion does not occur in the normal gland. Finally, the plasma contains many enzyme inhibitors that can inactivate trypsin, chymotrypsin, and elastase.

The pancreas also contains an enzyme called kallikrein, which catalyzes the release from plasma globulin of vasoactive peptides called kallidin and bradykinin. These kinins are potent vasodilators and mediate an inflammatory response causing leukocyte migration, increased capillary permeability, pain, and smooth-muscle stimulation. Tissue injury activates clotting factor XII, which sets off a process of activation of other proenzymes in the pancreas and plasma, eventually forming kallidin and bradykinin.

THE LIVER. This organ is probably the most important center in the body for biochemical synthesis and interconversions. The greater part of the amino acids, lipids, and sugars are processed here and used for the synthesis of endogenous materials, e.g., glycogen, proteins, phosphatides, and sterols.

Most amino acids liberated by digestion make their way to the liver via the portal system. Some are degraded further, their carbon skeletons being used for glycogen synthesis (liver glycogen is important as a nutritional reservoir and for regulation of blood sugar level) and their nitrogen appearing as urea. The rest go to form proteins, including several plasma components. The liver is in general a biochemical "clearinghouse," where structural components of the organism are degraded for the production of energy and new substances are formed to replace them.

Liver Structure. The liver consists of four discrete but interrelated structures that are physiologic-anatomic units (Figs. 12-15, 12-16).

CIRCULATORY SYSTEM. The liver has a dual blood supply (Fig. 10-24, p. 362; Fig. 12-17), consisting of the hepatic artery and the portal vein, both participating in the transport of oxygen and foodstuffs for assimilation. Sinusoids surrounding the liver cells merge to form central veins, which empty into sublobular veins (large collecting veins), then into hepatic veins, and finally into the inferior vena cava. The venous blood and the arterial blood to the liver usually enter at the periphery of the lobule, and the blood is mixed in the sinusoid into which they empty. The sinusoids are lined by a layer of discontinuous endothelium resting on a basement membrane, which is separated from the parenchymal cells by a potential space (of Disse). Also occupying the sinusoids and attached to the endothelium by "pseudopods" are the Kupffer cells or phagocytic histiocytes. The latter are bathed in sinusoidal blood as it passes through the lobule and remove particulate matter (including red blood cells). The blood vessels are accompanied by lymphatics and nerve fibers. Lymph channels carry liver lymph with its high protein content into the general circulation and help in the transport of material from the splanchnic bed to the systemic circulation. Nerve fibers transmit sensory impulses that help regulate intrasinusoidal vascular pressure and blood flow.

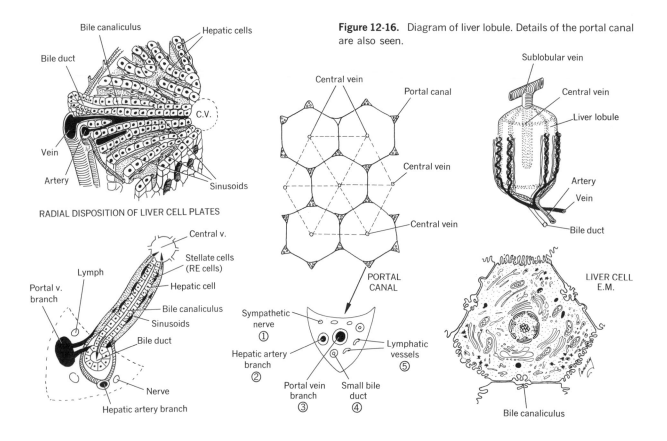

Bile canaliculus

Hepatic cells

Bile duct

C.V.

Vein

Artery

Sinusoids

RADIAL DISPOSITION OF LIVER CELL PLATES

Central v.

Stellate cells (RE cells)

Lymph

Hepatic cell

Portal v. branch

Bile canaliculus

Sinusoids

Bile duct

Nerve

Hepatic artery branch

Figure 12-16. Diagram of liver lobule. Details of the portal canal are also seen.

Central vein

Portal canal

Central vein

Central vein

PORTAL CANAL

Sympathetic nerve ①

Hepatic artery branch ②

Portal vein branch ③

Small bile duct ④

Lymphatic vessels ⑤

Sublobular vein

Central vein

Liver lobule

Artery

Vein

Bile duct

LIVER CELL E.M.

Bile canaliculus

Figure 12-17. **A.** The liver and its position in the circulation from the gastrointestinal tract. **B.** Normal bile pigment cycle. *1*, hemoglobin; *2*, reticuloendothelial cells convert hemoglobin to unconjugated bilirubin, which goes to liver; *3*, unconjugated bilirubin (indirect acting) goes to liver cells; *4*, liver cells conjugate bilirubin; *5*, conjugated bilirubin (direct acting) leaves liver via bile duct and goes to bowel; *6*, bilirubin is converted by bacterial action in bile to urobilinogen; *7*, small amount of urobilinogen is absorbed and excreted by kidney; *8*, fecal urobilinogen is variable and depends on hemolysis; *9*, urobilinogen is recirculated through liver (with obstruction of step 5, no urobilinogen is formed, and there is a marked increase in serum bilirubin). **C.** Alkaline phosphatase is delivered to the liver (*1*); it is excreted normally in the bile (*2*). In biliary obstruction serum alkaline phosphatase increases by virtue of impaired bile flow pathways and increased liver manufacture (*3*).

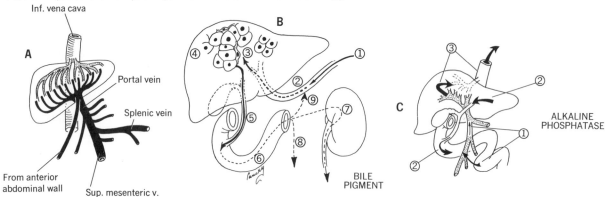

Inf. vena cava

A

Portal vein

Splenic vein

From anterior abdominal wall

Sup. mesenteric v.

B

BILE PIGMENT

C

ALKALINE PHOSPHATASE

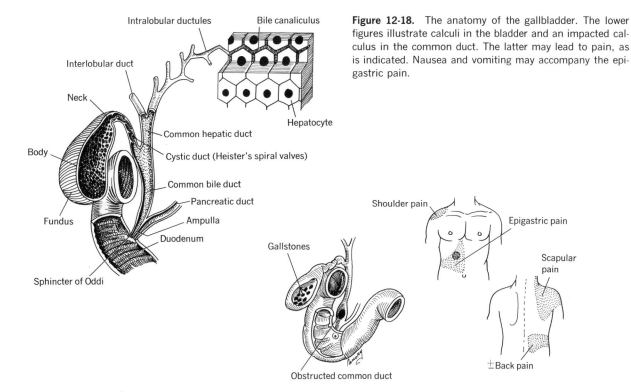

Intralobular ductules
Bile canaliculus

Interlobular duct

Neck

Body

Fundus

Sphincter of Oddi

Hepatocyte

Common hepatic duct

Cystic duct (Heister's spiral valves)

Common bile duct

Pancreatic duct

Ampulla

Duodenum

Gallstones

Obstructed common duct

Shoulder pain

Epigastric pain

Scapular pain

±Back pain

Figure 12-18. The anatomy of the gallbladder. The lower figures illustrate calculi in the bladder and an impacted calculus in the common duct. The latter may lead to pain, as is indicated. Nausea and vomiting may accompany the epigastric pain.

BILIARY PASSAGES. The biliary passages (Figs. 12-16, 12-18) consist of a series of thin-walled tubes into which conjugated bilirubin, cholesterol, certain drugs, and other materials are secreted by the liver cell. The bile canaliculi represent the biliary cell wall or plasma membrane of the liver cells. The canaliculi empty into bile ductules (Fig. 12-19), which lead to small interlobular bile ducts, then into septal bile ducts (medium-sized), into large intrahepatic bile ducts, and the major branches of the common bile duct.

The pigment bilirubin originates in the reticuloendothelial cells of the body and is derived mainly from the breakdown of old red blood cells, from the metabolism of heme in the formation of new red blood cells, and from the turnover of nonhemoglobin heme compounds. This is the so-called unconjugated bilirubin for transport, which is attached to serum albumin and is lipid soluble. It is carried to the parenchymal liver cells (Fig. 12-17), where microsomal enzymes convert it to water-soluble conjugates of glucuronide (75 percent), sulfate (15 percent), and other compounds. The now-conjugated bilirubin is excreted via the biliary tract to the intestine, where it is reduced by bacterial action into products such as mesobilirubinogen and stercobilinogen (both collectively called urobilinogen). The latter is largely excreted in the feces, but some is reabsorbed and reexcreted in the bile or appears in the urine.

Evaluating the state of the biliary network is of clinical importance, for it can establish the cause of jaundice and liver disease and suggest the possible treatment needed. One thus uses (1) direct biochemical analysis of bilirubin in the blood, urine, or feces; (2) measures of blood levels of various nonpigmented substances, other than bilirubin, that are excreted by the biliary system; (3) histologic studies of the network; (4) radiologic studies of the intrahepatic and extrahepatic radicles; and (5) disappearance of dyes from

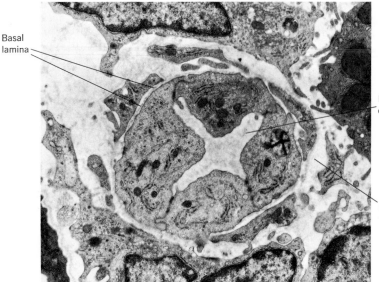

Basal
lamina

Bile
ductule

Portal
tract
area

Figure 12-19. Electron micrograph of the rat liver. Cross section of a bile ductule in the portal tract, showing epithelium, basal lamina, and intercellular association. ×12,000. (Courtesy of Dr. G. Colin Budd, Medical College of Ohio at Toledo.)

the plasma, which are handled by the liver circulation and excreted in bile.

RETICULOENDOTHELIAL CELLS OF THE LIVER. These cells are part of the reticuloendothelial system, which is widely distributed (60 percent of its elements however, are found in the liver, 5 percent in the spleen, and the rest in lymph nodes and other tissues). The reticuloendothelial (sinusoidal) cells in the liver are of three types: 25 percent are stellate-protruding phagocytic cells of Kupffer, 50 percent are flat phagocytic cells, and 25 percent are simply endothelial cells.

The Kupffer cells phagocytize foreign and endogenous substances. Some sinusoidal cells are of the immunocytic type and produce antibodies that contribute to resistance to infection or even lead to hypersensitivity conditions.

PARENCHYMAL CELLS. These are the liver cells themselves and have a polyhedral form (Fig. 12-16). Parenchymal cells contain a nucleus and cytoplasm rich in mitochondria, endoplasmic reticulum, lysosomes, and inclusions. The cells maintain a constant interchange with the biliary and vascular systems. The liver is a "biliovascular tree" in which the interspaces between the branches are filled with parenchyma.

Embryology. The liver (hepatic) diverticulum arises during the fourth week from the endodermal lining of the gut just caudal to the heart. This outgrowth of cells is destined to form the secretory tubules of the liver, its ductal systems, and the gallbladder. The growing hepatic tubules move between the layers of mesoderm of the ventral mesentery and spread them apart. The latter forms the fibrous capsule of the liver and the interstitial connective tissue of the liver lobules. Since both the parenchyma and ductal system have great potential for proliferation, they retain a remarkable power of regeneration even after birth.

Adult Liver. The mature liver (Fig. 12-15) is the largest gland in the body, weighing about 1500 gm (about $\frac{1}{15}$ of the body weight). It is extremely vascular; has many functions; and occupies part of the abdomen, chiefly on the right side just beneath the diaphragm. The right lobe of the liver occupies two thirds of its bulk, and the smaller left lobe, caudate lobe, and quadrate lobe occupy the rest. It is supplied by the hepatic artery. The portal vein carries blood to it from the digestive tract. Blood is drained from the liver via the hepatic veins, which open into the inferior vena cava. The liver is closely related to the diaphragm and is covered by the ribs, which protect it. In the infant, the abdominal bulge is probably due to the liver.

The falciform ligament of the liver is a fold of peritoneum lying between the liver and the anterior abdominal wall as far as the umbilicus. The round ligament (ligamentum teres) passes in its free edge. In the embryo, the latter carries the umbilical vein. The peritoneum that surrounds the liver disappears in the area where the liver grows against the diaphragm and is called the "bare area" of the liver. Where the peritoneum is reflected from the liver to the diaphragm anteriorly is called the anterior or upper coronary ligament; posteriorly, the posterior or lower coronary ligament. Where anterior and posterior reflections meet laterally, they are referred to as the right and left triangular ligaments.

The letter H is formed by structures on the liver's inferior surface. The left limb divides the surface into right and left lobes. The right limb of the H contains visceral structures, the fossa for the gallbladder anteriorly, and the inferior vena cava behind. The transverse part of the H is formed by the porta hepatis (transverse fissure) and contains the hepatic duct, hepatic artery, and branches of the portal vein, which make up the hepatic triad. The nerves of the liver and most lymph vessels (of which there is a rich network) are found in the porta. The hepatic duct lies to the right, the hepatic artery to the left, and the portal vein behind and between the two or behind the artery. The porta is bounded anteriorly by the quadrate lobe and posteriorly by the caudate lobe of the liver.

The hepatic artery (Fig. 10-24, p. 362) is one of the branches of the celiac artery and supplies arterial blood to the liver substance. Its origin is variable. It divides usually into a right and left branch to supply right and left lobes of the liver.

The portal vein also brings a great quantity of blood to the liver from the digestive tract. At the porta hepatis it divides into right and left branches.

The hepatic veins (right and left) carry blood from the liver to the inferior vena cava.

The nerves of the liver are derived from the right and left vagus and the sympathetic system and enter the liver at the porta hepatis, following the vessels and ducts to the interlobular spaces.

The lymph vessels of the liver terminate predominantly in a small group of lymph glands in and about the porta hepatis. The efferent channels from these glands go to the celiac lymph glands. However, some of the superficial vessels on the anterior liver surface pass to the diaphragm in the falciform ligament and reach the mediastinal glands. Another group follows the inferior vena cava into the thorax.

THE GALLBLADDER AND BILE DUCTS. *Embryology.* The hepatic diverticulum (outgrowth) arises from the foregut, and from it the gallbladder and extrahepatic biliary ducts develop. Initially, the bladder lies in the ventral mesentery, but at month two it becomes embedded in hepatic tissue and later it lies more superficial.

Adult Gallbladder. The gallbladder is a hollow, pear-shaped organ closely connected to the inferior surface of the right lobe of the liver (Fig. 12-18). It is about 7.5 to 10 cm (3 to 4 in.) long, holds about 45 ml ($1\frac{1}{2}$ oz) of bile, and forms the right boundary of the quadrate lobe of the liver. It is attached to the liver by peritoneum and consists of a fundus, a body, an infundibulum, and a neck. Its muscular mucosa is incomplete so that the mucosa rests on the smooth muscle with only a very loose narrow fibrous layer between. The mucosa has a honeycomb pattern since its secreting epithelium is thrown into ridgelike folds. There are no villi, and the mucosa has no glandular components.

The fundus usually projects beyond the liver. The body is the major part of the bladder and lies in the fossa on the inferior surface of the liver. The infundibulum (Hartmann's pouch) is found between the body and neck and looks like an overhanging pouch. The neck continues from the upper part of the infundibulum and narrows to become the cystic duct.

Ductal System. CYSTIC DUCT. The cystic duct is about 2.5 cm (1 in.) long and runs from the gallbladder neck to the porta hepatis, where it joins the hepatic duct to form the common bile duct. It is the only extrahepatic bile duct that is tortuous, due to the presence of Heister's spiral valve. The cystic duct may have many variables or anomalies.

COMMON HEPATIC DUCT. The common hepatic duct is about 2.5 cm (1 in.) long and is formed in the porta hepatis by the union of the right and left hepatic ducts, which emerge from the respective lobes of the liver. It is joined by the cystic duct to form the common bile duct.

COMMON BILE DUCT. The common bile duct (ductus choledochus) is from 7.5 to 10 cm (3 to 4 in.) long and about 0.625 cm ($\frac{1}{4}$ in.) wide. It begins near the porta hepatis, descends in the free margin of the lesser omentum, and continues behind the first part of the duodenum and behind the head of the pancreas. It terminates in the second part of the duodenum a little below the middle on its posteromedial surface. It is usually divided into supraduodenal, retroduodenal, infraduodenal, and intraduodenal portions. The last-named portion passes through the duodenal wall and is joined by the main pancreatic duct. A reservoir usually is formed by this junction in the duodenal wall, the ampulla of Vater. The latter opens into the duodenum on the summit of an elevation called the duodenal papilla, where the sphincter of Oddi is located and the union of bile duct and pancreatic duct is quite variable.

VESSELS. The vessels that supply the ducts (Fig. 10-24, p. 362) arise primarily from the cystic and posterosuperior pancreaticoduodenal arteries. The cystic artery usually arises from the right hepatic artery and then divides into a superficial and deep branch to the gallbladder. Its course is quite variable. The venous drainage is upward into the hepatic veins.

Lymphatics of the gallbladder drain into the lymph glands at the hilus of the liver and into the liver substance.

NERVES; HORMONES. The biliary tract is supplied predominantly by the vagus nerve, essentially motor and secretory. Intramural nerve plexi resemble those of the digestive tract but are sparse. Thus, contraction and relaxation of the organ and the biliary apparatus are under vagal influence, but hormonal factors play a major role. Cholecystokinin, which is secreted by the duodenal and jejunal mucosa, when in contact with HCl and fatty meals stimulates gallbladder contraction and release of stored bile.

PHYSIOLOGY OF THE LIVER AND ITS SECRETIONS (PARTICULARLY BILE). The major functional activity of the liver cell appears to take place in the endoplasmic reticulum, the mitochondria, the lysosomes, and other microsomes: (1) The endoplasmic reticulum carries out protein synthesis; (2) the mitochondria constitute the chief sites of oxidative phosphorylation; (3) lysosomes contain alkaline phosphatase and are involved in hydrolytic activities; and (4) other microsomes contain glucuronic acid and are concerned with bilirubin conjugation (Fig. 12-20).

A biliverdin precursor is liberated when hemoglobin breaks down in the reticuloendothelial system. This is reduced to bilirubin, which is quickly bound to albumin (1 gm of hemoglobin liberates about 34 gm of bilirubin—about 250 mg per day). The bilirubin-albumin complex is stable and cannot pass cell membranes. It is therefore not normally excreted in the urine and cannot be accepted by the hepatic cell until

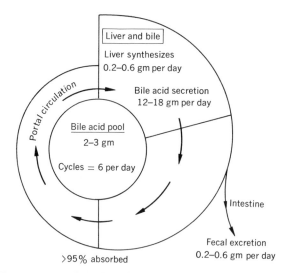

Figure 12-20. Enterohepatic circulation and kinetics of bile acids.

its protein bond is uncoupled at the plasma membrane interface. It enters the liver cell, where it is conjugated with glucuronic acid to form bilirubin-diglucuronide, which is water soluble; it then moves toward the secretory surface at the canaliculus to be secreted as the principal pigment of bile.

Bile helps digestion but contains no enzymes and is secreted by the liver cells at the rate of about 250 to 1000 ml per day into small ducts (biliary ducts), which converge to form the hepatic duct.

Bile consists of bile pigments (predominantly bilirubin) and bile salts as well as high concentrations of cholesterol, neutral fat, phospholipid, and inorganic salts mixed with water and mucus from the biliary tract.

Bile is secreted by the liver and enters the gallbladder to be stored and concentrated by the active reabsorption of salt and water across the gallbladder walls. Overdistention of the gallbladder is avoided in life by the absorptive capacity of the mucosa, which can remove up to one half of the fluid content of bile per hour, about 90 percent of its volume, and concentrate the residue up to ten times. It may become so concentrated that its cholesterol precipitates out of solution and forms crystalline gallstones, which can lodge in the ductal system, blocking bile flow and interfering with digestion of fat and absorption of fat in the intestine. Pain can result, due to gallstones being lodged in the ducts.

The bile salts are the most essential part of bile, since they are involved in digestion and absorption of fat. Note that the bile salts entering the intestinal tract are for the most part eventually reabsorbed and returned to the liver for reuse via the enterohepatic circulation. About 80 percent is recycled, and only 20 percent is normally lost in the feces. The bile pigments, however, are usually excreted through the gastrointestinal tract, and their breakdown products give feces their characteristic color. If the pigments cannot be excreted, they accumulate in the blood and tissues and give the skin its yellow or jaundiced appearance.

Bile enters the intestine, depending on the rate of hepatic secretion and the contraction of the muscle of the gallbladder wall. Stimulation of the vagus nerve to the liver increases bile secretion and causes the gallbladder to contract. This may be initiated emotionally as well as by eating. Food in the duodenum also increases bile secretion and gallbladder contraction and is mediated by the same duodenal hormones that control gastric and pancreatic secretion. Thus, fat and peptide fragments release cholecystokinin and pancreozymin, and these stimulate gallbladder contraction. Secretin and gastrin increase the volume of bile secreted by the liver but not the secretion of bile salts, nor do they affect gallbladder contraction. Bile salt concentration in the blood also stimulates bile secretion.

Bile breaks down the fat or emulsifies it, reduces the surface tension of fats in a watery medium, and increases the surface-to-volume ratio, facilitating attack by enzymes (like a detergent on greasy dishes). Bile also activates the fat-splitting enzymes from the succus entericus and pancreas so they can act on the droplets of fat. Bile, too, excretes the breakdown products of hemoglobin and provides a solvent for substances that are water insoluble, such as vitamin K.

Bacterial action in the intestine converts bilirubin (the bile pigment) into urobilinogen. Some of the urobilinogen is excreted in the stool and the rest is absorbed, a portion being excreted in the urine and the rest returned to the liver for reuse in bile synthesis. (If no urobilinogen is found in the urine in a jaundiced patient, one may assume some major bile duct obstruction.)

The liver is the major (not the only) site of cholesterol synthesis and is also the only real source of bile acid synthesis. The bile acids are derived from cholesterol in the form of cholic acid conjugated to glycine or taurine. If one ingests triglyceride, it is prepared for hydrolysis in the small intestine by bile salts, whereas cholesterol is esterified in the mucosa. In the liver cell, lipids are converted into free fatty acids, which eventually yield phospholipids, cholesterol, and reconstituted glyceride.

Most plasma proteins, such as albumin, fibrinogen, prothrombin, and globulin, are synthesized in hepatic parenchyma. Fibrinogen and prothrombin function in blood coagulation.

The normal liver, by a series of intracellular enzyme activities, is an important site of glycogen synthesis and storage. The exact site is unknown.

The embryonic liver is a site of blood cell formation, which continues in fetal life and for days after birth. A potential for the reestablishment of extramedullary (outside the narrow) blood cell formation remains throughout life.

The liver also serves as a catabolic or detoxifying organ. Both endogenous (serotonin, steroids) and exogenous (mineral and organic poisons, etc.) substances are degraded into innocuous forms.

The liver helps maintain homeostatic balance by the fact that it stores essentials such as glycogen, vitamin B_{12}, and vitamin K. It also synthesizes many enzymes, whose presence in reduced or elevated amounts in the blood is used as indications of abnormal hepatic function.

CONCLUSIONS OF INTESTINAL ABSORPTION. The various digestive juices (saliva, gastric juice, pancreatic juice, succus entericus, and bile) reduce the food materials to their final state and ready them for absorption through the cell lining of the remaining 5.7 m (19 ft) or so of small intestine (Tables 12-2, 12-3). The starches and sugars are in the form of glucose, the proteins are converted to amino acids, and the fats are changed to fatty acids and glycerol. The vast intestinal surface enables the nutrients to gain entry into the blood transportation system.

Gases diffuse passively across the lung lining, but nutrients need active transportation. The cells of the small intestine expend energy to transport the nutrients through the cell wall against a concentrated gradient, and because of this not all the nutrients reach the bloodstream.

The villi (Fig. 12-11) provide the areas of entry of digested food into the body. Each villus is covered with a single layer of epithelial cells through which the materials pass. Each villus contains a network of blood capillaries into which the amino acids, glucose, and some fats pass. These nutrients are then transported by veins and the hepatic portal system to the liver. In the center of each villus is a short, tubelike lymphatic vessel called a lacteal. Into this duct pass the glycerol and those fatty acids that were absorbed by the capillaries. These substances drain through the lymphatic system and eventually empty into the great veins of the neck and thus return to the bloodstream. At this point most of the nutrients are inside the body, but the alimentary tract has not completed its function. The food thus far has been drenched with water that has come to it from the blood supply, and this water must be returned to the body's interior to prevent its loss and assure constant and complete replacement.

Step VI: Intestinal Reabsorption. The large intestine forms the last 1.2 m (4 ft) of the gastrointestinal tract, is about 6.25 cm (2 1/2 in.) in diameter, and functions primarily to recover water. Its epithelial surface is only about 1/30 of the small intestine (has no villi and is not convoluted). It secretes no digestive enzymes and absorbs only about 4 percent of the total intestinal content per day. Thus, it stores and releases fecal material. Approximately every four hours, the contents of the small intestine are discharged into the large intestine through the ileocecal valve. About 500 ml of chyme enter the colon each day. Most of this material is composed of small intestinal secretions, since most of the ingested food has been absorbed before reaching the colon. The colonic secretions themselves are scanty and are mostly mucus. Food remains in the large intestine for a long time. Peristalsis here is very slow, allowing time for water to return to the bloodstream by osmosis. The residue remains behind and is called feces and consists of discarded

Table 12-3. Total Internal Body Secretions per 24 Hours

Saliva	1500 ml
Gastric secretions	2500 ml
Bile	500 ml
Pancreatic juice	700 ml
Intestinal secretions	3000 ml
Total	8200 ml

epithelial cells, indigestible cellulose, calcium, phosphate, iron, and other salts, many bacteria, and still about 70 percent water.

The accumulation of feces is passed to the terminal portion (last 12.5 cm [5 in.]) of the colon known as the rectum, which ends at the anus. The latter is closed off by both internal and external anal sphincters. The transfer of feces takes place through a remote-control reflex, the gastrocolic reflex, whenever food enters the stomach. Thus, it is normal and natural to defecate after breakfast; yet the hustle and bustle of modern living often force us to lose or overcome this habit, resulting in a very common cause of constipation. Diarrhea, on the other hand, is characterized by frequent defecation, usually of a high fluid content, and may be due to many conditions. The opening of the anal sphincters is partially under voluntary control, having been mentally conditioned since about the age of two years. Thus, to some degree, we control or cause our own constipation.

The bacteria in the digestive tract live in a relatively safe and stable environment with a sufficient food supply. In return, they suppress molds that manage to survive the digestive juices and that could damage the large intestine were they to multiply. This is a symbiotic relationship. Under normal conditions, these bacteria not only suppress molds but also manufacture one of the essential B vitamins that is not stored by the body.

Digestion is intermittent and even of an irregular nature since we eat at intervals during 12 to 24 hours and then a gap exists between the evening meal and breakfast. Yet the cells of the body demand and receive a steady supply of nutrients by way of the bloodstream. The liver regulates and measures out this supply.

The veins from the digestive system do not drain into the large systemic veins of the body. This prevents the entire blood system from being flooded with the digestive products of the latest meal. They drain instead directly into the liver by way of the hepatic portal vein.

The blood to the liver contains amino acids, glucose, and certain fats. Some glucose passes through without alteration to sustain the blood sugar levels. The excess glucose is stored by the liver cells. Since glucose is soluble, the liver, by linking glucose molecules together into chains, stores it as glycogen molecules. If glucose were stored, water would pass into the cell by osmosis and the cell membranes would burst open. Glycogen is an insoluble reserve and can be turned into glucose as the need arises.

Amino acids, on the other hand, cannot be stored. Any excess undergoes a process called deamination in the liver. Part of each molecule is made into urea, which is discarded through the kidneys, and the rest is turned into useful glucose.

Some fats are converted into a form that can be oxidized to provide energy, while other fats are changed into special forms used for structural purposes in the cell.

The liver also stores, regulates, transforms, and manufactures a variety of other things. It stores vitamins A and D as well as iron, whose distribution it regulates. It converts poisons in food into harmless compounds and helps to maintain the blood by synthesizing proteins such as globulin and fibrinogen. Thus, the liver, with its myriad of functions, not only is the largest "gland" in the body but also, because of its chemical activity, is the largest source of heat needed to maintain the body temperature.

Review Questions

1. What are the layers or coats of the wall of the digestive system?
2. What is said to regulate the gastrointestinal tract activity?
3. What organs make up the digestive or gastrointestinal tract?

4. In the mouth, identify or define the following:
   a. Vestibule
   b. Mouth proper
   c. Oral pharynx
   d. Gums
   e. Parts of the mouth floor visible when the tongue is raised
5. What is meant by deciduous or milk teeth? How many and which are they?
6. Compare the questions and answers of question 5 with similar ones related to the permanent teeth.
7. Describe the parts of a typical tooth.
8. Where are the soft and hard palates located? What are they made of?
9. Generally describe the dorsum (top) of the tongue. List the muscles of the tongue and their function. How are they innervated?
10. Are the motor and sensory nerves of the tongue alike? If not, how do they differ?
11. What are the salivary glands? Where are they? Which are they? What is their secretion like?
12. What is the pharynx composed of? What are its parts?
13. Where is the esophagus found? About how long is it? What is its function?
14. Describe the essential anatomy of the stomach, including its location.
15. From where does the stomach receive its blood supply? Name the major arteries to the stomach. How is blood drained from the stomach?
16. How does the histology of the stomach and small intestine differ?
17. What are the parts of the small intestine? Generally describe the blood supply of the portions of the small intestine.
18. How can you tell large intestine from small intestine? What are the named portions of the large intestine? How do they receive their blood supply?
19. Define the rectum and anal canal. What are the anal sphincters?
20. Why is the pecten of the anal canal so important?
21. Trace the digestion of proteins, carbohydrates, and fats through the digestive system, indicating the essential enzymes (catalysts) needed for the process.
22. What is a calorie? Why are they essential?
23. How do the accessory organs of digestion, such as salivary glands, liver, gall-bladder, and pancreas, contribute to food digestion?
24. Where are the organs mentioned in question 23 located? What is their essential anatomy?
25. What are the major functions of the large intestine, rectum, and anus? Does the large intestine secrete any digestive enzymes?
26. What is the difference between diarrhea and constipation?

References

Brooks, F. P.: *Control of Gastrointestinal Function.* Macmillan Publishing Co., Inc., New York, 1970.

Davenport, H. W.: *Physiology of the Digestive Tract*, 3rd ed. Year Book Medical Publishers, Inc., Chicago, 1971.

Ingelfinger, F. J.: Gastric Function. *Nutrition Today*, **6**:2–11, 1971.

Trier, S. S.: Structure of the Mucosa of the Small Intestine as It Relates to Intestinal Function. *Fed. Proc.*, **26**:1391–1404, 1967.

Truelove, S. C.: Movements of the Large Intestine. *Physiol Rev.*, **46**:457, 1966.

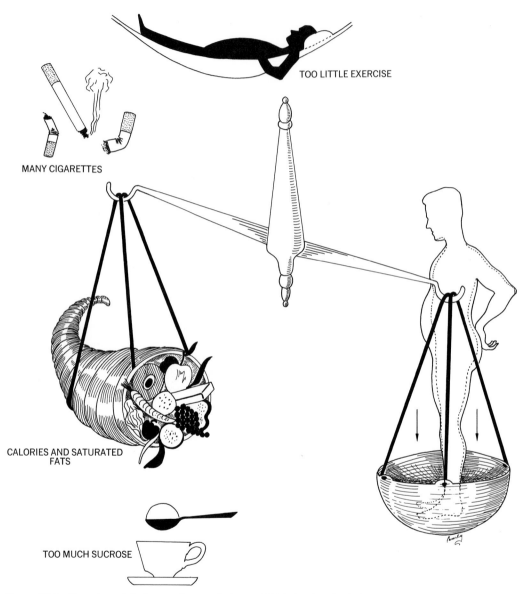

TOO LITTLE EXERCISE

MANY CIGARETTES

CALORIES AND SATURATED FATS

TOO MUCH SUCROSE

Figure 13-1. Organic metabolism and energy balance. A high intake of calories, saturated fats, and sucrose; physical inactivity; and cigarette smoking are all links in a chain leading to overweight, poor energy metabolism, and atherosclerosis.

Organic Metabolism and Energy Balance

CHAPTER 13

Intermediary Metabolism

Metabolism (Greek word meaning change) refers to the total collection of chemical reactions that take place in a living organism. It consists of two generalized classes of chemical reactions: those that result in the breaking down of a molecule into smaller and smaller parts, known as catabolic reactions; and those where small molecular fragments are put together to form larger molecules, known as anabolic reactions. Intermediary metabolism refers to the intermediate chemical steps in the intracellular transformation of foodstuffs in the body. In the adult organism, the rates of chemical synthesis and degradation are in balance and the composition of the body is said to be in a state of dynamic equilibrium.

Most of the molecules in the food we eat are utilized by the cells to provide chemical energy needed for cell structure and function (Fig. 13-1). Most cells, particularly liver cells, have the ability to convert one type of molecule to another, which enables the body to use a variety of molecules from different foods. Certain molecules are essential; for example, protein must be ingested to provide nitrogen for the synthesis of protein and other substances. Certain essential amino acids must be in the diet since the body does not form them. Other essential organic nutrients are certain fatty acids and vitamins. The latter are not used for energy or synthesis of structural parts but serve as cofactors or coenzymes in chemical reactions.

Cells, too, must have some mechanism for overcoming the tremendous energies needed for most organic reactions. The rates of chemical reactions in cells are accelerated by special protein molecules known as enzymes (see Chap. 1).

Vitamins

Nutrition is emerging as an exact science embracing many of the complicated and advanced aspects of biochemistry. Nutritional chemistry is concerned with the enzyme systems essential in the release of energy from foodstuffs and the utilization of this energy to carry out vital processes. These many enzyme systems require organic substances that the body cannot synthesize—the vitamins, substances whose lack in the diet causes deficiency diseases. Many vitamins have been purified and chemically identified, and their essential biochemical functions are being elucidated.

The importance of vitamins had been overlooked until almost 50 years ago, and research has only begun to uncover their primary physiologic actions (Table 13-1). It is still difficult for biochemists to define vitamins as a group. They make themselves known primarily by their lack, and their positive functions must be sought indirectly. Furthermore, there is so much overlapping among the various vitamins and other related groups of substances that even the classification is not perfect and the name *vitamin* (Latin *vita*, life; *amine*, an organic compound containing nitrogen) may not be appropriate.

Vitamins are organic regulators of metabolism, which are needed in the daily diet of all animals for normal growth and maintenance of life. Various animal

457

Table 13-1. The Vitamins

Group	Source	Function	Deficiency
Fat Soluble			
A	Yellow vegetables, milk, butter, eggs,	Maintenance of normal epithelium; synthesis of visual purple for night vision	Poor epithelial keratinization; susceptibility to night blindness
D	Whole milk, butter, egg yolk	Facilitation of absorption of calcium and phosphorus from intestine; utilization of Ca and P in bone development	Rickets in children; osteomalacia in adults
E (Tocopherol)	Spinach, lettuce, whole meat	No definite function known in humans; needed by rats in reproduction	No known human effects; sterility in rats
K	Cabbage, spinach, tomatoes, liver	Synthesis by the liver of prothrombin; needed for coagulation	Impaired mechanism of blood coagulation
Water Soluble			
B complex B_1 (thiamine)	Eggs, whole-grain cereals, bananas, apples, pork	Coenzyme in metabolism of carbohydrate (cocarboxylase); maintains normal appetite and absorption	Beriberi; polyneuritis
B_2 (riboflavin)	Meat, milk, eggs, liver, fruit	Coenzyme in metabolism (as flavoprotein)	Dermatitis; glossitis
B_6 (pyridoxine)	Milk, eggs, fish, liver, whole-grain cereals, yeast	Coenzyme in amino acid metabolism	Dermatitis
Nicotinic acid (niacin)	Milk, liver, tomatoes, leafy vegetables, peanut butter	Nicotinamide in metabolic processes such as energy release	Pellagra
B_{12}	Milk, eggs, liver, kidney, cheese	Maturation of red blood cells	Pernicious anemia
Pantothenic acid	Egg yolk, lean meat, skim milk	Necessary for synthesis of acetyl coenzyme A, metabolism of fats, synthesis of cholesterol, and antibody formation	Neurologic defects
Folic acid	Liver, fresh leafy green vegetables, yeast	Production of mature red blood cells; body growth; reproduction; tissue respiration	Macrocytic anemia
Biotin	Liver, eggs, milk; synthesized by intestinal tract bacteria	Coenzyme in amino acid and lipid metabolism	Not clearly defined in man
C (Ascorbic acid)	Citrus fruits, tomatoes, fresh green vegetables, potatoes	Production of collagen and formation of cartilage	Scurvy; susceptibility to infection; retardation of growth; poor wound healing; swollen and bleeding gums

species have different requirements. Ordinarily, or traditionally as far as man is concerned, metabolic regulators synthesized in the body by the ductless glands have been called hormones. What is called a "vitamin" for one species may not be a vitamin for another. We shall consider only those vitamins that man himself cannot synthesize. (The bacterial synthesis of vitamin K and most B vitamins in the intestine by bacteria or the formation of vitamin D on the skin through action of ultraviolet light should not be confused with synthesis by animal tissues themselves.) Some vitamins (e.g., riboflavin [one of the B vitamins] and niacin [also a B vitamin]) are essential parts of enzyme systems; others regulate in an unknown fashion. Vitamins probably furnish neither energy nor building units.

Initially, vitamins were called "accessory factors" when it was demonstrated that normal foods, in addition to the known nutrients of carbohydrates, proteins, fats, minerals, and water, contained small traces of other substances essential to well-being. Dietary alterations, and their effects on scurvy, rickets, beriberi, and night blindness, had been observed for centuries.

In 1911, Funk, who believed that the antiberiberi factor he had isolated from rice polishings was an amine, introduced the term *vitamines*—amines essential to life, not just accessories. He, too, formulated the vitamin hypothesis of deficiency disease. Osborne and Mendel, McCollum, and Davis in 1915 suggested the classification by solubility. Thus, fat-soluble vitamin A that cured nutritional eye disease was differentiated from water-soluble vitamin B that prevented beriberi in pigeons. They rejected the term *vitamine* because vitamin A lacked an amine group. In 1920, Drummond suggested a compromise terminology, using *vitamin* (without the "e") as the generic term.

Successive letters of the alphabet were given to new vitamins as they were discovered, and some letters were assigned out of order. Vitamin K, for example, refers to the Scandinavian term *Koagulation* and vitamin H to the German *Haut*, meaning skin. As time progressed, it was found that the original vitamin B was not a single vitamin but a group of vitamins and, therefore, subscripts were added for identification as they were discovered. Confusion was introduced when some were named as growth factors for test organisms and when subscripts were also used for fat-soluble vitamins merely to label closely related chemical compounds with identical physiologic actions rather than differentiating separate factors. Recently the trend has been to use chemical names; for example, vitamin B_1 is thiamine, B_2 is riboflavin, and C is ascorbic acid due to its antiscorbutic action. Vitamin nomenclature is still in its developmental stage and undoubtedly will undergo many future modifications.

General Observations

Vitamins and Stress. If we think of health as more than the absence of disease, the curative effects of vitamins in disease have been dwelt on too exclusively. Vitamins promote health and vigor as well as preventing manifest disease.

Health requires the ability to meet conditions of stress. Stress, it is believed, disturbs enzyme activity, and one of the physiologic manifestations of stress may be tissue hypoxia. William M. Govier and Margaret Grug have demonstrated that coenzyme breakdown does accompany tissue hypoxia, whether produced by hemorrhage, breathing high-nitrogen and low-oxygen mixtures, or ischemia.

Vitamins in Pregnancy. During pregnancy and lactation vitamin supplements are recommended for all women to help in the severe metabolic stresses encountered during this period. Vitamin supplementation has proved valuable in preventing premature labor and toxemia of pregnancy. Furthermore, the mother's nutrition is important to herself and her child since it has been shown that congenital abnormalities can be induced in the young of animals fed diets deficient in vitamins and other essential factors. Also to be noted is that nutritional deficiency of the fetus, resulting from metabolic disorders of the mother, placental disease, or vascular abnormalities, may exist in spite of the mother's adequate diet. Vitamins implicated in the production of congenital anomalies in experimental animals are vitamin A, folic acid, riboflavin, pantothenic acid, vitamin B_{12}, and vitamin D.

Vitamins and Growth. Growth is another form of metabolic stress, the ability to grow being a highly precise measure of bodily health. Growth failure accompanies the severe deficiencies of all vitamins thus far mentioned.

Vitamins and Resistance to Infection. Poorly nourished people have less than normal resistance to infectious diseases. Vitamin A deficiency, for example, lowers the body's natural resistance to bronchopneumonia by displacing normal epithelium with squamous epithelium. Furthermore, aside from its natural means of resistance, the body is capable of acquiring immunity to infection and responds to invading antigens with specifically modified molecules of globulins, called antibodies, that can inactivate the antigen. Experimentation has shown that vitamin deficiencies result in impairment in antibody synthesis.

Vitamins and Old Age. Age imposes its own cumulative stress on the organism and lessens its ability to cope with new stresses. Chronic vitamin inadequacies are quite common in later years. Factors predisposing to deficiencies include poor appetite, dietary problems, poor teeth, and changes in the vital organs. What effect vitamin supplementation has on a life-span is not certain, except to indicate that improved nutrition usually lengthens the active, useful life period and delays senility.

Relations Among Vitamins. A severe deficiency of any single vitamin will ultimately interfere with the functioning of other vitamins since all take part in the growth and maintenance of the organism. This is not fully understood, however. Since multiple deficiencies are common in man, such interactions should be considered in interpreting results of replacement therapy with a single vitamin.

Vitamins and Hormones. Although a relationship between hormones and vitamins appears to exist in the broad sense that both are metabolic regulators, no evidence exists to establish specific points of interaction. A few relationships have been suggested, between vitamin D and the parathyroid, thyroid, and pituitary glands; between ascorbic acid and the adrenal cortex; between folic acid and estrogen activity; and between the B vitamins and estrogen-androgen balance, since they are needed for liver inactivation of estrogens.

Organic Metabolism

Organic metabolism requires a dynamic anabolic (constructive) and catabolic (breakdown) steady state. Virtually all organic molecules are continually built up and broken down at a rapid rate (DNA is an exception). Thus, there is a dynamic state of molecular interconversion. Amino acids are derived from ingested protein and from the breakdown of body protein (see Chap. 12). These amino acid pools are used for resynthesis of body protein and other specialized amino acid derivatives, such as epinephrine, nucleotides, and so forth, although some is lost in the urine, skin, and hair. Amino acids can be converted into fat or carbohydrate by the removal of ammonia, which, in turn, is converted by the liver into urea, which is excreted by the kidneys as a major end product of protein metabolism.

Absorptive and Postabsorptive States

Absorption involves ingesting nutrients that enter the vascular system from the gastrointestinal tract, and postabsorption is a period when the digestive tract is empty and sustained energy is supplied by the body's own endogenous stores (Fig. 13-2). Carbohydrate (65 percent), fat (10 percent), and protein (25 percent) from an average meal enter the blood and lymph systems from the gastrointestinal tract predominantly as monosaccharides, triglycerides, and amino acids, respectively. All blood from the gastrointestinal tract goes to the liver before passing to the rest of the body. Fat, on the other hand, is absorbed into the lymph and not the blood.

In general, the major fate of carbohydrate and fat is catabolism to yield energy. Excess carbohydrate and fat are stored. Protein is broken down to amino acids, which are utilized by the cells or resynthesized to

Figure 13-2. Hormonal changes during absorptive (feeding) and postabsorptive (fasting) stages of digestion.

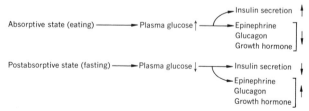

protein. However, excess amino acids cannot be stored as protein but are converted to carbohydrates or fat, and only then can they supply energy. Only a small amount of absorbed amino acids and fats is actually used directly for energy or immediate resynthesis of body protein and structural fat. Most of the amino acids, fat, and carbohydrate are transformed into adipose tissue fat. During absorption, glucose is the major source of energy.

During postabsorption, carbohydrate is synthesized in the body; however, its use for energy is reduced, fat and protein synthesis is cut down, and the oxidation of endogenous fat seems to provide most of the body's energy supply.

Glucose Metabolism

To some, sugar-containing foods mean a source of quick energy. To others, sugar means "sweet tasting." Sugar, the common name for sucrose (and the biochemical name for simple carbohydrates), is composed of a glucose molecule and a fructose molecule linked together. Starch, the major carbohydrate in the human diet (normally), is a polymer of many glucose molecules. Thus, before any carbohydrate can be used for energy, it must be converted to glucose. This breakdown of complex molecules to single units occurs in the gastrointestinal tract, while the conversion of fructose to glucose takes place in the liver.

Most carbohydrate, when reaching the liver, is built into glycogen or transformed into fat rather than oxidized for energy. Thus, glucose is a precursor of fat and provides the glycerol and fatty acid moieties of triglycerides. Even the glucose not entering the liver is carried to the adipose-tissue cells and transformed into fat; some is stored as muscle glycogen, and a large fraction is oxidized to CO_2 and HOH in the body cells for future energy needs (Fig. 13-3).

Glucose is the essential ingredient for energy metabolism in the living organism. Energy actually resides in the compound ATP (adenosine triphosphate) and not in the glucose molecule but is generated during the metabolic processes involving the breakdown of

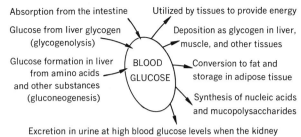

Absorption from the intestine   Utilized by tissues to provide energy

Glucose from liver glycogen (glycogenolysis)   Deposition as glycogen in liver, muscle, and other tissues

Glucose formation in liver from amino acids and other substances (gluconeogenesis)   BLOOD GLUCOSE   Conversion to fat and storage in adipose tissue

Synthesis of nucleic acids and mucopolysaccharides

Excretion in urine at high blood glucose levels when the kidney threshold is exceeded (about 170 mg % normally)

Figure 13-3. Processes adding or removing glucose from the blood.

glucose (Fig. 1-22, p. 30). ATP is the primary energy-storing compound in the cell. When it is hydrolyzed to adenosine diphosphate (ADP) during certain enzymatic reactions, there is a release of a remarkable amount of energy in the form of heat (about 8000 calories per molecule). This energy of ATP is used for such activities as the synthesis of macromolecules and muscle contraction. Since ATP cannot cross cell membranes, each cell produces its own supply.

Fortunately, most cells (except neurons) can shift from one fuel source to another. Immediately, when glucose is not available, fat is used. With the mobilization and loss of fat, ketosis and acidosis make their appearance. Also, but not as quickly, amino acids can be used to provide energy, with a resulting negative nitrogen balance.

Glucose Metabolism Pathway

When glucose enters the cell, it is first phosphorylated by the enzyme hexokinase to form glucose-6-PO_4 and ADP. This transformation requires energy, which is provided by the conversion of ATP (containing a high-energy bond) to ADP.

$$\text{Glucose} + \text{ATP} \rightleftharpoons \text{Glucose-6-}PO_4 + \text{ADP}$$

Glucose-6-PO_4 is a common intermediate for a variety of routes by which glucose can be metabolized; yet there are two major pathways. These are (1) glycolysis (Embden-Meyerhof pathway) (Fig. 1-20, p. 28; Fig. 13-4) and (2) the shunt pathway (pentose phosphate pathway) (Fig. 13-4). In the former, glucose-6-PO_4,

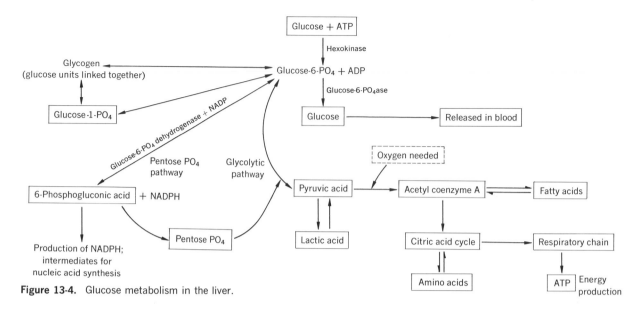

Figure 13-4. Glucose metabolism in the liver.

which is a six-carbon sugar, is converted (via a series of reactions) to two pyruvic acid molecules (each with three carbons). In the shunt pathway, it may be oxidized to 6-phosphogluconic acid and then to pentoses (which contain five carbons), or other compounds may be formed. A third possibility, taking place only in muscle and the liver, is the conversion of glucose-6-PO_4 to glycogen (Fig. 13-4).

In preparation for storage, the enzyme phosphoglucomutase transfers phosphate in the molecule to form glucose-1-PO_4 from glucose-6-PO_4. This is a completely reversible reaction and is why the balance between storage and release can be so quickly adjusted, especially in the liver (Fig. 1-20, p. 28). The next step is condensation with uridine triphosphate to form uridine diphosphoglucose. This prepares the glucose molecule for incorporation into the branched polymer, glycogen. This final reaction is stimulated by glycogen synthetase and possibly promoted by insulin (not positive). Glycogen is the storage form of glucose and is degraded as the need arises.

Glycogen is like a branching tree to which successive molecules of glucose are added. Thus, glycogen has no fixed molecular weight and its weight ranges widely.

In the liver, hydrolysis can occur, which frees glucose to be released into the circulation (liver is the principal organ of blood glucose regulation). It occurs in the liver since few tissues contain glucose-6-PO_4 ase (a phosphorylase) to perform this reaction. The action is promoted by epinephrine and glucagon, and this reconversion of glucose from glycogen constitutes an important method for correction of a low blood sugar in the face of increased insulin production or cellular demand.

Glycolysis, Citric Acid Cycle, and Respiratory Chain

Glycolysis, the primary metabolic pathway of tissues, takes place in the cell cytoplasm. Oxygen is not necessary for the glycolytic reactions. The first stage of glucose utilization involves the production of the three-carbon pyruvate, which involves many separate enzymes and eight different intermediates (Fig. 1-24, p. 32). The fate of pyruvic acid, however, depends on the presence of oxygen (Fig. 13-3). With little or no oxygen, it is converted to lactic acid. If O_2 is available, most pyruvic acid enters the mitochondria, where the enzyme systems of the oxidative cycle are located and packaged together; with the electron transfer machin-

Figure 13-5. Protein metabolism in the liver.

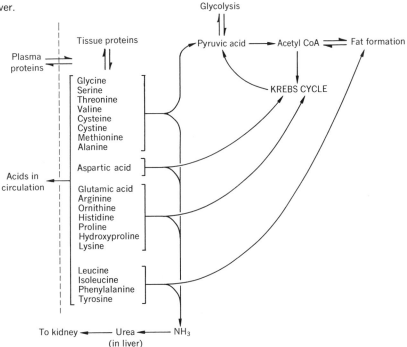

ery it is transformed to acetyl coenzyme A (acetyl CoA, a derivative of acetic acid) (Fig. 1-15, p. 24). Following complex steps, acetyl CoA is converted to CO_2 and HOH. These reactions involve the citric acid (Krebs) cycle (Fig. 1-21, p. 29) and a chain of oxidation-reductions involving several iron-containing respiratory enzymes called cytochromes (cytochrome oxidase). The last cytochrome in the series is oxidized by O_2, which itself is reduced to HOH (Fig. 1-19, p. 27). Oxygen lack, on the other hand, stimulates many mechanisms to compensate for inefficiency. It increases glucose entry into the cells which stimulates the effect of insulin.

Overall, 38 molecules of ATP are formed for each glucose molecule completely metabolized. Two ATP molecules form as a result of glycolysis, but most form in the oxidation-reduction reactions and thus energy generates. Since most of the energy reactions occur in the mitochondria, they have been referred to as the cell's "powerhouse."

The citric acid cycle also produces intermediates used in lipid and protein synthesis (besides the above-described energy metabolism) and in amino acid metabolism (Fig. 13-5). The reactions are reversible, and these intermediates can enter the cycle. Thus, carbohydrate, lipid, and protein metabolism are interrelated.

Pentose Phosphate Pathway

The pentose phosphate pathway (Fig. 1-20, p. 28; Fig. 13-4) does not generate ATP and is a secondary reaction in most tissues. It is particularly significant in adipose tissue, where the NADPH synthesized in the course of oxidation reactions is used in the formation of lipids. It is also very important in the red blood cells, since NADPH helps maintain the cell membrane, and this pathway provides ribose-5-PO_4 for the biosynthesis of nucleic acids.

Glucose needed by the tissues for either the glycolytic or pentose-PO_4 pathways is not stored in the cell but is obtained from the blood as needed. Although muscle has its own glycogen, it is not sufficient to meet metabolic needs, and muscle, too, gets its glucose from blood. The brain is almost completely dependent on glucose for its energy and, therefore, is highly sensitive to blood glucose changes. Thus, under normal conditions, levels of blood glucose are maintained within narrow limits by very complex hormonal systems.

Blood Glucose Regulation

The concentration of blood sugar after absorption is an indicator of carbohydrate metabolism balance. The level of blood glucose is maintained at a reasonably constant value (between 70 and 120 mg percent) despite the intermittent and irregular nature of supply of carbohydrate in the diet. This is essential for nutrition of individual tissues, like the brain, that lack an adequate reserve supply of carbohydrate. Glucose levels reflect the equilibrium between the release of glucose from the liver (glycogen being the most important reservoir for carbohydrate) and the metabolic needs of other tissues under the control of hormones and intracellular substances (Figs. 13-6, 13-7). If the concentration is to be constant, the glucose consumption and withdrawal of glucose from glycogen reservoirs must be balanced.

Five different hormones are known to be involved in the control of blood glucose levels: glucagon, epinephrine, glucocorticords, growth hormone, and insulin. Glucagon and insulin are secreted by the islet cells of the pancreas. Epinephrine is from the adrenal medulla, glucocorticoids from the adrenal cortex, and growth hormone from the pituitary.

Insulin. Of those mentioned only insulin lowers blood sugar; the other hormones raise it. Insulin, a protein hormone, is released from the beta cells of the islets of Langerhans of the pancreas (Fig. 15-13, p. 526). It is stimulated by a rise in blood glucose as well as by certain amino acids, fatty acids, and gastrointestinal digestive factors. Insulin lowers blood sugar primarily by facilitating the entrance of glucose into both skeletal and heart muscle as well as adipose tissue. Tissues such as the brain, liver, kidney, and mucosa of the gastrointestinal tract apparently do not need insulin for glucose transport across cell membranes. Insulin, on the other hand, may affect glucose utilization in other ways. For example, in the liver, it in-

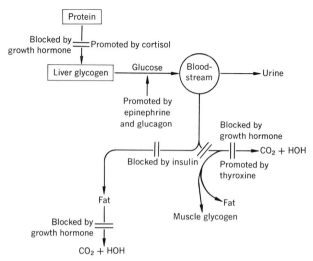

Figure 13-6. Glucose metabolism and homeostasis; hormones that affect utilization of glucose.

creases or decreases the activity of certain enzymes: augments the activity of an enzyme essential for glycogenesis (glycogen synthesis) and reduces the activity of one enzyme causing glycogenolysis (glycogen breakdown) (Fig. 13-7). Insulin enhances the activity of an enzyme that phosphorylates glucose to form glucose-6-PO_4, which is then mostly converted to glycogen. Thus, in the liver, insulin acts to increase glycogen formation and decrease the release of glucose into the blood. One should also keep in mind that insulin not only serves carbohydrate metabolism but also directly or indirectly increases protein and lipid synthesis and inhibits lipid breakdown.

Epinephrine (Adrenaline). Epinephrine is a hormone secreted by the adrenal medulla (Fig. 15-18, p. 531). In relation to organic metabolism it causes glycogen breakdown in the liver and muscle by stimulating the enzyme that catalyzes the reaction, and plasma glucose increases (Fig. 13-6). However, as muscle does not contain the phosphatase to catalyze the liberation of free glucose, muscle does not markedly contribute to a rise in blood glucose levels. Epinephrine is additionally related to carbohydrate metabolism by inhibiting the release of insulin from the pancreas. It also stimulates the breakdown of triglycerides in fat tissue

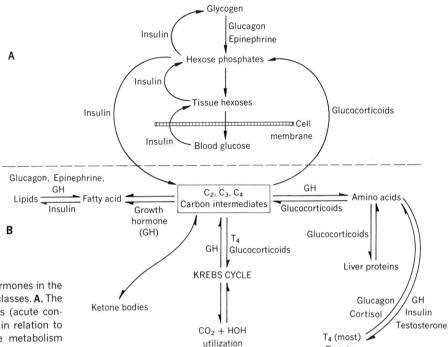

A

Glycogen

Insulin

Glucagon
Epinephrine

Hexose phosphates

Insulin

Tissue hexoses

Insulin

Cell
membrane

Insulin

Blood glucose

Glucocorticoids

B

Glucagon, Epinephrine,
GH

Lipids ⇌ Fatty acid

Insulin

Growth
hormone
(GH)

C₂, C₃, C₄
Carbon intermediates

GH

Amino acids

Glucocorticoids

Glucocorticoids

GH | T₄
Glucocorticoids

Liver proteins

KREBS CYCLE

Ketone bodies

Glucagon / GH
Cortisol / Insulin
Testosterone

CO₂ + HOH
utilization

T₄ (most)
Proteins

Figure 13-7. Overall effects of hormones in the regulation of the major metabolic classes. **A.** The key steps in glucose homeostasis (acute control of carbohydrate metabolism in relation to blood glucose). **B.** Carbohydrate metabolism and regulation.

by increasing enzyme activity there as well. Stimulating the sympathetic nerves to adipose tissue does the same thing, resulting in fatty acid increase in the plasma and facilitating cellular use of fatty acids, thereby sparing glucose and maintaining plasma glucose levels. Thus, epinephrine opposes the action of insulin.

Epinephrine release is controlled entirely by the preganglionic sympathetic nerves to the adrenal medulla. A lower plasma glucose reflexly stimulates epinephrine release, and the converse is true. The receptor initiating this response is a glucose receptor in the brain, probably in the hypothalamus. Epinephrine is also reflexly released in response to other stimuli, such as exercise, pain, hypotension, and stress.

Glucagon. Glucagon is a hormone produced by the alpha cells of the pancreatic islets (the beta cells produce insulin) (Fig. 15-12, p. 526). It is a small protein, which bears no structural relationship to epinephrine yet its metabolic effects are similar. It stimulates the breakdown of liver glycogen and adipose triglycerides,

resulting in a rise in blood glucose levels (Fig. 13-6). It also stimulates gluconeogenesis from amino acids.

Elevated blood glucose acts directly on the alpha cells to inhibit glucagon release, whereas the converse is true. It may be possible that excess glucagon plays a role in the etiology of diabetes mellitus.

Growth Hormone. People with growth hormone–producing tumors manifest not only excessive growth but also diabetes. Growth hormone exerts effects opposed to those of insulin. It increases adipose tissue breakdown of triglycerides, releasing fatty acids into the blood, and it inhibits the uptake and oxidation of glucose by many body tissues (Fig. 13-6).

Growth hormone secretion is directly controlled by a hypothalamic-releasing factor. The input leading to growth hormone–releasing factor is a decreasing plasma glucose, which stimulates glucose receptors in the hypothalamus, which in turn communicates neurally with secretory hypothalamic neurons. The converse is also true: growth hormone acts like glucagon and epinephrine in being a hyperglycemic agent.

Cortisol. Cortisol is an adrenocortical hormone (glucocorticoid) that also elevates blood glucose levels. Like growth hormone it is a powerful inhibitor of glucose uptake by many tissues. Cortisol is required for normal gluconeogenesis by the liver. It apparently facilitates both muscle protein breakdown and conversion of amino acids into glucose by the liver (Fig. 13-6). Cortisol is not secreted at a rate primarily determined by the state of glucose metabolism, but stress may play a role here.

Pathology

Diabetes Mellitus (Hyperglycemia). The major disorder of carbohydrate metabolism is diabetes mellitus (Greek *diabetes*, to siphon or run through, as with increased urine secretion; Latin *mellitus*, sweet). A tendency toward diabetes is inheritable. The condition may develop slowly, and some symptoms may be helped by appropriate measures, such as weight reduction or special diets. The cause is said to be a relative insulin deficiency. With a lowered insulin concentration there is a breakdown of triglyceride, with an elevation of fatty acids and ketones, since they must supply energy to cells that are prevented from taking up glucose due to insulin insufficiency. Thus, plasma glucose is high, since it cannot enter the cells. Only the brain is apparently saved, because its uptake of glucose is not insulin dependent. Consequently, due to the above-described cell deprivation and catabolic phenomena, there is a progressive weight loss, even despite the increased food intake created by constant hunger. The ingested protein and carbohydrate are converted to glucose, but an increased plasma glucose without insulin does not help.

Since normally we do not excrete glucose, inasmuch as all the glucose filtered at the glomerulus of the kidney is reabsorbed by the tubules, excess plasma glucose induces renal function changes that are serious. The filtered load of glucose may increase to the point where tubular reabsorption is exceeded and glucose is excreted in large amounts. Ketones in large amounts may also be present in the urine. Urinary excretion of these nutrients depletes the body; however, their loss also affects sodium and water excretion as well,

since the osmotic force exerted by unreabsorbed glucose and ketones holds water in the tubule, preventing its reabsorption, and sodium reabsorption is also retarded. As a result of this, there is a large excretion of water and sodium, leading to hypotension, brain damage, and death if not treated. Along with increased ketone bodies and fatty acids, which are moderately strong acids, there is an increase in hydrogen ion concentration by dissociation. The kidneys correspondingly excrete increased H^+ to maintain a body balance. This increased excretion of H^+ results in a marked increase in ventilation (breathing) in response to stimulation of the medullary respiratory centers by H^+ which in turn results in overexcretion of CO_2.

Arteriosclerosis, small-vessel and nerve disease, susceptibility to disease, and infection are but a few of the abnormalities associated with diabetes. Their cause is really unknown, and it appears certain that they involve more than an insulin deficiency since recent work has shown that many diabetics actually have normal or even elevated plasma insulin concentrations. There may very well be an insulin antagonist in the blood in these cases. However, it should be recognized that a number of hormones are involved in blood glucose control, although the common disorders center around insulin.

Diabetes is classified as either juvenile or maturity-onset. Juvenile diabetes is found in children and young adults, and little or no insulin appears to be present in the bloodstream. Thus, insulin injections are essential for therapy, and oral agents do not appear to function well. Maturity-onset diabetes is usually a milder form of diabetes and occurs most frequently in obese adults. Obesity and diabetes in later life appear to be related, and insulin does seem to exert marked effects on fat tissue. A study of this relationship may bear future discussion, since in some cases reduction in weight removes some chemical manifestations of the disease.

It has also been suggested that some of the circulating insulin in diabetes is in a form that cannot be properly used by the tissues. Drugs used in maturity-onset disease are taken orally and act by stimulating the release of insulin from the pancreas. Insulin injections are also used when needed under stress conditions.

It is apparent that in diabetes the biochemical manifestation is an impairment in regulation of blood glucose level, with hyperglycemia (elevated blood sugar) and glycosuria (spillage of glucose in the urine) being used as two major diagnostic features of the disease.

Since insulin facilitates glucose entry into many tissues, its deficiency or absence results in "starvation" of cells for glucose. To meet this need, glycogen stores are depleted and lipids (including fatty acids in their structure) are degraded and mobilized. These fatty acids undergo conversion to acetyl CoA, which enters the citric acid cycle, thus allowing fatty acids to be substituted for glucose in providing energy. In diabetes, however, fatty acids are excessively metabolized, leading to the formation of large amounts of side products called ketone bodies, such as acetone, acetoacetic acid, and beta-hydroxybutyric acid. Normally only a small amount of ketone bodies is formed. If the amount produced is too great for the tissues to metabolize, they accumulate in the blood, leading eventually to acidosis, an electrolyte imbalance and fluid loss that, if too severe, leads to diabetic coma. One sees this most often in untreated juvenile diabetes.

Hypoglycemia. Too much insulin will produce hypoglycemia (a decrease in blood glucose below normal levels). This condition may occur inadvertently when too much insulin is received or in individuals who respond to carbohydrates with an overproduction of the hormone. Tumors of the pancreatic beta cells have also been associated with the formation of abnormally large amounts of insulin. Other hormonal disorders, enzyme deficiencies, and liver disease may be involved.

Lipid Metabolism

Most ingested fat is absorbed into the lymph channels as fat droplets or chylomicrons and contains primarily triglycerides, which are stored in adipose-tissue cells. Fat-tissue-cell triglycerides come from ingested triglycerides, are synthesized from glucose in adipose tissue, and are synthesized in the liver and transported to the fat cells (Fig. 13-8). The free fatty acids are carried in the blood in combination with plasma albumin.

The fat cell was considered, at one time, to be metabolically inert, merely storing fat that had formed in the tissues. It is now known that these cells possess a very active glucose transport system, a well-developed hexokinase and glycolytic activity, and a high rate of fat formation from glucose. Fat storage also depends on an adequate supply of intracellular glucose (Fig. 1-22, p. 30). Fatty acids are stored after being neutralized by combining with phosphoglycerol to form a molecule of triglyceride or neutral fat.

Hormones affect lipid metabolism by controlling lipolysis and the release of fatty acids from lipid stores. Fatty acid levels in the blood regulate the rate of lipid oxidation by the tissues. Epinephrine, glucagon, and growth hormone stimulate lipid breakdown and utilization by activating lipases that catalyze the hydrolysis of triglycerides to fatty acids. The action on lipase may involve cyclic adenosine-3',5'-monophosphate (cyclic AMP) as an intermediate.

The primary balance in conserving triglycerides is performed by insulin, which promotes glucose uptake

Figure 13-8. Fat metabolism linked to carbohydrate metabolism.

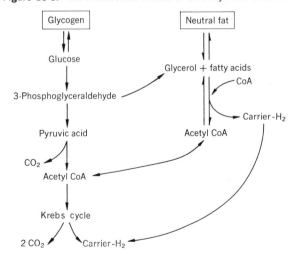

by adipose tissue and increases the supply and utilization of carbohydrates.

Protein Metabolism

There is a high rate of protein turnover in animal cells. It was once believed that amino acids absorbed from the gastrointestinal tract were used only to make up for "wear and tear." It is now known that protein synthesis, breakdown, and resynthesis are going on constantly (Fig. 1-22, p. 30). Protein survival time varies. Four hormones that have an anabolic effect and under proper conditions produce net protein synthesis and a positive nitrogen balance are growth hormone, insulin, testosterone, and thyroid hormone. Growth hormone spares amino acids from degradative metabolism, reduces carbon utilization and lipogenesis, increases amino acid transport, and increases protein synthesis by stimulating the translational activity of the ribosomes. Insulin action resembles that of growth hormone, and in fact they appear to act together. Insulin makes more carbohydrate available, and spares amino acids from gluconeogenesis and makes them available for protein synthesis. It also stimulates protein synthesis in the liver. Testosterone is also anabolic, affecting muscle and tissues involved in reproduction. Thyroid hormone levels are required for adequate protein synthesis, which is being mediated at the translational level in many tissues. An excess, however, leads to a general catabolic effect by increasing the metabolic demand for all metabolites, resulting in a reduced supply of amino acids and negative nitrogen balance. In thyroid deficiency, all metabolic interchanges proceed slowly.

Protein breakdown involves most significantly the glucocorticoids, which produce protein wasting and a negative nitrogen balance. Although they do stimulate hepatic protein synthesis, their catabolic effects on muscle and other tissues predominate.

Amino Acids

Much of the absorbed amino acids enters the liver cells and is converted into keto acids (carbohydrate) by removal of ammonia, which is converted into urea and excreted in the kidney (Fig. 1-22, p. 30). The keto acids enter the Krebs tricarboxylic acid cycle and are oxidized for liver cell energy or are converted into fatty acids. The amino acids not going to liver predominantly go to muscle cells, where they are synthesized to protein. Note, however, that most of the excess amino acids is not stored as protein but is converted into fat or carbohydrate, although a minimal amount is essential to maintain normal protein stores by preventing a protein breakdown in muscle and other tissues. It is the catabolism of adipose tissue triglycerides to fatty acids, which enter the Krebs cycle and are oxidized to CO_2 and HOH, that provides energy. The liver processes fatty acids by catabolizing them to acetyl CoA, which is processed into ketone bodies instead of entering the Krebs cycle. In this way the liver spares amino acid for glucose synthesis, and ketone bodies provide an energy source for tissues that can oxidize them via the Krebs cycle.

After a week's fast, even the brain begins to utilize ketone bodies, as well as glucose, for energy; thus, less protein need be broken down to supply the amino acids for gluconeogenesis, and the protein stores can survive for a long time.

Body Growth

Growth and development are complex processes. A gain in weight is not necessarily true growth but may represent water or adipose tissue excess. True growth involves lengthening of long bones, increased cell division, as well as increased synthesis and accumulation of protein. Two periods of rapid growth are usually described: during the first two years of life and during adolescence.

Determinants of Growth

Genetic makeup determines a person's growth capacity, but this can be influenced by food, disease, and other factors. Essential amino acids and fatty acids, vitamins, minerals, and protein are essential. No matter how much protein one eats, if the calorie intake is too low, normal growth cannot occur since the protein is just oxidized for energy. Nevertheless, one cannot stimulate growth beyond the genetically determined potential.

Hormones

The hormones most important for human growth are growth hormone, thyroxine, insulin, and sex hormones. In addition, adrenocorticotropic hormone (ACTH), thyroid-stimulating hormone (TSH), and prolactin influence the growth of the adrenal cortex, the thyroid gland, and the breasts. (These are all discussed in detail in Chap. 15.)

Growth hormone plays an essential role in facilitating growth, but it apparently does not direct its rate or course in time since average levels of this hormone are similar in adults and children. Its growth-promoting effects are due primarily to its ability to stimulate protein synthesis by increasing membrane transport of amino acids into cells and by stimulating the synthesis of RNA. It also causes large increases of mitotic activity and cell division. Growth hormone promotes the lengthening of bone by stimulating protein synthesis in both the cartilaginous center and bony edge of the epiphysial plate as well as by increasing the rate of osteoblast mitosis.

Thyroxine promotes the effects of growth hormone on protein synthesis. It plays an essential part in organ development, particularly of the central nervous system. Excess thyroxine does not cause excess growth (like growth hormone) but, rather, catabolism of protein and other nutrients.

Insulin is needed for normal growth since it is an anabolic hormone and affects amino acid uptake and protein metabolism.

As for the sex hormones, androgens stimulate protein synthesis in many body organs, and the adolescent growth spurt in both sexes is probably due in part to this anabolic effect. Androgens stimulate bone growth and eventually stop it by inducing complete conversion of epiphysial plates. Estrogen stimulates growth of the female sex organs and of sexual characteristics during adolescence but has less anabolic effect (unlike androgens) on nonsexual organs and tissues.

Total Body Energy Balance and Its Regulation

Basic Concepts

The source of energy utilized by cells in the performance of work (e.g., muscle contraction, active transport, molecular synthesis, etc.) is the breakdown of organic molecules, which liberates energy "locked" in the intramolecular bonds. The energy liberated takes the form of heat or is used for performing the body's work. In animal cells, most of the energy is seen as heat and only a small portion is used for work. The body, however, cannot convert heat to work and, therefore, the heat liberated is used to maintain body temperature. The energy for work is incorporated into molecules of ATP, which upon breaking down serve as a source for immediate energy to the body.

Body work can be classified as external (involving movement of external objects by using skeletal muscles) and internal (comprising all other biologic body activity, including muscle activity not related to movement). The real difference between these two forms of work is that the heat produced is, respectively, external and internal. The total energy liberated when organic nutrients are catabolized by cells is transformed sooner or later into heat. Only during periods of growth, when protein or fat synthesis exceeds breakdown, is the net energy stored in molecular bonds and this energy does not appear as heat. The total energy expended is thus equal to internal heat produced plus external heat produced plus energy storage during weight-gain periods. The units for energy are kilocalories, and the total energy expenditure per unit of time is the metabolic rate.

Metabolic Rate

One can measure metabolic rate directly or indirectly. Direct measurement involves placing a person in a calorimeter and directly measuring his internal and external heat production by the temperature changes recorded in the calorimeter. To indirectly measure the metabolic rate, one records the person's oxygen consumption in a given period of time (measurement of total ventilation and pO_2 of inspired and expired air). From this, one calculates heat production based on the fact that the energy liberated by the catabolism of foods in the body must be the same as when they are catabolized outside the body. Since we know how much heat is given off when 1 liter of O_2 is consumed in the oxidation of fat, protein, and carbohydrate outside the body, the same must be true inside. The value comes to an average of 4.8 kilocalories per liter of oxygen. A normal, fasting, resting adult male's rate of heat production is about equal to that of a single 100-watt light bulb. Since many factors affect the rate, all variables must be controlled as carefully as is possible. A basal metabolic rate (BMR) requires that the subject be mentally and physically at rest, the room temperature be comfortable, and no food be taken for at least 12 hours before the test. The value is then compared to normal values, taking into consideration sex, weight, height, and age. Thus, if 15.2 liters of oxygen are consumed per hour and we know that 1 liter of oxygen yields 4.8 kilocalories, then 15.2×4.8 equals 73 kilocalories per hour (calories produced or metabolic rate).

The growing child's resting metabolic rate is considerably higher than that of an adult since he expends more energy in the synthesis of new tissue. On the other hand, the metabolic rate gradually decreases with age. The resting metabolic rate of the female is generally less than that of the male but increases during pregnancy and lactation. Infection or disease generally increases total energy expenditure. The ingestion of food also increases metabolic rate. The energy expended in digestion and absorption is greatest for protein and less for carbohydrate and fat. Muscle activity, however, accounts for the greatest changes in metabolic rate. Even slight increases in muscle tone may increase the rate, and exercise can increase it 15 times. Many hormones, such as thyroxine and epinephrine, markedly increase the metabolic rate. Generally speaking, the kidneys play no role in the regulation of energy balance since almost all carbohydrates, amino acids, and lipids that are filtered at the glomerulus are reabsorbed by the tubules. However, in disease such as diabetes, urinary loss of organic molecules can be large and would greatly affect energy balance. Normally, too, very little loss occurs in feces or via hair and skin. In most adults, body weight remains relatively constant; however, this is variable, depending on man's own control since the amount of food intake is the dominant factor. Control mechanisms for heat production primarily regulate body temperature rather than total caloric balance. Thus, when you are cold, heat is produced by shivering even though you might be starving.

Control of Food Intake

Food intake seems to be controlled primarily by nerve cells in the hypothalamus. Destruction of the ventromedial hypothalamus in animals induces extreme overeating and obesity; thus, this center is a satiety center. Lesions in the lateral ventral hypothalamus cause inhibition of eating and food intake, and animals starve to death. The two areas are apparently interrelated. Thus, the lateral hypothalamus has a "feeding center" that stimulates the final motor acts of eating and associated food seeking, and the ventromedial area can inhibit the activity of this center. The hypothalamic centers serve only as integrating centers, processing sensory input and controlling motor output. Although food intake is stimulated by sight and smell, it is basically a process that continues unless turned off by an input signal. Complete denervation of the upper gastrointestinal tract apparently does not interfere with normal maintenance of energy balance; therefore, gastrointestinal sensory stimulation due to contractions of an empty stomach is not the major long-term regulator of food intake, although it may play some role of modification. The caloric content of food, too, may bear no relation to its bulk, and apparently "bulk" in the tract does not offer a good detecting system.

Theories. Three theories are presently in vogue as to the regulation and detection of total body energy content.

GLUCOSTATIC THEORY. This theory states that some form of glucose receptor exists in the brain, probably in the ventromedial nuclei of the hypothalamus (satiety center). These receptors probably initiate reflexes leading to the release of epinephrine and growth hormone when plasma glucose is reduced. Since the rate of cellular glucose utilization increases during and after eating (decreases in fasting), detection of greater use of glucose signals the center and inhibits eating (particularly if the hypothalamic centers are glucose-sensitive receptors themselves). Fasting signals decreased glucose use, removing the input satiety signal, and eating is promoted.

LIPOSTATIC THEORY. This theory suggests that a substance, possibly a blood-borne humoral agent, released from fat stores in direct proportion to their total mass would make a good indicator of body energy. A positive balance would increase the amount of adipose tissue, signaling satiety, and the converse would also be true.

THERMOSTATIC THEORY. This theory states that the increase in metabolic rate induced by eating raises the body temperature, and this elevation may constitute a satiety signal. Such a theory might explain why we tend to eat more in colder weather than in hot.

Obesity

Obesity is a common "disease." It often predisposes to illness and early death from many causes. The mortality rate is more than 50 percent greater in overweight people. "Too much eating" means consuming more than is needed to supply the body's energy needs, since beyond this point additional food is stored as fat. Thus, one's activity plays an essential role in obesity, and it may not be just food consumption or hormonal or glandular conditions. Interestingly, low levels of physical activity may actually induce or stimulate increased eating.

Most cases of obesity can be explained only in part by lack of physical activity. All obesity represents failure of normal food-intake control mechanisms. Only in rare cases does one see hypothalamic lesions, as observed in experimental animals. Cortisol and insulin may play a significant role where multiple metabolic disorders prevail. Psychologic factors, habit, and even environmental (social) factors may in reality relate to the hypothalamic maladjustments. The temporal pattern of eating may also be essential (eating large meals rather than nibbling all day). Meal eaters have relatively more body fat and less protein than do nibblers. Furthermore, for some reason, nibblers seem to have decreased susceptibility to diabetes and arteriosclerosis.

Body Temperature Regulation

The so-called cold-blooded animals cannot reflexly regulate their internal environment and cannot control the loss or retention of heat produced. Their body temperatures equilibrate with that of their external environment: if it is cold, they are cold; if hot, they are hot. Mammals and birds, on the other hand, regulate their body temperatures within very narrow limits and are called homeothermic. Temperature alterations affect both the rate of body chemical reactions and enzyme activity. Control must, however, be precise and operate in a limited range since even moderate temperature elevations result in nervous malfunction, protein denaturation, and even death. Convulsions occur at 106° to 107° F, with 110° F being the absolute limit for life. Most body tissues, however, can tolerate marked cooling, even below 45° F, the latter being used in cardiac and other surgery.

Enzymes make possible most of the complex chemical reactions of the body processes. They operate most effectively at about 98.6 F (37° C). Although the body temperature may vary normally between 96° and 100° F, it maintains this range fairly closely for ideal enzyme activity. A considerable rise above 100° F (high fever) results in a speedup of enzyme destruction, and the reactions they catalyze stop. A substantial drop below 96° F reduces enzyme activity and slows down all body processes, so that one finds great difficulty in action and even in thinking processes.

Generally speaking, internal temperature does fluctuate approximately 1° F in "normal" people in response to external temperature and activity. Oral temperature is about 1° F less than rectal, indicating that all areas of the body may differ slightly in temperature level. There is also a diurnal change, with levels being lower during sleep. Women have a slight temperature elevation during the middle of the menstrual cycle due to progesterone. The total heat content of the body is determined by the net difference between heat produced and that lost from the body. To maintain homeostasis, heat produced must equal heat lost.

There is, however, a difference in surface temperature and internal temperature. An attempt is made to keep surface temperature at 91.4° F, which is most comfortable. For this, an ideal air temperature is 68° F. At this temperature, the body surface temperature could be maintained without clothes. Unfortunately climate as is and the ideal temperature are often far apart, and we create for ourselves an artificial climate at the skin level with houses, heating, air conditioning, and clothes—all to maintain a surface environment at 91.4° F.

Heat Production. Heat is produced by all chemical reactions in the body, and body metabolism regulates and sets the basal level of heat production. It can be increased by muscle contractions or action of several hormones and decreased by sleep or inactivity.

Changes do occur, however, that upset the body temperature. There can be changes in the external environment, e.g., weather or clothing; and there can be internal changes, as the amount of heat generated can fluctuate according to cellular activity. Most of the heat to maintain the internal temperature at 98.6° F is acquired by the combustion of foods, mainly in the liver and muscle. By exercise we can voluntarily increase heat production if the temperature falls; and we shiver when we get cold, thus involuntarily contracting our muscles. This may even double our heat production. If these methods fail to elevate the temperature, man resorts to either artificial heating or more clothes.

Hormonal Control. Epinephrine and thyroxine can raise the basal metabolic rate (Fig. 13-9), but whether or not they normally contribute appreciably to temperature regulation in man is questionable. Experiments question whether their secretion actually does increase on exposure to cold. In any event, human hormonal changes in response to cold or heat are apparently of secondary importance. The changes in muscle activity are the major control of heat production for temperature regulation.

Muscle Changes. With cold, the first muscle reaction is a general increase in skeletal muscle tone, which results in shivering. The contractions are directly controlled by motor muscle neurons, which in turn are controlled by descending pathways under hypothalamic control. Cerebral cortex control can suppress this. Shivering is completely reflex, but voluntary heat production can occur by hand clapping, foot stamping, jumping up and down, and so forth. This can increase

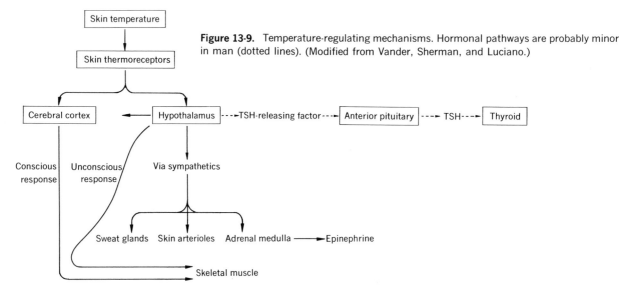

Figure 13-9. Temperature-regulating mechanisms. Hormonal pathways are probably minor in man (dotted lines). (Modified from Vander, Sherman, and Luciano.)

body heat production about seven times in a matter of seconds. Since there is no external work, all the liberated energy appears as internal heat.

With heat, muscle tone is reflexly decreased and voluntary movement reduced to a minimum. The ability to reduce heat, however, is limited since muscle tone normally is low, and man's ability to regulate temperature by decreasing metabolic rate is very limited because of the direct effect of a body temperature increase on metabolic rate.

Heat Loss. We lose heat by surface body exchange of heat with the surrounding environment in three ways: radiation, conduction, and convection.

RADIATION. Not only our body but all dense objects emit heat from their surfaces in the form of infrared rays, at all times regardless of the external temperature; this is by radiation, or a transfer by electromagnetic waves. The emission rate depends on the temperature of the radiating surface. The higher the skin temperature, the more heat is lost by radiation. It can only be minimized by reducing the flow of blood to the body surface (by constricting the blood vessels of the skin). On the other hand, if the body must lose heat, as during exercise, the vessels dilate and the skin temperature is raised. The sun is a huge radiator and exposure to it decreases heat loss by radiation.

CONDUCTION. Conduction is the exchange of heat by simple transfer of thermal energy from molecule to molecule. Thus, the body surface gains or loses heat this way only through direct contact with warmer or cooler substances (air included). Heat may go to one's clothing or to the air. Clothing serves to reduce heat loss by keeping a layer of relatively still air (a poor heat conductor) next to the skin. Cloth, too, conducts heat poorly.

CONVECTION. Convection takes place when air next to the body is heated, moves away, and is replaced by cool air, which then follows the same pattern. Loss of heat by convection is always happening, because warm air is less dense and rises. It is facilitated by things like fans and wind. Convection helps conduction by continuously maintaining a cool air supply. If there were no convection, little heat would be lost to the air and conduction could operate only by direct contact with objects or substances. Thus, for all practical considerations, convection and conduction are intimately related.

OTHER CONSIDERATIONS. Heat loss by all three processes is a result of the temperature difference between body surface and the environment around it. The body is surrounded by a layer of skin, with a varying insulation capability. The body temperature is kept at 99° F, whereas the skin changes continually. The skin is not a perfect insulator, and therefore, the temperature of its outer surface generally falls between that of the body and its environment. The skin's effectiveness as an insulator relies on physiologic control related to the changing blood flow to it. It is the vessels that carry heat to the surface and cut the skin's insulation. The vessels, in turn, are controlled by sympathetic vasoconstrictor nerves.

Exposure to cold increases the gradient between the environment and the body temperatures. Skin vasoconstriction increases skin insulation, cuts down skin temperature, and lowers heat loss. Heat, on the other hand, decreases the gradient; and to permit heat loss, skin vasodilation takes place, the gradient increases between the skin and the body, and heat is lost. Since even maximal vasodilation cannot establish a sufficient gradient between body temperature and environment to eliminate heat as fast as it is produced, we also rely on sweating.

Perspiration (sweating) is the only way heat can be lost if the outside temperature is greater than 91.4° F. It is so important to prevent the body from becoming overheated that sweating proceeds vigorously even though it means the loss of valuable, vital, and essential water. We normally lose about 600 ml per day by evaporation from the skin and lung lining. The sweat glands in the skin filter a watery solution from the blood, and the glands discharge the solution through ducts onto the skin surface, where it evaporates (conversion of fluid to vapor)—which absorbs heat. In this case, the heat comes from the skin itself, so the body is cooled. If, however, as in very humid climates, the sweat does not evaporate easily, almost no heat is lost and perspiration beads appear; although much water is lost, we are not relieved and feel uncomfortable.

Sweat contains an appreciable amount of salt, and when sweat evaporates the salt remains on the skin. In time, the solution on the skin's surface becomes more and more concentrated, which tends to raise the boiling point and leads to a reduced evaporation. Hence frequent showers tend to cool the body and wash away the salt, leading to better evaporation, at least for a short time.

Salt loss, however, upsets the osmotic balance of the body fluids and in extreme conditions may even result in irritability followed by convulsions. This is remedied by taking salt when the water loss is replaced by drinking. Without water and salt, man can survive only two or three days in tropical climates.

Other mechanisms that change heat loss by radiation, convection, or conduction are surface-area changes and clothing. Hunching up, pulling knees up, and hugging the body—in response to cold—cut down exposed surface area and decrease convection and radiation. Clothing is man's added insulator. Here the skin loses heat to the air space trapped by the clothes. The clothes, in turn, pick up heat from the inner air layer and transfer it to the outside. Thus, the thickness and type of clothing determine their insulating ability. Clothes may insulate against heat loss when it is cold and may insulate against external heat if they are loose enough to allow adequate air movement to permit evaporation. White clothing reflects radiant energy, while dark colors absorb it. Thus, wearing loose-fitting light-colored clothes is more cooling than being nude in the sun.

Control of Temperature Regulation. There are specific hypothalamic areas or temperature-regulating centers (thermoregulation): one for warmth, the other for cold. The major receptors that detect temperature changes are found in the skin and in the hypothalamus itself. In the skin are found both cold and warm receptors, stimulated by a lower and higher range of temperatures, which are transmitted to the hypothalamus, which in turn responds by motor output. These skin receptors are not sufficient, since they tell us of skin temperature but not of internal body temperature. Receptors for this are located in the hypothalamus. They are probably integrating neurons that are temperature sensitive. The neurons from these areas control the output of somatic motor nerves to skeletal muscle (tone and shivering) and sympathetic nerves to the skin arterioles (dilation and vasoconstriction), the sweat glands, and the adrenal medulla.

Emotional activity in the cerebral cortex affects the hypothalamus and may interrupt many autonomic functions. Thus, the mechanisms of temperature regulation are influenced by our emotional states. Embarrassment causes vasodilation in the skin, and we blush. Anxiety causes vasodilation and sweating, and we get hot and bothered. Anger causes an outpouring of epinephrine that causes vasoconstriction of surface vessels, and we get "white" with rage. Fear often produces a combination of vasoconstriction and sweating, resulting in a "cold sweat."

Fever. Fever or elevation of body temperature is due to a "resetting" of the hypothalamic thermostat rather than a breakdown of regulating mechanisms. With fever we regulate our temperature, in response to heat and cold, at a higher set point. When we feel a chill, it is due to the fact that the body thermostat has been raised and there is vasoconstriction, and shivering takes place. A combination of heat conservation together with "increased heat" raises the body temperature. When the fever breaks, the thermostat is reset to normal and we begin to feel hot and manifest vasodilation and sweating. Fever may be induced by chemical agents called pyrogens, the exact function of which is unknown.

Review Questions

1. Which are the major fat-soluble vitamins? Give at least one function for each.
2. List the major water-soluble vitamins. What do they do?
3. Can you name at least one source of the vitamins named in questions 1 and 2?

4. What is meant by metabolism? What is intermediary metabolism?
5. What are vitamins? How are they related to stress, growth, resistance, pregnancy, and old age; to each other; and to hormones?
6. What is the essential ingredient for energy metabolism in the living organism?
7. Relate the general steps of glucose metabolism in the cell. What are the two major biochemical pathways of glucose metabolism?
8. What is glycolysis?
9. How many molecules of ATP are formed for each glucose molecule metabolized?
10. Where is glucose most needed? How is blood glucose regulated?
11. What effect do the following have on blood glucose?
   a. Insulin   **c.** Glucagon   **e.** Cortisol
   b. Epinephrine   **d.** Growth hormone
12. What is diabetes mellitus? What are some of its effects? What are some of its diagnostic features?
13. What is the function of the fat cell?
14. What effect do epinephrine, glucagon, and growth hormone have on lipid metabolism?
15. What is the general fate of most excess amino acids?
16. How are growth hormone, insulin, testosterone, and thyroid hormone related to protein synthesis? What about glucocorticoids?
17. Which hormones are most essential for human growth?
18. What is meant by metabolic rate? How is it measured? Name several factors that might influence metabolic rate.
19. Where is the so-called feeding center located? How is it thought to function?
20. How might you explain obesity?
21. Why must man maintain a certain internal body temperature?
   a. How is body heat produced?
   b. How do hormones affect body heat?
   c. How is heat lost from the body?
   d. How is body temperature generally controlled?
   e. Why do we have a "fever"?

References

Carlson, L. D., and Hsieh, A. C. L.: *Control of Energy Exchange.* Macmillan Publishing Co., Inc., New York, 1970.

Changeun, J.: The Control of Biochemical Reactions. *Sci. Am.,* **212**(4):36–45, 1965.

Irving, L.: Adaptations to Cold. *Sci. Am.,* **214**(1):94–101, 1966.

Karlson, P. (translated by Doering, C. H.): *Introduction to Modern Biochemistry,* 3rd ed. Academic Press, Inc., New York, 1968.

Lentz, J.: The Age of the Enzyme. *Today's Health,* April, 1970, pp. 32–69.

Sebrell, W. H., Jr., and Harris, R. S. (eds.): *The Vitamins,* 2nd ed., Vol. 3. Academic Press, Inc., New York, 1970.

Yanofsky, C.: Gene Structure and Protein Structure. *Sci. Am.,* **216**(5):80–94, 1967.

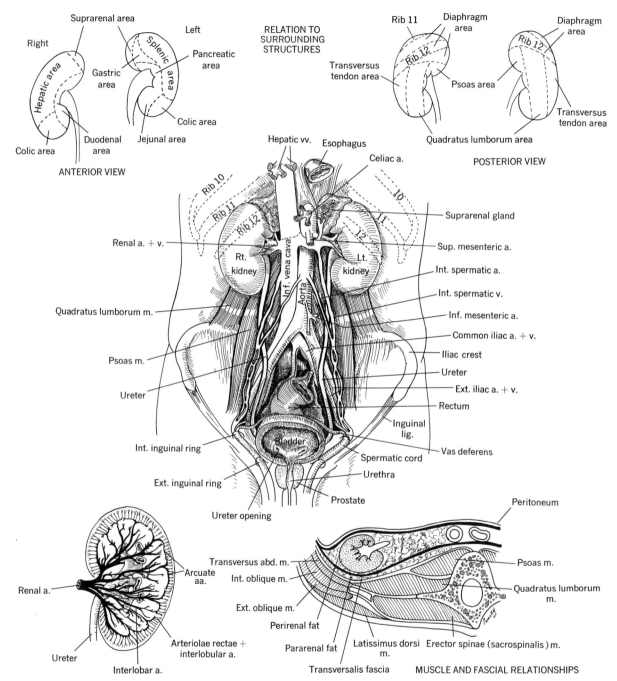

Figure 14-1. Relations of the urinary system, with special emphasis on the kidney.

Regulation of Extracellular Water and Electrolytes: The Urinary System

Most chemical reactions yield a useful product. Most reactions also manufacture by-products that are useless or even harmful—the excretory substances.

Carbon dioxide and water are products that are formed when solid food is "burned" to produce energy. Carbon dioxide is toxic in high concentrations because it increases the blood's acidity. Carbon dioxide is removed by the lungs (see Chap. 11).

Metabolic water, which is produced by reactions in living tissue, is excreted only when the body already has too much. Most excess water is eliminated by the kidneys (Fig. 14-1) and the rest by the skin and lungs. Thus, the skin and lungs must be considered as excretory organs although they also perform other functions (Fig. 14-2).

Many excretory substances exist, some taken in with food. Excess salts, the commonest being those with sodium, potassium, chlorine, bicarbonates, and phosphates, are excreted by the kidneys. Several toxic compounds in foods are also eliminated by the kidneys.

The water and carbon dioxide produced by combustion of proteins, fats, and carbohydrates present no excretory problem since they can accumulate for some time with little harm to the body. However, proteins, when converted, also yield ammonia, which is very soluble and 1 part in 25,000 is fatal. Proteins are broken down in the liver into amino acids and then into ammonia, but the latter is converted into the relatively harmless urea, which is excreted by the kidneys and to a slight extent by the skin in sweat.

Uric acid is another product of protein metabolism. Man is one of the few animals that makes this poorly soluble substance; with abnormal metabolism, an accumulation of uric acid results in gout, which leads to deposits of crystalline sodium urate in joints, particularly of the big toe, and causes severe pain.

The excretion of water, salts, and urea by the skin and carbon dioxide by the lungs is an unconscious, involuntary process and takes place regardless of the amount of water in the body. The only way to control excretion is via the kidneys, which are highly organized and specialized to perform this function.

The kidneys are versatile, and yet, although different from many organs, they have many similar units of simple cell structure. The refinement of action is at the cellular level with the complex activities taking place as a result of the passage of substances through a cell membrane by either diffusion, osmosis, or active transport.

Anatomy of the Kidneys, Ureters, and Urinary Bladder

Embryology

Three successive sets of renal corpuscles and tubules make their appearance in the developing kidney. The first, the pronephros, disappears and has no counterpart in the adult organ. The duct, however, persists and forms the duct for the second set, the mesonephros. Here, too, most of the tubules and all the corpuscles disappear. A few tubules remain and form parts of the male genital system, while the duct becomes

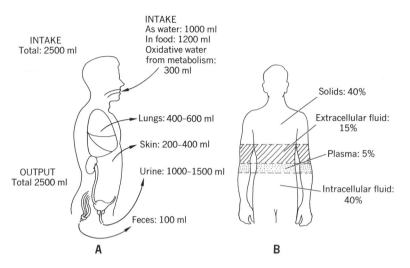

INTAKE
Total: 2500 ml

INTAKE
As water: 1000 ml
In food: 1200 ml
Oxidative water
from metabolism:
300 ml

Lungs: 400–600 ml

Skin: 200–400 ml

OUTPUT
Total 2500 ml

Urine: 1000–1500 ml

Feces: 100 ml

A

Solids: 40%

Extracellular fluid:
15%

Plasma: 5%

Intracellular fluid:
40%

B

Figure 14-2. Man's water balance. **A.** Average fluid intake and output in normal fluid balance. Man maintains the osmotic pressure of his body fluids by adjusting his water intake and output. The water of oxidation varies with the species. A kangaroo rat drinks no water, but his water needs are met by the moisture in his diet and that which his body manufactures. **B.** Fluid distribution in terms of body weight.

the duct of the epididymis and vas deferens (Fig. 14-3; Fig. 18-4, p. 592). The third set, the metanephros, becomes the fetal and definitive kidney. Thus, the nephrons are of metanephric origin, while the collecting ducts, renal pelvis, and ureter are of mesonephric origin.

Fascial Relations

The renal fascia (Fig. 14-1) is derived from the transversalis fascia, which, at the lateral border of the kidney, splits into an anterior prerenal and posterior retrorenal layer. Between this fascia and the kidney is the perirenal fascial space (of Gerota), which is filled with perirenal fat. External to the renal fascia is the pararenal fat. The anterior layer of fascia covers the kidney and its vessels, the aorta, and the vena cava and is continuous with the corresponding layer of the opposite side. The posterior layer attaches to the fascia of the deep back muscles. The two layers of renal fascia fuse at the upper pole of the kidney, leaving the suprarenal gland in its own chamber; however, they remain separate inferiorly, allowing for some mobility (floating) of the kidney.

The Kidney Proper

The kidney lies in the posterior wall of the abdominal cavity behind the peritoneum and encased in fascia

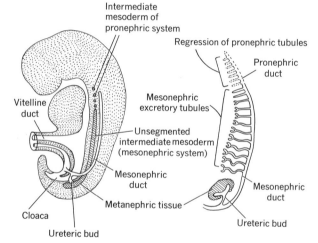

Intermediate
mesoderm of
pronephric system

Regression of pronephric tubules

Pronephric
duct

Mesonephric
excretory tubules

Vitelline
duct

Unsegmented
intermediate mesoderm
(mesonephric system)

Mesonephric
duct

Metanephric tissue

Cloaca

Ureteric bud

Mesonephric
duct

Ureteric bud

Figure 14-3. Diagram of excretory tubules of developing urinary duct system. Note relationship of intermediate mesoderm of pronephric, mesonephric, and metanephric systems.

(Figs. 14-1, 14-4). Besides the fascial layers just described, the kidney has its own "proper," closely fitting capsule, consisting of dense connective tissue that normally strips off easily from the kidney. Small blood vessels and capillaries are also present here.

The kidney is reddish-brown in color and fairly soft in consistency. It is about 10 to 12.5 cm (4 to 5 in.)

Figure 14-4. The kidney in cut section to demonstrate its parts and a diagram of a nephron with associated vessels.

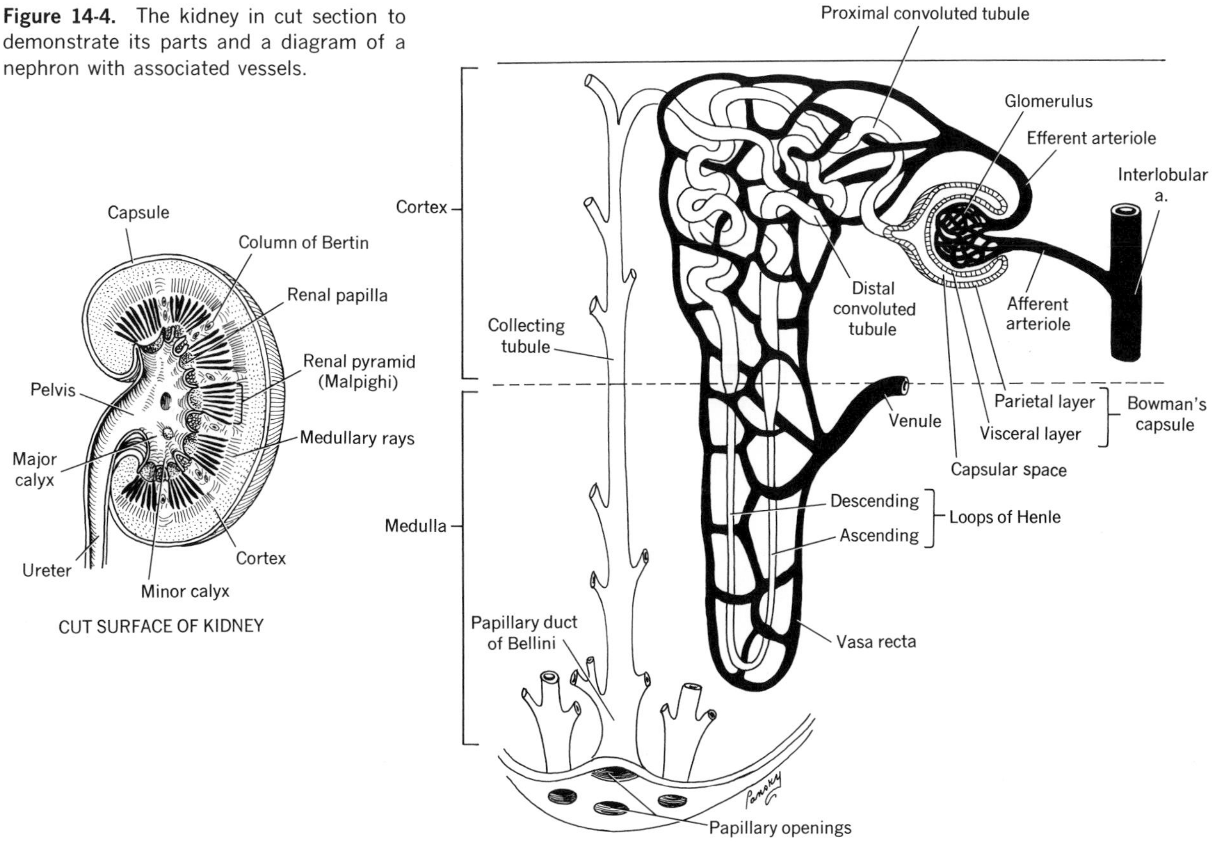

CUT SURFACE OF KIDNEY

long, 6.25 cm ($2\frac{1}{2}$ in.) wide, and about 2.5 cm (1 in.) thick (in the middle). It is bean shaped, and its surface is relatively smooth, not showing the fetal lobulations. It has an upper and lower pole, anterior and posterior surfaces, and medial and lateral borders. The hilum is an area on the medial border where the renal vessels and ureter enter and leave the kidney. The hilum leads into a wide space inside the kidney called the sinus, in which lie branches of the renal artery, tributaries of renal veins, and the dilated proximal part of the ureter, the pelvis. The pelvis is further divided into several primary divisions or major calyces, which in turn divide into several secondary minor calyces. The latter are associated with conelike projections of the kidney substance, the pyramids or papillae.

When viewing the cut, longitudinal surface of the kidney, one sees an outer, darker, red rim of tissue, the cortex, and an interrupted series of paler, flesh-colored pyramids, the medulla. In the area between the two are several larger vessels, the arcuate vessels.

The cortex is divided arbitrarily into a narrow inner zone and an outer zone. The inner zone is called the juxtamedullary zone, and the outer zone is termed the outer cortical zone. The medulla, too, is often divided into an inner zone (two thirds) and an outer zone (one third). One sees a large number of fine streaks radiating from the corticomedullary junction into the cortex, and we refer to these as the medullary rays, even though they lie in the cortex. (The medullary rays represent: [1] the straight portions of the kidney tubule and the loops of Henle and [2] the portions of the collecting system that pass straight down from the cortex to the

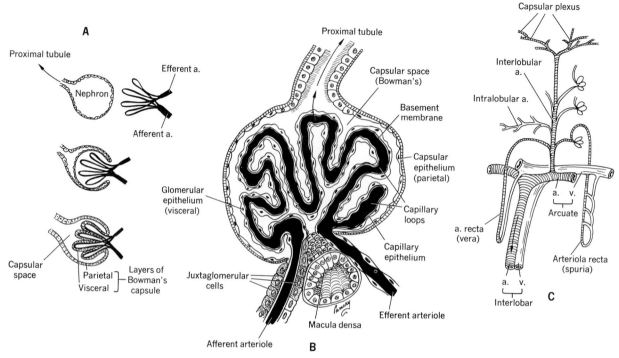

Figure 14-5. A. Bowman's capsule and its relation to the capillary loop during development. **B.** Diagram of a section of a glomerulus. **C.** Arterial supply to the glomerulus (including arteriola recta). (**A** and **B** modified from Ham.)

tip of the papilla. The cortical areas between the rays are filled with masses of coiled tubules.) The interlobar arteries occupy the renal (interlobar) columns between the pyramids.

The kidneys rest on four muscles: diaphragm, transversus abdominis, psoas, and quadratus lumborum. The twelfth rib, an important landmark to kidney anatomy, is separated from the kidney by pleura and diaphragm.

The kidneys are capped by the adrenal glands. The right kidney is related anteriorly to the liver (above), hepatic flexure of colon (below), and second part of duodenum near its hilum. Anterior to the left kidney is the stomach (above), spleen (lateral), and transverse colon (below), and the pancreas crosses its hilum. The lower pole contacts the jejunum.

Both kidneys lie above the level of the umbilicus and as high as the eleventh or twelfth ribs. The right is lower than the left due to the liver. The upper poles of the kidney lie closer together than do the lower.

The kidneys are kept in place by the attachment of their vessels, the pressure of surrounding organs, and fat and fascia. The lower poles of both kidneys may fuse across the midline to form a so-called horsehoe kidney.

Vessels and Nerves. The renal artery is a short branch of the aorta and may divide into many branches before entering the gland (Figs. 14-1, 14-4, 14-5). It may also give off a spermatic artery. The renal artery passes to the kidney in front or behind the inferior vena cava. It carries waste material to the kidney, while the renal vein carries more "purified" blood. After the artery enters the kidney hilum, it gives rise to interlobar arteries, which pass through the fatty tissue of the renal sinus and continue between the pyramids to the corticomedullary junction, where they lie in the connective tissue. The interlobar arteries terminate by dividing into the arcuate (arciform) arteries, which curve round (for a short distance) in the corticomedullary junction but do not anastomose with each other. From these, the straight arteries (arteriae rectae verae or interlobular arteries) pass into the cortex, being radially arranged in a straight course directed to and near the cortical surface. The interlobular arteries give off short side branches, the afferent arterioles of the renal corpuscles (Fig. 14-4).

In the renal corpuscle, the afferent vessel breaks up rapidly into wide capillaries and then they turn on themselves, join together, and form the efferent arteriole. Thus, the renal glomerulus of the corpuscle is formed.

The efferent arteriole turns toward the medulla and quickly divides up into many, very wide capillary channels, which run down into the medulla and gradually spread out between the loops of Henle, the collecting tubules, and the excretory ducts (Fig. 14-4). At points near the tip of the papilla, the individual capillaries turn abruptly up on themselves and pass back up the length of the medulla. As they near the corticomedullary junction, they join with each other to form venules and veins, which enter the arcuate vein in the junctional area. The latter become interlobar veins, then renal veins, which in turn join the inferior vena cava. The left renal vein is longer than the right.

The blood to the outer cortex divides up into a second series of capillaries (after having passed through the glomerulus and becoming an efferent arteriole), which form an anastomosing network around the convoluted tubules and medullary ray tubules. They form a complete peritubular network throughout the cortex. These vessels are associated with reabsorption of the tubules. From this network, capillaries also join to form venules, which unite to form the large stellate veins, which then turn toward the junctional zone to end in the arcuate veins. A few capillaries do occasionally anastomose with those of the capsule.

The lymph vessels follow the veins and drain into nearby glands in the region of the aorta and vena cava.

The nerves (Fig. 8-26, p. 277) are derived from the twelfth thoracic and first lumbar and run with the first rib. They lie in the pararenal fat for a distance.

The Nephron (Renal Structural and Functional Unit)

The nephron consists of the renal corpuscle, the proximal convoluted tubule, a descending limb, the loop of Henle, an ascending limb, and the distal convoluted tubule (Figs. 14-4, 14-6). The distal tubule empties into a collecting tubule, which eventually ends in the renal pelvis. A fluid called the glomerular filtrate is formed in the renal corpuscle as an ultrafiltrate of the blood plasma that circulates through the glomerular capillaries. As this fluid passes through the various segments of the nephron, it is modified by secretions into it and absorption of water and other constituents from it. The final product, urine, is drained through the collecting ducts into the renal pelvis.

Renal Corpuscle and Associated Structures. These are found only in cortical tissue. They extend from the corticomedullary junction to near the surface.

GLOMERULAR CAPSULE (BOWMAN). The glomerular capsule (Figs. 14-4, 14-5, 14-6) is a spherical, homogeneous membrane with two openings: one for the arterioles to enter and leave (the vascular pole) and the other at the place where the tubule exits from the capsule at the neck to form the proximal (first) convoluted tubule. The capsule develops around the tuft of glomerular capillaries as a double-walled cup composed of squamous epithelium (Figs. 14-4, 14-5, 14-6). The wall applied to the glomerulus is the visceral layer; the outer wall is the parietal layer or capsular epithelium. The thin, slitlike cavity between them is the capsular (Bowman's) space. At the vascular pole the visceral layer is continuous with the parietal, and at the urinary pole it is continuous with the cuboidal epithelium in the neck of the proximal convoluted tubule.

The visceral layer becomes extensively modified. Its cells, called podocytes (Fig. 14-6), have a number of processes, which give rise to small secondary processes called end-feet or pedicels. The foot processes become aligned on the outside of a continuous basal lamina (basement membrane), which they share with the glo-

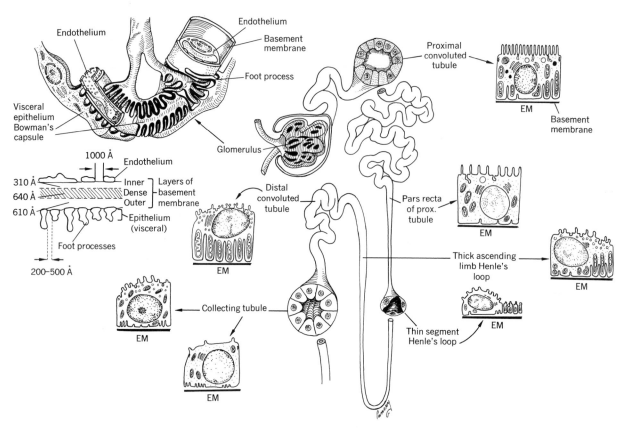

Figure 14-6. The normal histology and electron microscopy of the nephron. A detail of filtration layers is seen from the vascular compartment into Bowman's space (left figure).

merular endothelium on the inside. There are filtration slits or pores (about 250 Å wide) between adjacent foot processes. The endothelium of the glomerular capillaries is also extremely spread out and is perforated by circular pores (500 to 1000 Å wide). Since the basement membrane on the outside of the capillaries is fused with the visceral epithelium, it then is the only complete surface between capillary and Bowman's space and is the major filter that tends to hold back large molecules. Deep cells called mesangial cells are also found on the inner side of the capillaries.

AFFERENT ARTERIOLES. The afferent arterioles are end branches of the straight arteries. Near the renal corpuscle, the arteriolar adventitia disappears, the internal elastic lamina is lost, and the tunica media

changes from smooth muscle to epithelial-like epithelioid cells called the juxtaglomerular cells (Fig. 14-5), which contain granules. These cells seem to have some endocrine function and are believed to produce renin.

GLOMERULUS. The afferent arteriole gives rise to a number of wide capillaries, each of which supplies the glomerulus (Figs. 14-4, 14-5, 14-6). The capillaries have few points of anastomosis but are arranged in leaflets. The visceral epithelium is carried into the glomerulus and covers each leaflet of the tuft. The covering, however, is incomplete, as has been discussed.

EFFERENT ARTERIOLES. Efferent arterioles drain the glomerulus. They have the tunica media of typical arteriolar smooth-muscle cells, show no development of epithelioid cells, and have no internal elastic lamina.

Tubular Portion of the Nephron. The tubules (Figs. 14-4, 14-6) lie in a tangled mass in the cortex and in a regular parallel arrangement in the medulla. They are surrounded by reticular fibers and a variable amount of interstitial connective tissue and rest on a basement membrane. All the nephrons show the following:

1. A proximal (first) convoluted tubule with its neck joining the renal corpuscle. It is the most highly developed segment and the most easily influenced by disease or toxic materials. It has a striated or brush border associated with selective reabsorptive powers and is very granular, and its basal part has many mitochondria.

2. The nephric loop of Henle, a loop or U tubule in which one limb descends into the medulla and the other returns to the cortex. The cells are flattened and cuboidal with a slightly granular cytoplasm.

3. A second or distal convoluted tubule, which terminally connects directly with a collecting tubule, a tubule of the conducting system. It lies entirely in the cortex, the cells lack a brush border, and basal mitochondria are few.

The total area of filtration surface of all the glomeruli of both kidneys is about 1.5 m (5 ft), which is about that of the entire body surface.

The Excretory Duct System

The excretory duct system (Fig. 14-4) consists of straight tubules, collecting tubules, and excretory ducts. The straight tubules lie in the cortex and drain the distal convoluted tubules. Several may join one collecting tubule, and the cells are cuboidal with fine granulation. The collecting tubules lie in the medullary rays and outer zone of the medulla and gather the straight tubules; the cells are like those of the collecting tubules. The excretory or papillary ducts (Bellini) lie in the inner zone of the medulla and are formed by the junction, at very acute angles, of the collecting ducts. The cells are identical but increase in height near the mouth of the duct, which opens into the

minor calyces on the tips of the medullary pyramids and then into the major calyces and renal pelvis on the way to the ureter.

The Ureters

The ureters (Figs. 14-1, 14-7) are two muscular tubes about 25 cm (10 in.) long and approximately 0.625 cm ($\frac{1}{4}$ in.) in diameter. They carry a continual stream of urine down into the urinary bladder, where it is stored and voided at intervals through the urethra. The urine consists mainly of water in which the major solutes are urea and salt (NaCl). Creatinine, potassium, phosphate, and sulfate are also found, along with traces of uric acid, calcium, magnesium, glucose, protein, and bicarbonate. Each ureter has an abdominal and pelvic course and opens into the bladder at the upper angle of an area called the trigone. It has three constrictions: (1) near the kidney pelvis, (2) at the brim of the pelvis where it crosses the common iliac artery, and (3) just outside of the ureteral opening in the bladder. A kidney stone may be stopped at one of these constrictions as it passes downward. The ureter receives its blood supply from the renal, testicular, colic, vesical, and middle rectal arteries. Its nerve supply is derived from the renal, testicular, and hypogastric plexuses. Its lymphatics end in glands nearby, in the abdomen and the pelvis.

The Urinary Bladder

Embryology. The bladder originates from the cloaca (a part of the primitive hindgut) and the lower end of the wolffian duct.

The urachus (the allantois of the embryo) is a fibrous cord that connects the bladder apex to the umbilicus and becomes a fibrous cord. If it remains patent, urine can be discharged at the umbilicus.

At birth, the true pelvis is underdeveloped and allows little room for either bladder or intestines. Thus, in infancy, the bladder is an abdominal organ (full or empty) and is in contact with the abdominal wall. As the pelvis enlarges, the bladder sinks into it, so that at six years of age the pelvis accommodates the bladder (empty). Shortly after puberty, the bladder becomes an exclusive pelvic organ.

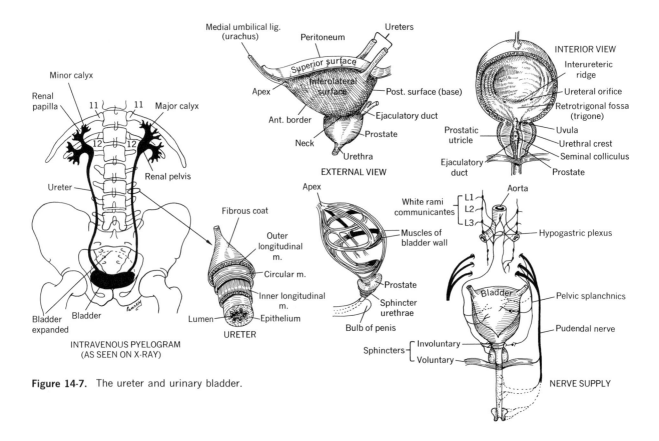

Figure 14-7. The ureter and urinary bladder.

Anatomy. The empty bladder has four angles, four surfaces, and four ducts for each angle (urachus, urethra, and two ureters) (Fig. 14-7). The urachus attaches to the apex (anterior angle) and becomes the median umbilical ligament, the ureters open into the posterolateral angles, and the urethra exits at the inferior angle. The surfaces are superior, two inferior lateral, and a posterior (base or fundus). In the female, the uterus lies on the superior surface. The two vasa deferentia and seminal vesicles lie on the posterior surface, which can be palpated through the rectum. The neck of the bladder is continuous with the urethra, and in the male it is embraced by the prostate gland. In the female it lies on the urogenital diaphragm. The lower part of the bladder is attached to the pubis via the pubovesical or puboprostatic ligaments.

As the bladder fills, the neck stays fixed and the upper wall and other surfaces become convex and it rises outside the pelvis. The empty bladder lies in the lower, anterior pelvis.

Histologically, the bladder has five coats: serous, fascial, muscular, submucous, and mucous. The serous coat is the outer peritoneal coat and is found only on the upper surface and part of the base (in the male). The fascial coat is beneath the peritoneum and sheaths the bladder. The muscular coat is strong, nonstriated, and arranged in interlacing bundles running in different directions (Fig. 14-7). These bundles pass obliquely and circularly around the bladder. At the bladder neck the bundles form a ring, the internal sphincter, which is continuous with the muscle wall of the urethra in the female and the muscle of the prostate in the male. The submucous coat consists of areolar tissue that lies between the muscle and mucous layers. The nerves and blood vessels ramify here before entering the mucosa, which is composed of transitional epithelium.

The mucosa of the bladder interior, when contracted and empty, is thrown into folds and appears wrinkled. When distended, the mucosa appears smooth. The trigone (Fig. 14-7) is a triangular area that occupies most of the inner surface of the posterior bladder wall. This area is thinner than the rest of the mucosa. The apex of the trigone is formed by the internal urethral opening, the base corresponding to a line that passes between the two ureteral orifices—the interureteric ridge. This ridge is mucous membrane raised by an underlying bar of muscle.

The bladder has three openings. Two are for the ureters, which constitute the inlets and appear as semilunar slits separated by the above-discussed ridge. The ureters pass obliquely for about 1.875 cm ($\frac{3}{4}$ in.) through the bladder wall, and in a sense this forms a valve that allows urine to pass into the bladder but prevents regurgitation as the bladder fills. The internal urethral orifice is at the trigone apex.

Vessels and Nerves. The arterial supply (Fig. 10-26, p. 364; Fig. 14-7) is predominantly from internal iliac artery via the superior and inferior vesicals and middle rectal. The veins form perivesical plexuses that drain into the inferior vesical veins and into the internal iliac. The lymph vessels of the anterior part pass to the external iliac glands, while from its posterior part they drain into the internal iliac glands.

The nerve supply (Fig. 14-7) follows the arteries and involves both sympathetic and parasympathetic nerves as well as voluntary control of the sphincters by way of the pudendal nerve. The sympathetic nerves (hypogastric plexus) are the "filling" nerves of the bladder since they inhibit the detrusor muscle that makes up the bladder wall. They cause increased tone in the internal sphincter by allowing the bladder to retain its contents. The parasympathetic nerves (pelvic splanchnic nerves) are the "emptying" nerves since they stimulate contraction of the detrusor muscle and relaxation of the internal sphincter. Then, by conscious relaxation of the external sphincter through the pudendal nerves, the urine is released.

Renal Physiology: The Kidney as a Dynamic Organ

The kidneys consist of two small organs, smaller than a man's head, that adjust the volume and composition of all extracellular fluids to the body's constantly changing needs. The kidney functions in two ways: there is a wholesale filtration process, which is then followed by a process of reabsorption of those materials necessary for the body's needs. These functions are carried on by about one million nephrons and depend on three major systems: (1) a complex of blood vessels, (2) a complex of tubes, and (3) an interstitium in the medulla.

The vessels in the cortex resemble a plexus and accommodate about 90 percent of the blood reaching the kidneys. The blood flows very rapidly in this area. The vessels in the medulla are arranged in loops and are meager, and the flow is relatively slow. With every heartbeat, one fourth of all the blood pumped by the heart passes through the renal artery, which in turn divides eventually into about a million arterioles.

The portion of the kidney confined to the cortex consists of Bowman's capsule and proximal and distal convoluted tubules. The parts in the medulla are Henle's loop and the collecting ducts.

In the medulla (Fig. 14-8) the concentration of salt and urea ranges from about 300 milliosmoles per liter (similar to blood) at the cortical boundary to about 1200 milliosmoles per liter in the papillary region. The interstitium is an essential portion of the mechanism for the production of a concentrated urine.

Each of the above-described systems is essential to the closely coordinated total body requirements. Therefore, in health, extracellular fluids are remarkably constant in both their composition and volume.

Blood enters the glomerulus at high pressure so that nonprotein components are forced through the capillaries into the cavity of the capsule. The total area of glomerular filtrate is about 1 sq m, and every day about 180 liters (190 qt) of filtrate are processed (equal to about two and one-half times the weight of an average man)—a huge volume that enables waste ex-

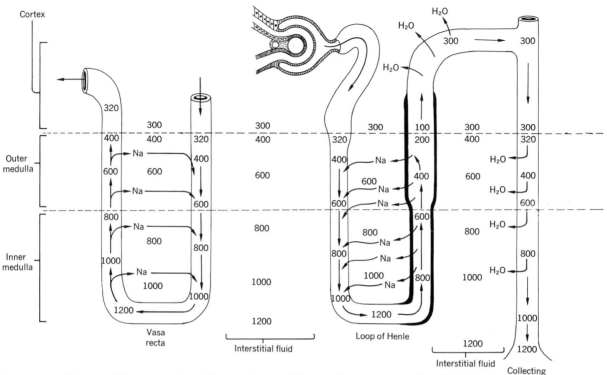

Figure 14-8. Diagram of the mechanism of the countercurrent theory of urine concentration. Flow in the descending limb is in the opposite direction of flow in the ascending limb of Henle. The ascending limb shows active sodium ion transport from lumen to surrounding medullary tissue but is impermeable to H_2O. This causes the medullary tissue to have a progressively higher concentration of sodium going from the convolutions to the bend of the loop of Henle. Thus, an osmotic gradient is created that will remove H_2O from the collecting duct (as shown). The numbers indicate the concentration of sodium in milliosmoles. (An osmole is the number of particles in solution that will depress the freezing point of water to $-1.86°$ C. One osmole is equal to 1000 milliosmoles.)

cretion and a homeostatic regulation of the internal environment.

The filtrate entering the capsule is like lymph, consisting of blood without red cells, platelets, or proteins. This first stage of excretion depends on the relatively high blood pressure, and thus the kidney is one of the first organs supplied from the aorta. Blood pressure is also maintained by the fact that blood not only leaves the glomerulus through a vessel of smaller diameter than the entering one but also has to pass through a circlet of muscle that adjusts the pressure in the glomerular capillaries. The renal arteries, too, can vary

in size to adjust to the hydrostatic pressure found, at a specific time, in the glomerular capillaries.

The arteriole leaving the glomerulus (efferent) winds around all parts of the tubule, into which it secretes several substances, while from the tubule reabsorption occurs into the arteriole. As much as 99 percent of the filtrate is reabsorbed, due primarily to the great total length of the kidney tubules—60 km (37 miles). Amino acids, sugars, fatty acids, and glycerol are totally reabsorbed by active transport, as are salts, depending on the amount of salts in the body. Water, on the other hand, is absorbed by osmosis, also in

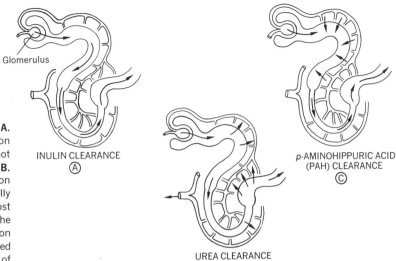

Figure 14-9. Renal function determinations. **A.** Inulin clearance measures glomerular filtration rate since inulin is normally filtered but not reabsorbed or excreted by the renal tubules. **B.** Urea clearance measures glomerular filtration and tubular reabsorption since it is normally filtered and partially reabsorbed. It is most useful in pathologic conditions that affect the glomeruli. **C.** PAH clearance measures filtration and tubular excretion since it is both filtered and excreted. It is useful as a measurement of renal plasma flow.

INULIN CLEARANCE
Ⓐ

UREA CLEARANCE
Ⓑ

p-AMINOHIPPURIC ACID
(PAH) CLEARANCE
Ⓒ

varying amounts. Potentially harmful substances like urea are excreted, and only 50 percent is reabsorbed. Of the substances actively secreted back into the tubules from the arterioles, one is creatinine, which is produced by a breakdown of worn-out muscle cells. Penicillin, too, is secreted in this manner and eliminated from the body.

The forces enabling filtration to take place from the glomerular capillaries into Bowman's capsule are hydrostatic pressure and diffusion, which each minute produce about 120 ml of an ultrafiltrate whose composition is very similar to that of blood plasma except that it is almost protein free (Fig. 14-8). The final urine that enters the renal pelvis is quite different from the glomerular one, because it is altered in its passage involving tubular reabsorption and secretion. The glomerular filtrate flows from Bowman's capsule into the very convoluted proximal tubule, where almost all of its glucose, amino acids, and other metabolites and much of its salts are reabsorbed into the bloodstream and general circulation by the copious blood supply of the cortex (Fig. 14-4). Since water passively accompanies these active transport processes, the filtrate volume is reduced by almost 80 percent. Since the average volume filtered is about 180 liters per day and the average person excretes about 1 to 2 liters of urine

per day, about 99 percent of the filtered water is reabsorbed into the peritubular capillaries. We measure this glomerular filtration by collecting the subject's urine over a 24-hour period and measuring a substance called inulin (a polysaccharide) in the urine, which is filtered by the glomerulus and not reabsorbed by the tubules (Fig. 14-9). Using a mathematical equation we can calculate the amount filtered by comparing the amount of inulin in the blood. This is called a clearance technique.

Although altered in composition and reduced in volume, the filtrate is still isotonic as it passes into the descending limb of Henle's loop and thus into the medullary portion of the kidney (Fig. 14-10).

The descending limb is surrounded by the interstitium of the medulla and, therefore, interstitial fluid, whose salt and urea concentration is progressively higher from cortex to papilla (300 to 1200 milliosmoles per liter). Furthermore, the descending limb is freely permeable to water. As the fluid inside the limb flows down the tubule, the osmotic pressure created by the salt and urea gradient causes the water in the limb to move out into the interstitium.

Thus, because of osmosis, the tubular (limb) fluid increases in concentration from 300 milliosmoles per liter at the cortex to 1200 milliosmoles per liter by the

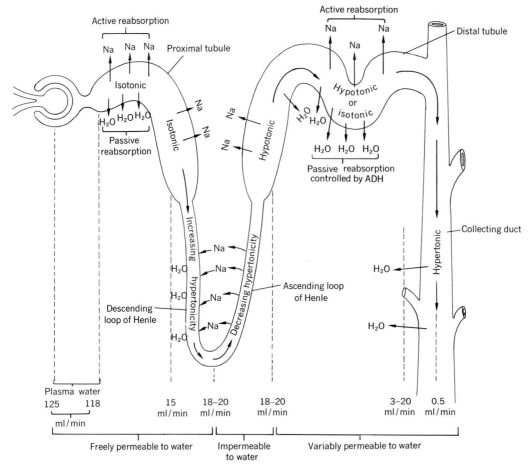

Figure 14-10. The functional organization of the nephron in relation to its reabsorption of sodium (Na) and water (H_2O) and the formation of a concentrated but dilute urine. (After studies by Pitts.)

time it reaches the turn of the loop in the thin segment.

When the greatly concentrated fluid makes the turn and enters the ascending limb, there also begins a countercurrent flow (flow against itself) and the fluid literally attempts to retrace its steps.

As the fluid continues to move up the ascending limb, the "sodium pump" transfers salt (actually sodium ions) out of the tubule into the medullary interstitium. Note that, in contrast to the descending limb, the ascending limb is impermeable to water (Fig. 14-10).

Because of the countercurrent flow phenomenon,

the salt concentrations at corresponding levels of tubular fluid and interstitium equalize or are roughly similar. Thus, "pumps" of limited capacity are sufficient to move sodium ions (Na^+) from the tubule into the interstitium and maintain a small concentration difference. The sum of these increments from papilla to cortex results in a concentration gradient whose range is far greater than could be produced by any one-step pumping mechanism known in biologic systems.

Vasa recta is the name given to the blood supply of the medulla. The loop arrangement about the tubules and the slow blood flow permit the removal of both salt and water at a rate that keeps them from accumulating, but they do not destroy the osmotic gradient (Fig. 14-8).

By the time the fluid in the ascending loop reaches the distal tubule it is relatively hypotonic (has a decreased osmotic pressure). It is here that it is finally restored to isotonicity.

In the distal convoluted tubule, active sodium transport continues and is controlled by the juxtaglomerular apparatus (Fig. 14-5) in the presence of an antidiuretic hormone (ADH) from the pituitary gland, which controls the permeability of the distal tubule to water. The sodium is accompanied by passive water loss. In addition, excretion of potassium (K^+) occurs here. Note that H^+ is secreted throughout the system.

ADH increases water reabsorption in the distal tubule and collecting duct, although normally 85 percent of it has been absorbed as it passes from proximal to distal tubule. NaCl is reabsorbed to the extent of seven eighths in the proximal tubule and one eighth in the loop of Henle, and distal tubule aldosterone aids this absorption. Other salt-regulating hormones include the cortisols, which assist in reabsorption of Na^+ and depress that of K^+. The parathyroid hormone decreases reabsorption of phosphates and calcium. If the level of salt in the tissue fluid falls, the kidney's juxtaglomerular cells secrete renin (Fig. 14-11), which is an enzyme that acts on the substance in the plasma called angiotensinogen to form angiotensin. The latter then stimulates aldosterone secretion by the adrenal cortex. Aldosterone influences sodium retention by the kidney tubule to overcome the salt loss in the tissues.

The renal blood pressure normally stays fairly constant. However, during shock it may alter due to epinephrine, causing constriction of the arterioles going to the glomerulus.

When the distal tubule fluid, now finally restored to isotonicity, crosses the corticomedullary boundary, it enters the system of collecting ducts. The collecting ducts are confined to the medulla; hence the fluid is also exposed to the range of osmotic forces from cortex to medulla, previously mentioned (Figs. 14-8, 14-10). In the presence of ADH, which in the collecting ducts also controls water permeability, water is lost to the

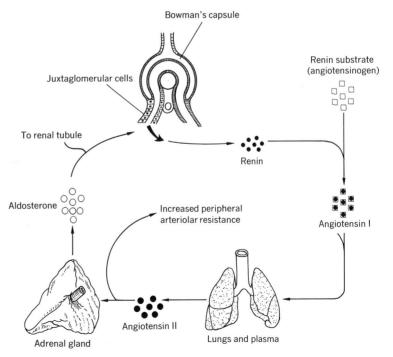

Figure 14-11. The biochemical pathway from renin to aldosterone. When renin is secreted by the juxtaglomerular cells into the circulation, it acts (enzymatically) on renin substrate to produce angiotensin I. The latter is converted to angiotensin II in the lungs and to a degree in the plasma. Angiotensin II acts on the adrenal cortex to cause the secretion of aldosterone. It also acts on the peripheral arterioles to increase peripheral resistance. The two mechanisms help to maintain the circulation, since increased resistance and aldosterone-mediated sodium and water retention are activated by this process. (After studies by Davis.)

Table 14-1. Some Major Urinary Constituents

	Grams per 24 Hours
Water	600–2500 ml, 600–2500 gm
Urea	33.0
Uric acid	0.6
Creatinine	1.0
Sulfate as H_2SO_4	2.0
Phosphate	1.7
Chlorine	7.0
Ammonia	0.7
Potassium	2.5
Sodium	6.0
Calcium	0.2
Magnesium	0.2
Glucose	0–trace

medullary interstitium by osmosis. This produces a urine many times more concentrated than the plasma from which it originated (Table 14-1).

Thus, the kidneys respond rapidly and with precision to the many alterations that take place in the internal environment of the body and maintain this inner world remarkably constant in both health and disease. The regulatory processes are complex but essential, for the kidneys not only function as organs of waste disposal but also are regulators of the composition of blood and, via the blood, the body tissue fluids.

Urination

Urine flows to the bladder via the ureters, propelled by peristaltic contractions of the smooth muscle of the ureteral wall (Fig. 14-7). The bladder, as noted, has walls of thick, smooth muscle layers and on contraction squeezes inward, increasing the pressure of the urine. At the base of the bladder is a muscle layer often spoken of as the first part of the urethra and sometimes referred to as the internal urethral sphincter. It is not distinct, and in reality the last part of the bladder itself (when relaxed) functions as a sphincter. When the bladder contracts, this "sphincter" is pulled open due to a change in bladder shape.

The urethra, too, is surrounded in part by a circular layer of skeletal muscle. When contracted, it can hold the urethra closed and thus has been called the external urethral sphincter.

Urination (micturition) is a spinal reflex but can be influenced by the brain. The bladder is well supplied by parasympathetic fibers that cause it to contract. The external sphincter, on the other hand, is innervated by somatic motor nerves that can be voluntarily controlled. Many stretch receptors from the bladder wall enter the spinal cord and synapse with the parasympathetic and somatic neurons to stimulate the former and inhibit the latter. Higher brain centers both facilitate and inhibit these motor pathways. In the adult one needs about 300 ml of urine in the bladder to initiate the reflex of bladder contraction.

Other Kidney Functions

Osmoregulation

Osmoregulation (maintenance of the osmotic pressure of body fluids) is even more important than excretion, and survival depends on it. The osmotic pressure depends on two variables: the amount of water and the concentration of dissolved substances (particularly salts).

Loss of water must be balanced against intake (see discussion of water later in this chapter). A water shortage increases the osmotic pressure of body fluids, whereupon this is registered in the hypothalamus and we begin to feel thirsty. After we drink, the kidneys serve to adjust what water we retain or excrete and the correct amount remains in the tissue fluids and in the blood.

The salt concentration in the body fluids is the other major factor in osmoregulation. Beside helping to maintain osmotic pressure, some salts are needed for bone building and some for coenzymes. Many metabolic body processes work only in the presence of exact ion concentrations of salts such as potassium or magnesium; the heart, for example, will stop with too great an increase of potassium concentration in the blood, and a small increase of magnesium puts one to sleep. The balance of salt ions is essential, and ions of one kind can be toxic unless counteracted by the correct amount of ions of another kind. The kidneys operate continuously to maintain the total salt concentration (Fig. 14-12).

Acid-Base Balance

The kidneys also regulate the acid-base balance of the blood, which must be controlled within very nar-

Figure 14-12. Sodium and water balance as well as the various factors and organs involved in their regulation.

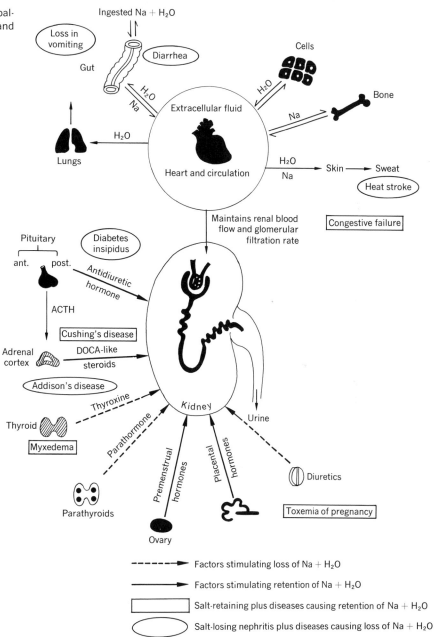

row limits if cells are to survive. An alkaline or basic substance is one that tends to counterbalance the effects of an acid, but not always. Acidity is due to the presence of hydrogen ions continuously being produced in the body. Any excess hydrogen ions are removed by the kidney and the urine becomes acid, but this permits the blood to maintain its slight alkalinity.

The autonomic nerves partially help to regulate kidney function. During certain states of shock, the blood may be shunted away from the kidneys by nerves constricting the afferent arterioles of the glomerulus. Epinephrine has the same effect. However, the kidney normally functions by filtering a constant amount of blood and then varying the amount reabsorbed, and the major influence on the kidney is hormonal. ADH from the posterior pituitary gland regulates water reabsorption; aldosterone from the adrenal gland regulates salt absorption. The secretion of both of these is controlled by the hypothalamus, which has receptors sensitive to the degree of blood dilution. Similar receptors undoubtedly exist elsewhere in the body. Other hormones also affect the kidneys; thyroxine and growth hormone have their effects, while the parathyroid glands influence excretion of calcium and phosphates.

Metabolism

The kidney, usually regarded as an excretory organ, is also very important in metabolism. In addition to the usual enzymes of oxidative metabolism, it contains high levels of amino acids and amine oxidases responsible for free ammonia formation. The latter is excreted as NH_4^+ in urine.

Figure 14-13. In cardiac failure there is decreased renal perfusion pressure and renal blood flow **(1)**. This initiates renin release **(2)** and activates the angiotensin-aldosterone system **(3, 4,** and **5)**. In addition to direct effects on the kidney, alterations are present in circulating dynamics in the liver **(6** and **7)** and periphery **(8)**. Thus, there is increased sodium and water retention, leading to edema and ascites **(9)**.

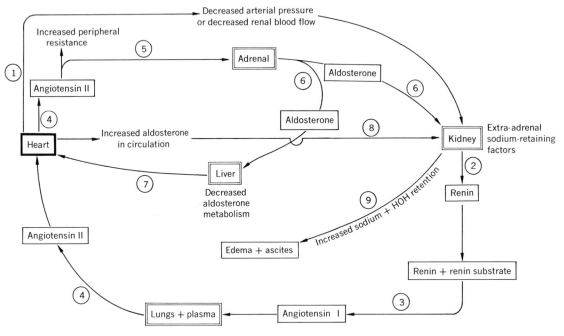

The major metabolic waste product excreted in urine is urea. Ammonium ion, uric acid, and others may be present (Table 14-1). Certain materials signify diseases: Protein in urine is invariably associated with kidney damage and sugar with diabetes (level of blood sugar overloads reabsorption system).

Renal Pathology

Pyelonephritis, Glomerulonephritis, and Others

Pyelonephritis is a kidney infection caused by bacteria. Glomerulonephritis, a relatively common renal disease, may result secondarily from an allergic response in a throat infection caused by specific bacteria (mostly streptococci). Congenital defects, kidney stones, tumors, and toxic chemicals all alter the kidney and create damage. Obstruction of the terminal excretory ducts (ureter or urethra) may cause injury as a result of pressure built up behind the obstruction. It may also predispose to infection.

Uremia

The end stage of a progressive disease usually results in a nonfunctioning kidney (Fig. 14-13). The symptoms and signs of renal malfunction are often independent of the causative agent and are collectively known as uremia (urine in the blood). The latter may result as follows: Kidney destruction markedly reduces glomerular filtration rate (assuming a normal diet). The substances that reach the tubule by filtration are filtered in smaller amounts. In addition, the excretion of potassium, hydrogen ions, and other substances is impaired since there is diminished renal tubular secretion in a diseased kidney (Fig. 14-14). The buildup of all these materials (sodium, chloride, water, calcium, wastes, potassium, hydrogen ion, etc.) in the blood causes the symptoms and signs of uremia.

The artificial kidney is an apparatus used to eliminate the excess ions and wastes that accumulate in the blood when the kidneys fail. The other hope for patients with permanent renal failure is kidney transplantation.

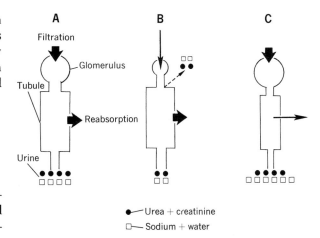

Figure 14-14. Glomerulotubular balance. **A.** Normal, where a balance between filtration and reabsorption is present. **B.** Glomerular disease, such as acute glomerulonephritis, diabetic glomerulosclerosis, or amyloid nephropathy. There is a decrease in water and solute filtered, but tubular reabsorption is unaltered. Thus, less urea, creatinine, sodium, and water are excreted. **C.** Tubular disease, such as pyelonephritis, polycystic kidney disease, and hydronephrosis. The relatively large filtered load exceeds the reduced reabsorptive ability of the damaged tubules, and excess sodium and water are lost in the urine.

Water

Distribution of Water

Most of the body consists of water, and most body processes occur in the aqueous environment (Fig. 14-15). Nothing can live without water. Water is always moving in and out of the cells. All living things appear to require about the same concentration of water. In the normal adult human of average weight, water accounts for 45 to 75 percent of body weight (varies with the amount of fat; fat is essentially water free, as the leaner the individual, the greater the proportion of water to total body weight). Another water trait that all animals have in common is that none of us can tolerate a body salt concentration greater than 0.9 percent. For life to persist, this metabolic equilibrium is essential, since no matter where the animal is or how far removed from his normal habitat he main-

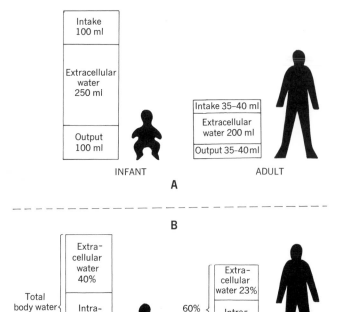

Figure 14-15. **A.** Estimate of daily water exchange in infant and adult (milliliters per kilogram of body weight). **B.** Total body water in infant and adult (percent of body weight) and major distribution.

tains this crucial water balance. Water in the cells (intracellular water) is usually 30 to 40 percent of body weight. Water in the spaces between the cells (extracellular water) consists of blood plasma (4.5 percent of body weight), interstitial fluid that bathes the cells (16 percent of body weight), and lymph (about 2 percent of body weight). Thus, extracellular water totals about 22.5 percent of body weight. Until 12 months of age the infant has the highest proportion of body water (Fig. 14-15). In young adults the average is 63 percent for males and 52 percent for females, with a decrease in total body water in the elderly.

Water Loss

Water makes its way out of the body through the kidneys, lungs, skin, and gastrointestinal tract (Fig. 14-2). In the normal adult, the loss from the gastro-

intestinal tract is very small. The average loss per day through the kidneys is about 1500 ml and through the lungs and skin (insensible loss) about 1000 ml, amounting to a total of 2500 ml per day. The loss through the lungs and skin increases when there is fever, an increase in respiratory rate, a hot and dry environment, and an injury to the skin (e.g., burns). The loss by the kidneys varies (1) with solute load (an increase occurs in diabetes mellitus or after ingestion of large amounts of food), which obligates the kidneys to excrete enough urine to carry the solutes into the bladder, and (2) with levels of antidiuretic hormone (ADH) from the pituitary gland. ADH controls the reabsorption of water in the kidney's distal convoluted tubules. Increased ADH results in increased water reabsorption and a more concentrated urine.

Water Replacement

Body water is normally replaced (1) by ingestion of liquids (about 1000 ml), supplemented by the water content of meat, fruit, and vegetables (food about 1200 ml), which contain 60 to 97 percent water; and (2) by metabolism or combustion of foodstuffs, which yields water on oxidation (about 300 ml). Metabolism of every 100 kilocalories of fat, carbohyrate, or protein releases about 14 ml of water. Thus, intake is about 2500 ml.

Water Balance

Water balance is determined best by daily body weight. A weight change of 1 kg (2.2 lb) is equal to a fluid gain or loss of about 1000 ml (1 liter).

Nature, Distribution, and Measurement of Electrolytes

Chemical compounds in a solution may react in one of two ways. They may remain intact (undissociated) and are called nonelectrolytes. Examples of these in the body water are urea, dextrose, and creatinine. On the other hand, they may dissociate in solution to form ions. This process is called ionization, and the compounds are referred to as electrolytes. Ions, which are the dissociated particles of an electrolyte, carry an electrical charge. For example, sodium chloride (NaCl) dissolved in water provides sodium ions (Na^+) and chloride ions (Cl^-) carrying positive ($+$) and negative ($-$) charges, respectively.

Faraday was the first to use the term *electrolyte* and demonstrated that, if electrolytes are placed in a solution, an electric current is carried through the solution by electrically charged atoms or molecules. These molecules were called ions (Greek word meaning going). Those ions carrying a positive charge migrated downward to Faraday's negative electrode or cathode and were thus called cations (Greek *kation*, down), while those with a negative charge migrated upward to Faraday's positive electrode and were called anions (Greek *anion*, up). We find electrolytes in all body fluids (Table 14-2).

The major cations in the body are sodium (Na^+), potassium (K^+), calcium (Ca^{++}), and magnesium (Mg^{++}). Sodium and potassium are the predominant cations. This is true of virtually all living cells. The two are kept apart by semipermeable membranes but,

Table 14-2. Electrolyte Composition of Body Fluids

		Intracellular	Extracellular		Interstitial
			Vascular		
		mEq/liter H_2O	mg/liter plasma	mEq/liter plasma	mEq/liter fluid
Cations	Na^+	10	3260	142 (137–147)	145
	K^+	150	200	5 (4–6)	4
	Ca^{++}	—	100	5 (4–6)	2.5
	Mg^{++}	40	36	3 (1–4)	2
Total cations		200	3600	155	153.5
Anions	Cl^-	0–3	3657	103 (98–106)	114
	HCO_3^-	10	1647	27 (25–29)	30
	Phosphate	150	106	2 (1.7–2.6)	2
	SO_4^-	—	16	1 (0.2–1.3)	1
	Organic acid anions	—	175	6 (4–8)	5.5
	Proteins	40	65,000	16 (15–20)	1
Total anions		200	72,000	155	153.5
Total mEq/liter fluid		~400		~310	
Total millimoles particles/liter fluid		~310		~310	~310

most important, by active secretory processes. Sodium is the chief extracellular cation and potassium the major cation of the intracellular space. This exclusion of sodium from the cell is due to some form of active transport that selectively removes the cations against a concentration gradient.

The most important anions are chloride (Cl^-), bicarbonate (HCO_3^-), phosphate (HPO_4^{--}), and sulfate (SO_4^{--}). Chloride and bicarbonate predominate extracellularly, while in the cell one finds phosphate, sulfate, and bicarbonate (Table 14-2). Chloride is almost entirely absent from within the cell.

Normally, the volume and osmolality of the extracellular fluid are maintained within very narrow limits. Since sodium and its associated anions form almost all of the osmotically active solute of the extracellular fluid, the control of these particular ions as well as water is a very essential mechanism to the organism. None of us can tolerate a body salt concentration greater than 0.9 percent, and thus this metabolic equilibrium is essential.

The chemical and physiologic activity of electrolytes are proportional to (1) the number of particles found per unit volume (moles or millimoles) and (2) the number of electrical charges per unit volume (equivalents or milliequivalents per liter). One more frequently uses milliequivalents than milligrams per 100 ml, since the latter gives no direct information as to the number of ions or electrical charges but merely the weight of the electrolytes per unit volume.

The electrolytes of body fluids are expressed in terms of chemical activity or "equivalents," with the suitable term being milliequivalents per liter. The number of milliequivalents per liter (mEq/L) is derived as follows (Table 14-4):

$$mEq/L = \frac{mg/100 \text{ ml} \times 10 \times \text{valence}}{\text{atomic or molecular weight}}$$

In a healthy person, the number of milliequivalents per liter for plasma electrolytes will vary only in a very narrow range (Table 14-4).

Calculations are simple. For example, there are 10 mg of calcium in 100 ml or 100 mg per liter of

Table 14-3. Atomic Weights and Balances of Important Ions

Material	Electrolyte	Valence	Atomic Weight
Sodium	Na^+	1	23
Potassium	K^+	1	39
Magnesium	Mg^{++}	2	24
Calcium	Ca^{++}	2	40
Chloride	Cl^-	1	35.5
Sulfate	SO_4^{--}	2	32
Phosphate	HPO_4^{--}(80%)	1.8	31
	H_2PO_4 (20%)	1.8	31

Table 14-4. The Ranges of Normal Values of Important Ions in Milliequivalents per Liter (mEq/L)

Sodium	136–145
Potassium	3.5–50
Calcium	4.3–5.3
Magnesium	1.5–2.5
Chloride	100–106
Phosphates	2.6–3.2
Bicarbonate	21–32

plasma (10 mg/100 ml $\times$ 10 = 100 mg/1 liter). Calcium has an atomic weight of 40, and its valence is 2. Thus, using the above formula:

$$\frac{100 \text{ mg} \times 2}{40} = 5 \text{ mEq/L of calcium}$$

Calculate the milliequivalents of sodium or chloride per liter in a 0.9 percent (0.9 gm/100 ml) solution of sodium chloride.

$$\frac{900 \text{ (mg/100 ml)} \times 10 \times 1 \text{ (valence)}}{23 + 35.5 = 58.5 \text{ (molecular weight of NaCl)}}$$
$$= 154 \text{ mEq/L of sodium or of chloride}$$

Since milliequivalents measure the number of electrovalent bonds present in solution, 100 mEq of cations are always balanced by the exact number of milliequivalents of anions with which the cation is combined.

The intervascular plasma and interstitial fluids are grouped as extracellular fluid. The major difference between plasma fluid and interstitial fluid is the higher concentration of protein in plasma (they are otherwise very similar). Cellular fluid (intracellular) is characterized by a high concentration of potassium and phosphate and much protein. (See Fig. 14-15 for average normal values of fluid in the various body compartments.)

Water is not an inert carrier of biologic materials, but has a controlling influence not only on biochemical materials but also on the molecular form assumed by proteins and other biopolymers as well as on cellular architecture. The maintenance of the correct distribution of water and electrolytes between cells, tissue spaces, and plasma is essential for body function. The metabolism of the cell is highly sensitive to changes in volume as well as the composition of the fluid in the cell. These, in turn, depend on the extracellular fluid, which resembles primordial dilute seawater. Thus, the ability of higher animals to control the composition of the extracellular fluid is a great asset and an important evolutionary advance, for this permits man to become independent of his environment.

The lungs allow gaseous exchange with the external environment; the kidneys regulate the water and electrolyte composition with the external environment. The latter regulate the water and electrolyte composition of the intravascular fluid (blood) directly and the interstitial and intracellular fluids indirectly. Nutrients and metabolites exchange between cells and intravascular fluids by diffusion through the interstitial fluids, so that the fluid environment of the cells remains relatively unaltered. Routine laboratory procedures, however, measure only electrolytes in the blood; they do not give a picture of measurements in the cellular space.

The distribution of water between blood and interstitial fluid is governed by the balance between blood pressure, which tends to force liquid into the tissue spaces, and plasma osmotic pressure, which tends to withdraw fluid from the interstitial spaces. At the arterial end of the capillary, blood pressure in the vessel is greater than osmotic pressure and drives water into the tissues. At the venous end of the capillary, osmotic pressure is greater than blood pressure and fluid is drawn into the vessel (Fig. 14-16).

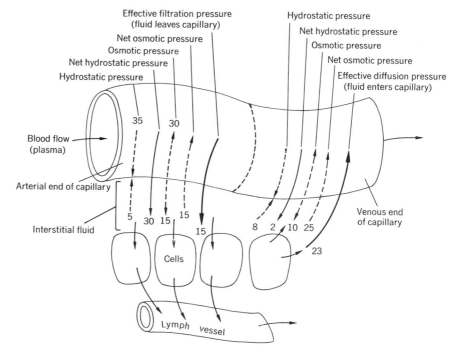

Figure 14-16. Diagram showing how substances pass between capillaries and cells. Blood pressure at arterial end of capillary tends to force some blood out of capillaries, while the osmotic pressure tends to suck fluid back. Since blood pressure is higher than osmotic pressure, fluid (lymph) and electrolytes (and even some protein) are forced into interstitial spaces and O_2 can diffuse into cells. At the venous end of the capillary, blood pressure has fallen while osmotic pressure is relatively stable. Since osmotic pressure is now higher, some lymph passes back into the capillaries together with waste CO_2 and nitrogenous matter that has diffused out of the cells. The excess is drained into the lymph vessels.

The distribution of water between blood and interstitial fluid is also very important, since movement of too much water from the plasma to the tissues can cause a flooding of those tissues, creating edema, or a reverse situation can create dehydration. However, because cell membranes are freely permeable to water, no permanent difference in water concentration normally occurs between the various body spaces.

The amount of water needed by the body is generally determined by the quantity required to maintain the volume and osmotic pressure of the body fluids and to replace water lost by excretion through kidney, skin, and lungs.

If there is water deprivation, the primary deficiency is seen in the plasma, which is directly connected with the excretory channels. As the plasma concentrates and its osmotic pressure rises, water is withdrawn from interstitial fluid of tissues. Water, in turn, is lost from the cell to the interstitial fluid. With loss of cellular water (cell dehydration), potassium ions are lost to the extracellular fluid and excreted in urine. If a person loses more than 20 percent of body water, he usually does not survive. Sodium ions also play an important role in osmotic pressure of extracellular fluid. A drop in sodium ions reduces plasma volume, and this cannot be corrected by water alone but needs additional ions.

When one drinks a great deal of water, it is absorbed from the gastrointestinal tract into the blood and from there spreads into the extracellular fluid, making the latter area lower in osmolarity than the cells it bathes. Water then diffuses into the cells until the osmolarity is equal, even though it is lower than normal. On the other hand, if one were to take in a fair amount of isotonic sodium chloride (a salt solution equal to the osmolarity of the extracellular fluid), this too would be taken up by the gut and distributed to the extracellular fluid. However, in this case since only the volume had been altered and not the osmolarity, the water would not move and would remain in the extracellular space as long as the salt also stayed there. Thus, it is evident that the total extracellular fluid volume really depends on the amount of extracellular sodium found there. Conversely, when the extracellular volume is decreased, bone liberates sodium

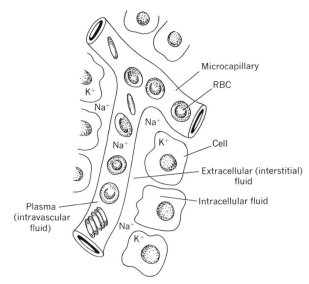

Figure 14-17. Extracellular fluid is the "vehicle" for exchange of nutrients between the microcapillaries and the cells. Sodium (Na^+) is the major cation of the interstitial fluid and plasma; potassium (K^+) is the major intracellular cation.

into the extracellular fluid to help maintain its volume and this sodium helps in pulling water out of the cells (Fig. 14-17). Note, however, that these movements of sodium and water do not eliminate excesses or deficiencies but merely distribute them around temporarily. Long-term regulation requires the control of total body sodium and water, and the most important mechanism for the regulation of extracellular volume and osmolarity constitutes the renal excretion of both.

Understanding Osmotic Pressure to Understand Water Balance

Most of the membranes of the body are semipermeable membranes that permit free passage of water molecules and many other uncharged molecules, but partially or completely prevent the easy passage of large molecules and charged ions.

Suppose a solution containing a large number of nonpermeable dissolved particles is placed on one side of a semipermeable membrane and another solution containing a relatively small number of nonpermeable dissolved particles is put on the other side. A difference in pressure develops between the two solutions on either side of the membrane, and only water will pass across the semipermeable membrane. The water will move from the less concentrated to the more concentrated solution until the concentration of dissolved particles is equal on both sides.

If the solution having the greatest concentration of solute is enclosed and connected to a vertical capillary, the pressure exerted is indicated by a rise of liquid in the tube. If one completely closes the chamber, the pressure is shown by an increase in tension upon the membrane or case enclosing it. This pressure is called osmotic pressure.

Thus, osmotic pressure reflects an effort by the system to eliminate the unfavorable state of nonuniform solute concentration. The solvent (usually water) thus diffuses across the membrane until there is equilibrium. Since the concentration of a solvent can never be precisely the same, this condition can only be attained if the free energy change resulting from this concentration difference is compensated for by the work done against a pressure difference. Thus, the osmolality of a fluid is a measure of the concentration of individual solute particles dissolved in it. Osmotic pressure depends only upon the number of solute particles in unit volume, regardless of size. This is an inverse measure of water concentration, since the higher the osmolality, the lower the water concentration. To repeat, the milliequivalent unit measures the chemical and physiologic activity of an electrolyte and is based on the number of ionic charges present in an electrolyte solution. The unit of measurement of osmotic activity of a solution is the milliosmole (mOsm), which is a measure of the amount of work that dissolved particles can do in drawing fluid through a semipermeable membrane. Thus, osmotic activity depends on the number of actual particles in solution, regardless of what charge they may carry (Fig. 14-18). Electrolytes in either nonionized or ionized forms or even nonionizable substances, such as glucose, urea, and creatinine, exert an osmotic effect. Sodium and chloride exert the major osmotic forces in the extracellular fluid.

Serum osmotic pressure is primarily related to the concentration of sodium bicarbonate and chloride,

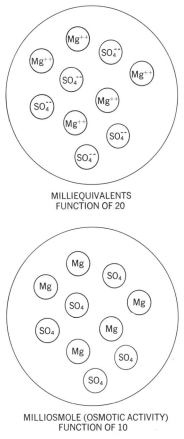

MILLIEQUIVALENTS
FUNCTION OF 20

MILLIOSMOLE (OSMOTIC ACTIVITY)
FUNCTION OF 10

Figure 14-18. The total number of available electrovalent bonds is the chemical activity, measured in milliequivalents (mEq). The total number of particles present is the osmotic activity, measured in milliosmoles.

with sodium in the dominant role. There is little osmotic effect from protein in the serum; however, high levels of urea or glucose can elevate serum osmotic pressure, with glucose being more effective.

A great share of the body's store of electrolytes and water serves in the secretions of the digestive tract (Fig. 14-19). In a healthy person, the water and electrolytes in these various secretions are not lost from the body (except for a small amount via feces or a larger amount during diarrhea or vomiting) but are ultimately reabsorbed.

	Minimum Need	Maximum Tolerance (without need to adapt)	Usual Dietary Range
Water Liters per 24 hours	2	20	2.5–4
Sodium (Na) mEq per 24 hours	15	400	90–250
Potassium (K) mEq per 24 hours	30	400	80–200
Magnesium (Mg) mEq per 24 hours	5	400	15–30

Figure 14-19. Twenty-four-hour needs and variations of water, sodium, potassium, and magnesium in a 70-kg healthy male.

Osmoreceptors are found in the hypothalamus and respond to changes in extracellular osmolality (lower or higher water concentration). The hypothalamic cells receive neural input, which regulates their rate of secretion of antidiuretic hormone (ADH).

Two afferent (sensory) pathways control hypothalamic ADH secretion. One responds to baroreceptors (pressure receptors) located in some major vessels and sensitive to stretching of the wall, and the other responds to osmoreceptors located in the hypothalamus. The latter are tiny vesicles that swell and shrink depending on the osmolarity of the fluid around them. A high extracellular osmolarity (i.e., lower water concentration) draws water out of the vesicles and they shrink. The shrinking stimulates the hypothalamic cells that secrete ADH and they increase their rate of secretion. Conversely swelling inhibits the cells. Thus, a simultaneous increase in extracellular volume and decrease in extracellular osmolarity cause maximum ADH inhibition (the converse is also true). Alcohol is a powerful inhibitor of ADH release and may account for the large amount of urine flow.

Diabetes Insipidus. Diabetes insipidus is characterized by the constant excretion of huge volumes of very dilute urine. ADH administration helps enormously. The disease is probably due to the inability to produce ADH as a result of hypothalamic damage. This results in poor tubular permeability to water, regardless of extracellular volume or osmolarity.

Thirst. Thirst is stimulated both by lower extracellular volume and by a higher plasma osmolarity. These circumstances stimulate ADH production. The thirst centers, too, are located in the hypothalamus. One speculates that the osmoreceptors and baroreceptors that initiate ADH-controlling reflexes may also affect the nearby thirst centers.

Electrolyte Balance

The various body fluids tend to maintain osmotic equilibrium despite a varied electrolyte distribution. The regulation of volume and ionic composition of extracellular fluid is necessary for control of intracellular fluid. The central nervous system is especially sensitive to changes in intracellular fluid, and death can occur from too little or too much water.

Sodium chloride and water are freely filterable at the glomerulus and undergo (as previously described) tubular reabsorption but no tubular secretion. The reabsorption is an active process and requires an energy supply. Chloride, on the other hand, is reabsorbed by passive diffusion and depends on the movement of sodium. Water is also moved by passive diffusion or osmosis, and it, too, depends on the movement of sodium. Thus, sodium plays the key role.

Sodium. As the major cation of the extracellular fluids, sodium is very essential in fluid balance and also is important in the therapy of fluid and electrolyte disturbances. Changes in sodium concentration stimulate an appropriate response by the pituitary gland via ADH. Essential facts to know are as follows: (1) 1 mEq of sodium weighs 23 mg; (2) a normal 70-kg (154-lb) man has a total body content of exchangeable sodium of about 2700 to 3800 mEq; (3) about 200 mEq per liter is outside the cells, with the concentration in the extracellular fluid ranging from 136 to 148 mEq per liter; (4) there is only a small amount of sodium in the cell itself, about 10 mEq per liter of intracellular water; (5) bone contains as much as 1500 mEq of sodium, but only about one third of this is available for exchange with sodium in other parts of the body; and (6) normally most of the ingested sodium is found in the urine. Since the kidney can conserve sodium, normal requirements are maintained by the average diet, which usually has at least 100 mEq per day. Sodium balance, however, can be maintained on an intake of as little as 12 to 15 mEq per day.

SODIUM EXCRETION AND CIRCULATION. Under normal conditions, sodium loss from the skin is negligible. As temperatures rise, with fever and with muscular exercise, sodium loss can be marked. The concentration of sodium in sweat is usually about 25 to 30 mEq per liter (it may be as high as 100). Sodium loss in sweat is actually a product of temperature regulation. It is essential to maintain normal body temperature even though a loss of sodium is undesirable from the standpoint of electrolyte balance. Under adverse conditions, one may lose a full intake of sodium in sweating within six to eight hours.

As stated above, under conditions of normal health and moderate environmental temperature, the kidney is the major regulator of sodium. Total urinary excretion of sodium per 24 hours usually equals sodium intake and varies with the dietary intake (Fig. 14-19). When the body must conserve sodium, the kidney is able to reabsorb it and excrete a relatively sodium-free urine.

Sodium balance is influenced by a hormone produced by the adrenal cortex called aldosterone, which is the most potent, naturally occurring inhibitor of renal sodium excretion since it stimulates tubular sodium reabsorption (Fig. 14-20). The antidiuretic hormone of the pituitary (ADH) promotes water reabsorption in the distal kidney tubules and, therefore, also has an effect on the extracellular sodium concentration.

Aldosterone secretion is controlled by reflexes originating in the kidney itself. Cells lining the arterioles in the kidney synthesize and secrete a protein into the blood called renin, which is an enzyme that catalyzes the reaction in which angiotensinogen (a large plasma protein) is broken down into smaller polypeptides called angiotensin. Renin is the "rate-limiting factor" in angiotensin formation, and this in turn depends on the rate of its secretion by the kidneys. Angiotensin stimulates aldosterone secretion and thus controls this hormone's release from the adrenal cortex (Fig. 14-11).

Angiotensinogen is synthesized in the liver and is always found in the blood. The cells that secrete renin are intimately associated with the epithelial cells lining one portion of the tubules and somehow detect the

Figure 14-20. Effects of aldosterone on sodium and potassium reabsorption and secretion.

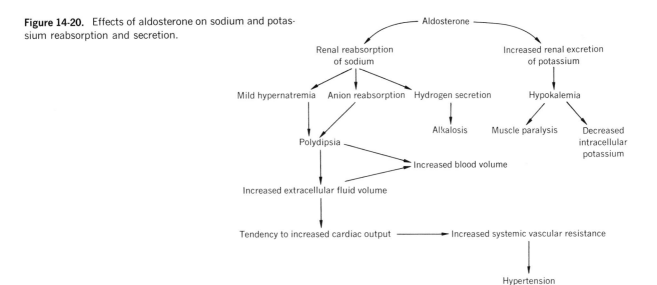

amount of sodium in the tubule and regulate their secretion of renin accordingly (in inverse proportion to the sodium amount). This forms a feedback system, so that any reduction in extracellular volume reduces intratubular sodium by reducing the glomerular filtration rate, which sets off the chain of hormonal events to restore normal volume. An expansion of extracellular volume, on the other hand, lowers renin, angiotensin, and aldosterone levels and results in decreased sodium reabsorption and increased sodium excretion. Other hormones besides aldosterone may serve a similar purpose.

Changes in sodium excretion must be accompanied by similar changes in water excretion. As mentioned previously, the ability of water to follow sodium depends on ADH. Thus, a decreased extracellular volume reflexly creates increased ADH production as well as increased aldosterone secretion. This mechanism is due to a group of hypothalamic neurons that end in the posterior pituitary, where ADH is released into the blood. The hypothalamic cells receive sensory input from vascular baroreceptors, which have been stimulated by varying atrial blood pressures. Increased atrial pressure inhibits ADH production (the converse is also true).

Not only does sodium pass in and out of the body in a continuous and well-regulated balance, but also it is maintained in a steady, constant internal circulation. Large amounts of this ion are produced in the gastrointestinal tract—almost 8200 ml per 24 hours. Sodium moves rapidly between the vascular and interstitial compartments, an exchange closely related to the fluid exchange between the two compartments, but enters the cells from extracellular fluid fairly slowly. Yet the concentration of sodium in the cells can quickly change as a result of rapid movement of water in and out of the cell.

If there is interference with reabsorption of intestinal secretions, marked alterations in fluid and electrolyte balance can be expected. In the case of gastrointestinal function, the most common disturbance is diarrhea and vomiting. Even though the gastric juice has a low sodium content, the gastric mucus has a high sodium concentration. Vomiting due to irritation can induce increased production of mucus, and, therefore, much sodium can be lost. As the total daily amount of all intestinal secretions is about twice that of plasma volume, loss from vomiting or abnormal stools (diarrhea), together with a reduced intake, can lead to dehydration and a sodium deficit. Loss of such water and sodium from the plasma results in a transfer of sodium and water from the interstitial fluid to attempt to replace the loss. If this continues, water is drawn from the cells, increasing dehydration; if the plasma volume is not sustained, the general circulation fails.

Potassium. Potassium is the major cation of the intracellular fluids, and thus its balance plays an essential role in body function as well as in fluid and electrolyte therapy. Since potassium plays an essential role in the excitability of nerves and muscle, its regulation is closely monitored and very important. Membrane potentials depend on the ratio of intracellular and extracellular potassium concentration. Therefore, if one raises the extracellular concentration, the resting-membrane potential is lowered and this increases all excitability. On the other hand, lowering extracellular potassium reduces excitability. Essential facts to remember with reference to this cation are as follows: (1) 1 mEq of potassium weighs 39 gm; (2) a normal 70-kg (154-lb) man has a total body content of exchangeable potassium of about 3200 mEq (about 2300 mEq in women); (3) most of the potassium (about 2300 mEq) of the body is inside the cell (the actual intracellular concentration is about 125 mEq per liter); (4) in a healthy adult, the serum potassium ranges between 3.5 and 5.0 mEq per liter; (5) the daily intake and output are well balanced, and the normal turnover per day is about 50 to 150 mEq or about 1.5 to 5 percent of the total body potassium content; (6) the average diet normally meets daily requirements and amounts to 80 to 200 mEq per day; and (7) the actual minimum daily need is about 30 mEq, whereas one can tolerate, under an acute load, a maximum of about 400 mEq (only for a short time).

POTASSIUM EXCRETION AND CIRCULATION. Potassium is predominantly an intracellular ion. Each day one excretes potassium in an amount equal to that ingested. The kidney function is the major mechanism of regulation. Unlike sodium, the kidney does not conserve potassium very well or quickly and potassium is completely filtered at the glomerulus; also unlike sodium, it is not reabsorbed by the tubules but is secreted. When a diet is high in potassium, its concentration in the body cells increases, including the renal tubular cells. The latter facilitate the secretion of potassium into the tubule and increase its excretion. A low-potassium diet (one with sodium and little or no potassium) or loss from diarrhea lowers tubular cell potassium and decreases secretion, but there can be a marked depletion in a normal person in one week.

Aldosterone, too, helps not only in sodium reabsorption but also in the case of potassium (Fig. 14-20). This is reflexly initiated by an excess or deficit of potassium, due not to the renin-angiotensin process but, rather, to the fact that the adrenocortical cells that secrete aldosterone are themselves sensitive to potassium changes in their extracellular fluid. Thus, increased potassium results in increased extracellular potassium, which stimulates hormone production by the adrenal, which in turn increases tubular secretion and excretion of the ion. Potassium and hydrogen ions compete for exchange with sodium ions in the kidney tubules, and steroids from the adrenal glands increase the loss of urinary potassium when sodium is given. Hence a patient with normal kidneys who becomes stressed by disease or during a postoperative period may excrete as much as 60 to 120 mEq of potassium in 24 hours. If there is tissue injury, the urinary excretion may rise to 200 mEq per 24 hours.

The total amounts and concentrations of potassium in the body are relatively constant and in a dynamic state; yet this cation does continually move in and out of cells. The potassium gradient between the extracellular and intracellular compartments is influenced by adrenal steroids, testosterone (male sex hormone), pH changes, glycogen formation, and decreased blood sodium. Thus, for each 0.1-unit fall in blood pH, plasma potassium increases by 0.6 mEq per liter and the converse is also true. One always evaluates plasma concentration in terms of concomitant blood pH and base status.

Potassium is an essential part of electrolyte replacement. Up to one third of patients (except for those in complete renal failure) can exhibit hypokalemia (low blood potassium) and need supplemental potassium. One finds an increase in potassium loss in conditions such as kidney diseases, metabolic alterations, and diarrhea in infants. Potassium deficiency may also be the result of intensive therapy using diuretics, particularly with a lowered intake of potassium without supplementing the loss.

A patient in the hospital receiving intravenous fluids needs potassium daily. Potassium balance is essential, and one must keep in mind that normal intake and turnover are high. Although potassium and sodium are similar in terms of dietary intake, body content, and maximum tolerance, one must keep in mind that, whereas sodium can be rigorously conserved, potassium is poorly conserved, particularly in a stressful condition.

Calcium. Bone contains more than 90 percent of the body's calcium in the form of phosphate and carbonate. When serum calcium drops, the bones act as a source of available calcium. Calcium is not usually needed in routine intravenous therapy due to the tremendous reservoir in bone. Intravenous calcium, furthermore, does not prevent the decalcification of bone from disuse in the case of patients lying in bed for long periods of time. Extracellular calcium concentration affects neuromuscular activity and thus must also be precisely controlled. Calcium serves in bone formation, blood clotting (see Chap. 16), and muscle contraction. A low extracellular calcium leads to severe muscle twitching, even convulsions.

CALCIUM EXCRETION. Three areas take part in the extracellular calcium regulation: bone, the kidney, and the gastrointestinal tract. Bone, as mentioned above, contains most of the body's calcium and is available for either deposit or withdrawal of calcium from extracellular fluid. The gastrointestinal tract regulation of calcium is under precise hormonal control, and the kidneys filter and reabsorb calcium as well as phosphate, which plays a role in calcium regulation.

Parathormone from the parathyroid gland (see Chap. 15) is controlled by calcium concentration in the extracellular fluid surrounding the cells of the gland. Low calcium levels stimulate production and secretion; high calcium levels inhibit them. Thus, there is a direct action on the gland, with no intermediates. Parathormone itself (1) increases the movement of calcium and phosphate from bone into the extracellular fluid; (2) increases gastrointestinal absorption of calcium by stimulating the active transport system to move the ion into the blood, elevating plasma calcium levels; and (3) increases renal tubular reabsorption of calcium and reduces renal tubular reabsorption of phosphate, resulting in an increase of extracellular calcium levels. The converse is also true.

It should be noted that ionic calcium and phosphate are related, so that if extracellular phosphate increases it forces the deposition of some extracellular calcium into bone, lowering the serum calcium concentration and keeping the calcium phosphate production constant. If, however, parathormone causes bone breakdown, both ions are released.

Deficiency of vitamin D, a vitamin produced by the skin in the presence of sunlight, creates an abnormal deposition of calcium in bone. Vitamin D is apparently needed for parathormone effectiveness. Thyrocalcitonin, a thyroid hormone, increases bone uptake of calcium and acts in opposition to parathormone. Its secretion apparently depends on calcium concentration in the extracellular fluid surrounding the cells of the thyroid gland. Increased extracellular calcium induces this hormone release.

Magnesium. Magnesium is essentially an intracellular ion like potassium. It appears to be very important in many enzymatic systems that involve protein, carbohydrate, and lipid metabolism. Essential facts to remember are the following: (1) 1 mEq of magnesium weighs 12 mg; (2) 1 gm of magnesium sulfate supplies 16 mEq of magnesium; (3) in the normal, average adult, the total body content is about 2000 mEq; yet only about one half is rapidly exchangeable (bones contain a large amount of nonexchangeable magnesium); (4) most of this element lies in the cells; (5) the plasma level in healthy adults ranges between 1.5 and 2.5 mEq per liter, whereas the intracellular amount is from five to ten times greater; (6) most of the magnesium in the diet is excreted in the feces; yet the exact

regulation of its absorption is not well known; and (7) about 5 to 10 mEq are excreted in the urine per day under normal conditions. The minimum daily need is about 5 mEq, but in acute cases one can tolerate up to about 400 mEq.

MAGNESIUM CONTROL AND CLINICAL SIGNIFICANCE. Normal dietary intake ranges between 15 and 30 mEq per 24 hours. Deficiency can occur after prolonged periods of intravenous therapy with magnesium-free solutions. Deficiency is also known to occur in severe malabsorption in the digestive tract, prolonged nasogastric tube suction, acute pancreatitis, primary aldosteronism, and chronic alcoholism, just to name a few. Deficiency is characterized by hyperirritability of the central nervous and neuromuscular systems, resulting in muscle tremors, tetany, convulsions, or delirium. On the other hand, excess magnesium, which can occur in uremia, may result in cardiac arrest.

Chloride. This was the first electrolyte that could be measured easily. The so-called chloride shift was emphasized because the ion was easy to measure. Hypochloremia (deficiency of blood chloride) usually indicates an alkalosis due to increased levels of bicarbonate (HCO_3^-). This is one of the major problems in persistent vomiting and other conditions resulting in potassium loss. Thus, the chloride ion is very essential in hypokalemic (low-potassium) alkalosis. The condition may persist if one supplies only potassium. Yet one does not give too much chloride. The 154 mEq of chloride in a liter of normal saline far exceeds any routine use. Normal saline is very useful, however, when a condition of severe hypochloremic alkalosis exists.

Acid-Base Balance

Hydrogen Ion Control (pH). The acidity or alkalinity of a solution depends on the concentration of the hydrogen ions (H^+). The H^+ concentration (the concentration of free H^+ in solution is also referred to as pH) of the extracellular fluid is critical and is carefully regulated since H^+ markedly influences enzyme activity. The weight of H^+ in water is about one ten-millionth (10^{-7}) of a gram per liter. For convenience however, H^+ concentration is more commonly expressed as the negative logarithm, using the symbol pH. Thus, a H^+ concentration of one ten-millionth of a gram per liter equals pH 7. This use of negative logarithms means that as the H^+ concentration increases, the pH becomes lower and the solution becomes more acidic. As the H^+ concentration decreases, the pH rises and the solution becomes more basic. A pH of 7 represents ten times the amount of H^+ as a pH of 8 (10^{-7} vs. 10^{-8}).

A solution of pH 7 is neutral, because at this concentration the number of protons (H^+) is exactly balanced by the number of hydroxyl ions (OH^-). Thus, an acid solution has a pH lower than 7, and an alkaline solution has a pH greater than 7. Extracellular fluid has a pH ranging between 7.35 and 7.45 (slightly alkaline). Complex processes operate to maintain this narrow range of proper physiologic function.

The addition of an acid or base to water produces a change in pH and, in equivalent concentrations, the stronger the acid or base, the greater is the resultant change in pH. Little is known of cellular pH, but it is felt that it varies within wide limits (e.g., pH of 4.5 for prostate cells to about 8.5 for osteoblasts), whereas the pH of blood plasma is always 7.0 to 7.8. Furthermore, the pH of any cell is itself not uniform, but varies according to the ultrastructural spaces in the cell structure.

Many of the components of a cell (inorganic and organic) have the ability to acquire and lose electrical charge. One way in which molecules can gain and lose electrical charge is by the gain or loss of protons. A proton is a hydrogen atom devoid of its electron. Thus, when a single electron of the hydrogen atom is completely transferred to another atom, only the positive proton of the hydrogen nucleus remains and this positive proton is the hydrogen ion (H^+).

$$H \text{ (atom)} \longrightarrow H^+ \text{ (+ proton or ion)} + e^- \text{ (electron)}.$$

The gain and loss of protons constitute acid-base chemistry.

An acid is any substance that can lose or form a hydrogen ion (a proton or H^+ donor). Thus, the acidity of a solution depends on the H^+ concentration. A base is any substance that can gain or accept a hydrogen ion (a proton or H^+ acceptor). The pH of a solution is thus inversely proportional to the concentration of H^+ in solution, for an increase in the concentration of dissociated H^+ in solution means that the pH is going down.

An acid-base reaction is any reaction involving the exchange of protons (H^+). Acids are called "strong" or "weak" depending on how easily they lose their protons (H^+) when placed in water. Hydrochloric acid (HCl) is a strong acid because in water (HOH) it dissociates almost 100 percent into H^+ and Cl^-. Acetic acid is a weak acid because in water it dissociates only about 1 percent to give H^+ and acetate ions. Thus, in water H^+ and Cl^- have no affinity for each other and exist independently and will not regenerate. Under physiologic circumstances, $HCl \longrightarrow H^+ + Cl^-$. On the other hand, the acetate ion has a great attraction for the proton (H^+) and, when the ions are added independently to a solution, they react with each other: $Ac^- + H^+ \rightleftharpoons HAC$. The double arrow indicates that the reaction can proceed in both directions until equilibrium is reached. From the above one can see that since acetate accepts H^+, it is a base.

The body has several major sources of hydrogen ions: (1) the dissociation of phosphoric and sulfuric acids into hydrogen ions and the anions phosphate and sulfate; (2) the liberation of hydrogen ions by many inorganic acids, such as fatty acid and lactic acid, on dissociation; and (3) the liberation of hydrogen ions from metabolically produced carbon dioxide: $CO_2 + HOH \longrightarrow H_2CO_3 \longrightarrow H^+ + HCO_3^-$ (bicarbonate). The lungs normally eliminate CO_2 as rapidly as it is produced since as blood flows through the lung capillaries, CO_2 diffuses into the alveoli: $HCO_3^- + H^+ \longrightarrow H_2CO_3 \longrightarrow CO_2 + HOH$. Thus, the hydrogen ions generated by carbonic acid here are reincorporated into water molecules. If lung disease occurs, however, CO_2 is retained in the body and some of the hydrogen ion generated in the venous blood from the reaction of this CO_2 with water is also retained and raises the extracellular hydrogen ion concentration, which has to be eliminated by the kidneys to maintain a balance.

The Buffer System. Most physiologic reactions occur in water solutions and involve a variety of weak acid systems. Certain combinations of chemicals found in the extracellular fluids act as a "buffer" against sudden changes in the hydrogen ion concentration. A buffer can be likened to a "chemical sponge." Thus, depending on circumstances, it is able to "soak up" surplus hydrogen ions (protons) or release them in order to prevent marked alterations and fluctuations of the pH, which if uncontrolled would prove to be lethal to the cells and tissues. A "buffer system" consists of a "weak" acid in combination with one of the salts of that acid.

The H^+ concentration of the extracellular fluid is very small. If an excess occurs, much is bound (buffered) by other ions and eventually the kidneys eliminate the excess. It is the buffering that minimizes H^+ concentration changes until excretion takes place. The most important body buffers are (1) bicarbonate (HCO_3^-), (2) large anions such as plasma protein, (3) intracellular phosphate complexes, and (4) hemoglobin (Hb). The buffers all act by binding hydrogen ions (buffer$^-$ + H^+ = H buffer). The bound complex is a weak acid and can exist as the undissociated molecule or can dissociate into H^+ and buffer$^-$. Note that none of the buffering systems actually eliminate hydrogen ions from the body but merely cause their binding to some molecule, thereby preventing an increase in free hydrogen ion concentration.

The substances that become components of buffer systems are both organic and inorganic. The major buffers in the body fluids, described as anion-acid pairs are distributed as follows:

1. Blood:
   a. Plasma: $\dfrac{HCO_3^-}{H_2CO_3}$; $\dfrac{protein^-}{H\ protein}$; and $\dfrac{HPO_4^{--}}{H_2PO_4^-}$

   b. Red blood cells: $\dfrac{HCO_3^-}{H_2CO_3}$; $\dfrac{Hb^-}{HHB}$; $\dfrac{HbO_2^{--}}{HHbO_2}$;

   and $\dfrac{HPO_4^{--}}{H_2PO_4}$

2. Interstitial fluid: is the same as plasma except for a lower concentration of protein (Table 14-2).

3. Intracellular fluid (Table 14-2): is the same as in the red blood cells except that there is no hemoglobin (Hb). Quantitatively, the protein and phosphate systems are three times greater in the cell than in the plasma.

The most important proton (H^+) acceptor in the extracellular fluid (blood) is the carbonic acid-sodium bicarbonate system. Bicarbonate's available concentration is about 20 to 30 mEq per liter. In the hemoglobin system there is available 10 mEq per liter of proton acceptors. Thus, together, the 5-liter total of blood in a normal human adult has enough buffer capacity to absorb almost 150 ml of normal HCl before the pH of the fluids becomes dangerously acidic. Furthermore, the major buffer capacity of the body is attributed to the proton acceptors found in other tissues, especially muscle. They can neutralize about five times as much acid as do blood buffers. Clinically speaking, disturbances of acid-base balance usually involve imbalances in the above discussed system.

Normally, the extracellular fluid contains a ratio of 1 part of carbonic acid (H_2CO_3) to 20 parts of bicarbonate (HCO_3^-). The addition of hydrogen ions (addition of an acid) will result in the formation of more H_2CO_3 in the following manner:

H^+Cl^- (hydrochloric acid) +
$\qquad Na^+HCO_3^-$ (sodium bicarbonate) $\longrightarrow$
$\qquad\qquad\qquad\qquad\qquad Na^+Cl^- + H_2CO_3$

The neutralization or buffering of the H^+Cl^- that has been added is brought about by the formation of H_2CO_3. The latter is, in turn, excreted from the lungs, through increased depth and rate of breathing, as CO_2 (carbon dioxide). This occurs as follows:

$H_2CO_3 \longrightarrow H_2O$ (water) $+ CO_2$ through the action
$\qquad\qquad\qquad$ of the enzyme carbonic anhydrase

After the above reactions have taken place, the ratio of bicarbonate (HCO_3^-) to H_2CO_3 is altered since bicarbonate has been decreased proportionally more than the H_2CO_3; thus, the absolute value of the ratio falls, giving an acid pH. To compensate for this, the bicarbonate is replaced by kidney mechanisms causing

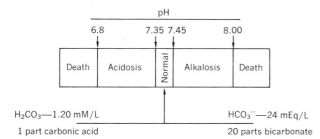

<figure>
pH

6.8   7.35 7.45   8.00

| Death | Acidosis | Normal | Alkalosis | Death |

H_2CO_3—1.20 mM/L HCO_3^-—24 mEq/L

1 part carbonic acid   20 parts bicarbonate
</figure>

Figure 14-21. Acid-base balance. Twenty parts of bicarbonate salt balance 1 part of carbonic acid for normal body pH. Any change in proportion changes the pH. If bicarbonate goes down to 10 parts to 1, the pH drops to 7.10 and acidosis results. If the bicarbonate increases to 40 parts to 1, the pH rises to 7.70 and alkalosis results.

acidification of urine and excretion of the fixed anion (Cl^-). Finally, as these processes occur, the buffer ratio returns to its original value of 1:20.

Bicarbonate. The diagnosis of metabolic acidosis and alkalosis is usually made from the bicarbonate level in the blood (Fig. 14-21), which is normally 22 to 26 mEq in arterial blood and 24 to 28 mEq in venous blood.

Although the blood constitutes only a fraction of the body's fluids and although the blood carbonic acid system constitutes only about one tenth the total available body buffer, this system is used to assess the status of all the other buffers of the body. This is because changes in the acid-base balance at both intracellular and extracellular levels must be reflected ultimately in the blood's buffer system.

The two major routes for elimination of protons (H^+) and for maintaining a normal ratio of anion to weak acid buffer (HCO_3^-/H_2CO_3) are the lungs and the kidneys. Under normal conditions, 20 to 40 liters of 1 N acid are eliminated by the lungs per day and about 50 to 150 ml of 1 N acid are excreted via the kidneys. Thus, the carbon dioxide excreted by the lungs accounts for the greatest portion of acid eliminated continuously from the body. However, there are other acids having no avenue of elimination except by the kidneys. Although these are produced metabolically in smaller amounts than is carbonic acid, it is just as vital to have them excreted as to eliminate CO_2 from the

lungs. Renal regulation of acid-base equilibria involves (1) reabsorption of Na^+ and HCO_3^- (conservation of one of the major bases of the body), (2) excretion of weak acids, and (3) formation and excretion of NH_4^+ (ammonia ions).

The concentration of bicarbonate in the blood plasma is regulated by the kidneys through acidification of the urine, when there is a relative or absolute deficit of bicarbonate, or by alkalinization of the urine, when there is a relative or absolute excess of bicarbonate. Thus, in the first case, the kidney makes "new" bicarbonate by the reverse reaction of carbonic anhydrase, $H_2CO_3 \rightleftharpoons H^+ + HCO_3^-$, sending H^+ into the urine and HCO_3^- back into the body fluids. In the second case, HCO_3^- is excreted in the urine with Na^+ and K^+.

In respiratory acidosis and alkalosis (see below) primary changes in carbonic acid are reflected by secondary changes in bicarbonate. Respiratory acidosis is characterized by accumulation of CO_2 due to hypoventilation (decreased breathing), which results in an increase of H_2CO_3 so that the normal ratio of H_2CO_3/HCO_3^- of 1:20 is decreased. This leads to a drop in pH and the characteristic acidosis. Conversely, in respiratory alkalosis there is a deficit of H_2CO_3 due to hyperventilation (increased breathing), with a resultant increase in the ratio of 1:20 with bicarbonate and a rise in pH characteristic of alkalosis (Fig. 14-22).

Thus, using the equation $H^+ + HCO_3^- \rightleftharpoons H_2CO_3 \rightleftharpoons HOH + CO_2$, one notes that an increased extracellular H^+ concentration drives the re-

Figure 14-22. Alterations in acid-base equilibrium.

	Respiratory Component (plasma pCO_2)	Metabolic Component (plasma HCO_3^-)	Blood pH
Respiratory Acidosis	↑↑	↑	↓
Respiratory Alkalosis	↓↓	↓	↑
Metabolic Acidosis	↓	↓↓	↓
Metabolic Alkalosis	↑	Normal or ↑	↑

action to the right and H$^+$ is removed from solution. The converse drives it to the left, releasing H$^+$. This buffer mechanism is important because extracellular bicarbonate concentration is normally high and is regulated by the kidneys. Note, however, that for the reaction to continue to proceed to the right, CO_2 must be removed. This actually does happen since a greater extracellular H$^+$ concentration stimulates the respiratory center to increase alveolar ventilation and CO_2 is eliminated by the lungs. On the other hand, a lower H$^+$ concentration drives the reaction to the left, CO_2 and HOH combine to generate H$^+$ and HCO_3^-, and all returns to normal. The lower H$^+$ concentration decreases alveolar ventilation and CO_2 elimination by inhibiting the medullary respiratory center. Thus, the added dimension of gas elimination and normal lung function allows HCO_3^- to function efficiently as a buffer. Lung disease itself can cause an increase in H$^+$ concentration and impair the efficiency of the HCO_3^- buffering system. Other buffers, however, attempt to handle the added load.

Hemoglobin. Most metabolically produced CO_2 is carried from the tissues to the lungs as HCO_3^-. The bicarbonate ions are generated by CO_2 hydration to form H_2CO_3 and dissociated to HCO_3^- and H$^+$. As blood passes through the lungs, the reaction reverses and H$^+$ and HCO_3^- recombine. Hemoglobin buffers the H$^+$ as it passes from tissues to the lungs since reduced hemoglobin has a greater affinity for H$^+$ than does oxyhemoglobin. Thus, as blood passes through the tissues, a part of the oxyhemoglobin loses its oxygen and is changed to reduced hemoglobin. At the same time CO_2 enters the blood and in the red blood cell undergoes reactions to generate HCO_3^- and H$^+$. Most of the H$^+$ is bound to hemoglobin as a result of strong affinity. This maintains the acidity of the venous blood only slightly greater than that of the arterial. The reactions are reversed in the lungs since hemoglobin becomes saturated with O_2 and its ability to bind H$^+$ is reduced; the H$^+$ are released and react with HCO_3^- to give CO_2, and the latter diffuses into the alveoli to be expired. Thus, the presence of O_2 in the lungs is a very important factor in hemoglobin's ability to bind H$^+$ (Fig. 14-23).

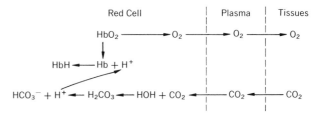

Figure 14-23. Hemoglobin buffering of hydrogen ions as blood passes through tissue capillaries.

Renal Hydrogen Ion Regulation. The amount of organic, phosphoric, or sulfuric acid formed in the body depends on the type of food eaten. A high-protein diet results in considerable protein breakdown and a large amount of sulfuric acid. These are but one source of hydrogen ions (one that liberates about 40 to 80 milliosmoles of H$^+$ per day). To maintain a homeostatic equilibrium one must eliminate approximately what one ingests. This is done primarily by the kidneys, which must be able to alter H$^+$ excretion, regardless of the source of H$^+$ variation, as well as compensate for gastrointestinal variations and H$^+$ loss or gain.

Practically all the H$^+$ excreted enters the renal tubules by tubular secretion; a high extracellular H$^+$ concentration induces greater secretion (and the converse is also true). The controlling mechanism involves the tubule and extracellular concentration levels, but there seems to be no neural or hormonal control. The renal capabilities depend on tubular H$^+$ secretion and buffers in the urine. If no buffers are present, free H$^+$ concentration of the tubule fluid gets so high that H$^+$ actually diffuses back into the surrounding blood capillaries as fast as it is secreted. The major urinary buffers are HPO_4^{--} and NH_3 (ammonia).

$$HPO_4^{--} + H^+ \longrightarrow H_2PO_4^-$$
$$NH_3 + H^+ \longrightarrow NH_4^+$$

The HPO_4^{--} found in the tubule fluid arrives there by filtration and is not reabsorbed. The ammonia, however, is formed in the tubular cells by the deamination of certain amino acids, which are transported into the tubular cells from the surrounding capillary plasma. The amount of ammonia that remains in the tubular lumen depends on the rate of H^+ accumulation, since the tubular cell membrane is permeable to NH_3 but not NH_4^+. Thus, the more H^+ present, the greater the conversion to NH_4^+ and increased H^+ secretion.

Acidosis and Alkalosis. In order to define the terms acidosis and alkalosis consider the following equilibrium:

$$HA \text{ (weak acid)} \longrightarrow H^+ + A^- \text{ (anion of weak acid)}$$

The ratio of $[H^+]$ to $[A^-]/[HA]$ is a constant. At a given $[H^+]$, which can be called normal or physiologic, the ratio of $[A^-]/[HA]$ will have a characteristic value. If we add HA or $[H^+]$ to the system or remove $[A^-]$, the condition is defined as acidosis since the $[H^+]$ is now greater than normal or the pH is less than normal and the ratio of $[A^+]/[HA]$ has decreased. If, on the other hand, HA or $[H^+]$ is removed or A^- or any other base added to the system, the result is termed alkalosis. Now the $[H^+]$ is less than normal, the pH is higher, and the ratio of $[A^-]/[HA]$ has a higher-than-normal value.

In order to characterize blood acid-base status, three variables are needed: (1) the respiratory component, (2) the metabolic component, and (3) the blood pH. Knowing any two will enable one to calculate the third. The respiratory component is represented by the blood or plasma pCO_2. The metabolic one can either be described in terms of all blood buffers in 1 liter of whole blood or be presented by a single buffer in plasma, usually plasma bicarbonate concentration. Simple disturbances of acid-base equilibrium usually involve a single primary cause and are classified as follows (Fig. 14-22):

RESPIRATORY ACIDOSIS. Respiratory acidosis occurs where there is an increase in plasma pCO_2 due to a decrease in lung alveolar ventilation. This can be due to pulmonary edema (fluid in the lungs), suppression or injury of the respiratory center, obstructive emphysema, diseases of the chest wall interfering with breathing, or muscle disorders involving respiratory muscles.

RESPIRATORY ALKALOSIS. Respiratory alkalosis occurs where there is a decrease in plasma pCO_2 due to an increase in lung ventilation. This can be due to emotional hyperventilation or anything that affects the respiratory center directly.

METABOLIC ACIDOSIS. Metabolic acidosis is associated with a rise of nonvolatile acids, as seen in uncontrolled diabetes; lactic acid accumulation (primary or secondary), as in starvation; or loss of bicarbonate from the extracellular fluid, as seen in severe diarrhea and kidney tubule dysfunction.

METABOLIC ALKALOSIS. Metabolic alkalosis is associated with a loss of strong acid from the body and is seen most often in severe, persistent vomiting (with a loss of hydrochloric acid). It may also occur from loss of hydrogen ions in kidney dysfunction, as a result of potassium deficiency or diuretic therapy, or because of excessive intake of bicarbonate.

PROBLEMS OF HOMEOSTASIS. Two major clinical conditions relating to problems of homeostasis (the control of composition or concentration and volume of fluid compartments of the body) are shock and burns.

Shock. Shock is characterized by low blood pressure, poor perfusion of the tissues caused by microcirculatory stasis, and decreased blood volume due to a shift of fluid from the extracellular to intracellular compartments. A decreased effective blood volume underlies all forms of shock. One treats by restoring circulatory volume, replacing extracellular fluid, correcting acidosis, providing oxygenation, relieving pain, maintaining hydration, replacing bicarbonate, preventing heat loss, and the like.

Burns. Burns result in shock, with a shift of extracellular fluid into the interstitial space from the plasma, which is aggravated by loss of fluid into the burned area. Fluid is also lost from the burned surface, and denuded areas create marked edema. There is a danger of kidney damage due to released hemoglobin, arising from red-cell destruction at the time of the burn and cells broken down within two or three days. Smoke and heat damage to the lungs is serious as the capacity of the lungs to function in bicarbonate regulation is impaired. The reabsorption of large amounts of edema on days two and three also occurs, and overtreatment with fluids, in victims with lung damage, can lead to pulmonary edema. Large amounts of sodium enter the edema in the burned areas, and intracellular potassium is exchanged for sodium. With water loss due to reabsorption of edema fluid, more sodium is lost. Because of the damage to many cells, release of potassium is common; so one has to be careful not to give large amounts of potassium.

One bases estimates of fluid losses on area burned, which is crude because fluid losses depend on the depth of the burn as well as area.

Review Questions

1. Where are the kidneys located?
2. Describe the general anatomic parts of the kidney.
3. Trace the renal artery into the kidney. Retrace the venous flow.
4. What is a nephron? List its parts. Describe what generally takes place in each portion.
5. Describe the anatomy of the bladder.
6. What is the difference between ureter and urethra?
7. What controls emptying and filling of the bladder?
8. Do you understand water reabsorption by the kidney? What happens to other materials? Does the kidney excrete as well as secrete? Is glucose excreted normally?
9. What is inulin? How does it test kidney function?
10. What is meant by the "sodium pump"?
11. Where is the juxtaglomerular apparatus located? How does it function?
12. How do ADH and parathyroid hormone affect kidney function?
13. Generally speaking, what are the functions of the kidney?
14. How is water distributed in the body? How is it lost and replenished?
15. List the essential electrolytes of body fluids. What is an electrolyte? How do we determine the milliequivalents per liter of an electrolyte?
16. What is meant by edema?
17. Define osmotic pressure. What does serum osmotic pressure depend on?
18. How is electrolyte balance controlled? What role does the hypothalamus serve? What is diabetes insipidus?
19. What is the major extracellular cation?
20. What is the normal major regulator of sodium? How else is it influenced?
21. Where is potassium predominantly found?
22. What control the body's calcium levels?
23. What other ions are essential? Why?
24. What is meant by pH? What is an acid? What is a base?
25. What is a buffer? How does it function? What are the major buffers of the body? How do buffers act?
26. What is meant by metabolic acidosis and alkalosis?
27. What is meant by respiratory acidosis and alkalosis?
28. How do shock and burns relate to acid-base balance?

References

Davenport, H. W.: *The ABC of Acid-Base Chemistry,* 5th ed. University of Chicago Press, Chicago, 1969.

Hills, A. G.: *Acid-Base Balance, Chemistry, Physiology, Pathophysiology.* The Williams & Wilkins Co., Baltimore, 1973.

Pitts, R. F.: *Physiology of the Kidney and Body Fluids,* 2nd ed. Year Book Medical Publishers, Inc., Chicago, 1968.

Smith, H. W.: The Kidney. *Sci. Am.,* **188**(1):40–48, 1953.

Solomon, A. K.: Pumps in the Living Cell. *Sci. Am.,* **207**(2):100–108, 1962.

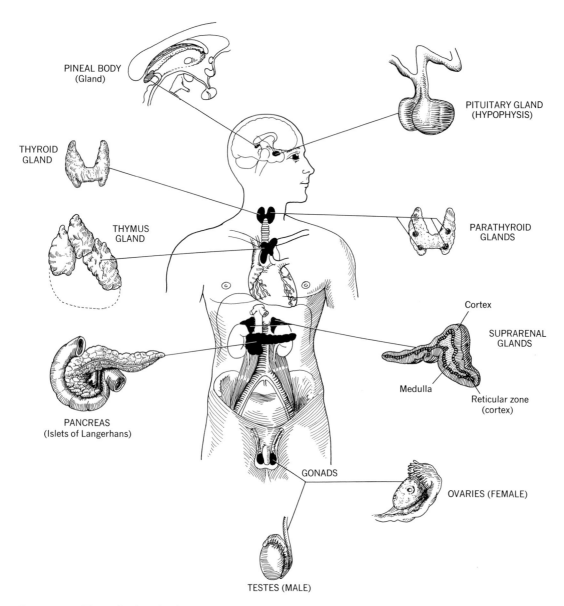

Figure 15-1. The endocrine glands and their general locations in the body.

PINEAL BODY
(Gland)

PITUITARY GLAND
(HYPOPHYSIS)

THYROID
GLAND

PARATHYROID
GLANDS

THYMUS
GLAND

Cortex

SUPRARENAL
GLANDS

Medulla

Reticular zone
(cortex)

PANCREAS
(Islets of Langerhans)

GONADS

OVARIES (FEMALE)

TESTES (MALE)

Chemical Messengers: The Endocrine System

An elaborate system of controls governs the many biochemical reactions taking place in the body tissues. In the cell, regulation is maintained by inducing or repressing protein synthesis and metabolic feedback. In addition, in higher organisms, there is a hormonal system composed of cells that discharge their synthesized contents into the circulatory system. In developing organisms, hormonal control is manifested through regulation of metamorphosis and later through growth regulation. In adults, hormones are responsible for the integrated activity of organ systems and subsystems by (1) altering cellular function in response to variations in the external environment, (2) inducing sustained activity by cells, and (3) changing the level of activity of tissues and organs and thus maintaining a homeostasis in the internal environment.

The glands of the body are organs that produce or secrete a variety of substances, such as saliva, pancreatic juice, bile, epinephrine, and so forth. There are two kinds of glands. The first type is the kind seen, for example, in the digestive system, which transports its secretions to where they are needed by means of a tube or duct; these are called exocrine glands. The other type has no duct, and its cells secrete directly into the vascular system (Fig. 15-1); these are referred to as endocine glands (Greek *endon*, within; *krino*, to separate), and their secretions are called hormones. Certain other organs, e.g., the kidneys, secrete hormones from time to time, in addition to performing their major function. Specific substances attributed to a particular endocrine gland have been traced to specific cells in that gland. The pancreas, for example, contains groups of clearly demarcated cells called islets that produce pancreatic hormones. Furthermore, endocrine cells may store secretions either as inactive granules in their cytoplasm (as in the pituitary gland) or as clear, proteinaceous material in a cavity surrounded by the secretory cells (as in the thyroid gland). Some secretions are derived from neural crest cells, which also produce neural tissue. The cells making up the neural tissue have the ability not only to conduct an electrical impulse but also to secrete neurohumors at their terminals that excite adjacent cells. The secretory cells from the embryonal neural tissue lose their ability to conduct impulses and are found as the chromaffin cells of the adrenal medulla, where they produce catecholamines, norepinephrine, and epinephrine. Neurosecretory cells have the ability to conduct impulses and to secrete. Their secretions are called neurohormones when they pass into the body fluids or circulation. They can be found in the hypothalamus.

The endocrine glands provide an alternative communication and regulatory system to the nervous system. The two systems supplement each other or perform separately. Hormones also have an effect on the central nervous system. Hydrocortisone, for example, strongly affects our mood and awareness of reality. Many emotional disturbances are "decoded" in the hypothalamus and thereby affect the pituitary and may interfere with normal pituitary function. In other words, the total behavior of man is controlled by the interplay between hormones and nerve impulses, and over long periods of time extreme conditions can result in illnesses that may not appear to be stress or hormonally oriented. The nervous system permits us to make rapid responses, but when the stimulus stops so does the reaction. Hormones, on the other hand, are not suitable for quick responses, since they take time to be conveyed throughout the body; however, their effects are long lasting.

Certain hormones affect every cell (like the growth hormone, which increases protein synthesis), while others affect only certain tissues. There are two conditions under which the endocrine system operates: (1) gland cells must make hormones and (2) target cells must respond to certain hormones and not others. Thus, hormones influence every cell in the body, and both physical health and mental health are dependent on the function of this system. Unlike vitamins, the hormones are produced by the organism itself. They are remarkably effective at very low concentrations, being produced in small amounts ranging from nanograms to milligrams per day. Their blood concentrations are low and their tissue concentrations even lower. In the normal animal, the secretion rate of a hormone is determined by the need for that hormone, and the effect on tissues is determined largely by the capacity of the tissues to respond and the amount of hormone present.

The control mechanism is complex and relatively specific for each endocrine gland. Hormones seldom have any direct effect on the actual organs secreting them. Hormones act as trigger substances that initiate biochemical reactions that persist after the hormones are no longer present. Although hormones have specific effects in cells, many have numerous effects, such as insulin, which not only changes cellular permeability to glucose but also increases the biosynthesis of protein in muscle and other tissues. As to how hormones act, some interact directly with key enzymes to alter their kinetics, others influence membrane permeability, and still others intervene in protein synthesis by interacting with specific gene loci, thereby stimulating or blocking the formation of messenger RNA and the synthesis of specific enzymes. Regulation of high efficiency also controls the removal of excess hormone.

Other hormones are produced at irregular intervals.

When we eat, the stomach is stimulated by the prospect of food and secretes a hormone called gastrin, which passes through the bloodstream and back to the stomach, which in turn is stimulated to secrete gastric juices. The stomach therefore secretes a substance that eventually stimulates itself. However, most hormones act at a distant site and are produced in cycles or intervals, and the amount depends on the body's needs. Blood pressure is controlled by hormones since regulation is essential; if it is too high there is tissue damage, and if it is too low the kidneys will not function adequately.

No hormone is believed to start a chemical reaction. Like enzymes, they only accelerate the speed of chemical reactions that otherwise would move too slowly. Every reaction in a cell needs a substrate and an enzyme to act upon it. A reaction may also need a metallic coenzyme and, of course, the energy source in the form of adenosine triphosphate (ATP). Hormones may influence any one of the above stages: affect the enzyme, influence the availability of substrate, or affect the inhibitor or activator. How the hormone affects many of the stages is unknown. Hormones do make the cell more permeable to specific compounds. For example, growth hormone stimulates growth of the body's cells by speeding up amino acid transport across cellular membranes. Antidiuretic hormone (ADH) from the posterior pituitary influences cell membranes by preventing water loss from the body by making the kidney less permeable to water.

Like enzymes, hormones are needed in very small amounts. They are derived from three classes of chemical substances (Table 15-6, p. 535): (1) the smallest group is derived from two amino acids, tryptophan (which is converted to serotonin and melatonin) and tyrosine (which is the source of the catecholamines and thyroid hormones); (2) a large group of hormones originate from cholesterol and are converted to adrenocortical and gonadal steroids; and (3) the largest group of hormones are peptides or proteins and vary in length from 8 to more than 180 amino acids and may have carbohydrate groups attached to them. Hormones differ from enzymes in that they are quickly destroyed after doing their job. If they went on producing, their effects would be too long and thus it is easier to replace them as needed.

No hormone is secreted at a continuous steady rate. Some are secreted in a daily rhythm; others have longer cycles. The female reproductive hormones are made to cycle in roughly 28 days and are altered by pregnancy. These hormones control all the stages of menstruation and of pregnancy from the moment the egg leaves the ovary until it is either discarded or with pregnancy grows into a new being. Each hormone is produced at a specific time to enable all stages to follow systematically. Each hormone also, like the pituitary, has control on an organ system far distant from it.

In general, hormones comprise a regulatory system superimposed on existing controls of cellular metabolism. Hormones appear to act directly on their target cells and apparently affect the rate at which metabolites and nutrients reach the cells. Hormones also affect membrane transport (cellular permeability) and can cause changes in transport of ions, water, amino acids, and sugars. They also tend to affect other endocrine glands in subtle ways not completely understood. A disturbance in one may have a tremendous effect on other processes, which is often difficult to relate to the initial deficiency.

Endocrine Glands and Their Secretions

Pituitary (Hypophysis) and Hypothalamus
Embryology. The pituitary gland consists of two portions embryologically: (1) an anterior glandular portion (adenohypophysis), derived from the roof of the mouth; and (2) a posterior nervous part (neurohypophysis), derived from the floor of the third ventricle of the brain. The former is an outgrowth (Rathke's pouch) from the roof of the primitive pharynx (stomodeum).

Anatomy. The pituitary gland (Fig. 15-2) is an unpaired organ located on the floor of the skull in a bony

Figure 15-2. Pituitary gland and its divisions.

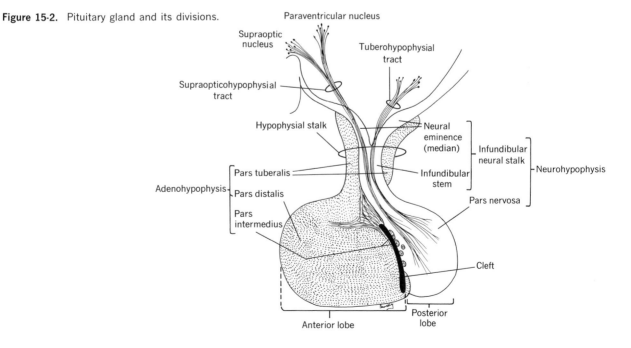

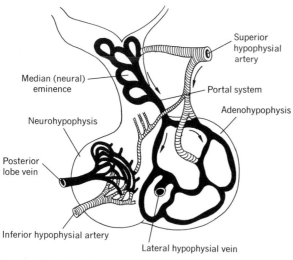

Figure 15-3. Blood supply of pituitary gland.

pocket called the sella turcica (Fig. 5-13, p. 142). It is approximately 13 × 8 mm in size and has a "stalk" or infundibulum that connects it to the brain. It also has a "skin" or covering of dura that ensheathes it. It is one of the major glands of the body since it secretes ten hormones that directly or indirectly regulate the activity of many endocrine glands. The anterior lobe (the larger of the two) consists of the pars distalis, the pars tuberalis, and a portion of the pars intermedia. This region is functionally referred to as the adenohypophysis. The pars tuberalis and pars distalis are the source of seven hormones. The pars intermedia produces another hormone (MSH) and in some species is separated from the pars distalis by a narrow cleft. The anterior lobe is highly vascular and extensively supplied by capillaries, which reach it along the infundibulum from an arterial circle (Fig. 15-3). The pars intermedia contains a few blood vessels, some finely granulated cells, and small masses of colloid material.

The posterior lobe (neurohypophysis) consists of a portion of the pars intermedia, the pars nervosa, and the infundibular stalk and contains no nerve cells, even though it develops as a downgrowth from the brain.

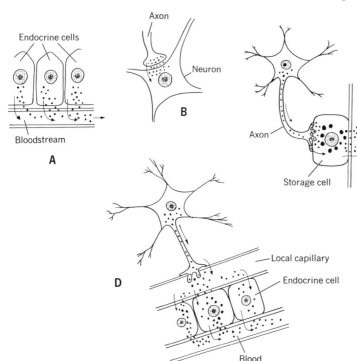

Figure 15-4. Neurohumoral secretions. **A.** In a classic hormone secretion the endocrine cell secretes directly into the bloodstream. **B.** At a classic synapse the axon of a nerve cell locally releases a transmitter substance to activate the next cell. **C.** In neurosecretion of oxytocin, for example, the hormones are secreted by nerve cells and pass through their axons to storage cells in the posterior pituitary to be eventually secreted into the bloodstream. **D.** Hypothalamic releasing factors (hormones) go from the neurons that secrete them into local capillaries, which carry them through portal veins to endocrine cells in the anterior pituitary lobe, whose secretions they in turn stimulate.

It does have numerous neuroglial cells (nonneural connective tissue cells of the central nervous system) and fibers, as well as small collections of colloid material. The nerve fibers project from the hypothalamus into the neurohypophysis, form the hypothalamic-hypophysial tract, and terminate on capillaries (Figs. 15-2, 15-4). Thus, neurosecretory granules are formed in the cell bodies of the hypothalamus (supraoptic and paraventricular), are transported via the nerve fibers into the posterior pituitary, and are stored in the nerve terminals resting on the capillaries. Functionally, the neurohypophysis consists of the pars nervosa, infundibular stem, and median eminence. The latter two are anatomically related to the hypophysial stalk.

THE ADENOHYPOPHYSIS: HYPOTHALAMIC RELATION. This portion of the pituitary gland (Fig. 15-5) is found to contain at least six different types of cells, which are identified by special stain and fluorescent-antibody techniques, size and composition of granules, location, and activity. The terms *basophil, acidophil,* and *chromophobe* refer to the cells' granules that stain with basic, acid, or neutral dyes. Each has its own secretion (all polypeptides) and has been characterized in relation to pathologic states, pregnancy, and the effects of removal of other endocrines (adrenalectomy, thyroidectomy, etc.). The only hormone not yet assigned a "cell type" is lipotropin.

The activity of the adenohypophysial cells is regu-

Figure 15-5. The hypothalamic-hypophysial pathway and secondary influences on the endocrine glands and other organs.

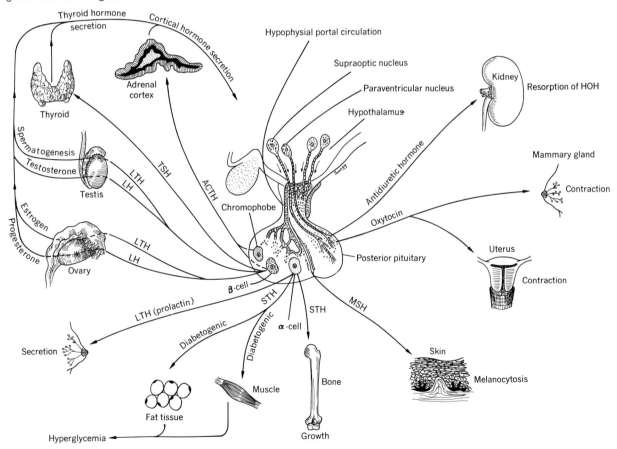

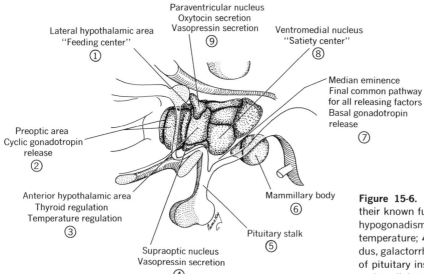

Padaventricular nucleus
Oxytocin secretion
Vasopressin secretion
⑨

Lateral hypothalamic area
"Feeding center"
①

Ventromedial nucleus
"Satiety center"
⑧

Median eminence
Final common pathway
for all releasing factors
Basal gonadotropin
release
⑦

Preoptic area
Cyclic gonadotropin
release
②

Anterior hypothalamic area
Thyroid regulation
Temperature regulation
③

Mammillary body
⑥

Pituitary stalk
⑤

Supraoptic nucleus
Vasopressin secretion
④

Figure 15-6. Diagram of hypothalamic nuclei and their known functions. Lesions in **1** cause aphagia; **2**, hypogonadism; **3**, disturbances of metabolism and temperature; **4**, diabetes insipidus; **5**, diabetes insipidus, galactorrhea; **6**, precocious puberty; **7**, syndromes of pituitary insufficiency; **8**, obesity and hyperphagia; and **9**, diabetes insipidus.

lated by hypothalamic hormones delivered to the cells by way of a specialized vascular pathway, and thus one should really consider the hypothalamus and pituitary as a unit rather than speak of the so-called pituitary master gland. Two sets of primary capillaries have been described, which are connected by blood vessels that pass over the surface of the pituitary stalk. One "bed" is in the median eminence (Fig. 15-3) and the other in the adenohypophysis. Blood flowing into the bed of the median eminence is collected by a "portal system" of long and short portal vessels and distributed to a secondary bed around cells of the adenohypophysis. The portal system apparently is essential for regulation of pituitary secretion, and there is indication of a highly specific, point-to-point neurovascular relation. A recent development was the discovery of hypothalamic releasing hormones that regulate the synthesis and secretion of such secretory cell types in the adenohypophysis (Fig. 15-6).

Regulation of the cells of the adenohypophysis is under the control of blood-borne materials introduced into the capillaries of the hypothalamic-hypophysial portal system at the median eminence. Each pituitary cell is influenced by at least one regulator. See Table

Table 15-1. Mammalian Hypothalamic Neurohormones

Hormone	Nature
Corticotropin-releasing hormone (CRH)	Polypeptide (?)
Luteinizing hormone–releasing hormone (LRH or LH-RH)	Not a polypeptide (?)
Follicle-stimulating hormone–releasing hormone (FRH or FSH-RH)	Polypeptide (?)
Thyroid-stimulating hormone–releasing hormone (TRH)	Not a simple polypeptide
Growth hormone–releasing hormone or somatotropin-releasing hormone (GRH or SRH)	Probably a polypeptide
Prolactin release–inhibiting hormone (PIH or PRIH)	?
MSH release–inhibiting hormone (MIH or MRIH)	Not a polypeptide

15-1 for the seven regulators known (five activators and two inhibitors). Others undoubtedly exist. Little is known about these hypothalamic releasing hormones. They have been classified as neurohormones since they release into the bloodstream and activate distant cells (neurohumors are released at nerve terminals and activate adjacent nerve bodies). The cells producing neurohormones serve as intermediaries between the nervous and endocrine systems and have been called transducer cells (transforming neural to hormonal stimuli). This allows environmental stimuli (afferent impulses) that are perceived in the central nervous system to affect the endocrine system and can serve as an elaborate control and integrating mechanism of body function.

Somatotropin (STH GH, Growth Hormone). STH (Fig. 15-5, Table 15-6) has a molecular weight of about 21,500 (man) and contains 188 amino acid residues and two disulfide bridges. The hormone stimulates the formation of specific antibodies, and immunochemical reactions with the sera produced by these antibodies have been used as a basis for its assay. The factors controlling the secretion of STH by the pituitary are too complex to be discussed here (see endocrinology textbooks listed at the end of this chapter).

EFFECTS. The stimulatory action of STH on all organs (visceral enlargement) is obvious in adult animals. The hormone is associated with a marked positive nitrogen balance, causes a reduction in plasma amino nitrogen, increases amino acid transport across membranes, and increases protein synthesis. It also causes mobilization of fatty acids from deposits (ketogenic), thus playing a role in providing cellular fuel. It inhibits glucose utilization by muscle tissues and decreases the sensitivity of hypophysectomized animals to insulin (diabetogenic). Recently it has been reported to stimulate transfer RNA and messenger RNA formation. Thus, it stimulates protein synthesis, fat metabolism, and bone growth.

THYROTROPIN (THYROID-STIMULATING HORMONE, TSH). Thyrotropin (Fig. 15-5, Table 15-6) has a molecular weight of about 28,000. Its secretion is under dual control: (1) thyroid hormone acts directly on the pituitary gland and (2) the existence of a hypothalamic feedback pathway has been suggested by the isolation of a neurohormone (TRH) that induces TSH release.

EFFECTS. The hormone acts on two different sites: (1) the thyroid gland; and (2) the fat deposits in the body, inducing release of fatty acids. The effect on the thyroid causes an increased blood flow through the gland and then an increased rate of breakdown of the colloid thyroglobulin, resulting in an increased discharge of thyroid hormone and a reduction in follicle colloid. Thyroid hormone, in turn, causes an increased metabolic rate, growth, and development (Table 15-6).

LATS (LONG-ACTING THYROID STIMULATOR). LATS is a substance found in the plasma of hyperthyroid patients. It is not present in the pituitary and may be an autoantibody against a thyroid component.

Adrenocorticotropin (Adrenocorticotropic Hormone, ACTH). ACTH (Fig. 15-5, Table 15-6) consists of 39 amino acid residues and has a molecular weight of about 4500. Its molecule is the smallest of the hormones of the adenohypophysis. It causes depletion of adrenal ascorbic acid (vitamin C) and induces steroid synthesis by the adrenal. Many factors affect the secretion of ACTH. Stress can provoke an increased secretion of ACTH (acting on the hypothalamus via the central nervous system). CRH (hypothalamic hormone) regulates ACTH secretion by the pituitary. ACTH, in turn, can inhibit CRH secretion. Glucocorticoids appear to have a direct effect on ACTH secretion as well as on CRH secretion. It is now believed, however, that the primary site of feedback action is the hypothalamus and other nervous areas.

EFFECTS. ACTH acts on three major sites: (1) The hormone acts on the two innermost zones of cells in the adrenal cortex (zona fasciculata and reticularis), which are the sites of glucocorticoid synthesis and secretion (Fig. 15-18, p. 531); the outer cortical cell layer (zona glomerulosa), which secretes aldosterone (Fig. 15-18), is hardly affected by ACTH. (2) ACTH affects fat depots, exerting a lipolytic action like STH, thyrotropin, and catecholamines. (3) It stimulates melanocytes, causing skin darkening, just like MSH from the hypothalamus. Thus, the adrenal cortex responds to ACTH by reducing its ascorbic acid content, converting cholesterol to glucocorticoids, and increasing its secretion rate, which raises the level of activity of the adrenal by increased oxygen and glucose utilization; it stimulates cell division of the inner layers and cortex growth and, by increasing adenylcyclase activity, causes cyclic AMP synthesis to enhance steroid synthesis.

Prolactin (Luteotropic Hormone, LTH). Prolactin (Fig. 15-5, Table 15-6) has a molecular weight of 23,000. Human prolactin is difficult to distinguish from human growth hormone. It does differ in other species. Evidence indicates that the hypothalamus continually inhibits prolactin secretion (prolactin release–inhibiting hormone, PIH).

EFFECTS. In nonmammalian vertebrates, prolactin serves as other than a lactogenic hormone. In the rat it helps maintain the corpus luteum, in the pigeon causes the crop to secrete "crop milk," in the tadpole inhibits tail resorption, and in some fish makes survival possible even after hypophysectomy. Generally it functions in milk preparation in the breast (after complete development of the gland through ovarian hormone secretions).

Gonadotropins (LH [or ICSH in the Male] and FSH). There are no pure gonadotropins (Fig. 15-5, Table 15-6) available, but five preparations (three of non-pituitary origin) are in use. FSH and LH are made from pituitary tissue. Human menopausal gonadotropin (HMG) is isolated from urine of menopausal women and has both FSH and LH activity. The serum from pregnant mares (PMS) is predominantly FSH-like with a little LH action. Human chorionic gonadotropin (HCG) is found in urine of women about seven days after conception and peaks at six weeks. It has an LH effect, with some FSH activity with increased doses. The gonadotropins have a molecular weight in the range of 28,000 to 31,000.

Functionally LH and FSH complement each other. Spermatogenesis is possible only if they are secreted together, since FSH operates in the intermediate stages of sperm development and ICSH stimulates androgen secretion needed for sperm maturation. In the female, ovulation requires that FSH and LH are sequentially used. Testosterone-blood levels regulate the activity of the neurosecretory cells that secrete the gonadotropin-releasing hormones (FRH and LRH)—a negative feedback mechanism. Regulation in the female is more complicated. The secretion of FSH and LH is

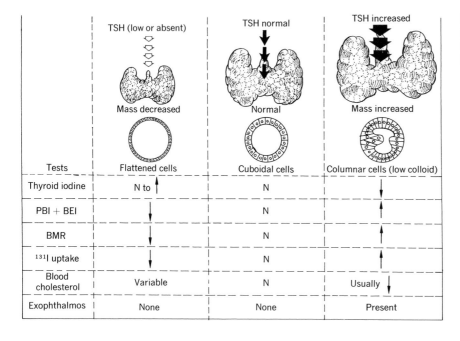

Figure 15-7. Effect of thyrotropic hormone on the thyroid and thyroid function tests. (After studies by Rawson and Netter.)

Tests	TSH (low or absent) Mass decreased Flattened cells	TSH normal Normal Cuboidal cells	TSH increased Mass increased Columnar cells (low colloid)
Thyroid iodine	N to ↑	N	↓
PBI + BEI	↓	N	↑
BMR	↓	N	↑
¹³¹I uptake	↓	N	↑
Blood cholesterol	Variable	N	Usually ↓
Exophthalmos	None	None	Present

at a steady base level. A surge of secretion takes place at the fourteenth day (middle of the menstrual cycle).

EFFECTS. The function of FSH in males is not clear. However, with ICSH it maintains testicular function and probably involves stimulation of androgen secretion since final maturation of sperm requires androgens. In females, FSH acts on growing follicles to stimulate their maturation and growth.

In males, ICSH causes the interstitial cells to secrete testosterone. In females, LH is related to the final maturation of the follicle, estrogen secretion, ovulation, corpus luteum formation, and progesterone secretion.

Lipotropins (Lipotropic Hormone, LPH). Two lipolytic polypeptides have been described, an alpha and a beta, and function as fat-mobilizing substances.

Melanocyte-Stimulating Hormone (MSH, Intermedin). This hormone (Fig. 15-5, Table 15-6) is also present in an alpha and beta form, with the alpha having greater biologic activity. All the MSH molecules have a common heptapeptide core (seen in ACTH and the lipotropins). Apparently the secretion of MSH is dependent on the release of an MSH release–inhibiting hormone (MIH) produced in the supraoptic area of the hypothalamus. Melatonin (produced by the pineal gland) may act on the hypothalamus to stimulate MIH secretion as well as act directly on the melanophore to cause pigment concentration around the nucleus and blanching of the skin (see below).

EFFECTS. There are two types of melanin-containing cells. In cold-blooded animals the material is contained in melanophores. Skin color is determined by pigment distribution in the cytoplasm, and the skin is pale when the pigment is concentrated around the nucleus and dark when it is scattered throughout the cytoplasm. In mammals (and birds) melanin is found in melanocytes, which synthesize and store their pigment and pass it on to nearby cells and structures. The alpha form of MSH causes increased melanin synthesis in both cell types. The exact role of MSH in man is not clear.

Clinicopathologic Considerations. Hyperfunctioning

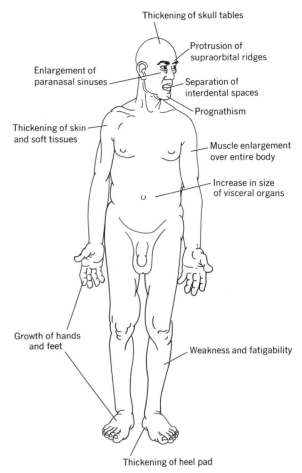

Figure 15-8. Acromegaly—its physical manifestations.

disorders (Fig. 15-7) include gigantism and acromegaly. Gigantism is increased production of growth hormone during the prepubertal period, resulting in excessive skeletal and somatic growth. Acromegaly results from an overproduction of growth hormone in the adult. The face, feet, and hands gradually enlarge from overgrowth of connective tissue and bone; the features become coarse, with thickened skin and prominence of the facial bone (mandible and frontal). Weakness, fatigue, headache, and joint pain develop (Fig. 15-8).

Hypofunctioning disorders include partial and complete hormone deficiencies, with symptoms depending on the hormones involved. Deficiency of growth hormone is the most common and results in dwarfism. The child is of normal size at birth, but in time the normal increase in height does not occur; there is a pudgy appearance with abnormal fat deposits around the abdomen, hips, and pectoral areas.

THE NEUROHYPOPHYSIS. Secretions associated with this part of the pituitary gland (Fig. 15-5) are produced in the hypothalamus. Seven natural cyclic nonpeptides have been discovered in neurohypophysial extracts— oxytocin, vasotocin, mesotocin, echthyotocin, glumitocin, arginine vasopressin, and lysine vasopressin— which apparently originate in the supraoptic and paraventricular nuclei in mammals (Figs. 15-2, 15-6). They are transmitted via axons of neurons that make up the nuclei to the capillary beds in the neurohypophysis. The hormones are stored there in granules accumulated at the axon terminals and must be released from these granules, called neurophysins, before being secreted. The neurohypophysis may also serve to transport and store the hormones, since three hormone-binding proteins (neurophysin I, II, and III) have been discovered. Each hormone is secreted independently from its own hypothalamic nucleus: (1) the supraoptic nucleus for vasopressin secretion and (2) the paraventricular for oxytocin secretion.

Oxytocin and Arginine Vasopressin. These hormones are found in the posterior pituitaries of most mammals. The other oxytocins and vasopressins are found in other animal species. The hormones differ in amino acid composition, have similar qualitative activities, but differ remarkably in their quantitative effects.

Regulation of vasopressin secretion is initiated by many different pathways converging on the hypothalamus, with changes of plasma osmotic pressure being the prime mechanism. However, changes in blood volume or blood pressure also contribute to control of vasopressin secretion. Pain also affects the nervous system, induces stress, and inhibits secretion of vasopressin.

Oxytocin in pregnant animals induces parturition and is important in the lactating female since, in its absence, no milk is released from the mammary gland.

EFFECTS. Both oxytocin and vasopressin can cause similar responses of different magnitudes. Vasopressin exerts an antidiuretic effect physiologically by affecting water and urea permeability of renal nephrons. An absence of it causes a condition called diabetes insipidus, in which large volumes of water are consumed and lost in the urine. Oxytocin is apparently important only in the pregnant and lactating female and causes contraction of uterine muscle and milk ejection.

Thyroid Hormones

The thyroid gland produces two different kinds of hormones. The first is actually a group of substances, the iodinated derivatives of thyroxine (see structure below) called thyroxine and triiodothyronine; the second is a polypeptide called calcitonin, which involves regulation of blood calcium levels (Table 15-6).

The thyroid gland uses iodine in the biosynthesis of the thyroxines. Seven iodinated amino acids have been isolated as constituents of the normal thyroid gland. However, the term *thyroid hormone* usually refers to either 3, 5, 3′-triiodothyronine (T_3) or thyroxine (T_4).

T_3—three iodines

T_4—four iodines

The thyroid gland by the combination of inorganic iodine with the amino acid tyrosine forms T_3 and T_4,

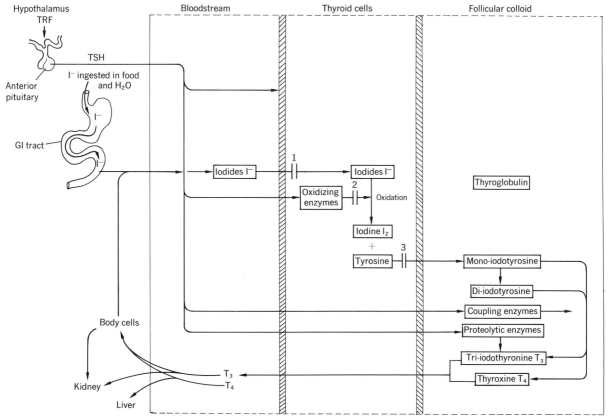

Figure 15-9. Formation, secretion, transport, and excretion of thyroid hormones. Iodide (I⁻) is trapped by thyroid-follicle cells and is converted by oxidizing enzymes in the cells into iodine (I_2), which combines with tyrosine to form the thyroid hormones. The latter are stored attached to thyroglobulin. When the need arises, they are freed by proteolytic enzymes and pass into the blood. TSH stimulates hormone production by acting to abet iodide trapping and the activity of three sets of enzymes. A variety of substances tend to interfere with hormone production. Thiocyanates (KSCN) and perchlorates ($KClO_4$) block the iodide trap **(1)**; thiouracil, the oxidizing enzymes **(2)**; and sulfonamides, the combination with tyrosine **(3)**.

the thyroid hormones (Fig. 15-9). Iodine is trapped in the gland and is concentrated there. What is not utilized is excreted by the kidneys. T_3 and T_4 are stored in the thyroid gland bound to thyroglobulin. Thyroglobulin is a glycoprotein with a molecular weight of 660,000. It is completely digested prior to release of its amino acids into the bloodstream, and thus the organic iodide of the blood is predominantly T_4 and

some T_3 until released into the bloodstream by enzymatic action. In the blood, both hormones are bound to plasma proteins, chiefly to thyroid-binding globulin (TBG) and, to a lesser degree, to thyroid-binding prealbumin (TBPA). Over 90 percent of the total hormone in peripheral blood is T_4, which is more strongly bound to TBG than is T_3. The amount of free T_3 and T_4 is minimal, but the free fractions are responsible for the

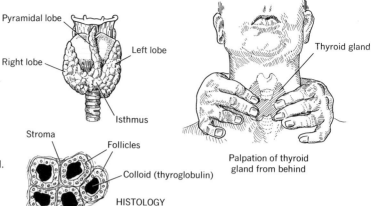

Pyramidal lobe

Left lobe

Right lobe

Isthmus

Thyroid gland

Palpation of thyroid
gland from behind

Stroma

Follicles

Colloid (thyroglobulin)

HISTOLOGY

Figure 15-10. The human thyroid gland.

physiologic effects of the hormones. Although T_3 is present in smaller amounts, it is more active than T_4 and equal to it physiologically.

Thyroid-stimulating hormone (TSH) from the pituitary affects all stages of thyroid hormone formation, including iodide trapping, hormone synthesis, storage, and release. High levels of thyroid hormone in the blood inhibit TSH secretion by the pituitary, and low levels act in reverse.

Anatomy. The thyroid gland (Fig. 15-10) consists of two lobes, often connected by a narrow isthmus across the front of the trachea. It normally weighs about 30 ± 10 gm and measures about $4 \times 3 \times 2$ cm. It is composed mainly of spherical follicles lying in a highly

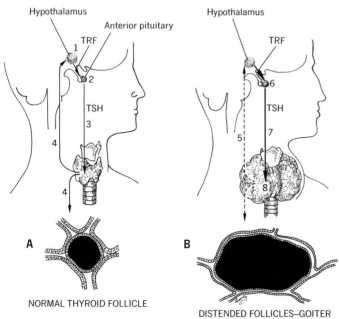

Hypothalamus

Anterior pituitary

TRF

TSH

A

NORMAL THYROID FOLLICLE

Hypothalamus

TRF

TSH

B

DISTENDED FOLLICLES—GOITER

Figure 15-11. Feedback mechanism to control production of thyroid hormones. Neurosecretion begins in hypothalamus **(1)** with secretion of thyrotropin-releasing factor (TRF), which goes directly to the pituitary **(2).** The latter releases thyrotropin or thyroid-stimulating hormone (TSH) **(3)** to bring about the synthesis and secretion of thyroid hormones T_3 and T_4 into the circulation **(4).** The amount of T_3 and T_4 reaching the hypothalamus, in turn, controls TSH secretion, completing the negative feedback loop. If there is an iodine deficiency, not enough hormone is produced **(5)** to turn off the system and, therefore, excess TRF **(6)** and TSH **(7)** are secreted, stimulating the iodine-depleted thyroid tissue to grow **(8),** produce more colloid, distend, and become a "goiter."

vascular matrix. Each follicle consists of a single layer of cuboidal epithelial cells enclosing a colloidal mass of thyroglobulin. The follicle size and the height of the epithelial cells depend on the functional condition of the gland, which in turn reflects the concentration of TSH in the blood. An enlarged thyroid gives rise to a goiter due to excessive stimulation of the thyroid by TSH. The latter may be caused by inadequate production of T_3 or T_4 as a result of dietary iodine deficiency, poor iodine absorption, or excess production of TSH. If too much TSH is secreted, the enlarged thyroid secretes great amounts of T_3 and T_4, resulting in a hyperthyroid condition (note that the long-acting thyroid stimulator, LATS, an abnormal protein found in plasma, may also cause this condition). Goiter appears in an endemic form in areas where there is iodine deficiency and can be cured by supplementation. This goiter of deficiency results in a thyroid with large follicles, much colloid, and shortened, squamous epithelial cells (Fig. 15-11). The goiter of TSH excess is characterized by small follicles, little colloid, and an epithelium of columnar cells.

Effects. The thyroid hormones regulate the metabolism of most of the adult tissues, such as kidney, liver, heart, and skeletal muscle, but they do not normally affect the lungs, lymphatic system, gonads and accessory organs, nervous tissue, skin, smooth muscle, or the thyroid gland itself. They have a depressant action on the pituitary. T_3 appears to act sooner than T_4 since it is bound less firmly to serum proteins and its effects are of shorter duration, but no true qualitative differences occur.

Thyroid hormones play an essential role in normal growth and development. Thyroid deficiency resulting from iodine shortage or congenital thyroid insufficiency during infancy leads to a condition called cretinism. The cretin is a mental and physical dwarf with inhibited growth and maturation of both the skeletal and nervous systems. Gonadal development is juvenile as well. Mental retardation prevails due to both poor brain development and inadequate nerve growth. The condition can be reversed, if treated early in the growth period with T_3 or T_4.

The thyroid hormones seem to be exclusively involved in the maturation of the brain, bone, and skin. Growth hormone apparently stimulates normal growth, but T_4 enhances the rate of growth.

Excessive doses produce an effect on catabolic and oxidative metabolism, with an increase in oxygen use, negative nitrogen balance, and weight loss—all considered a thyrotoxic effect.

Thyroid secretion is also related to environmental temperature (thermogenesis). Thus, thyroidectomized animals are more sensitive to cold, and in human myxedema (adult thyroid deficiency) the heat-regulating mechanism is impaired.

The major signs and symptoms of thyroid states are as follows (there are many other signs and symptoms that exist in either state) (Table 15-7):

HYPOTHYROIDISM. Hypothyroidism is characterized by coarse and dry skin, lethargy, sensitivity to cold, decreased or absent sweating, facial edema, increased appetite, and even deafness.

HYPERTHYROIDISM. Clinical manifestations of this disorder include warm and moist skin, five-finger tremor, sensitivity to heat, increased sweating, palpitations, exophthalmos, insomnia, familial history of thyroid disease, and even eyelid retraction.

Pancreatic Hormones

The pancreas functions (1) in the regulation of carbohydrate metabolism by the endocrine secretion of two hormones that control the concentration of blood glucose (Table 15-6) and (2) in the synthesis of digestive enzymes. About 50 years ago Banting and Best showed that extracts of pancreas caused hypoglycemia in diabetic dogs. Abel crystallized insulin in 1926, and in 1955 Sanger determined its amino acid sequence. Insulin was synthesized by Katsoyannis and associates in 1964.

Anatomy. The adult human pancreas (Fig. 15-12) is a compact, lobulated organ in man. It weighs from 50 to 70 gm, of which the islets of Langerhans, which are distributed throughout the gland but most numerous in the tail, represent about 1 to 2 percent of pancreatic volume. The islets develop from the ends of

Figure 15-12. The pancreas, gross and microscopic details.

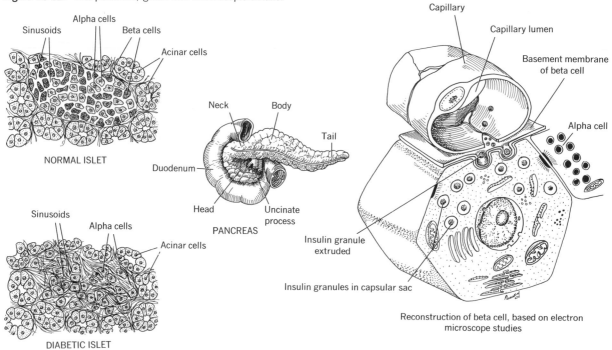

NORMAL ISLET

Sinusoids
Alpha cells
Beta cells
Acinar cells

DIABETIC ISLET

Sinusoids
Alpha cells
Acinar cells

PANCREAS

Neck
Body
Tail
Duodenum
Head
Uncinate process

Capillary
Capillary lumen
Basement membrane of beta cell
Alpha cell
Insulin granule extruded
Insulin granules in capsular sac

Reconstruction of beta cell, based on electron microscope studies

Figure 15-13. Electron micrographs of several adjoining beta cells **(A)** and a portion of a beta cell at higher magnification **(B)**. The beta granules of man and several other species are membrane-bound spherical vesicles containing dense crystals of varying configuration. **A,** $\times 5,400$; **B,** $\times 11,448$. (Courtesy of Dr. Dale E. Bockman, Medical College of Ohio at Toledo.)

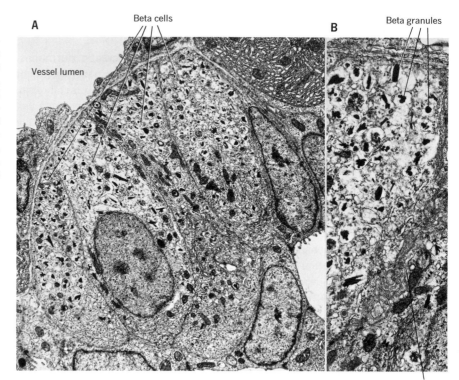

A
Beta cells
Vessel lumen
B
Beta granules
Cell membrane

primitive ducts and are first seen in the fetus at the end of the third month of gestation. The beta cells constitute 60 to 80 percent of the cells of the islet.

The islets are uniform in size, rounded, and arranged in clusters or cords separated by capillaries. Special staining techniques are required to distinguish between the alpha and beta cells of the islet. With correct fixation and staining, beta cells show deep-purple granules and occupy about two thirds of the islet-cell population. The rest are alpha cells, c cells, and delta cells. The latter two are rare. The beta cell produces insulin (a hypoglycemic factor), the alpha cell releases glucagon (a hyperglycemic factor), and the delta- and c-cell function is unknown.

The beta cells, as viewed by electron microscopy (Figs. 15-12, 15-13), contain one or more granules lying free in membranous sacs. Two basement membranes separate the beta cells from the capillaries: one next to the epithelial cell, and the other adjoining the endothelium. Fibroblasts and nerve fibers lie between the paired membranes. The beta granule is a preinsulin, insulin-protein complex, and its shape varies in different species. The granules are apparently formed in the ribosomes of the endoplasmic reticulum. The release of insulin is accomplished by the movement of the granule-containing sacs to the cytoplasmic membrane, where they are expelled into the extracellular space. Here they dissolve in the extracellular fluid and enter nearby capillaries. Alpha granules are round and completely fill their enveloping sacs. Delta granules look like the alpha but are less dense.

Insulin. Bovine insulin has a molecular weight of about 6000 and consists of two polypeptide chains (A and B) with amino acids connected by disulfide bonds. As mentioned above, it probably is formed by the cleavage of a macromolecule called proinsulin.

Several stimuli induce pancreatic insulin secretion. However, the beta cell is apparently not dependent on outside trophic support, since cutting the vagus nerves or removing the anterior pituitary gland does not cause marked changes in the cells even though the storage and secretion of insulin are somewhat reduced. An excess of growth hormone, glucocorticoids, or thyroid hormone will stimulate and eventually exhaust the beta cells. All of these hormones produce hyperglycemia, which demands more insulin.

The major stimulus for insulin secretion is blood sugar concentration (Fig. 15-14). Thus, ingestion of food stimulates secretion. Food in the stomach and duodenum causes secretion of three gastrointestinal hormones (gastrin to increase gastric hydrochloric acid, secretin to stimulate secretion of watery pancreatic juice, and pancreozymin to stimulate pancreatic enzyme secretion). All three increase insulin secretion. Glucagon in small doses stimulates secretion. Norepinephrine and epinephrine, on the other hand, inhibit secretion, even if glucose is given at the same time. Thus, insulin is predominantly regulated by the interaction of hormones with the beta cells and by the action of blood sugar as a negative feedback stimulus.

EFFECTS. Insulin affects the metabolism of glucose, fatty acids, and amino acids, since it is a general metabolic hormone modifying the use of nutrients. It aids in cell permeability, particularly for glucose and other monosaccharides. Muscle cells, fibroblasts, and adipose tissue cells do not allow rapid, free entry of sugars since they have a special transport system to do so. This system requires insulin, for without it sugar cannot enter the cell unless the blood level is very high. Neurons and red blood cells, however, do not need the hormone to transport sugar across their cell membranes. Even the intestine and the kidney tubule seem to function without insulin. The liver, too, can transport sugars freely without it.

Hyperglycemia (increased blood sugar) induces insulin secretion (the converse is true). With hypoglycemia (reduced glucose levels) glucagon is released and apparently leads to liver glycogenolysis.

DIABETES MELLITUS. This disorder is not fully understood. It is essentially due to either complete or partial insulin deficiency, with increased blood sugar (hyperglycemia) as an invariable manifestation. It is a chronic systemic disease characterized by (1) disorders in metabolism of insulin and carbohydrate, fat, and protein; and (2) disorders of structure and function of blood vessels. The early symptoms are related to the metabolic problems; the late complications are related to the circulatory problems.

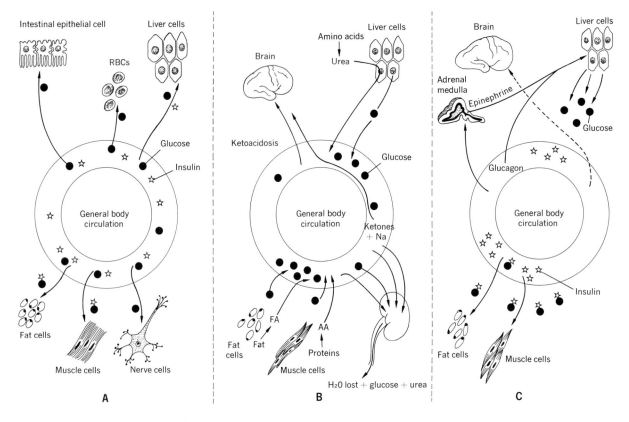

Figure 15-14. Insulin and its regulatory function in the body under various conditions. **A.** Normal cellular uptake of glucose. In fat and muscle cells insulin promotes glucose uptake. The uptake and release of sugars in the liver are unclear. Insulin has no effect on glucose uptake in the intestinal cells, red blood cells, or nerve cells. **B.** Hyperglycemia (insulin depletion). Stored glucose is released by liver and urea output also increases. Glucose is unable to enter fat or muscle cells without insulin. Thus, glucose level in general circulation rises and H_2O plus glucose and urea are secreted by the kidney. There are polyuria, glycosuria, ketoaciduria, and mineral loss. Ketoacidosis may lead to coma and death. **C.** Hypoglycemia (insulin excess). Excess insulin promotes rapid glucose uptake by muscle and fat cells. Epinephrine and glucagon promote release of stored glucose in liver, leading to hepatic depletion. Rapid decline in blood glucose stimulates adrenal medulla to produce epinephrine (causes anxiety, tremors, weakness, etc.) and also deprives the brain of glucose, leading to shock and convulsions. (After studies by Jackson.)

Two types are known: an early-onset or juvenile type, which appears before age 20; and a late-onset or maturity type, generally seen after 40 years of age (see Table 15-2 for a comparison of the characteristics of both types). In the juvenile there is an absolute insulin deficiency, whereas in the adult there is more often a delayed release of internal insulin in relation to a carbohydrate challenge or even a subnormal ability to synthesize or release it.

The pancreas in diabetes rarely shows gross changes. Interestingly, in one third of the cases, conventional histology reveals no microscopic changes; however, when present, it is the islets that are affected. Changes are seen in the kidneys, where there is an early appear-

Table 15-2. Characteristics of Adult-Onset and Juvenile Diabetes

	Juvenile Type	Adult Type
Age of onset	Usually but not always during childhood or puberty	Mostly over 40
Type of onset	Abrupt	Usually gradual
Family history	Frequently positive	Commonly positive
Nutritional status (at onset)	Usually undernourished	Obesity usual
Thirst and hunger	Polydipsia, polyphagia, and polyuria	May be none
Hepatomegaly	Common	Uncommon
Stability	Brittle—blood sugar fluctuates widely in response to small changes in insulin dose, exercise, and infection	Blood sugar fluctuations
Control	Difficult	Easy with proper diet
Ketosis	Frequent (if treatment insufficient)	Uncommon except in stress or sepsis
Plasma insulin	Negligible to zero	Response may be delayed but adequate or diminished (not absent)
Vessel complications	Infrequent until present for 5 years	Frequent
Diet	Mandatory	If used, insulin may not be needed
Insulin	Needed by all	Needed in 20 to 30% of cases
Oral agents	Rarely useful	Useful

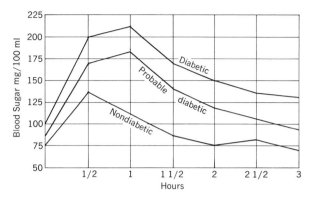

Venous Blood Sugar (mg per 100 ml)					
	Fasting	1/2 hour	1 hour	1 1/2 hours	2 hours
Normal	<100	<160	<160	—	<100-110
Probable	<110-120	130-159	160-180	>135	110-120
Diabetes	>110	>150-160	>160	>140	>120

Figure 15-15. Criteria used for interpretation of glucose tolerance tests.

ance and an increased incidence of arteriosclerosis; degenerative changes appear in the nervous system; diabetic retinopathy (eye changes) with increased frequency of cataracts is seen; and infant mortality in children born of diabetic mothers is high.

The disease is generally characterized by polyuria (increased urination), polydipsia (frequent drinking), and polyphagia (excess food consumption) as well as hyperglycemia (increased blood sugar) and glycosuria (sugar in the urine). The normal blood glucose concentration in fasting animals is 70 to 120 mg percent. In diabetes it is over 150 mg percent and may reach 300 to 400 mg percent or higher.

Although many tests are used in the diagnosis of diabetes, the most commonly accepted is an oral glucose tolerance. The standard test is one in which 100 gm of carbohydrate are given and blood glucose is determined at 30, 60, 90, 120, and 180 minutes. (See Fig. 15-15 for criteria used to interpret the resultant curves.)

Glucagon. Glucagon is a single polypeptide chain of 29 amino acid residues with a molecular weight of 3485, and thus it does not resemble insulin. The hormone is secreted by the alpha cells of the pancreas, which are stimulated by hypoglycemia. Although blood sugar concentration is its most potent regulator, pancreozymin is also effective.

EFFECTS. Glycogenolysis and hyperglycemia are produced by glucagon action on the liver. Thus, regulation of blood glucose is its primary role.

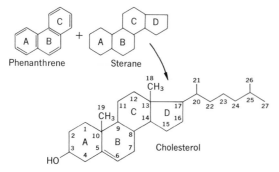

Figure 15-16. Cholesterol is the parent sterol and accepted precursor of the five major classes of steroid hormones. Cholesterol is composed of three fused cyclohexane rings **(A, B,** and **C)** attached to a terminal cyclopentane ring **(D)** with a saturated, branched side chain at C17.

Steroid Hormones

Steroids are derivatives of a hydrogenated phenanthrene core to which a cyclopentane or a sterane

is attached. The number of biologically active ingredients of this basic structure is high, but the naturally occurring derivatives form a definite pattern with a hydrocarbon side chain at position 17 and methyl groups at 10 and 13 (Fig. 15-16).

The most commonly occurring animal steroid is cholesterol, whose name is derived from the fact that it is a common constituent of human gallstones and is deposited in the bile (chole) duct. Cholesterol is found in all normal tissues. It is the major component of esterified lipids, myelin sheaths, and most cell membranes as well as a metabolic precursor of most of the steroid hormones (Fig. 15-17). The hormones° to be considered are as follows: sterol (cholesterol), found in all tissues and cells; estrogens (estradiol, estrone, and estriol), found in the ovarian follicles; androgens (testosterone), produced by the interstitial cells of the

°Each hormone group is followed by a parenthetical indication of its principal compound.

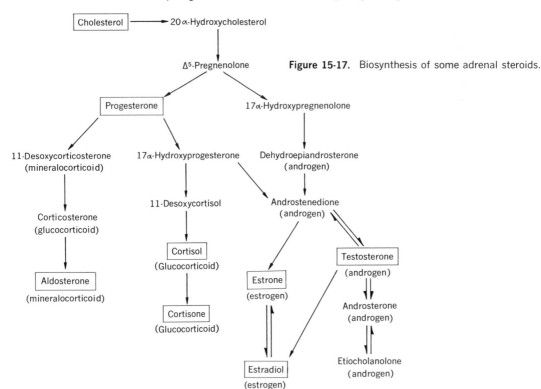

Figure 15-17. Biosynthesis of some adrenal steroids.

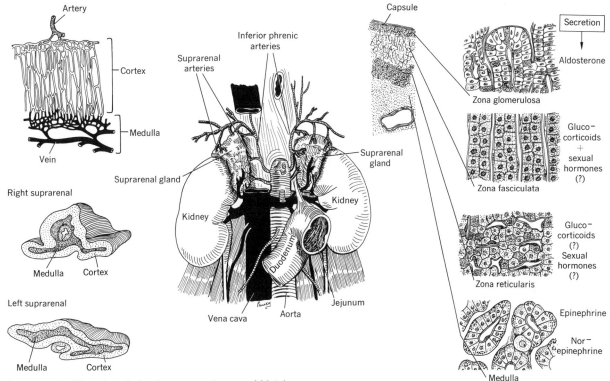

Figure 15-18. The adrenal gland, gross anatomy and histology.

Labels in figure:
Artery
Cortex
Medulla
Vein
Right suprarenal
Medulla   Cortex
Left suprarenal
Medulla   Cortex
Suprarenal arteries
Inferior phrenic arteries
Suprarenal gland
Kidney
Duodenum
Vena cava   Aorta   Jejunum
Suprarenal gland
Kidney
Capsule
Secretion
Aldosterone
Zona glomerulosa
Gluco-corticoids + sexual hormones (?)
Zona fasciculata
Gluco-corticoids (?) Sexual hormones (?)
Zona reticularis
Epinephrine
Nor-epinephrine
Medulla

testes; progesterone (progestogen), produced by the ovarian corpus luteum; mineralocorticoid (aldosterone), synthesized by the zona glomerulosa cells of the adrenal cortex; and glucocorticoid (cortisol), synthesized by the zona fasciculata of the adrenal cortex (Fig. 15-18).

The Sex Hormones. The gonads of both sexes serve to form gametes and produce hormones responsible for growth and maintenance of accessory sexual structures (see Chaps. 17 and 18). In both sexes, these hormones are responsible for sexual behavior and body configuration (Table 15-7).

ANDROGENS. The androgens, of which testosterone is the most prominent, are produced mainly by the interstitial cells of the testes. Although the testes are the primary source, the adrenals also synthesize and secrete small amounts of androgens. These hormones serve in the maintenance of the testes, reproductive

tract, and secondary sexual characteristics of the male. Testosterone is necessary for spermatogenesis, continued activity of the glands of the reproductive tract, growth of the testes and scrotum, growth and distribution of body hair, and body configuration. Metabolically androgens stimulate protein metabolism and induce protein synthesis in muscle.

Androgens given to females may cause sex reversal. Castration in prepubertal animals results in failure of the reproductive tract to mature as well as failure of secondary sexual characteristics to develop. Castration after puberty results in atrophy of both reproductive tract and secondary sexual characteristics.

ESTROGENS. The estrogens are produced during follicular growth by the developing ovarian follicle, and during the luteal phase of the ovarian cycle by the corpus luteum (along with progesterone). Estradiol is the most active naturally occurring estrogen. The two other major estrogens are estrone and estriol. The high concentrations of estrogens in the urine of stallions are assumed to come from the testes (also found in the urine of normal men). Estrogen secretion by the ovary is controlled by pituitary hormones (FSH and LH). Estrogens, in a "loop circuit," act on the hypothalamus and pituitary to inhibit LH secretion. They are important in the development of the gonadal duct system of the female; they are also important in mammary gland duct development. Estrogens effect metabolism of both organic and inorganic substances, including bone, fat, and protein. Administration also results in elevations of serum calcium and redistribution of bone calcium.

PROGESTINS. The progestin predominantly formed by the corpus luteum is progesterone (progestogen). Progesterone plays a number of roles in the reproductive life of mammals. It speeds the movement of the ovum through the oviducts; it prepares the uterine mucosa for implantation of the fertilized egg; and it inhibits uterine muscle contractility, permitting pregnancy to persist, and in its absence pregnancies terminate quickly. Progesterone is also essential for the development of the mammary glands, and in the presence of estrogens (which cause duct growth) it is responsible for the development of the glandular alveolar portions. Milk production, however, needs prolactin, ACTH, growth hormone, and thyroid hormone.

RELAXIN. Relaxin has been found in the placenta, uterus, and corpus luteum. It is a polypeptide with a molecular weight of 9000. It causes the ligaments of the symphysis pubis to distend. Relaxin also increases dilation of the cervix in pregnant women at parturition. It helps maintain gestation, along with estrogen and progesterone.

Table 15-3. Acute Adrenal Insufficiency

1. Hypotension
2. Nausea and vomiting
3. Collapse
4. Hyperthermia
5. Hypoglycemia
6. Hyponatremia and hyperkalemia

The Adrenal Hormones. Many hormones operate under daily requirements; however, the body is subject to periodic stresses, such as the strains of living, tiredness, or disease, that demand special body reactions. Sudden stresses of only a few minutes may produce hormone secretions (particularly epinephrine). Removal of the adrenal glands therefore affects the metabolism of animals (Table 15-3), since the adrenal steroids permit the animal to make adjustments to metabolic stresses. The most crucial metabolic disturbance after gland removal is the inability to retain sodium, leading to a variety of ion and water imbalances that arise from the loss of water from the circulation. These eventually lead to death if not corrected. A disease of adrenal insufficiency is Addison's disease, which results in muscle weakness, hypotension, decreased heart output, acidosis, and a general inability to adapt to major chemical and physical stresses.

ANATOMY. The adrenal glands (Fig. 15-18) are small glands weighing about 3 to 6 gm each. They lie on the right and left kidneys. Each gland is firmly attached and surrounded by renal fascia. Multiple arteries supply the glands, and many veins drain them. The gland is divided into two portions, a cortex and a medulla.

ADRENOCORTICAL HORMONES. There are three outer layers of cortical tissue that secrete about 40 to 50 steroids. The cells of the cortex develop from embryonic mesoderm into the three layers (Fig. 15-18): (1) the first layer is a thin, cortical layer under the capsule, the zona glomerulosa, consisting of columnar cells where aldosterone, a mineralocorticoid, is synthesized and secreted (ACTH does not influence this secretion); (2) the next layer is the zona fasciculata, which merges gradually with (3) the zona reticularis, the deepest layer surrounding the inner medulla. The cells of the inner two layers synthesize glucocorticoids under the control of ACTH (typical negative-feedback control system). The major glucocorticoids are cortisol (hydrocortisone) and corticosterone. After hypophysectomy, layers two and three atrophy while layer one hypertrophies.

As stated, the glucocorticoids are under ACTH control (Fig. 15-5). The feedback mechanism involves the

Table 15-4. Actions of Cortisol

Muscle wasting
Resorption of bone matrix
Calcium resorption
Increased antibody release
Decreased antibody production
Increased neural excitability
Anti-inflammatory action
Antiallergic action
Increased insulin output (beta cells exhausted)
Increased renal secretion of calcium
Increased glomerular filtration
Potassium loss, sodium retention
Fat deposition
Gluconeogenesis
Increased gastric acidity

hypothalamus, where inhibition occurs to the release of CRH with a reduction of ACTH secretion by the pituitary. Control of aldosterone secretion involves a different regulatory mechanism.

Control of Aldosterone Secretion. Stimulation of renin secretion by the kidney juxtaglomerular apparatus takes place in a declining blood volume or drop in pressure or with a decrease in sodium ion concentration. Renin causes cleavage of angiotensin I (decapeptide) from angiotensinogen (a plasma globulin). A converting enzyme in plasma hydrolyzes angiotensin I to angiotensin II (an octapeptide). Angiotensin II exerts the stimulus on the adrenal cortex that induces aldosterone secretion, which causes an increase in sodium retention, which in turn regulates blood pressure and volume.

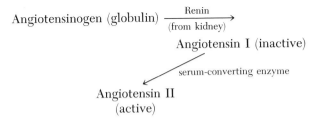

Effects of Adrenal Steroids. An excess secretion of the steroid cortisol due to an adrenal tumor results in Cushing's disease (adrenogenital syndrome). This causes marked localization of fat deposits (moonface, buffalo humpback), muscle weakness, disturbances of electrolyte balance, hypertension, and often diabetes (Table 15-4). If the tumor produces aldosterone, there is extensive disturbance of salt distribution and the development of edema, hypertension, heart failure, and muscle weakness. There may be some masculinization, with steroid hypersecretion leading to virilism in females (atrophy of breasts, inhibition of ovulation, and hirsutism). On the other hand, glucocorticoids administered therapeutically have been used in the treatment of rheumatoid arthritis, rheumatic fever, hypersensitivity, and many allergic and inflammatory diseases.

Thus, aldosterone regulates sodium and potassium balance (electrolyte balance), but most of the metabolic actions of adrenocortical extracts are attributed to the glucocorticoids. Glucocorticoids play a role in stimulating the biosynthesis of enzymes involved in glucogenesis, which divert carbohydrate precursors to glucose formation. The interplay and antagonism of glucocorticoids and insulin in regulating the level of these enzymes are at both the molecular and the genetic level. Insulin, in contrast to the adrenal steroids, acts as a suppressor of the glucogenic enzymes and induces glycogenic and glycolytic enzymes.

Glucocorticoids appear to affect oxidation, synthesis, mobilization, and storage in the process of fat metabolism. Cortisol, for example, induces fat transport from subcutaneous tissues to the liver and serum, resulting in ketonuria and ketonemia.

Adrenal Medulla Hormones

The principal hormone of the adrenal medulla is epinephrine (identified in 1904). Since it is the primary product of the medulla, it is also the most prevalent form of the circulating hormone. Norepinephrine is also present in smaller amounts in the adrenal glands but functions mainly as the neurotransmitter in the sympathetic nervous system and acts locally on effector cells. These two hormones (norepinephrine and epinephrine) along with their precursor, dopamine, make up a group of substances called catecholamines, which chemically share a common orthodihydroxy structure (Fig. 15-19).

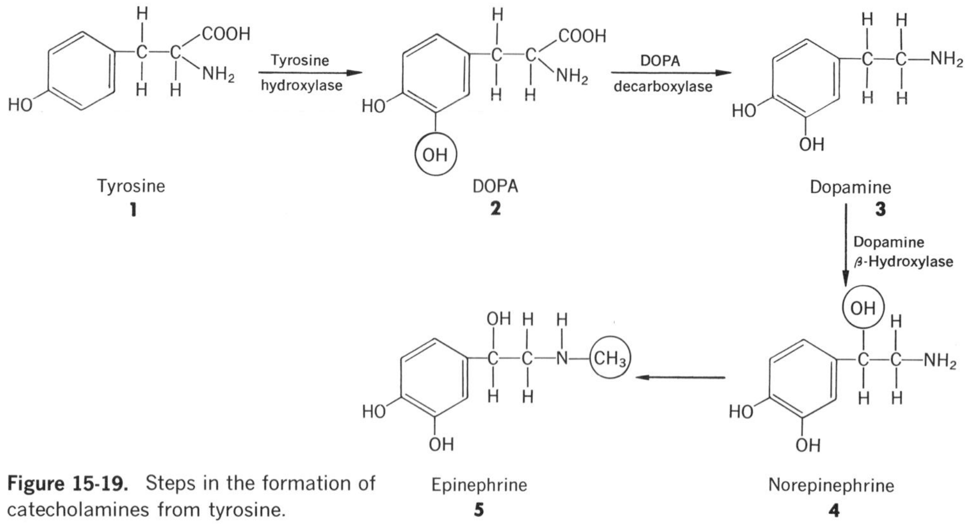

Figure 15-19. Steps in the formation of catecholamines from tyrosine.

The catecholamines are "emergency" or "stress" hormones that enable an animal to mobilize and meet emotional and physical changes.

Release of epinephrine is under direct neural control of hormone secretion. Stimulation of the splanchnic nerves causes secretion of hormones, and cutting the nerves or chemically interfering with them (nicotine) reduces secretion. The adrenal medulla cells are embryologically derived from neural crest tissue and, therefore, are modified postganglionic cells that maintain a connection with the preganglionic fibers of the sympathetic system; secretion from these cells is under splanchnic nerve control. Catecholamines are stored in cytoplasmic granules of the chromaffin cells and are discharged by nerve stimulation. It has been suggested that epinephrine and norepinephrine are formed by different cells of the medulla and may be released separately depending on the stimulus.

Effects. The response to the catecholamines differs widely in smooth muscle and the cardiovascular system. There are apparently two kinds of tissue targets: alpha receptors and beta receptors. The blood vessels of the skin, mucosa, and kidneys constrict due to effects of the alpha receptors, while the vessels of skeletal muscles dilate due to effects of the beta receptors. Apparently low doses of epinephrine affect beta receptors, lowering blood pressure, whereas high doses

Table 15-5. Hormonal Responses Involving Cyclic AMP

Hormone	Response
Catecholamines	Phosphorylase activation in liver
Epinephrine, etc.	Phosphorylase activation in heart Positive inotropic response in heart Lipolysis (adipose tissue)
Glucagon	Phosphorylase activation (liver) Insulin release (pancreas)
ACTH	Steroidogenesis (adrenal cortex)
LH (ICSH)	Steroidogenesis (corpus luteum)
Angiotensin	Steroidogenesis (zona glomerulosa)
Vasopressin	Permeability changes
TSH	Thyroid hormone production
MSH	Melanocyte dispersion
Serotonin	Phosphofructokinase activation
Gastrin	HCl production

activate both receptors, resulting in an increase in peripheral resistance and a rise in blood pressure.

The effects of epinephrine are similar to those produced by stimulating the sympathetic system. Heart rate and blood pressure rise so that more blood reaches the tissues; the arterioles of the muscles and heart

Table 15-6. Summary of Hormones, Their Targets, and Their Actions

Gland and Hormone(s)	Target	Action
I. *Amino Acid Derivatives*		
1. Thyroid		
a. Thyroxine (T_4)	Most body cells	Increased metabolic rate, increased
b. Triiodothyronine (T_3)		oxygen consumption, growth and development
2. Adrenal medulla		
a. Norepinephrine	Most body cells	Increased cardiac activity, elevated
b. Epinephrine	Most body cells	blood pressure, glycolysis, hyperglycemia, vasoconstriction
3. Pineal gland		
a. Melatonin	Melanophores	Depresses melanin
4. Argentaffin cells, Platelets, and nerves	Arterioles, central nervous system	Vasoconstriction
II. *Steroid and Lipid Derivatives*		
1. Testis		
a. Androgens (testosterone)	Most body cells	Development and maintenance of maturation and male sexual characteristics
2. Ovary		
a. Estrogens (estradiol)	Most body cells	Development and maintenance of maturation and female sexual characteristics as well as female sex cycle
3. Corpus luteum		
a. Progesterone (progestogen)	Uterus and mammary gland	Maintains uterine endometrium, stimulates mammary duct formation, helps maintain pregnancy
4. Adrenal cortex		
a. Hydrocortisone	Most body cells	Balances carbohydrate, protein, and
b. Cortisone	Most body cells	fat metabolism; anti-inflammatory action
c. Aldosterone	Kidney	Reabsorbs sodium ions from urine and thus helps in water and electrolyte metabolism
5. Prostate, seminal vesicles, brain, and nerves		
a. Prostaglandins	Uterus	Contraction of smooth muscle
III. *Peptide and Protein Derivatives*		
1. Pituitary gland (hypophysis)		
a. Adenohypophysis (anterior lobe)		
(1) Pars distalis		
(a) Somatotropin (STH, GH, Growth hormone) (acidophil)	All body cells	Growth of tissues, particularly bone and muscles; metabolism of protein; mobilization of fat

Table 15-6 (Continued)

Gland and Hormone(s)	Target	Action
(b) Adrenocorticotropin (ACTH, adrenocorticotropic hormone, corticotropin) (chromophobe)	Adrenal cortex	Synthesis and release of glucocorticoids
(c) Thyrotropin (thyroid-stimulating hormone, TSH) (basophil)	Adipose tissue, thyroid gland	Lipolysis, synthesis, and secretion of thyroxine and triiodothyronine
(d) Male follicle-stimulating hormone (FSH) (basophil)	Seminiferous tubules	Production of sperm
(e) Interstitial cell-stimulating hormones (ICSH, LH, or luteinizing hormone) (basophil)	Testes	Secretion and synthesis of androgens, development of interstitial tissue
(f) Female follicle-stimulating hormone (FSH) (basophil)	Ovarian follicles	Maturation of the follicles
(g) Female LH (basophil)	Ovary (interstitial cells)	Final follicle maturation, estrogen secretion, ovulation, formation of corpus luteum, progesterone secretion
(h) Prolactin (luteotropin luteotropic hormone, LTH) (acidophil)	Mammary glands (alveolar cells)	Milk production in prepared gland
(2) Pars intermedia		
(a) Melanocyte-stimulating hormone (α- and β-MSH, intermedin)	Melanophores	Pigmentation (darkening of the skin)
b. Neurohypophysis (posterior lobe)		
(1) Oxytocin	Uterus, mammary gland	Parturition (contraction of uterus), milk ejection
(2) Vasopressin (antidiuretic hormone, ADH)	Kidneys, arteries	Reabsorption of water and urine concentration, contraction of smooth muscle, blood pressure elevation
2. Pancreas		
a. Insulin	All body cells	Carbohydrate, fat, and protein metabolism; blood sugar level (hypoglycemia)
b. Glucagon	Liver	Glycogen mobilization (hyperglycemia)
3. Ovary		
a. Relaxin	Pelvic ligaments	Relaxation of ligaments and separation of pelvic bones
4. Thyroid		
a. Thyrocalcitonin (CT)	Bones and kidneys	Excretion of calcium and phosphorus (inhibited calcium release from bones and decreased calcium levels)

Table 15-6 (Continued)

Gland and Hormone(s)	Target	Action
5. Parathyroid		
a. Parathyroid hormone (para-thormone, PTH)	Bones and kidneys	Elevated blood calcium and phosphorus levels, mobilization of calcium from bones, inhibited calcium excretion by the kidneys
6. Kidneys*		
a. Erythropoietin	Bone marrow	Increases erythrocyte production
b. Renin	Adrenal cortex	Aldosterone synthesis and secretion

*The renal hormones are enzymes that appear to activate plasma substrates, which in turn act on the target organs.

Table 15-7. Hormones and Syndromes

Organ	Cell	Abnormality Benign (B) or Malignant (M)	Syndrome
Pituitary	Eosinophil	B	Gigantism, acromegaly
	Basophil	B	Cushing's syndrome
	Chromophobe (rare)	B	
Thyroid	Follicular cell	B	Hyperthyroidism
	Parafollicular (?)	Medullary calcium (M)	Hypercalcitoninemia (?)
Parathyroid	Chief cell	B	Hyperparathyroidism
Thymus	?	?	Myasthenia gravis
Adrenal Cortex	Fetal cortex	B	Adrenogenital syndrome
	Zona fasciculata	B	Aldosteronism
	Cortical cells forming nodules (?)	B	Hypertension (?)
Medulla Chromaffin tissue	Chromaffin	B	Pheochromocytoma
	Paraganglioma	B	Epinephrine overproduction
Islets of Langerhans	Beta cell	B	Hyperinsulinism
	Alpha and delta cells	M	Zollinger-Ellison syndrome
	Alpha cell	B	Glucagonemia
Argentaffin tissue	Kulchitsky cell	M	Carcinoid syndrome
Testis	Interstitial cell	B	Hyperandrogenism
	Stertoli-Leydig cell	M	Masculinization
Ovary	Granulosa theca	M	Precocious-puberty hyperestrinism
	Gynandroblastoma	M	Usually androgenic
Testis, ovary, uterus	Trophoblast	Chorionic gonadotropin	Estrogenic stimulation

dilate to allow extra blood to flow through them. In contrast, the arterioles of the gastrointestinal tract constrict to divert blood to more important places, and the liver secretes extra glucose (promotes mobilization of glycogen stores and elevation of blood glucose) to provide more energy to the body. The bronchi and bronchioles also dilate to allow more oxygen to reach the lungs.

Norepinephrine, on the other hand, affects the alpha receptors and causes vasoconstriction and a rise in blood pressure. It is the major chemical mediator at the postganglionic adrenergic nerve terminals.

Little is known of a hypofunctioning medulla, but hyperfunction caused by pheochromocytomas or chromaffin tissue tumors is known. They result in persistent hypertension, elevation of metabolic rate, hyperglycemia, and lipemia, all responses to epinephrine action.

Epinephrine, like glucagon, increases the rate of formation of cyclic adenosine-3′,5′-monophosphate (cyclic AMP), which in turn increases the rate of formation of active glycogen phosphorylase, facilitates neuromuscular transmission, and affects alpha-receptor activation, resulting in vasoconstriction (Table 15-5).

Parathyroid Gland Hormones

The parathyroid glands develop as two pairs of small yellowish bodies on the posterior surface of the thyroid gland but may actually be embedded in the thyroid tissue. The parathyroid glands are composed of two cell types: (1) a small principal or chief cell; and (2) an oxyphil cell, characterized by its purple-staining granules. The chief cell is the source of parathyroid hormone (PTH), which is the hypercalcemic hormone, consisting of a single polypeptide chain with a molecular weight of about 9500 (Tables 15-6, 15-7).

Calcium ion concentration in the plasma acts via a negative-feedback mechanism to regulate the secretion of PTH (Fig. 15-20). Thus, a decline in plasma calcium concentration causes increased PTH secretion. In contrast, an elevation of plasma calcium provokes a secretion of thyrocalcitonin (CT), which causes a decline in plasma calcium concentration. CT acts quickly and its action persists for a brief time; PTH acts more slowly and lasts longer. Thus, PTH and CT

regulate plasma calcium and phosphate concentration over a very narrow range of variation. Both hormones act on the bone cells, which are in chemical equilibrium with calcium and phosphate ions of the plasma.

Effects. PTH acts on the osteoclasts of bone to promote dissolution of bone minerals, which releases calcium into the plasma and produces hypercalcemia. PTH also acts on the kidney by decreasing calcium salt secretion. The hormone acts on the intestinal mucosa to facilitate calcium absorption. Thus, PTH results in hyperglycemia and hypophosphatemia of short duration.

Calcium and Phosphate. Calcium is found in the plasma in both an ionized and a nonionized state bound to protein (Figs. 15-20, 15-21; see Chap. 13). The concentration of free calcium ions depends on the concentration of plasma proteins. The concentration of free ions in the plasma is about one half of the total calcium concentration. Calcium is biologically active only in the ionic state, when it is involved in the regulation of cellular activity by controlling cellular irritability. In a hypocalcemic state, nerves fire spontaneously, muscles react to lower-threshhold stimuli, heart contraction is increased, and relaxation is retarded. These symptoms are seen in hypoparathyroidism and are called tetany.

Lack of calcium results in the formation of rachitic bones lacking in minerals. Since calcium is important in blood clotting, this mechanism may be interfered with.

Phosphorus (as phosphate) concentration is inversely related to calcium concentration (Figs. 15-20, 15-21; see Chap. 13). Phosphates are important in bone mineral; in addition, they form the backbone of nucleic acids, are components of nucleotides, are involved in membrane structures, and are important in phosphorylation (essential in intermediary metabolism) and in buffer systems.

Calcium metabolism is also influenced by vitamin D from the diet (regulates the synthesis of a calcium-binding protein in gastrointestinal tract mucosa, which facilitates calcium absorption), growth hormone from the pituitary (increases the utilization of calcium during growth), estrogens from the ovary and adrenal

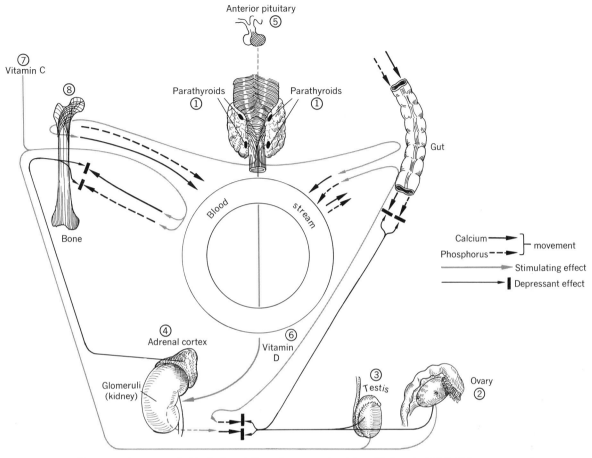

Figure 15-20. Diagram of some probable factors influencing normal calcium (Ca) and phosphorus (P) metabolism. **1.** The principal action of parathormone is disputed. Some believe it is on the kidney to increase urinary excretion of phosphorus; others feel it is extrarenal, probably on bone, stimulating calcium resorption. There is secondary minor action, possibly on gut, to augment calcium absorption. **2.** The ovarian estrogens appear to influence a decrease in urinary P and Ca excretion; appear to decrease fecal P and Ca; and may stimulate bone deposition. **3.** The testicular androgens act on bone, where they stimulate production of bony matrix and facilitate Ca and P deposition; they decrease fecal and urinary Ca and P. **4.** The adrenal cortex appears to decrease Ca and P deposition in bone by antianabolic effect on protein in bony matrix. **5.** The anterior pituitary acts by gonadal and adrenal stimulation and possibly directly on parathyroids. **6.** Vitamin D's principal action is to stimulate absorption of Ca from intestines. It may stimulate urinary excretion of P and possibly stimulate bone resorption. **7.** Vitamin C seems to play a role in the formation of adequate bony matrix to facilitate Ca and P deposition. **8.** Physical stress and strain on bone stimulate Ca and P deposition where needed.

(causes closure of epiphyses, thus arresting bone growth, and prevents calcium resorption, thus stabilizing calcium metabolism), and thyroid hormones (play an important role in calcium metabolism and regulation of bone maturation in both growing animals and adults).

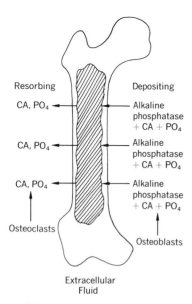

Figure 15-21. Calcium and phosphorus exchange in bone. Resorption is stimulated by parathormone (and possibly vitamin D) and functional inactivity. Deposition is stimulated by local stress and strain and possibly estrogens. Matrix growth needs protein and vitamin C and is stimulated by androgens, inhibited by 11-oxysteroids. Arrows indicate direction of movement of calcium and phosphorus.

Other Hormones

Other substances sometimes classified as hormones are serotonin, the prostaglandins (see later discussion), histamine, and the kinins. However, they act locally and are not dependent on transport to target sites. Still others are certain gastrointestinal and renal hormones.

Gastrointestinal Hormones. Hormones of the gastrointestinal tract (see Chap. 12) are (1) gastrin, which causes an increase in secretion of acid-containing gastric juice; (2) secretin from duodenal mucosa, which causes the flow of a watery pancreatic juice high in salt and low in enzymes; (3) cholecystokinin from intestinal mucosa, which causes the gallbladder to contract; (4) pancreozymin from intestinal mucosa, which causes the pancreas to secrete a juice high in enzymes; and (5) enterogastrone from duodenal and jejunal mucosa, which inhibits gastric motility and secretion.

Renal Hormones. The kidney is the source of two proteins, the enzymes renin and erythropoietin. These proteins exert different effects, but both have hormonal activity. Renin is the specific hormone secreted by the kidney, but its ultimate biologic effect of producing persistent hypertension (high blood pressure) is exerted by angiotensin II, a serum or blood protein.

Erythropoietin is probably produced by the kidneys. It induces hyperplasia of the bone marrow and subsequently an increase in the numbers of circulating reticulocytes and eventually red blood cells.

Pineal Gland

The pineal gland (Fig. 15-22) is a small, ovoid body located centrally in the brain, with its stalk attached to the anterior aspect of the posterior wall of the third

Figure 15-22. Light and the pineal gland. Light stimuli reach the pineal gland via the sympathetic nervous system, as is seen in the diagram. Only after being carried along the inferior accessory optic tract into the spinal cord and through the pre- and postganglionic sympathetic fibers do light impulses enter the gland, where they are taken directly to the pineal cells involved in the synthesis and secretion of the hormone melatonin. (After studies by Wurtman.)

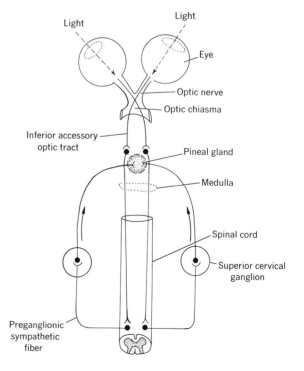

ventricle. It weighs 40 mg in the infant and 150 mg in the adult. In man, the function of the gland is not fully known; however, its apparent role is one of light mediation for the neuroendocrine system. Light input (or absence) controls pineal synthesis and secretion of the hormone melatonin. Little information is available as to which photoreceptor cell of the eye (rod, cone, etc.) mediates the nonvisual effects of light. We do not know whether all visible and ultraviolet wavelengths are equally effective. Since humans spend much of their lives under artificial light sources, which often bear little similarity to natural sunlight, no data are available to evaluate the biologic consequences of such living. Excess exposure to artificial light sources or inadequate exposure to natural light may have still-unknown, harmful biologic effects. In other animals, the gland apparently serves as a photoreceptor containing melatonin. Constant exposure to light causes the pineal gland to shrink and the ovaries to enlarge. Synthesis of melatonin is regulated by the amount of light to which an animal is exposed. Pinealectomy causes the gonads of rats to enlarge. Melatonin, formed from serotonin, is an antagonist of melanocyte-stimulating hormone (MSH) and causes bleaching of the frog or tadpole skin by causing a contraction of pigment granules in melanocytes. Few pathologic alterations of the pineal gland are common.

Thymus

The thymus (Fig. 16-10, p. 559) is located in the anterior upper mediastinum. It has two flat, ovoid lobes with a thin, tough capsule that separates it from neighboring tissues. It is gray-white and has a lobulated structure. At birth the thymus is about 15 gm, is 30 gm at puberty, and in the adult decreases to about 15 gm. It is not certain whether its lymphocyte content arises from elements in the gland or migrates into it from nearby mesenchyme or from precursors in a distant site (such as bone marrow).

The function of the thymus is not entirely known. Its immunologic importance with regard to cellular immunity has been clearly established. There is also evidence of participation in humoral and cellular immune mechanisms.

The thymus frequently is abnormal in myasthenia gravis. However, the relationship of thymoma or thymic hyperplasia to this disease has not been definitely established.

Cyclic AMP

Enzymes catalyze the chemical reactions that proceed in the living cell. Many mechanisms exist that,

Figure 15-23. An example of the role of cyclic AMP in glucose release from the liver. The release of blood sugar by a storage cell for glycogen in the liver is mediated by cyclic AMP. Epinephrine (#1 messenger) arrives at the liver membrane and activates the enzyme adenylate cyclase, causing it to convert some ATP in the cytoplasm into cyclic AMP (#2 messenger). The latter activates the protein kinase, which activates a second kinase. The second kinase triggers a series of sequences that eventually converts glycogen to glucose, which then enters the bloodstream.

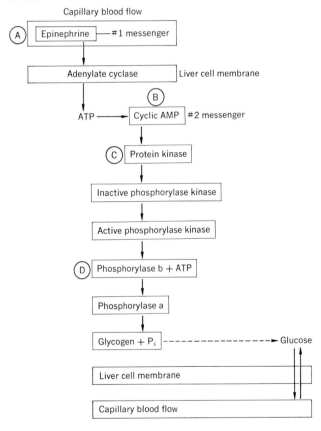

in turn, control the speed at which these enzymes operate. A small molecule that appears to play an important role in regulating the speed of chemical processes in organisms is cyclic adenosine-3′,5′-monophosphate, or cyclic AMP (Figs. 15-23, 15-24).

The term *cyclic* refers to the fact that the atoms in the single phosphate group of the molecule are arranged in a ring. Cyclic AMP serves as a chemical messenger that regulates the enzyme reactions in cells that store sugars and fat, and it controls gene activity.

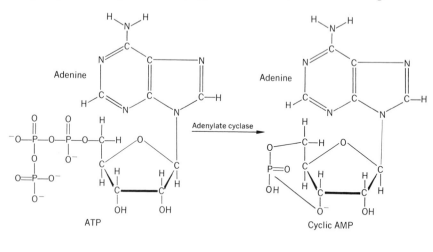

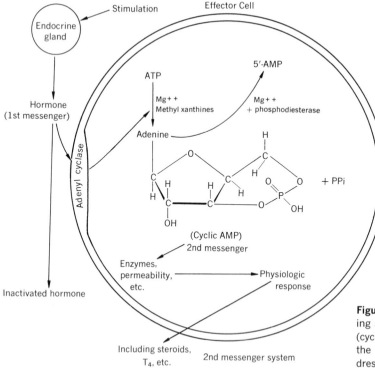

Figure 15-24. The second messenger system involving adenyl cyclase. The second messenger nucleotide (cyclic AMP) picks up hormone-delivered messages at the cell membrane and finds the intracellular "addresses."

A precondition for one kind of uncontrolled cell growth (cancer) has been shown to be related to an inadequate supply of cyclic AMP. Cyclic AMP is present in a wide variety of cells. It is about 1000 times less abundant than adenosine triphosphate (ATP). However, although found in very small amounts, it is present in virtually every organism from bacteria to man and in almost every type of body cell.

The study of the initial steps that led to the discovery of cyclic AMP in 1958 gave Dr. Earl W. Sutherland, Jr., the 1971 Nobel Prize in physiology and medicine. The cyclic AMP molecule is formed from ATP by the action of a special enzyme, adenylate cyclase. The enzyme, which is a target for a number of hormones, is located in the cell membrane. The activity of the enzyme is normally low, and the transformation of ATP into cyclic AMP occurs at a slow rate. When a hormone that is the "first messenger" enters the bloodstream, it travels to the target cell and binds to specific receptor sites on the outside of the cell wall. Note that thyroid cells have receptors that "recognize" thyrotropin, adrenal cells have receptors that recognize ACTH, and so forth. The binding of the hormone to the receptor site increases the activity of the adenylate cyclase in the cell membrane and, using the abundant supply of ATP on the inner side of the cell membrane, cyclic AMP is produced. The cyclic AMP can then diffuse throughout the cell, acting as a "second messenger" to instruct the cell to respond characteristically. Thus, a thyroid cell would respond to cyclic AMP by secreting thyroxine. In the liver, the instruction results in conversion of glycogen to glucose.

The cellular control of cyclic AMP, which is a very strong regulator of cell function, is accomplished by controlling the synthesis of cyclic AMP and by the action of enzymes called phosphodiesterases, which degrade cyclic AMP into an inert form of adenosine monophosphate. The level of cyclic AMP is also controlled by diffusion through the cell wall.

Cyclic AMP binds to still another enzyme, protein kinase, which is inactive until cyclic AMP is present. Once activated, the kinase activates phosphorylase in the liver cell, for example, which results in the breakdown of glycogen. A special enzyme mediates the synthesis of glycogen—glycogen synthetase, which like phosphorylase has an active and inactive form. While some molecules of cyclic AMP are initiating the chain of events leading to glycogen breakdown, other molecules are converting glycogen synthetase from the active to inactive form, and thus glycogen synthesis is kept controlled and energy is not wasted to create too much glucose.

The fat cells supply the need for fatty acids, as do the liver cells for glucose. Fatty acids are stored as triglyceride. It is cyclic AMP that initiates the breakdown of stored fat. Thus, in response to a number of hormonal stimuli, the level of cyclic AMP begins to rise. This activates a protein kinase, which in turn activates a second enzyme, triglyceride lipase, which when converted to active form begins to degrade the stored triglyceride into the needed fatty acids.

Cyclic AMP has been established as an intermediate in a number of hormonal effects besides the cases cited (Table 15-5). In addition to its effects in the cell, cyclic AMP is said to stimulate the expression of genetic information. Here, cyclic AMP is able to stimulate the transcription of many genes. The details of this work are too complex to be discussed here, and the reader is referred to the list of references at the chapter end.

Most of the actions of cyclic AMP, however, still remain to be explored. It has been suggested that abnormalities in the metabolism of cyclic AMP may explain the nature of many diseases. In cholera, the responsible bacteria produce a toxin that stimulates the intestine to secrete much water and salt, resulting in dehydration. Apparently, the toxin initially stimulates the cell to accumulate cyclic AMP, which in turn instructs the cell to secrete the salty fluid. In the case of cancer, it seems possible that some abnormality of the cancer cell may be due to its inability to accumulate normal amounts of cyclic AMP, and these abnormal cells may contain abnormally low levels of cyclic AMP. Research in this area is in progress.

The role of cyclic AMP as a "second signal" leads to the suggestion that endocrine specificity depends on the structure of the membrane receptor and the enzymatic profile of the target tissue.

If one considers a "second messenger," one must

weigh the possibility of an "anti-second-messenger" effect. The so-called antilipolytic effect of insulin depends on its ability to inhibit the formation of cyclic AMP in fat tissue. Prostaglandin was first isolated from the prostate gland but is now known to be present in many tissues and antagonizes cyclic AMP action in a variety of tissues. It also is known to block the lipolytic effects of epinephrine without affecting metabolic or cardiovascular responses.

Prostaglandins

The prostaglandins are a family of substances showing a wide diversity of biologic effects. They resemble other known hormones in their effects, but chemically are a different class of compounds. They are basically fatty acids, and the body produces them in very small quantities (a man normally synthesizes only about 0.1 mg per day of two of the most important prostaglandins). Furthermore, they are rapidly broken down by catabolic enzymes. Bergstrom and Sjovall in Sweden crystallized two of these substances called prostaglandin E and prostaglandin F-alpha. (See Fig. 15-25 for some important prostaglandins.)

The prostaglandins are 20-carbon carboxylic acids and are synthesized in the body from certain polyunsaturated fatty acids by the formation of a five-member ring and the incorporation of three oxygen atoms at certain positions (see Fig. 15-25 for one example). The enzymes involved in their synthesis are widely distributed in many tissues, and their synthesis in many tissues has been demonstrated. It also seems that the cell membrane (as with adenyl cyclase) is a prime site for their formation and that they are produced there as needed. Investigators have found only about $1\,\mu g$ per gram of wet tissue, except for the seminal fluid of man or an extract from a seminal gland, which usually contains 100 times this concentration of prostaglandins. It was this relatively large amount that led to their discovery. Recently, chemists have developed methods of total synthesis of prostaglandins and they are now available for research.

Prostaglandins are among the most potent of all known biologic materials, producing marked effects in very small doses. However, their existence in the body is very ephemeral, since they are acted upon readily by catabolic enzymes. Some of the effects of prostaglandins (demonstrating their versatility and wide range of function) are as follows: (1) one prostaglandin (E_2) lowers blood pressure; (2) prostaglandin F_2-alpha, on the other hand, raises blood pressure; and (3) the effects are generally based on broad powers relating to the regulation of the activity of smooth muscles, secretion, and blood flow. Of particular note is their effect on the female reproductive system, where prostaglandin E_2 or F_2-alpha stimulates uterine contraction. Since they are found in amniotic fluid and venous blood of women during labor contractions, they un-

Figure 15-25. All prostaglandins are variants of a basic 20-carbon carboxylic (COOH-bearing), polyunsaturated fatty acid incorporating a five-member cyclopentane ring and oxygen atoms at certain positions. Small structural changes are responsible for different biologic effects.

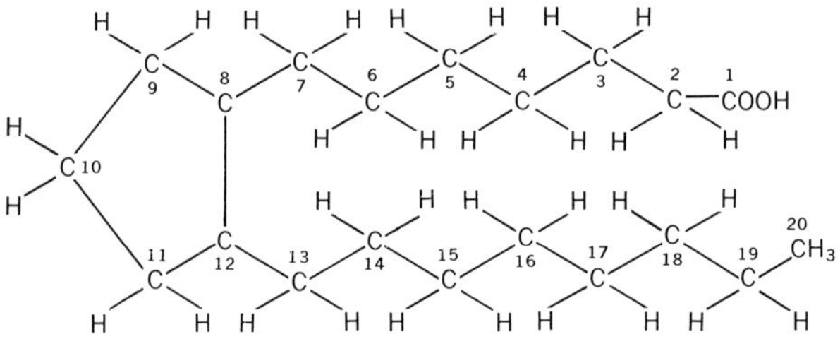

Prostanoic acid

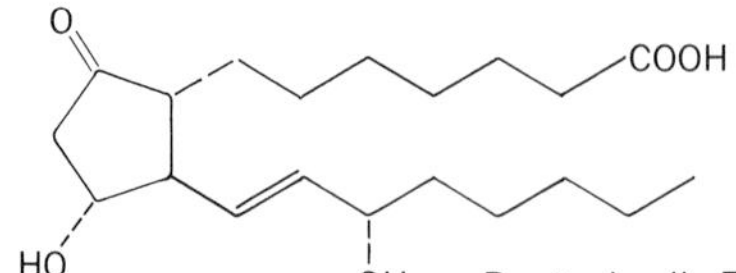

Prostaglandin E_1

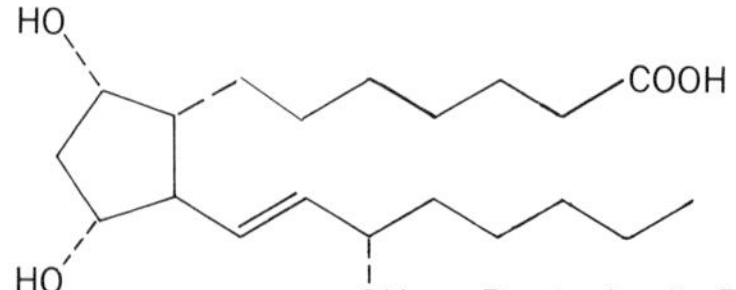

Prostaglandin F_1—alpha

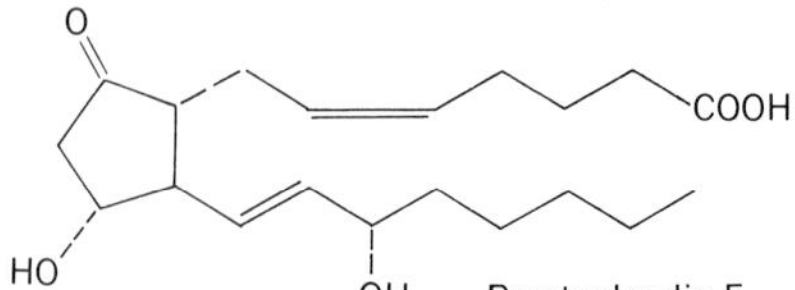

Prostaglandin E_2

doubtedly play a role in parturition. Infusion of prostaglandin E_2 can induce delivery in a few hours, as does oral administration. The use of these substances as agents for causing abortion and inducing menstruation is under investigation. Infusion of prostaglandin F_2-alpha in monkeys, after mating, has been shown to produce a reduction in progesterone secretion and it is felt the hormone may even induce regression of the corpus luteum.

A number of other biologic activities of prostaglandins have been noted: (1) they seem to hold promise for the prevention of peptic ulcers, since prostaglandins E_1 and E_2 have been shown to inhibit gastric secretion in dogs, probably by changing the chemical activity in the parietal cells; (2) experimentally, in asthmatic subjects, an aerosol preparation of type E has improved airflow by relaxing the bronchial smooth muscle; (3) infusion has been shown to help regulate blood pressure; (4) topically applied, prostaglandin E has been shown to clear nasal passages by widening them by vessel constriction; and (5) prostaglandin E_1 has been shown to counteract the effects of many hormones in stimulation of metabolic processes (breakdown of lipids in fatty tissue) and thus help regulate metabolism.

Prostaglandins appear to play an important role in animal physiology. Production and release of these substances can be evoked by nerve stimulation, and there appears to be a clear interrelation between prostaglandins and substances involved in nerve impulse transmission since infusion of type E_2 markedly inhibits the release of norepinephrine in response to nerve stimulation. The prostaglandin acts like a brake on the nerves' action in inciting norepinephrine release. They may play a role in the mechanism that normally controls transmission in the sympathetic nervous system.

To complicate the entire picture and to illustrate the wide distribution of these substances one only has to note that they are formed by the lungs during anaphylaxis, by the kidney when its blood vessels are constricted, by the brain when sensory nerves are stimulated, by the skin in allergic contact eczema, and in certain inflammatory conditions. Thus, they occur normally and in certain pathologic conditions.

Generally, since the cell membrane consists predominantly of phospholipids, proteins, and other components and since the former supply the material for fatty acids, they lead to prostaglandin formation and, therefore, a cell-membrane functional relationship. Here is a medium where endocrine and nervous systems meet and where the translation of messages of specific hormones, as well as the regulation of growth and cell differentiation, occurs. Here, too, in the membrane is where the prostaglandins probably act to influence the formation of cyclic AMP.

Review Questions

1. What is the difference between an exocrine gland and an endocrine gland?
2. What are some of the characteristics of hormones?
3. How do hormones work?
4. Describe the anatomy of the pituitary gland.
5. Name the hormones of the anterior and posterior lobes of the pituitary gland. What do they do?
6. Name the conditions resulting from a hyperfunctioning growth hormone secretion before and after puberty. What about hypofunctioning?
7. Generally speaking, what is the role of the hypothalamus in hormone secretion?
8. What two different kinds of hormones are produced by the thyroid gland? (Where is the thyroid gland located?)
9. What do the thyroid hormones do?

10. What is insulin? What is its major stimulus? What does insulin do?
11. What is diabetes mellitus? What are its symptoms? How does one know when one has this disease? What is the difference between the juvenile and adult types? Can they be treated?
12. What portion of the pancreas is its endocrine part? What does this endocrine part secrete?
13. What is a steroid? Write the formula of the parent sterol. What hormones fall under this classification and where are they produced?
14. What is the function of the following:
    a. Androgen
    b. Estrogen
    c. Progestin
    d. Relaxin
15. Where is the adrenal gland located? What are the layers of the adrenal cortex? What hormones do each of these layers secrete?
16. Where is aldosterone secreted? What controls its secretion? What does it do?
17. What do the glucocorticoids do? What happens when they hyperfunction?
18. What is the function of the adrenal medulla? What are catecholamines? How do they affect us?
19. Correlate parathyroid hormone function with calcium ion concentration. How does calcitonin relate to calcium? What role does phosphate play?
20. Where is the pineal gland located? What is its function?
21. What is cyclic AMP? Why is it important?
22. What are prostaglandins? How are they said to function?

References

Arnaud, C. D.; Rasmussen, H.; and Anast, C.: Further Studies on the Inter-relationship Between Parathyroid Hormone and Vitamin D. *J. Clin. Invest.*, **45:**1955–64, 1966.

Davidson, E. H.: Hormones and Genes. *Sci. Am.*, **212**(6):36–45, 1965.

Forsham, P. H., and Melmon, K. L.: The Adrenals. In Williams, R. H. (ed.): *Textbook of Endocrinology*, 4th ed. W. B. Saunders Co., Philadelphia, 1968, pp. 287–404.

Foster, G. V.: Calcitonin (Thyrocalcitonin). *N. Engl. J. Med.*, **279:**349–60, 1968.

Guillemin, R., and Burgus, R.: The Hormones of the Hypothalamus. *Sci. Am.*, **227**(5):24–33, 1972.

Hermann, P.: Ultrastructure of the Human Thyroid. *Acta Endocrinol. (Kbh.)*, **53**(Suppl. 110):1–102, 1966.

Lacy, P. E.: The Pancreatic Beta Cell: Structure and Function. *N. Engl. J. Med.*, **276:**187–95, 1967.

Leak, A., and Coggins, C. H.: The Neurohypophysis. In Williams, R. H. (ed.): *Textbook of Endocrinology*, 4th ed. W. B. Saunders Co., Philadelphia, 1968, pp. 85–103.

Lloyd, C. W.: The Ovaries. In Williams, R. H. (ed.): *Textbook of Endocrinology*, 4th ed. W. B. Saunders Co., Philadelphia, 1968, pp. 459–536.

Lowy, C., and Williams, E. D.: The Pancreas and Diabetes Mellitus. *Postgrad. Med. J.*, **43**:50–60, 1967.

McCann S. M.; Dharrival, A. P. S.; and Porter. J. C.: Regulation of the Adenohypophysis. *Annu. Rev. Physiol.* **30**:589, 1968.

McCann, S. M., and Porter, J. C.: Hypothalamic Pituitary Stimulating and Inhibiting Hormones. *Physiol. Rev.*, **49**:240, 1969.

McGowan, G. K., and Sandler, M.: Symposium on the Thyroid Gland. *J. Clin. Pathol.*, **20** (Suppl):309–412, 1967.

Parker, M. L.: Juvenile Diabetes, a Deficiency in Insulin. *Diabetes*, **17**:27–37, 1968.

Pastan, I.: Cyclic AMP. *Sci. Am.*, **227**(2):97–105, 1972.

Reklin, R.: The Pineal Gland. *N. Engl. J. Med.*, **274**:944–50, 1966.

Robinson, G. A.; Butcher, R. W.; and Sutherland, E. W.: *Cyclic AMP.* Academic Press, Inc., New York, 1971.

Shields, T. W.: The Thymus Gland. *Surg. Clin. North Am.*, **49**:61–70, 1969.

Tepperman, J.: *Metabolic and Endocrine Physiology*, 3rd ed. Year Book Medical Publishers, Inc., Chicago, 1973.

Turner, C. D., and Bagnara, J. T.: *General Endocrinology*, 5th ed. W. B. Saunders Co., Philadelphia, 1971.

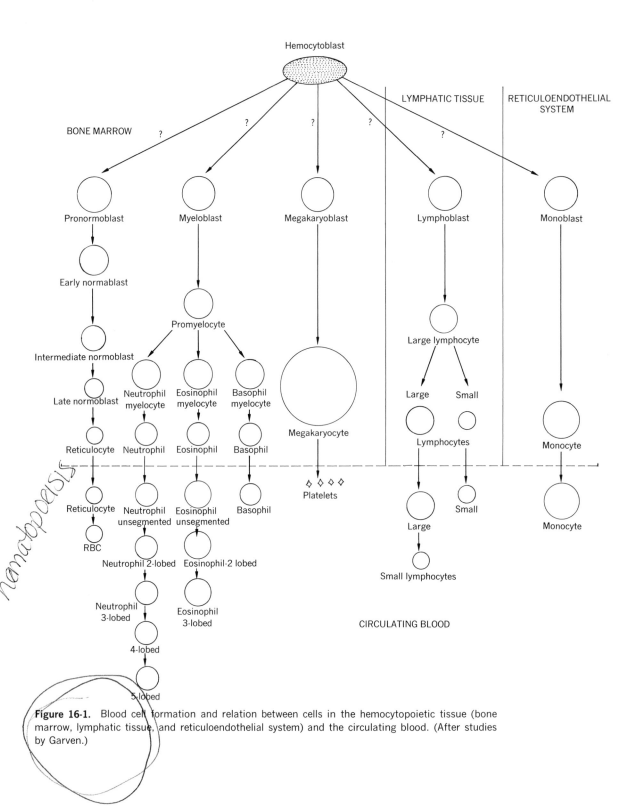

Figure 16-1. Blood cell formation and relation between cells in the hemocytopoietic tissue (bone marrow, lymphatic tissue, and reticuloendothelial system) and the circulating blood. (After studies by Garven.)

Defense Mechanisms of the Body

One of the obvious basic features of man's existence is the fact that his external environment is often hostile and unfavorable from the moment of birth and abounds with potentially dangerous microorganisms: bacteria, viruses, and other unicellular and multicellular parasites. These microorganisms (1) cause cellular damage directly by releasing enzymes that break down cell membranes and organelles, (2) give off toxins that act throughout the body to disrupt organs and tissues as well as the neuromuscular system, and (3) constitute a steady drain of the body's energy supplies since some (viruses in particular) often take hold of the metabolic and reproductive mechanisms of the cell they inhabit.

The body's first line of defense against infection is the anatomic barrier of the surfaces exposed to the environment. Most of our vulnerable organs and tissues are also protected by the skeletal system: the heart and lungs lie in the thoracic cage and the central nervous system in the bony vertebral column. The kidneys are embedded in a mass of shock-absorbing fat, and the pelvis protects other essential organs. Furthermore, the central nervous system with its many sensory receptors and motor endings helps us avoid innumerable hazards. The experience and learning stored in the higher centers of the central nervous system (i.e., the highly developed cerebral cortex) allow us to choose a particular course of action.

Few microorganisms can penetrate intact skin. Furthermore, sebaceous and sweat glands secrete chemicals that are toxic to certain bacteria, and the mucous membranes contain antimicrobial chemicals as well as having a sticky mucus where ciliary action sweeps the organisms away or where phagocytic cells can engulf them. Thus, our major defense is the "unbroken" cover or shield of skin and mucous membranes of the throat and digestive system. If the skin is penetrated or burned, there is a serious loss of body fluids and the body works rapidly to close the openings. The blood provides its own "plug" by forming a clot there, and later the surrounding tissues repair themselves. The mucous membranes are delicate and moist and, therefore, more vulnerable than the skin. However, they do protect themselves with their own secretions: the eye surface is bathed with antiseptic tears from the lacrimal glands; the salivary glands produce antibiotics; and the acid secretion of the stomach does do some sterilization on our food intake. Parts of the digestive tract also contain lymphoid follicles, "patches" of defensive white cells that deal with invading microbes that have managed to get beyond the acid of the stomach.

In spite of the body's external defense surfaces, small pathogenic (disease-producing) microorganisms do enter. They cause illnesses either by disrupting body cells or by producing poisons called toxins. Their effects are often specific and affect a limited area in a certain way. For example, the bacteria causing diphtheria produce a toxin that acts on heart muscle and some motor neurons; therefore, they must be completely destroyed or their toxins neutralized. Several defense mechanisms are in operation, the most important being the chemical substances referred to as antibodies and complement and a group of specialized phagocytic cells (Fig. 16-1). These mechanisms tend to function interdependently.

One of the unique aspects of the body is its determination to destroy any substance recognized by it as a foreign protein. Thus, the body rejects skin grafts or a heart implant from another because it contains proteins foreign to those inherent in the body itself (some exceptions in the tissues and organs of identical twins do exist).

The Cells of Defense

If one examines a drop of blood microscopically after appropriate staining, one notes that over 99 per-

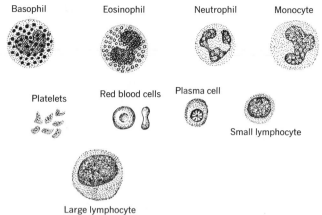

Figure 16-2. Diagram of blood cells seen in a blood film.

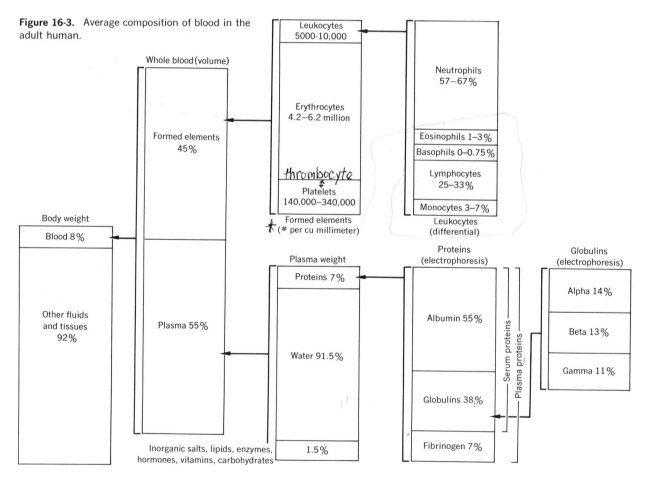

Figure 16-3. Average composition of blood in the adult human.

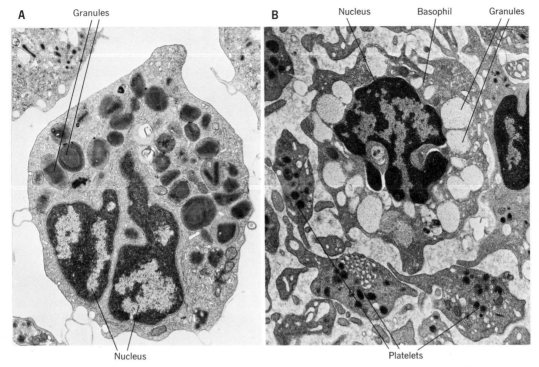

A   Granules   B   Nucleus   Basophil   Granules

Nucleus   Platelets

Figure 16-4. A. Electron micrograph of an eosinophil. × 10,000. **B.** Electron micrograph of a basophil; also seen are several platelets. × 10,000. (Courtesy of Dr. Dale E. Bockman, Medical College of Ohio at Toledo.)

cent of the cells are erythrocytes or red blood cells (see Chap. 11 for function) and the rest are leukocytes or white blood cells, classified according to structure and their reaction to staining dyes (Figs. 16-2, 16-3). There are polymorphonuclear granulocytes with lobulated nuclei and abundant cytoplasmic granules: the eosinophil with red granules (Figs. 16-2, 16-4); the basophil with blue-purple granules, which are few in number and whose function is unknown (Figs. 16-2, 16-4); and the neutrophil (Figs. 16-2, 16-5), whose granules show no dye preference. Another type of leukocyte is the monocyte (Figs. 16-2, 16-6), which is larger and has a single oval or horseshoe-shaped nucleus (mononuclear) with few granules in the cytoplasm. The last class of leukocytes is the lymphocyte (Figs. 16-2, 16-7), which is smaller than the monocyte and has a large nucleus (also mononuclear) with little cytoplasm. Normal human blood contains approxi-

mately five million red cells and seven thousand white cells per cu mm of blood. The latter consist of 50 to 70 percent neutrophils, 1 to 4 percent eosinophils, 0.1 percent basophils, 2 to 8 percent monocytes, and 20 to 40 percent lymphocytes.

The erythrocytes, polymorphonuclear granulocytes, and a few monocytes and lymphocytes are produced in the bone marrow (Fig. 16-1). The lymphoid organs (lymph nodes, spleen, and other tissue) produce only monocytes and lymphocytes. The monocyte may be a derivative of the lymphocyte, as is the plasma cell.

Two other cell types involved in resistance to infection are the plasma cells and the macrophages. The plasma cells are found in the lymph nodes and other lymphoid tissue and appear to be derived from lymphocytes or lymphocyte precursors; they are not normally found in blood and thus are not categorized as white cells. The macrophages are found scattered throughout

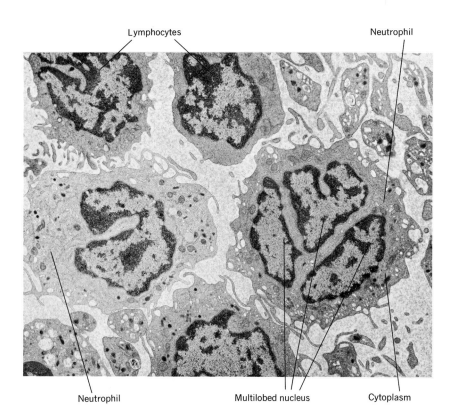

Lymphocytes   Neutrophil

Neutrophil   Multilobed nucleus   Cytoplasm

Figure 16-5. Electron micrograph of a neutrophil; two lymphocytes are also seen. ×7,000. (Courtesy of Dr. Dale E. Bockman, Medical College of Ohio at Toledo.)

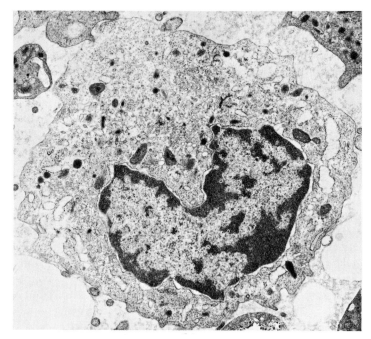

Figure 16-6. Electron micrograph of a monocyte. ×12,500. (Courtesy of Dr. Dale E. Bockman, Medical College of Ohio at Toledo.)

Figure 16-7. Electron micrograph of a lymphocyte. ×12,000. (Courtesy of Dr. Dale E. Bockman, Medical College of Ohio at Toledo.)

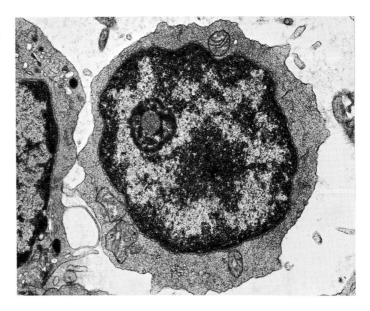

the tissues of the body (in the liver, spleen, lymph nodes, and bone marrow) and actually form part of the lining of the lymphatic and vascular channels. They have the ability to ingest foreign particles. In response to an infection, not only are they capable of mitotic activity but monocytes and lymphocytes can transform into macrophages. The macrophage itself and the plasma cell are "dead-end" cells and cannot convert to another form. The mechanism of controlling the production of these cells is virtually unknown.

The functions of the cells described are concerned with antibody formation and phagocytosis. Antibody results from an interplay of macrophages, lymphocytes, and plasma cells. Phagocytosis, or the ingestion of a particle or cell by another cell, is performed by the neutrophilic and eosinophilic granulocytes, the monocytes, and the macrophages. Note, too, that the phagocytic cells phagocytize nonmicrobial material, such as old red cells (in the liver and spleen).

Antigens

An antigen is a chemical substance that, when introduced into an animal, results in a specific immune response. This response is usually recognized by the appearance of antibodies (see below) with which the antigen reacts specifically. Proteins are generally regarded as the best antigens, but other materials such as nucleic acids and carbohydrates may be antigenic. Factors thought to be important in antigenicity may be molecular size, structural complexity, molecular rigidity, and the structural relationship of the antigen to the host's own tissue components. Each active region of the antigen that combines with antibodies (even multiple antibody molecules) is called an antigenic site. Antigenic sites may be considered to be complementary in structure to the portion of antibodies with which they combine.

Antibodies, Complement, and Phagocytosis

An antibody is a plasma glycoprotein (belonging to the group gamma globulin) that is capable of combining chemically with the specific antigen (usually also a protein) that stimulated its production. The term *immunoglobulins* has been applied to antibodies.

Antibody Structure and Function

Antibodies are found in the gamma globulin portion of the plasma proteins. There are three major fractions of gamma globulin (each in itself very heterogeneous):

1. Gamma (γ) M (IgM, 19s, macroglobulin)
2. Gamma (γ) A (IgA, B_2A, 6.6 to 13S)
3. Gamma (γ) G (IgG, 7S)

Gamma G is the most abundant of the three fractions and is one of the most well characterized of the antibodies; 70 percent of the total are gamma G. The designation 7S means a sediment coefficient of 6.65. The molecular weight is about 150,000.

Recent work has shown that each molecule of the antibody is made up of subunits of two identical polypeptide chains with a molecular weight of 50,000 (called H, A, or heavy chains) and two with a molecular weight of about 20,000 (L, b, or light chains). The chains are bound by disulfide bridges (Fig. 16-8). The H or heavy chains are class specific—gamma G, gamma M, and gamma A—and each is antigenically distinct. Although the light (L) chains are also distinguishable because of antigenic differences, their antigenic classes—K and g delta—are not class specific for immunoglobulins.

Gamma A represents about 20 percent of the total gamma globulins in man and is also heterogeneous.

Like gamma G it has a sediment coefficient of 6.65 but tends to aggregate to a variety of forms with centrifugation (95, 115, and 135).

Gamma M is of relatively high molecular weight, with a sediment coefficient of its main component of 195 (295 and 325 can occur). Evidence seems to show that the M molecular unit is like the other two, consisting of polymers of a basic unit joined by disulfide bridges.

Antigenic substances react only with the type of antibodies elicited by their own kind or related kinds of molecules. The body fights off these "foreign" antigens (or proteins) by producing antibodies. The plasma cells and perhaps other cells have the capability of making antibodies. An essential determinant of antigen capacity is molecular weight (size), and most are greater than 10,000. Note, however, that many small molecules become antigenic by attaching themselves to one of the host's proteins to form a larger complex that induces antibody formation. Furthermore, low-molecular-weight substances can combine with antibodies already induced by other antigens and, since the original antigen is similar or identical to the small molecule, they may induce allergic reactions. Components of almost any foreign cell or molecule not normally present in the body can act as an antigen (even penicillin); for that matter, under certain circumstances even normal body components can induce

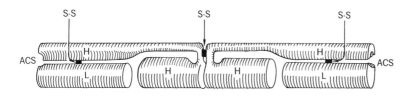

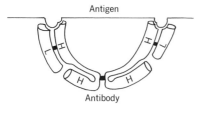

Figure 16-8. Shown is a diagrammatic hypothetic model of the gamma G globulin (top drawing), indicating relations between heavy **(H)** and light **(L)** chain components. Both H and L chains contribute to the two antibody-combining sites **(ACS).** Three interchain disulfide bonds **(S—S)** are noted. Since the chain is flexible, it permits bivalent combination with a single antigen (lower drawing).

antibody formation. Proteins, nucleic acids, lipids, carbohydrates, and many synthetic materials can elicit antibody synthesis. Most cellular membranes (even bacterial) contain antigenic material. The antigens match up with the antibodies somewhat like a key fits a lock, and the combination renders the antigen harmless. Since most foreign proteins that enter the body are from bacteria or viruses, the ability to produce antibodies is a major line of man's defenses.

When disease-causing organisms attack the body in sufficient numbers to make us ill, during the initial stage of infection neutrophils (white cells) swarm and converge to the infected part, engulf the antigen (by phagocytosis), and break it up. Soon they are followed by lymphocytes and monocytes, which engulf both antigen and neutrophils and turn into macrophages. Meanwhile, basophils disintegrate, liberating enzymes. The blood supply to the digestive system is reduced so that we lose our appetite. There is also a rise in temperature—we call it fever. These changes speed up the chemical reactions that are necessary for recovery, although our enzymes can survive this pace only for a short time. There is also little energy to spare for muscular activity, and we are further incapacitated because the brain is confused by toxins or poisons in the blood and the elevated temperature. After the infection is past, the macrophages return to the lymphoid tissues carrying antigen particles. Future cells, resulting from the division of these same macrophages, also contain the antigen and, therefore, a second infection by the same antigen is recognized by the cells. During a second infection, eosinophils are more numerous than neutrophils and they penetrate the macrophages and disintegrate them, which attracts more lymphocytes and macrophages from other areas to continue the destruction of the antigen. Plasma cells are also attracted; they secrete antibodies, which cover the antigen with a substance that makes it easier for other white cells to engulf it.

At birth, there are few lymphocytes in the lymphoid tissue but many are stored in the thymus gland. They are unable to multiply or attack antigens. However, after birth the thymic lymphocytes migrate from the thymus to the lymphoid tissues, and a thymic hormone apparently also migrates to the lymphoid tissues and stimulates the lymphocytes to become effective—to multiply and turn into plasma cells that produce antibodies. The hormone also stimulates lymphoid tissue to produce its own lymphocytes.

When we suffer a local infection (a wound or boil), we can observe what happens in the inflammatory process, which is a relatively nonspecific response to tissue damage depending on the injurious agent (heat, cold, trauma, bacteria, etc.). The first symptom is redness (rubor) and heat (calor) as local blood vessels dilate from carrying more blood, and thus more white blood cells are brought to the area of infection. Then as poisons begin to circulate, the bone marrow and lymph nodes are stimulated to increase their production of white cells. Since germs multiply at a rapid rate, the white blood cells and antibodies must destroy them at a faster rate than they are produced and the mass of germs and white blood cells builds up to form a visible swelling. The pressure of the swelling, along with the extra heat of cell activity, stimulates receptors to relate messages to the brain, which are interpreted as sensations of pain, heat, and throbbing. Connective tissue cells also are multiplying around the area in an effort to help localize the infection.

Meanwhile, as the battle progresses there are millions of cellular and bacterial casualties, including some of the connective tissue cells. The debris of shattered cells is called pus, which drains into the lymph-carrying vessels and passes to the lymph glands. Here it is broken down into harmless substances, and in the process the glands become swollen and tender. If the germs are winning, the glands themselves may become infected. The ability to multiply and produce toxins is the major concern of the germs, and the body counteracts this by its ability to act quickly and with sufficient strength. We speak only of the body's natural abilities, but modern medicine has given us many drugs and antibiotics to help tip the scales in our favor.

Many diseases occur only once in a lifetime (diphtheria, measles, etc.) even though we may come in contact with the causative microorganisms many times. The first infection produces specific antibodies; if they deal successfully with the infective organisms, we develop an immunity to further infections of that organism, since once antibodies are produced to destroy an organism they remain in the blood plasma permanently and repel any subsequent invasion of the same germ.

Immunity is an index of a host's ability to resist the damaging effects of microorganisms and their products. However, the outcome of an infection depends on the virulence of the microorganism as well as the host's defensive mechanisms. The defense mechanisms can be divided into those that require specific activation, called acquired immunity, and those that do not need prior stimulation, called natural immunity. (See p. 557 for a discussion of acquired active or passive immunity.)

The relative importance of cellular and humoral factors in both natural and acquired immunity has been argued. Cellular immunity operates chiefly through phagocytosis by polymorphonuclear leukocytes and either fixed or circulating mononuclear cells. Phagocytosis can occur at the injury site or in a region apart from the injury (lymph node). Phagocytosis is a short-range phenomenon that depends on direct contact with an invading agent and on chemotaxis to direct the cell to the agent (most bacteria are good chemotactic agents). On the other hand, antibodies, complement, and other human plasma humoral factors function as opsonins (characterized by their ability to "coat" the microorganisms and permit them to adhere to the phagocytic cells).

One of medicine's great triumphs has been its work with antibodies and the antibody mechanism in order to confer immunity to diseases before the infection occurs. One method used is the preparation of weakened bacterial strains or viruses of diseases, such as typhoid, cholera, diphtheria, or bubonic plague. These strains of weakened material are called vaccines, which are too weak to produce dangerous symptoms when injected but strong enough to stimulate the body to produce antibodies that can remain in the body and fight future infections. One can also inoculate or vaccinate with blood serum that contains ready-made antibodies, but this works only if done shortly after one is infected and even then results in only a short-term immunity.

Function of Antibodies. Antibodies combine chemically with the toxins and viruses (antigens) to neutralize them (prevent attachment to the host cell membranes and prevent cell entry). They also neutralize bacterial toxins by chemically combining with them to prevent their interaction with susceptible cell-membrane sites. Antibodies do have more than one possible site for combining with an antigen and, therefore, usually form chains or aggregates of complexes that are phagocytized.

Complement

The plasma proteins called complement are also involved in resistance but are not induced by antigens and are distinct from antibodies. They are heat labile, have enzymelike properties, and are normally present in plasma (production site unknown). Complement has the capacity, in cooperation with the antibody and cellular elements, to destroy a variety of pathogenic organisms and other foreign substances. The commonest functions of complement are bacteriolysis, bactericidal action, hemolysis, and acceleration of opsonic actions (an opsonin is a substance, normally found in blood serum, that is necessary to prepare bacteria for phagocytosis). Complement (C') has been fractionated into four components: $C'1$, $C'2$, $C'3$, and $C'4$. Serum complement levels have been shown to be depressed in serum sickness, acute glomerulonephritis, and certain other conditions, indicating that complement may serve a role in tissue injury. Still other conditions, such as rheumatoid arthritis, are characterized by a rise in serum complement levels.

Often the combination of an antibody with its specific antigen in the bacterial cell wall does not damage the cell. However, with the attachment of complement to the complex, the combination becomes lethal due to its catabolic enzyme activity. Thus, initially the cell membrane or wall antigen combines with the antibody, an essential step for complement effectiveness.

Complement then coats the antigen-antibody complex. Thus, complement plays a dual role in immunity—it can be both helpful and harmful.

Phagocytosis

Phagocytosis is apparently the only way of disposing of many bacteria and other microorganisms not sensitive to plasma factors. However, the plasma factors usually prepare the bacteria for the process. Thus, the bacteria are "coated" with antibody (with or without complement), which facilitates phagocytosis (ingestion of foreign or other particles). Other plasma factors apparently also perform in this manner. Thus, antibodies can neutralize bacterial toxins and viruses, participate with complement in the killing of microorganisms, and prepare microbial cell surfaces for phagocytosis (Table 16-1).

Production of Antibodies

Circulating antibodies are synthesized by plasma cells found in the spleen, lymph nodes, and lymphoid tissue. These cells, in turn, apparently are derived from primitive lymphoid tissue and lymphocytes. An antigen triggers the differentiation of lymphocytes, replication of plasma cell precursors, and plasma cells themselves. The antibody is released from the plasma cell endoplasmic reticulum into the interstitial fluid, lymph, or blood.

The antigen is carried from its entry site by way of the lymphatics or bloodstream (free in solution or in a phagocyte). It is, however, phagocytized either at its entry or at its arrival in the lymphoid organs. During catabolic breakdown of the antigen by the

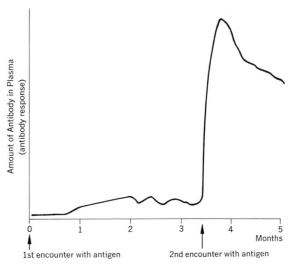

Figure 16-9. Antibody production with initial antigen contact and a subsequent encounter with the same antigen.

phagocyte, the RNA of the latter is altered (coded) to reflect the nature of the antigen. This information is transferred to surrounding lymphocytes or plasma-cell precursors to trigger their differentiation, thereby leading to mature plasma cells and antibody synthesis.

Active and Passive (Acquired) Immunity

Active Immunity. The response of the antibody phenomena depends on time—whether the body has been exposed to that antigen before. Antibody response to a first contact of antigen occurs slowly and takes several days. However, a subsequent infection elicits an immediate and marked antibody response and confers a greater enhanced resistance toward subsequent infection with a particular microorganism (Fig. 16-9). This resistance, as a result of actual contact, is known as active immunity and is now accomplished by the use of vaccines or microbial derivatives (living or weakened microbes). Note that not all microorganisms induce this form of immunity, and in those cases the second infection reacts like the first.

Passive Immunity. Passive immunity is the transfer of actively formed antibodies from one person (or

Table 16-1. Function of Cells in Resistance to Infection

Cells	Phagocytes	Antibody Producers
Macrophages	+	+
Monocytes	+	−
Polymorphonuclear granulocytes	+	−
Plasma cells	−	+
Lymphocytes	−	+

animal) to another, with the recipient receiving pre-formed antibodies. This occurs between fetus and mother across the placenta during the early months of development or when specific antitoxin or pooled gamma globulin is given for measles, hepatitis, or tetanus. Note, however, that this type of immunity is short lived and may last only a few weeks. It may even be dangerous, since the very antibody injected could serve as an antigen in the recipient and create allergic responses.

Cell-Bound Antibody

Apparently certain antigens in every cell of an individual are unique to that person (except an identical twin), and they induce antibody formation when the cells are transplanted into another person and yet specific antibodies cannot be found in the plasma. The antibodies may, however, be bound to the lymphocytes rather than roaming "free" in the plasma. Thus, the combination at the transplant site of antigen and cell-bound antibody plus complement destroys the tissue cells. These considerations have become very important with the recent advent of organ and tissue transplantation.

Self-Recognition

It appears that any tissue antigens present during embryonic and early neonatal life are recognized by the body as "self" and normally no antibodies are formed against them later in life. However, under certain conditions antibodies can be a "two-edged sword" and the body does produce antibodies against its own tissues, a process attributed to a phenomenon called autoimmunity. This has been explained by the fact that cells of certain tissues, which have never before (during normal embryonic existence) been exposed to whatever cells must recognize as "self"-antigens, are encountered in diseases in which proteins leak out from their usual and proper location. If these cells appear in the blood postnatally as a result of injury or infection, the body's defense mechanism treats them as foreign cells and induces antibody formation to neutralize and destroy them. The latter can also attack normal tissue and create disease or damage.

Thus, in these cases, the antibodies have done more harm than good.

The Thymus and Antibody Relationship

The thymus (Fig. 16-10; see Chap. 15) is a lymphoid organ, but it is incapable of producing antibody. It does seem to play a critical role (at least in animals) in conferring antibody-producing capacity on the lymphoid tissue cells during the early postnatal period. One theory is that the thymus "seeds" the lymph nodes and spleen with cells that can produce additional lymphocytes by mitosis. It apparently also produces a "hormone" that induces the final differentiation of lymphoid cells into immunologically competent cells; however, this does not induce antibody formation but, rather, confers (on the cells) the capacity to do so. These hypotheses, unfortunately, are based predominantly on animal research.

Blood Groups and Types

The safe administration of blood from donor to patient requires the typing and cross-matching of both patient and donor blood. This ensures the patient of getting only "agreeing" blood.

The compatibility of blood and the classification of systems are based on the presence or absence of specific polysaccharide antigens (agglutinogens) in the membranes of the red blood cells and the presence or absence of antibodies (agglutinins) in the plasma or serum.

ABO Blood Groups

This is the main classification of blood and the most important of the erythrocyte membrane antigens. The blood groups are inherited, with A and B being dominant. Agglutinogens and agglutinins are identified in the accompanying illustration (Fig. 16-11). Note that the blood group is named for the antigenic substance (agglutinogen) in the red blood cells.

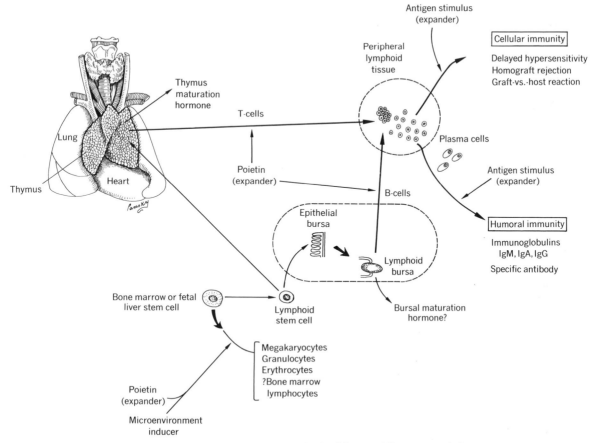

Figure 16-10. Arising in early life from a common stem cell (in fetal liver and bone marrow), two separate sets of lymphoid cells differentiate and populate the peripheral lymphoid organs during late gestation. One set is differentiated in the thymus (T cells). These represent the majority of circulating small lymphocytes in blood and lymph, nodes and spleen. When they contact an antigen, they proliferate and initiate cellular immune responses. The second set of lymphoid cells (B cells) differentiate in birds in the bursa of Fabricius and possibly in the gut-associated lymphoid tissue in mammals. These cells synthesize and secrete antibody. When they meet an antigen, they proliferate and produce specific antibody (IgG, IgA, IgM, IgD, IgE) in blood and a special form of IgA in body secretions.

Group A	The red cells contain the agglutinogen A
	The plasma or serum contains the isoagglutinin anti-B
Group B	The red cells contain the agglutinogen B
	The plasma or serum contains the isoagglutinin anti-A
Group AB	The red cells contain the agglutinogens A and B
	The plasma or serum contains no isoagglutinins for anti-A or anti-B
Group O	The red cells contain no agglutinogens for A or B
	The plasma or serum contains both isoagglutinins anti-A and anti-B

Figure 16-11. Summary of ABO blood groups (agglutinogen-antigen; isoagglutinin-antibody).

The blood group of any given blood sample is determined by the addition of specifically prepared serums, which contain either anti-A antibodies or anti-B antibodies. Thus:

1. When anti-A serum is added to the blood to be typed, clumping of the red cells indicates the sample blood is group A.

2. When anti-B serum is added to blood to be typed, clumping of the red cells indicates the sample blood is group B.
3. When the red cells of the blood to be typed are clumped by both anti-A serum and anti-B serum, the blood is group AB.
4. When the red cells of the blood to be typed are not clumped by either anti-A serum or anti-B serum, the blood is group O.

If a type A person receives type B blood, the following can be expected: the recipient's anti-B antibody will attack the transfused (B person's) cells and the anti-A antibody in the donor's plasma will attack the recipient's cells. The latter is of little importance because the transfused antibodies are diluted in the recipient's plasma and are practically ineffective. The destruction of the transfused cells, however, causes the "transfusion reaction" and clotting. Thus, in transfusions, one must check on the donor's red blood cells (not plasma) and the recipient's plasma (not red blood cells). Often one needs to use only the plasma portion of blood in transfusion. The advantage of this is that plasma keeps better and no red blood cells are present, relieving one of worry about blood types and agglutination. Thus, plasma from many donors can be mixed. Type O people do not encounter any anti-O antibody and, therefore, have been called the universal donors; on the other hand, type AB people have no plasma antibodies to A or B and have been referred to as the universal recipients. This classification, however, can be dangerous since it is now known that there are many other incompatible red cell antigens and plasma antibodies beside the ABO type. In short, careful matching is necessary (Figs. 16-12, 16-13).

Rh Blood Types

The symbol Rh refers to an agglutinogen (antigen) first found in red blood cells of the rhesus monkey. As with the A and B agglutinogens of the ABO system, the Rh factor is found in the red cell membrane and occurs in about 85 percent of the white population of the United States, regardless of the ABO grouping. The Rh system adheres to the general immunity pat-

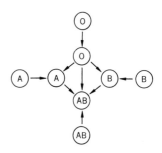

Figure 16-12. Simplified diagram to show which ABO group is acceptable for transfusion without causing clotting. The arrows indicate permissible direction of blood group donors.

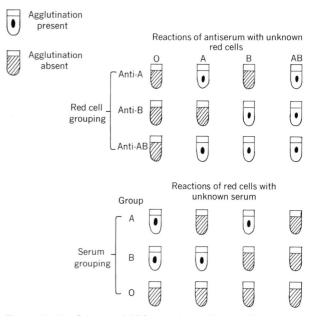

Figure 16-13. Scheme of ABO grouping, both red-cell grouping and serum grouping.

tern. When the red cell contains this property, the blood is called Rh positive; if lacking, the blood is designated as Rh negative. Serum isoagglutinins (antibodies) do not occur naturally in the blood. Instead, they develop only in Rh-negative blood when Rh-positive blood is injected and may give rise to severe (even fatal) reactions. The initial transfusion may only "sensitize" the recipient and cause the development of antibodies without the occurrence of severe symptoms. However, once sensitized, the recipient may experience severe reactions to subsequent infusions of Rh-positive blood.

The above-listed events can occur in pregnancy when the fetus inherits Rh-positive blood from an Rh-positive father but is carried by an Rh-negative mother. As the Rh-positive red cells from the fetus cross the placenta into the mother, antibodies are produced in the mother's blood due to this "invasion." There may be little effect on the mother; however, the effects on the infant may be severe, especially in subsequent pregnancies, causing severe anemia or even brain damage when the built-up maternal antibodies cross the placenta and attack the fetal red cells and create hemolysis (destruction). Fortunately only about 5 percent of Rh-negative mothers produce such antibodies while carrying an Rh-positive child. The danger, as stated, increases with subsequent pregnancies due to the "memory component" of immune mechanisms, but the first child is usually safe.

Other Subfactors

As stated above, transfusions require at least the determination of ABO group and Rh type. There are other subfactors, unfortunately, that require a final cross-matching of the recipient's blood with the blood to be given, in spite of the fact that the two bloods are of the same major group and type. The following blood groups and types are not all of the various groups and types present, but they are the ones most commonly determined for clinical purposes.

1. ABO blood groups: A, B, AB, O, A_1, A_2, A_3, A_1B, and A_2B
2. Rh blood types: $RH_0(D)$, $RH_0'(CD)$, $Rh_0''(DE)$ rh'(C), rh''(E)—all are Rh positive and have subtypes rh (Rh negative)

Systemic Response to Injury or Infection

Inflammation is local tissue response. The most common and striking systemic response is fever. The substance primarily responsible for resetting the hypo-thalamic "thermostat" is a protein pyrogen released by the neutrophils (perhaps other cells) that take part in the inflammatory reaction. Certain bacteria release proteins that may also act directly on the brain to cause fever, but they may function strictly to damage neutrophils and thus release the pyrogen into the blood or lymph. Fever is not always or necessarily protective, nor does it enhance body resistance; indeed, fever may be harmful in its effects on the central nervous system, and convulsions may result. Furthermore, by increasing body chemical reactivity, fever burdens the circulatory system.

The Resistance of the Body

The body resists infection by altering the ability of the microbe to injure tissue; by creating resistance mechanisms of its own, such as humoral factors, anatomic barriers, and phagocytes; or by altering the inflammatory response. These processes can be altered by man himself.

Cortisol, in large doses, may inhibit inflammation by preventing leukocyte emigration and increasing capillary permeability, but it can also lower resistance. An existing disease may predispose to infection (e.g., diabetes). Any injury to a tissue reduces its resistance by altering the chemical environment or blood supply. At times, the very basic mechanism is deficient (e.g., deficiency of plasma gamma globulin or complement), and thus there is inability to synthesize antibodies. A decrease in leukocyte production due to drugs makes one prone to disease. The converse is also true in that a large number of leukocytes may create problems if these are immature or abnormal (as in leukemia). Antibiotics are harmful to microbes but relatively innocuous to body cells; yet they, too, cannot be used indiscriminately and may alter body resistance. Antibiotics exert many effects; however, in general, they interfere with the synthesis of one or more of the bacteria's own essential macromolecules. Such drugs are relatively effective against viruses, but their toxicity is not selective; thus, antiviral chemotherapy is not effective as yet.

Antibodies formed in response to a viral infection manifest a defense mechanism. A chemical substance called interferon also plays a role. Interferon is a protein that inhibits viral growth and multiplication, and it is produced by many different body cells but only in response to a viral infection of a particular cell. Unlike antibodies, interferon is not specific, and all viruses induce the same kind of interferon synthesis. On the other hand, interferon can inhibit the multiplication of many different viruses. The interferon system is a rapid-reacting one and begins within hours of an infection. Antibodies cannot enter intact human cells, whereas interferon functions inside the infected cell to prevent further viral growth, spread, and multiplication from cell to cell.

Allergy

Allergy refers to an acquired hypersensitivity to a particular substance and is simply an antigen-antibody reaction resulting in cell damage. It usually involves nonmicrobial antigens. An initial exposure to an antigen (pollen, weeds, etc.) leads to the formation of antibodies and a "memory" storage similar to the one that characterized immunity. On reexposure, the specific antibodies join the antigens and with complement trigger cell damage by initiating the release of histamine, thereby causing an inflammatory reaction. Unfortunately the inappropriate response by the body may be more violent and damaging than the antigen triggering it. Antihistamines offer an effective but incomplete therapy. A complete discussion of hyposensitization and blocking of antibody as other means of dealing with this problem is too involved to be considered in this textbook.

Blood and Hemostasis (Prevention of Loss)

Constituents of Blood

Blood constituents (Fig. 16-3) fall into three categories. The first category includes nutritional sub-stances (substances absorbed from the gastrointestinal tract, the products of digestion, and substances formed in the body as metabolites or secretory substances [hormones] for use in other parts of the body). The second category consists of excretory substances (substances transported to the liver, kidneys, lungs, and skin for elimination). The third major group of constituents includes functional or intrinsic substances (involved in maintaining the functional integrity of the blood) and consists of red blood cells, white blood cells, platelets, albumins, globulins, factors of the coagulation mechanism, electrolytes, enzymes, lipid, and water. All of these, except the factors of the coagulation mechanism, have been discussed previously. In man, blood coagulation (clotting) is only one mechanism to minimize blood loss as a consequence of vessel damage. The hemostatic mechanism used also depends on the vessels damaged and the location of injury. Venous bleeding is less dangerous than arterial because of the vein's low hydrostatic pressure. Furthermore, since venous bleeding is into the tissues, a hematoma (blood accumulation) may increase interstitial pressure and eliminate the pressure gradient and stop the bleeding.

In man, the platelets (Fig. 16-4) play an essential role in the hemostatic mechanism. They are colorless corpuscles, smaller than red cells, have many granules but lack nuclei, and originate from large cells called megakaryocytes found in bone marrow. It has been estimated that there are about 250 thousand platelets per cu mm of blood.

The events prior to clot formation consist of (1) contraction of the smooth muscle in the wall of injured vessels, (2) a sticking together of the endothelial surfaces of the injured vessel, (3) the clumping of platelets to form a plug, (4) a humoral facilitation of vasoconstriction (local release of vasoconstrictor substances such as serotonin and epinephrine from the aggregated platelets), and (5) blood coagulation (formation of a fibrin clot and retraction of the clot).

Blood Coagulation

The mechanism of coagulation is a very complicated process, and as many as 12 factors found in the blood

Table 16-2.   The Nomenclature and Function of the Blood-Clotting Factors

Factor	Function
Factor I (fibrinogen)	Constitutes 7% of the plasma proteins and is important in the third stage of coagulation, in which it is converted to fibrin. Remove fibrinogen from plasma and you get serum.
Factor II (prothrombin)	Is a stable glycoprotein that is important in the second stage of coagulation, where it is converted to thrombin. It is synthesized in the liver and requires the presence of vitamin K.
Factor III (platelet factor)	Refers to a group of substances called thromboplastins, derived from many sources including the platelets, lungs, brain, and other tissues. Practically all tissues contain thromboplastin, which may be released when the tissues are injured. Tissue thromboplastin differs from plasma thromboplastin in that the latter is formed during the actual process of clotting through the interaction of a group of circulating protein coagulation factors and a phospholipid released from platelets. Tissue thromboplastin is an "incomplete" thromboplastin, requiring the presence of factors V, VII, and X in order to convert prothrombin to thrombin. Plasma thromboplastin, on the other hand, is "complete" in that it is capable of converting prothrombin to thrombin directly.
Factor IV (calcium)	All of the other coagulation factors are proteins; calcium is an inorganic ion. Its presence is essential for the first stage of coagulation, and it acts as a catalyst in the conversion of prothrombin to thrombin in the second stage of coagulation.
Factor V (labile factor)	Is called the labile factor or accelerator and is derived from plasma globulin. It speeds the conversion of prothrombin to thrombin in the presence of tissue thromboplastin. It is consumed during blood clotting and is thus not found in serum. A deficiency of this factor causes a decrease in prothrombin activity. A congenital deficiency is called parahemophilia, a bleeding disorder.
Factor VI (formerly accelerin; no name at present)	Has not been assigned to any coagulation factor.
Factor VII (stable factor)	Is called the stable factor, since it is stable to heat and storage. It has a high serum concentration. Like factor V it is considered to be an accelerator in the conversion of prothrombin to thrombin. Severe deficiency also creates parahemophilia. It is not consumed during blood clotting and is found in the serum after normal coagulation.
Factor VIII (antihemophilic globulin)	A deficiency causes hemophilia A, which is a sex-linked characteristic transmitted by females but in which hemorrhagic problems occur exclusively in males. Bleeding may occur spontaneously or after minor injury. It is essential in the first phase of clotting since it is an important precursor of thromboplastin. It is almost completely used in the clot formation. Its activity originates in the globulin fraction of normal plasma.
Factor IX (Christmas factor or plasma thromboplastin component, PTC)	Is another factor essential for the formation of intrinsic blood thromboplastin, but it influences the amount of thromboplastin formed rather than the rate of its formation. Deficiency is inherited (sex-linked recessive trait), as is the deficiency of factor VIII, and is called hemophilia B or Christmas disease. It occurs in normal plasma or serum associated with globulins. It is not consumed in the clot and is seen in serum.
Factor X (Stuart-Power factor)	A deficiency is apparently inherited and occurs in hemorrhagic disease of the newborn, liver disease, and vitamin K deficiency. A deficiency results in nosebleeds, as well as bleeding into joints or soft tissues. Plasma with this deficiency shows defective generation of thromboplastin, impaired use of prothrombin, reduced prothrombin activity, and prolonged recalcified clotting time.
Factor XI (plasma thromboplastin antecedent, PTA)	Plays a role in the formation of plasma thromboplastin. Deficiency is congenital and causes a mild hemophilia (hemophilia C). It is transmitted (in contrast to the other two) as a mendelian dominant and thus is seen in males and females. It is seen in serum and plasma.
Factor XII (Hageman factor)	Is a coagulation factor. It is apparently important in initiating coagulation outside of the body. It is an inherited recessive trait characterized by minor bleeding episodes.
Factor XIII (fibrin-stabilizing factor)	Thrombin catalyzes the conversion of factor XIII into its active form, which cross-links subunits of the fibrin clot to form insoluble fibrin.

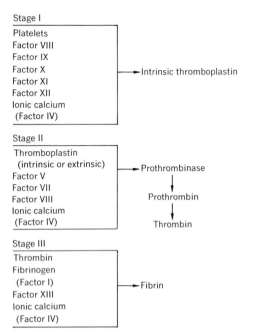

Stage I

| Platelets
Factor VIII
Factor IX
Factor X
Factor XI
Factor XII
Ionic calcium
(Factor IV) | ──►Intrinsic thromboplastin |

Stage II

| Thromboplastin
(intrinsic or extrinsic)
Factor V
Factor VII
Factor VIII
Ionic calcium
(Factor IV) | ──►Prothrombinase
↓
Prothrombin
↓
Thrombin |

Stage III

| Thrombin
Fibrinogen
(Factor I)
Factor XIII
Ionic calcium
(Factor IV) | ──►Fibrin |

Figure 16-14. Intrinsic coagulation system.

have been implicated (Table 16-2). In addition to protein plasma factors, calcium is a required cofactor for several steps; yet calcium deficiency is never, apparently, a cause of clotting defects in man since so little of it is needed. The mechanism of coagulation occurs in three phases (stages) (Fig. 16-14). It is the dominant hemostatic defense and is a dynamic process acting in concert with other mechanisms to achieve and maintain hemostasis.

Phase I. Phase I involves the formation of thromboplastic activity. It begins with injury to a blood vessel, which provides a contact surface and a site of accumulation of platelets previously described. Plasma factors involved in the formation of thromboplastin are factor IV (calcium), factor VIII (antihemophilic globulin), factor IX (PTC), factor X (Stuart-Power factor), factor XI (PTA), and factor XII (Hageman). Factors VIII and XII are thromboplastin precursors.

Phase II. Phase II involves the conversion of prothrombin to thrombin. In actual coagulation of the blood, the plasma thromboplastin (from phase I) is capable of converting prothrombin to thrombin di-

rectly. It is believed, however, that an accessory mechanism exists with the release of tissue thromboplastin. Conversion of prothrombin (which has been secreted in an inactive form by the liver and is in the blood normally) to thrombin is under the influence of this type of thromboplastin and requires factors V, VII, and VIII and calcium. Although fibrinogen is always present in the blood, thrombin is not and requires vessel injury to be generated.

Phase III. Fibrinogen (factor I) is converted to fibrin by the thrombin formed in phase II. It is accelerated by calcium. Thrombin acts like an enzyme to break down the fibrinogen molecule into molecular fibrin and other substances. The molecules of fibrin then polymerize and combine end to end and side to side to form the fibrous filamentous network or gel recognized as a fibrin clot. Red cells and platelets adhering to it give the clot its characteristic appearance.

Figure 16-15. Intrinsic mechanism for clotting. Note that activation of V and VIII requires phospholipids, which are obtained from platelets. What starts the chain of events with factor XII is uncertain, but it may be contact with foreign surfaces.

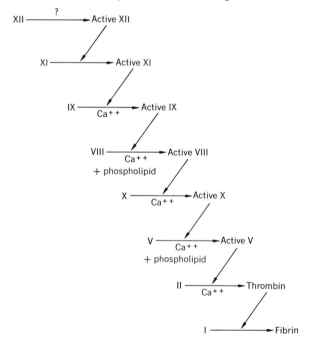

Clotting Pathways. INTRINSIC CLOTTING PATHWAY. One of the two clotting pathways, the intrinsic one, reguires only factors present in plasma (Fig. 16-14). These components are still poorly characterized. The process is stepwise; each step is activated by the preceding one in the chain and, in turn, activates that which follows (Fig. 16-15).

EXTRINSIC CLOTTING PATHWAY. This pathway begins with factor X. It also involves factor VII, which is not found in plasma but in the tissue (Fig. 16-16). Both pathways end with fibrin (Fig. 16-17), but their relative importance is still questionable.

Role of Liver. The liver plays several indirect roles in the functioning of the clotting mechanism. It is the site of production of many of the plasma factors (prothrombin and fibrinogen), and the bile salts produced there are needed for normal gastrointestinal absorption of the fat-soluble vitamin K, which is an important cofactor in the liver synthesis of prothrombin and plasma factors. Thus, patients with liver disease or poor fat absorption in the gastrointestinal tract often have bleeding abnormalities.

Other Considerations. Tissue (rather than blood) factors may contribute to clotting in that they may contain substances that can substitute for both platelet phospholipids and several plasma factors. These factors, however, do not appear to play a major role in

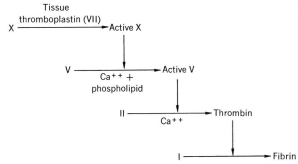

Figure 16-16. Extrinsic mechanism for clotting.

intravascular clotting; yet they may have a role in the response to bacterial infections by initiating interstitial fibrin clots to block bacterial spread.

Blood in itself is extremely complex, and the "scheme" of coagulation is no exception. To more fully appreciate the mechanism, a more complete view of the various clotting factors is presented in Table 16-2.

Retraction of the Clot

After the clot is formed, the first thing that happens is syneresis (clot retraction), which causes consolidation of the fibrin clot and is due to the platelets. When blood is collected in a tube, clotting occurs in five to eight minutes and in the next 30 minutes it retracts and squeezes out the fluid to form a "hard clot." As

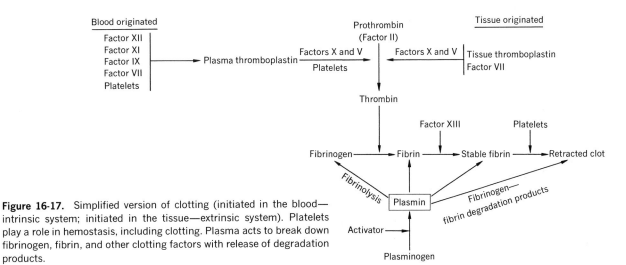

Figure 16-17. Simplified version of clotting (initiated in the blood—intrinsic system; initiated in the tissue—extrinsic system). Platelets play a role in hemostasis, including clotting. Plasma acts to break down fibrinogen, fibrin, and other clotting factors with release of degradation products.

fibrin strands form around the platelets, the agglutinated mass of platelets sends out pseudopods, which contract to pull the fibrin fibrils together and squeeze out the clear yellow serum (plasma without fibrinogen).

The second postclotting event is fibrinolysis, or dissolution of a fibrin clot. The source of activity is the blood itself, which contains an inactive enzyme called plasminogen. In the presence of activators from the tissue, the inactive plasminogen becomes plasmin, an active proteolytic enzyme that dissolves fibrin clots.

The platelets in adhesion and agglutination thus function in hemostasis to aid in (1) the formation of the platelet plug; (2) the discharge of serotonin, resulting in vasoconstriction; (3) the discharge of phospholipid, resulting in a cofactor for coagulation; and (4) the adherence to fibrin and contraction, resulting in clot retraction.

Anticlotting System

It is evident that a substance(s) exists that can decompose or lyse fibrin and dissolve a clot or prevent its formation (Fig. 16-17). It has been called fibrinolysin. Its effects are as complicated as the clotting mechanism. A second naturally occurring anticoagulant is heparin, found in many cells of the body, especially mast cells, and this substance acts by interfering with the ability of thrombin to split fibrinogen. It is widely used as an anticoagulant clinically; yet its natural role in the body is unproven.

Clot Formation and Thrombosis

Clots formed in intact vessels are pathologic. The blame has been placed on a "hyperactive" clotting mechanism or on alterations in the lining of the blood vessels, creating lipid and connective tissue deposits in the wall. Excessive clot formation or the growth of a clot may occlude the entire vessel, leading to damage of the tissue supplied. Fragments (emboli) may break off and plug the microcirculation of organs as well as induce cardiovascular reflexes, resulting in hypotension, abnormal cardiac rhythm, and even death.

Hematopoiesis

The hematopoietic tissue is widely dispersed and includes the myeloid and lymphoreticular tissues of the bone marrow, spleen, and lymph nodes and the "fixed" reticular tissue found in almost every organ. It also includes various lymph aggregates, such as the tonsils, adenoids, Peyer's patches of the intestinal tract, and lymphoid tissue of the bronchi.

Disorders of the hematopoietic system primarily represent changes in cellular kinetics and abnormalities in the function as well as the character of the cells produced. One may see normally increased production of cells (hyperplasia), defective production of varying degrees (sickle cells, fragmentation, etc.), decreased production (hypoplasia), or abnormal hematopoiesis (dysplasia). In the peripheral blood, the cells may be subjected to extreme abnormalities with an increased rate of cell destruction and a shortened lifespan. Because of its rapid cell turnover, no other tissue is more vulnerable to modifications by chemicals, therapeutic agents, or deficiency conditions. Some of the major proliferative disorders are leukemia, lymphoma, Hodgkin's disease, myeloma, and proliferation on an abnormal immunologic basis. (For general cell development, see Fig. 16-1; for greater details, consult any standard textbook of histology or hematology.)

Bone Marrow

The bone marrow functions in an organlike manner, is responsive to stimuli, and is sensitive to toxic material. The hematopoietic elements are distributed in an organized fashion in a reticular network in association with fat. The latter varies with age (increases) but is related to the marrow's cellularity. From the second to seventh decades hematopoiesis is active in the marrow of the trunk, skull, and upper ends of the femur and humerus.

Erythropoiesis

Anemia, one of the most common disorders seen, is characterized by a significant reduction in either the total number of red blood cells or the amount of hemoglobin in the peripheral circulation. Red cell production and loss are normally maintained as a very sensitive feedback mechanism. Thus, anemia results from loss due to hemorrhage or cell destruction (hemolysis), failure to produce red cells, or both.

An increase in the number of red blood cells occurs less frequently than anemia and is called polycythemia.

Granulopoiesis

The importance of granulocytes in inflammatory reactions is well known. An increase in the numbers of granulocytes in the peripheral blood (leukocytosis) has been relied upon as a good indicator of acute inflammation. Variations in the number of peripheral blood granulocytes are characteristic of certain diseases. A very fine control mechanism for granulocyte kinetics exists and is apparent from the general constancy of the granulocyte count. Granulopoietin, a substance recently identified, controls granulopoiesis via a feedback mechanism and regulates the number of cells and the size of the granulopoietic mass.

Lymphopoiesis

The lymphocyte has been assigned an ill-defined function in chronic inflammation. There is not a single lymphocyte population or function; rather, there are several lymphocyte systems with highly specialized functions, distributions, and regulating mechanisms. Lymphocytes play an important role in cellular immune reactions (including graft rejection), reaction to neoplasms, response to viruses, and delayed hypersensitivity. Exactly how this works is uncertain, but it is known that lymphocytes produce several factors called lymphokinins that are active immunologic mediators and their release at the site of antigen introduction promotes the immune response. Some lymphocytes are also capable of producing immunoglobulins and specific antibodies.

Two central lymphoid tissues have been shown in animals: the thymus and, in birds, the bursa of Fabricius. They arise embryologically from a marrow reticular stem cell via a migratory lymphocytoid cell. These two organs are responsible for the development and regulation of two functionally different lymphoid tissue components. The thymus is responsible for the development of cellular immunity (including hypersensitivity), graft rejection, resistance to neoplasms, and response to viral infections. The bursa or its mammalian equivalent probably controls the development of humoral immunity, such as immunoglobulin and antibody production.

Lymphopoiesis is widely distributed throughout the body and occurs principally outside the bone marrow. The circulating lymphocyte is a dormant primitive cell rather than a mature element. Most lymphocytes also have a prolonged life-span and leave the circulation but recirculate periodically. Several immune deficiency disorders result from developmental abnormalities of the thymus or bursa.

Lymphopoiesis is activated under the influence of antigen, particularly where this antigen is introduced but especially in areas of lymphoid concentration, such as lymph nodes and lymphoepithelial tissue (e.g., Peyer's patches of the intestine). The extent of proliferation depends on the type of antigenic stimulus. Thus, if it is extensive, it may involve not only regional lymph nodes but also the whole lymphoreticular system. Under the stimulation of antigen, the small lymphocyte is transformed into the stem cell with changes in its morphology and the stem cell can produce more lymphocytes. There is decreased transformation in immunologic deficiencies such as leukemia or thymic deficiency. Other forms of lymphocyte transformation also take place: development of plasma cells from lymphocytes, and formation of medium and large lymphocytes. Thus, because of its potential multidirectional transformation, the small lymphocyte is capable of bringing into an inflammatory area a variety of immunologically functioning cells. Lymphocytes are recirculated in the system. The major pathway is from the blood through the lymph nodes into the lymph, followed by reentry into the blood via the thoracic lymph ducts.

Lymph Nodes

The normal lymph node has cortical and medullary areas, distributed in a reticular stroma with secondary cortical reaction or follicular centers and encompassed by a prominent collagen capsule (Fig. 10-36, p. 375). Afferent lymphatics are seen entering from several directions through the capsule and emptying into a subcapsular sinus, from which the lymph follows a sinusoidal route through the lymphoid tissue to emerge at the hilus of the node into the efferent lymphatic. The node varies with age, prior immunization, and general body nutrition as well as with time following exposure to stimuli.

Bacterial infections are associated with lymphoid reactions. Lymphoid hyperplasia is generally a response to antigens without evidence of infection of the node, but lymphadenitis is an infection of the node with evidence of infection. Nodes are generally small and discrete but enlarge with stimulation. The lymphoid tissue of the malpighian corpuscles of the spleen differs from lymph nodes in that it principally participates in systemic reactions to antigens and is composed of small lymphocytes without reaction centers.

The Spleen

The spleen is the largest lymphatic organ of the body. It is composed of fibrous trabeculae, running through it from the capsule; white pulp between the

Figure 16-18. Anatomy and histology of the spleen and its circulation. **A.** Visceral surface of spleen. **B.** Transverse section of spleen showing trabecular tissue and splenic vein and its tributaries. **C.** Diagram of splenic structure. **D.** Scheme of splenic circulation. **E.** Circulation through spleen (three dimensional). **(C** and **D,** after studies of Moore; **E,** after studies of Garven.)

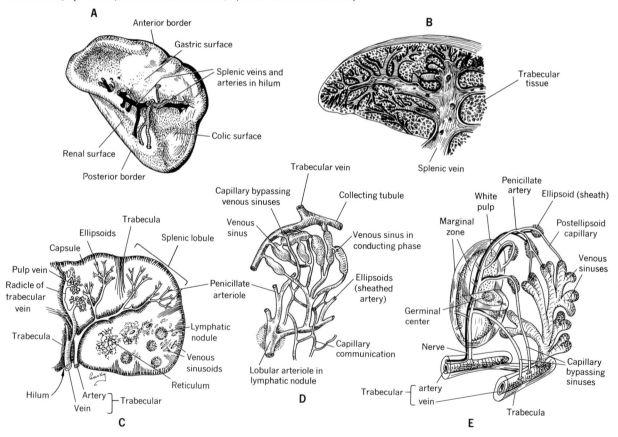

trabeculae, consisting of ensheathed branches of the splenic artery and cords and nodules of lymphatic tissue; and red pulp, made up of atypical lymphatic tissue (the splenic cords) that fills the spaces between venous sinuses.

The spleen functions to remove most of the worn-out red blood cells of the body and makes bilirubin from their hemoglobin. It also extracts iron from the hemoglobin; produces antibodies, lymphocytes, and monocytes; and can liberate "stored" blood into the circulatory system when needed. Yet the spleen is not essential to life, since other hematopoietic tissues can take over its function.

The adult spleen (Fig. 16-18) is completely covered by peritoneum except at its hilum. It is in contact with the diaphragm on the left side at the level of the ninth, tenth, and eleventh ribs. It rests on the splenic flexure of the colon and is in contact with the left kidney. The vessels and nerves enter and leave at the hilum.

The splenic artery (branch of the celiac) does not enter the spleen as a single vessel but breaks up into from five to eight branches. The vein is formed at the hilum and is joined by the left gastroepiploic and other veins from the greater curvature of the stomach. It runs along the back of the pancreas and joins the superior mesenteric vein to form the portal vein. In its course, it receives the inferior mesenteric vein.

Splenectomy (spleen removal) has been performed for injuries to the spleen, hemolytic jaundice, and a number of other diseases.

Review Questions

1. What is the body's first line of defense? Give several examples.
2. A drop of blood after being spread on a slide, stained, and viewed microscopically reveals red blood cells and polymorphonuclear granulocytes.
   a. What general percentage of these cells is seen?
   b. What forms of granulocytes are seen? How are they differentiated? Where are they formed? How do they function?
3. What is an antibody? Of what is it composed? How does it work?
4. What is an antigen? Cite several examples.
5. What is complement?
6. What is meant by acquired immunity, natural immunity, and cellular immunity?
7. What is meant by the ABO blood group classification of blood? List the groups and their cross-reactivity.
8. What is the universal recipient? Or universal donor? Why are they so called?
9. Discuss briefly the Rh factor and its significance.
10. Why do we become allergic?
11. In man, which cells play the "key" role in the hemostatic mechanism?
12. What are the 12 factors utilized in blood clotting?
13. Outline the factors and the order of sequence of blood clotting.
14. Define:
    a. Hematopoiesis
    b. Bone marrow
    c. Anemia
15. What is a lymph node?
16. What is the spleen? Where is it located? What is its function?

References

Benacerraf, B.: Cell Associated Immune Reactions. *Cancer Res.*, **28**:1392–98, 1968.

Elves, M. W. (ed.): *The Lymphocytes.* Lloyd-Luke, LTD., London, 1966.

Franklin, E. C., and Frangione, B.: Immunoglobulins. *Annu. Rev. Med.*, **20**:155–74, 1969.

Fudenberg, H. H.; Pink, J. R. L.; Stites, D. P.; and Wang, A.: *Basic Immunogenetics.* Oxford University Press, New York, 1972.

Gowans, J. L., and McGregor, D. D.: The Immunological Activity of Lymphocytes. *Prog. Allergy*, **9**:1–78, 1965.

Harker, L. A. (ed.): *Hemostasis Manual.* University of Washington, Seattle, 1970.

Jerne, N. K.: The Immune System. *Sci. Am.*, **229**(1):52–60, 1973.

Lerner, R. A., and Dixon, F. J.: The Human Lymphocyte as an Experimental Animal. *Sci. Am.*, **228**(6):82–91, 1973.

Levine, B. B.: Immunochemical Mechanisms of Drug Allergy. *Annu. Rev. Med.*, **17**:23–28, 1966.

Mayer, M. M.: The Complement System. *Sci. Am.*, **229**(5):54–66, 1973.

Metcalf, D.: Regulation of Lymphopoiesis. In Gordon, A. S. (ed.): *Regulation of Hematopoiesis*, Vol. II. Appleton-Century-Crofts, New York, 1970, p. 1383.

Miescher, P. A., and Jaffe, E. R. (eds.): The Lymphocyte. *Semin. Hematol.*, **6**:1–403, 1969.

Pearsall, N. N., and Weiser, R. S. (eds.): *The Macrophage.* Lea & Febiger, Philadelphia, 1970.

Porter, R. R.: The Structure of Antibodies. *Sci. Am.*, **217**(4):81–90, 1967.

Samter, M. (ed.): *Immunological Diseases*, 2nd ed., Vols. I and II. Little, Brown & Co., Boston, 1971.

Uhr, J. W.: Delayed Hypersensitivity. *Physiol. Rev.*, **46**:359–419, 1966.

The Life Cycle

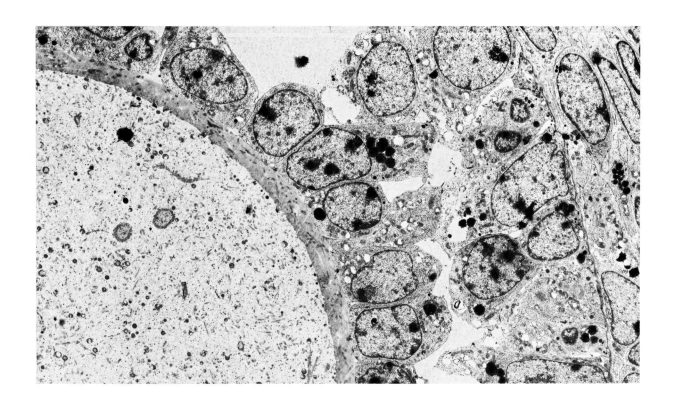

UNIT V

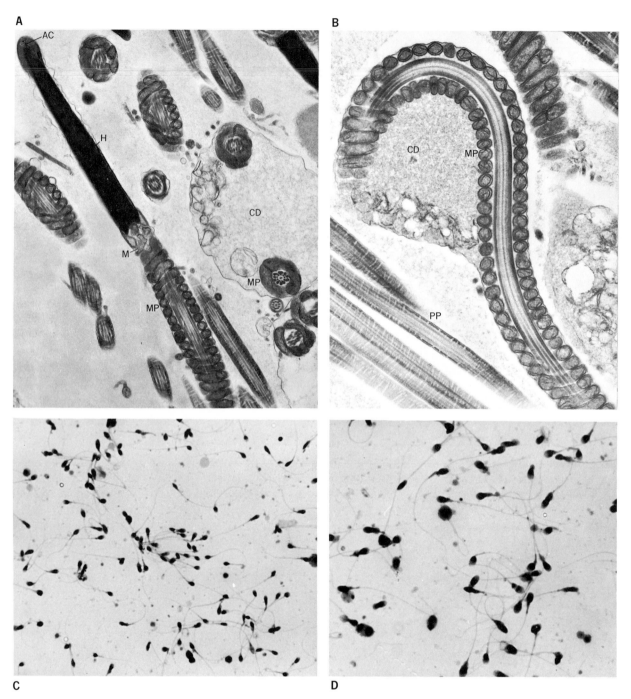

Figure 17-1. **A** and **B.** Sections of mouse sperm, showing the major anatomy. *AC*, acrosome; *H*, head; *M*, mitochondria; *MP*, middle piece; *CD*, cytoplasmic droplet; *PP*, principal piece. **A** demonstrates both longitudinal and cross sections of sperm; ×12,500. **B** is a longitudinal section of the sperm flagellum, showing the middle and principal pieces; ×25,000. **C** and **D.** Histologic sections of human sperm, magnified ×450 and ×720, respectively. (Courtesy of Drs. Jyotsna Chakraborty and Leonard Nelson, Medical College of Ohio at Toledo.)

The Male Reproductive System

CHAPTER 17

The Reproductive Cycle

Mammalian reproduction is characterized by a series of cycles, consisting of immaturity, puberty, maturity, menopause, senility, and death. The reproductive portion of this cycle varies from species to species and individual, and in man a remarkable sex difference exists in the length of the reproductive life.

The menstrual cycle in the female gives a fairly clear indication of her reproductive life, but there is no such clear indication in the male since the onset of puberty and climacteric is less overt. The male retains his fertility long after it has vanished in the female of similar age. Men of 70 or 80 years of age have been fathers, but only rarely do women of 50 or more years become mothers.

The primary organs of reproduction in mammals, the gonads, are called ovaries in the female and testes in the male. They are organs of both external and internal secretion. The ovaries produce ova and sex hormones, estrogenic and progestational hormones (mainly estradiol and progesterone) that are concerned with the development and maintenance of female attributes, including sexual receptivity and the inception of pregnancy (see Chap. 15). The testes produce spermatozoa (Fig. 17-1) and sex hormones, androgenic hormones (predominantly testosterone) that induce and maintain the attributes of maleness (see Chap. 15).

Accessory Organs of Reproduction

The accessory organs of reproduction consist of the system of ducts, and glands lining them, through which the sperm or ova are transported from the gonads. They play an essential role in the reproductive process. In the female, these organs consist of the reproductive tract leading from the ovaries to the exterior, namely, the fallopian tubes (oviducts), uterus, cervix, vagina, vulva with its associated labia, and clitoris. In the female, the mammary glands or breasts are also included, although artificial feeding has rendered the breasts superfluous in the human and they may be destined to degenerate into only a psychosexual symbol. The gonads and accessory organs undergo varying changes in the reproductive life cycle of the female.

In the male, the accessory organs are the scrotum (containing the testes), the epididymis (which collects the spermatozoa from the testes), the vas deferens (which conducts the sperm to the urethra), the seminal vesicles and prostate gland (whose secretions dilute the sperm mass), and the penis (the organ that transmits spermatozoa from the male to the female) (Fig. 17-6, p. 579; Fig. 18-4, p. 592).

Secondary Sexual Characteristics

The secondary sexual characteristics are also typical of many animals, including man, and although they are gonad dependent they demonstrate marked changes during life and play a minor role in the reproductive process. The secondary sexual characteristics are virtually absent until puberty, at about 10 to 14 years of age. Puberty refers to sexual maturity or the age at which conception becomes possible. Adolescence has a broader meaning and includes the total period of transition from childhood in all respects, not just sexual. The secondary sexual characteristics are better developed in the male than in the female and include such things as facial hair, a different pubic hair pattern, a deeper voice in the male, and a larger, more muscular male body. They depend on the testicular androgenic substances in the male. In their absence, sexual differentiation does not appear. In the female, on the other hand, pubic hair pattern is not gonad dependent but, like axillary hair, depends on body maturation at puberty due to androgenic substances from the adrenal glands. Voice changes at puberty, resulting in deepening of pitch, are androgen dependent and tend to regress in old age. Genetic factors influence the amount and distribution of head hair, so baldness is not necessarily seen in a eunuch (one who lacks functional testes) and probably results from some action of the male hormones. Skin changes at puberty and congestion of the sebaceous glands are androgen dependent, leading to a degree of acne.

The smaller female size and her subcutaneous fat accumulation, creating "rounded" contours, are due in part to the bone-ossifying and fat-mobilizing effects of estrogen. Together with her wider pelvis these are also some of the secondary female sexual characteristics.

Male External Anatomy

Embryology

The digestive tract, which originates from endoderm, ends as the blind cloaca, which also has an anterior diverticulum, the allantois or urachus. The mesonephric duct becomes the adult vas deferens and grows caudally into the cloaca. The ureter develops as an outpouching of the mesonephric duct. Originally the ureter and vas end in one terminal duct, which is subsequently absorbed into the bladder wall and the prostatic urethra. As a result, the ureter and vas deferens eventually have separate openings. A mesodermal urorectal septum (Fig. 17-2) divides the cloaca into an anterior urogenital and posterior intestinal portion. The cloacal sphincter anteriorly becomes the superficial muscles of the perineal region associated with

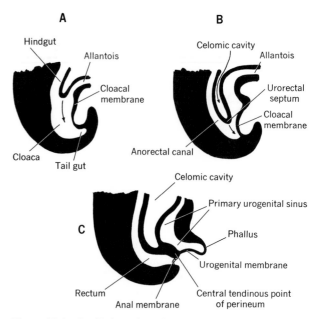

Figure 17-2. Sagittal sections through the cloacal area of the embryo (about 30 days), indicating the separation of the anterior urogenital area from the posterior rectoanal and descent of the urorectal septum.

the penis and the urogenital diaphragm, which closes off the pelvis anteriorly (see below). Posteriorly the cloacal sphincter forms the external anal sphincter (see Chap. 12).

Perineum

The perineum (Fig. 17-3, Table 17-1) is the area between the thighs that extends from the coccyx to the pubis and below the pelvic diaphragm. It is bounded by the pubic symphysis, rami of the pubes and ischia, ischial tuberosities, sacrotuberous ligaments, edges of the gluteus maximus muscles, and coccyx. A line drawn between the ischial tuberosities divides the perineum into an anterior urogenital triangle and a posterior anal triangle.

Urogenital Triangle. The urogenital triangle (Fig. 17-3) is the urogenital portion of the perineum in the male (and female). The superficial fascia beneath the skin has an outer fatty layer and contains cutaneous vessels and nerves. Over the scrotum, however, the fat

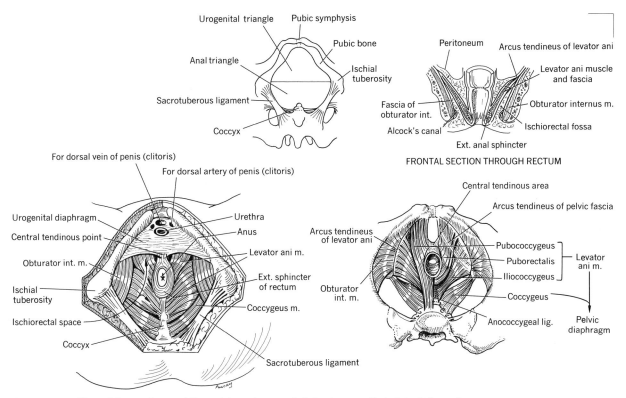

Figure 17-3. The pelvic opening and the pelvic and urogenital diaphragms that close it (seen from below).

disappears from the superficial layer and gives rise to a thin layer of involuntary or dartos muscle, which creates the rugosity of the scrotal skin. The deep layer of superficial fascia is membranous and here is called Colles' fascia and is found only over the urogenital area. It is prolonged over the penis and scrotum.

Superficial Perineal Space. The superficial perineal space (pouch) (Fig. 17-4) is a compartment in the urogenital part of the perineum, bounded below (floor) by Colles' fascia and above (roof) by the urogenital diaphragm. Its contents consist of the roots of the bodies of the penis and urethra (two corpora cavernosa penis and the corpus cavernosum urethrae), their overlying muscles, the superficial transverse perineal muscles that lie behind the penis, and branches of the internal pudendal vessels and nerves that pierce the urogenital diaphragm to reach this area.

Deep Perineal Pouch. The deep perineal pouch (urogenital diaphragm) (Figs. 17-3, 17-4; Table 17-1) is actually the space occupied by the urogenital diaphragm itself and is thus located between the inferior and superior fascial layers of the urogenital diaphragm. The urogenital diaphragm is a musculomembranous diaphragm, stretched across the pubic arch and attached to the ischiopubic rami, that separates the perineum from the pelvis. It consists of the two facial layers just mentioned and two muscles, the deep transverse perineal and sphincter urethrae membranaceae. Both muscles are supplied by the perineal nerve. The pouch also contains the membranous urethra, as it passes from the bladder to the penis, the bulbourethral glands (Cowper's), the internal pudendal artery, the artery to the bulb of the penis, and the dorsal nerve of the penis.

Table 17-1. Muscles of the Pelvis

Muscle	Origin	Insertion	Action	Innervation
1. Obturator internus	Internal surface of obturator membrane and margin of obturator foramen	Only structure to exit from lesser sciatic foramen to insert on inner surface of great trochanter of femur	Rotates thigh laterally	Nerve of obturator internus (sacral plexus, L5, S1, S2)
2. Piriformis	Margins of anterior sacral foramina and greater sciatic notch of ilium	Exists from greater sciatic notch to insert on upper border of great trochanter of femur	Rotates thigh laterally and abducts flexed thigh	Sciatic plexus (S1, S2)
3. Pelvic diaphragm				
a. Levator ani	Back of pubis, pelvic fascia, spine of ischium	Central point of perineum, externus sphincter ani, sides of lower part of sacrum and coccyx	Chiefly draws anus upward in defecation and aids in support of pelvic floor	Pudendal, S3, S4
b. Coccygeus	Spine of ischium and sacrospinous ligament	Sides of lower part of sacrum and upper part of coccyx	Assists in raising and supporting pelvic floor	S3, S4
4. Urogenital diaphragm				
a. Sphincter urethrae	Ramus of pubis	With its fellow, in a median raphe behind and in front of urethra	Constricts membranous urethra	Pudendal (perineal)
b. Deep transverse perineal	Ramus of ischium	With its fellow, in a median raphe	Assists sphincter urethrae	Pudendal (perineal)
c. Superficial transverse perineal	Ramus of ischium	Central point of perineum	Draws back and fixes central point of perineum	Pudendal (perineal)

Ischiorectal Fossa (Anterior Prolongation). The anterior prolongation of the ischiorectal fossa (Figs. 17-3, 17-4) is the space between the urogenital diaphragm and the pelvic diaphragm.

Central Tendinous Point of the Perineal Body. This point (Figs. 17-3, 17-4) is a fibromuscular mass lying between the anal canal and the urogenital diaphragm.

Male External Genitalia

Penis. The penis (Fig. 17-5) is fixed posteriorly at its root and is mobile anteriorly at its body. It consists of three fibroelastic "cylinders" (bodies) filled with erectile tissue. The three bodies are the right and left corpora cavernosa penis and the corpus cavernosum (or spongiosum) urethrae. Each of the two roots or crura of the corpora cavernosa arises from the ischiopubic rami laterally and is covered by the ischiocavernosus muscle. The corpora are united in the body of the penis but are separated at its root. The corpus cavernosum urethrae lies on the lower or ventral surface of the united corpora cavernosa and through it passes the longest of the three divisions of the urethra. The corpus cavernosum urethrae expands anteriorly to form the glans penis, in which the anterior ends of the corpora cavernosa are embedded. Posteriorly, the

Figure 17-4. Muscles of the penis and general relationship of the male external genitalia to the pelvic floor.

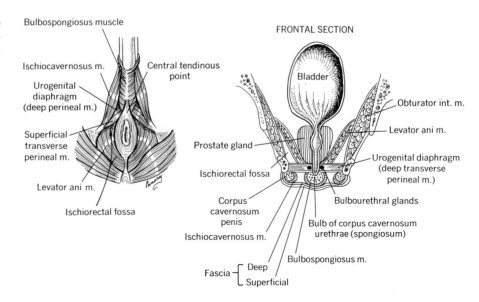

Bulbospongiosus muscle

Ischiocavernosus m.

Urogenital diaphragm (deep perineal m.)

Superficial transverse perineal m.

Levator ani m.

Ischiorectal fossa

Central tendinous point

FRONTAL SECTION

Bladder

Obturator int. m.

Levator ani m.

Prostate gland

Ischiorectal fossa

Corpus cavernosum penis

Ischiocavernosus m.

Urogenital diaphragm (deep transverse perineal m.)

Bulbourethral glands

Bulb of corpus cavernosum urethrae (spongiosum)

Bulbospongiosus m.

Fascia — Deep
— Superficial

Figure 17-5. The male external genitalia. The prostate and bulbourethral glands are also shown.

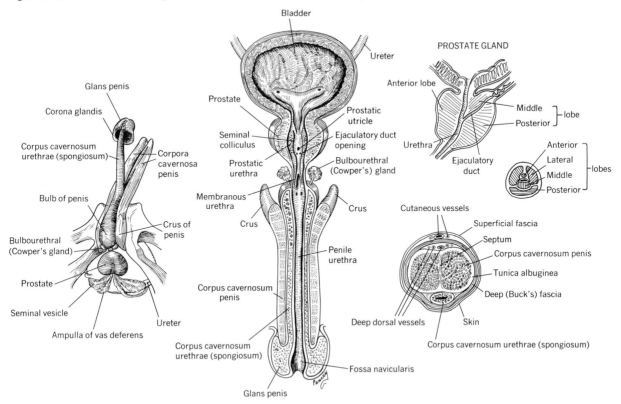

Bladder

Ureter

PROSTATE GLAND

Glans penis

Corona glandis

Corpus cavernosum urethrae (spongiosum)

Corpora cavernosa penis

Bulb of penis

Bulbourethral (Cowper's gland)

Prostate

Seminal vesicle

Ampulla of vas deferens

Crus of penis

Ureter

Prostate

Seminal colliculus

Prostatic urethra

Membranous urethra

Crus

Corpus cavernosum penis

Corpus cavernosum urethrae (spongiosum)

Glans penis

Prostatic utricle

Ejaculatory duct opening

Bulbourethral (Cowper's) gland

Crus

Penile urethra

Fossa navicularis

Anterior lobe

Urethra

Ejaculatory duct

Middle — lobe
Posterior —

Anterior —
Lateral
Middle — lobes
Posterior —

Cutaneous vessels

Superficial fascia

Septum

Corpus cavernosum penis

Tunica albuginea

Deep (Buck's) fascia

Deep dorsal vessels

Skin

Corpus cavernosum urethrae (spongiosum)

corpus cavernosum urethrae terminates in a large, bulbous sac, the bulb of the penis.

The skin of the penis is continued forward as the foreskin or prepuce. It is devoid of hair, and the fascia has no fat. The prepuce forms a "cuff" over the glans, and within the cavity of the prepuce are modified sebaceous glands that secrete smegma. Another fold of skin on the under aspect of the gland, the frenulum, passes forward and is attached to the prepuce. The subcutaneous tissue (continuation of Colles') is loose and contains the superficial dorsal vein (drains skin and prepuce), which is accompanied by superficial lymphatics that pass to the subinguinal lymph nodes. Under the subcutaneous layer is the fascia of the penis (deep fascia or Buck's fascia), which covers the body and corona of the glans to the root. The suspensory ligament is a fibroelastic band from the linea alba and forms a sling for the penis near the symphysis pubis. Beneath Buck's fascia are the deep dorsal vein, arteries, lymph vessels, and nerves. The veins drain into the plexus surrounding the prostate. The erectile tissue is finally surrounded by an elastic, distensible, fibrous tissue layer, the tunica albuginea (Fig. 17-5).

Urethra. The urethra (Fig. 17-5) is a fibroelastic structure about 20 cm (8 in.) long and consists of three parts: prostatic (through the prostate), membranous (through the urogenital diaphragm), and spongy or cavernous (through the corpus cavernosum urethrae of the penis).

PROSTATIC PART. The prostatic part is the widest part and about 2.5 cm (1 in.) long. It has a vertical ridge on the posterior wall, the urethral crest, surrounded by two gutters, the prostatic sinus (prostatic ducts open here). The summit of the crest is called the colliculus, upon which the prostatic utricle (blind sac homolog of the female uterus and vagina) opens, and there are ejaculatory ducts (union of vas deferens and duct of the seminal vesicle).

MEMBRANOUS PART. The membranous part is about 1.25 cm ($\frac{1}{2}$ in.) long and on each side of it (in the urogenital-diaphragm) lie the bulbourethral glands (note that these do not secrete here). It is surrounded by the voluntary muscle, the sphincter urethrae (compressor urethrae), supplied by the pudendal nerve.

SPONGY OR CAVERNOUS PART. The spongy part is the longest portion, lying in the dorsal aspect of the corpus cavernosum urethrae and glans of the penis and expanding terminally in the glans as the navicular fossa. Cowper's glands open here proximally. Its external opening is the narrowest part.

The internal and external urinary sphincters govern urine flow. The internal is derived from the outer and middle muscle layers of the bladder. The external comes from the sphincter urethrae membranaceae, a part of the deep perineal muscle of the urogenital diaphragm.

Scrotum. The scrotum (Fig. 17-6) is a bag of skin and fascia in which the testes are found. Embryologically it arises from the right and left labioscrotal folds of the abdominal wall. In the female the folds remain separated as the labia majora, but in the male they fuse behind the penis to form the scrotum. The median raphe is a vestige of the fusion line. The skin is a single pouch of delicate texture, dark in color and distensible. Its rugosity varies with temperature and is due to the dartos muscle immediately beneath it. The superficial fascia has no fat and is mostly dartos. This superficial layer separates the two testes by a median partition.

Lining the wall of the scrotal "chamber," and thereby forming coverings for the testes beside skin and dartos, are three complete layers derived from the anterior abdominal wall (Fig. 17-6). From without inward, they are the external spermatic fascia, derived from the aponeurosis of the external oblique; the cremasteric fascia and muscle derived from the internal oblique muscle which by its contraction can draw the testis toward the subcutaneous ring and produce the cremasteric reflex; the internal spermatic fascia, derived from transversalis fascia; and the tunica vaginalis testis, a layer originally derived from serous peritoneum but which has descended into the scrotum with its development from the abdominal wall. The last-named layer is the deepest lining and has a visceral layer that covers the epididymis and testis intimately and a parietal layer closely adhered to the internal spermatic layer. A potential space filled with a film of fluid separates the visceral and parietal layers and keeps surfaces moist.

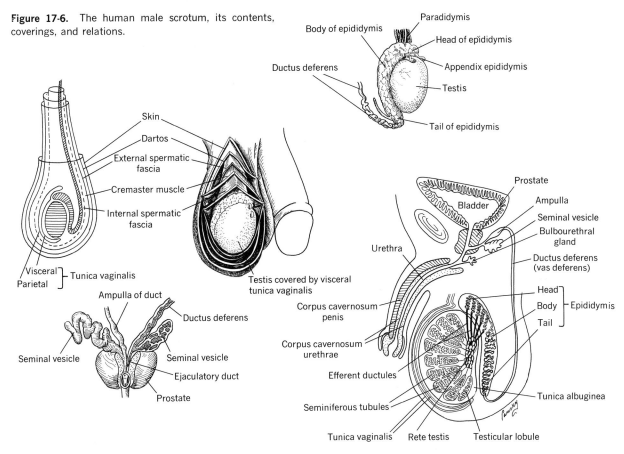

Figure 17-6. The human male scrotum, its contents, coverings, and relations.

Paradidymis

Body of epididymis

Head of epididymis

Ductus deferens

Appendix epididymis

Testis

Tail of epididymis

Skin

Dartos

External spermatic fascia

Cremaster muscle

Internal spermatic fascia

Visceral
Parietal ⎱ Tunica vaginalis

Testis covered by visceral tunica vaginalis

Ampulla of duct

Ductus deferens

Seminal vesicle

Seminal vesicle

Ejaculatory duct

Prostate

Prostate

Bladder

Ampulla

Seminal vesicle

Bulbourethral gland

Ductus deferens (vas deferens)

Urethra

Head ⎱
Body ⎱ Epididymis
Tail ⎱

Corpus cavernosum penis

Corpus cavernosum urethrae

Efferent ductules

Seminiferous tubules

Tunica albuginea

Tunica vaginalis

Rete testis

Testicular lobule

Testis. The testis (Fig. 17-6) is one of a pair of oval bodies lying in the scrotum, variable in size but approximately 3.75 cm ($1\frac{1}{2}$ in.) in length and about 2.5 cm (1 in.) in other dimensions. The vas deferens and epididymis are applied to its dorsal border. It is enveloped by tunica vaginalis (see above). Deep to the latter and forming a complete fibrous covering for the testis is the tunica albuginea, which becomes thick at the posterior glandular border to form the mediastinum testis, from which incomplete fibrous septa pass into the testis proper to divide the gland into lobes. Within these lobes lie the convoluted seminiferous tubules, which conduct sperm from the testis. The combined length of these tubules is about 225 m (750 ft). They pass into a series of straight tubules into the mediastinum to unite and form the network of the rete

testis. From this network, about 20 efferent ductules, the vas efferentia, arise and emerge from the upper pole of the testis and end in the epididymal head. The epididymis is divided into a head (near the upper and posterior border of the testis), a body, and the tail (lower end). The vas or ductus deferens emerges from the tail end and ascends out of the scrotum in the spermatic cord, conveying the spermatozoa from testis to urethra. The spermatic cord consists of the vas deferens and its associated vessels and nerves. The arteries of the cord are the testicular artery (from aorta) and the artery of the vas deferens (from inferior vesical). The veins form a plexus, the pampiniform plexus, which empties into the testicular veins. The lymphatics follow the cord. The genital branch of the genitofemoral nerve (L1 and L2) supplies the cremaster muscle and a sensory branch to the area. Autonomics are also present.

Male Pelvic Viscera

General Considerations

As one looks down into the pelvis, one sees that the rectum lies dorsally in the concavity of the sacrum and coccyx and that the bladder lies anteriorly behind the pubic bone. Both ureters lie at the side of the pelvis on their way to the bladder. The seminal vesicles lie on the back of the bladder below the terminal part of the vas deferens (Figs. 17-5, 17-6). The prostate lies below the bladder and encloses the prostatic urethra. The pelvic peritoneum covers all the pelvic structures. Laterally, as peritoneum covers the side walls of the rectum, are the pararectal fossae, and between bladder and rectum is the rectovesical pouch.

Vas Deferens

The vas deferens (ductus deferens) (Fig. 17-6) is a duct with a scrotal, inguinal, and pelvic course. It begins at the tail of the epididymis, runs up over the back of the testis, and ascends to the superficial inguinal ring in the abdominal wall. It passes through the inguinal canal as a part of the spermatic cord, which it leaves at the deep inguinal ring (deep in the abdominal wall). From the latter it runs to the base of the prostate, crossing over the ureter and lying on the medial side of the seminal vesicle. It is always covered by fascia and never enters the abdominal or peritoneal cavity proper. It expands slightly to form the ampullae (elongated sacs that are semen reservoirs) or the duct, where a secretion of the mucous membrane of the vas is added. Each duct converges and unites with an excretory duct of the seminal vesicles to form the two ejaculatory ducts. These, in turn, pass through the prostate and open into the prostatic urethra. The artery to the vas comes from either the superior or the inferior vesical artery, a branch of the internal iliac.

Seminal Vesicle

The seminal vesicle (Fig. 17-6), which lies on the back of the bladder, is an offshoot of the vas deferens. It is one of a pair of lobulated sacs, about 5 cm (2 in.) long but can be teased apart to unfold to almost 15 cm (6 in.) long and 1.25 cm ($\frac{1}{2}$ in.) wide. Histologically the seminal vesicle looks like the ampullae of the vas. It not only serves as a reservoir for spermatozoa but also produces a secretion that forms part of the seminal fluid. Large amounts of the carbohydrate fructose are found here and may be used by the sperm for energy. Its duct joins that of the vas deferens to form the ejaculatory duct. The arteries to the seminal vesicle are branches of vesical arteries and middle rectal. The veins correspond to the arteries, and the lymph vessels drain into the external and internal iliac glands.

Prostate Gland

The prostate gland (Figs. 17-5, 17-6) is an organ that surrounds the urethra just as it leaves the bladder. It consists of glandular elements, smooth muscle, and fibrous tissue and is about 3.125 cm ($1\frac{1}{4}$ in.) long and 3.75 cm ($1\frac{1}{2}$ in.) broad. It adds a secretion concerned with the vitality of the spermatozoa. The prostate looks like a chestnut, with its base or superior surface under the bladder and its apex pointing downward. The posterior part is flattened and rests against the lower 2.5 cm (1 in.) of the rectum and is the part felt by digital examination. It is this surface, too, that is pierced by the ejaculatory ducts. The gland has a true capsule of its own and another surrounding layer of connective tissue called its false capsule. Between the two there is a massive venous plexus that receives the deep vein of the penis.

Lobes. In fetal life, depressions appear on the urethral walls just below the bladder. These depressions form buds that grow and penetrate the surrounding muscle and connective tissue to form five lobes.

ANTERIOR LOBE. The anterior lobe at birth has no glandular elements; therefore, tumors rarely occur here, and it seldom affects the urethra.

POSTERIOR LOBE. The posterior lobe is inferior to the ejaculatory duct and grows up behind the duct, and thus it is posturethral and postspermatic. It lies behind the middle lobe and is the lobe felt in the rectal examination. Primary cancers do develop here.

LATERAL LOBES. The two lateral lobes grow laterally, anteriorly, and posteriorly. Hypertrophy of these can cause urinary obstruction by compressing the urethra.

MEDIAN LOBE. The median (middle) lobe is clinically the most important. It lies behind the urethra, in front of the spermatic cord, and below the neck of the bladder and is very glandular. This lobe normally projects into the urethra, causing a prominence, the seminal colliculus (verumontanum or crista urethralis). This is the lobe in which tumors often grow and may block the internal urethral opening.

Vessels. The blood supply is from the inferior vesical and middle rectal arteries. The venous plexus around it continues backward to terminate in the internal iliac vein. The lymph vessels end in the internal iliac and sacral lymph glands.

Male Maturation

Developmental and Growth Changes and Secretion

The male external genitalia undergo five stages of development: (1) infantile stage, (2) scrotal enlargement, (3) penis enlargement, (4) sculpturing and darkening of the penis, and (5) the adult condition. Stage 2 begins at about 12 years of age, and stage 5 is reached at about 17 years of age. The epididymis and accessory glands are active at birth, being subjected to maternal hormones during late fetal life. Postnatal regression leads to quiescence until puberty, when development begins at around ten years and keeps pace with the testes.

The testes normally are descended into the scrotum early in life, but they increase very little in size during the first ten years. They grow rapidly thereafter and reach their full size at about age 20. There is apparently little decrease even with old age. The secreting interstitial cells of the testes develop with the testes but contribute little to their bulk. These cells are predominantly responsible for most (if not all) of the production of androgen (the adrenal gland produces the rest). This hormone is secreted in the urine and shows a clear rise and fall during reproductive life. Functionally, however, androgen is adequate for many years, and the loss of androgen control of the pituitary gland seems to begin only around 60 years of age. The seminiferous tubules begin to show activity at about ten years of age and active spermatogenesis begins about three years later, at 13, so that at 14 and 15 years of age the male is producing spermatozoa in number and quality. Thus, the male is potentially "fertile" at about 14 or 15 years of age.

Once begun, spermatogenesis (production of germ cells) proceeds right through adulthood and into old age. The process of spermatogenesis requires testosterone, but testosterone production does not depend on spermatogenesis. Thus, testosterone deficiency may produce sterility by interrupting spermatogenesis, but interfering with seminiferous tubular function does not change the production of testosterone by the interstitial cells. Thus, vasectomy (cutting the vas deferens) prevents sperm passage but does not alter hormone secretion or sexual drive or "maleness." However, there is a natural great variation in the number of spermatozoa produced by different individuals, and this varies with the frequency of ejaculation. Twenty million spermatozoa in one ejaculation is considered a low count and marginal for fertility, whereas 500 million is high.

The physical signs of sexual capacity are erection, orgasm, and ejaculation, but not all are necessarily connected, for erection and orgasm can take place even in infants whereas the ability to ejaculate appears later in life and may occur without erection (e.g., a nocturnal emission). The capacity for erection and ejaculation is retained into old age; yet ejaculation is much less vigorous in old men.

Spermatogenesis

The seminiferous tubules of the testis have a complex lining of cells, the vast majority of which are in

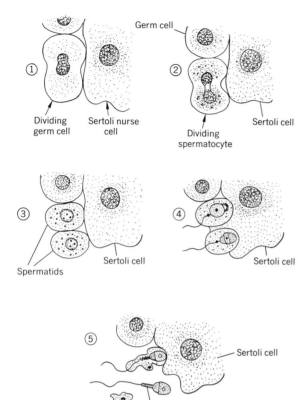

Dividing germ cell

Germ cell

Sertoli nurse cell

① ②

Dividing spermatocyte

Sertoli cell

③ ④

Spermatids

Sertoli cell

Sertoli cell

⑤

Sertoli cell

Cytoplasmic remnant   Spermatozoon

Figure 17-7. The germ cells in the tubules of the testis form the spermatozoa. The germ cells divide to form spermatocytes, which divide to form spermatids with one half the chromosome number. The spermatids embed in Sertoli (nurse) cells and grow tails. The cytoplasm contracts away, and the sperm become free swimming.

various stages of division and giving rise to the sperm cells (Fig. 17-7). Before puberty, the seminiferous tubules are lined by special "nurse cells" as well as germ cells from which the spermatozoa develop. At puberty, the undifferentiated germ (or stem) cells, called spermatogonia (the outer layer of dividing cells in the periphery of the tubules), begin to multiply by a series of mitotic divisions, and they differentiate to form primary spermatocytes as they move away from the membrane lining the tubule. The germ cells form a reservoir of stem cells for future spermatozoa. Each spermatogonium contains a diploid number of chromosomes (44 somatic plus an XY), and each undergoes repeated mitotic divisions to produce spermatogonia (continuing the stem line) and primary spermatocytes. Thus, the process occurs continuously from puberty to senescence; yet during this period some tubules are active and others temporarily quiescent. It is of interest that, in any small region of the seminiferous tubules, the entire process of spermatogenesis proceeds in a regular sequence, so that at any one time all the primary spermatocytes in one area of the tubule are undergoing division and in another area the secondary spermatocytes may be dividing.

Each primary spermatocyte (with the same chromosome number as the spermatogonia) divides twice to form two daughter cells or secondary spermatocytes. Each of these again divides to produce two spermatids. The latter differentiate to form mature spermatozoa. The transformation of the spermatids to mature spermatozoa involves no cell division. The two divisions that result in spermatids (primary to secondary) are reduction divisions or meiosis, during which the chromosome number of the diploid primary spermatocyte (46) is halved to give the haploid (23) that is found in the spermatozoon (Fig. 17-8). As this takes place, the spermatid is engulfed by the cytoplasm of the Sertoli cells, which are found in the wall of the seminiferous tubule. The sperm cluster around the tips of Sertoli cells, where they gain nourishment. They are then released and float in the lumen of the tubule. By their accumulation, "growth pressure" develops and the sperm are forced through the rete testis and efferent ducts into the ductus epididymis, where they undergo physiologic maturation and obtain the ability for movement and fertilization. The tail of the epididymis acts as a storehouse for sperm, and they may survive here for several weeks to months. They are carried through the vas deferens to the ampulla of the vas by the "pressure" as well as the peristaltic contractions of the vas, where they wait to be ejaculated or die (and are absorbed). If ejaculated into the acid vagina, the survivors move at a rate of 15 mm per minute into the uterus and uterine tubes. They lose their ability to fertilize after 48 hours and die in 96 hours. A redistribution of genes between the homologous chro-

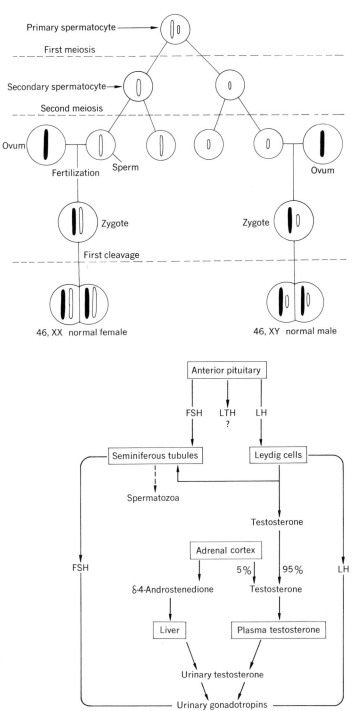

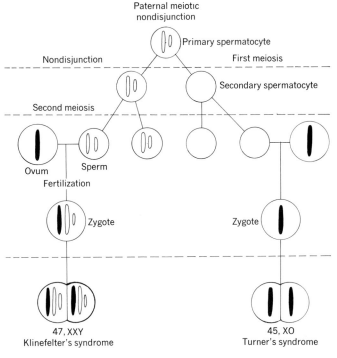

Figure 17-8. Normal meiosis of spermatocytes produces equal numbers of X and Y sperm that lead to normal offspring.

Figure 17-9. Nondisjunction during meiosis produces either XXY or XO child.

Figure 17-10. The pituitary-gonadal axis. The testes serve as hormonal and reproductive organs. FSH acts on the germinal epithelium of the testes to promote spermatogenesis. LH acts on the Leydig cells, in which testosterone matures and is secreted. The function of luteotropin (LTH) or prolactin in the male is unknown. The contribution of the adrenal cortex to both plasma and urinary testosterone is small. (After studies by Paulsen.)

mosomes also occurs (Fig. 17-9). Thus, crossing-over of chromosome parts may allow for genetic exchange and occurs between homologous chromosomes (see Chap. 1). The above-described process from the first division of the sperm cell to the production of the spermatozoa in the ejaculate takes about 72 days.

Sperm reproduction is under the control of the anterior pituitary gland (Fig. 17-10). Removal of the latter results in infantile testes. In man and most other mammals, the production of spermatozoa can take place only if the testicular temperature is maintained below body temperature. Thus, when the temperature is high, the scrotum relaxes; in the cold, it contracts close to the body. Furthermore, should the testes fail to descend into the scrotum (they normally do so in the month before birth), no spermatozoa are produced; however, if the testes are carried down surgically, sperm production commences. Altitude may also affect sperm production, probably due to oxygen lack rather than temperature. Other factors resulting in damage to spermatozoa are irradiation and radiomimetic agents that act like x-rays.

The Sperm Cell (Spermatozoon—Definitive Germ Cell)

A man normally produces about 100 to 300 million sperm cells in one ejaculation. Under favorable circumstances, one of these will fertilize the single egg cell normally shed by a woman at ovulation. The spermatozoon not only stimulates the egg to divide and develop but also brings to it a genetic contribution from the male, which joins that of the female in a fertilized egg. The spermatozoa° are made in the testes, mixed with secretions from several glands (prostate, seminal vesicles, and glands lining the genital

---

° Spermatozoa were first identified in semen in 1677 by the Dutch medical student John Hamm and confirmed by Antonj van Leeuwenhoek in 1679. "Ovists" thought they merely nourished or stimulated the female seed, whereas "animists" thought each one contained the entire miniature embryo or homunculus. K. E. von Vaer discovered the human egg in 1827, and M. Barry first described spermatozoa penetrating the egg in 1843. Details of sperm production in the testis were described by R. Kolliker in 1841, and the role of spermatozoa in fertilization was established by E. van Benedeen in 1883.

ducts) to form semen, and ejaculated into the female vagina during intercourse. They reach the fallopian tubes aided by muscular contractions of the uterus, where, in its distal end, the egg is fertilized.

Histology of Sperm. The head of the mature spermatozoon (Fig. 17-11) consists almost entirely of a nucleus that carries DNA, which at fertilization transmits the genetic contribution of the father. The head is covered in front by a cap, the acrosome (a protein-filled vesicle derived from the Golgi apparatus), which contains certain lytic enzymes that permit the sperm to penetrate the ovulated egg. The nucleus and acrosome vary in shape from species to species. The middle piece of the sperm contains a sheath of mitochondria, where enzymes exist that provide for the transfer of energy (derived from sugar metabolism) to the tail fibers for active movement.

The sperm tail articulates with the head by what appears to be a sort of ball-and-socket joint, where the capitulum of the connecting piece is the ball that articulates with the basal plate, which lines the concavity at the base of the spermatozoon head. The tail is composed of actomyosinlike contractile fibers, which are arranged in a pattern consisting of a central pair of fibers with an outer ring of nine coarse fibers and an inner ring of nine fine fibers. This pattern also exists in flagellae of both animal and plant cells. Contraction of the fibers produces a whiplike movement of the tail, enabling it to propel the sperm at 1 to 4 mm per second.

Male Sexual Act

Erection

The male sexual act produces erection of the penis, which permits entry into the female vagina and ejaculation of the semen (containing sperm) into the vagina. Erection is a vascular phenomenon. The penis consists of three cylindric cords of erectile tissue or vascular spaces. The arterioles supplying these spaces are normally constricted; so they contain little blood and the penis is flaccid. During sexual excitation, the arterioles

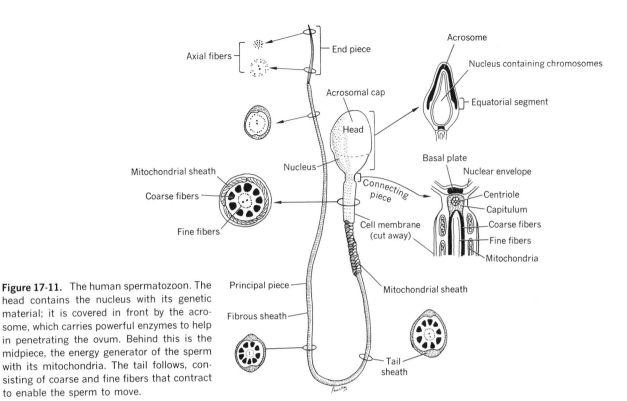

Figure 17-11. The human spermatozoon. The head contains the nucleus with its genetic material; it is covered in front by the acrosome, which carries powerful enzymes to help in penetrating the ovum. Behind this is the midpiece, the energy generator of the sperm with its mitochondria. The tail follows, consisting of coarse and fine fibers that contract to enable the sperm to move.

dilate, the spaces fill with blood, and the penis becomes hard and rigid. As the erectile tissue expands, the emptying veins are compressed and hinder vascular outflow. The entire process may take only five to ten seconds. The vascular dilatation is made possible by stimulation of the parasympathetic nerves and inhibition of the sympathetic nerves to the arterioles. Furthermore, the parasympathetic nerves stimulate the urethral glands to secrete a mucuslike material that helps in lubrication. The major sensory input is the highly sensitive mechanoreceptors found in the tip of the penis. Afferent fibers relay impulses and synapse in the lower spinal cord to trigger efferent outflow. Higher brain centers, via descending tracts, may exert facilitative or inhibitory influences on the efferent neurons and, therefore, emotions and thoughts may cause erection without actual mechanical stimulation or there may be impotence (no erection) due to some psychologic phenomena.

Ejaculation

Ejaculation is predominantly a spinal reflex, with the afferent pathway the same as that of erection. When a peak (climax) is reached, a patterned automatic sequence of efferent discharge is relayed to both smooth muscle of the genital ducts and skeletal muscle at the base of the penis. Thus, the genital ducts contract, emptying the semen into the urethra, and it is then expelled from the penis into the vagina by rapid muscular contraction. During this process, the sphincter at the base of the bladder is closed so that neither urine is expelled nor sperm travel back into the bladder. A simultaneous skeletal muscle contraction throughout the body is then followed by rapid muscle and psychologic relaxation.

Semen

In man, the average volume of semen is about 3 ml (varying from 0.5 to 10 ml) and contains about 300 million sperm. A boar, in contrast, ejaculates up to 500 ml. In man, liquid semen coagulates rapidly due to a clotting enzyme in the prostate that acts on a substance produced in the seminal vesicle. This clot liquefies again in a few minutes due to another prostatic enzyme. Why human semen clots and then liquefies is not clear. The components of semen are not necessarily ejaculated as a homogeneous mixture. The semen of the first part of the ejaculate more closely resembles the contents of the vas deferens than do the succeeding fractions.

When spermatozoa penetrate the mucus at the neck of the womb (uterus), they become arranged in a direction determined by the orientation of long-chain molecules of the mucus. This latter orientation is determined by simple physical forces. In women, the consistency of the mucus at the womb's neck varies with the menstrual cycle. It is most fluid at midcycle, when conception is most likely to occur if intercourse takes place.

Other Considerations

Immune Properties

Spermatozoa may act as antigens. It is suggested that they meet antibodies in the female genital tract and that binding of these to the sperm may have some significance. Apparently, mobile living sperm do not bind the antibody but immobile sperm do, and the latter may also play a role in the removal of sperm from the female tract by white cells in the blood.

Sex Determinant

Two different kinds of spermatozoa determine the sex mechanism. Half carry the female-determining X chromosome and half carry the male-determining Y chromosome. The X male chromosome gives rise to a female offspring, the Y to a male.

Although many of the characteristics of sperm cells are determined by the genetic constitution of the parent cells from which they are derived, there is evidence that the gene complement of the sperm itself can determine some of its characteristics. Thus, some researchers have claimed that the X and Y spermatozoa differ in specific gravity and other characteristics, and other investigators have stated that the two types can be separated (this has not been completely verified).

Artificial Insemination

In recent years, methods for semen preservation outside of the body have improved and extended the use of artificial insemination. Semen may be preserved at normal refrigeration temperature ($+4°$ C) for several days, but after freezing to $-196°$ C semen has been kept up to ten years. Artificial insemination by the husband is sometimes used where the husband has a very low fertility or is impotent. Artificial insemination by donor is used when the husband and wife prefer it due to hereditary defects carried by the husband or in cases of severe rhesus (Rh) incompatability.

Review Questions

1. What is meant by secondary sexual characteristics?
2. Where is the perineum of the male?
3. Describe the component parts of the male penis.
4. Differentiate between the various portions of the male urethra.
5. What makes up the wall or coverings of the male scrotum?
6. Describe the passage of sperm from testes to the tip of the penis.
7. Where is the prostate located? What are its parts?
8. Where is the seminal vesicle? What does it accomplish?

9. What is spermatogenesis? What does it accomplish? What are the steps of spermatogenesis?
10. Describe the essential anatomy of a sperm.
11. What enables an erection to take place? How is it controlled? Define ejaculation.
12. What is semen?
13. What is meant by artificial insemination?

References

Ford, C. S., and Beach, F. A.: *Patterns of Sexual Behavior.* Methuen, Ltd., London, 1965.

Leblond, C. P.; Steinberger, E.; and Rossen Runger, E. C.: *Mechanisms Concerned with Conception.* Pergamon Press, Ltd., Oxford, 1963.

Lloyd, C. W., and Wersy, J.: Some Aspects of Reproductive Physiology. *Annu. Rev. Physiol.,* **28**:267, 1966.

Masters, W. H., and Johnson, V. E.: *Human Sexual Response.* Little, Brown & Co., Boston, 1966.

————: *Human Sexual Inadequacy.* Little, Brown & Co., Boston, 1970.

Mittwoch, V.: Sex Differences in Cells. *Sci. Am.,* **209**(1):54–62, 1963.

Tepperman, J.: *Metabolic and Endocrine Physiology,* 3rd ed. Year Book Medical Publishers, Inc., Chicago, 1973.

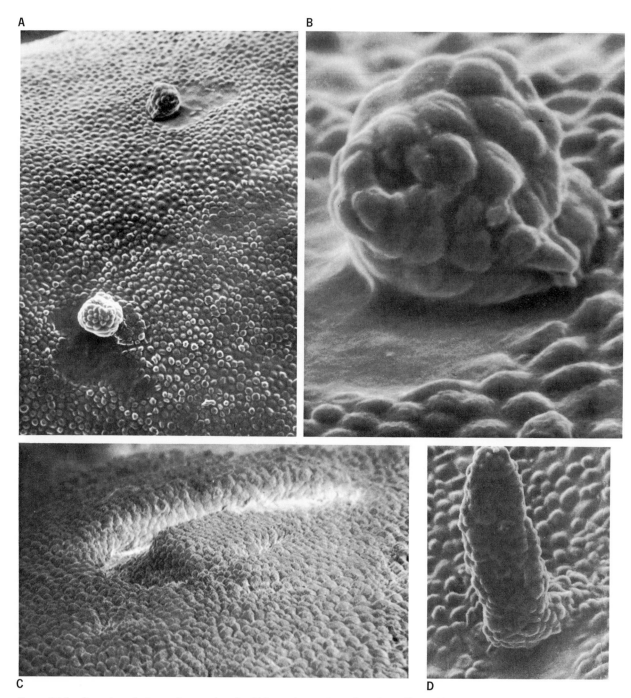

Figure 18-1. Scanning electron micrographs of rabbit ovarian surface. Seen is a villus, which may be normal, or an adenoma. **A,** ×400; **B,** ×6000; **C,** ×840; **D,** ×2000. (Courtesy of Prof. Pietro Motta, University of Rome.)

The Female Reproductive System

CHAPTER 18

Female External Anatomy

Perineum

The perineum (Fig. 17-3, p. 575) is the region between the thighs and the lower part of the buttock. It is shaped like a diamond and is bordered by the inferior pubic ligament anteriorly, the tip of the coccyx posteriorly, and the ischial tuberosities at each side. As in the male, a line between the tuberosities divides the region into an anterior urogenital triangle and a posterior anal triangle. The anal triangles in male and female are essentially similar, while the urogenital ones differ. On the surface of the latter, in the female, are found the external genitalia.

Female External Genitalia

The female external genitalia (Fig. 18-2) consist of the structures listed below.

Mons Veneris. This is the cushionlike eminence on the front of the pubes and consists of fat under the skin and covered with hair.

Vulva. The vulva (pudendum) includes the labia majora, labia minora, body and glans of the clitoris, and the vaginal opening. It extends from the pubes to a point in front of the anus.

LABIA MAJORA. The labia majora are homologous to the male scrotum, with the vestibular cleft between the two corresponding to the scrotal raphe. In nulliparae (women who have never borne children) the labia are approximated. Each labium is a rounded cutaneous fold covering a long process of fat. The two are united below the mons pubis (veneris) to form the anterior commissure; they connect posteriorly to form the posterior commissure. Each labium is covered by skin and hair and has many sebaceous glands; its inner surface is more delicate with large sebaceous follicles.

LABIA MINORA. The labia minora (nymphae) are also skin folds but medial to the labia major. They close the vestibule and are concealed by the majora in the young, but with age and births they are frequently pendulous and even externally visible. Anteriorly they split to meet above the clitoris as the prepuce and a layer below the glans clitoris to form the frenulum. In front of the posterior commissure of the majora they are connected by a fold of skin, the fourchet (frenulum of the labia). The fossa navicularis is a shallow depression in front of the fourchet. These labia have no hair and look like mucous membrane

VESTIBULE. The vestibule of the vagina is the cleft between the minora into which the vagina and urethra open. On each side are the openings of the Bartholin glands.

CLITORIS. The clitoris is the homolog of the penis and is found at the apex of the vestibule. It consists of a body and a glans.

URETHRAL ORIFICE. The urethral orifice (external urinary meatus) is about 2.5 cm (1 in.) below the clitoris and is 4 to 5 mm (0.16 to 0.20 in.) in diameter. A group of small glands (homolog of male prostate) are found at its sides near the lower end. They group together and use a common paraurethral or Skene's duct, which opens at the sides of the urethral opening.

VAGINAL ORIFICE. The vaginal orifice is near the posterior end of the vestibule and is partially closed,

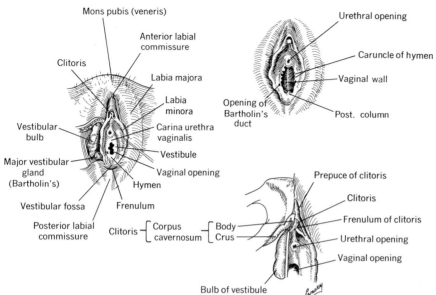

Mons pubis (veneris)

Anterior labial commissure

Clitoris

Labia majora

Labia minora

Vestibular bulb

Carina urethra vaginalis

Major vestibular gland (Bartholin's)

Vestibule

Vaginal opening

Vestibular fossa

Hymen

Frenulum

Posterior labial commissure

Clitoris { Corpus cavernosum { Body Crus

Urethral opening

Caruncle of hymen

Vaginal wall

Opening of Bartholin's duct

Post. column

Prepuce of clitoris

Clitoris

Frenulum of clitoris

Urethral opening

Vaginal opening

Bulb of vestibule

Figure 18-2. Female external genitalia.

in the virgin, by the hymen, which is a thin, vascular duplication of mucous membrane that varies in shape and thickness. After rupture it is seen as irregular projections into the vaginal orifice, the hymenal caruncles. It usually, even when intact, still presents a central opening that can be stretched sufficiently to allow one examining finger. Whether torn or intact, the hymen persists anteriorly as a membranous band as it nears and joins the tissues forming the urethral floor at the meatus.

Superficial Perineal Fascia. The superficial perineal fascia divides into a superficial fatty layer and a deeper membranous one (like the abdominal wall). Where the fatty layer covers the labia, the fat is replaced by a thin layer of involuntary muscle (homolog of the male dartos of the scrotum). The deep layer is called Colles' fascia here and is only found in the urogenital area.

Superficial Perineal Pouch. The superficial perineal pouch is a space between the deep layer of superficial fascia and the perineal membrane (fascia of urogenital diaphragm). It contains the crura of the clitoris, bulb of the vestibule, greater vestibular or Bartholin glands, and superficial perineal muscles, nerves, and vessels.

CRURA OF THE CLITORIS. The crura of the clitoris

are like those of the penis and consist of the corpora cavernosa clitoris, only in miniature, and are covered by the ischiocavernosus muscle. They retard the outflow of blood and assist in production of erection of the clitoris. The corpora fuse anteriorly to form the small unpaired body of the clitoris. It, too, is covered by a glans.

BULB OF THE VESTIBULE. The bulb of the vestibule is similar to that of the male but bilateral; i.e., it consists of two oblong masses of erectile tissue covered by the bulbocavernosus (or bulbospongiosus) muscles, which act as constrictor muscles of the erectile tissue. The masses unite near the body of the clitoris and in front of the urethra.

PERINEAL BODY (CENTRAL POINT). This is a musculotendinous point found in the midline of the perineum anterior to the anus and posterior to the perineal compartment. The superficial muscles attach here (similar to the male).

GREATER VESTIBULAR OR BARTHOLIN GLANDS. These correspond to the male Cowper's glands but differ in position. They lie on the side of the posterior part of the vagina under cover of the bulbs. The ducts open into the vestibule. They are palpable only when diseased and enlarged.

Muscles of the Perineum and Pelvis

The muscles of the perineum and pelvis (Table 17-1, p. 576) are arranged in three layers. The superficial layer in the superficial space consists of the superficial transverse perineal, bulbocavernosus, and ischiocavernosus muscles, all of which are supplied by the perineal branch of the pudendal nerve. The middle layer (urogenital diaphragm) consists of the urethral sphincter and the deep transverse perineal muscles. The deep layer is the levator ani and coccygeus (which together form the pelvic diaphragm).

Urogenital Diaphragm. The urogenital diaphragm (Fig. 17-3, p. 575; Fig. 18-3) consists of an inferior fascia or perineal membrane and a superior fascia with the above-mentioned muscles between. Above the superior fascia (between it and the pelvic diaphragm) is the anterior recess of the ischiorectal fossa (which is fat filled).

Pelvic Diaphragm. The pelvic diaphragm (Fig. 17-3, p. 575; Fig. 18-3) consists of the levator ani, the coccygeus muscles, and their superior and inferior fascias.

Figure 18-3. Frontal section through uterus to illustrate the relations of the internal and external female genitalia and pelvic and urogenital diaphragms.

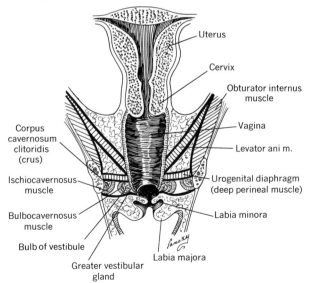

Uterus
Cervix
Obturator internus muscle
Corpus cavernosum clitoridis (crus)
Vagina
Levator ani m.
Ischiocavernosus muscle
Urogenital diaphragm (deep perineal muscle)
Bulbocavernosus muscle
Labia minora
Bulb of vestibule
Greater vestibular gland
Labia majora

Vessels and Nerves

Internal Pudendal Artery. The internal pudendal artery (Fig. 10-26, p. 364) leaves the pelvis to enter the gluteal region through the greater sciatic foramen, turns around the spine and sacrospinous ligament, and reaches the anal perineum by passing through the lesser sciatic foramen. It passes along the lateral wall of the ischiorectal fossa (lying in Alcock's canal) on the obturator internus muscle fascia, and here gives off the inferior rectal artery to the anus, anal canal, superficial fascia, and skin. The main vessel continues to the urogenital area to supply the skin and tissue of the labia. It continues as the perineal artery to supply the muscles of the superficial space, and the deep branches go to the clitoris (deep and dorsal) as well as the erectile tissue of the superficial space.

Dorsal Vein. The dorsal vein of the clitoris joins the venous plexus on the vaginal wall as well as the pudendal vein. The latter drains the same area as the artery supplies.

Pudendal Nerve. The pudendal nerve is the chief source of muscle and skin innervation; is derived from S2, S3, and S4; and follows the pudendal artery and vein.

Anal Triangle

See the discussions on muscles of the anus and rectum (Chap. 12) and pelvic diaphragm (Chap. 7).

Female Pelvic Viscera

Embryology

Prior to sex determination, four parallel tubes grow caudally in the retroperitoneal tissue of the posterior abdominal wall (Fig. 18-4): the right and left müllerian (or paramesonephric) ducts and the right and left wolffian (or mesonephric) ducts. They terminate in the part of the cloaca called the urogenital sinus. In the male, the wolffian duct predominates to form the spermatic duct, epididymis, vas deferens, and ejaculatory duct. In the female, the müllerian duct predominates and becomes the fallopian tube (oviduct), uterus, and vagina. The cephalic ends of the two müllerian ducts

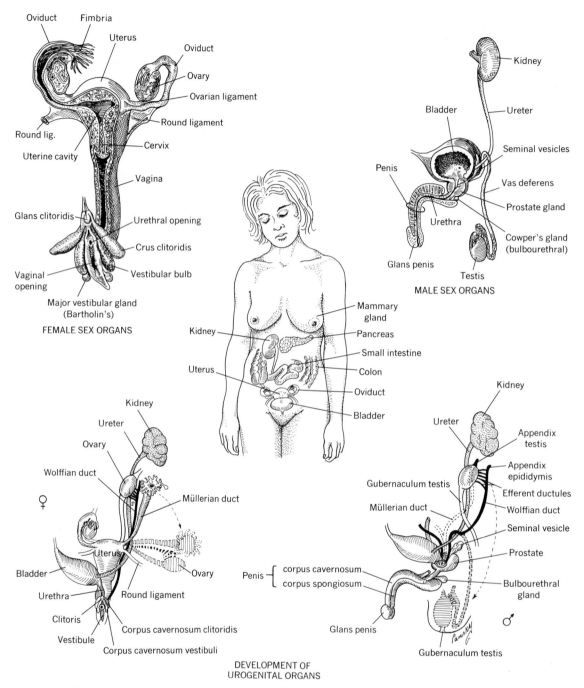

Oviduct　Fimbria

Uterus

Oviduct

Ovary

Ovarian ligament

Round ligament

Round lig.

Uterine cavity

Cervix

Vagina

Glans clitoridis

Urethral opening

Crus clitoridis

Vestibular bulb

Vaginal opening

Major vestibular gland (Bartholin's)

FEMALE SEX ORGANS

Kidney

Bladder

Ureter

Seminal vesicles

Penis

Vas deferens

Prostate gland

Urethra

Cowper's gland (bulbourethral)

Glans penis

Testis

MALE SEX ORGANS

Kidney

Mammary gland

Pancreas

Small intestine

Colon

Uterus

Oviduct

Bladder

Kidney

Ureter

Ovary

Wolffian duct

Müllerian duct

Kidney

Ureter

Appendix testis

Appendix epididymis

Gubernaculum testis

Efferent ductules

Müllerian duct

Wolffian duct

Seminal vesicle

Uterus

Ovary

Prostate

Bladder

Round ligament

Urethra

Penis { corpus cavernosum, corpus spongiosum

Bulbourethral gland

Clitoris

Vestibule

Corpus cavernosum clitoridis

Corpus cavernosum vestibuli

Glans penis

Gubernaculum testis

DEVELOPMENT OF UROGENITAL ORGANS

Figure 18-4. Female and male pelvic visceral embryology. The mature male and female reproductive tracts are also seen.

open into the body cavity, whereas the caudal ends fuse with each other and form a single tube that opens into the urogenital sinus. The cephalic ends become the fallopian tubes; the intermediate parts fuse to form the upper vagina. The intermediate part of one duct pulls away from the sides of the pelvis to meet its fellow of the opposite side and thus pulls a fold of peritoneum with it, the so-called broad ligament of the uterus, which is really the mesentery of the müllerian duct. In the female, each of the wolffian ducts only persists at its extremities, forming the tubules of the epoophoron and paroophoron. (In the male, each of the müllerian ducts disappears except at its caudal end, where it becomes the prostatic utricle, and at its cephalic end, where it becomes the appendix testis.)

The ovary develops as a retroperitoneal abdominal organ (as does the testis) but becomes pelvic in the adult. The gubernaculum of the ovary is attached caudally to the skin, which is later to become the labium majus, and in its course it attaches to the side of the uterus. The gubernaculum passes through the inguinal canal of Nuck (a prolongation of peritoneum). As the ovary descends into the pelvis, it takes its nerve and vascular supply along. The gubernaculum becomes the ligament of the ovary and the round ligament (teres) of the uterus. (In the male, the gubernaculum of the testis passes through the inguinal canal into the scrotum, pulling with it a processus vaginalis and the testis.) The ovary may descend abnormally and come to lie outside of the pelvis, even following the gubernaculum into the labium. This is spoken of as an ectopic ovary.

Female Visceral Support

The female viscera (Figs. 18-5, 18-6) are supported by six ligamentous structures: the broad ligaments, the

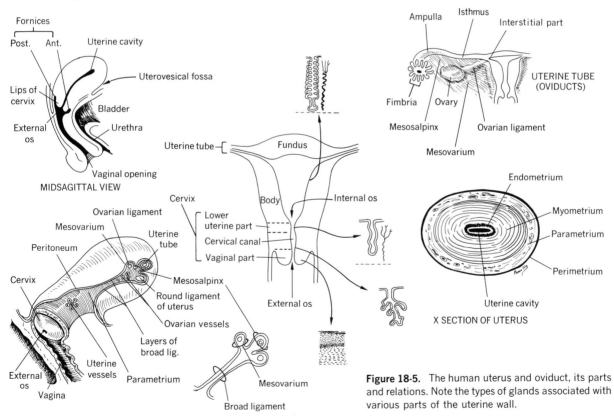

Figure 18-5. The human uterus and oviduct, its parts and relations. Note the types of glands associated with various parts of the uterine wall.

Figure 18-6. Female pelvis, as seen from several views.

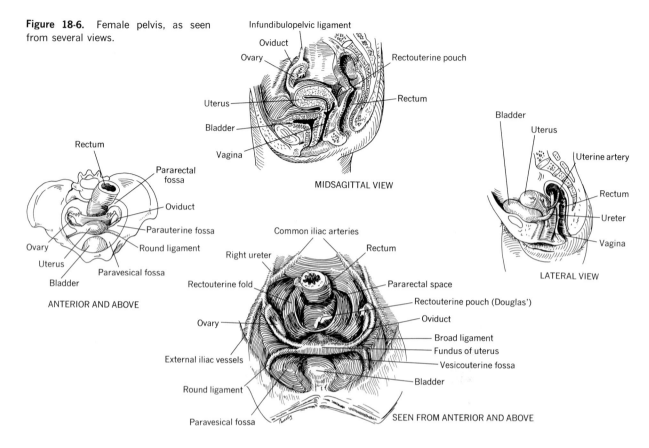

infundibulopelvic ligaments, the round ligaments, the uterosacral ligaments, the cardinal or Mackenrodt's ligaments, and the ovarian ligaments.

Broad Ligaments; Infundibulopelvic Ligaments. The broad ligaments of the uterus are peritoneal (mesenteric) folds that cover the uterus and pass laterally from it to the lateral wall of the pelvis. Its two layers (anterior and posterior) envelop the uterus and laterally cover the fallopian tube (here called the mesosalpinx) and suspend the ovary as the mesovarium. Its most lateral aspect, from the fimbriated end of the fallopian tube to the wall of the pelvis, is called the infundibulopelvic ligaments. Since it is actually peritoneum, it becomes continuous with the peritoneum on the pelvic wall and floor. Between the two layers of ligament, to the sides of the uterus, is the extraperitoneal connective tissue or parametrium, which con-

tains, the oviduct, round ligament of the uterus, ovarian ligament, epoophoron, and the uterine and ovarian vessels.

Round Ligaments. The round ligaments are not peritoneal but true ligamentous fibromuscular cords running from the upper lateral angle of the uterus to the internal inguinal ring, where they enter the inguinal canal to terminate in the labium majus. These ligaments draw the uterus forward and, with intra-abdominal pressure, help maintain the normal position of the uterus. If they fail to function, the uterus may prolapse.

Uterosacral Ligaments. The uterosacral ligaments are fibromuscular cords that pass from the back of the upper end of the cervix on each side of the rectum and end in the sacrum. They create ridges, called rectouterine folds, as the peritoneum covers them.

They function to hold the cervix up and back and help maintain the antiflexon (forward bending) of the uterus. If they fail, the cervix is displaced down and forward, permitting backward displacement of the body of the uterus, and intra-abdominal pressure causes uterine prolapse.

Cardinal Ligaments. The cardinal (transverse cervical or Mackenrodt's) ligaments are located lateral to the cervix and vagina and are continuous on each side with the uterosacral ligaments. They form the bases of the broad ligaments and lie within them. Blood vessels (predominantly veins) are the major component. They are really a condensation of the parametrial tissue that helps suspend the cervix and uterus from the pelvic walls.

Ovarian Ligaments. Each ovarian ligament is a fibromuscular cord that is found inside the broad ligament and extends from the ovary to the lateral side of the uterus, where it attaches between the oviduct and round ligament. It is a portion of the embryonic gubernaculum (the round ligaments form the rest).

Uterus

The uterus (Figs. 18-5, 18-6) is a muscular, pear-shaped organ located between the bladder and rectum. It measures 7.5 cm (3 in.) by 5 cm (2 in.) by 2.5 cm (1 in.), and it is flat anteriorly where it faces the bladder and convex posteriorly. The uterus is lined with mucous membrane and covered partially by peritoneum. The fallopian tubes enter at its widest part.

The uterus has three parts: the body or corpus, the main portion; the base or fundus, which lies above the opening of the oviducts and forms its upper dome-shaped portion; and the cervix, which projects into the vagina. The upper portion of the corpus is covered by peritoneum that is reflected onto the bladder in front of it, forming a potential space between them, the uterovesical pouch. The posterior surface is covered by peritoneum that extends below the body to cover the cervix and is then reflected onto the rectum, also creating a space, the rectouterine space (the pouch of Douglas). Laterally, the broad ligament covers the parametrium, the vessels and nerves, and the ligaments of the uterus.

The cervix (Fig. 18-5) is the narrower, cylindric part of the uterus that enters the vagina via the external os and lies at right angles to it. It thus has a supra-vaginal portion and an intravaginal portion. The vaginal portion is covered with mucous membrane reflected from the vaginal wall. In the nulliparae, this is a transverse slit with smooth and rounded lips; however, in the multiparae (women having had several children), it is wider and its lips are irregular.

The uterine cavity is small and in sagittal section is an elongated, narrow cleft that becomes continuous with the cervix. Its mucous membrane is velvety and smooth. The mucous membrane of the cervix, on the other hand, is raised into folds, secondary folds or the arbor vitae present in the nulliparae.

The body of the uterus is the most movable part, while the cervix, although movable, is firmly held by the cardinal ligaments. In the virgin, the fundus of the uterus is tipped forward (anteversion). Furthermore, the uterus is also bent forward on itself (anteflexion). This position varies and is subject to change, depending on the posture and condition of the bladder and rectum.

The uterus consists of three layers: an outer serous layer, or perimetrium; a middle muscle layer, or myometrium; and an inner mucous layer, or endometrium. The serous coat is actually visceral peritoneum (mesothelial cells). The muscle coat makes up the greater share of the organ and consists of interlacing bundles of smooth-muscle fibers united by connective tissue. These fibers increase in size (probably in number) during the first half of pregnancy and never return to virginal size. The inner lining is smooth and from 2 to 5 mm thick. The epithelium is columnar and firmly attached to the muscle. The endocervix is also columnar but contains many compound racemose (grapelike) glands that secrete a clear mucus.

Vagina

The vagina (Figs. 18-5, 18-6) is a flattened, distensible, musculomembranous canal that extends from the vulva, through the urogenital diaphragm, to encircle the uterine cervix. Its anterior wall is 7.5 cm (3 in.) long and its posterior wall 9.25 cm ($3\frac{1}{2}$ in.) long, and the walls are normally in contact with each other. Thus, the vagina, in transverse section, appears as a transverse slit. The space that exists between the intravaginal part of the cervix and the vaginal wall is divided into anterior, lateral, and posterior fornices (cul-de-sacs). The normal nulliparous vagina is held in position by its surrounding fascia: between it and the bladder and pubis anteriorly, and between it and the rectum (rectovaginal fascia or fascia of Denonvillier) posteriorly. When these fasciae are torn or stretched, rectoceles (prolapse into rectum) and cystoceles (prolapse into bladder) develop. One other relation is important: laterally, the vagina is crossed by the ureters and the uterine arteries are also nearby.

The major vessel of the vagina is the vaginal artery, but it also receives branches from the uterine and middle rectal arteries. The veins form networks in the submucous coat and on its surface. The lymph vessels to various portions of the vagina drain into the external and internal iliac glands, the sacral and common iliac nodes, and, from its lowest part, the superficial inguinal glands.

The vagina has four layers: an external fascial coat with a plexus of veins; a coat of smooth muscle, which is next; a deeper submucous layer of elastic areolar tissue, also with a dense venous plexus; and an inner mucous coat, which is covered by stratified epithelium but with no mucous glands (it is kept moist by mucus from the uterus).

Fallopian Tubes

The fallopian tubes (oviducts or uterine tubes) (Figs. 18-5, 18-6) are paired and about 10 cm (4 in.) long but only about 0.625 cm ($\frac{1}{4}$ in.) wide. They lie in the broad ligaments and emerge from the uterus near the junction of its body and fundus. They curve horizontal, outward, and then backward to reach the ovaries, which they overlap. Each tube is usually divided into four parts: (1) the interstitial part, which is short and passes through the uterine wall (about 1.25 cm [$\frac{1}{2}$ in.] thick); (2) the isthmus, which is the medial 2.5 cm (1 in.) that is near the uterus and joins the next portion; (3) the ampulla, which is the widest and longest division and relatively thin walled and dilatable; and (4)

the outer, expanded, trumpetlike infundibulum, whose extreme surface is broken up by numerous fingerlike processes called fimbriae. The entire tube is enveloped by a fold of peritoneum called the mesosalpinx.

Histologically the tube contains the outer serous mesosalpinx, a subserous layer, a muscular layer continuous with that of the uterus, and a mucosa, which is arranged in longitudinally directed folds. This mucous membrane is covered with ciliated epithelium, and the cilia "beat" toward the uterine cavity.

The arterial supply is from tubal branches of the ovarian and uterine arteries. The nerve supply is via the uterine and ovarian plexuses (T11, T12, and L1). The lymph vessels drain into aortic glands along with those of the ovary and uterine fundus.

Ovaries

The ovaries (genital glands of the female) (Figs. 18-1, 18-5, 18-6, 18-7) are paired organs that consist of two areas, a central medulla and an outer cortex. The medulla is characterized by many blood vessels in connective tissue. The cortical zone, besides containing connective tissue, is composed of glandular tissue in the form of follicles in various stages of development.

Each ovary looks like a large almond. It projects from the posterior layer of the broad ligament by the mesovarium. It is covered by cuboidal epithelium (not peritoneum). The gland is smooth and pink in the nulliparae but gray and puckered in the multiparae and elderly women (due to repeated discharge of ova from its surface and resultant scars). It eventually becomes shrunken, wrinkled, and atrophic.

Each ovary lies in a depression on the side wall of the pelvis. During pregnancy the ovary is carried with the uterus into the abdomen, but after childbirth the organ returns to the pelvis but not to its original position—it is found anywhere in the dorsal part of the pelvis. It still remains anchored to the uterus by its ovarian ligament and to the pelvic wall by its infundibulopelvic ligament.

The blood supply is by the ovarian artery (a direct branch of the aorta), which also supplies the oviduct and upper uterus and anastomoses with the uterine

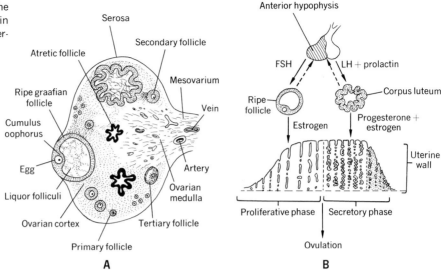

Figure 18-7. A. Histology of the human ovary. **B.** Menstrual cycle in the human female (simplified version).

Labels for A: Serosa, Atretic follicle, Secondary follicle, Ripe graafian follicle, Mesovarium, Cumulus oophorus, Vein, Egg, Artery, Liquor folliculi, Ovarian medulla, Ovarian cortex, Tertiary follicle, Primary follicle

Labels for B: Anterior hypophysis, FSH, LH + prolactin, Ripe follicle, Corpus luteum, Estrogen, Progesterone + estrogen, Uterine wall, Proliferative phase, Secretory phase, Ovulation

artery (see Chap. 10). Several veins from the hilum unite to form a single (or double) ovarian vein. The right empties into the inferior vena cava; the left joins the renal vein. The nerve supply is via the aortic and renal plexuses. The lymph vessels follow the veins and end in aortic lymph nodes.

The Breast (Mammary Gland)

The breast (mammary gland) is both ectodermal and mesodermal in origin. While it is the last gland to function in the adult, it is the first of the epidermal glands to appear. It is actually an analog of the sweat gland and arises along the so-called milk line or ridge, which extends from the neck through the middle of the clavicle to the thigh. Although seven papillae are seen along this line embryologically, only one remains at birth. The female breast is an organ subject to many disorders, the majority being benign. Because of its physiologic change during menstruation, pregnancy, lactation, and postmenopausal atrophy, the breast is subject to many alterations in gross and microscopic structure during life. The breast is also one of the two most frequent sites of cancer in women.

Anatomy

The anatomy of the breast (Fig. 18-8) centers around its glandular and ductal structure. Ectodermal thickenings grow inward from the nipple. As they radiate peripherally, they subdivide into one, two, and three ducts, the last receiving the milk secretions during lactation. The circular pigmented area surrounding the nipple is the areola, which contains from 12 to 15 elevations arranged in a circular fashion. These mark the site of Montgomery's glands, which resemble sebaceous glands. They secrete sebum during pregnancy for lubrication.

The glandular tissue is firm in texture, and there is more tissue centrally than at the periphery. The exocrine glands of the breast are evident in the active state, but in the resting phase no alveoli are seen and they do not secrete. The upper lateral (outer) part of the breast extends almost into the axilla in a segment called the tail of Spence.

The breast is subdivided into lobes (15 to 20), which are subdivided into lobules, which in turn are subdivided into acini. All are enmeshed in areolar tissue interspersed with fat. The fibrous and fatty connective tissue is called the stroma. Between the lobules also course numerous fibrous strands called Cooper's ligaments (Fig. 18-8), which support the breast and run

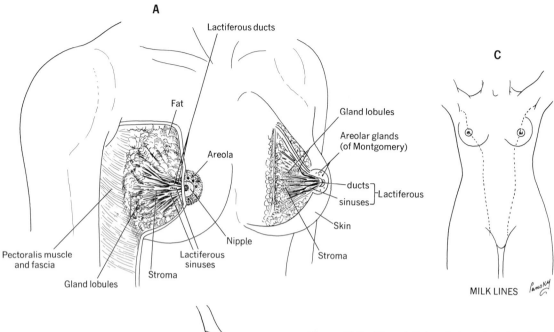

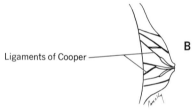

Figure 18-8. The adult female breast. **A.** The breast has been dissected to show its detailed anatomy. **B.** The breast is anchored by Cooper's ligaments as well as by general attachment to the pectoral fascia. **C.** The milk lines are ridges that consist of potential mammary tissue in the fetus. Failure to disappear may result in accessory breasts.

between the skin and the deep fascia. The lobes, like the spokes of a wheel, converge on the nipple, and each lobe is drained by a lactiferous duct. The ducts branch all through the breast tissue and terminate in saclike structures called alveoli. The latter secrete the milk. Near the ducts and alveoli are special contractile cells, the myoepithelial cells.

The size, shape, and structure of the breast are very variable. Each of the two breasts is found in the superficial fascia on the front of the chest. It extends from the sternum to the midaxillary line and vertically from the second to the sixth ribs. There is no capsule.

Before puberty, the breast is small with little structure. With the onset of puberty, the female breast is subject to alternate hyperplasia (increased size due to increased cells) and involution. The hyperplasia consists of an increase in both stroma and duct structures

as a result of the actions of estrogen and progesterone. In addition, normal breast development needs prolactin and growth hormone. The cycles of this process are coordinated with the menstrual periods, with the phases of hyperplasia exceeding those of involution until the breast assumes adult size. With pregnancy, there is a wave of marked proliferative activity and many alveoli develop. Regressive changes occur as lactation ends. With menopause, there is a gradual reduction in stroma and glandular material until all that remains are a few dilated ducts and a small amount of fatty, fibrous connective tissue.

The normal male breast does not significantly change during life and consists of a few ductal rudiments behind a vestigial nipple.

Vascular Supply. The arteries are derived from the internal thoracic and branches of the axillary (Fig.

10-23, p. 361). The supply reaches the breast predominantly by its medial and lateral sides. This permits incisions inferiorly as well as plastic surgery to take place with minimal interference with blood supply.

The major veins follow the arterial pattern; yet many form venous plexi beneath the areola.

Lymph Supply. The lymph drainage (Fig. 10-37, p. 376) consists of three parts: a cutaneous supply, which flows into the axillary glands but may cross to the opposite side as well; the areolar lymphatics, which drain nipple and areola and pass to a subareolar plexus, which also drains to the axilla; and the glandular supply, which also drains to the axilla.

Lymphatics from this region may pass to the axillary glands of the same or opposite side; or they may pass deep to glands inside the chest cavity on each side of the internal thoracic vessels; or they may even go down to the inguinal region (epigastric) and invade the abdominal wall or follow inguinal drainage.

Lactation

During each menstrual cycle, breast changes occur in association with alterations of blood concentrations of estrogen and progesterone. These are minor when compared to changes taking place during pregnancy as a result of hormonal action on both the ducts and alveoli.

Even though the breast enlarges markedly as pregnancy progresses, there is no milk secretion. A clear fluid called colostrum may be produced in small amounts. True milk production takes place several days after birth, at which time there is an increased secretion of prolactin and cortisol. This secretion begins during labor and is maintained throughout the nursing period. Apparently the placenta exerts an inhibitory effect due to its secretion of progesterone, which in large amounts stimulates the release of a hypothalamic prolactin-inhibiting factor. With birth, the placenta is removed and inhibition ceases. A reflex initiated by mechanoreceptors in the uterus during labor apparently also initiates prolactin and cortisol secretion. Last, prolactin and cortisol secretion is reflexly maintained by input to the hypothalamus from afferent receptors in the nipples, as a result of stimula-

tion by suckling. When nursing ceases, milk production ceases. Suckling also inhibits the release of follicle-stimulating hormone (FSH) and luteinizing hormone (LH) from the pituitary gland and thus blocks ovulation. This inhibition, however, is of short duration, and about 50 percent of women ovulate despite nursing.

Milk is secreted into the alveolar lumen and is moved into the ducts by a process called milk letdown, which is accomplished by the contractions of the myoepithelial cells around the alveoli. This contraction is under control of oxytocin, also released reflexly as a result of suckling. Apparently higher brain centers also influence oxytocin release since nursing mothers may even "leak" milk when the infant cries. Thus, psychologic factors play a strong role in nursing. Milk contains water, protein, fat, and the carbohydrate lactose, all built up by the alveolar cells with the help of insulin, growth hormone, cortisol, and other hormones.

Female Maturation and Physiology

The female, like the male, undergoes somatic changes as well as gonad and accessory reproductive organ development. In the female, there is one clearly defined state, namely, menarche (appearance of the first menstruation), that serves as a reference point for female puberty and obscures other stages, such as first ovulation, breast development, and body changes. At puberty the vagina enlarges and its secretions become more acid, owing to the formation of acid from the glycogen of the epithelium. The clitoris becomes erectile, the Bartholin glands secrete mucus, and the labia enlarge.

The effects of estrogen are analogous to those of testosterone in the male and exert a major control over all accessory sex organs and secondary sex characteristics. Estrogen maintains the entire genital tract (uterus, oviducts, and vagina), the glands lining the tract, the external genitalia, the pelvis, and the breasts. It is important in female body hair distribution, general body shape, and fat deposition in many areas (hips, abdomen, etc.). Estrogen also contributes to general body growth at puberty. Last, as will be noted, it is important for follicle and ovum maturation and, with the gonadotropins, in ovulation and menstruation. During most of a woman's reproductive life, estrogen supports the ovaries, accessory organs, and secondary characteristics. Its action appears to relate to protein synthesis. Note, however, that it is not uniquely a female hormone, since small quantities are secreted by the male adrenal and testicular interstitial cells.

Progesterone, on the other hand, is present in large amounts only during the luteal phase of the menstrual cycle, and its effects are less profound than those of estrogen. During pregnancy, however, it does affect the breasts and maybe the uterine muscle.

Menstrual Cycle

Onset. During early childhood the uterus develops rapidly; however, beginning at age ten it increases in size many times under ovarian estrogen stimulation. Then a change in hormonal and neural stimuli causes a breakdown of the endometrial uterine lining (with or without a preliminary ovulation), and there is a pouring out of blood—the first menstrual bleeding. The age of the first menses varies from person to person, occurring between 13 and 15 years in about 60 percent of females. The age of puberty has changed substantially in modern times, having decreased by about four years in the last century, probably owing to improvement in environmental conditions. Note that even in the male better physical development has been associated with earlier puberty.

Length of Cycle. The length of the menstrual cycle is quite variable, not only between females but in the same individual. The median length ranges from 24 to 29 days. The length of the cycle does not vary in any regular manner during the major portion of a female's reproductive life, although it is often irregular in years after menarche and in menopause.

Basic Phases of Cycle. The menstrual cycle (Fig. 18-9) normally has two basic phases: (1) the follicular phase (before ovulation) and (2) the luteal phase (after ovulation).

The follicular phase is characterized by the production and release of folliculin or estradiol by the developing follicle, which is found in the liquor folliculi of the vesicular follicle and produced by the actions of theca interna cells and follicular cells. It causes the uterine mucosa to repair (regenerate and proliferate) and to begin the proliferative phase of the cycle, and there is growth of an ovum(egg)-bearing follicle in the ovary.

The luteal phase begins at ovulation (when the ovum leaves the ovary), which is near midcycle (about day 14 of a 28-day cycle), and extends to the onset of the next menstruation. During this period, the endometrium develops and becomes glandular, under the control of progesterone, a hormone secreted by the corpus luteum of the ovary. This period of uterine growth prepares it for receiving the fertilized ovum. If fertilization does not occur, the corpus luteum regresses, progesterone levels fall, the uterine lining breaks down, menstruation begins, and a new ovarian cycle begins. If the egg is fertilized, received by the uterus, and implanted in the uterine wall, the corpus persists and no menstruation occurs. Ovulation is thus associated physiologically not with the preceding menstruation but, through the corpus luteum, with the one that follows.

Hormonal Influence. The changes in uterine morphology in the menstrual cycle are due to the effects of estrogen and progesterone. Estrogen stimulates growth of the smooth muscle (myometrium) and the glandular epithelium (endometrium) that lines the inner surface of the uterus. Progesterone acts on this "prepared" endometrium to convert it to a secreting tissue, so as to provide a suitable environment for the fertilized ovum; under its influence, the glands become coiled and filled with glycogen, the blood vessels become numerous and spiral, and various enzymes accumulate in the glands and connective tissue of the linings.

Endometrial Changes. The endometrial changes in a normal (nonpregnant) menstrual cycle are summarized below (Fig. 18-9).

Menses. As the corpus luteum regresses, the fall in blood progesterone and estrogen deprives the active

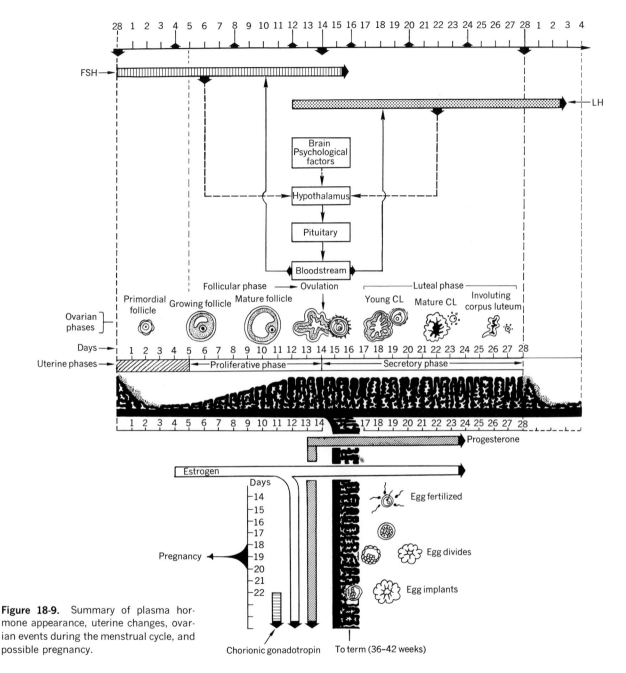

Figure 18-9. Summary of plasma hormone appearance, uterine changes, ovarian events during the menstrual cycle, and possible pregnancy.

endometrium of its hormonal support so there is marked constriction of the uterine blood vessels, leading to a decrease in oxygen and nutrients. A breakdown begins and the entire innermost layers of the uterine mucosa begin to slough as the menstrual flow, marking day 1 of the cycle. After initial vasoconstriction, the endometrial arterioles dilate, resulting in hemorrhage through the weak capillary walls. Thus, the menstrual flow consists of not only blood but also endometrial debris. The menstrual phase lasts about three to five days, and during this time blood estrogen levels are low.

Repair or Proliferative Phase. The menstrual flow stops as the endometrium begins to repair and reepithelialize its denuded mucosa and thicken and establish straight uterine glands under the influence of a rising blood estrogen level. This process is initiated in an area as soon as the "old" mucosa is shed and, therefore, can be seen in one area while the shedding is still taking place in another. Repair is complete by day 6, and proliferation continues for nine to ten days after menstrual flow ceases. Time of this phase varies with the time of the entire cycle.

Secretory (Luteal, Premenstrual) Phase. This phase is characterized not only by an increase in mucosal thickness due to an increase in interstitial fluid but also by changes in the uterine glands with increased tortuosity and production of a glycogen-rich secretion. It extends from about 14 days after menses begins until just before the onset of the next menses, a total of about 13 days. It is the most constant period, as time goes. It occurs after ovulation when the corpus luteum is formed, and then progesterone and estrogen induce this secretory type of endometrium.

Ischemic Phase. Just prior to menses the corpus luteum breaks down when the spiral arterioles to the innermost uterine mucosa contrict and cause an ischemia, which makes the functional zone of the mucosa essentially bloodless. Anoxia, as a result of the ischemia, leads to death of the endometrial tissue and its dissolution and fragmentation. Discharge of that tissue leads to the beginning of menstruation, as the spiral arterioles dilate, rupture, and allow for extravasation of blood with necrotic tissue. Thus, the uterine changes merely reflect the effects of alterations in levels of blood estrogen and progesterone, and the pattern of the latter reflects the hypothalamic-pituitary-ovarian interreactions.

LENGTH OF PHASES. The two basic phases of the cycle—the follicular phase and the luteal phase—are variable in length. Thus, an attempt to calculate ovulation time may be difficult. The best that can be done is to try and ascertain that ovulation did indeed occur a couple of days or so previously, either by using basal body temperature, which is slightly elevated by the action of progesterone, or by determining the level of pregnanediol, the excretion product of progesterone that is found in the urine.

Failure of Ovulation. A failure of ovulation results in an anovular cycle. The biphasic nature seen in the normal cycle is absent. This does not result in a failure to menstruate, although bleeding may be less; nor are anovular cycles necessarily shorter in length. Thus, neither the rhythm of the menstrual cycle nor the mechanism causing endometrial breakdown is dependent on the development and regression of the corpus luteum.

Ovaries, Ova, and Twins

The ovaries show less absolute increase in size during sexual development than do the testes. They increase to about ages 20 to 40 years and then decrease to almost half size. Of note is that their oocyte (egg) population is laid down very early in life and is not increased thereafter. Originally the ovaries probably contain about 500 thousand oocytes, whose number even before puberty is reduced by follicle decay and continues to decrease throughout reproductive life. At menopause the supply of eggs may be entirely exhausted. The loss by ovulation itself is negligible, for if a woman loses one egg each 28 days for 30 years, that amounts to only about 400 eggs. Due to this laying down of ovarian ova very early in life, it means that during the reproductive life the ova are between 15 and 45 years old and that an older woman, say of 45 years, who produces a child has done so with a relatively old ovum. This may account for the great number of abnormal fetuses produced by older women.

The usual rule for women is the release of a single ovum during each cycle, and the two ovaries usually ovulate alternately. However, a single egg can result in twins by splitting of the embryo in early development. These become monozygotic (single-egg) twins (identical twins) owing to their identical genetic makeup. They are of the same sex and look much alike. The embryos are in a single chorionic sac but may have separate amniotic sacs or share a single one. If, in the case of monozygotic twins, longitudinal splitting of the embryonic disk is delayed or incomplete, fusion of the paired embryos (to variable degrees) results in conjoined or siamese twins. The majority of twins, however, result from ovulation of two eggs to produce dizygotic (fraternal) twins, where siblings of the same or opposite sex resemble each other but no more so than siblings from separate births. They have separate extraembryonic membranes. They have separate births. The total twinning rate is about 1 in 80 of all Western births. There appear to be more dizygotic twins in older women.

Oogenesis

Early in embryonic development special primordial germ cells are formed, and from these the future eggs (and spermatozoa) are derived. These primordial sex cells have a diploid number of chromosomes (44 plus 2X) and serve as stem cells in fetal life, and by mitosis they give rise to daughter cells incorporated into the primordial follicles.

Cells that form eggs gather just below the surface of the developing ovary, and their multiplication ceases just after birth. The germ cells become progressively involved in a process called meiosis, which results in halving of the chromosome number (Fig. 2-5, p. 62; Fig. 2-7, p. 66). As meiosis begins in the germinal cells, the nuclei assume a distinctive spherical form—the germinal vesicle—and the cells at this stage are called primary oocytes, which stay quiescent until puberty, when they are stimulated to follicle development by pituitary FSH and the primordial germ cells become primary follicles.

At birth, all the germ cells in the developing ovary are at the stage of being primary oocytes. They remain so until just before ovulation (some oocytes reach this period in 13 to 15 years, at puberty, whereas others wait until later in the reproductive cycle, at menopause). Once an oocyte enters meiosis, no further multiplication can occur and, therefore, the human female is born with her complete quota of potential eggs.

After the germ cells become primary oocytes (Fig. 18-7), they are covered by a single layer of cells derived from the ovarian surface and are called primordial follicles. No other changes occur until puberty. At this time and during the reproductive life, follicles with their contained oocytes undergo growth. The oocyte (egg or ovum) swells and the follicle cells multiply, and both processes proceed at an equivalent rate to a point when the oocyte reaches its maximum size. The ovum is separated from the follicle cells by a relatively thick membrane, the zona pellucida, which is probably formed by the follicle cells. The follicle continues to grow, mostly by the formation of a fluid-filled space, the antrum, between the cells but also by the addition of new cell layers from mitosis of the original follicle cells and by the growth of specialized ovarian connective tissue cells. When the antrum forms, the ovum has reached full size and the follicle is referred to as a graafian or vesicular follicle (Fig. 18-7). The antrum holds the follicular liquor, which contains estrogenic hormones. A basement membrane is found between the follicular cells and the cells of the theca externa and the theca interna (from the stromal cells around the follicle). The interna and follicular cells elaborate estrogenic hormones. During early development the follicle had moved to the center of the ovary; however, with maturity it moves to the surface, where the egg cell is extruded at ovulation. In the vesicular follicle, the ovum is surrounded by follicular cells that form the cumulus oophorus. After ovulation, the same cells carried with the oocyte are called the corona radiata.

During reproductive life many follicles develop, become vesicular, and remain near the ovarian surface but never mature, yet during each cycle one follicle does reach the degree of maturity for ovulation before the others. When the mature one releases its ovum at ovulation as a result of LH, the secretion of FSH is inhibited and the large vesicular follicles still present in the ovary degenerate.

Just before ovulation, the events of meiosis (which had stopped at birth) resume and the chromosomes become visible as the envelope of the germinal vesicle disappears (see Chap. 1). As a result of two meiotic

Figure 18-10. A. The egg develops in the ovary. As the follicle matures, the egg ruptures from the ovarian surface and enters the uterine tube, where it is fertilized in the outer one third of the tube. The stages of fertilized egg development take place as shown. The blastocyst finally implants in the uterine wall. **B.** Electron micrograph of an ovulated rabbit ovum. ×8000. (Courtesy of Dr. David D. Cherney, Medical College of Ohio at Toledo.)

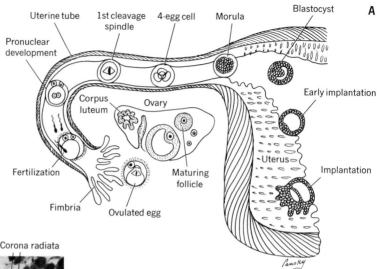

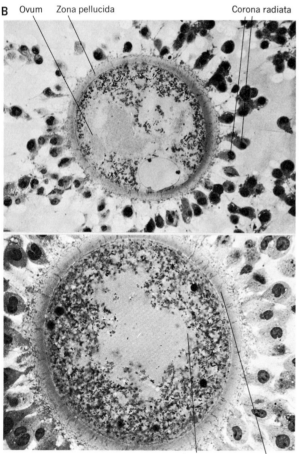

divisions, the number of chromosomes is halved (haploid) and the ovum contains 23 tissue cell chromosomes and a sex chromosome (X) (Fig. 2-5, p. 62). Note that in the divisions the ovum has lost half of its chromosomes but almost none of its cytoplasm, since the division is unequal and the polar bodies given off are rudimentary. The same is true in spermatozoon formation and is necessary if the combination is to lead to the diploid number and reestablish the normal human number. Errors in meiosis may lead to an embryo with three sets of chromosomes (triploidy), or failure of separation of a pair of chromosomes may result in an embryo with one too many or one too few chromosomes. Both of the last-named anomalies are lethal or lead to abnormal embryos. The polar bodies have no known function except a passive one to eliminate chromosome material (Fig. 2-5, p. 62).

Ovulation

The egg is ovulated from the ovary under the hormonal influence of FSH and LH (Fig. 18-10). Just prior to ovulation the follicle has a final growth spurt, and then it opens to release the egg in a "gentle" fashion. The follicle at this time projects above the ovarian surface, a slit develops in the exposed part, and it widens enough to allow the egg to escape along with a surrounding group of cells. The thinning of the follicle wall may be due to proteolytic (protein-digesting) enzymes that are induced to form in the follicle by the pituitary hormones.

There is no reliable means of detecting the occurrence or estimating the exact time of ovulation, but a sign that the event is occurring is the so-called *mittelschmerz*, or middle pain, which happens in the middle of the cycle and which is assumed to represent abdominal irritation induced by the entry of follicular contents; however, this concept may be erroneous. Other objective approximate indications are a sharp rise in the amount of LH in circulating blood before ovulation; a rise in basal body temperature, which often shows a sudden rise of 0.5° C halfway through the cycle; and an increase in the number of vaginal wall cells with deeply stained nuclei, which occurs at about the middle of the cycle. Thus, vaginal smears (cells and mucus) can be examined. In addition, vaginal pH has been determined with variable success, as have cervical glucose levels.

On ovulation, the egg with its surrounding cells and tightly adhering zona pellucida is carried out of the ovary by the antral fluid and is caught in a stream created in the peritoneal fluid by the beating movement of the fine hairlike cilia that line the inside of the oviduct. It is thus carried into the oviduct opening and passes along largely due to cilia action aided by muscular contraction of the oviduct wall. The egg comes to rest (temporarily) in the outer one third of the oviduct, the ampulla, where fertilization usually takes place. Ovulation has been induced artificially by the injection of appropriate hormones. These have led to multiple births.

Note, however, that the egg may fail to leave the follicle or may fail to enter the oviduct and enter the peritoneal cavity instead. The egg may also fail to be fertilized in the oviduct and pass back into the cavity. Such ectopics rarely last long but often require surgical intervention due to the high risk of hemorrhage or other complications. Occasionally more than one egg may develop and ovulate, and we get multiple births. When ovulation is blocked by oral contraceptives, maturation of follicles may still continue and follicles may accumulate in the ovary. Thus, multiple ova may be available for release during the "nonblocked" cycle, enhancing the multiple-birth possibility.

Physiology of Coitus

Scientific knowledge of intercourse is increasing rapidly because of recent advances in hormone estimation and the advent of electronic devices, making it feasible to conduct experiments in an inoffensive way. The work of A. C. Kinsey, W. B. Pomeroy, C. E. Martin, and P. H. Gebhard in the late 1940s and 1950s as well as the more recent investigations of W. H. Masters and V. E. Johnson have carried the study of sexual physiology from the realm of obscurity into one with a scientific basis.

Masters and Johnson have demonstrated that sexual intercourse has four distinct phases: excitement, plateau, orgasmic phase, and resolution. Of note is the fact that Sigmund Freud also postulated a ladder or steps in dreams symbolizing the sexual act.

During intercourse, changes in the shape of the vaginal "barrel" and corrugation of the vaginal wall (which may act as a frictional aid to the penis) have been observed, as has the process of uterine erection or elevation. Normally, the uterus is directed upward and forward, but during coitus it shifts into a position pointing upward and backward. Simultaneously, a tenting effect at the vault or top of the vagina takes place, and it is here that the pool of semen collects. After female orgasm, the mouth of the uterus has been seen to gape and the womb found to undergo contractions very similar in nature to those observed during the first stage of labor but lacking the subjective feeling of pain.

The environment of the vagina may also play an important role in affecting the sperm and their survival. The vagina and its secretions are acidic with a pH of about 3.8, while the seminal plasma in which the sperm are suspended is slightly alkaline with a pH of about 7.3. The acid apparently immobilizes the sperm and tends to hinder their progress, but there is a buffering effect of the seminal plasma that neutralizes the acid. Thus, the volume of the ejaculate and its buffering ability may be just as important as or more important than the actual sperm count in cases of infertility. Furthermore, the vagina may contain a lethal factor in certain individuals who would be fertile by laboratory standards but remain infertile because the sperm are immediately immobilized on deposit in the vagina. Apparently rhythmic contractions of the vagina and uterus during copulation assist the progress of the sperm.

Oxytocin, a hormone produced in the hypothalamus and stored and released from the posterior part of the pituitary gland, causes uterine contraction, especially during labor, and is also responsible for ejection of milk during breast feeding. Since oxytocin causes uterine contraction, it has been suggested that its release during coitus acts as an aid to sperm transport. Testosterone, the male sex hormone (produced in the testis), is also apparently increased in the circulation during copulation and plays an essential role.

Parameters such as blood pressure, heart rate, and respiration have been measured by Masters and Johnson, so that one can advise the cardiac or asthmatic patient with regard to future sexual activity. In the human male, there is a rise in blood pressure with a peaking at the point of ejaculation. There is also a similar peak at orgasm in the female. The blood pressure rises from a norm of 120 to 175 mm Hg at the moment of ejaculation in the male, while the female reaches a peak of 200 mm Hg at orgasm. After orgasm, the pressure falls sharply to a little below the resting level and then returns to normal. The peak lasts only about one minute on either side of the climax.

Noticeable hyperventilation (heavy breathing) in the male occurs at orgasm. The breathing rate is faster before the climax in the male, but the volume of air breathed is greater after ejaculation. The female pattern is a bit different. Periods of apnea or breath holding are seen before regular hyperventilation at climax. Thus, the breathing rate and the volume of air breathed are low before climax but are increased after orgasm. Muscular tension, increased heart rate, and hormonal factors (besides the exertion of intercourse) must all play a role. Pressure changes, too, do occur

in the uterus and vagina during intercourse and orgasm.

Besides oxytocin (in the female) and testosterone (in the male) investigations are underway concerning the prostaglandins present in human semen. They appear to exert a powerful effect on the womb, and it has been suggested that they play a significant role in the process of sperm transport by causing smooth muscle contractions. Apparently the prostaglandins are absorbed from the vagina or affect the uterus directly. They are so powerful that they have been used to initiate labor (see Chap. 15).

Fertilization

Fertilization (Figs. 18-10, 18-11) is the fusion of the egg (female germ cell) and a spermatozoon (male germ cell). At intercourse the spermatozoa, in semen, are deposited in the upper vagina and under their own

Figure 18-11. Egg undergoing fertilization. The chromosomes on the second meiotic spindle will separate and one half will be emitted in the second polar body, leaving the ovum with one half its normal number. The sperm contains a similar number of chromosomes so that fusion to form a zygote will return the number to its full diploid complement.

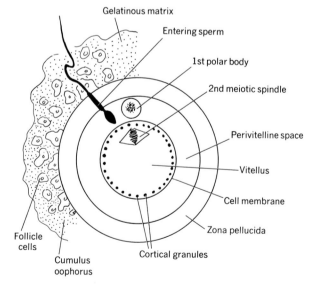

power (by movements of their tails) pass into the uterine cervix, where conditions are favorable for them. Passage through the uterus and oviduct is brought about mainly by the marked contractions in the walls of these very organs, in addition to the currents and countercurrents created by the beating cilia. The transport time is fairly rapid, about 5 to 15 minutes from the time of deposit into the vagina. During passage through the genital tract, the spermatozoa undergo a marked reduction in number, owing to dilution in the tract secretions and by general inefficiency. Thus, only a few thousand, out of the hundreds of millions deposited, enter the oviduct and only a few dozen actually come near the egg at the end of the oviduct.

The egg at fertilization looks like a spherical mass of cytoplasm and contains a nucleus and other cell structures, such as mitochondria, ribosomes, Golgi apparatus, and cytomembranes. It is limited by a cell membrane, which is surrounded by a thick, transparent coat, called the zona pellucida, that is separated from the membrane by a fluid-filled perivitelline space. Outside of the zona lie follicle cells that have accompanied the egg. The cells are embedded in a gelatinous matrix that binds them to the egg, and together they are called the cumulus oophorus.

The spermatozoon must penetrate the cumulus and zona pellucida to fertilize the egg. It has been suggested that this occurs as follows: On leaving the male tract, the spermatozoon cannot fertilize an egg but undergoes a physiologic change called capacitation, which occurs while it is in the female tract. The spermatozoon undergoes "destabilization" by removal of a protective coating, and it then experiences the so-called acrosomal reaction whereby small perforations are produced in the acrosome wall permitting enzymes to escape. The spermatozoon is now able to digest a path through the layers surrounding the egg.

The spermatozoon, after taking a curved path through the zona pellucida, comes to rest in the perivitelline space with its head flat on the surface of the egg. The tail is still protruding through the slit it made in the zona. The spermatozoon head then attaches firmly to the egg, a fusion takes place involving union of the cell membranes of the egg and spermatozoon, and the two cells become enclosed in the same membrane. Thus, a single cell is formed from the two gametes. Shortly after, the spermatozoon nucleus moves into the egg and enlarges to form the male pronucleus (Fig. 2-5, p. 62). The tail apparently serves best for egg penetration and not passage up the female tract.

The egg responds to sperm contact, and changes in the zona pellucida and egg surface take place to prevent the entry of more than one spermatozoon. This mechanism to prevent polyspermy may be related to a change stimulated by sperm entry—namely, the evacuation of a number of small bodies, the cortical granules, which are found just under the egg surface. They may release material that passes across the perivitelline space and induces zona reactions. This process is not perfect, and at times two or more do participate in fertilization (polyspermy) and such an event usually dooms the embryo to an early death. At this time, too, the second meiotic division resumes and ends with the formation of the second polar body, and the chromosomes in the egg form the female pronucleus (Fig. 2-5, p. 62).

The female and male pronuclei now grow and develop synchronously. Nuclear envelopes are seen around the swollen sperm head, and egg chromosomes and small nucleoli appear in each pronucleus. As the pronuclei grow, the number and size of nucleoli increase. When the pronuclei reach their full development, the two come together in the center of the egg. At this time they shrink, the nucleolar number diminishes, the nuclear envelopes fade away, and the chromosomes finally reappear and arrange themselves at the center of another spindle, the cleavage spindle. Fertilization is now complete, having taken about 12 hours from sperm entry, and the hereditary materials of egg and sperm have united. The egg is now called a zygote (Fig. 2-5, p. 62; Fig. 2-7, p. 66).

The next stage involves division of the zygote into two equal parts, each with an equal share of hereditary material. Embryonic development begins with the formation of the two-cell embryo. A few hours later the first two cells divide to make four. Further division leads to the formation of a hollow sphere, or blastocyst.

Just after the blastocyst stage, the inner cell mass (found at one pole of the blastocyst) begins to assume the form of a fetus, which then becomes a child. At this phase if the inner cell mass splits into two distinct halves and each reorganizes itself and grows, we have true or identical twins. They are necessarily of the same sex and resemble each other closely.

At fertilization, equal parts of DNA (genetic material) are contributed by the egg and the spermatozoon. The haploid cells contribute an amount equivalent to one half that of the normal (diploid) body cell, and thus the zygote contains only the same amount of DNA as a normal body cell. When the zygote divides, the two cells of the early embryo would receive on this basis only half the amount of DNA. This is avoided by the synthesis of DNA during fertilization (important function of the pronuclei). The DNA is doubled so that each cell of the two-cell embryo receives an amount of DNA equal to that in a normal body cell.

The fertile life of the human egg may be around 24 hours. However, after ovulation the egg deteriorates progressively, and even though it can be fertilized it may lose its ability to give rise to a normal embryo. Embryos with anomalies usually die during pregnancy, but if they live they may have mental or physical abnormalities.

Menopause

The menopause is variable in onset and has no definite end point; thus, the process is difficult to time accurately. It is influenced by ethnic group, age at menarche, socioeconomic status, occupation, urban or rural environment, marital status, and maternal history. It usually occurs between 40 and 50 years of age, and there appears to be a trend toward later menopause (as there is toward earlier menarche). Menopause is characterized by irregularity of menstruation, increased ovulatory failure, and increased secretion of pituitary gonadotropins in response to a marked reduction in ovarian estrogen secretion. What affects ovarian egg exhaustion, overall hormonal changes, or

the whole mechanism controlling reproduction is not certain.

Fertility

Fertility involves the combined operation of coitus, whose physiology, effectiveness, and frequency change during the life cycle. In the male, reaction time increases with age and the ejaculatory process changes gradually. In the female, sexual response is more dependent on experience (than in the male), and interest in coitus and capacity for orgasm may increase rather than decrease. However, the frequency of coitus seems to decrease with increasing age, probably largely determined by the male. Effectiveness also apparently decreases, and the number of acts of coitus needed to produce a pregnancy increases with age. The probability that conception will occur after a single act of coitus is small, and the advent of effective contraceptive means is further disassociating sexuality and reproduction.

Some Additional Thoughts on Hormones of Reproduction

The gonadotropins in the female are released in a discontinuous cyclic manner and, therefore, there is a regular periodic discharge of releasing hormones from the hypothalamus. Men, on the other hand, secrete their pituitary gonadotropins continuously and apparently do not experience this periodic function from the hypothalamus. Thus, animals are born with this sex difference imprinted on their hypothalamus. After a dose of the male hormone testosterone, the cyclic properties of the hypothalamus are obliterated in a newborn female rat; when the same animal becomes sexually mature, gonadotropins are secreted continuously and she goes into a state of permanent estrus and is always willing to accept the male.

If fertilization takes place, the small speck of growing cells of the fertilized ovum must elaborate a "biologic signal" to change the normal events of menstruation. This turns out to be a hormone, chorionic gonadotropin, secreted by the fertilized egg itself and essential for the survival of the egg. It can be detected in the blood and urine of a pregnant woman only a few days after the fertilized egg has implanted in the uterus. Its presence is the basis of nearly all pregnancy tests since it is produced only by a newly conceived embryo.

If the pituitary hormones were available for administration as pills or injections, ovulation could be turned "on" or "off." Unfortunately, neither FSH nor LH—each consisting of thousands of amino acid units in long chains—has been synthesized. On the other hand, chorionic gonadotropin from human placentae can be extracted and, although not the same as LH, it shares the property of LH in causing ovulation. However, the fully developed follicle requires FSH, and for this there is no substitute. On the other hand, a partially inactivated form of pituitary FSH is excreted in female urine and is concentrated in the urine of menopausal women; therefore, this urine is an alternative source to using pituitaries from corpses. Treatment with this FSH, followed by chorionic gonadotropin when the follicles are grown, has resulted in multiple births due to overdosage, multiple ovulation, and lack of a feedback principle to keep secretion within limits. (See Chap. 15 for detailed discussion of the hormones of reproduction.)

Review Questions

1. What differences, if any, are there between the male and female perineum?
2. What is meant by the vulva (or pudendum)?
3. What are the male homologous structures of the following?
    a. Labia majora
    b. Clitoris
    c. Paraurethral glands
    d. Involuntary muscle in the fascia of the labia
4. Compare the structure of the male penis and the female clitoris.
5. List the six ligamentous supporting structures of the female viscera. Which are truly ligaments and which are peritoneal reflections?
6. Describe the following:
    a. Uterus
    b. Vagina
    c. Fallopian tubes
    d. Ovaries
7. What is breast tissue? Where is the breast located? What are its major blood supply and lymph drainage?
8. Describe the process of lactation, indicating the hormones involved.
9. What is meant by menstruation? What is the menstrual cycle?
10. Relate the changes in the menstrual cycle with:
    a. Uterine changes   b. Ovarian changes   c. Hormonal changes
11. What is the difference between:
    a. Identical twins   b. Siamese twins   c. Fraternal twins
12. Describe the process called oogenesis. Why does it occur?
13. What do ovulation and fertilization accomplish?

References

Austin, C. R.: *Fertilization*. Prentice-Hall, Inc., Englewood Cliffs, N.J., 1965.

Inhorn, S. L.: Spontaneous Human Abortions (Chromosomal Studies). In Woollam, D. H. M. (ed.): *Advances in Teratology*, Vol. II. Academic Press, Inc., New York, 1967.

Page, E. W.; Villee, C. A.; and Villee, D. B.: *Human Reproduction*. W. B. Saunders Co., Philadelphia, 1972.

Patten, B. M.: *Human Embryology*, 3rd ed. McGraw-Hill Book Co., New York, 1968.

Wood, C.: *Sex and Fertility*. Thames & Hudson, London, 1970.

Development and Aging

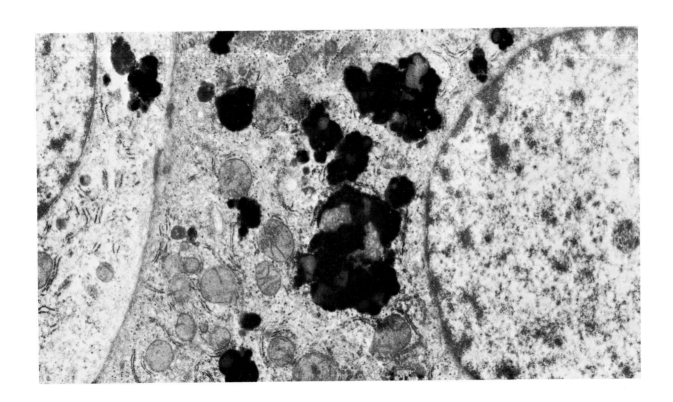

UNIT VI

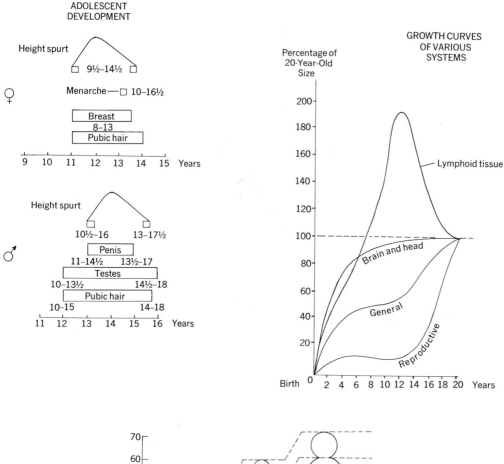

ADOLESCENT
DEVELOPMENT

Height spurt

♀
Height spurt ⌂ 9½–14½ ⌂

Menarche—□ 10–16½

Breast
8–13
Pubic hair

9   10   11   12   13   14   15   Years

♂
Height spurt
10½–16   13–17½

Penis
11–14½   13½–17
Testes
10–13½   14½–18
Pubic hair
10–15   14–18

11   12   13   14   15   16   Years

GROWTH CURVES
OF VARIOUS
SYSTEMS

Percentage of
20-Year-Old
Size

200
180
160
140   — Lymphoid tissue
120
100
80   Brain and head
60
40   General
20   Reproductive

Birth   0   2   4   6   8   10 12 14 16 18 20   Years

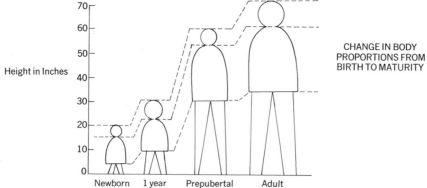

70
60
50
Height in Inches   40
30
20
10
0
Newborn   1 year   Prepubertal   Adult

CHANGE IN BODY
PROPORTIONS FROM
BIRTH TO MATURITY

Figure 19-1. Growth and its many manifestations and variations.

Human Development

CHAPTER 19

Review of Early Development, External Body Form, and Growth of Embryo and Fetus

Early Development

The menstrual age or the length of time following the last menstrual period (LMP) may provide an index of prenatal age, owing to the assumption that ovulation takes place precisely two weeks after the LMP. This, of course, is unjustified, and delayed ovulation may account for instances of so-called prolongation of gestation.

An egg has an inherent ability to develop into an individual similar to that from which it is derived. The active processes of development involve two events: the growth of the developing embryo and the differentiation of the embryo into different parts. The ability to do these things is determined when the egg and spermatozoon fuse. However, although genetically equipped for its task, the egg depends on the external environment for its development and thus it is placed in the uterus, a constant warm-temperature environment, where the development is best favored. Thus, the egg genetic potential, or genotype, also is dependent on environment and what it becomes. Its structural differentiation and metabolic capacities are designated as the phenotype. Mammalian eggs are simple and small and contain just sufficient protoplasm to carry them from ovary to uterus (about five days), where they then obtain nourishment from the tissue fluids around them until they reach the state of embryonic

development of having their own circulating system. This system comes into close association with that of the mother and, although the two blood systems never actually mix, they are separated by only thin layers of cells that permit nutritive materials (O_2 and CO_2) to pass across from the side of greater concentration to that of lesser concentration.

The initial step in development is the division of the fertilized egg. From a single cell, it divides into two cells or blastomeres (Fig. 4-2, p. 98; Fig. 18-10, p. 604); which in turn divide into four cells, and so on until there is a solid ball of cells called the morula, present by day 4 after fertilization and still covered by the zona pellucida. This cell division or cleavage is such that the single cell divides into smaller and smaller units, with initially no increase in total volume. However, the two daughter cells, even at the first division, are not identical. Rather, one may be larger than the other, and the larger may divide sooner than the smaller; thus, initially, there appears to be a difference in cellular structure and behavior. After a few divisions, the differences between the cells become more marked. By the 20-cell stage, the cells that are to form the implantation mechanism and eventually the placenta and fetal membranes have already separated off from those that are to become the embryo proper (Fig. 19-2).

In the morula, areas of clear fluid collect among the cells in the center. These changes take place as the uterine cavity is approached. The areas coalesce to form a central reservoir, and the egg is transformed into a distended, thin-walled vesicle, the blastocyst

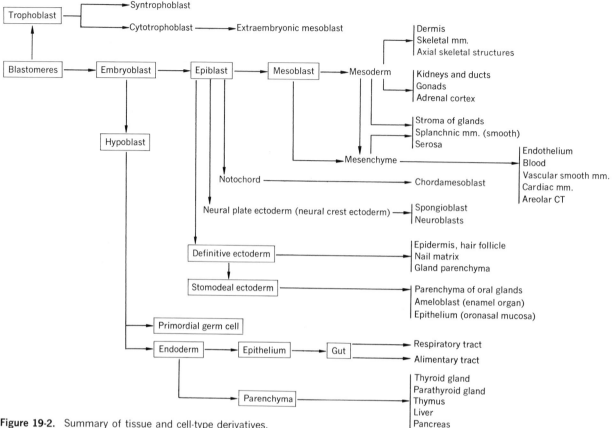

Figure 19-2. Summary of tissue and cell-type derivatives.

(Fig. 18-10, p. 604; Fig. 19-3), with a thickened area of larger cells at one pole, the inner cell mass. This pole is also referred to as the embryonic pole, because the inner cell mass forms the embryo and the remaining peripheral cells form the trophoblast (Fig. 19-3), which invades the uterus and forms the placenta. In the blastocyst stage, all the mammals are alike and it is difficult to differentiate one animal from another. At this stage, too, the embryo is still about the size of the original egg but now begins to grow and swell in proportion to the distending amniotic fluid.

Placentation

Trophoblast. The trophoblast (Fig. 19-3) is the most precocious part of the egg. Its cells divide more rap-

idly, are thus smaller than the other cells, and soon specialize in structure. The inner cell mass divides more slowly and appears more primitive. The food and oxygen supply of the free blastocyst is met to a small degree by substances present in the surrounding uterine secretions, but as it grows rapidly its needs increase. Thus, the outer layer (trophoblast), which is the source of all extraembryonic membranes except for the vitelline sac endoderm, comes into contact with the uterine lining (stratum compactum) (Fig. 19-3). The embryo attaches to a restricted part of it (usually its upper, posterior wall), where it implants and becomes embedded about six or seven days after ovulation (blastocyst implantation or nidation), and this is completed by the eleventh day.

Figure 19-3. Early stages in the development of the human embryo. The blastocyst attaches to the uterus on about day 6.

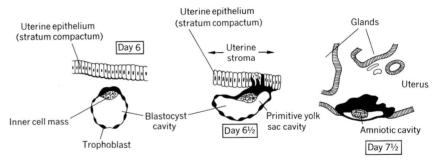

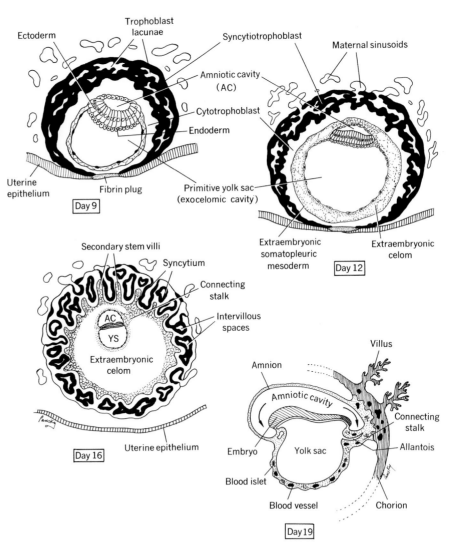

The preparation of the uterus to receive the egg is hormonally controlled, and the endometrial (inner uterine lining) becomes swollen, fluffy, well supplied with blood, and actively secreting (in secretory phase). The secretory endometrium consists of (1) stratum compactum, which is the zone near the uterine cavity consisting of a layer of simple columnar epithelium supported by compact connective tissue stroma (uterine gland epithelium penetrates this zone); (2) stratum spongiosum, which lies beneath stratum compactum and consists of dilated, tortuous uterine glands with intervening connective tissue stroma and spiral arteries and veins; and (3) stratum basale, which is the deepest layer, lying on the uterine muscle. The stroma of the last-named layer contains the basal portions of the glands, and its epithelium serves as a reservoir of cells that epithelialize the denuded uterine surface after menses; this basal layer receives blood from basal arterioles, which maintain nutrition of the layer during the ischemia before menstruation. The first two layers are lost with each menstrual flow, and during pregnancy they become the decidua (Latin *decidua*, falling off) of the extraembryonic membranes at birth.

This preparation takes place in women regularly every month, but if a fertilized egg is present and attaches, certain connective tissue cells at the implantation site enlarge until they look like gland cells. They are called decidual cells since they form the decidua, which appears to have both a nutritional and a protective role. With blastula implantation there are related endometrial changes—a swelling of the stromal cells (a "decidual" reaction). Since both fetal chorion and maternal endometrium are shed at birth, the term *decidua* applies to all endometrium during pregnancy. This tissue is divided into (Fig. 19-4): (1) decidua basalis, which is the endometrium between conceptus and uterine myometrium that becomes the maternal part of the placenta; (2) decidua capsularis, which is a thin endometrial layer that covers the implanted chorionic vesicle and, with expansion, comes into apposition to and fuses with the remaining uterine endometrium, obliterating the uterine lumen; and (3) decidua parietalis, which includes all endometrium except the two types mentioned above and that part

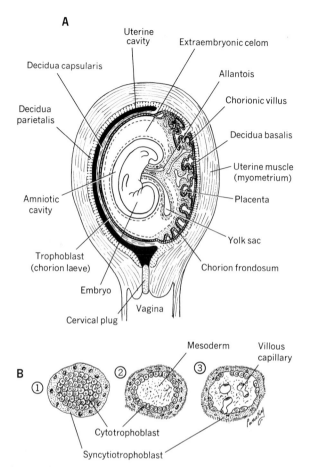

Figure 19-4. **A.** Relation of fetal membranes and uterine wall at about two to three months of age. **B.** Scheme of villus development. **1.** Transverse section of a primary-stem villus, showing core of cytotrophoblast covered by syncytiotrophoblast (syncytium). **2.** Secondary-stem villus with a core of mesoderm and a single layer of cytotrophoblast. **3.** Mesoderm of secondary villus with capillaries; cytotrophoblast breaking down.

in the cervical canal. This part, with development of the fetus, fuses with the decidua capsularis. At this time, too, further ovulations by the ovary are suppressed. This is accomplished by the secretions of the corpus luteum as well as by lactation.

Steps of Implantation. The steps of implantation (Fig. 18-10, p. 604; Fig. 19-3) are as follows:

1. The zona pellucida disappears, allowing the out-

ermost blastomeres (trophoblast) to contact the endometrium.

2. The blastocyst attaches to the uterine stratum compactum, with the cells near the inner cell mass (embryoblast) attaching first.

3. Penetration of the epithelium and superficial layers of the stratum compactum is followed by uterine vessel, stroma, and gland disruption, and the debris and extravasated blood form the embryotrophe, which serves to nourish the conceptus during implantation.

4. The defect at the implantation site is closed by a coagulum and then on the eleventh day by epithelial repair.

5. The invasion into the uterine mucosa by the expanding blastocyst continues, and its outer layer, the trophoblast, proceeds to elaborate by forming two layers in early implantation: (a) an outer syntrophoblast (syncytium), consisting of a multinucleated, protoplasmic mass without clear intercellular divisions in a single, continuous plasma membrane; and (b) an internal cytotrophoblast, consisting of well-defined epithelioid cells and the source of the extraembryonic mesoblast and amniotic epithelium. It is this layer that is believed to be the source of the placental gonadotropic hormones.

6. The syntrophoblast (in contact with the maternal tissue) extends into the uterine stroma by digestion and regression of the stroma. It has no cellular boundaries, and its surface has microvilli. It is thought to be a source of the steroid hormone progesterone. Lacunae appear in the syntrophoblast, which are filled with blood from maternal vessels opened by the invading trophoblast, and swelling (decidual reaction) takes place in the nearby endometrial cells.

7. With capillary and vessel erosion and blood filling the lacunae, the uterine-embryo relationship is referred to as being of the hemochorial type.

Abnormal Implantation. Abnormal implantation sites, called ectopic (usually terminated by rupture from the site and hemorrhage), consist of (1) tubal—in the uterine tube; (2) interstitial—in the uterine tube as it passes through the uterine wall; (3) abdominal—in the peritoneal cavity (usually the pelvis); and (4) ovarian—in the ovary itself or nearby ligaments.

The trophoblast protects the embryo, provides an attachment to the mother, and facilitates exchange of nutrients and elimination of waste. The blastocyst formation is characteristic of the development of all placental mammals. Why it takes place is uncertain. In any event, the trophoblast always forms not only an isolating covering for the embryo but also an expanding membrane that establishes a large contact area with the maternal tissues for subsequent exchange of fluid and dissolved nutrients between mother and developing child.

Further Development

Two-Layered Vesicle. Once the egg is implanted, the inner cell mass begins the active assertion of its prerogatives of segmentation, readjustment in position, and differentiation. In human embryos, between the seventh and ninth days, these cells arrange themselves as the amniotic vesicle and yolk sac vesicle. Next, the two vesicles become flattened against each other to form a two-layered plate of ectoderm and endoderm. It is this embryonic disk that forms the embryo; once formed, development speeds up. The yolk sac is so called because it has the same relation to the intestinal system of the mammalian embryo as it has in the yolk-fed embryos of lower vertebrates. It has no yolk but plays an essential role in the early nutrition of the embryo as well as in the transmission of immunity from mother to offspring.

Three-Layered Vesicle. Just after the two-layered vesicle is established, a third intermediate layer is seen between the other two and is called the mesoderm. The portion of the original wall of the blastocyst that makes up its outer covering, after the endoderm and mesoderm have been established, is called the ectoderm. Thus are created the three germ layers of the embryo. The different germ layers contain cells with different developmental potentialities, and the embryologic origin of different body parts depends on the growth, division, and differentiation of these three germ layers (Fig. 19-2).

The mesoderm grows peripherally and soon extends beyond the boundaries of the embryonic disk into the spaces of the original blastocyst not occupied by the embryo and yolk sac. It is originally a uniform sheet of cells but later splits into two layers enclosing the primitive body cavity. The outer layer of mesoderm is closely associated with the ectoderm and with it forms the chorion. The blastocyst is now called the chorionic vesicle. The inner layer closely invests the yolk sac wall. This extensive differentiation of the extraembryonic layers of the mesoderm foreshadows an early development of the membranes derived from them.

The chorionic vesicle floats in the maternal blood lake, although syntrophoblastic processes extend to and contact maternal tissues (at various points) around the margin of the lake, thereby anchoring the vesicle (Fig. 19-3). The cytotrophoblast penetrates the syntrophoblastic processes to form the anchor. Fibrinoid material appears between fetal and maternal tissues at this point and may serve to "cement" the union. Substances in the maternal blood lake that bathe the chorionic vesicle include breakdown products of the invaded maternal tissues, blood, uterine gland secretions, and tissue fluid (the embryotrophe). The conceptus is maintained by nutrients from the embryotrophe that reach it by simple diffusion, active transport, and pinocytosis by the syntrophoblast. As the vesicle enlarges, the chorion expands to form small, fingerlike processes—chorionic villi—which increase the surface area of the chorionic vesicle available for fetal-maternal exchange. Three stages of development are seen (Fig. 19-4B): primary villus, when the formed syntrophoblast is invaded by a core of cytotrophoblast; secondary villus, when the core is invaded by an additional core of mesoderm; and tertiary villus (definitive or functional), which is established by blood vessels in the mesodermal villus core. Mid-to-late gestation is characterized by thinning of syntrophoblast, disappearance of cytotrophoblast, pigmentation of syntrophoblast, and appearance of fibrinoid material in the villi. Thus, late in gestation, the placental barrier consists of syntrophoblast, mesenchymal tissue, and capillary endothelium, and increased permeability of the barrier is now related directly to the structural attenuation or thinning of the syntrophoblast.

CHORION LAEVE. At first the villi cover the chorionic sac completely. Later that portion of the sac associated with the decidua capsularis smooths out and becomes devoid of villi (smooth chorion or laeve) (Fig. 19-4).

CHORION FRONDOSUM. Villi on the basal side of the vesicle increase in size and number and are modified to form the fetal part of the placenta. With placental growth, generations of villi are formed by the branching of preexisting villi (Fig. 19-4).

OTHER CONSIDERATIONS. The mesodermal network appears to arise before the endoderm can spread around the interior of the blastocyst; thus the primary yolk sac has endoderm on its roof only, whereas the rest of it is formed by a thin sheet of mesodermal cells.

Formation of Membranes. The first part of the body cavity of the embryo to be established is the region where the future heart and its blood vessels are to develop (the pericardial region). This is apparently correlated with the lack of yolk as a reserve food supply in the mammalian egg. Thus, the system demands membranes capable of establishing metabolic interchange with the maternal circulation as well as a fetal circulation able not only to transport and distribute food material throughout the developing embryo but also to remove wastes to be disposed of into the maternal blood.

Cleavage of the egg leads to the formation of membranes that enclose the embryo, before birth, and separate the part forming the embryo from that of the trophoblast. The membranes are not incorporated into the embryonic body but are discarded at birth and thus are called extraembryonic membranes (Figs. 19-3, 19-4). The first to form is the amnion, which arises by the folding of the chorion (mesoderm plus ectoderm) over the front and hind ends of the future embryo. The folds fuse, and the embryo becomes enclosed and protected from shock in a liquid-filled space, the amniotic cavity.

Amniotic Fluid. The amniotic fluid is initially a diffusate from the embryotrophe. Later it includes fetal urine, desquamated epithelial cells, and fetal hair. Excessive fluid (polyhydramnios) is related to cranial defects; oligohydramnios is marked by reduced fluid. Amniocentesis (fluid withdrawal) enables one to examine cells for sex chromatin and sex determination. Karyotype analysis can also be made by cell culture. While the later type of amniotic fluid is forming, the inner cell mass becomes flattened into a disk, the germ or embryonic disk, along the top of the yolk sac. Although the disk has no semblance of body form, there is an early development of a longitudinal groove on the surface of the extoderm, the ventral groove, which foreshadows the future body axis. Soon the head of the embryo becomes more clearly demarcated and then we speak of the embryo as having a head and tail region.

Fetal Circulation. The first blood vessels form in the part of the mesoderm that has condensed on the yolk sac wall and in the area where the future heart will form. It is this part of the yolk sac with its vascular network, together with the intervening trophoblast (chorion), that forms the primary maternal connection. It is a temporary structure whose functions are gradually taken over by the allantois (Fig. 19-4) with its blood vessels. The latter is an outgrowth of the hind end of the gut canal of the embryo, which expands in all directions, displacing the yolk sac and crowding the amnion. This tiny endodermal tube enters the body stalk (a bridge of mesoderm connecting the embryo to the chorion) as if anticipating the arrival of the endodermal component. It acquires its own investment of vessel and blood-forming mesoderm at an early stage, and its functional significance is limited to its accompanying vessels (allantoic and then umbilical). Its intraembryonic part contributes to the formation of the urachus and urinary bladder, while the part in the body stalk is incorporated into the umbilical cord and undergoes obliteration (Fig. 19-5).

Umbilical Cord. The umbilical cord is a tubular sheath of amnion with surrounding structures. In the amnion covering the cord are (1) a remnant of extraembryonic celom and (2) extraembryonic mesoderm of the body stalk, specialized as Wharton's jelly. The cord contains the allantoic diverticulum, allantoic and umbilical vessels, and the vitelline sac and vessels. Certain portions of the allantoic sac now fuse with the adjacent chorion and enter into varying placental relationships with the uterine wall. The allantoic vessels take over the transportation of substances from mother to embryo and embryo to mother. The basic plan is a fetal circulation covered by trophoblast, bathed in maternal blood that is circulating within a trophoblast-lined space. Once created, the placenta is essential to fetal growth and development.

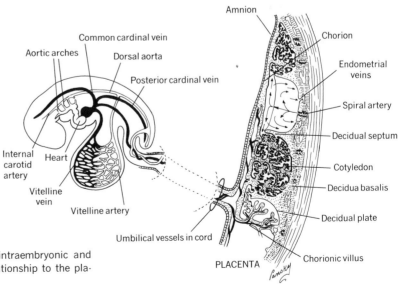

Figure 19-5. Schematic drawing of both intraembryonic and extraembryonic blood vessels and their relationship to the placenta.

Summary of Placental Circulation

Placental circulation (Fig. 19-5) involves the vessels in the chorion and villi conducting blood to and from the embryo (or fetus) and the maternal blood bathing the villi.

Fetal or Embryonic Portion. Vessels are formed in the embryo, vitelline sac, body stalk, and chorionic villi simultaneously. They join and form a primitive vascular system linking embryo to villi. With the onset of cardiac activity, blood passes through vessels in the villi that are bathed by maternal blood, and this is where exchange occurs.

Maternal Portion. The maternal portion of circulation is derived from the blood lake, which divides into lacunae between syntrophoblastic processes. The lacunae then covert into intervillous spaces (when villi form). The lacunae receive blood from maternal capillaries opened by the invading trophoblast.

Intervillous Spaces. Intervillous spaces contain blood entering from the eroded ends of spiral arterioles due to implantation depth. They are drained primarily by veins similarly opened, although some blood flows peripherally into a marginal sinus (an extension of the intervillous space), which in turn is drained by the uterine veins at the placental periphery.

The spiral arterioles opening into the spaces are narrowed at their orifices to increase the velocity of blood escaping into the intervillous space, and thus blood spurts toward the chorionic plate and then filters back to the space for venous drainage.

The Fetus

Before the end of the second month, the embryo has turned itself into something recognizable. The principal external features of the body are clearly visible, and the developing young acquires the status

Figure 19-6. The figure in the center indicates the relative size of the uterus during various months of pregnancy. The fetuses indicate development during early growth period to the point of a recognizable "human" fetus.

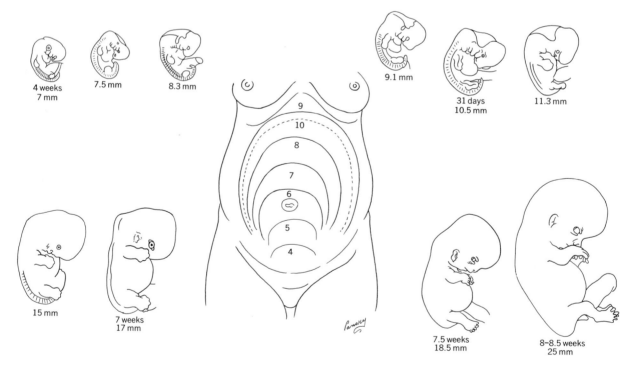

4 weeks
7 mm

7.5 mm

8.3 mm

9.1 mm

31 days
10.5 mm

11.3 mm

15 mm

7 weeks
17 mm

7.5 weeks
18.5 mm

8–8.5 weeks
25 mm

of a fetus (Fig. 19-6). The factors responsible are changes in the curvature of the body and the loss of dorsal convexity so that the head becomes erect and the body straight. The face develops, and the external features of the eye, ear, and nose appear. The limbs become distinguishable with their digits outlined. The prominent tail clearly seen in the embryo in the fifth week becomes inconspicuous. The umbilical cord becomes a definite entity, and the head, which earlier formed the chief ventral prominence, now recedes. The fetus becomes recognizable due to the settling of the heart and the effacement of the branchial arches and gill clefts.

During the second week of development, the extra-embryonic mesoderm in the cavity of the blastocyst is thinned out but a little stalk of this tissue, the body stalk, remains, connecting the amniotic and yolk sac vesicles to the inside of the trophoblast covering. This ensures that when the first blood vessels of the embryo begin to develop, as they soon do in the mesoderm of the yolk sac and the body of the stalk, they have direct access to all parts of the trophoblast as well as to the cores of the villi and thus establish the foundation for vascularization of the fetal portion of the placenta.

As the embryonic disk takes the form of an embryo and later a fetus, the amniotic cavity becomes greatly distended with fluid and then completely fills the exocelomic space, so that its expanding wall compresses the neck of the yolk sac against the body stalk forming the umbilical cord. Simultaneously, the expanding of the amnion brings it into contact with the chorion, with which it fuses. Note that the yolk sac never acquires a functional connection with the chorion.

The most striking feature of the second week of pregnancy of early human development is the great increase of the trophoblast. At first the entire chorion is uniformly covered by villi. These eventually thin out and disappear except at the base of the implantation, where optimal contact exists between the chorionic surface and maternal blood (Fig. 19-4).

The human embryo occupies the uterus for about 40 weeks and comes to weigh about 3 kg ($6\frac{1}{2}$ lb), or about $\frac{1}{20}$ of its mother's weight (Fig. 19-7). The anti-

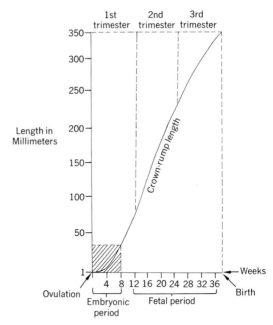

Figure 19-7. Growth curve for developing embryo and fetus. (After studies of O'Rahilly.)

bodies derived from the mother's blood protect the baby from infection and bridge the gap between birth and the time when the offspring gradually begins to produce its own active immunity. Antibody transfer from mother to fetus passes across the placenta, and the infant is born with a full complement of antibodies. For many reasons, the immune mechanisms of the mother seldom reject this fetal "foreign graft." The embryo lives as a human parasite, making indirect use of the maternal organs of digestion, respiration, and excretion while its own organs are growing and developing for a future independent existence. At the same time, the placenta also produces hormones, acting like an endocrine gland. Thus, the hormones of the placenta, pituitary, and ovary prepare the mother for the infant's birth as well as for suckling of the newborn.

Placental Fate

At term, rupture of decidua and fetal membranes (amnion and chorion) is followed by fetal expulsion. The decidua, placenta, and associated membranes (the

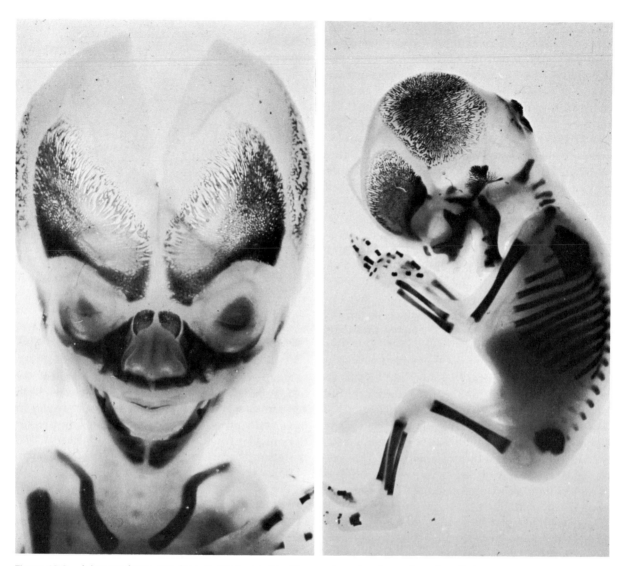

Figure 19-8. A human fetus seen from in front and side. The specimen has been cleared so that the developing bones are visible through the transparent skin and soft tissues. (Courtesy of Dr. Bruce A. Gastineau, Medical College of Ohio at Toledo.)

afterbirth, collectively) are delivered about 15 to 20 minutes later. Decidual and placental detachment from basal endometrium starts by the formation of a cleavage plane that splits the spongy layer and permits the afterbirth to be delivered by efforts of uterine contractions. These myometrial contractions also par-

tially constrict arterial inflow and prevent excessive bleeding of the denuded endometrium. In the case of placenta praevia the placenta develops over the internal os of the uterine cervix instead of on its back wall, and at labor the placenta is aborted before the fetus and can result in fetal anoxia (even death).

The Embryo and Heredity

The nine-month period between fertilization and birth presents more hazards than any similar span up to the ninth decade and includes abnormalities in the germ cells (ovum or sperm) as well as in the environment (as a result of maternal illness).

Every embryo is genetically different from the mother's tissues in which it develops. It contains genes from the father that lead to production of proteins foreign to the mother, and yet the uterus does not reject the embryo as "foreign material." However, if the conceptus is somehow abnormal, the uterus often rejects it. Thus, for every normal child born, at least five are lost in spontaneous abortions or stillbirths. Miscarriage is the layman's term for loss of pregnancy before the baby is able to exist independently. Medically, all are abortions. If the pregnancy ends naturally, it is a spontaneous abortion.

The fetus is capable of independent existence at about the twenty-eighth week of pregnancy, yet some do survive born before that time. Pregnancy lasts about 40 weeks, and at 20 weeks the fetus weighs about 400 gm. About 15 percent of all pregnancies are lost before the fetus is viable at 20 weeks. Stillbirths are about one tenth as common, with 1 to 2 percent of pregnancies being lost during 20 to 40 weeks. Of interest is the fact that about 30 percent of all fertilized ova are lost without the woman even being aware of pregnancy. Thus, of all eggs fertilized, about 50 percent fail to produce a living child.

Inherited characteristics are transmitted from parent to offspring through the chromosomes. These are contained in the nucleus of the cell and are clearly visible as separate entities only when the cell is dividing.

Heredity is the process by which both physical and mental traits are transmitted from parents to offspring, and it is a very uncertain process. It was once thought that in the process of procreation the father implanted a tiny replica of himself in the female womb and the female protected and nourished it until birth, and there were theories that characteristics were passed onto the child through the mixing of blood of mother and father (royal blood, full blooded, etc.). Not until the midnineteenth century were any advances in the principles of heredity made, and then by an Augustinian monk named Gregor Johann Mendel (born in 1822). The results of his work with plants were published in a short paper in 1865 (about the end of the American Civil War). In brief, he determined that a hereditary characteristic is inherited as a complete unit. (Cross a tall pea with a dwarf pea and the first-generation offspring are either tall or short—no haphazard mixing, no intermediate size.) There is also a predominance of certain characters. (The first-generation offspring are more likely to be tall than short, more likely to be purple than white, etc.) Furthermore, hereditary characteristics of both parents are contained in each child, although not always visible, and these "concealed" characteristics can be transferred to the grandchildren without any change in their nature. Finally, although heredity contains an element of uncertainty, certain things are predictable. In any strain there is a certain mathematical proportion that is predictable by well-defined laws of mathematics.

Every species of living thing has its own characteristic number of chromosomes. Man has 46, white rats 42, pea plants 14, etc. The highest is more than 1500, found in a one-celled creature called a rhizopod.

Around 1915, a scientist concluded that a chromosome did not resemble a solid threadlike structure but was more like a necklace with many beads. Since every living thing has hundreds of characteristics, and if only 46 chromosomes carry hereditary material from generation to generation, then each chromosome must transmit these characteristics. These "beads"—as many as 1250 on one human chromosome—are the actual hereditary sites and are called genes. The study of heredity is known as genetics.

An organism is homozygous if it has two identical copies of a gene and heterozygous if two different forms of the gene are present. Genetic maps have been constructed on the basis of quantitative studies of the distribution of progeny resulting from crosses between individuals of different genetic makeup.

Few characteristics of living things are transmitted by only one gene. For example, skin color is affected by at least eight different genes, accounting for the considerable variety; facial features, body structure, and intelligence depend on hundreds of genes. The odds of two individuals being exactly alike are astronomic (except for identical twins, which come from one fertilized egg and, therefore, have the same gene assortment). There are more than eight million ways that the chromosomes (23) of the female and the chromosomes (23) of the male can combine. The odds of any two children having the same chromosome complement are 1 out of 70 trillion, and since each chromosome has 1250 genes, the number is beyond calculation. Furthermore, genes are subject to change. A normal mother and father may occasionally produce an abnormal child. The defect may be slight (e.g., an extra toe or finger, a small blemish) or may be severe (e.g., mental defect, gross physical defect, or absence of limbs). Of course, the defect can be due to an accident during pregnancy or a disease process (e.g., German measles), but often it is linked to an altered gene or a mutation (which can be caused by x-rays and rays of natural and artificial radioactive material).

In 1940, a link between enzymes and genes was made by E. L. Tatum and G. Beadle of Stanford University. The discovery between 1941 and 1944 that the gene was composed of nucleic acid and the experiments with bacteriophage in 1952 established DNA as "the stuff of heredity" (see Chap. 1).

Variation in structure is common yet is usually slight and within the limits of normality. Variations beyond this limitation are termed anomalies, abnormalities, malformations, terata, or birth defects. These may be compatible with life or so extensive that they are incompatible with life during either prenatal or postnatal existence.

Teratology refers to the study of defective development. It regards not only the actual malformation or anomaly but also its mechanism of origin. About 8 percent of bodies coming to autopsy have anomalies, but only 2 percent of live infants demonstrate obvious external defects. Yet within, and including, the first year of postnatal life, the latter figure rises to about 4 percent.

Abnormalities can occur as a result of genetic or chromosomal disorders. The recognition that genes control the development of particular bodily components (cells, structures, etc.) has enabled us to understand the effect of altered genes on the mechanism of abnormal development. Some gene changes are recognizable only by their effects on the chemical material in the cell, altering its metabolism. Other gene mutations are recognizable only in the whole organism and not in single cells. Genetic changes can physically alter the organism, such as by the presence of extra fingers or toes, a cleft palate, or even a clubfoot. Genetic changes may produce abnormalities in several body systems, but at present all one can see is the effect of genetic mutation and not the actual alteration of molecular structure. Unfortunately many abnormalities may be produced by other factors, such as environmental agents (in the case of cleft palate).

It is impossible to identify all the chromosomes in the cell. Therefore, we must cut out each chromosome from a photograph, pair them up like a jigsaw puzzle, and create what is known as a karyotype. Karyotype analysis (chromosome analysis) in cultured cells enables us to visualize certain chromosomal abnormalities (Fig. 2-3, p. 57). Some structural and functional abnormalities have been linked to chromosomal alterations. Chromosomal abnormalities affecting the entire organism (arising during germ cell formation, especially in meiosis) include absence of a chromosome (e.g., an X chromosome or allosome in Turner's syndrome), presence of extra chromosomal material attached to another chromosome (translocation), or an additional separate chromosome (trisomy 21, Down's syndrome, or mongolism) (Fig. 17-9, p. 583).

If during cleavage, or later in development, faulty chromosome segregation takes place, then mosaicism is the result and only a part of the cells of the organism carries the defect.

Genetic mutation usually applies to changes in, or deletion of, specific individual genes, which are passed on to succeeding generations as a known cause of infant abnormalities. The gene defect may be carried on either an autosome or a sex chromosome. It may behave as a dominant or recessive. The former produces its abnormal effects in the person who carries it and so is manifested. The recessive genes carried on autosomes rarely produce obvious effects in their carriers since the normal chromosome of the pair prevents expression of the recessive one. If two carriers of recessive mutant genes marry, their offspring may receive a recessive gene from each parent (statistically 1 in 4 offspring) and thus with two recessives would show evidence of the disease.

The cause of mutation is variable, including radiation on parental germ cells. The latter can alter gene structure and thus affect development of structure(s) that depend on that gene. Chemicals, too, like caffeine and LSD (lysergic acid diethylamide) can cause chromosome fragmentation and perhaps gene mutation.

Recessive mutant genes may be carried on the X chromosome. If the carrier is a male and since he has only one X chromosome, the abnormality is inevitably manifested. If the carrier is a female, she appears normal because the dominant normal gene or her other X chromosome suppresses the recessive one. Hemophilia and muscular dystrophy are common X-linked inherited diseases. Genetic lethals, on the other hand, are serious enough to cause death of the embryo, abnormalities, and abortion.

There is statistical evidence that certain types of mating are associated with increased pregnancy loss and sterility. This is based on studies of marriages between people of certain A, B, and O blood groups. These incompatabilities are associated with spontaneous abortion. Rh blood group incompatibility is not, since Rh hemolytic disease is a cause of late fetal loss but not of abortion.

Chromosomal abnormalities are found in about 1 of every 200 liveborn infants. The most common is that associated with mongolism or Down's syndrome, which occurs in every 600 or 700 infants and is associated with the presence of an extra small chromosome in group G (set 21 trisomy). Most chromosomal abnormalities, however, are found in spontaneous abortions.

Human Teratology

About 85 percent of all clinical congenital malformations have no precise etiology. About 12 percent can be traced to known hereditary factors, genetic disturbances, or chromosomal abnormalities. About 1 or 2 percent are undoubtedly the result of environmental insults. The rest can only, at present, be said to be due to a subtle interaction between environmental factors and the genetic milieu.

Environmental Factors

The placental barrier protects the fetus, especially during early stages of development, and yet a number of factors (extrinsic) may bypass the placenta and exert an influence on the developing embryo. The nature of that effect depends on the teratogenic agent as well as the developmental time period in which it acts.

Infectious Agents. Infectious agents can affect the developing organism adversely, particularly in its embryonic period. Viruses (e.g., the rubella virus, causing German measles) have been shown to have adverse effects either by competing with the embryonic cells for nutrients or by producing a toxin that inhibits cellular activity by blocking normal metabolic pathways. Bacterial infections (not clearly identified as having an effect on the developing structure) such as syphilis have been shown to affect and damage fetal tissues directly.

Physicochemical Agents. Radiation of the embryo may have an adverse effect, ranging from complete cellular destruction, due to ionization, to damage of the cell (especially the nucleus). Differentiation between radiation of germ cells prior to conception and radiation of the developing embryo might be noted. In the former, the effect would result in germ cell mutation and each cell of the conceptus would carry the defect in its genes. In the latter, only a part of the cells of the embryo might be involved and the resulting defect would be limited to the degree of cell injury and the role those cells played in development.

Any factor that alters or modifies the physical status of the cell (changes in O_2 or CO_2 or even pH) creates a situation whereby the affected cells would or could be limited in participating in the developmental processes and lead to an abnormal development.

Chemicals or drugs with antimetabolic actions are able to produce abnormalities by inhibiting (wholly or partially) participation of the developing cell in the developmental process. For example, the tranquilizer thalidomide plays such a role by its ability to block or inhibit growth of mesodermal tissues (particularly that of the limb buds).

Immunologic Factors. Best known of the immunologic factors is the Rh factor, in which incompatibility of maternal and fetal blood results in the development of maternal antibodies. If these antibodies enter the fetal circulation, they result in clumping and destruction of fetal red cells.

Immunologic suppression of organs or tissues by the development of other organs or tissues may result in antibody formation. The question of maternal-fetal incompatabilities allowing for immunologic suppression and resulting in abnormal development has not as yet been answered.

Nutritional Factors. Failure of a cell to obtain adequate nutrients may be a result of competition with infectious microorganisms. Inability to use available nutrients may be due to physiocochemical agents blocking metabolic pathways. Faulty maternal diet, per se, as a cause of abnormal development has not been demonstrated; if diet is sufficient to maintain the mother and permit conception and implantation, it can supply the embryo. Yet mineral deficiencies (e.g., lack of iodine) may result in infantile cretinism due to failure of thyroid gland function (this can be reversed by iodine substitution following the embryonic period).

Essentials of Postnatal Development

Birth

Birth is a dramatic moment in the cycle of life. For about nine months the organism has been a parasite, attached to the mother and remaining warm and protected in her womb. Suddenly, the fetus is projected into the air and deprived of oxygen from the placenta and must regulate its own temperature, get its own nourishment, and use its own organs for digestion, respiration, and excretion. Its most immediate need at birth is the need for oxygen.

Infancy

Several factors may control a newborn baby's efforts to take its first breath of air: (1) the stimulation of the skin's sensory nerves by cooler air or (2) the handling or cutting of the umbilical cord. The lungs, which up to this point have been relatively solid and collapsed, fill almost at once (delay could be fatal). The newborn baby's tissues contain large stores of glycogen, which can be converted to glucose and then to lactic acid to provide energy for immediate needs. Heat, needed to counteract the drop in temperature suddenly met in the external environment, is obtained by the combustion of a special type of "brown fat" found only in the newborn.

The newborn, as far as we know, has no consciousness since both the sensory and motor areas of the cerebral cortex are undeveloped. Yet the brain is almost the size of that of an adult, with ten billion neurons, many of which will degenerate over the years.

The spinal cord and lower brain centers are well developed at birth, enabling the infant to perform some reflex movements. It can suck, is sensitive to touch, can grasp small objects, can taste slightly, and has sensations of hot and cold. The only response is crying and vague arm and leg movements. After several months, these basic reflexes disappear and are gradually replaced by conditioned reflexes associated at first with feeding. Certain noises are distinguished and the eyes can fix on objects, but the brain is not sufficiently developed for any meaning to be attached to either sights or sounds. The infant, recognizing or hearing a familiar person, may smile or move his arms. At first his upper body moves awkwardly, since weight supporting is not possible as yet owing to the disproportionately large head.

By about the fifteenth postnatal month, the visual cerebral cortex and the eye muscles are better developed. The infant responds more directly to sounds, but this lags behind his visual ability. Many visual and auditory sensations have been acting on the cerebral cortex and are retained for future use. By the age of 18 months the child is curious and learns by mimicry and trial-and-error experimentation. At this time, the child usually develops the ability to speak and during the following year begins to build monosyllables into simple, coherent sentences.

From the age of one year, possibly earlier, until five or seven years, the building of cortical patterns will influence the child's future behavioral phenomena. From unawareness, the infant enters a world where constant adjustment is a necessity; and from his early impressions, during which his mother was most influential, the world now becomes agreeable or hostile, fair or unreliable. An anxious environment may result in inhibitions and lack of concentration. These years are very essential in the formation of a child's personality.

Locomotion proceeds in stages. First one learns to position the head in space, then to creep on all fours, and later to move about while sitting upright; finally the bones and muscles of the legs are strong enough to support the body weight, and the biped stance (on two feet) is achieved. The inner ear and cerebellum are rapidly developing, and the infant can balance for awhile with some support. Once the child is standing and walking, the leg muscles pull the infantile bones into their correct shape.

In the first six months the infant approximately doubles its birth weight, but the rate diminishes thereafter. By the age of five years, his height is about doubled and he has altered in all proportions (Fig. 19-1). The head, which accounted for one fourth of the length at birth, now is only one eighth of the body length. As body proportions alter, the internal organs also grow since they must cope with the demands of increasing size. These growths are hormone influenced, especially by thyroxine and growth hormone. The stages of maturation are not usually affected by short-term malnutrition or normal childhood illnesses. If a child is exceptionally small, the cause may be hereditary or a hormonal disorder and not necessarily a lack of feeding.

Childhood

By the age of five years, understanding, sensation, and voluntary motor control are at a high control level, and speech provides communication and speeds mental development so that reading and writing are feasible. Between the ages of five and eight the child's abilities and emotions increase in complexity, and after eight the child develops new mental abilities—he systematizes facts and reasons objectively. However, he is not as yet capable of abstract thought. After eight, however, the greatest physiologic change (since birth) takes place. In about 50 percent of females between the ages of 8 and 12 (and in others a bit later) and in boys a year or so older there is a new type of endocrine development, particularly in the sex organs and the pituitary gland. This hormonal change prepares both for producing the next generation.

Adolescence and Puberty

To the age of 11, girls and boys have about the same physique; however, the new hormone balance causes a rapid increase in growth—the adolescent spurt—which gives males and females their distinctive shape and increases their height (Fig. 19-1). The adolescent is also more capable and skillful and able to make rapid judgments. The spurt usually begins earlier in females than in males so that the female may be stronger than the male at 12 or 13 years of age. The male, however, soon overtakes her and is usually much stronger and about 10 percent bigger.

This rapid growth slows down after two or three years and only then is sexual maturity reached. This phase of maturity is called puberty. At this time, usually at ages 12 to 14, girls experience the onset of menstruation.

Up to a certain age, 25 in boys and 18 in girls, the body's resistance and strength increase and reach a maximum due to rises in blood pressure and number of red blood cells. The efficiency of the muscle cells also increases, and development of the central nervous system enables the adolescent to adopt new skills and perform them quickly. The peak of intelligence is said to be reached at about 18 years of age. With cerebral development, the development for abstract thought has increased, which can bring about emotional problems as all of the world's contradictions are examined.

Early Adulthood

During the ages of 20 to 30 years, early adulthood, the bones and muscles continue to develop in the male as does his ability to run up an oxygen debt. In females, these changes are slight. It is at this age that we begin to cope with the problems of life, and what we are capable of doing now depends on how successfully our inherited characteristics have been blended with the experience gained during earlier development.

Maturity

From here on physiologic life for male and female remains about the same. For women, however, another marked change takes place during her 40s. Endocrine changes occur that affect most parts of her body, especially the ovaries. Ovulation ceases so she no longer is fertile, and she is said to be in menopause.

After 30 years of age, both sexes begin a gradual process of aging (see Chap. 20) that affects every cell of the body. Yet the age at which this occurs varies with each individual. Connective tissue forms around muscle fibers, interfering with oxygen supply. The blood has less oxygen, because in the lungs the alveolar walls thicken and become less permeable. The blood tends to be more sluggish and moves more slowly through the narrowing blood vessels. The number of muscle fibers diminishes as does the ability to cope with the oxygen debt, and the loss of elasticity and the deposit of substances in the blood vessels produce a rise in blood pressure, which can be dangerous.

Old Age

After 70, the cells undergo more drastic changes. The cartilage of the joints begins to degenerate, causing annoying stiffness, and the cells of the body repair themselves very slowly and with difficulty. Since degenerative changes have been going on in the central nervous system throughout life, in later years these changes have an important effect. Alertness, speed of decision, and thought slow down. The level of intelligence, however, remains about the same in those who remain mentally active throughout middle age, and increased experience often offsets slight mental lapses or lethargy.

Yet man gains a kind of immortality in his offspring, and the cycle of life continues through the living cell.

Review Questions

1. Differentiate between genotype and phenotype.
2. What is meant by:
   a. Blastomere   b. Morula   c. Blastocyst   d. Trophoblast
3. What is placentation? Where does it usually take place?
4. What are decidual cells? Define the decidua basalis, capsularis, and parietalis.
5. Describe the general steps of implantation. What is an ectopic pregnancy? Why is it dangerous?
6. Differentiate between the primary, secondary, and tertiary villi.
7. What is the amniotic cavity?
8. What are some of the functions of the placenta? What eventually happens to it?
9. What is a karyotype? Of what significance is it?
10. List some of the essential steps and stages of postnatal development.
11. What is meant by:
    a. Adolescence   c. Menstruation   e. Menopause
    b. Puberty   d. Early adulthood   f. Aging

References

Beck, L. F.: *Human Growth.* Harcourt, Brace & World, Inc., New York, 1969.

Beer, A. E., and Billingham, R. E.: The Embryo as a Transplant. *Sci. Am.,* **230**(4):36–46, 1974.

Carr, D. H.: Chromosomal Errors and Development. *Am. J. Obstet. Gynecol.,* **104**:327, 1969.

Dawes, G. S.: *Foetal and Neonatal Physiology.* Year Book Medical Publishers, Inc., Chicago, 1968.

DeBusk, A. G.: *Molecular Genetics.* Macmillan Publishing Co., Inc., New York, 1968.

Hurlock, E. B.: *Adolescent Development,* 4th ed. McGraw-Hill Book Co., New York, 1973.

Joselyn, I. M.: *Adolescence.* Harper & Row, Publishers, New York, 1971.

Michael, R. P.: *Endocrinology and Human Behavior.* Oxford University Press, New York, 1968.

Timiras, P. S.: *Developmental Physiology and Aging.* Macmillan Publishing Co., Inc., New York, 1972.

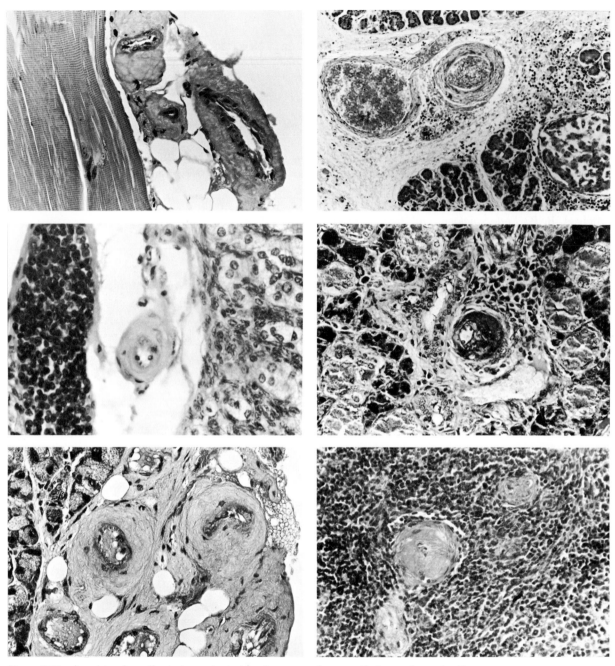

Figure 20-1. A variety of small arteries and arterioles from several organs that are atherosclerotic (a common occurrence in old age) with narrowed lumens. They also have fatty, fibrous perivascular alteration. ×250.

Consciousness, Behavior, and Aging

CHAPTER 20

Consciousness

Behavior is an individual's response to his external and internal environment. Consciousness is both a state and an experience (thoughts, ideas, perceptions, dreams). One's consciousness is defined by his behavior (from coma to extreme alertness) as well as by the pattern of brain neuron activity, which can be recorded by means of electrodes to the scalp. The graphic record is called the electroencephalogram (EEG).

Variations in behavior or consciousness change the frequency and amplitude of the EEG patterns (Fig. 20-2). The EEG corresponds to cortical neuron activity, but its specific components are unknown. It has, however, been used as a tool in medicine to determine the degree and state of diseased or damaged brain areas.

The state of consciousness is the result of interplay between three neuronal systems, one causing arousal and the other two sleep, and all three are related to different areas of the reticular formation (see Chap. 8). The latter is found in the center of the brain stem, ascending and descending between the spinal cord and brain. The neurons of this formation receive a continuous pattern of information from higher cerebral areas (cortex, basal ganglia, limbic system, etc.), spinal cord, cerebellum, and other brain stem structures. Its output projects (the reticular activating system) to the cord, the cerebellum, and subcortical and cortical areas of the cerebrum. There are also numerous synaptic endings in the brain stem, and, therefore, the reticular formation influences itself as well as all areas of the central nervous system. It helps, in general, to smooth out and coordinate muscle activity, contains the primary respiratory and cardiovascular control centers, monitors many impulses ascending and descending through the central nervous system, and plays a role in determining the state of consciousness.

Wakefulness

When one is awake and relaxed with eyes closed (decreased level of attention), the prominent EEG wave pattern is a slow, large, regular oscillation of 8 to 13 cps called the alpha rhythm. When one pays attention to an external stimulus or even thinks hard about something, this pattern is replaced by lower but faster oscillations and we speak of EEG arousal. The latter is related to "paying attention" to stimuli, even to opening one's eyes and trying to see. As already stated, the neural structures associated with the above EEG patterns lie in the brain stem reticular formation. Chemical or electrical stimulation of this area causes EEG arousal, and destruction produces coma and an EEG similar to that of sleep. The activating system does not relay specific afferent information but only arouses the brain and facilitates information reception. Specific sensory inputs are selectively picked and others tuned out, like one hearing a child cry yet paying little attention to a truck's motor. One reason for such selectivity, of course, is habituation. If one hears a bell ring, one may respond and be aroused; however, were that bell to continue to ring monotonously, one's response would progressively decrease

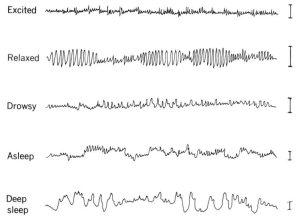

Excited

Relaxed

Drowsy

Asleep

Deep sleep

Figure 20-2. Electroencephalograms during different states of sleep and wakefulness. The excited or alert state shows a low amplitude; the fast wave indicates activity. When the subject is relaxed, a larger, slower wave (alpha) is seen. A mixture of slow and fast waves is seen in drowsiness. In sleep, periods of fast and slow waves are seen. In deep sleep, large-amplitude, slow waves are typical.

(habituation). Habituation is not due to receptor fatigue or adaptation but is also mediated by the reticular formation.

Other parts of the brain beside the reticular formation also play a role in wakefulness. The cortex is necessary for prolonged wakefulness, and the state of alertness involves a relationship between cortex and the reticular formation. Hypothalamic areas have also been implicated in behavioral aspects of the wakeful state.

Sleep

We spend about a third of our lives sleeping. Sleep is actually an active process involving two distinct states characterized by different behavioral patterns and EEGs (Fig. 20-2).

As one gets drowsy, the alpha rhythm is replaced by irregular, low-voltage potential differences; with deepening sleep, the EEG waves become more irregular, larger, and slower—the slow-wave sleep. The latter is periodically interrupted by episodes of paradoxical sleep, when one seems asleep but the EEG pattern is similar to that of arousal (awake and alert). One can be awakened easily from slow-wave sleep, and one rarely dreams in this state but little physiologic change is noted. In paradoxical sleep, however, postural muscle tone is inhibited and periodic twitching of limbs and facial muscles occurs as well as rapid eye movements behind closed lids. Thus, paradoxical sleep has also been called rapid-eye-movement sleep, or REM. Respiration and heart rate are irregular, and blood pressure varies. One tends to dream in REM.

Slow-wave sleep occupies about 75 percent of total sleeping time and paradoxical sleep about 25 percent. If a person is deprived of paradoxical sleep by being awakened continually, for several nights, he spends a greater-than-usual amount of time in paradoxical sleep the next time he sleeps.

The major control of sleep is exerted by two neuronal systems that differ from the action of the reticular activating system. One is responsible for slow-wave sleep and is found in the midline of the lower brain stem and requires cerebral cortex. The second is responsible for paradoxical sleep and is located in the pons. The areas involved may be known, but how they operate is unclear. The fact that sleep-wakefulness is cyclic, as are slow-wave and paradoxical sleep, appears to indicate that there are biologic clocks or pacemakers present in some of the neurons of the reticular formation. The reasons for this rhythmic nature of sleep are unclear; yet experiments have shown that there is a buildup of chemical transmitters such as serotonin, acetylcholine, and norepinephrine during sleep that may account for the cycling.

But why sleep? The brain does not rest during sleep, and there is no generalized inhibition of cerebral neuron activity. In fact, during slow-wave sleep there is a good deal of neuronal activity, and areas of the brain are even more active during paradoxical sleep than during waking. Even blood flow and oxygen use do not decrease! There is, however, a change in the distribution of neuronal activity, with some pathways being less active than while awake. Thus, sleep may represent a period of rest for certain specific elements to replenish substrates. It has also been suggested that the value of sleep is not in short-term recovery but in the relatively long-term chemical and structural changes that the brain must undergo to make memory and learning possible. The answers are very vague because we really do not know.

The Mind

The workings of the mind somehow involve an entity imposed between afferent and efferent impulses, implying that it is something more than action potentials and synapses. We really have no idea of the mechanisms that give rise to conscious experience. All that can be said is that the mind involves the interaction of many components and is the result of coordinated interaction of many areas of the nervous system.

Behavior: Emotion and Motivation

Motivation has been defined in behavioral terms as the process responsible for the goal-directed quality of behavior. Much of behavior is related to homeostasis or the maintenance of a stable internal environment and is usually sustained until the needs are satisfied. This is difficult to relate to physicochemical terms, however, and is undefinable in neurophysiologic terms.

Emotion is related to motivation and is a complex phenomenon whose most common component is subjective feeling—fear, love, hate, etc. Physical components autonomically mediated (blushing, sweating, change in heart rate, etc.) are also present, as are somatic responses (laughing, crying, etc.). The integration of all these components confers on behavior its goal-directed quality. The strong interrelation of somatic and autonomic activity, along with subjective feelings, often leads to long-standing emotional states and psychosomatic disease.

Motivation is also tied to reward and punishment. The former is what one works for and it strengthens behavior, while the latter does the reverse. Rewards and punishments may constitute the incentives for learning and become crucial factors in directing behavior. Much of man's behavior is thus influenced by both rewards and punishments, many of which he is

unaware of. Little is known of the mechanisms that underlie the subjective components of these phenomena (emotion and motivation).

We do know that the brain area most important for the integration of motivated behavior related to homeostasis is the hypothalamus, where the centers for food intake, temperature regulation, thirst, and many others are located. The limbic system, an interconnected group of brain structures in the cerebrum, is also involved in emotion and motivated behavior. It includes part of the frontal lobe cortex, cells in the temporal lobe, and parts of the thalamus. All are connected together and to the hypothalamus as well. Damage to the limbic system leads to a great variety of behavioral changes, particularly associated with emotion. A temporal lobe lesion can make a savage animal docile, while damage to an area deep in the brain may do the reverse. A lesion in part of the hypothalamus can result in a rage response as well as bizarre sexual behavior in which animals attempt to mate with other species, and so forth.

The subjective aspects of feeling are assumed to involve the cortex, particularly the frontal lobes, since damage here leads to alterations in mood such as fear, aggressiveness, rage, euphoria, and depression. There may be both facilitatory as well as inhibitory frontal regions. The frontal cortex does have connections to the hypothalamus that suggest an interrelationship.

Memory and Learning: Molecular and Morphologic Theories

Learning is defined as the increase in the likelihood of a particular response to a stimulus as a result of experience. Rewards and punishments are important ingredients of learning, as is our environment. The processes involved in setting down memory occur in minutes or hours, and they are called short-term memory. After this formative period, the memory is stored as long-term memory. The former memory involves the cerebral cortex, the latter the limbic system. There is, however, no specific site for memory storage.

Memories do not fade or decay with time but, rather, may be changed or suppressed by other experiences. Long-term memory may be due to the fact that large molecules in neurons are changed during learning and information is stored in the configuration of these molecules (most frequently implicated are RNA and protein). We are uncertain about the mechanism of learning, but it appears possible that there are chemical processes that in some way represent the storage of learned material in a coded form, like the coding of genetic information by the nucleotide sequences in chromosomal DNA and its transfer to RNA and proteins. Many experiments have tried to prove this, but most are controversial. Some theories suggest that changes in the relationship between cells take place during the formation of a memory trace. It has also been postulated that modifications of glial cells around neurons provide the storage of this trace. Other theories indicate that new synaptic relationships are established or old ones become more efficient when new information is sent to the brain. Relationships between cells could also change with the interposition of small, newly formed nerve cells between the input and output elements of the nervous system.

Thus, molecular and morphologic theories are not the entire answer since structural changes of relationships occur only through changes in macromolecules such as protein and RNA. How the latter occur is obscure. We do know, however, that learning can be rapid, can be translated from an initial form involving action potentials to a permanent form of storage, and can survive trauma, shock, and the like; that information can be retained for long periods of time and, therefore, that memory and learning must reside in systems that do not turn over rapidly; that information can be recalled from memory stores after long periods of time—memory does not "atrophy" with disuse; and that, after an initial period of rapid learning, the process slows down and proceeds at a rate offering diminished returns. It seems unlikely that there is only one process underlying "learning and memory."

Language

Man differs from other animals in the fundamental process of language. No experimental animal has man's highly developed language skills, although there are no simple anatomic differences between man's brain and that of other mammals. Subtle differences between the two hemispheres of man's brain exist that are related to the fact that, in adults, language appears to function predominantly in the left hemisphere.

One speaks of aphasias, which are specific language deficits not due to mental defects but related to conceptualization or expression (see Chap. 8). In other words, the patient cannot understand spoken or written words, even though he sees them or hears them clearly (conceptualization). In the expressive form the patient has difficulty performing coordinated oral and respiratory movements for language, even though he understands the spoken word and knows what he wants to say. He may also not be able to write.

Various cortical areas have been related to language. Areas in the frontal lobe near the motor cortex are associated with articulation of speech, and areas in the temporal and parietal lobes are involved in sensory functions and language interpretation. These cortical-area specializations develop in childhood as one acquires language skills.

If the left hemisphere is injured in childhood, language can develop in the intact right side or may even develop with a later injury after a period of time. Prognosis gets worse with age (after early teens, interference is permanent). The capacity for language is characteristic of man and develops under stimulation and opportunity. It is partially inherited but strongly influenced by environmental situations.

In general, behavioral functions are not solely controlled by any one area of the nervous system, such as the cerebral cortex, but are shared or influenced by many areas. The cerebral cortex and subcortical areas, particularly the limbic and reticular systems, form a unique interconnected system that contributes to the expression of behavioral performance. Thus, the use of language and thinking are interdependent processes.

Aging and the Aging Process

Human aging begins with fertilization of the egg, and initially it is a function of the chemical properties of genetic material. Life is programmed for growth, cell division, and differentiation. When these cease or are inhibited, generalized deterioration takes over.

Some of the effects of heredity on life-span are due to the transmission of susceptibility or resistance to disease. There are many known or suspected hereditary disorders. Certain diseases are clearly genetic in origin (autosomal dominant inheritance) and easy to recognize since the affected individual usually has a clinically affected parent or other family members. Other diseases are more difficult to recognize because the parents are normal clinically, but the siblings may be affected (autosomal recessive inheritance; e.g., sickle-cell anemia). Still other diseases affect only males (sex-linked inheritance; e.g., hemophilia). Common diseases in which hereditary susceptibility has been transmitted are coronary atherosclerosis, rheumatoid arthritis, diabetes mellitus, gout, hypertension, and even cancer (to name a few).

Practically all genetic diseases can be modified or altered by treatment and environment. Even the genetic information for life-span can be modified. Manipulations of diet have resulted in prolongation of life in animals and also in the delay of onset of major diseases.

Death is the universal outcome of life in free-living cells as well as in multicellular organisms. Its probability, in any species, increases with the passage of particular periods of time. The time periods show extreme variations in different species, ranging from days to centuries. However, the changes of aging are remarkably similar—there is a period of growth and development, which reaches a peak and is maintained for a relatively short time, which is then replaced by retrogressive changes leading to old age and death.

Aging is a biologic process that increases susceptibility to disease. Aging is associated with skin wrinkling, thinning and graying of hair, decreased motor ability and power, decreased auditory and visual acu-

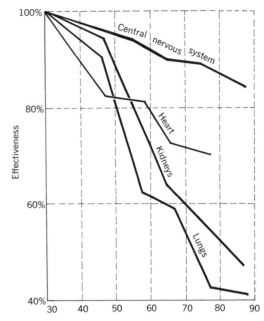

Figure 20-3. Some of the vital systems and how their function alters with age.

ity, involution of sex organs and secondary sex characteristics (such as breast atrophy and uterine atrophy), decreased agility, poor healing of broken bones, and a variable loss of mental faculties (Fig. 20-3; Table 20-1).

Microscopic examination of the tissues of aged indi-

Table 20-1. The Organs and Changes Resulting from Aging

Organ	Effect
1. Hair	Turns gray, thins out
2. Brain	Loses 10,000 cells per day
3. Ears	Hearing capacity declines
4. Eyes	Sight dims
5. Heart	Pumps less blood
6. Lungs	Take in less oxygen
7. Kidneys	Function diminishes sharply
8. Bladder	Capacity and control diminish
9. Endocrine glands	Output of hormones decreases
10. Skin	Dries out, wrinkles
11. Muscles	Shrink and become weaker
12. Joints	Stiffen and swell

viduals reveals atrophic changes in skin, bone, muscle, brain, kidneys, and sexual organs, as well as the sex target organs. Along with the atrophy of parenchymal cells one sees increased collagenization. In addition, degenerative changes in blood vessels are almost universal. Vessel changes involve the large as well as the small arteries, arterioles, and veins. The degenerative changes take the form of hyaline degeneration and calcification of the aorta and large vessels, atherosclerosis, and arteriosclerosis. Associated with such changes are variable degrees of secondary calcification and thrombus formation (Fig. 20-1).

Generally speaking, the changes with age are predominantly those of diffuse vascular and connective tissue disease. Note, however, that changes in aged tissues are not diagnostic since changes similar to those of aging can be produced by a variety of disease processes as well as physical and chemical agents. Thus, ionizing radiation produces almost identical changes in the skin and connective tissue, dietary changes cause similar vascular changes in experimental animals, and collagen disease produces alterations in the connective tissues and in many organs (spleen, kidneys, and skin). There is also a similarity between the functional and structural changes in aging and those seen in starvation and panhypopituitarism.

Body composition, in terms of fat, water, cell solids, bone mineral, and specific gravity, varies with nutritional and pathologic states, sex, and exercise habits. Apparently, body water decreases with age. The decrease in specific gravity with age reflects, in part, the increase in fat content and loss of lean body mass. Cell solids (mostly protein) decrease remarkably, with the greatest change occurring in muscle. Note that certain body constituents change little, namely, serum protein, total protein, total globulin, albumin, gamma globulin, and antibodies.

Although changes in aged tissue can be mimicked by various causes not connected with age, it is true that the total alterations in the body are unique to aging. Although aging is a generalized biologic process, not all people die in the same manner nor do all tissues show the same degree of change in all aged individuals. The causes of death in old age usually are

cardiovascular disease (cardiac failure, myocardial infarction), cerebrovascular disease (hemorrhage or cerebral thrombosis), renal disease (uremia secondary to arteriosclerotic vascular disease), combinations of the preceding, and intercurrent disease (cancer, pneumonia, etc.).

The tissue changes in different individuals also vary in involvement (Table 20-1). Some tissues are remarkably unaltered in aged people. The thyroid, adrenal, and pituitary glands are apparently structurally intact when contrasted to the gonads, which hyalinize and atrophy. The bone marrow and peripheral blood are generally intact, but there is a tendency for fewer and less prominent lymph nodes with age. Nevertheless, an aged person can respond to antigenic stimulation either by antibody production or by cell proliferation. Thus, there is no evidence that age leads to immunologic incompetence or a depressed circulating gamma globulin. Although the kidneys alter, the liver is not particularly changed from that of a young adult.

In the female, beginning after menopause, there are atrophic changes in the breast and uterus, which proceed to involute with age. Yet in spite of the atrophic changes, each of these organs can be rejuvenated by the use of female sex hormones. Thus, it appears that, at least in some tissues, the atrophy of old age is secondary to the loss of tropic hormones. In contrast, the ovaries and testes (gonads) of senile people do not respond to gonadotropins. Of interest too is the fact that age-associated changes in regard to senile osteoporosis can be reversed by the use of estrogens, calcium, vitamin D, and fluoride. Thus, apparently, the aging process does not affect all cell types in a similar fashion or to the same degree.

The aging process demonstrates the interrelationship between the endocrine, vascular, and connective tissues. The "senile" changes are the aggregate effects of the accumulated residues of past illnesses, the secondary effects of circulatory alterations and connective tissue changes, and the direct effects of the aging process on cellular structure and function.

Aging takes its toll over a protracted period of time, and with time the changes that occur strongly tax homeostatic compensation and there exists a vicious cycle of cause and consequence. The decline in physical and even mental prowess begins at a time when outward appearance, microscopic examination, and laboratory tests indicate no abnormal findings or any indication of old age. Consider the athlete, who is "old" at 30, and the very brief span of his prime years.

The exact pathogenesis of the aging process, although studied at various levels, still cannot be defined. Many hypotheses do exist and two major ones are considered here: the wear-and-tear theory and the somatic mutation hypothesis.

The wear-and-tear theory states that aging is associated with the accumulation of degradation products of cells, which eventually "poison" the cells. Examples cited are the wear-and-tear pigment that is found in heart muscle cells and the increased amount of collagen that occurs with age. The "age" pigments accumulate over a period of time and are seen as intracellular bodies or lipofuscin deposits. It has even been suggested that lysosomes are involved in the formation of the pigment. However, it is not known whether the lipofuscin "age" pigments represent a causative factor in the aging process at the cellular level.

The increased incidence of arteriosclerosis with age also favors the wear-and-tear theory. Yet most body cells show no such accumulations with age, and it may very well be that the age-related accumulations are a consequence rather than a cause of age.

A negative version of the wear-and-tear theory presumes that cells use up their supply of preformed enzymes or other limiting substances and then die. It might explain the behavior of red blood cells, white blood cells, and other cells, but why, then, do thyroid cells not appear to age?

The somatic mutation theory receives experimental support from observations that ionizing radiation can shorten the life-span. Such shortening effects are associated with an increased incidence of abnormal mitoses and abnormalities in chromosomes. Chromosomal abnormalities in the blood of apparently normal humans also increase with age. In general, experimental investigation has shown that aging is associated with nonlethal chromosomal changes and that such changes are presumed to compromise the DNA control of cell function and eventually lead to cell death. Cells with a low mitotic rate would be most affected since the effect of the genetic change would be cumulative.

There is evidence also of differences in biosynthesis and structure of DNA. The rate of DNA synthesis with different templates by tissue extracts from young animals varies considerably from those prepared from older animals. The DNAase activity of young and old animals differs, and the stability of certain DNA-protein complexes is not the same in all ages. An interrelationship among these several manifestations of age-related alterations in DNA metabolism appears clearly indicated.

In general, changes in the metabolism of nucleic acids and proteins that occur with aging suggest that they play a major role in the process. The exact nature of the relationship among these variations is, however, not readily apparent.

The somatic mutation theory also does not explain all. For example, why are so many of the changes in aging related to degenerative changes in the vascular system rather than to primary parenchymal cell damage? It does not explain why such a high incidence of somatic cell mutation is not associated with observable changes in cellular function or differentiation. Arteriosclerosis is not produced by known mutagens but can be produced by decreased estrogen levels or dietary manipulations.

Thus, much remains to be learned about the aging process in man's complex system. In his homeostatic, interrelated, multicellular system there are many vulnerable sites in which damage might initiate a cycle of decompensation. In a multicellular system, too, all tissues are not aged nor are all age-connected changes merely a result of time passage. The process of aging and death is a vital component of the evolutionary continuum.

One may summarize this discussion by presenting several other theories on aging that have received some attention. It is obvious that the more theories are presented, the less we apparently seem to know. Under general theories that, regardless of their philosophies, offer little aid to experimentation, we consider

(1) the depletion of irreplaceable matter due to the rate of living, (2) vitality depletion, (3) converging of physiologic variables to cause aging, (4) accumulation of stresses and consequent stress damage, and (5) aging as an adaptive effect to increase the survival value of the species by eliminating older individuals. Intermediate theories having a firm experimental basis, relating to one or more areas of aging syndromes, and resting on chemical causes are as follows: (1) the progressive decrease in the number of many cells with aging seems to play a role; (2) the slowing down of transmission rate of neural impulses results in a lack of coordination, which finally disorganizes the body systems beyond repair; (3) since hormones exert such a powerful controlling influence, any imbalance is bound to affect the functioning of the entire organism; (4) enzyme deterioration is a major factor in aging since the content of most, although not all, enzymes of cells and tissues decreases with progressing age; (5) autoimmunity develops with aging and, consequently, so does age-dependent pathology (as age progresses, very large molecules emerge, which are altered so that they are no longer self-recognized by the body, resulting in their being attacked and leading to pathology); (6) somatic mutations cause formation of inferior cells by spontaneous change in nucleic acid templates; (7) continuous radiation damage may cause aging; and (8) toxin formation in the digestive tract may affect the defense mechanisms. Last, there are the so-called basic theories that relate to experimentally verifiable causes. These are (1) the "clinker" theory, which is self-explanatory; (2) the protein hysteresis theory, which refers to protein coagulation and the effects of fluctuations of pH with age; (3) the denaturation theory, which suggests that irreplaceable molecules are rendered nonfunctional by thermal denaturation; (4) the calcification theory, which states that uncontrolled deposits or reactions of calcium cause many of the symptoms of aging and senility; and (5) the cross-linkage theory, which reveals that the large molecules necessary for life processes are progressively immobilized in all cells and tissues by cross-linkage. Thus, cell death or even mutations can be caused particularly by cross-linkage of DNA.

When the age problem itself has been solved, one might hope that the age-dependent diseases, such as cardiovascular disease and cancer, will also lend themselves to treatment and cure.

Review Questions

1. What is meant by consciousness?
2. What is an electroencephalogram? How does it differ in sleep and wakefulness?
3. What is REM? What is sleep? Why sleep?
4. How would you define emotion? Motivation?
5. Differentiate between long- and short-term memory.
6. What are some of the morphologic changes associated with aging?
7. What major general body tissues are affected by aging?
8. What are some tissues (and organs) relatively unaltered by age?
9. What are two general theories given for aging?
10. Discuss other aging theories.

References

Agranoff, B. W.: Memory and Protein Synthesis. *Sci. Am.*, **216**(6):115–22, 1967.
Bakerman, S. (ed.): *Aging Life Processes*. Charles C Thomas, Publisher, Springfield, Ill., 1969.

Curtis, H. J.: *Biological Mechanisms of Aging.* Charles C Thomas, Publisher, Springfield, Ill., 1966.

Eccles, J. C. (ed.): *Brain and Conscious Experience.* Springer-Verlag, New York, 1966.

Gazzaniga, M. S.: The Split Brain in Man. *Sci. Am.,* **217**(2):24–29, 1967.

Hayflick, L.: Aging Symposium, Summarized by Krohn, P. L. *Science,* **152:**391, 1966.

Lenneberg, E. H.: On Explaining Language. *Science,* **164:**635, 1969.

Penfield, W., and Roberts, L.: *Speech and Brain-Mechanisms.* Atheneum Publishers, New York, 1966.

Pribram, K. H.: The Neurophysiology of Remembering. *Sci. Am.,* **220**(1):73–86, 1969.

Riley, M. W.; Riley, J. W.; and Johnson, M. E.: *Aging and Society.* Russell Sage Foundation, New York, 1969.

Sinex, F. M.: Biochemistry of Aging. *Perspect. Biol. Med.,* **9:**208, 1966.

Strehler, B. L.: *Time, Cells and Aging.* Academic Press, Inc., New York, 1962.

Timiras, P. S.: *Developmental Physiology and Aging.* Macmillan Publishing Co., Inc., New York, 1972.

Walford, R. L.: *The Immunologic Theory of Aging.* Williams & Wilkins Co., Baltimore, 1969.

Prefixes, Suffixes, and Combining Forms

a-   not, without (apnea)
ab-   from, away (absorption)
ad-   to, toward (adrenals)
ag-, af-, as-   see **ad-**
-algia   pain, complaint (neuralgia)
amylo-   starch (amylopsin)
an-   not, without (anemia)
ana-   up (anabolism)
anti-   opposite, opposed to (antitoxin)
-ase   termination denoting an enzyme (amylase)
auto-   self (automatism)

bi-   two, twice (biceps)
bili-   pertaining to bile (bilirubin)
bio-   life (biology)

calci-   calcium, lime (calcification)
calor-   heat (calorimeter)
cata-   down (catalysis)
cerebro-   pertaining to the large brain
chole-   bile (cholecystokinin)
chrom-   color (chromosome)
-cidal   killing (bactericidal)
co-, com-, con-, cor-   with, together
coll-   glue (colloids)
contra-   opposite (contralateral)
corpus   body (corpuscle)
-cyt-   cell (leukocyte)

-dermic, -dermis   skin (hypodermic)
di-   two, twice (dichromatic)
dia-   through, apart (diaphragm)

dis-   negative (disinfect)
dys-   bad (dyspepsia)

-ectomy   to cut out (tonsillectomy)
em-, en-, endo-   in, into (embolus)
-emia   blood (anemia)
entero-   intestine (enterokinase)
epi-   on, above, upon (epidermis)
erythro-   red (erythrocyte)
ex-   out (expiration)

-fer   to carry (afferent)
-fract   break (refraction)

-gastric   stomach (pneumogastric)
-gen or **gen-**   producing (glycogen)
-glosso-   tongue (hypoglossal)
glyc-   glucose, sugar (glycosuria)
-gnosis   knowledge (diagnosis)
-gog or **gogue**   leading (secretogog)
-graph   to write (kymograph)

hemo-   blood (hemorrhage)
hetero-   other, different (heterogeneous)
hydro-   water (hydrolytic)
hyper-   over, above measure (hypertrophy)
hypo-   under, less than (hypotonic)

in-   in, into (insertion)
in-   not, without (insufficiency)
inter-   between, together (intercostal)
intra-   within (intrathoracic)

ir- not (irreciprocity, irregular)
-itis inflammation (tonsillitis)

-ject to throw (injection)

kata- down (katabolism)
kin- to move (kinesthetic, kinetic)

-lac- milk (lactose)
leuco- or leuko- white (leukocyte)
-logy doctrine, science (physiology)
lympho- pertaining to lymph (lymphocyte)
-lysin or -lysis or -lytic dissolving, destruction (hemolysis)

macro- large (macrophages)
mal- bad (malnutrition)
-meter measure (manometer)
micro- small (microorganism)
mole- mass, body (molecule)
mono- one (monosaccharide)
myo- muscle (myoglobin)

nephr- kidneys (nephritis)
neur- pertaining to nerves (neurasthenia)

-oid like (ameboid)
-ole small (bronchiole)
-oma swelling, tumor (sarcoma)
-opia sight (myopia)
-osis a condition (cyanosis); a process (phagocytosis)
oste- or osteo- bone (osteology)
ovi- or ovo- egg (oviduct)

para- near, by, beside (parathyroid)
patho- suffering, disease (pathology)
peri- around, near (pericardium)
phago- to eat (phagocyte)
-phil loving (hemophilia)

-plasm form (cytoplasm)
-plegia or -plexy stroke (apoplexy)
-pnea breathing (dyspnea)
pneumo- air, lungs (pneumonia)
poly- many (polysaccharide)
post- behind, after (postganglionic)
pre- before, in front of (precentral)
pro- before, giving rise to (proenzyme)
proprio- one's own (proprioceptive)
proto- first (protoplasm)
pseudo- false (pseudopod)
psycho- mind (psychology)
pulmo- lung (pulmonary)

re- back, again (regurgitation)
-renal kidney (adrenal)
-rrhea flow (diarrhea)

sarco- flesh, muscle (sarcoplasm)
-sclero- hard (sclera, sclerosis)
semi- half (semicircular)
-some body (chromosome)
-sthenia strength (asthenia)
sub- under, below (subnormal)
supra- above

-tax, -taxis change in location in response to stimuli
  (thermotaxis)
thermo- heat (thermogenesis)
-thrombo- clot, coagulation (thrombin)
-tome or -tomy to cut (tonsillectomy)
trans- beyond, through (transudation)
-tropic to turn (chemotropic)

-ule small (saccule)
-uria pertaining to urine (glycosuria)

vaso- pertaining to blood vessels (vasodilation)

Atlas of Regional Anatomy

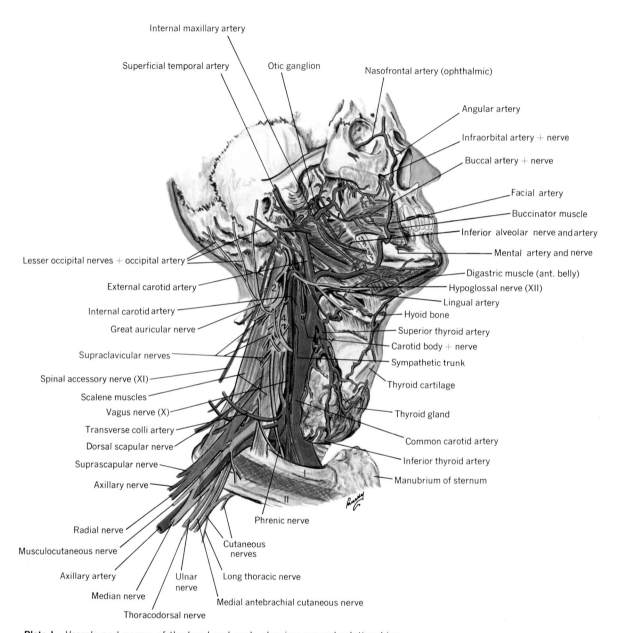

Internal maxillary artery

Superficial temporal artery

Otic ganglion

Nasofrontal artery (ophthalmic)

Angular artery

Infraorbital artery + nerve

Buccal artery + nerve

Facial artery

Buccinator muscle

Inferior alveolar nerve and artery

Mental artery and nerve

Lesser occipital nerves + occipital artery

Digastric muscle (ant. belly)

External carotid artery

Hypoglossal nerve (XII)

Lingual artery

Internal carotid artery

Hyoid bone

Great auricular nerve

Superior thyroid artery

Supraclavicular nerves

Carotid body + nerve

Sympathetic trunk

Spinal accessory nerve (XI)

Scalene muscles

Thyroid cartilage

Vagus nerve (X)

Transverse colli artery

Thyroid gland

Dorsal scapular nerve

Common carotid artery

Suprascapular nerve

Inferior thyroid artery

Axillary nerve

Manubrium of sternum

Radial nerve

Musculocutaneous nerve

Cutaneous
nerves

Axillary artery

Ulnar
nerve

Long thoracic nerve

Median nerve

Phrenic nerve

Thoracodorsal nerve

Medial antebrachial cutaneous nerve

Plate I. Vessels and nerves of the head and neck, showing general relationships.

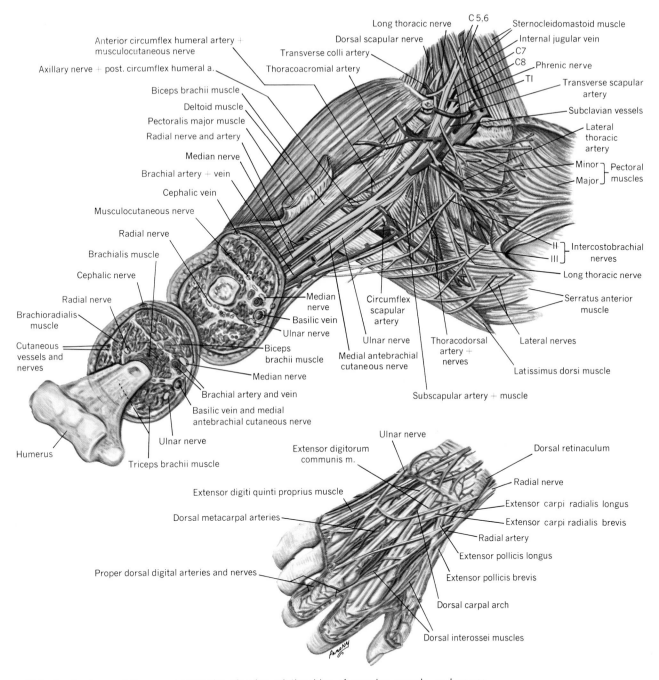

Plate II. Anatomy of the upper extremity, showing relationships of muscles, vessels, and nerves.

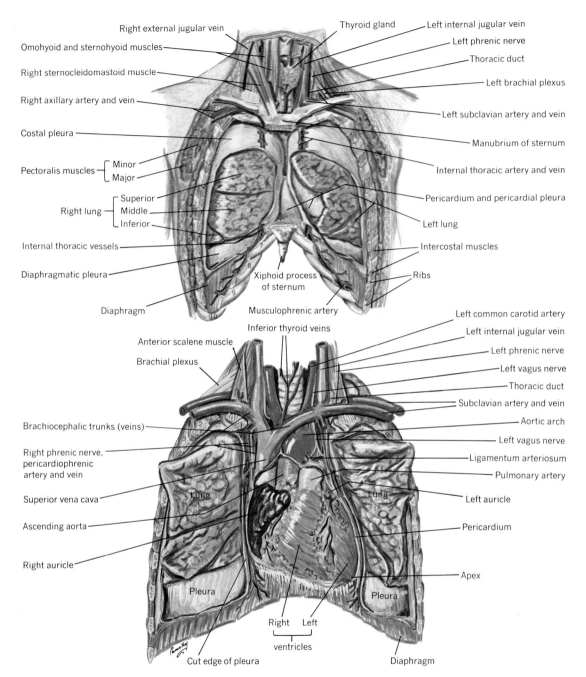

Right external jugular vein

Thyroid gland

Left internal jugular vein

Omohyoid and sternohyoid muscles

Left phrenic nerve

Thoracic duct

Right sternocleidomastoid muscle

Left brachial plexus

Right axillary artery and vein

Left subclavian artery and vein

Costal pleura

Manubrium of sternum

Pectoralis muscles — Minor / Major

Internal thoracic artery and vein

Right lung — Superior / Middle / Inferior

Pericardium and pericardial pleura

Left lung

Internal thoracic vessels

Intercostal muscles

Diaphragmatic pleura

Xiphoid process of sternum

Ribs

Diaphragm

Musculophrenic artery

Inferior thyroid veins

Left common carotid artery

Anterior scalene muscle

Left internal jugular vein

Brachial plexus

Left phrenic nerve

Left vagus nerve

Thoracic duct

Subclavian artery and vein

Brachiocephalic trunks (veins)

Aortic arch

Left vagus nerve

Right phrenic nerve, pericardiophrenic artery and vein

Ligamentum arteriosum

Pulmonary artery

Superior vena cava

Left auricle

Ascending aorta

Pericardium

Right auricle

Lung

Lung

Apex

Pleura

Pleura

Right Left
ventricles

Cut edge of pleura

Diaphragm

Plate III. Anatomy of the thorax (chest). In the top view the lungs and pleurae are seen overlapping the heart and pericardium.

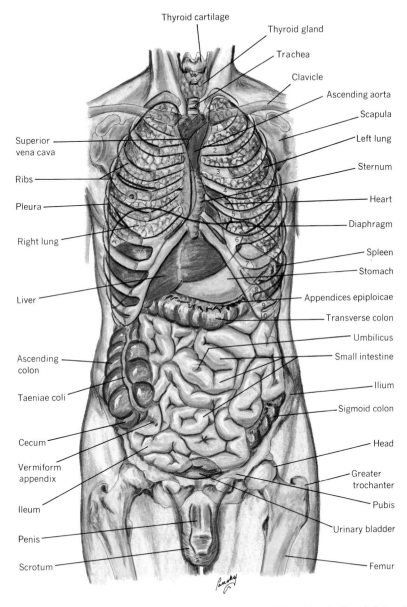

Thyroid cartilage

Thyroid gland

Trachea

Clavicle

Ascending aorta

Scapula

Superior vena cava

Left lung

Sternum

Ribs

Heart

Pleura

Diaphragm

Right lung

Spleen

Stomach

Liver

Appendices epiploicae

Transverse colon

Umbilicus

Small intestine

Ascending colon

Ilium

Taeniae coli

Sigmoid colon

Cecum

Head

Vermiform appendix

Greater trochanter

Pubis

Ileum

Urinary bladder

Penis

Scrotum

Femur

Plate IV. Thoracic and abdominal viscera, showing normal relationships to the skeleton (anterior view).

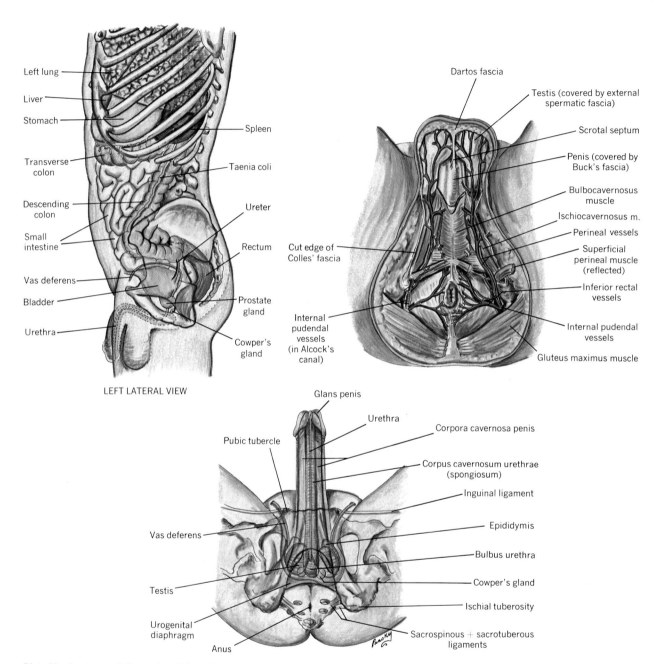

Left lung

Liver

Stomach

Spleen

Transverse colon

Taenia coli

Descending colon

Ureter

Small intestine

Rectum

Vas deferens

Bladder

Prostate gland

Urethra

Cowper's gland

LEFT LATERAL VIEW

Dartos fascia

Testis (covered by external spermatic fascia)

Scrotal septum

Penis (covered by Buck's fascia)

Bulbocavernosus muscle

Ischiocavernosus m.

Perineal vessels

Cut edge of Colles' fascia

Superficial perineal muscle (reflected)

Inferior rectal vessels

Internal pudendal vessels (in Alcock's canal)

Internal pudendal vessels

Gluteus maximus muscle

Glans penis

Urethra

Corpora cavernosa penis

Pubic tubercle

Corpus cavernosum urethrae (spongiosum)

Inguinal ligament

Vas deferens

Epididymis

Bulbus urethra

Testis

Cowper's gland

Ischial tuberosity

Urogenital diaphragm

Sacrospinous + sacrotuberous ligaments

Anus

Plate V. Anatomy of the male pelvis and related structures; perineum and external genitalia.

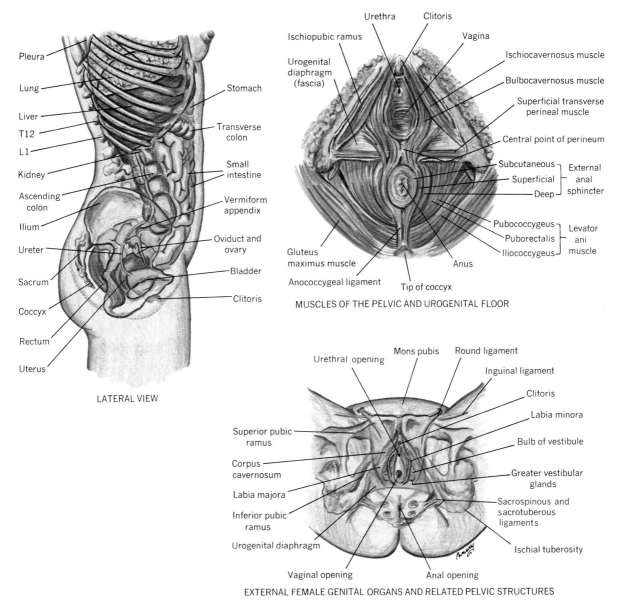

LATERAL VIEW

Pleura
Lung
Liver
T12
L1
Kidney
Ascending colon
Ilium
Ureter
Sacrum
Coccyx
Rectum
Uterus

Stomach
Transverse colon
Small intestine
Vermiform appendix
Oviduct and ovary
Bladder
Clitoris

MUSCLES OF THE PELVIC AND UROGENITAL FLOOR

Urethra
Clitoris
Ischiopubic ramus
Vagina
Urogenital diaphragm (fascia)
Ischiocavernosus muscle
Bulbocavernosus muscle
Superficial transverse perineal muscle
Central point of perineum
Subcutaneous
Superficial
Deep
External anal sphincter
Pubococcygeus
Puborectalis
Iliococcygeus
Levator ani muscle
Gluteus maximus muscle
Anus
Anococcygeal ligament
Tip of coccyx

EXTERNAL FEMALE GENITAL ORGANS AND RELATED PELVIC STRUCTURES

Urethral opening
Mons pubis
Round ligament
Inguinal ligament
Clitoris
Labia minora
Bulb of vestibule
Greater vestibular glands
Sacrospinous and sacrotuberous ligaments
Ischial tuberosity
Superior pubic ramus
Corpus cavernosum
Labia majora
Inferior pubic ramus
Urogenital diaphragm
Vaginal opening
Anal opening

Plate VI. Anatomy of the female pelvis and related structures; perineum and external genitalia.

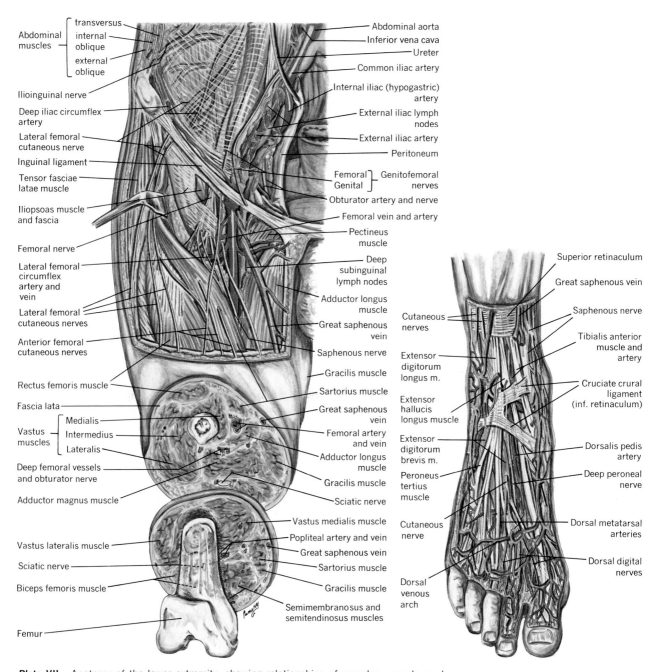

Abdominal muscles
- transversus
- internal oblique
- external oblique

Ilioinguinal nerve

Deep iliac circumflex artery

Lateral femoral cutaneous nerve

Inguinal ligament

Tensor fasciae latae muscle

Iliopsoas muscle and fascia

Femoral nerve

Lateral femoral circumflex artery and vein

Lateral femoral cutaneous nerves

Anterior femoral cutaneous nerves

Rectus femoris muscle

Fascia lata

Vastus muscles
- Medialis
- Intermedius
- Lateralis

Deep femoral vessels and obturator nerve

Adductor magnus muscle

Vastus lateralis muscle

Sciatic nerve

Biceps femoris muscle

Femur

Abdominal aorta

Inferior vena cava

Ureter

Common iliac artery

Internal iliac (hypogastric) artery

External iliac lymph nodes

External iliac artery

Peritoneum

Femoral } Genital } Genitofemoral nerves

Obturator artery and nerve

Femoral vein and artery

Pectineus muscle

Deep subinguinal lymph nodes

Adductor longus muscle

Great saphenous vein

Saphenous nerve

Gracilis muscle

Sartorius muscle

Great saphenous vein

Femoral artery and vein

Adductor longus muscle

Gracilis muscle

Sciatic nerve

Vastus medialis muscle

Popliteal artery and vein

Great saphenous vein

Sartorius muscle

Gracilis muscle

Semimembranosus and semitendinosus muscles

Superior retinaculum

Great saphenous vein

Saphenous nerve

Tibialis anterior muscle and artery

Cruciate crural ligament (inf. retinaculum)

Dorsalis pedis artery

Deep peroneal nerve

Dorsal metatarsal arteries

Dorsal digital nerves

Cutaneous nerves

Extensor digitorum longus m.

Extensor hallucis longus muscle

Extensor digitorum brevis m.

Peroneus tertius muscle

Cutaneous nerve

Dorsal venous arch

Plate VII. Anatomy of the lower extremity, showing relationships of muscles, vessels, and nerves.

Index

Figures in **boldface** type refer to pages on which illustrations appear.

glucose, 464
groups, 558, **559, 560**
hemostasis, 562
oxygen content, 399
  saturation, 399
oxygenation, 399
pH, 506
plasma, 344, **344**
pressure, **340,** 343, **346**
  in arteries, **340,** 341
  in arterioles, **340,** 341
  in capillaries, **340,** 341
  diastolic, 339
  gradient, **340,** 341
  heart action, 339
  mean arterial, **340,** 341
  normal, **340,** 341
  peripheral resistance, 341
  respiration, 341
  of skin, 302
  systolic, 339
  in veins, **340,** 341
  in venules, **340,** 341
  viscosity of blood, 3
  volume of blood, 341
proteins, 344, **344**
rate of flow, **340**
reservoirs, 352
salts. *See* Electrolyte(s)
serum, 344
sinuses, 259, **260**
subfactors, 561
sugar, 464
types (or groups), 559, 560
typing, 559, 560, **560**
velocity, 339, **340,** 341
venous, 352
vessels, functions, 346
  kinds, **340**
  nerve supply, 351
  nutrition, 351
  structure, **340**
volume, **340,** 341, 344
BMR, 469
Body(ies), cavities, 127
  directions, 125, **126**
  framework and movement, 123,
    126

growth, 468
Nissl, 114, 118
organization and direction, 125
planes, 125, **127**
restiform, **256**
of rib, **150,** 152
skeleton, **124**
of sphenoid, **141, 142**
of sternum, **150,** 152
of stomach, **423,** 424
of uterus, **593,** 595
of vertebra, **131,** 132
Body water, balance, 494
  distribution, 493
  loss, 494
  replacement, 494
Bolus, 437
Bone(s), age changes, 103
  ankle, 159
  anvil (incus), **314,** 315
  arm, 145
  astragalus. *See* Talus
  atlas, 133, **133, 173**
  axis, 133, **133, 173**
  blood vessels, 104, **105,** 130
  breakdown and resorption, 103,
    104, 105
  calcaneus, 160, **160, 161**
  canaliculi, 102, **102**
  canals, 102, **102, 103**
  cancellous (spongy), 102, **103**
  capitate, 148, **148,** 149
  carpal, 148, **148**
  cells, 101
  centers of ossification, 106
  cervical, **131**
  of cheek, 139
  classification, 128
  clavicle, 103, **144,** 145
  coccyx, 131, **133,** 153
  collar. *See* Clavicle
  compact, 102, **102**
  conchae, 139
  cranial, 134
  cuboid, 160, **160,** 161
  cuneiforms, 160, **160, 161**
  definition, 127, 128
  ear, **314,** 315

elbow, 146, **147**
epiphysis, 105, **105,** 106, 107
epistropheus, 133, **133**
ethmoid, **136,** 139
face, **136,** 138
femur, 156, **156,** 157
fibula, **157,** 158, **159**
finger, **148,** 149
flat, 103, 128
foot, 159, **160**
formation, 103, 130
  endochondral, 103, 104, **105**
  intramembranous, 103, **103**
fresh, 128
frontal, 103, 136, **136,** 139, **141**
function, 128
greater multangular, 149
gross anatomy, 101
growth, 102, **103**
hammate, **148,** 149
hammer (malleus), **314,** 315
hand, **148,** 149
heel, 160, **160, 161**
hips, 153, **153**
histology, 102, **102**
humerus, 145, **146**
hyoid, **140, 198**
ilium 153, **153**
incus, **314,** 315
inferior nasal conchae or
    turbinates, 139
irregular, 128
ischium, 153, **153,** 154
jaw, **136,** 139
kneecap (patella), **156,** 158
lacrimal, 139
lacunae, 102, **102**
lamellae, 102, **102**
lesser multangular, 149
long, 103, 128
lower extremity, 156
lumbar vertebrae, **131**
lunate, **148,** 149
malar, 139, **141**
malleus, **314,** 315
mandible, **136,** 140, **140, 141**
marrow, 106, 127, 129, **129, 548,**
    566

Bone(s) [*cont.*]
    maxillary, **136**, 139, **141**
    medullary cavity, 106, 127, 129, **129**
    metabolism, 130
    metacarpal(s), **148**, 149
    metatarsal, 159, **160**, 161, **161**
    multangulars, **148**, 149
    names, 131
    nasal, **136**, 139, **141**
    navicular, 160, **160**, 161
    number. *See* individual bones
    occipital, 136, **136**, **141**, 143
    of orbit, **136**
    os coxae, **133**, 134
    os innominatum, 153
    palatine, 137, **137**, 139
    parietal, 103, 136, **136**, **141**
    patella, **156**, 158
    pathology, 130
    pelvic, **153**, 153
    phalanges, **148**, 149, **160**, 161, **161**
    pisiform, 148, **148**, 149
    pneumatic, 128
    primary, 104
    properties, 128
    pubic, 153, **153**
    radius, 146, **147**
    ribs, **150**, 151
    sacrum, **131**, **133**, 153, **155**, 174
    scaphoid, **148**, 149
    scapula, 144, **144**
    sculpturing, 106
    sesamoid, 161
    shaft (diaphysis), 105, **105**
    shin, 158
    short, 128
    of skeleton, **125**, 127
    sinuses, **383**, 384, **385**
    skull, 134, **134**
    sphenoid, 139, **141**, 142, **142**
    stapes (stirrup), **314**, 315
    sternum, **150**, 152
    structure, 101, **102**, 127, **129**
    talus, 160, **160**, 161
    tarsal, 159, 160
    temporal, **134**, 140, **141**, 143
    thoracic, **131**

    of thorax, 149, **150**
    tibia, **157**, 158, **159**
    tissue, 101
    toe, **160**, 161
    trapezium, **148**, 149
    trapezoid, **148**, 149
    triquetrum, **148**, 149
    turbinates, **383**, 384
    types, 128
    ulna, 146, 147, **147**
    upper extremity, 144
    vertebra(e), 131, **131**
    vomer, **137**, 138
    wrist, 148, **148**
    zygomatic, **136**, 139, **141**
Boutons terminaux, **115**, 119
Bowlegs (genu varum), 179, 180
Bowman's capsule, **479**, **480**, 481
Brachia conjunctivum cerebelli, **256**
    pontis, **256**
Brachial artery, **359**, 360
Brachial plexus, 276, **277**
Brachial vein, **368**
Brachialis muscle, **182**, **211**, **212**, 217
Brachiocephalic artery, 355, **356**
Brachiocephalic veins, 366, **367**
Brachioradialis muscle, **211**, **214**, **218**
Bradykinin, 445
Brain, 248, 294
    arterial supply, 258
    association areas, **251**
    attention coordination area, **251**
    auditory area, **251**, 252
    autonomic center, 254
    circulation, 258, **358**
    convolutions, 249
    coverings, **260**, 261, **261**
    divisions, 248
    eye movement area, **251**
    fissures, 249, **250**
    fluid spaces, **260**, 261, 262
    functions, **251**
    growth, 248
    hemispheres, 249
    lobes, **249**, 250, 251, 252
    motor area, 250, **251**, 252
    nuclei, **255**, **257**
    olfactory area, **251**, 253, 255

    psychoauditory area, **251**
    psychomotor area, **251**
    sensory area, 251, **251**, 252
    speech area, 250, **251**
    stem, **257**, **274**
    tactile perception area, **251**
    tracts, **257**, 283, 285
    veins, 258
    ventricles, 248, **248**, **254**, 263
    visual area, **251**, 253
    weight, 248
Breasts (mammary glands), 597, **598**
    lymph supply, 599, **599**
    vascular supply, 598, **598**
Breathing. *See also* Respiration
    abdominal, 149
    acclimation, 401
    amounts of air exchanged, 396
    blood flow in veins, 396
    chest, 397
    costal, 397
    deep (diaphragmatic), 397
    external (lung), 396, 407, 408
    intercostal, 149
    internal (cell), 396, 408
    mechanism, 396, 408
    movements and circulation, 396
    normal, 396, 407
    process, 407
    shallow (costal), 397
        hypoxia, 400
    transport of gases, 396, 408
Bregma, **136**, 136
Broad ligament of uterus, **593**, 594, **594**
Broca's area, 250, **251**
Brodmann's map of cerebral cortex, **251**
Bronchi, 392, **393**
    histology, 395
Bronchial arteries, **356**, 395
Bronchial veins, **369**
Bronchiectasis, 410
Bronchioles, 394, **394**
    histology, 396
Bronchitis, 410
Bronchopulmonary segments, **391**, **393**, 394

parietal, 76, **76, 422,** 425
permeability, 81
phagocytosis, 84
photoreceptor(s), 89
physiology, 3
pigment, 99
plasma, 73, 76, 99
procaryotic, 9
for protection, 86
pyramidal, of Betz, **116**
red blood, 84
reproduction, 584, 602, 613
reticular, **98,** 99
reticuloendothelial, 99, 448
satellite, 35, **116,** 117
Schwann's, 35, **35,** 117
secretion, 73
shape and size, 9, 10
significance, 60
stem, 548
structure and function, 3
summary, **2,** 51
supporting, 86
surface, 30
tendon, 99
theory, 3
zymogenic, 75, **75**
Cellular energy, 25
  morphology, 30
Cellulose, 25, **25**
Celom (ventral cavity), 127
Cement of tooth, 418, **418**
Center(s), appetite, 470
  auditory, **251**
  cardiac, **334**
  feeding, 470
  heat-regulation, 474
  higher autonomic, **289**
  inspiratory, **395**
  lower autonomic, **274**
  in medulla, **257**
  in midbrain, **257**
  in pons, **257**
  pupillary, **289**
  reflex, 295
  respiratory, **395**
  satiety, 470
  of smell and taste, **251,** 319

in spinal cord, **275**
swallowing, **272**
vasomotor, **340**
visual, **251**
vital, **251**
vomiting, 441
Central canal of spinal cord, 272
Central nervous system, 121, 247
Central sulcus of cerebrum, **249,**
  250, **250,** 251
Central tendinous point, 576, **577,**
  590
Centrioles, 2, 44, **44,** 60
Centromere, 60
Centrosome, 44, **44**
Cephalic vein, 366, **368**
Cerebellum, 248, 295
  functions, 193, 258, 294
  structure, **249, 256,** 258
Cerebrospinal fluid, 261, 263
Cerebrum, 294
  aqueduct, **248**
  areas, 249, **249**
  arteries, 258, 358, **358**
  circulation, 263
  convolutions, 249, **249**
  cortex, 249
  fissures, 249, **249, 250**
  functions, **251**
  hemisphere, 250, 294
    inferior surface, **250,** 254, **255**
    lateral surface, 249, **249,** 250
    medial surface, **249, 250,** 253
  lobes, **249,** 250, 251, 252
  nuclei, 249, **251**
  peduncle, 248, **248,** 256, **256**
  sulci, 249, **249, 250**
  tracts, **252, 253**
  vascular supply, 258
  veins, 259
  ventricles, 248, **248, 254,** 263
Cerumen, 313
Cervical canal, **593,** 595
Cervical curve, **131,** 132
Cervical ganglia, 289, **289**
Cervical glands, **593**
Cervical lymph nodes, **373**
Cervical plexus, 276, **277**

Cervical spinal nerves, **277**
Cervical sympathetics, **289**
Cervical vertebrae, 133, **133**
Cervix of uterus, **593,** 595
Chalazion, 305
Cheeks, 416
Chemical changes, composition of
    blood, **550**
  in digestion, 435
  in metabolism, 457
Chemoreceptors, 409
Chest cavity, 127, 327
Chewing muscles, 194, **196**
Chiasm (chiasma), optic, **310**
Chiasmata formation, 67
Chief cells, of pineal gland, 117
  of stomach, 425
Childhood, 627
Chloride, **13, 495,** 496, 504
  cell relationship, 186
  shift, 504
Chloroplasts, 89
Choanae, **137,** 138
Cholecystokinin, 451, 540
Cholesterol, 24, 530, **530**
  esterase, 444
  synthesis, 451
Choline acetylase, 186
Cholinergic fibers, 287, **288**
Cholinesterase, 293
Chondrocytes, 101, **101**
Chondroitin sulfate, 49
Chorda tympani nerve (VII), 197,
    198, 315
Chordae tendineae, **326,** 330
Chorion, **615,** 616, 618
Choroid coat of eye, **304,** 307
Choroid plexus, 261
Christmas factor (IX), **163**
Chromaffin cell(s), 117
Chromatid, **59**
Chromatin, 46
Chromosome(s), 15, 18, **57,** 623
  number, 57, 62, 623
  X, 64, 586
  Y, 64, 586
Chyle. *See* Chylomicrons
Chylomicrons, 80, 467

Extensor(s), carpi, radialis, **211, 214**
    ulnaris, **211, 214**
  digitorum, communis, **214**
    longus, **235, 239**
  hallucis longus, **234**
  indicis, **215, 218**
  pollicis, brevis, **215, 218**
    longus, **215, 218**
  retinaculum, 219, **221, 222**
  tertius, **235**
External auditory meatus, **137, 141,**
    142, 313, **314**
External carotid artery, 357
External ear, 313, **314**
External genitals, female, 589
  male, 574
External iliac artery, 363
External iliac vein, **371**
External intercostal muscles, **204,**
    205
External jugular vein, 366, **367**
External laryngeal nerve, **201**
External oblique muscles, **182,** 206,
    207, 208, 209, **209**
External occipital protuberance, **137**
External os of cervix, 595
Exteroceptors, 283, 299
Extracellular electrolytes, **495**
Extracellular fluid or substances, 48,
    496
Extrapyramidal system, 286
Extremity. *See* Limb
Extrinsic eye muscles, **307,** 309
Extrinsic factor (EF), 441
Eye, 303, **304**
  accessory structures, 303
  accommodation, 311
  anatomy, 303, **304**
  arterial supply, 305, **306**
  chambers, **304,** 308
  lymphatics, 305
  nerves, **266,** 309
  photoreceptor, 89
  physiology of vision, 310
  refractory media, 306
  veins, 305, **306**
  vessels, **306,** 309
Eyeball, **304, 305,** 306

Eyeglasses, 311
Eyelashes, 305
Eyelids, 303, **305**

F

Facial artery, **356,** 357
Facial bones, **136,** 138
Facial expression, muscles, 193, **194,**
    195
Facial nerve, **255, 265,** 268, **269**
Facial vein, 366, **367**
Factor, blood clotting, **563, 564**
Falciform ligament, 449
Fallopian tubes, **593, 594,** 596
False pelvis, 153
False vocal cords, **387,** 389
Falx cerebelli, **261,** 262
Falx cerebri, **260, 261,** 262
Faraday, Michael, 495
Farsightedness, **310,** 311
Fascia, Buck's, 578
  cervical, 201, **202**
  Colles', 578
  cribriform, 232
  deep, of abdomen, 206
    of ankle, 241
    of arm, 217
    of foot, 242
    of forearm, 219
    of hip, 230
    of leg, 237
    of thigh, 232
  endothoracic, **204**
  of kidneys, **476,** 478
  lata, 232
  lumbodorsal, **207, 225**
  muscle, 186
  orbital, 309
  pretracheal, 201, **202**
  prevertebral, 201
  Sibson's, of lung, 393
  spermatic, 209
  superficial, of neck, 201
  transversalis, 206, 209
  visceral, of neck, **202**
Fasciculus, cuneatus, **275,** 284

  gracilis, **275,** 284
  interfascicular, **275**
Fat(s), 23
  absorption, 80, **81**
  brown, 85
  cell, 85, 467
  digestion, 467
  emulsified, 451
  metabolism, 467, **467**
  mobilization, 451
  neutral, **23,** 24, **467**
  polyunsaturated, **23**
  saturated, **23**
  storage, 85
    in liver, 467
  unsaturated, **23**
Fat-sparing action of carbohydrates,
    467
Fatty acids, 23, **23,** 24, **442,** 445, 467
Fauces, **416,** 420
Faucial tonsils, 420
Feces, 453
Feeding center, **518**
Felon, 225
Female, chromosomes, 64
  external genitalia, 589, **590**
  gametes, 62, 602
  gonads. *See* Ovary; Testes
  hormones, 600, **601**
  maturation and physiology, 599
  pelvic viscera, 591
    embryology, 591
    support, 593, **593, 594**
  reproductive organs, 595, 596
  sexual cycles, **597,** 600
  urethra, 589
Femoral artery, 364, **365**
Femoral nerve, 232, **277, 282, 283**
Femoral triangle, 232
Femoral vein, **370**
Femur, 156, **156, 157,** 175
Fenestra, ovalis, 315
  rotundum, 315
Fertility, 608
Fertilization, 62, **62,** 606, **606**
Fetal blood formation, 619
Fetal circulation, **329,** 332, 619, **619**
Fetus, 620, **620**

Hamulus, **137**, 138
Hand, intrinsic muscles, **215, 220,**
    221, 222, **222, 223**
Haploid, number of chromosomes,
    63, 582, **583**
Hard palate, **383**, 419
Hardening of arteries. *See*
    Arteriosclerosis
Harvey, William, 5
Haustra, **426**
Haversian canal, 102, **102**
    lamellae, 102, **102**
    system, 102
Head, muscles, 193, 196
Hearing, anatomy, 313
    physiology, 317
Heart, **326, 328**, 330, **332**
    anatomy, 330
    beat, 335
    block, 335
    blood supply, 332, **333**
    boundaries, **328**
    cardiac cycle, 337
    cavities, **326**, 330
    conduction system, 335, **336**
    course of blood through, 338
    covering, 327
    flow velocities, 339
    impulse conduction, 335
    lining, 331, **336**
    murmurs, 339
    muscle, 112, **113**
    nerve supply, 333, **334**
    nodes, 335, **336**
    output, 337, 341
    pacemaker, 335
    pressures, **337**
    rate, 338, 341, 342
    role in blood pressure, 341
    sounds, 334, **338**, 339
    strength of beat, 341
    structure, 331
    valves, 330
    volumes, 339
Heartburn, 438
Heat, hormonal control, 472
    loss, 473
    of muscle contraction, 186, 472
    production, 470, 472

receptors, 303, **303**
    regulation, 474
    variations, 472
Heel bone, 160, **160, 161**
Helix, double, 16, 18
Hematocrit (packed cell volume,
    PCV), 344
Hematopoiesis, 566
Heme, 400
Hemiazygos vein, **369**, 370
Hemispheres, of cerebellum, **249**,
    295
    of cerebrum, 249, **249, 250**
Hemocytoblasts, **548**
Hemoglobin, 85, 403, 508, **508**. *See*
        also Gases, transportation
    A (adult), 402
    F (fetal), 402
    function, 400
    synthesis, 400
Hemolysis, 402
Hemophilia, 625
Hemopoiesis, 403
Hemopoietin, 406, 566
Hemorrhage, 377
Hemorrhoids, **431**, 432
Henle, limb of, **479, 482**, 483
Heparin, 566
Hepatic artery, **362**, 363, 445, **446**
Hepatic duct, **443**, 447, **447**, 450
Hepatic lobes, **443**, 449
Hepatic veins, **369**, 445, **446**, 449
Heredity, 623
Hernia, inguinal, 210
    umbilical, 210
Hertz (Hz) (cps), 317
Hexose monophosphate shunt, **462**
Hiatus, of facial canal, **142**, 143
High blood pressure. *See*
        Hypertension
High-energy bonds, 25, 26
High-energy compounds, 25, 26
Hilton's (intersphincteric) white line,
    431
Hilum, of kidneys, 479
    of lungs, **391**, 393
    of spleen, **568**, 569
Hindbrain, 248, **248**
Hinge joint, **164**, 166, 170

Hip, bones, 153, **153**
    deep fascia, 230
    joint, 166, 175, **175**
    muscles, 230, **238**
His, bundle of, 335, **336**
Histamine, 540
Histiocytes, 99
Histogenesis, 94
Histology, arteries, 347, **347**, 350
    veins, 352
Hodgkin's disease, 566
Homeostasis, 299, 509
    blood pH, 504
    blood sugar, 461
    fluid distribution, 494
    red blood cells, 403
    role of kidneys, 508
    water balance, 498
Hooke, Robert, 3
Horizontal (transverse) plane, 125, **126**
Hormone(s), 73, 513
    ACTH, or adrenocorticotropic,
        517, 519, 532, **536**
    ADH (antidiuretic), 489
    adrenal, 532
    aldosterone, **489, 530**, 531, 533
    androgens, 530, **530**, 531
    calcitonin, 522, **536**
    catecholamines, 534, **534**
    cholecystokinin, 540
    corticoids, **530**, 531, 532
    corticosterone (compound B), **530**,
        532
    cortisol, **530**, 531, 532, **533**
    cortisone (compound E), **530**, 532
    diabetogenic, 519, 529
    DOC, **530**, 532
    enterogastrone, 540
    epinephrine, 77, 464, 533, **534**
    erythropoietin, **537**, 540
    estradiol, 530, **530**, 532
    estrogens, 530, **530**, 531
    estrone, 530, **530**, 532
    female, 530, **530**. *See also*
        Hormone(s), estrogens;
        progesterone
    follicle-stimulating (FSH), **517**,
        520, **536**
    gastrin, 441, 527, 540

Masseter muscle, **182,** 194, **196**
Master gland. *See* Adenohypophysis
Masters and Johnson, 605
Mastication, 436
  muscles, 194, **196**
Mastoid air cells (sinuses), 315
Mastoid foramen, **137,** 138
Mastoid process, **137,** 138, **141,** 142
Matrix, 86, 97, 99
  osteoid, 103
Matter, 12
  gray, of central nervous system,
    116
  white, of central nervous system,
    116
Maturation of gametes, 581, 603
Maturity, 628
Maxilla, **136,** 139, **141**
Maxillary air sinus, 384
Maxillary artery, **357,** 358
Maxillary bone, **136,** 139, 141
Meatus, external acoustic, **137, 141**
  inferior, 139
  internal acoustic, **142,** 143
  middle, 139
  nasal, 139
  superior, 139
Meckel's diverticulum, **426,** 429
Medial, 125, **126**
Medial lemniscus, **257,** 284
Medial malleolus, 158, **159**
Median basilic nerve, **218,** 219, **279,**
    281, 283
Median basilic vein, **368**
Mediastinum, 327, **328**
  veins, **369**
Medulla oblongata, 248, **249, 255,**
    257, 294
  of adrenal gland, **531,** 532
  of kidney, 479, **479**
Medullary cavity of bone, 129, **129**
Medullary sheath, 35, 276
Medullated nerve fibers, **35,** 118, 119
Megakaryocytes, 130
Meiosis, 47, 63, **63,** 603
  genetic significance, 67, 603
  stages, 64
  summary, **63, 64,** 65

Meissner's corpuscles, 303, **303**
Meissner's plexus, **414,** 427
Melanin, 521
Melanocyte, **300,** 521
Melanocyte-stimulating hormone
    (MSH), **517,** 521, **536**
Melanophores, 521
Melatonin, 521, 541
Membrane(s), alveolar-capillary, 396
  arachnoid, 260, 261, **261**
  basement, **48,** 49, **49**
  basilar, **316,** 318
  of brain. *See* Meninges
  capillary, 350
  cell, **2,** 30, **31,** 46, **46**
  Descemet's, 307
  fascial, 100
  fetal, 618
  intercostal, **204**
  interosseous, 146, 158, **159**
  mucous, 384, 416
  of muscle, 186
  obturator, **175**
  plasma (cell), **2**
  polarization, 185
  postsynaptic, 186
  Reissner's, **316,** 318
  selectively permeable, 33
  semipermeable, 33
  serous. *See* Peritoneum; Pleura
  of spinal cord. *See* Meninges
  synovial, **164,** 165
  tectorial, **173**
  tympanic, 313, **314**
  vestibular, **316**
Membranous labyrinth, **314,** 316
Memory, 296, 633
Mendel, Gregor Johann, 623
Meningeal arteries, **137, 357,** 358
Meninges, cranial, **260,** 261, **261**
  spinal, **260, 261,** 272
Menisci, 177, **177,** 179
Menopause, 608, 636
Menstrual cycle, 600, **601**
Menstruation, 600
Mental foramen, **136, 140**
Mental protuberance, **136,** 140, **140**
Merocrine (eccrine) secretion, 75, **75**

Meromyosin, **110,** 112
Mesaxon, **35**
Mesencephalon, 248, **248**
Mesenchyme, 98, **98**
Mesentery, **424**
  arteries, **362,** 363
  omenta, **424**
  veins, **372**
Mesoderm, 107, **614, 615,** 617
Mesothelium, **92,** 95
Metabolic acidosis, 507, **507,** 509
Metabolic alkalosis, 507, **507,** 509
Metabolism, 457
  amino acid, 463, **463**
  basal rate, 469
  calories, 434, 470
  carbohydrates, **28, 30,** 461, 464
  cell, 25
  definition, 457
  fats, **30, 464, 465,** 467, **467**
  glucose, 461, **461,** 464
  mineral salts, 494, 495, **495**
  pathways, 29, **28–30, 464, 465**
  proteins, **30, 464, 465,** 468
  synthesis of ATP, 25, 26, **28,** 29, 463
  vitamins, 457
  water, 477
Metacarpal bones, **148,** 149
Metaphase, 59, **59, 61**
Metatarsal arch, 161
Metatarsal bones, **160,** 161, **161**
Metencephalon, **248**
Methemoglobin, 400
Metopic suture, 135, **135**
Microglia, **116,** 117
Micrometer (micron), 6
Microscope, electron, **4,** 6, **7**
  light, **4,** 5, **7**
  scanning, **4,** 8
Microsome fraction, 36
Microtubules, 45, **45**
Microvilli, **34,** 35, **45, 76,** 79
Micturition, 490
Midbrain, 248, **248**
Milk, cow's, 434, **434**
  maternal, 434, **434,** 599
  secretion, 599
  sugar, 434, **434**

Muscle(s) [*cont.*]
  hamstrings, 232, **234, 239, 241**
  of hand, **215, 220, 221, 222, 223**
  heart. *See* Cardiac muscle
  hyoglossus, **198, 199, 200, 201**
  iliacus, **238**
  iliocostalis, **226, 227, 230**
  iliopsoas, 232, **238**
  inferior oblique, **307,** 309
  infrahyoid, **200**
  infraspinatus, 210, **211, 212**
  infratemporal, 194, **196**
  intercostals, **204, 205**
  internal oblique, 206, **207, 208,**
      209, **209**
  interossei, of feet, **237, 243**
      of hands, **215, 218**
  intrafusal, 189
  involuntary. *See* Cardiac muscle;
      Smooth muscle
  laryngeal, **386, 388,** 389
  latissimus dorsi, **182, 208,** 225, **225,**
      226
  of leg, **234, 241**
  levator, anguli oris, **194, 195**
      ani, **575, 576**
      costarum, **204, 230**
      labii superioris, **194, 195**
          alaque nasi, **194, 195**
      palpebrae superioris, 303, **305,**
          309
      scapulae, **189, 201, 229**
      veli palatini, 197, **197, 201**
  longissimus capitis, cervicis, and
      dorsi, **226, 227**
  longus capitis and colli, **228, 229**
  of lower extremity, 230
  lumbricales, of foot, **236, 243**
      of hand, **215,** 219, **220, 221, 224**
  masseter, **182**
  of mastication, 194, **196**
  membrane, 186
  mentalis, **194, 195**
  multifidus, **228, 229, 230**
  musculus uvulae, 197, **197**
  mylohyoid, 194, **196, 198, 199, 200**
  nasalis (compressor naris), **194, 195**
  of neck, 197, **198, 199, 200,** 225,
      226, 229

  nerve supply, **115,** 184
  nonstriated, **107,** 108, **110**
  oblique, of eye, **307,** 309
  obturator, externus, **233, 238**
      internus, **238, 576**
  occipitalis, **194, 195**
  opponens, digiti V, **215, 218**
      pollicis, **215, 220, 222**
  orbicularis, oculi, **182,** 193, **194,**
      195
      oris, **182,** 193, **194, 195**
  palatoglossus, 197, **197, 199**
  palatopharyngeus, 197, **197**
  palmaris, brevis, **211, 221, 224**
      longus, **211,** 213, **221,** 222
  papillary, **326,** 330
  pectineus, 232, **233, 239, 240**
  pectoralis, major, **182, 203, 204,**
      208
      minor, **203, 204, 208**
  of pelvic floor, **575, 576**
  perineal, **576**
  peroneus, brevis, **235, 239,** 241
      longus, **235, 239,** 241
  pharyngeal constrictors, **200, 201**
  piriformis, **238, 576**
  plantaris, **235, 239**
  platysma, **194, 195**
  popliteus, **235**
  procerus, **194, 195**
  pronator, quadratus, **214, 218**
      teres, **211, 212, 218**
  properties, 89, 107
  psoas major, **229,** 230, **238**
  pterygoids, 194, **196**
  pyramidalis, **207, 209**
  quadratus, femoris, **233, 238**
      lumborum, **229,** 230, **238**
      plantae, **236, 243**
  quadriceps femoris, 232, **234**
  rectus, abdominis, **182, 203,** 206,
      207, 208, 209
      of eye, **307,** 309
      femoris, **234, 239, 240**
  relaxation, 188
  rhomboids, **229**
  risorius, **194, 195**
  rotatores, **230**
  sacrospinalis, **207, 226, 227, 230**

  sartorius, **182,** 232, **234, 239, 240**
  scalene, **201, 203, 229**
  of scalp, 194, **195**
  semimembranosus, 232, **234, 239,**
      240
  semispinalis capitis, cervicis, and
      thoracis, **226, 228, 229**
  semitendinosus, 232, **234, 239, 240**
  sense, 283, 284
  serratus, anterior, **182, 203, 205**
      posterior, inferior, 225, **226**
          superior, 225, **226**
  shoulder, 210, **211, 212**
  skeletal (striated), **107,** 108
  sliding filaments, 112
  smooth (plain), **107,** 108, **110**
  of soft palate, 197, **197**
  soleus, **235**
  sphincter, pupillae, 308
      urethrae, **576**
  spinalis, **226, 227, 230**
  spindles, 189, **189**
  splenius, capitis, 199, **201,** 227
      cervicis, **227, 229**
  stapedius, 315
  sternocleidomastoid, **182,** 198, **199,**
      200
  sternohyoid, **198, 199, 200**
  sternothyroid, **198, 200**
  striated, 109, **110**
      triad, 111
  styloglossus, **198, 199**
  stylohyoid, **199, 201**
  stylopharyngeus, **198, 199, 202**
  subcostal, **205**
  suboccipital, **227, 229**
  subscapularis, **203,** 210, **212**
  superior oblique, **307,** 309
  supinator, **214, 218**
  supraspinatus, 210, **211, 212**
  temporalis, 194, **196**
  tensor, fasciae latae, **208, 233,**
      238
      veli palatini, 197, **197, 202**
  teres, major, **208, 211, 212**
      minor, **208,** 210, **211, 212**
  thigh, 232, **233**
  of thorax (torso), 201, **203, 204**
  thyrohyoid, **199, 200, 201**

Phosphorus, 13, 538, **539, 540**
Phosphorylation, 26
Photoreception, 89
Photosynthesis, 89
Physiology, of sight, 310
Pia mater, **260**, 261, **261**
Pigmentation, of hair, 301
   of skin, 300
Piles (hemorrhoids), **431**, 432
Pillars of fauces, **416**, 420
Pineal gland, 248, **248**, 254, 540, **540**
Pinna, 313
Pinocytosis, 33, **34**
Piriformis muscle, **238, 576**
Pisiform bone of wrist, **148**, 149
Pitressin. *See* Vasopressin
Pituitary gland, 515
   portal venous system, 516, **516**
Pivotal joint, **164**
Placenta, **616, 619**
   circulation, **619**, 620
   fate, 621
Placentation, 614, **615, 616, 619**
Planes of body, 125, **128**
Plantar aponeurosis, 242, **243**
Plantar arteries, **365**, 366
Plantar flexion of foot, **126**
Plantar surface, **126**
Plantar venous arch, 370, **370**
Plasma, 344, **344**, 496
   cells, **550**, 551
   membrane, 30, **31**
   proteins, 344, **344**
   thromboplastic antecedent, **563**
Plasmin, 566
Plasminogen, 566
Plate, cribriform, of ethmoid, 142, **142**
   pterygoid, **137, 141**
   tarsal, 305, **305**
Platelet(s), **551**, 562
   factor (III), **563**
Platysma muscle, **194, 195**
Plethysmograph, 353
Pleura, 390, **390, 391**
   cavity, 390, **390, 391**
   fluid, 390
   space, 390

Plexus, autonomic, 290
   brachial, 276, **276**
   cardiac, 290
   celiac, 290
   cervical, 276, **277**
   choroid, 258
   coccygeal, 276, **278**
   hypogastric, 290
   intramuscular of Auerbach, **414,** 427
   lumbar, 276, **277**
   pulmonary, 290
   sacral, 276, **278**
   solar, 290
   of spinal nerves, 276, **277**
   submucosal of Meissner, **414,** 427
Plica, circularis, 430
   semicircularis, of colon, **426**
   of eye, 303
Pneumogastric nerve, 270, **271**
Pneumonia, 410
Podocytes, 481, **482**
Polar bodies, **62**, 604
Polarization, 185, 186
Poliomyelitis, **274**
Polycythemia, 567
Polymorphonuclear leukocytes. *See* Neutrophil
Polynucleotide, **16**
Polysaccharides, 25
Polyuria, 529
Pons, 248, **255, 257**, 258
Popliteal artery, 364, **365**
Popliteal fossa, 232
Popliteal line, 158, **159**
Popliteal surface, of femur, **156**, 158
Popliteal vein, 370, **370**
Pores, **2**
Portal circulation, 371, **372**, 432
Postabsorptive state of metabolism, 460, **460**
Postcentral convolution, **250**, 251
Posterior, 125, **126, 127**
Posterior commissure, **249**, 254
Posterior ligament of uterus, 594
Posterior lobe of pituitary gland, 516
Posterior rami, 275
Posterior root of spinal nerve, 275, 276

Postganglionic fibers, 279, 289
Postganglionic neurons, **284**, 289, **289**
Postnatal development, 626
Postural reflexes, 285
Postural tone, 285
Posture, 188, 192, 285
Potassium, **13**, 495, **495**, 502
   cell relationship, 186, 502
   excretion and circulation, 502
Potential, action, 186
   difference, 186
   excitatory, 186
   inhibitory, 186
   postsynaptic, 186
   resting, 186
Pouch, rectouterine (of Douglas), **594**, 595
Poupart's ligament. *See* Inguinal ligament
Precentral convolution, 250, **250**
Precuneus, **250**, 254
Prefrontal functions, 250, **251**
Preganglionic (presynaptic) neurons, 279, 290
Pregnancy, 623
   breast changes, 598
Premenstrual period, 600
Premolar teeth, 418, **418**
Prepatellar bursa, **177**, 179
Prepuce, 578
Pressoreceptors, 303, **303**, 352
Pressoreflexes, 352
Pressure, alveolar, **398**
   atmospheric, 397, **398**
   blood, 339, **340**, 343, **497**
   capillary, **398**, 399
   cardiac, **337**
   difference, **497**
   glomerulocapsular, 486
   gradient, 497, **497**
   hydrostatic, **497**
   intrapleural, **398**, 399
   intrapulmonic (alveolar), **398**, 399
   intrathoracic, **398**, 399
   osmotic, 343, 497
   partial, of CO_2, **398**, 399, **399**
     of O_2, 399, **399**
   pathways, 284

Salivary digestion, 436
Salivary glands, **417**, 420
Salivary nuclei, **255**
Salts. *See* Electrolytes
Santorini, duct of (accessory duct), 443
Saphenous veins, 370, **370**
Sarcolemma, 108, **110**, 111, **113**
Sarcomere, 109, **110**, 111, 185
Sarcoplasm, 109, 111
Sarcoplasmic reticulum, 109, **110**, 111, 186
Sarcostyles, 108
Sartorius muscle, **182**, 232, **234**, 239
Satellite cell, 35, **116**, 117
Satiety center, **518**
Saturation, oxygen, 399, **399**
Scala, tympani, **314**, 315, **316**
    vestibuli, **314**, 315, **316**
Scalp, muscles, 194, **195**
Scaphoid (navicular) bone, **148**, 149
Scapula, 144, **144**
    angles, **144**, 145
    borders, **144**, 145
    processes, **144**, 145
    surfaces, **144**, 145
Scapular notch, **144**, 145
Scarpa's fascia, 206
Schleiden, Matthias, 3
Schwann, Theodor, 3
Schwann cells, 35, **35**, 117, **276**
Sciatic foramina, 155
Sciatic nerve, 232, **278**, **282**, **283**
Sciatic notches, **153**, 154, 155
Sclera, **304**, 306, **306**
Sclerosis, **274**
Scoliosis, 174
Scrotum, **207**, 578, **579**
Sebaceous glands, **300**, 302
Secondary sex characteristics, 573
Secretin, 441, **442**, 527, 540
Secretion(s), of acid, 76
    of bile, 450
    of cell, 73
    endocrine, 77
    gastric, 438, **439**
    granules (zymogen), 20

internal. *See* Hormone(s)
intestinal, 441
merocrine (eccrine), 75, **75**
pancreatic, 73
of protein, 73
of saliva, 420
of urine, 490
Segmentation of ovum, 603
Self-recognition, 558
Sella turcica, **142**, 143
Sellar joints, 166
Semen, 585
Semicircular canals, **314**, 316, 317
Semilunar notch, **147**, 148
Semilunar valves, 330, **376**
Semimembranosus muscle, 232, **234**, **239**, **240**
Seminal ducts, 580
Seminal fluid (semen), 580, 585
Seminal vesicles, **579**, 580
Seminiferous tubule(s), 579, **579**, 581
Semipermeable membrane, 33
Semitendinosus muscle, 232, **234**, **239**, **240**
Sensations, 299
    exteroceptive, 299
    proprioceptive, 299
Sense organs, 299
Sensory areas of brain, 251, **251**, 252
    neuron, **276**
    pathways, 283
    receptors, **276**
    tracts, 275, 283
Septum, interatrial, 330
    interventricular, **326**
    nasal, 139
    pellucidum, **249**, 253
Serosa, **414**, 415
Serotonin, 540
Serous membrane. *See* Peritoneum; Pleura
Serratus, anterior, **182**, **203**, **205**
    posterior, inferior, 225, **225**, **226**
    superior, 225, **226**, **229**
Sertoli (nurse) cells, **582**
Serum, 344
Sesamoid bones, 158

Sex, chromosome, 586
    determination, 586
    glands. *See* Ovary; Testis
    hormones. *See* Estrogen(s); Progesterone; Testosterone
Sexual characteristics, secondary, 573
Shinbone, 158, **159**
Shivering, 472
Shock, 509
Shoulder, blade, 144, **144**
    joints, **168**, 169
    muscles, 210, **211**
Sickle cell hemoglobinemia, 402, 566
Sigmoid colon, **426**, 430
Sigmoid sinus, 260, **260**
Signet ring cartilage. *See* Cricoid cartilage
Simple sugars, 25
Sinoatrial (SA) node, 335, **336**
Sinus(es), air, of skull bones, 384
    blood, of skull, 259, **260**
    cavernous, **260**, 366
    ethmoid, 385, **385**
    frontal, 142, **383**, 384, **385**
    maxillary, 384, **385**
    paranasal, **383**, 384, **385**
    sagittal, **142**, 144, **367**
    sigmoid, **142**, 144
    sphenoid, **383**, 385, **385**
Sinusoids, 445, **446**
Skeletal muscle, 109, **110**
    red and white, 109
    system, 125
    triad, 111
Skeleton, **124**, 127
    age changes, 636
    appendicular, **125**, 127
    axial, **125**, 127
    differences between male and female, 153, **154**
Skene's (paraurethral) glands, 589
Skin, 86, 299
    of abdomen, 206
    appendages, 301, 302
    blood supply, 302
    characteristics, functions, 86, 299
    creases at wrist, 221, **223**

Synovial membranes, **164,** 165
  knee joint, 179
Synovitis, 170, 179
Synthetases (ligases), **20**
Syntrophoblast, 617
System, definition, 127
  vestibulocochlear, 313
  visual, 303
Systemic circulation, 355
Systolic, 338, **338**
Systolic blood pressure, 339
Systolic discharge, 339

T

T tubule (sarcotubule), 110, **110,** 111,
    113
T wave, **336,** 342
Taenia coli, **426,** 430
Talipes equinus, 180
Talus, 160, **160, 161**
Tarsal bones, 160, **160, 161**
Tarsal glands, 305, **305**
Tarsal plates, 305, **305**
Taste buds, 319, **320**
Taste sensations, 269, 270, 319, **320**
Tear ducts, 303, **305**
Tear glands, 303, **305**
Tectorial membrane, **173**
Tectum, 248, **248**
Teeth, 417, **418**
Tegmen tympani, 143, 314, **314**
Tegmentum, 256
Telencephalon, 248
Telodendrites (nerve endings), 118, 119
Teloglial cell, **115**
Telophase, **59,** 60, **61**
Temperature, influence on heart
        rate, 342
  normal, 471
  regulating center, **518**
  regulation, 471, **472**
  tracts, 283
Template, 18
Temporal bone, **141**
Temporal fossa, **136,** 140, **141**
Temporal lines, 140, **141**
Temporal muscle, 194, **194, 196**

Temporal petrous portion, **141, 142,**
    143
Temporal squamous portion, **141**
Temporal zygomatic process, 142
Temporomandibular joint, 166, 167,
    167
Tendon(s), 100
  conjoined, **207,** 209
  organ, **189**
"Tennis elbow," 170
Tension, 191
Tensor, fasciae latae, **208, 233, 238**
  veli palatini, 197, **197**
Tentorium cerebelli, **260, 261,** 262
Teratology, 624, 625
  environmental factors, 625
  immunologic factors, 626
  infectious agents, 625
  nutritional factors, 626
  physicochemical agents, 625
Teres, major, **208, 211, 212**
  minor, **208,** 210, **211, 212**
Terms of direction, 125, **126**
Testicle. *See* Testis
Testis, 579
  cells, 581
  descent, 581
  function, 581
  structure, 579, **579**
Testosterone, 24, **24, 530,** 531
Tetanus, 184
Tetany, 538
Thalamus, 248, 254, **255,** 294
Thebesian valve, 330
"Thermostat," physiologic, 471, 474
Thiamine, **458**
Thigh bone. *See also* Femur
  deep fascia, 232
  muscles, 232, **232**
Thirst, 500
Thoracic aorta, **356,** 361
Thoracic cavity, 127, 390, **390, 391**
Thoracic duct, **373,** 374
Thoracic inlet, 150, 390
Thoracic outlet, 150, 390
Thoracic vertebrae, **131,** 133
Thoracolumbar. *See* Parasympathetic
    nervous system

Thorax, 149
  bones, 149, **150**
  changes in size, 149
  muscles, 201, **204**
  nerves, **278,** 279
Thrombi, 410
Thrombin, 564
Thrombocytes. *See* Platelets
Thrombogen. *See* Prothrombin
Thrombophlebitis, **354,** 355
Thromboplastin, **563,** 564
Thumb, muscles, **216, 220**
Thymine, 16
Thymus gland, 541, **559,** 567
  and antibody relationship, 558,
    559
  involution, 541
  lymphocytes, 555
  veins, 369
Thyrocalcitonin (TCT, calcitonin),
    522, **536**
Thyroglobulin, 523, **523**
Thyroid, arteries, **356,** 357
  cartilage, 385, **386, 387**
  gland, 524, **524**
  hormone, 522, **523, 535**
Thyroid-stimulating hormone (TSH),
    517, 519, **535**
Thyrotropic hormone, **517,** 519, **520,**
    535
Thyroxine, 522, **523, 535**
Tibia, **157,** 158, **159**
  artery, **365,** 366
  collateral ligament, 177, **177**
  nerve, 241, **278, 282, 283**
  tuberosity, **157,** 158, **159**
  vein, **370**
Tibialis muscle, anterior, **234, 239**
  posterior, **236, 241**
Tidal volume, 407, **408**
Tissue(s), 93
  adipose, 100
  areolar, 100
  bone, 101
  cartilage (gristle), 100, **101**
  classification, 94
  connective (supporting), 97
    adult supporting, 99

Vestibular system, 313, 319
Vestibule, of ear, 316
  of female genitals, 589, 590, **590**
  of larynx, **387**, 388
  of mouth, 416
Villus(i), 427, **427**, 452
  chorionic, **615 616,** 618
Virchow, Rudolph, 3, 58
Viruses, 68
Viscera, female pelvic, 591, **592, 593**
  male pelvic, **577,** 580
Visceral, 125
Visceral muscle, **107,** 108, **110**
Visceroreceptors, 290
Viscosity of blood, 343
Vision, binocular, 313
  color, 311
  daylight, 312
  far, **310,** 311
  near, **310,** 311
  night, 312
  pathways, 310, **310**
  physiology, 310
  receptors, **304,** 308
Visual acuity, 308
Visual area of brain, **251,** 253
Visual cones, **304,** 308
Visual fields, **310**
Visual hallucinations, 253
Visual purple, 312
Visual rods, **304,** 308
Visual system, 303
Vital capacity, 407, **408**
Vital centers, **334, 395,** 409
Vitamin(s), 457, **458**
  deficiency, **458**
  definition, 457
  function, **458**
  groups, **458**
  and growth, 459
  and hormones, 459
  interrelationship, 460
  and old age, 460
  in pregnancy, 459
  and resistance to infection, 459
  source, **458**
  and stress, 459

Vitreous humor, **304,** 308
Vocal cords, **386,** 388, **388**
Vocal folds, **386,** 388, **388**
Vocal muscles, **386, 388,** 389
Voice, box, **386, 387, 388, 388**
  production, 409
Voiding, 490
Volar arch(es), **359**
Volar plate, 172
Volkmann's canals, 102, **102**
Volume, percent O$_2$, **398,** 399, 400
Voluntary movements, 188, 286
Voluntary muscle (striated), 188
Vomer bone, **137,** 138
Vomiting, 441
  center, 441
Von Economo's map of cerebral
          cortex, **251**
Vulva, 589, **590**

W

Waking state (wakefulness), 631
Waldeyer's ring, 419
Water, 12
  absorption, **488,** 498
  balance, **478,** 494, 498
  distribution, 493
  functions in cell, 497
  intake, **478, 494**
  loss, **478,** 494, 498
  metabolism, 493
  movement, **488** 489, 498
  output, **478, 494**
  replacement, 494
  retention, **488**
Watson, James D., 18
Wave, of negativity, 342
  of sound, 317
Wax (cerumen), ear, 313
Weight control, 469, 470, 471
Wharton's duct, **416, 417,** 420
White blood cells, **550,** 551
  count, **550**
White matter, **273,** 275

White rami, **284,** 289
Willis, circle of, **358,** 360
Window, oval, **314,** 315
  round, **314,** 315
Windpipe. *See* Trachea
Wirsung, duct of, 443
Wisdom teeth, 418
Wolffian duct, 591, **592**
Wrist, bones, **148,** 148
  muscles, **214, 218, 221,** 221, **222**

X

X chromosomes, 64, 586
Xiphoid process, **150,** 153

Y

Y chromosome, 64, 586
Yeast, 21
Yellow marrow, 129
Yolk sac, **615,** 618

Z

Z line (or disk), 110, **110,** 111, 112,
          113, 186
Zinc, **13**
Zona, fasciculata of adrenal, **531**
  glomerulosa of adrenal, **531**
  orbicularis of hip joint, **175**
  pellucida, **606,** 607
  reticularis of adrenal, **531**
Zygomatic arch, **137,** 138, 141
Zygomatic bones, **136, 141**
Zygomatic process, **136, 137,** 138,
          139, 142
Zygomaticus muscles, **182, 194, 195**
Zygote, 62, 94
Zymogen (secretory) granules, 20, **75,**
          445
Zymogens, **442,** 445